# Agricultural Economics

# Agricultural Economics

## Second Edition

**S Subba Reddy**
*Former* Professor and Head
Department of Agricultural Economics
SV Agricultural College, Tirupati

**P Raghu Ram**
*Former* Professor and Head
Department of Agricultural Economics
SV Agricultural College, Tirupati

**TV Neelakanta Sastry**
*Former* Professor and Head
Department of Agricultural Economics
SV Agricultural College, Tirupati

**I Bhavani Devi**
Professor and Head
Department of Agricultural Economics
SV Agricultural College, Tirupati

**Oxford & IBH Publishing Co. Pvt. Ltd.**
New Delhi
*( A Unit of* **CBS Publishers & Distributors** Pvt Ltd **)**

CBSPD

**CBS Publishers & Distributors** Pvt Ltd

New Delhi • Bengaluru • Chennai • Kochi • Kolkata • Lucknow • Mumbai
Gujarat • Hyderabad • Jharkhand • Nagpur • Patna • Pune • Uttarakhand

# Agricultural Economics
## Second Edition

ISBN-13: 978-81-204-1786-1
ISBN-10: 81-204-1786-0

## OXFORD & IBH
New Delhi
( A Unit of CBS Publishers & Distributors Pvt Ltd )

Published by **Satish Kumar Jain** and produced by **Varun Jain** for

## CBS Publishers & Distributors Pvt Ltd

4819/XI Prahlad Street, 24 Ansari Road, Daryaganj, New Delhi 110 002, India.
Ph: 011-23266838, 23289259      Website: www.cbspd.com
e-mail: delhi@cbspd.com

*Corporate Office:* 204 FIE, Industrial Area, Patparganj, Delhi 110 092
Ph: 011-4934 4934      Fax: 011-4934 4935
e-mail: publishing@cbspd.com; publicity@cbspd.com

### Branches

- **Bengaluru:** Seema House 2975, 17th Cross, KR Road, Banasankari 2nd Stage, Bengaluru 560 070, Karnataka, India
  Ph: +91-80-26771678/79      Fax: +91-80-26771680      e-mail: bangalore@cbspd.com
- **Chennai:** 18/8B, Subbaraya Street, Shenoy Nagar, Chennai 600 030, Tamil Nadu, India
  Ph: +91-044-42032115, 044-26681266      e-mail: chennai@cbspd.com
- **Kochi:** 42/1325, 1326, Power House Road, Opp KSEB, Power House, Ernakulum Kochi 682 018, Kerala, India
  Ph: +91-484-4059061-65, 67      Fax: +91-484-4059065      e-mail: kochi@cbspd.com
- **Kolkata:** 147, Hind Ceramics Compound, 1st Floor, Nilgunj Road, Belghoria, Kolkata-700056, West Bengal, India
  Ph: +033-25633055, 033-25633056      e-mail: kolkata@cbspd.com
- **Lucknow:** Basement, Khushnuma Complex, 7 Meerabai Marg (Behind Jawahar Bhawan), Lucknow-226001, UP, India
  Ph: +0522-4000032      e-mail: tiwari.lucknow@cbspd.com
- **Mumbai:** PWD Shed, Gala no 25/26, Ramchandra Bhatt Marg, Next to JJ Hospital Gate no. 2, Opp. Union Bank of India, Noorbaug, Mumbai-400009, Maharashtra, India
  Ph: 022-66661880/89      e-mail: mumbai@cbspd.com

*Representatives*

- Gujarat     0-9879558667
- Nagpur      0-8692091830
- Uttarakhand 0-9716462459
- Hyderabad   0-9885175004
- Patna       0-9334159340
- Jharkhand   0-9811541605
- Pune        0-9664372571

*Printed at* Chaman Enterprises, Daryaganj, New Delhi, India

**Dr. A. PADMA RAJU**
Vice-Chancellor

**ACHARYA N.G. RANGA AGRICULTURAL UNIVERSITY**
Rajendranagar Hyderabad - 500 030
Phone : 040-24015035 (O)
040-24015031 (F)
Email: angrau_vc@yahoo.com
Grams "AGRIVARSITY"

# Foreword to the Second Edition

The revised edition of the book AGRICULTURAL ECONOMICS is well organized and presented under seven sections and each section is divided into chapters. It is a big voluminous one containing fifty eight chapters in all. besides appendices. Over the years the book has received wide spread recognition and popularity due to the demand generated by the students, teachers, researchers belonging to economics and applied economics at all India level and abroad. The students and teachers of State Agricultural Universities, Horticultural Universities and Veterinary Universities besides traditional Universities in the country are evincing keen interest in the book for their academic needs. The popularity of the book is evident by its reprinting by the publishers annually since its first publication in 2004.

The first section of the book is devoted to present the basics of micro economics. The second section of the book deals with macro economics. The third section provides the contents of agricultural production economics. The key concepts of farm management economics are furnished in fourth section. Section five embodies the subject of agricultural finance. The area of agricultural marketing is provided in sixth section. Finally, the seventh section includes the topics concerned with the role of agriculture in Indian economy.

The chief merits of the book lie in providing absolute clarity on the key concepts of agricultural economics, an outcome of the painstaking efforts of the authors, besides identifying the farmers" decision problems and finding solutions to the same at the field level. In the process they used sophisticated programming models, production function techniques and time series analysis. Particularly a detailed account has been given for presenting price forecasting models. Another notable feature of the book is objective questions which are appended for each section.

All the four authors of the book have served the Acharya N. G. Ranga Agricultural University at different places and cadres. Their vast experience in teaching and research has reflected in the preparation of such a quality book. Though there arc a number of books on Agricultural Economics, this book has its own place among all.

I congratulate and appreciate all the four authors individually by name for bringing name and fame to ANGRAU through such a valuable piece of work. I appreciate the efforts of Oxford & IBH Publishers. New Delhi for publishing and reprinting the book annually and undertaking the publication of second edition of the book during the year 2015.

A. Padma Raju

(A. Padma Raju)

# Foreword to the First Edition

There are many known textbooks written by several authors dealing separately with various branches of agricultural economics during the past. But, the contents of all the branches of agricultural economics were not furnished in one book. As a result, the students of agricultural economics might not have an easy access to them. The present book by the authors is probably the first one that bridges this gap. The merit of this book lies in presenting the major fields of agricultural economics that are structured for academic use of undergraduates in various State Agricultural Universities for easy comprehension. The authors have made earnest efforts in publishing this book by including all the latest and relevant information on different fields of agricultural economics. Infact such a book is long pending and is very much required by the user agencies, students, staff and researchers. The book is presented in seven sections covering the various areas of agricultural economics, *viz.*, microeconomics, macroeconomics, agricultural production economics, farm management, agricultural finance, agricultural marketing, economic problems of Indian agriculture and WTO. For example, at the very outset the book commences with the field of microeconomics followed by macroeconomics, which form the basis for any student studying agricultural economics. These two sections provide a good foundation for the students before they start learning the areas of production economics, farm management, finance, marketing, etc. Agricultural production economics, which includes all the basic economic relationships, is very lucidly presented in the book using geometric, algebraic and mathematical approaches. Farm management principles are clearly elucidated with related agricultural data. The complex Operations Research techniques are clearly explained with their applications to the decision problems in agriculture. The authors highlighted the significant need of the credit for the farmers to generate income, employment and required capital base in agriculture. The book provides adequate knowledge and study of various financial institutions in public and cooperative sectors including the tests of farm credit proposals. Studying and identifying the elements of risk and uncertainty of farming are given good account covering both the theory and their applications to agriculture. The exigency of an efficient marketing of farm commodities and its related concepts are clearly highlighted. The subject matter of agricultural marketing includes the topics on the marketing functions, classification of markets, marketing functionaries, marketing efficiencies, defects in the existing marketing systems and their remedial measures, marketing channels, marketing institutions and the various aspects related to marketing risks. All these topics are well presented with clarity. In the last section of the book the most pressing economic problems of Indian agriculture including WTO and its implications are well discussed and furnished with the help of relevant and recent agricultural data. On the whole, I feel extremely happy for bringing out this book in the present form.

I personally congratulate the authors for their painstaking and sincere efforts in bringing out such a comprehensive book on agricultural economics. I am very confident that the book would get wider acceptance by all State Agricultural Universities and other related Universities and institutions concerned with agricultural economics.

Dr. N. Sreerama Reddy
Dean of Agriculture
Acharya N.G. Ranga Agricultural University
Hyderabad

# Preface to the Second Edition

We have great pleasure in presenting to the students, teachers and researchers this second edition of the book. This edition provides a new look with the additional feature of objective type ot questions. It takes care of recent data wherever it is required. During the last decade, the students and the teachers in India and abroad showed much interest and few of them offered suggestions to incorporate which have been duly included in this revision. We earnestly express our heartfelt gratitude to all patrons of the book and are confident that this revised edition will find even more welcome reception. We all also express our gratitude to the Oxford & IBH Publishing Co. New Delhi, for undertaking the reprinting of the book for the past 11 years and also for bringing out the second edition of the book.

Suggestions for further improvement of the book are most welcome.

Tirupati
1 June, 2015

S. Subba Reddy
P. Raghu Ram
T.V. Neelakanta Sastry
I. Bhavani Devi

# Preface to the First Edition

The basic purpose of this book is to assist the reader to develop a thorough understanding of the subject of agricultural economics. To accomplish this objective an attempt has been made to present the subject lucidly to suit the requirements of undergraduates in Indian Agricultural Universities. The book presents a comprehensive treatment of microeconomics; macroeconomics, agricultural production economics, farm management, agricultural finance, agricultural marketing and economic problems of Indian agriculture. For convenience the book is furnished in seven sections. In section I, the field of microeconomics contains basic concepts of economic theory of consumer behaviour, theory of production, of distribution, market structure etc. adequate to provide a foundation for the graduates in the subject.

Section II includes key chapters on national income, money, public expenditure, public revenue, unemployment, business cycles and inflation to provide requisite insights on fundamental macroeconomic aspects.

Section III describes simple input-output relationships *viz.*, including factor-product, factor-factor and product-product relationships and their implications to farming.

Section IV focuses on the role of management in the organization of the farm business. The topics that appear in this section are economic principle, types of farming, types of farm business organization, planning, techniques of planning, farm records and accounts, acquisition and management of farm resources and risk and uncertainty.

There is no denying the fact that farm finance strengthens farm business and helps to increase the productivity of scarce resources. Therefore an attempt is made to present the most important topics of agricultural finance. These are problems of agricultural finance, classification of credit, institutional agencies supplying credit, tests of farm credit proposals, tools of farm financial analysis and agricultural project analysis.

From the welfare point of view of the farmers, consumers and middlemen, agricultural marketing plays a crucial role. Therefore there is a need to analyse the agricultural marketing from all dimensions. The topics that figure here are the processes of agricultural marketing, problems in agricultural marketing, suggestions to improve agricultural marketing, marketing channels, agricultural marketing efficiency, agricultural prices and role of Government in agricultural marketing.

In section VII, economic problems of Indian agriculture are highlighted including a brief note on agricultural price policy.

Many colleagues in the Department and University have helped us in this endeavour. We would like to express our gratitude to them individually. Any errors and omissions found in this book are entirely those of the authors. We welcome suggestions from the readers for the improvement of the book.

Tirupati                                      S. Subba Reddy

1 June, 2015                              P. Raghu Ram

T.V. Neelakanta Sastry

I. Bhavani Devi

# Contents

# Agricultural Economics

The word agriculture comes from the Latin words *ager*, referring to the soil and *cultura*, to its cultivation. Agriculture, in its widest sense can be defined as the cultivation and/or production of crop plants or livestock products. It is synonymous with farming: the field or field—dependent production of food, fodder and industrial organic materials.

Having known the meaning of agriculture, let us know what economics is. Economics is the science that studies as to how people choose to use scarce productive resources to produce various goods and to distribute these goods to various members of society for their consumption. Now having defined agriculture and economics, we look into the field of agricultural economics.

## Definitions

Agricultural economics is an applied field of economics in which the principles of choice are applied in the use of scarce resources such as land, labour, capital and management in farming and allied activities. It deals with the principles that help the farmer in the efficient use of land, labour and capital. Its role is evident in offering practicable solutions in using scarce resources of the farmers for maximization of income.

Prof. Gray has defined agricultural economics as *"The science in which the principles and methods of economics are applied to the special conditions of agricultural industry"*.

According to Prof. Hibbard *"Agricultural economics is the study of relationships arising from the wealth-getting and wealth-using activity of man in agriculture"*.

Jouzier defined that *"Agricultural economics is that branch of agricultural science which treats that manner of regulating the relations of different elements comprising the resources of the farmer, whether it be the relations of each other, or to human beings in order to secure the greatest degree of prosperity to the enterprise"*.

Snodgras and Wallace defined agricultural economics as *"an applied phase of social science of economics in which attention is given to all aspects of problems related to agriculture"*.

Accordingly to Goodwin *"Agricultural economics as a social science is concerned with human behaviour during the process of producing, processing, distributing and consuming the products on farms and ranches"*.

*Agricultural economics is an applied social science dealing with how humans choose to use technical knowledge and the scarce productive resources such as land, labour, capital and management to produce food and fibre and to distribute it for consumption to various members of the society over time (Cramer and Jensen).*

These definitions indicate that the field of agricultural economics deals with the problems of the farms as the units of industry, the income earning and spending activities of the farmers and also the management of farm business and bringing necessary changes according to the situation so as to bring stability to farm income.

## IMPORTANCE OF AGRICULTURAL ECONOMICS

Akin to economics, the field of agricultural economics finds to seek relevance between cause and effect using the most advanced methods *viz.*, production functions and programming models. It uses theoretical concepts of economics to provide answers to the problems of agriculture and agri-business. Initially earnest efforts were made by the economists to use the economic theory to agricultural problems. Now the subject of agricultural economics is enriched in many directions and fields taking the relevant tools of sciences particularly mathematics and statistics. Agricultural depression which occurred in the last quarter of 19th century and middle of 20th century brought about increased attention and concern to find out plausible causes and solutions for world agricultural depression. Here in this context the contribution made by Agronomists, Economists, Horticulturists, *etc.*, is noteworthy. Agriculture is the integral part of the world food system, having the foundation links between crops and animal production system. Agricultural Economists here have to play a major role in understanding the intricacies involved in the foundation systems. The students of agricultural economics should have a clear insight and understanding of the influence of climatic conditions in determining as to how the commodities are produced and marketed in line with the consumption needs. Knowledge regarding problems in production, finance, marketing and Government policies and their impact on production and distribution is very essential to find out suitable solutions for the farm problems. The students of agricultural economics are taught the subject disciplines *viz.*, microeconomics, macroeconomics, agricultural production economics, farm management, agricultural marketing, price analysis, *etc.*, to fulfill their requirements. The present book on **Agricultural Economics** is presented in the following sections.

Section   I   :   Microeconomics
Section   II  :   Macroeconomics
Section   III :   Agricultural Production Economics
Section   IV  :   Farm Management
Section   V   :   Agricultural Finance

# SECTION I
## MICROECONOMICS

# CHAPTER 1

# Economics–its Subject Matter

Nature has blessed the humans with abundant natural wealth to live on this earth. Humans would have been contended with what nature provided, had they been able to peg their wants (requirements) at a given level. But it is not so, man being born in this world is influenced by biological, physical and social needs, which keep him always busy in searching out the means to keep him satisfied. To fulfill his requirements arising out of various needs, he involves in an activity called economic activity. Conversely, man would have freed himself from economic activity, had there been no resource scarcity and also what humans want can be satisfied without limit. But neither of the possibilities being found because of the scarcity of resources imposed by the nature, humans always deeply engage in arriving at an equation, which balances their unlimited wants and limited means. By engaging themselves in the economic activity people aim at maximizing their satisfaction from their scarce resources. Thus, scarcity is the pivot for the economic activity of the people representing consuming and producing segments leading to the origination of a field of study called economics. The field of economics keeps on going as long as the human race exists on the earth with their toiling to satisfy their ever-new and ever-emerging wants and satisfying the same through their efforts. Thus, the field of economics constitutes wants, efforts and satisfaction. These also form the subject matter of economics. The subject matter when examined further reveals the participation of two different economic agents *viz.,* consumers and producers in the economic activity. Consumers, given their income levels would attempt to satisfy their wants which have been short-listed based on the degree of urgency from unlimited wants. In this process the consumers fulfill the goal of maximization of satisfaction by consuming the goods and services that can be purchased given the limitation of money. The other participants in the economic activity are the producers. They employ all the necessary factors of production (inputs) and bring out goods required by the consumers.

## DIVISIONS OF ECONOMICS

### Traditional Approach

Traditionally, the subject matter of economics can be studied under four divisions. These are consumption, production, exchange and distribution.

## Consumption

It means the use of wealth to satisfy innumerable wants. It also means the destruction of utility. All the goods that are produced are consumed immediately or some time in future. Through the consumption activity, we use utilities hence consumption represents using up of utilities.

## Production

It is an activity that helps to create utility. It simply means the addition of utilities. Hence production is defined as the creation of utility. Through the process of production one set of goods is transformed into the other. In the economic sense mere creation of utilities is not treated as production, and in fact the goods that are produced should have value too. In the production process, inputs or resources are transformed into products.

## Exchange

The word exchange of goods implies transfer of goods from one person to the other. Exchange of goods takes place among groups of individuals, countries, markets, regions and so on. The exchange of goods leads to an increase in the welfare of the individuals through creation of higher utilities for goods and services.

## Distribution

If refers to the sharing of wealth produced by the community among the agents of production. Proper distribution of wealth and resources leads to growth and economic welfare of the people in the nation.

# MODERN APPROACH

## Microeconomics and Macroeconomics

As against the traditional approach, modern approach treats microeconomics and macroeconomics as the two basic divisions of economics.

The term "microeconomics" has been derived from the Greek word "*MICROS*" meaning small. In other words micro means a millionth part. It is otherwise known as price theory. It focuses on price determination. Microeconomics fundamentally deals with economic behaviour of individual economic units such as consumers, resource owners and business firms. It is concerned with the flow of goods and services from business firms to consumers and also the flow of resources or their services from resource owners to business firms. Microeconomics covers theory of consumer behaviour, theory of value (product pricing and factor pricing) and theory of economic welfare. Microeconomics is somewhat abstract because it cannot include all the economic activities of real world.

The term macroeconomics has been derived from the Greek word "*MAKROS*" meaning large. Macroeconomics otherwise is called income theory. It treats the economic system as a whole, rather than treating the individual economic units of which it is composed. Macroeconomics is concerned with the value of the overall flow of goods and the value of the overall flow of resources. Thus, it covers, theory of income and employment, theory of money and prices, banking, theory of economic growth, macro theory of distribution, general equilibrium analysis, policy formulation and analysis, *etc*. Thus, it is concerned with the study of aggregates.

Though distinction exists between microeconomics and macroeconomics, both are essential for a thorough understanding of the economy.

## DEFINITIONS OF ECONOMICS

The word economics has been derived from the Greek word *"OIKONOMICAS"* with *"OIKOS"* meaning a household and *"NOMOS"* meaning management. It is understood that the beginning was made by the Greek Philosopher, Aristotle who in his book *"Economica"* focussed that the field of economics deals with household management. The economists in defining the term, economics followed several approaches and concepts. The concepts on which various definitions of economics given are 1) Wealth 2) Welfare 3) Scarcity and 4) Growth.

### 'Wealth' Definition of Economics

Adam Smith (1776) who is regarded as *"Father of Economics"* in his book, entitled *"Wealth of Nations"* defined economics as *"An enquiry into the nature and causes of the wealth of nations"*. J.S. Mill, another classical economist* defined economics as *"The practical science of production and distribution of wealth"*.

Here, the term, economics is defined as the field of science concerned with wealth. Wealth in this context refers to abundant supply of money or affluence. These definitions invited the criticism of Carlyl as he called economics as a 'dismal science'.

### 'Welfare' Definition of Economics

Alfred Marshall came out with a new concept in the definition of economics towards the end of 19<sup>th</sup> century. He defined economics as *"A study of mankind in the ordinary business of life: it examines that part of individual and social action which is most closely connected with the attainment and with the use of the material requisites of well-being"*.

This definition of economics has received the acceptance of a number of economists, as it considered that wealth was not the end itself but a means of achieving welfare, *i.e.*, it is a source of attaining human welfare. However, the idea of Marshall was condemned by Lionel Robbins on the points that economics was treated as a social science rather than a human science; and the term "human science" is more apt on the ground that any individual either as a member of a society or in isolation does have economic problems. Secondly, emphasis on material requisites of well-being was objected, as demarcation was created between material and non-material activities, when both are equally important. Thirdly, his approach of economics is limited to the study of mankind in the ordinary business of life, leaving the human life in the extra-ordinary business of life such as famines, wars, *etc.*, is another point of criticism.

### 'Scarcity' Definition of Economics

Lionel Robbins opposing the Marshall's definition putforth the following definition based on the scarcity concept. *"Economics is the science which studies human behaviour as*

---

\* Classical school: The economists who wrote on economics during the period 1750-1850, often known as the classical economists and who formulated a systematic body of economic principles for the first time. The principal members of the school were Adam Smith, Malthus, Ricardo, Nassau Senior, James Mill, John Stuart Mill and Bentham.

*a relationship between ends and scarce means which have alternative uses"* according to Robbins. In this definition 'ends' indicate human wants. 'Means' are the resources with which wants are fulfilled. Though the resources are scarce, they have numerous alternative uses. This definition examines as to how an individual either as a consumer or a producer or a businessman, *etc.*, shortlists the unlimited wants in the light of limited available resources. Indeed, this is an important aspect of human behaviour.

Robbin's definition is superior to 'wealth' and 'welfare' definitions because 'welfare' aspect is embodied in the definition and 'wealth' is represented as means which is always scarce. However, some limitations were pointed out with Robbin's definition. 'Scarcity' is not the problem always and 'abundance' also gives rise to problems, which was not recognized by Robbins. His definition treats economics as a positive science only without touching normative aspects of economics. It has not taken into account the growth aspect of the economy, the dynamic nature of adjustment, *etc.*

### 'Growth' Definition of Economics

According to Samuelson, *"Economics is the study of how men and society choose, with or without money, to employ scarce productive resources which could have alternative uses, to produce various commodities over time, and distribute them for consumption now and in future among various people and groups of society".*

Keynes defined economics as *"The study of the administration of scarce resources and of the determinants of employment and income".*

## IS ECONOMICS A SCIENCE OR AN ART?

By definition science is a systematized body of knowledge having an empirical correspondence. Analogous to science, an art is also systematized body of knowledge. It directs through a system of procedures to attain a given objective or goal. It tells us how to do a thing.

Treating economics as a science, a given theory is formed through conduct of experiments, recording observations, analysis of data recorded, drawing the conclusions and finally testing them. In economics also the same procedure is followed to present any principle or theory. Hence economics is as good as any science. Only the question is regarding precision. The scientific experiments are conducted under laboratory conditions, while economic theories are subjected to several causal factors that influence human behaviour. The situation of controlled experiments in economics is not a possibility, since it deals with human behaviour, which is unpredictable. This indicates the fact that the degree of precision of economics as a science is less, when compared with the pure sciences, but nonetheless economics is a science.

As an art, economics shows solutions to the problems. It helps us how to do a thing. The role of economics as an art can be found in any sphere of economic activity. For example, it advocates how to maximize the profits of a firm given the resource constraints. Given a problem, the field of economics guides us to solve the same. Thus, the field of economics has the attributes of science and art. Economics therefore is a science as well as an art.

## ECONOMICS–A SOCIAL SCIENCE

Economics studies human beings as members of the society participating in the economic activities. It does not study humans as isolated individuals. He is interdependent. Thus economics is a social science.

# ECONOMICS–POSITIVE SCIENCE OR NORMATIVE–SCIENCE

In economic theory, we make an effort to explain the nature of economic activity and predict the events in the economy as facts change. Such an effort helps us to know the environment in which we live and what part is related to others and what causes what. Economists differentiate between positive economics and normative economics. Positive economics is completely objective and is limited to the cause and effect relationship of economic activity. It is simply concerned with the way the economic relationships are present in different economic activities (what they are).

Normative economics studies the way that economic relations ought to be. Normative economics evaluates. Policy making, a conscious intervention in the economy for the welfare of the people is essentially a normative in character.

# METHODS OF ECONOMIC INVESTIGATION

Every field of science follows certain methods to formulate the laws or principles. There are two methods followed in the scientific study of economics. They are: 1) Deductive method and 2) Inductive method.

## Deductive Method

Economists belonging to classical school viz., J.S. Mill and Bacon advocated this method. This method is called analytical or abstract or apriori method. Here the economists proceed from general to particular. It proceeds from certain fundamental assumptions or truths. Through this method theoretical abstractions are derived from the real world. We further analyse these abstractions and hypotheses with the data and finally make logical conclusions about the objectives of the research studies.

## Inductive Method

This method is known as historical or empirical or posterior method. The economists belonging to historical school like Roscher, Frederick, etc., advocated this method. In this method facts are generated through surveys. Here the investigator moves up from particular to general. It is a realistic method because, it is based on facts. We collect data from sample units and analyse the data and infer or draw conclusions relevant to the study. Here we give importance to the predictive power of models.

Though distinction does exist between deduction and induction methods both are needed for scientific investigation, as neither deduction nor induction alone is suffice to formulate and test theories, hypotheses and economic laws.

# ECONOMIC LAWS

Economic laws are the principles that govern the actions of the individuals in their economic activities. Just like any law of science, economic laws too are conditional i.e., applicable when certain conditions are fulfilled. What economists do is that they consider the basic factors into account while developing a theory, keeping other factors influencing the theory as constant. This implies that for developing a theory in economics some kind of abstraction is necessary. There is an important role for assumptions.

"Economic laws are statements of uniformities, which govern human behaviour concerning the utilization of limited resources for the achievement of unlimited ends (Robbins)".

## Characteristics of Economic Laws

1. *Economic Laws are not the Governmental Laws:* The laws of Government are very stringent and any violation of these laws amounts to punishment. Economic laws, on the other hand, are applicable, only if certain conditions are satisfied.
2. *Economic Laws are Merely the Statements of Tendencies:* These are based on the tendencies of humans who behave in a particular way to a given phenomenon. This is the expected behaviour. This expected behaviour, however may not be found, for certain reasons. This leads to unpredictable character of economic laws. Certainty is one thing, which is not guaranteed with regard to economic laws.
3. *Economic Laws are Hypothetical:* These hold good under the assumption of a number of things. Economic laws are characterized by the phrase *ceteris paribus* (other factors are held constant).
4. *Economic Laws are Positive but not Normative:* They only describe the economic phenomenon but do not prescribe how it should be.
5. *Some Economic Laws are Axiomatic in Character:* It means that they are self-evident as that of law of diminishing marginal utility and generalizations drawn are universally valid.
6. *Economic Laws Lack Exactness of the Laws of Science:* This prompted Marshall to compare the economic laws to the laws of tides rather than the simple laws of gravitation.

# CHAPTER 2

# Basic Concepts in Economics

Basic concepts in the field of economics are discussed below:

## GOOD

Anything that satisfies a human want is called good. Goods are tangible and material outcome of production. *Examples*: Foodgrains, pulses, oilseeds, machinery, implements, seeds, fertilizers, cloth, book, pen, *etc.*

## SERVICE

A service is any act or performance that one party can offer to another *i.e.*, essentially intangible and does not result in the ownership of any thing. Services are intangible, non-material, inseparable, variable and perishable. The services rendered by doctors, teachers, lawyers, engineers, labourers, *etc.*, are the examples.

## CLASSIFICATION OF GOODS

Goods are categorized based on four criteria *viz.*, supply, transferability, consumption, and durability as given below:

### Based on Supply

Based on supply, goods are classified into economic goods and free goods. Economic goods are those goods, which are produced through human efforts and are to be purchased at a given price. Supply is less than demand. They have value in use and value in exchange. Buildings, machinery, furniture and a host of other goods of our daily use are the economic goods. Free goods are the free gifts of nature. Their supply is more than demand and one can get to the extent they need. No efforts are needed to be putforth by humans to secure free goods. Since these are freely available in nature, no price needs to be paid. They have value in use but no value in exchange. Examples are, air, sunshine, rainfall, *etc.*

Though a clear distinction is made between free goods and economic goods based on their distinct characteristic features, the distinction between the two is lost under certain situations. Water, which is a free good near the canals and rivers for the consumers, becomes an economic good in the water scarcity places. Similarly, sand which is a free good in riverbeds becomes an economic good in the places of house construction activities.

## Based on Transferability

As given by Marshall, goods are classified as follows based on transferability.

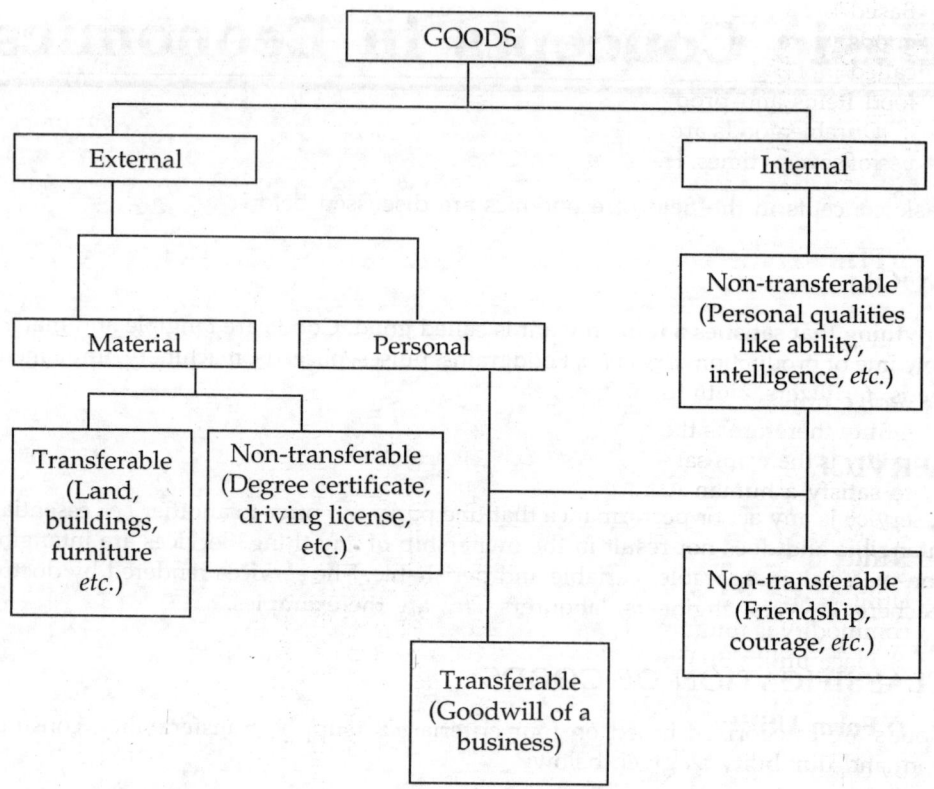

## Based on Consumption

Based on consumption, goods are classified into consumer goods and producer goods.

### Consumer Goods

Consumer goods are those from which consumers directly derive the satisfaction using the goods. These are otherwise known as goods of first order in view of their ability to give direct satisfaction. Food, cosmetics, clothes, books, pens, *etc.*, are the examples of consumer goods.

### Producer Goods

Producer goods are those that help to produce other goods. They can be used by consumers or producers or both, because it depends upon how the good is used. From the consumers' point of view they give satisfaction indirectly. The examples are machines, factory buildings, raw material, *etc.* The chance of a producer good to become a consumer good is possible based on its usage. For instance, electricity when used at home, it becomes a consumer good and the same becomes a producer good when used in industry.

## Based on Durability

Based on durability goods are classified into mono period goods and poly period goods (durable goods). Those goods which are used only once to satisfy a need are called mono period goods. They cease to exist once their use was over. *Examples*: All food items and productive resources like seeds, fertilizers, *etc.*

Durable goods are those goods, which are used time and again. They can be made use of several times. Here the relevant examples are, machinery, implements, buildings, *etc.*

# UTILITY

In the process of economic activity, consumers exhibit their desire to possess a good and/or service. Desire for a good arises in view of the utility the consumers derive from the purchase of that particular good. If the consumers feel that they can not derive utility from the possession of a good, they pay least attention for the same. Utility therefore is the capacity of a good that satisfies a human want. In other words, utility is the want satisfying power of a good. Also it means the power of a commodity to satisfy a human want.

## Kinds or Types of Utility

Utilities are classified into different kinds. A particular kind or type of utility for a commodity is found in a particular situation. The kinds of utilities are 1) Form utility 2) Place utility 3) Time utility and 4) Possession utility.

## 1. Form Utility

By changing the form of a good, greater utility is created. It does not mean that before change of form of good, there was no utility. It means that change in the form offers greater utility to the good. *Examples*: Processing of paddy into rice, wheat into flour, coffee seeds into coffee powder, butter into ghee, cotton into cloth, *etc.* The goods or commodities, which assume new forms thus will have certainly high utility and high value.

## 2. Place Utility

By virtue of its position in an area, the commodity will have different utilities. Such utility is called place utility. Spatial movement of the goods *i.e.,* moving a good from one place or market to another place or market increases its utility. Mostly the goods are transported from the places of production down to the places of consumption and also from the places of surplus production to the places of scarcity. For example, cement is transported from the places of production to the areas of consumption, similarly fertilizers are made available to the farmers by transporting them from the factories. Apples from Himachal Pradesh, the area of abundant production are transported to Southern and Western parts of the country (non-producing areas), thereby increasing the utility of apples. For scarce commodities in two different places depending upon demand and supply, utilities vary. Similarly, at production centres and consumption centres, utilities change for different commodities. Production does not necessarily mean conversion of raw materials into products. It also implies creation of place utility, since goods are made available in the areas of scarcity.

## 3. Time Utility

Any time lag between production and consumption of commodities creates time utility. Through storage over time, greater utility is created for the products. Storage helps to create time utility. Agricultural production being season bound with consumption spreading throughout the year, the commodities need to be made available to the consumers as and when required. The storage helps us to perform this function. Agricultural commodities like paddy, wheat, oilseeds, pulses, *etc.*, are stored to make them available for the regular use of consumers throughout the year. In doing so, higher utility for commodity is imparted which we call, time utility.

## 4. Possession Utility

Commodities in the transaction process, change the hands from one person to another person. Commodities in the hands of producers have some utility and by the time they reach consumers through the traders their utility is increased. Such utility due to possession or transfer of ownership of the commodity is called, possession utility. For example, paddy in the hands of producers, *i.e.,* farmers are having less utility compared to that of consumer in the form of rice. Similarly any other commodities like fruits, vegetables, livestock products, *etc.,* would have higher utility when these goods change hands from farmers. Such a rise in the utility of commodities due to possession is called possession utility.

## Characteristics of Utility

1. *Utility is Subjective:* Utility is not satisfaction by itself, it is subjective to the interest of an individual. It depends on the individual's frame of mind. Hence a given commodity need not bring the same utility to all the consumers. Utility varies from person to person. For example, non-vegetarian food to a vegetarian, alcohol to a teetotaler, motorcycle to a child, *etc.,* give zero utility. Utility varies with a regular smoker when compared with an occasional smoker and similarly for a regular non-vegetarian to an occasional non-vegetarian. Utilities derived by the occasional users of a good are greater than those of regular users.

2. *Utility Varies with Purpose:* Depending upon the purpose for which a commodity is used, utility of the same varies. For example, utility derived from water varies from its use as drinking water against its use as irrigation or for power generation or for industrial use, *etc.* Similarly, rice bran as cattle feed against using it for extracting oil and cotton for beds against cloth, *etc.,* have different utilities.

3. *Utility Varies with Time:* A particular commodity gives different utilities for the same person in different time periods. Cool drinks and ice-creams provide greater utility to the same individual in summer than in winter. Availability of tractor gives more utility to the same farmer during sowing season against off-season. The wine stored over time in scientific warehouses would have higher utilities and fetch higher price.

4. *Utility Varies with Ownership:* Ownership of a good creates far greater utility from a good than that when it is hired. For a farmer, land ownership brings higher

utility over leasing in the land. Similarly, owing farm machinery offers greater utility than hiring in, for a farmer.

5. *Utility Need not be Synonymous with Pleasure:* Utility derived from a commodity need not be associated always with pleasure for the consumer. Consuming items of our taste brings in the utility one expects. Similarly, buying an asset also gives the desired utility. These are all the pleasures enjoyed by the consumer from the goods. As against this, consumption of a good is a painful feeling for an individual though it possesses utility. For example, for a sick man, bitter medicines are difficult to swallow. Similarly, restricted dietary items on the health grounds are certainly not those which bring pleasure, though they possess utility.

6. *Utility does not Mean Satisfaction:* Utility is not satisfaction by itself. Utility is the quality of a good by virtue of which it gives satisfaction to an individual. The question of obtaining satisfaction from a good depends on the consumer's choice for the same.

## VALUE

It is the capacity of a good to command other things in exchange. It is the rate of exchangeability. In reality, value can be used as value-in-use as well as value-in-exchange. Free goods have value-in-use and not value-in-exchange. Economic goods possess both value-in-use and value-in-exchange. But in economics always the term, value means value in exchange.

**Attributes of Value:** Following are the attributes, which make the goods possess value.
1. Goods must possess utility
2. Goods must be scarce; and
3. Goods must be transferable or marketable.

## PRICE

When the value of a good is expressed in terms of money, it is called price. Price expresses value in terms of money.

## WEALTH

In ordinary language wealth is synonymously used to mean riches. Wealthy man is the one who is blessed with riches. But in the economics, the term, wealth conveys a different meaning. Anything, which has value is wealth. When one says value, only economic goods possess value. Consequently economic goods represent wealth.

*Wealth consists of all potentially exchangeable means of satisfying human wants needs (J.M. Keynes).*

### Characteristics of Wealth

1. *Wealth should Possess Utility:* Wealth must be capable of satisfying human wants.
2. *Wealth Must be Scarce:* Scarcity is the binding factor for exchange. Economic goods are exchangeable. Since wealth represents economic goods, it must be scarce.
3. *Wealth Must be Transferable:* Transferability implies changing the ownership of a good from one to another. It is nothing but exchangeability.
4. *Wealth Must be External to Person:* This quality of wealth enables for exchange.

Individual competencies do not represent wealth, for these form a source of wealth. Skill of a surgeon, architect, artist, *etc.*, are individual qualities which help them to earn wealth. These do not satisfy all the characteristics of wealth.

**Types of Wealth:** These are 1) Individual wealth 2) Social wealth 3) National wealth 4) Cosmopolitan wealth and 5) Negative wealth.

1. *Individual Wealth:* Individual wealth consists of all tangible and intangible possessions of the individuals, besides loans due to them. Land, buildings, vehicles, shares, bonds, deposits, commodities for sale, cash, *etc.*, are the tangible possessions. Goodwill of a business, copyrights, patents (non-material external goods), *etc.*, are intangible possessions. From the total value of all these possessions, loans owed to others are to be deducted to arrive at the individual wealth.

2. *Social wealth:* It is wealth, which is collectively used by all the people in a nation. Hence it is also termed as collective wealth or communal wealth. Common properties of the community like roads, railways, public parks, libraries, Government hospitals, Government colleges, *etc.*, represent the social wealth.

3. *National Wealth:* National wealth is an aggregate of all individuals' wealth and collective wealth of the country. To this amount the loans due to people and loans due to the Government from foreign countries should be added. From this amount so arrived at, the debts that the people and the Government owed to foreign countries should be deducted. Economists like Marshall feel the inclusion of free gift of nature like mountains, rivers, *etc.*, in national wealth.

4. *Cosmopolitan Wealth:* Here the wealth belongs to the world but not to one country. Rivers, oceans, forests, *etc.*, extending over the nations are the examples of cosmopolitan wealth. It is the sum total of wealth of all nations.

5. *Negative Wealth:* Negative wealth is the exclusive debts owed by the individuals and the nation.

## Wealth and Welfare: A Comparison

Wealth and welfare apparently appear to carry the same meaning, but they are two different aspects having relationship. Wealth as defined earlier, is used in the sense of economic goods. Free goods have no place while representing wealth. Once we say economic goods the question of good and bad from social point of view is ignored. As long as they are economic goods, it hardly matters whether they are harmless or harmful. Therefore, economic goods represent both harmless and harmful goods. Wealth is a path for welfare, as most of our desires are satisfied by means of wealth.

Welfare is the well being of individual or community. It refers to the condition of mind. Here, any good whether it is free or economic, is counted as long as it causes well being of an individual or community. Both economic and free goods lead to welfare. Welfare is one, which is never associated with any unacceptable things. Welfare is subjective as it differs from individual to individual.

# CHAPTER 3

# Consumer Behaviour and Demand

The theory of demand begins with the analysis of the behaviour of consumer, since the market demand is the summation of the individual consumers. Consumption decisions are made by individual or household groups. Consumer is assumed to be rational. Based on his income and the market prices of various goods and services, he plans spending of his income with the aim of attaining maximum possible utility or satisfaction.

## HUMAN WANTS

Economic activity of the humans is directed towards the satisfaction of their wants. The fulfillment of human wants generally is considered as the goal of economic activity. Wants of different sections of the people in the society *i.e.*, wage earners, salaried class, businessmen, political leaders, *etc.*, may vary. Similarly importance attached by each group for a given want also varies.

### Characteristics of Human Wants

1. *Wants are Unlimited:* Wants are continuously cropping up in the minds of the humans. If one want is satisfied, immediately another want emerges. They go on multiplying. There is no end to human desires, hence unlimited.
2. *Wants Recur:* If one want is satisfied at a point of time, the same want again repeats in the future. It can be satisfied for that moment only and it is again ready to be fulfilled. Wants like hunger, recreation, *etc.*, need to be satisfied time and again because of their recurrence.
3. *A given Want is Satiable:* Though human wants are unlimited, a given want at particular point of time is completely satisfied. For example, for a man who is thirsty, adequate potable water completely satisfies his thirst.
4. *Wants are Complimentary:* Goods are required in pairs or groups to satisfy human wants. For example, bread and butter, tractor and driver, fertilizer and irrigation water, pen and paper, *etc.*, can serve as complimentary goods to satisfy human wants.
5. *Wants are Competitive:* Wants are unlimited but means to satisfy them are limited. Wants compete among themselves and hence given preference. This compels the

consumer to choose the most urgent wants keeping in view of the limited income, and

6. *Wants have Alternative means:* There is more than one way of satisfying a certain want. A given want can be satisfied with alternative goods. If a man is thirsty, he can be satisfied with water or cool drink or coconut water, *etc.* Similarly, hunger can be satisfied with rice or bread or fruits, *etc.* The desire to have entertainment may be satisfied by watching a movie in a cinema theatre or television.

## Origin of Wants

The basic need of humans *i.e.,* food is the most obvious reason for the origination of wants. Two other desires which also arise from necessity are shelter and clothing. Both are essential for the humans to protect themselves from the vagaries of weather. Individual's habits and tastes give rise to demand for wide variety of goods. The customs of the society in which we live in, influence the origination of wants. Differences in age of the individuals, climate, education, social position, *etc.,* will give rise to a variety of wants. Finally the satisfaction of existing wants results in the generation of new wants. Thus as the economy develops, new wants may crop up adding to the existing wants.

## Classifications of Wants

Wants are broadly divided into three categories *viz.,* necessaries, comforts and luxuries.

### 1. Necessaries

Necessaries are those wants which must be satisfied. The goods which are used to satisfy basic needs of humans are called necessaries. They are further classified into necessaries of existence, necessaries of efficiency and conventional necessaries.

a) *Necessaries of Existence:* These are the necessaries which are essential for living. Human existence is not possible without fulfilling the necessaries of existence. *Examples:* Food, water, clothes, shelter, *etc.*

b) *Necessaries of Efficiency:* These are not as essential as those of necessaries of existence, but at the same time essential in improving the efficiency of an individual. *Examples*: Nutritious diet, table and chair to a student, class rooms with good ventilation, *etc.*

c) *Conventional Necessaries:* These are the necessaries, which arise out of customs or habits. The customs prevailing in a society influence the individuals to follow them. *Examples*: Customs like celebration of functions and habits like smoking, drinking, gambling, *etc.*

### 2. Comforts

Comforts are those which fall between necessaries and luxuries. Man is not satisfied with fulfilling necessaries only. The comforts also increase the efficiency. *Examples*: Cushion chairs in a classroom, revolving chair in the saloons, fans in house/office, *etc.*

### 3. Luxuries

Luxuries are those which satisfy superfluous wants of individuals. These are neither essential for life nor increase the efficiency. Luxuries represent wasteful expenditure of the individuals. Luxuries are further classified into harmless luxuries, harmful luxuries and defense luxuries. Harmless luxuries are those, the expenditure on which will not cause any harm to the individuals. For example, well furnished bungalow, expensive food habits, *etc.*, fall under harmless luxuries. Harmful luxuries on the other hand are injurious to the health of the users. *Examples*: Alcohol, smoking, *etc.*

Defence luxuries are those which protect the users during the period of crisis. Expenditure on gold ornaments, jewellery, *etc.* though appears to be luxurious, at the same time it would help the individuals during the periods of crisis.

Though demarcation is made among necessaries, comforts and luxuries, in reality these are interchangeable. What is necessary for an individual may turn out to be a comfort for another person and luxury for someone else. Air travel is a luxury for a common man but an absolutely necessary for a busy business executive. Due to development of the economy the wants and desires may change over time. For example TV and telephone were once considered as luxuries, now because of their abundance and less price, we are using them as necessary goods. Changing consumerism is going to give a new dimension to the human wants.

## LAW OF DIMINISHING MARGINAL UTILITY

Two techniques are used in the analysis of consumer's behaviour. *viz.*, (1) utility analysis or the Marshallian approach through which utility can be measured and (2) the indifference curve technique. The former approach is called *'cardinal approach'* and the later is called *'ordinal approach'*. The law of diminishing marginal utility, the law of equimarginal utility, consumer's surplus, *etc.*, fall under cardinal approach.

Having understood the concept of utility, an attempt is made here to examine a related law of utility *viz.*, the law of diminishing marginal utility. This law was initially formulated by German economist H.H. Gossen. The law says that more a commodity that an individual possesses, the less is the utility that is derived from it. More emphatically, the law is based on the observation that each successive unit of a commodity brings in lesser and lesser utility until the point of satiation is reached. After the point of satiation is reached, any further consumption results in disutility *i.e.*, the consumer derives negative satisfaction. More and more we have a thing, the less and less we want it. It is evident through this law that though some goods have greater utility than others, subsequent units of the same good, give diminishing satisfaction. Marshall defined the law as *"The additional benefit a person derives from a given increase of his stock of any thing diminishes with the growth of the stock he has"*.

### Assumptions of the Law

1. The units of commodity consumed in succession should be identical in size, colour, taste, freshness, maturity, *etc.* Homogeneity of the commodity is the basic requirement of the law.
2. Consumption of the units of the commodity should be continuous without interval, *i.e.*, there should not be any gap during the consumption process. The law becomes in-operative, if such a condition is violated.
3. During this particular activity, it is also assumed that tastes, preferences, incomes, *etc.*, remain unchanged.

## Explanation of the Law

The law is explained through a table and also with the help of a graph. Before looking into the law, two concepts need to be understood *viz.*, marginal utility and total utility.

*Marginal Utility:* It is the additional utility derived by an individual, by the consumption of one more unit of the commodity.

*Total Utility:* It is the sum amount of satisfaction derived from the consumption of different units of the commodity.

The case of an individual consuming apples at a point of time and the related utilities are presented in Table 3.1.

**TABLE 3.1 Utility Schedule for Apples.**

| Number of apples | Total utility | Marginal utility |
|---|---|---|
| 0 | 0 | — |
| 1 | 20 | 20 |
| 2 | 36 | 16 |
| 3 | 48 | 12 |
| 4 | 56 | 8 |
| 5 | 60 | 4 |
| 6 | 60 | 0 |
| 7 | 56 | -4 |
| 8 | 50 | -6 |

As per the table, consumption of first apple gives a marginal utility of 20 units, while the second 16 units, the third 12 units and so on. The consumption of successive units of apples results in declining marginal utilities but remain positive till marginal utility became zero. Any further consumption of apples makes the marginal utility negative.

With regard to the total utility, it goes on increasing, right from the consumption of first unit till the point of satiety (maximum satisfaction) is reached. At this point, the total utility is maximum. This is attained at the 6$^{th}$ unit of the apple. Consumption of further units of apples results in declining total utility. The relation between marginal utility and total utility is that as the marginal utility is falling, total utility is increasing at a decreasing rate. When marginal utility is zero the total utility is maximum, and finally the negative marginal utility results in declining total utility.

The graphical representation of the law in Figure 3.1 indicates that marginal utility is falling throughout, while total utility is increasing at a decreasing rate till marginal utility becomes zero. Total utility starts falling, when marginal utility becomes negative.

## Importance of the Law

It is the basic law on which various other laws like law of demand, law of substitution, *etc.*, rely upon. It is something like a pendant. Marshall has built up the theory of taxation and public expenditure by applying this law to money.

## Exceptions to the Law

Inspite of it being a basic law of economics, it has limitations in its application. This law does not apply for those persons with abnormal qualities *i.e.*, those who are misers and those having drinking, gambling habits, *etc.* Similarly, this law does not apply for persons who enjoy their hobbies.

## LAW OF EQUI-MARGINAL UTILITY

This law is also known as the law of substitution. *"The law implies that if a person has a thing which he can put to several uses, he will distribute it between those uses in such a way that it has the same marginal utility in all"* (Marshall). A consumer generally is confronted with the problem of buying from among several goods and services, given his limited income. He is left with the choice making, regarding purchase of commodities and their quantities so that the purchases that are decided upon ensure him maximum satisfaction. Here, the consumer aims at maximizing total utility by consuming possible goods and services given the income constraint. In this process the consumer substitutes the goods having greater utility for those which have lesser utility. This process is continued till the marginal utilities of the commodities purchased are equalized. Hence the name, the law of equimarginal utility.

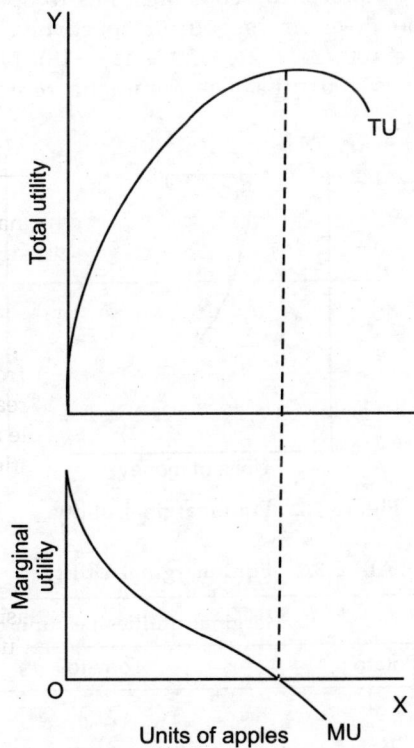

**Figure 3.1** Diminishing Marginal Utility.

## Assumptions

Before explaining the law, the following assumptions are made.
1. The consumer behaves rationally.
2. He has full knowledge about the commodities, their attributes, prices, *etc.*, in the market.

3. Utility is measurable cardinally in terms of utils.

4. Commodities that are choosen are divisible and substitutable.

### Explanation of the Law

Given the income constraint, the consumer makes prudent decisions in his purchases such that the allocation so made ensures him maximum satisfaction. Let us assume that the consumer has got Rs. 25 to spend. He has the options of spending this amount on three vegetables *viz.*, potato, tomato and ridge gourd. The marginal utilities that are derived from the consumption of these vegetables and the amounts of money spent are presented in Table 3.2. The marginal utilities are derived from the consumption of three vegetables by spending a unit of money (each unit of money is equal to Rs. 5). First unit of Rs. 5 on potato gives a marginal utility of 19 utils, second unit 16 utils, third 14 utils and so on. Now to maximize the satisfaction from the three vegetables the consumer has to spend Rs. 25 in such a way that the marginal utility of the last unit of money is equal in all uses. Given the marginal utilities derived from the three vegetables, the consumer has to spend first three units on tomato, one unit each on potato and ridgegourd respectively. The total utility through this combination would be 100 (22 + 21 + 20 + 19 + 18). No other combination of vegetables gives as high as 100 utils. Diagrammatic representation is shown in Figure 3.2.

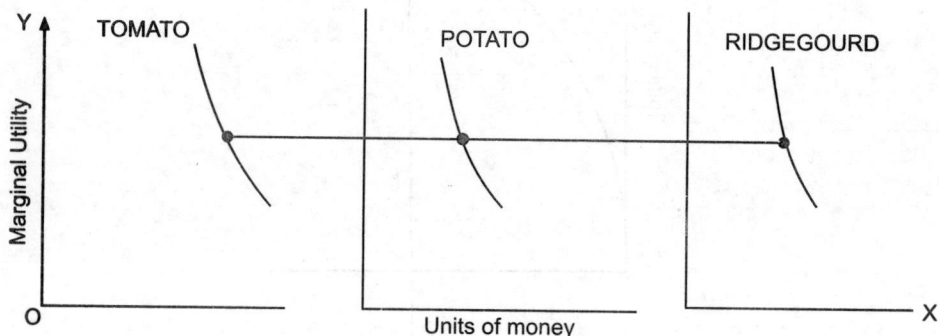

**Figure 3.2**  Equi-marginal utility.

**TABLE 3.2  Equi-marginal Utility.**

| No. of units (Rs. 5 per unit) | Marginal utilities per unit of Rs. 5 | | |
| --- | --- | --- | --- |
| | Potato | Tomato | Ridge gourd |
| 1 | 19 | 22 | 18* |
| 2 | 16 | 21 | 17 |
| 3 | 14 | 20 | 15 |
| 4 | 10 | 16 | 12 |
| 5 | 5 | 14 | 11 |

This combination is going to hold good till some changes occur either in the prices or his income or his tastes. This law is also called as the principle of proportionality, as the consumer allocates his expenditure in such a way that the marginal utilities of the goods purchased would be in proportion to their prices. Consumer attaining equilibrium by spending his limited money is shown below:

$$\frac{\text{MU of X}}{\text{Price of X}} = \frac{\text{MU of Y}}{\text{Price of Y}} = \frac{\text{MU of Z}}{\text{Price of Z}} = K$$

### Practical Importance of Law of Equi-marginal Utility

1. *Consumption:* A rational consumer follows this law, while planning his expenditure. He spends in such a way that marginal utility derived from each unit of money gives nearly equal utility in the various goods he purchases.
2. *Production:* A rational producer allocates his limited resources among various possible enterprises in such a way that the marginal value product derived from each unit of resource on various enterprises is the same.
3. *Marketing:* The consumer should keep in mind that marginal utility of the commodity and price of the commodities should be equal in purchasing the commodities from the markets. Thus this law guides the consumers to spend the given amount efficiently on different goods which provide utilities.
4. *Distribution:* The share of each factor of production is determined on the basis of marginal value productivity.
5. *Prices:* When the price of a commodity goes up in view of shortfall in supply, consumer prefers that commodity which is relatively less scarce. This preference of consumer brings down the price of the commodity.

### Limitations of the Law

It needs a very careful scrutiny by the consumer regarding prices of various commodities and their substitutes. More often, the expenditure pattern is influenced by the habits, except in the case of very high priced goods for which certain amount of thinking goes before taking buying decisions. Another limitation for the operation of the law is indivisibility of certain goods which would not permit the equalization of marginal utilities. Further, it is possible that buying decisions are influenced by advertisement in which case rationality criterion may not be followed.

### Limitations of the Cardinal Utility Approach

The cardinal approach of Marshall has been criticized by modern economists. Utility being a psychological phenomenon, it is not amenable to measurement by numbers, weights, *etc*. The marginal utility of money is taken to be constant by Marshall which is not the common experience of the consumer, as the marginal utility of money would be more when he is left with less and less money. Another shortcoming is that, it does not help to separate the income and substitution effects.

## CONSUMER SURPLUS

The concept of consumer surplus was first introduced by Alfred Marshall and according to him, it is measured in monetary units and is equal to the difference between the amount of money that a consumer actually pays for buying certain quantity of commodity and the amount he would be willing to pay for the same quantity rather than go without it. When a consumer is prepared to buy a commodity, he always calculates the utility he is going to derive from its consumption. Every rational consumer compares the utility he derives from the consumption of a commodity, against the price he has to pay. If the utility is more than the price paid, he prefers it and if it is *vice-versa*, he does not buy the same good. The surplus of utility he derives is consumer's surplus. In a nutshell, consumer's surplus is the difference between what the consumer is willing to pay and what he actually pays.

## Assumptions of the Consumer Surplus

1. Marginal utility of money for the consumer is assumed to be the same throughout the process of exchange.
2. Commodity does not have substitutes
3. In the market at the given point of time there are no differences of income, tastes, preferences and fashions among the consumers, and
4. Each commodity is considered independent of others.

## Explanation of the Consumer Surplus

In explaining the consumer's surplus, the law of diminishing marginal utility is recapitulated here. The particulars presented in Table 3.3 reveal that when a person prefers for purchase of mangoes at a price of Rs. 5 each, the marginal utility derived is equivalent to Rs. 20 from the first mango, Rs. 15 from the second mango, Rs. 10 from the third mango and Rs. 5 from the fourth mango, at which marginal utility and price are equal. Consumer will not go beyond the 4th mango because at this level the price of a mango is equal to marginal utility obtained by the consumer.

From the consumption of first three mangoes, the consumer enjoys a surplus utility of Rs. 30. This is the total consumer's surplus. There is no consumer's surplus for the fourth unit. The diagrammatic representation of consumer's surplus is shown in Figure 3.3.

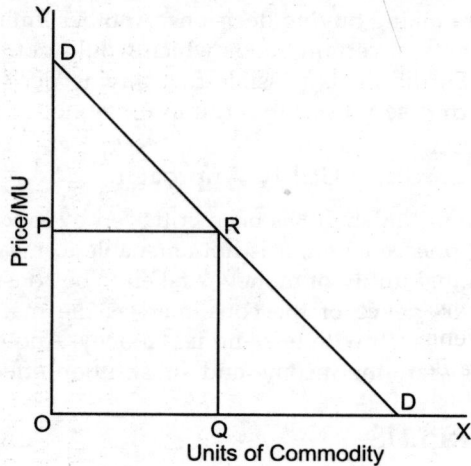

**Figure 3.3**  Consumer's surplus.

**TABLE 3.3  Estimation of Consumer Surplus.**

| Units of mangoes | Marginal utility (price willing to pay) | Total utility | Market price (Rs.) | Consumer's surplus |
|---|---|---|---|---|
| 1 | 20 | 20 | 5 | 15 |
| 2 | 15 | 35 | 5 | 10 |
| 3 | 10 | 45 | 5 | 5 |
| 4 | 5 | 50 | 5 | 0 |

Along X-axis units of commodity are measured and along Y-axis marginal utility in terms of money is measured, DD is the demand curve. At OP Price the number of mangoes purchased is OQ. At price OP, the total amount paid for OQ units of commodity (mango) is OP × OQ and in graph the area is represented as OQRP. But the actual amount the consumer is willing to pay for OQ units is OQRD. The consumer's surplus therefore is OQRD – OQRP *i.e.*, DRP. Increase in price causes a reduction in consumer's surplus, while a fall in price leads to an increase in consumer's surplus.

## Difficulties in Measuring Consumer Surplus

1. A complete list of demand prices is not available. What we know as consumers is the knowledge of part of demand schedule and what we are prepared to pay for certain commodities is only the guess work.
2. For necessaries of life and conventional necessaries, the consumer's surplus is immeasurable and indefinite.
3. Variation in the incomes of the consumers also poses a problem. Whether one is rich or poor, he has to pay the same price for an unit of commodity. Naturally a person with low means has to make a greater sacrifice to buy a commodity. This situation of the consumers makes it difficult to measure the consumer's surplus.
4. Marginal utility of money changes. As we keep on spending the money on a commodity, we will be left with less and less of money. The marginal utility of each unit of money therefore increases, as we keep on spending money in succession. If we cannot consider this phenomenon, the estimation of consumer's surplus does not stand to rationality and reasoning and
5. The presence of substitutes always makes the consumer to go for X commodity instead of Y commodity, if its price rises. In the absence of substitutes, the utility of X commodity would have been much greater.

## Importance of Consumer Surplus

1. *Conjunctural Importance:* When the people enjoy larger consumer's surplus, it does not indicate that they are better off. Thus it serves as an index of economic betterment.
2. *Useful to the Monopolist:* The monopolist can freely raise the prices of the goods if they bring in higher consumer's surplus, without any fear of foregoing the sales.
3. *Helps in Public Finance and Taxation:* More taxes can be imposed by the Government to get more revenue, on those goods for which consumer's surplus is high.
4. *Helps to Measure Benefits from International Trade:* In the international trade, those commodities which are cheaper in the foreign markets are imported. Before their imports the consumers were paying higher price. With the availability of imported goods which are cheaper, the consumers get surplus of satisfaction. Greater surplus indicates, larger benefits from international trade.

# CONSUMER SURPLUS AND PRODUCER SURPLUS— ALGEBRAIC APPROACH

$$Q_d = 26 - 2p \qquad (1.1)$$

$$Q_s = 6 + 2p \qquad (1.2)$$

$$\overline{P} = \frac{26+6}{2+2} = \frac{32}{4} = 8$$

Here $Q_d$ and $Q_s$ denote demand and supply functions

$$\overline{Q_d} = 26-6 = 10; \quad \overline{Q_s} = -6 + 16 = 10$$

$\overline{P}$ is equilibrium price $\overline{Q_d}$ and $\overline{Q_s}$ are the equilibrium demand and equilibrium supply. The assumed P values given in Table 3.4 when substituted in equation (1.1) and (1.2) we get values of quantity demanded and quantity supplied. Using the values of demand and supply, the demand and supply curves are drawn to estimate consumer's surplus and producer's surplus (Figure 3.4).

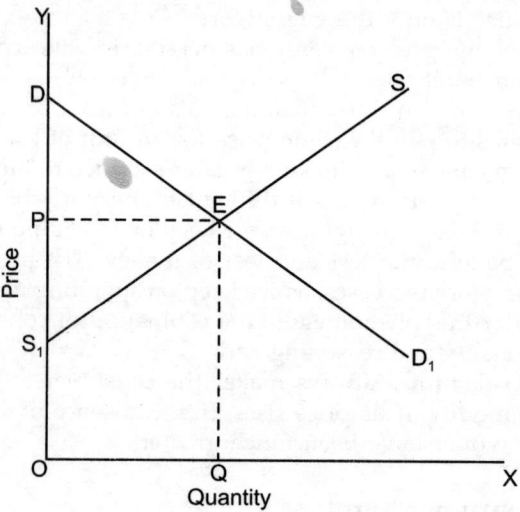

**Figure 3.4** Consumer's surplus and producer's surplus.

**TABLE 3.4 Demand and Supply and Price of Commodity.**

| $Q_d$ | $Q_s$ | P |
|---|---|---|
| 20 | 0 | 3 |
| 14 | 6 | 6 |
| 10 | 10 | 8 |
| 6 | 14 | 10 |
| 0 | 20 | 13 |

We know from economic theory that consumer's surplus is the triangle area of DE$\overline{P}$ and producer's surplus is the area of triangle $S_1$E$\overline{P}$. This means consumer's surplus is the area under demand curve less rectangle area O$\overline{Q}$E$\overline{P}$.

Producer's surplus is the area above supply curve and below equilibrium price. Algebraically this is obtained by subtracting the area O$\overline{Q}$E$S_1$, from rectangle area O$\overline{Q}$E$\overline{P}$.

Now to work out producer's surplus and consumer's surplus we will have to follow the steps below:

1. Let us assume the demand and supply functions of a commodity as

$$Q_d = 26 - 2p \qquad (1.1)$$
$$Q_s = -6 + 2p \qquad (1.2)$$

2. The inverse demand and supply functions can be worked out as

$$P = -\frac{D}{2} + \frac{26}{2}$$

$$P = \frac{S}{2} + \frac{6}{2}$$

For demand, $P = 13 - \frac{1}{2} D$                   (1.3)
For supply, $P = \; 3 \; + \frac{1}{2} S$                 (1.4)

3. $Q_d = Q_s$                                                  (1.5)

Equation (1.5) is market-clearing condition. Using this equation we can work out equilibrium price, equilibrium demand and equilibrium supply.

$$26-2P = -6 + 2p$$

$$\overline{P} = \frac{26+6}{2+2} = \frac{32}{4} = 8$$

Substitute $\overline{P} = 8$ in equation (1.1) and (1.2) we see equilibrium demand

$$= 26 - 2 \times 8 = 10 \; ; \text{Equilibrium supply} = -6 + (2 \times 8) = 10$$

4. Work out consumer's surplus as shown below:
   a) We know that by taking definite integral of function with respect to real numbers (here D for demand and S for supply are the real numbers), we can find the area under demand and supply curve. We require to define ordinates of demand function and supply function. The ordinates of demand function are 0 and D *i.e.*, 0 and 10. The ordinates of supply function are 0 and S *i.e.*, 0 and 10.
   b) Now let us proceed to work out consumer's surplus at first. This is done by taking out definite integral of inverse demand function with respect to D and then subtracting rectangle area $\overline{OQEP}$ from area under demand curve.

$$\text{C.S.} = \int_0^{10} (13 - 1/2)^{dD} - \overline{P}.\overline{Q}_D$$

$$= \left[ (13D - \frac{1}{2} \times \frac{1}{2} D^2) \right]_0^{10} - 8 \times 10$$

$$= [13 \,(10) - 1/4 \,(10)^2 \,] - 0 - 80$$

$$= 130 - 25 - 80$$

$$= \text{Rs. } 25$$

∴ The consumer's surplus is Rs. 25

**Producer Surplus (P S):** It is the difference between market price and minimum variable cost per unit of output.

This is worked out by taking definite integral of inverse supply function with ordinates 0 and 10 and this value is subtracted from rectangle area, $O\overline{Q}E\overline{P}$ i.e.,

$\overline{P} \times \overline{Q}$

$$PS = \overline{P} \bullet \overline{Q} - \int_0^{10} \left(3 + \frac{1}{2}S\right)^{d \bullet s}$$

$$= 8 \times 10 - \left[\left(3 \times S + \frac{1}{2} \times \frac{1}{2}S^2\right)\right]_0^{10}$$

$$= 80 - \left[3 \times 10 + \frac{1}{4}(10)^2\right]$$

$$= 80 - [30 + 25]$$

$$= 80 - 55 = Rs. 25$$

Thus we can find out the consumer surplus and producer surplus by using definite integral process with the given ordinates.

5. We now define economic surplus (ES) as

$$ES = CS + PS$$
$$= 25 + 25$$
$$= Rs. 50$$

# INDIFFERENCE CURVE

It is assumed that the consumer is rational and attempts to maximize utility (the want satisfying quality in a good) from his spendable income. It is a fact that consumers make choice from among many goods so as to maximize satisfaction from their limited money income. In view of the limitations of cardinal utility approach, modern economists developed an alternative technique viz., indifference curve technique to examine the consumer's behaviour.

The indifference curve analysis is based on the assumption of ordinal utility. It explains the behaviour of consumer in terms of his preferences or rankings for different combinations of two goods. Since the satisfaction derived from the consumption of various goods cannot be measured objectively, it is enough if the consumer indicates his preference for the various combinations of commodities.

An indifference curve is derived from an indifference schedule. A list of various combinations of two goods, arranged in such a way that the consumer is indifferent to the combinations preferring none of any other is called indifference schedule. An hypothetical indifference schedule is given in Table 3.5.

TABLE 3.5   Indifference Schedule.

| Combinations | X | Y |
|---|---|---|
| A | 1 | 20 |
| B | 2 | 15 |
| C | 3 | 11 |
| D | 4 | 8 |
| E | 5 | 6 |
| F | 6 | 5 |

In the above schedule, the consumer is indifferent whether he buys 'A' combination of 1 unit of X + 20 units of Y or 'D' combination of 4 units of X + 8 units of Y or any other combination. All the above combinations of X and Y commodities yield the same satisfaction.

If the various combinations are plotted on a graph and are joined by a line, it becomes indifference curve.

In Figure 3.5 the curve labelled I is an indifference curve. All combinations on the indifference curve are equally satisfactory to the consumer.

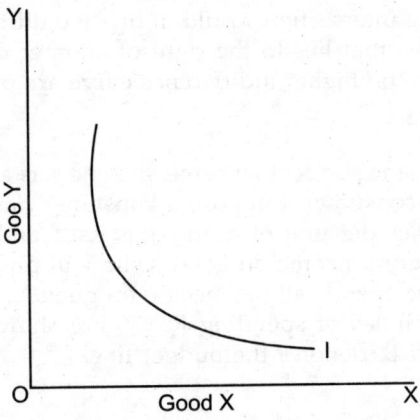

**Figure 3.5** Indifference curve.

An indifference curve is the locus of the combinations of two goods that are equally satisfactory to a consumer, or to which, the consumer is indifferent.

A diagram showing a set of indifference curves is called indifference map Figure 3.6. An indifference map is graphic device that shows the taste of an individual consumer. The combinations of two goods that lie on a higher indifference curve always preferred to those that lie on a curve below it.

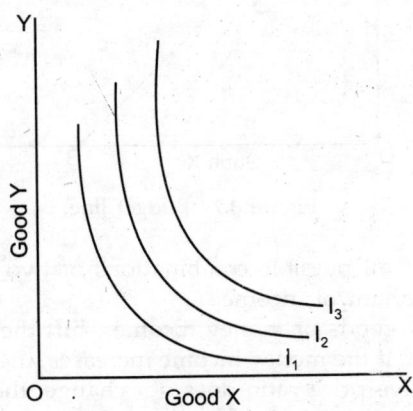

**Figure 3.6** Indifference map.

## Properties of Indifference Curves

1. *Convex to the Origin:* The absolute slope of an indifference curve declines as we move along the curve from left downwards to the right. This means that marginal rate of substitution of goods* is diminishing.
2. *Negatively Sloped:* An indifference curve has a negative slope which implies that if the quantity of one good increases, the quantity of other good must be decreased. This is to keep the consumer on the same level of satisfaction.
3. *Non-intersecting:* Indifference curves do not intersect with each other. If they intersect, then the point of intersection would imply two different levels of satisfaction.
4. An indifference curve that lies to the right of another denotes higher utility and combination of goods on higher indifference curve are preferred by the consumer.

## The Budget Line

Consumer choices are made subject to income. Income acts as a constraint to maximize the satisfaction by the consumer. Suppose a consumer has Rs. 50 to spend on two goods X and Y. The price per unit of X and Y is Rs. 5 and Rs. 2 respectively. If the consumer spends his entire income on good X, he will purchase 10 units of X good. On the other hand, if he spends all his income on good Y, he will purchase 25 units of Y. These two possibilities of spending Rs. 50 are shown in Figure 3.7. The line connecting point A and B, denotes the budget line.

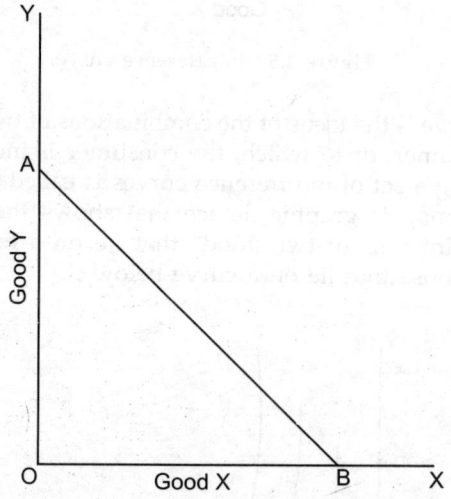

**Figure 3.7**  Budget line.

A budget line shows all possible combinations of two goods that could be purchased with a given amount of income.

Changes in prices of goods or money income, shift the budget line. Keeping the prices of goods constant, if the money income increases, the budget line shifts parallel to itself. Since the relative price ratio does not change, the slope of the budget line remains the same. The slope of the budget line indicates the price ratio of two goods.

---

* Marginal rate of substitution is the rate of exchangeability between the two goods which are equal'y preferred. It indicates the amount by which one good is reduced to gain another good by one unit. Marginal rate of substitution is negative and the sign is ignored.

## Consumer's Equilibrium

Having known the consumer's preference as indicated by indifference map and the information on income and prices of goods as given by the budget line, now we can determine that combination of goods, which maximizes the satisfaction. The rational consumer wants to choose the highest indifference curve given the budget constraint.

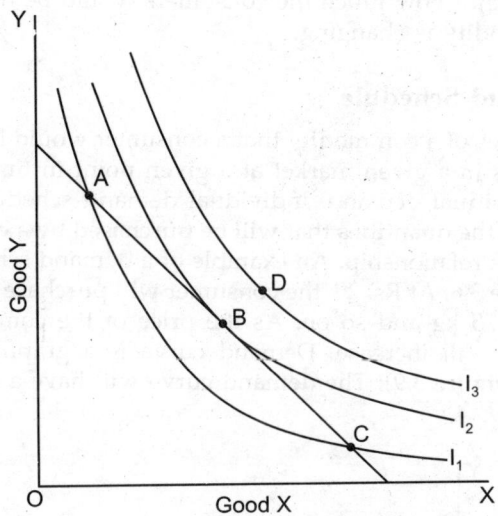

**Figure 3.8**  Consumer's equilibrium.

From Figure 3.8, we can observe that the optimal position of the consumer is at point B where the budget line is just tangent to the indifference curve $I_2$. Combination D is preferred to B but is beyond budget limitation. Combinations A and C are on the budget line, but are inferior to combination B, because they are on lower indifference curve ($I_1$). Any indifference curve to the right of $I_2$ is beyond the reach of the consumer's income and any to the left of $I_2$ is less preferable than $I_2$. Thus, the combination B indicated by the tangency of indifference curve $I_2$ and the budget line is the best combination of goods that the consumer can purchase. At the point of tangency, the slopes of indifference curve and the budget line are equal. Thus,

$$\frac{\Delta Y}{\Delta X} = \frac{P_x}{P_y}$$

## DEMAND

Demand normally means the desire or willingness for a good. But in economics simple desire or willingness for a good alone may not represent demand. Apart from the desire or willingness, consumer should be able to buy the good. Demand is therefore an effective desire. Thus, desire and ability to buy are the key components of demand. More specifically demand is defined as a schedule that shows the amounts of a product or service the consumers are willing and able to purchase at each price in a set of possible prices during some specified time in a specified market. Consumers like to possess a particular commodity but without ability to pay, in which case it is not a demand. Apart from these, two more requisites are essential, *viz.*, time and market, for demand is likely to vary over time and also among the markets. The conditions hence imposed are a specific time and a specific market to measure the demand. One condition that governs here is that all other factors, other than price influencing the demand remain the same. According to Bowden, demand means "propensity of the consumers to buy different quantities of a particular good at different unit prices". It indicates how much the consumers would be buying when the price per unit of a commodity is changing.

## Individual Demand Schedule

The various quantities of a commodity that a consumer would be willing to purchase at all possible prices in a given market at a given point in time, other things being equal is called individual demand. Individual demand schedule is merely a list of prices together with the quantities that will be purchased by a consumer. It is pairing of quantity and price relationship. An example of a demand schedule for a consumer is illustrated in Table 3.6. At Rs. 20, the consumer will purchase 0.25 kg, at Rs. 16, 0.50 kg and at Rs. 12, 0.75 kg and so on. As the price of the commodity decreases the quantity demanded will increase. Demand curve is a graphical representation of demand schedule (Figure 3.9). The demand curve will have a downward slope. DD is the demand curve.

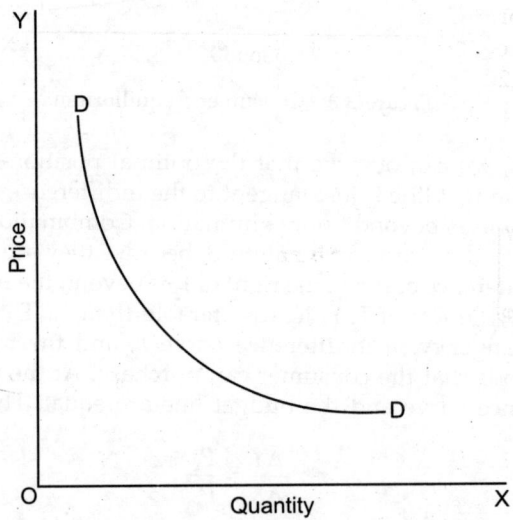

**Figure 3.9** Demand curve.

**TABLE 3.6 Hypothetical Demand Schedule for Tomatoes.**

| Price (in Rs./kg) | Quantity demanded per week (in kg) |
|---|---|
| 20 | 0.25 |
| 16 | 0.50 |
| 12 | 0.75 |
| 8 | 1.00 |
| 4 | 1.25 |
| 2 | 1.50 |

## Market Demand

Market demand is the sum of the demand of all the consumers in a market for a given commodity at a specific point of time. Assume that in a market there are only three consumers, viz., A, B and C, with individual demand schedules as presented in Table 3.7. A look at the table indicates different pairs of quantity and price relationship.

**TABLE 3.7 Market Demand Schedule for Tomatoes.**

| Price (Rs/kg) | Individual demand schedule/week (kg) | | | Market demand (A + B + C) |
|---|---|---|---|---|
| | A | B | C | |
| 20 | 0.25 | 0.5 | 0 | 0.75 |
| 16 | 0.50 | 1.0 | 0 | 1.50 |
| 12 | 0.75 | 1.5 | 0 | 2.25 |
| 8 | 1.00 | 2.0 | 1 | 4.00 |
| 4 | 1.25 | 2.5 | 2 | 5.75 |
| 2 | 1.50 | 3.0 | 3 | 7.50 |

For example, consumer C is not willing to buy tomatoes for any price higher than Rs. 8 kg. Given the individual demand schedules, market demand schedule can be worked out at each price level as indicated in the last column of the table. It is horizontal summation of the demand of individual consumer at each unit price. The market demand at Rs. 20 kg is 0.75 kg, at Rs. 16, 1.5 kg and at Rs. 12 it is 2.25 kg and so on. The particulars in Table 1.7 are graphically depicted to represent the demand curve of consumers A, B and C and the entire market (Figure 3.10). The market demand curve for tomatoes is drawn based on the first and last columns of the table.

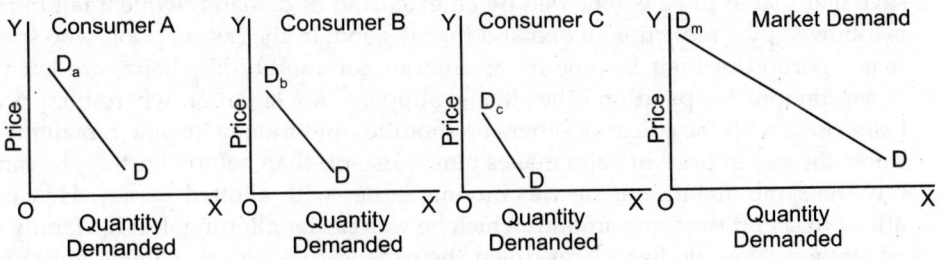

**Figure 3.10** Market demand curve.

## Autonomous Demand and Derived Demand

The goods, whose demand is not linked with the demand of other goods are supposed to have autonomous demand. Consumer goods are the examples here. The demand for certain goods is related with the demand for other goods, which is called derived demand. The demand for fertilizers, pesticides, etc., is a derived demand, for it is linked with the demand for agricultural products. Thus the goods which are demanded for their own sake have autonomous demand, while the goods that are required to produce other goods have derived demand.

## KINDS OF DEMAND

1. *Price Demand:* It refers to various quantities of a good or service that a consumer would be willing to purchase at all possible prices in a given market at a given point in time, *ceteris paribus.*
2. *Income Demand:* It refers to various quantities of a good or service that a consumer would be willing to purchase at different levels of income *ceteris paribus.*
3. *Cross Demand:* It refers to various quantities of a good or service that a consumer would be willing to purchase not due to changes in the price of the commodity under consideration, but due to changes in the price of related commodities.

## LAW OF DEMAND

The law of demand explains the functional relationship between the quantity demanded of a commodity and its unit price, *i.e.,* a rise in the price of a commodity or service is followed by a reduction in the quantity demanded and a fall in the price is followed by an extension in demand, with other conditions remaining the same. Increased prices tend to contract demand and falling prices extend it. Demand various inversely with the price, other things being equal. The tendency of the consumer is to buy more quantities of a commodity at a lower price than what he buys at a higher price.

### Exceptions to the Law of Demand

Following are the exceptional cases, where, the law of demand does not hold good.

1. *Giffen Goods (Inferior Goods):* This phenomenon which is explained below was given by Sir Robert Giffen. It was named after him as Giffen paradox. This phenomenon says that rise in price is followed by an extension of demand, while a fall in price is followed by a reduction in demand for the good. In the case of poor, who spends major portion of their income on an inferior commodity like bajra, are left with lesser amount to spend on other items. Suppose in a situation where the price of bajra rises, with the prices of other commodities and money income remaining the same, the rise in price of bajra makes him worse-off than before. Further, he cannot buy the same quantity as he was buying earlier with allotted money. He cannot afford to spend the same amount which he was earlier allotting, lest his family will be starved. Now he has to cut down the expenditure on other items not only to maintain but also to increase the quantity of bajra he bought per unit of time. On the other hand, if the price of bajra falls, the increased real income enables him to spend on superior foodgrains. This leads to a contraction of demand for bajra, even though there is a fall in its price.
2. *Prestigious Goods:* When the possession of a good brings in social distinction, consumer would go for the same even if its price is higher. An example to be cited here is the diamonds that the rich people purchase, as the possession of the same is prestigious to them.
3. *High Priced Commodities:* When the consumers view that those products which are superior are sold at higher prices, they do not mind to buy more of the same at higher prices.
4. *Fear of Shortage:* If the existing price is higher and it is expected to increase further, consumers would buy more of it even at higher price, fearing for the shortage.

## Movement Along the Demand Curve

It refers to change in the quantity demanded due to change in price. It can be either extension or contraction of demand.

Extension of demand means buying more quantity of commodity at a lower price, while contraction of demand indicates buying less at a higher price. As given in Table 3.7 for consumer A, when the price of tomatoes falls from Rs. 20 kg to Rs. 16 kg the quantity demanded increased from 0.25 kg to 0.50 kg. It is a case of extension of demand. For the consumer A, when the price increased from Rs. 12 to 16 the quantity demanded decreased from 0.75 kg to 0.50 kg. It is case of contraction of demand. The terms extension and contraction refer to the movement on the same demand curve (Figure 3.11). The downward movement from A to B is extension, while the upward movement from B to A contraction. Extension and contraction of demand represent the "change in quantity demanded". This is purely a price resulted phenomenon of demand changes for a commodity, while other factors influencing demand are assumed to be at fixed level.

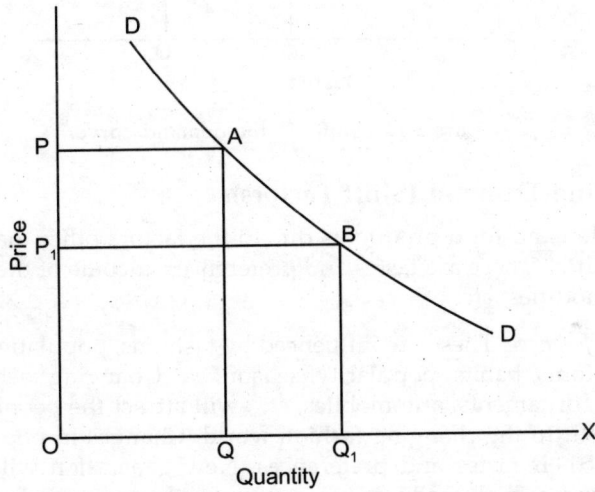

**Figure 3.11** Movement along the demand curve.

## Shifts in the Demand Curve

It refers to change in demand not due to change in price but due to change in the values of other variables influencing demand. It can increase or decrease in demand. As against extension and contraction of demand, increase and decrease in demand result in the shifting of the demand curve. Increase in demand means more demand at the same price or same demand at higher price. On the other hand, decrease in demand means less demand at the same price or same demand at lower price. This is shown in the Figure 3.12. When there is an increase in demand, the demand curve shifts upwards to the right side of the initial demand curve DD. $D_1D_1$ is the new demand curve representing increase in demand. At 'OP' price, the quantity demanded is OQ. Increase in demand indicates the purchase of same quantity of a commodity (OQ) at a higher price of $OP_1$ or purchase of $OQ_1$ quantity at the same price of OP. The decrease in demand is indicated by the shift of the demand curve towards left downwards to the initial demand curve. $D_2D_2$ is the demand curve representing decrease in demand. Decrease in demand

36

indicates purchase of the same quantity of a commodity (OQ) at a lower price of $OP_2$ or purchase of $OQ_2$ quantity at the same price of OP. Increase and decrease denote 'change in demand'.

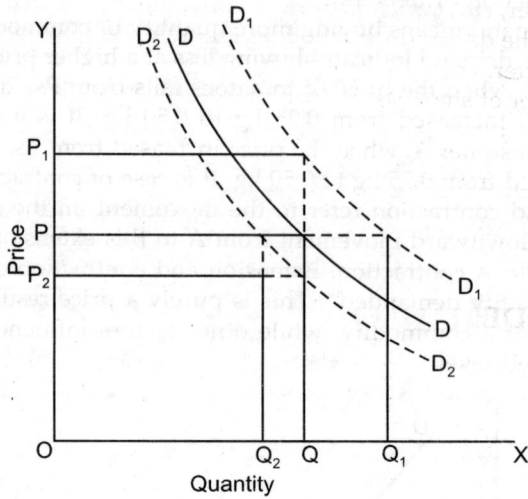

**Figure 3.12** Shifts in the demand curve.

## Factors Affecting Demand (Shift Factors)

The change in demand for a product is due to the factors other than the price factor of the same product. These are tastes and preferences, income of the consumer, prices of related commodities, *etc.*

1. *Tastes and Preferences:* These are influenced by fashions, population changes, advertisement, customs, habits, popularity, season, *etc.* Changing fashions in the men/women wear, ornaments, automobiles, *etc.*, will attract the people to buy them, so that they are fit to the changing fashion world. Changes in population also bring variations in their tastes and preferences. New generation will have something different attitudes in all walks of life, consequently new tastes emerge. Advertisement has become an important instrument of sales promotion/selling strategies of the manufactures. With the growing access to several mass media sources, people are beholden by their impact, consequently tastes and preferences are ever changing. Customs of the society enforce the people to change their life styles, as a result of which new preferences add to their long list of wants. People are habituated to certain aspects of life and develop new preferences. Popularity of a product attracts the new consumer towards it. This implies that we have new preferences for products. Popularity of the product is one, the influence of which cannot be avoided by the consumer. Season-bound requirements compel the people to have a new set of preferences each season. Thus, the changes in tastes and preferences make the consumer more satisfied than before.
2. *Income:* Changes in the income of the consumer leads to a shift in the demand curve. The proportion of income spent on food and other necessaries decreases with increase in the income level of the consumers.
3. *Price of Other Related Goods:* The change in price of one good influences the consumption of other good, depending upon the relationship between the goods.

If two commodities are supposed to be substitutes such as meat and chicken, coffee, tea, *etc.*, the rise in price of one good results in increase in demand for the other good. There are certain goods which need to be used together *viz.*, bread and jam, bread and butter, *etc.*, which are called complements. Increase in price of one good brings down the demand for the other good. A rise in price of bread reduces the demand for butter.

4. *Habits:* Large number of smokers in a region influence the demand for tobacco and its products.
5. *Region:* In high altitude regions demand for wine, woolen clothing, meat, *etc.*, would be persistent.
6. *Season:* Demand for eggs decreases in summer season and increases in winter season.

## ELASTICITY OF DEMAND

The law of demand says that demand varies inversely with the price, other things being equal. From the law of demand, we know the direction in which quantity and price are moving. What is not known is the extent by which quantity demanded is responsive to changes in price. This is the information, which is precisely needed by businessmen and policy-makers. Alfred Marshall developed the concept of elasticity of demand which measures the responsiveness of quantity demanded to changes in price. Elasticity of demand indicates the degree of relation between quantity demanded and price. In fact elasticity of demand is the rate at which quantity demanded changes because of change in price. To be more precise, elasticity of demand is defined as *"the relative change in the quantity demanded to the relative change in the price"*.

### Types of Elasticities of Demand

There are three types of elasticities of demand *viz.*, i) Price elasticity of demand, ii) Income elasticity of demand, and iii) Cross elasticity of demand

### (i) Price Elasticity of Demand ($E_d$)

This shows the responsiveness of quantity demanded of a commodity, when price of that commodity changes, with other factors being constant.

$$\text{Price elasticity of demand } (E_d) = \frac{\%\text{ change in quantity demanded}}{\%\text{ change in price}}$$

(or)

$$= \frac{\text{Proportionate change in quantity demanded}}{\text{Proportionate change in price}}$$

$$= \frac{\dfrac{\text{Change in quantity}}{\text{Initial quantity}} \times 100}{\dfrac{\text{Change in price}}{\text{Initial price}} \times 100}$$

### (ii) Income Elasticity of Demand ($E_I$)

It measures the responsiveness of demand due to changes in the income of the consumers in terms of percentage, when other factors influencing demand *viz.*, price of the commodity, price of substitutes, tastes, preferences, *etc.*, are kept at constant level.

$$\text{Income elasticity demand } (E_1) = \frac{\%\text{ change in quantity demanded}}{\%\text{ change in income}}$$

(or)

$$= \frac{\text{Proportionate change in quantity demanded}}{\text{Proportionate change in income}}$$

$$= \frac{\dfrac{\text{Change in quantity}}{\text{Initial quantity}} \times 100}{\dfrac{\text{Change in income}}{\text{Initial income}} \times 100}$$

## Cross Elasticity of Demand ($E_{xy}$)

Demand for one good (X) is also influenced by the price of other related good (Y). These may be substitutes or complements. It is the ratio of percentage change in quantity demanded of commodity (X) and percentage change in price of related commodity (Y).

$$\begin{array}{l}\text{Cross elasticity of} \\ \text{demand } (E_{xy})\end{array} = \frac{\%\text{ change in quantity demanded of commodity (X)}}{\%\text{ change in price of related commodity (Y)}}$$

(or)

$$= \frac{\text{Proportionate change in quantity demanded of commodity (X)}}{\text{Proportionate change in price of related commodity (Y)}}$$

$$= \frac{\dfrac{\text{Change in quantity (X)}}{\text{Initial quantity (X)}} \times 100}{\dfrac{\text{Change in price (Y)}}{\text{Initial price (Y)}} \times 100}$$

## DEGREES OF ELASTICITY OF DEMAND

Based on the magnitudes of elasticity of demand, it can be categorized into five degrees, *viz.*, perfectly elastic demand, perfectly inelastic demand, relatively elastic demand, relatively inelastic demand, and unitary elastic demand.

### Perfectly Elastic Demand

A slightest change in price of a commodity leads to an infinite change in quantity demanded. The demand in such a situation is said to be perfectly elastic. The demand is hypersensitive and the elasticity of demand is infinite. Here the demand curve will be a horizontal line parallel to X-axis (Figure 3.13). It is mostly a theoretical concept.

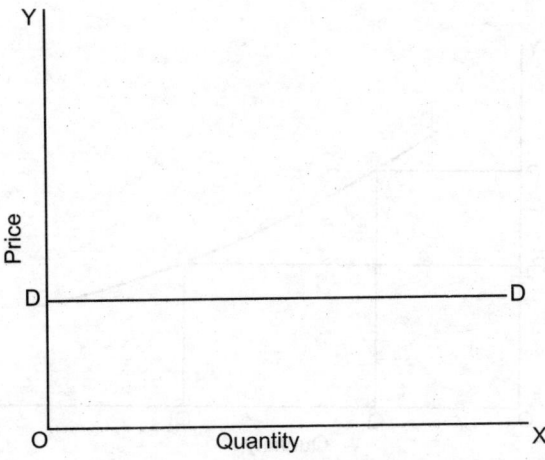

**Figure 3.13** Perfectly elastic.

## Perfectly Inelastic Demand

It is situation in which change in price of a commodity leaves the demand unaffected. The price of the commodity may increase or decrease, but the quantity demanded remains the same. The demand here is insensitive. Elasticity of demand is zero. The demand curve is vertical to X-axis (Figure 3.14). The case of perfectly inelastic demand is also a theoretical concept.

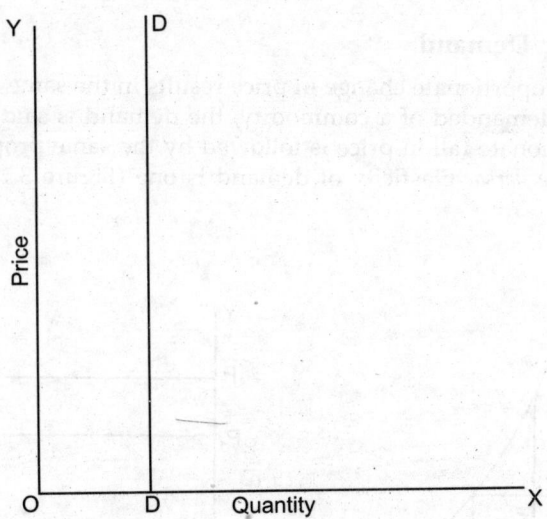

**Figure 3.14** Perfectly inelastic.

## Relatively Elastic Demand

It means that lesser proportionate change in the price of a commodity is followed by a larger proportionate change in the quantity demanded. A small proportionate fall in the price is accompanied by a larger proportionate increase in demand and *vice versa*. Elasticity of demand is greater than unity (Figure 3.15).

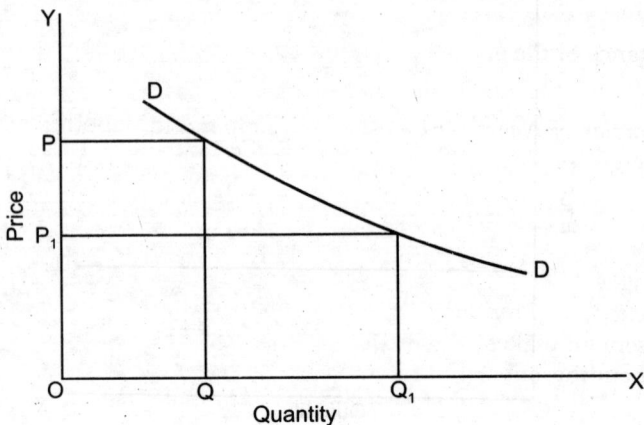

**Figure 3.15** Relatively elastic.

## Relatively Inelastic Demand

It means that large proportionate change in the price of a commodity is followed by a smaller proportionate change in the quantity demanded. This is to say that a large proportionate fall in the price is followed by a smaller proportionate increase in the quantity demanded and *vice versa*. Elasticity of demand is less than unity (Figure 3.16).

## Unitary Elastic Demand

When a given proportionate change in price results in the same proportionate change in the quantity demanded of a commodity, the demand is said to be unitary elastic. A given proportionate fall in price is followed by the same proportionate increase in demand and *vice versa*. Elasticity of demand is one (Figure 3.17).

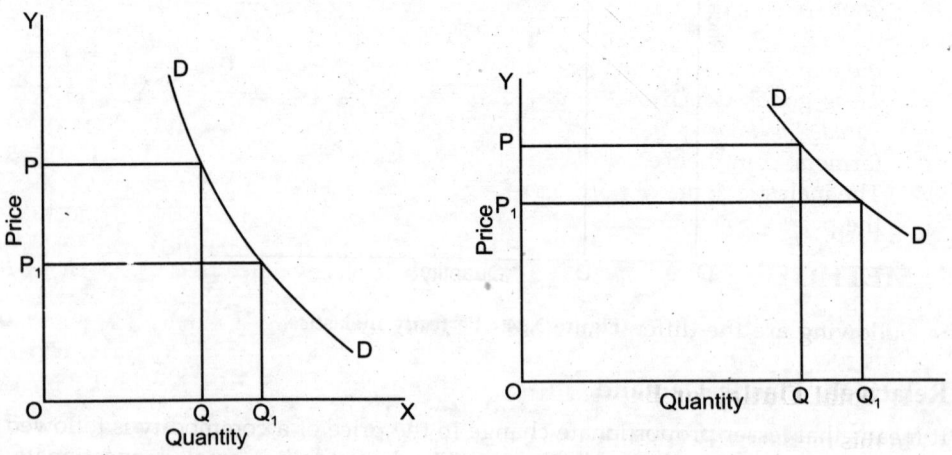

**Figure 3.16** Relatively inelastic.                **Figure 3.17** Unitary elastic.

## Practical Importance of Elasticity of Demand

The importance of the elasticity of demand can be visualized from the points as given below:

1. *Determination of Wages:* The elasticity of demand for labour plays an important role in the determination of wages. If the demand for labour is elastic, any pressure put up by labour in the form of strikes to get higher wages would be unsuccessful. On the other hand, if the demand for labour happens to be inelastic, even a threat of strike would help the workers to get the approval of their employers in raising their wages.

2. *The Elasticity of Promotional Activity:* The producers are well convinced that the advertisement makes the demand for a product less elastic. Hence, they would not mind spending substantial amount of money on advertising. The price increase therefore will not reduce the sales.

3. *Determination of Monopoly Price:* The monopolist considers the nature of demand for his product before fixing the prices. If the demand is elastic, a lower price would help him to realize more profits. On the other hand, if the demand is inelastic, he is in a position to fix a higher price. The monopolist while practicing price discrimination also takes into account the elasticity of demand. He fixes a lower price for the product in the market in which the demand for the product is elastic and he charges a higher price for the same product in the market in which the demand is less elastic or inelastic.

4. *Undertaking the Public Utilities:* The Government itself runs some enterprises or industries in the interest of people, otherwise they are subjected to exploitation by the private people. In the case of electricity, the demand is inelastic and it is very essential item in the lives of humans. In the interest of the public, electricity boards are run by the Government to supply power at reasonable rates.

5. *Taxation:* The nature of demand for a good helps the Government, looking into the possibilities of raising the revenue. If the demand is less elastic for a good, by levying more indirect taxes on that good, the Government would get larger revenue.

6. *International Trade:* The country gains in the international trade by exporting those goods, for which the demand in the export market is less elastic and by importing those goods for which the demand is elastic. The less elastic nature of demand in the export market helps the nation to charge a higher price and pay less price for those goods which are imported, and

7. *Paradox of Poverty in Plenty:* A bumper crop instead of bringing prosperity to the farmers, ruin their economic position. It is a common phenomenon in agriculture. The inelastic demand for the products in the years of bumper harvest brings down the prices, consequent to which the farmers fail to realize prices of normal years.

# METHODS OF MEASURING PRICE ELASTICITY

Following are the different methods of measuring price elasticity of demand.

## 1. Total Outlay or Expenditure Method

In this method we compare total expenditure of the consumer before and after change in price. The elasticity of demand is unity when the total expenditure remains unaltered eventhough, there is price change. The demand is said to be elastic when the total expenditure increases with fall in price and decreases with rise in price. Inelastic demand is observed when the total amount spent on the commodity by the consumer increases with increase in price and decreases with fall in price. An hypothetical example of total expenditure method is shown in Table 3.8.

Between 1 and 2, the elasticity is greater than unity (relatively elastic) because the total expenditure increases with fall in price and decreases with rise in price. Between 3 and 4, the elasticity is unitary as the total outlay remains the same, though there is variation in the price. Between 5 and 6, the fall in price is followed by a fall in the total outlay and rise in the price is accompanied by an increase in total expenditure. Therefore, the elasticity is less than unity (relatively inelastic).

TABLE 3.8 Total Expenditure Method.

| S. No. | Price (Rs./unit) | Quantity (units) | Total expenditure (Rs.) |
|--------|------------------|------------------|-------------------------|
| 1 | 5.00 | 6 | 30.00 |
| 2 | 4.50 | 10 | 45.00 |
| 3 | 4.00 | 15 | 60.00 |
| 4 | 3.00 | 20 | 60.00 |
| 5 | 2.00 | 25 | 50.00 |
| 6 | 1.00 | 30 | 30.00 |

## 2. Point Elasticity of Demand

It is a geometrical method for measuring elasticity at a point on the demand curve. Point method is used when price and quantity changes are extremely small. This method is applicable only when we possess information on the minutest changes in price and quantity. Measurement of point elasticity is shown in Figure 3.18.

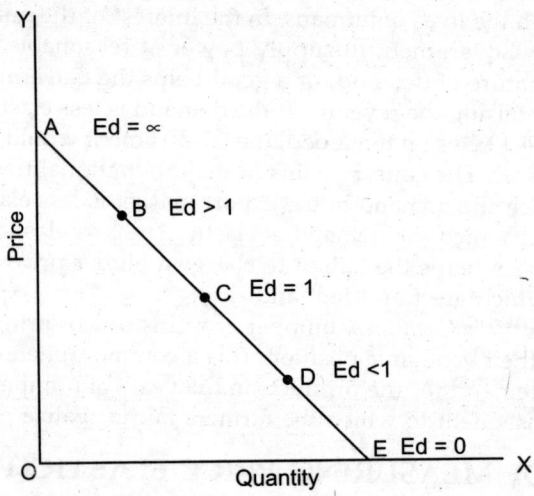

Figure 3.18   Point Elasticity of Demand.

A straight-line demand curve joining the two axes is considered to show the measurement of point elasticity. Elasticity at any point on the demand curve is the ratio of lower part of the straight line to the upper part. It is important to note that point elasticity of demand on straight line is different at every point. At any point to the right of mid point (C) the point elasticity is less than unity, at any point to the left of mid point the point elasticity is more than unity, and at mid point (C) the elasticity is unity. At the point where the linear demand curve intersects Y-axis (A) the point elasticity is infinite, while at the point where the demand curve intersects X-axis (E), the point elasticity is zero.

## 3. Arc Elasticity of Demand

The price elasticity of demand measured between two distinct points on a demand curve is called arc elasticity of demand. This method uses the mid points between the old and new data in the case of price and quantity. It studies a portion of the demand curve between two points. An arc is a segment or a portion of a curved line. Arc elasticity is employed in order to compute price elasticity coefficient from the discrete data. The following formula is used to compute arc elasticity.

$$E_d = \dfrac{\dfrac{Q_1 - Q_2}{(Q_1 + Q_2)/2}}{\dfrac{P_1 - P_2}{(P_1 + P_2)/2}} = \dfrac{\Delta Q/(Q_1 + Q_2)}{\Delta P/(P_1 + P_2)}$$

Where,

$Q_1$ stands for the first quantity observed and $Q_2$ for second quantity. Similarly, $P_1$ is the first observed price and $P_2$ is the second price.

## Factors Determining Elasticity of Demand

1. *Type of Goods:* The demand is inelastic or less elastic for necessaries. It is obvious because the price changes do not influence the consumption of these goods. On the other hand, in respect of comforts and luxuries, the demand is elastic or relatively elastic.
2. *Goods Having Several Uses:* The demand for those goods which can be put to several uses is elastic. For example, electricity when it is cheap, it can be put to uses like cooking, apart from its regular use in production of industrial goods, transport, lighting, *etc.* In case it is dearer, its use is limited and demand automatically declines.
3. *Existence of Substitutes:* The demand for those commodities which have good substitutes is elastic. The price variations of such goods would have a bearing on the quantities demanded. If the price of groundnut oil increases, with the prices of other edible oils, say sunflower remaining constant, the consumers tend to shift to sunflower oil. Conversely a decrease in price of groundnut oil makes consumers to switch over back to groundnut oil.
4. *Possibilities of Postponement:* For such goods the purchase of which can be post-poned by the consumers, the demand is elastic. The consumers prefer to buy the same when their prices are cheap. Televisions, cars, ornaments, *etc.*, are the relevant examples here.
5. *Range of Prices:* When the price range of goods is very high or very low the demand is inelastic. The demand for goods like salt, matches, *etc.*, with low price range, the demand is inelastic. Analogously for goods like diamonds, luxurious cars, *etc.*, the demand is inelastic as consumers cannot afford to the changes in the price of such a high range.

# CHAPTER 4

# Theory of Costs

Knowledge regarding various relationships existing between costs and output is necessary to comprehend the concepts of equilibrium conditions of different firms under different market situations. Here we require to know how the fixed cost (FC), variable cost (VC), total cost (TC), average fixed cost (AFC), average variable cost (AVC), average cost (AC), and marginal cost (MC) are related to different output levels of the firms under different levels of technology used by the firms. Here we have assumed that every business firm has a certain state of technology or know how *i.e.*, how they are producing various levels of outputs with the given prices of inputs and outputs. With data on those variables, we can understand the relationships between costs and output levels. Basically we require data on output, fixed costs, variable costs and the prices of inputs and outputs. From this data we finally derive all the seven cost concepts, *viz.*, TFC, TVC, TC, AFC, AVC, AC and MC. These cost concepts would have implications for output expansion of the firms and equilibrium position of the firms in different time periods. In the cost theory, economists use different names for cost concepts under different contexts. They are money costs or nominal costs, real costs, opportunity costs, economic costs, implicit costs, explicit costs, deflated costs, social costs, short run costs, long run costs, separable costs, *etc.*

## Nominal Costs or Money Costs

These are usually expressed in money terms at current prices. Firms incur costs in producing different products. Every level of output is associated with a given level of costs under a given level of technology, prices and input levels at a particular point of time. Generally the cost of production refers to costs incurred per an unit of output produced by a firm under a given technology. Nominal costs of production refer to per unit cost of production of output at current market prices.

## Real Costs

When the costs of inputs, and input services are expressed at constant prices they become real costs.

## Opportunity Cost

Opportunity cost is the value of return sacrificed or foregone from the next best alternative activity. In farming farmers don't have to pay for their owned resources, *viz.*, family labour, owned bullock labour, owned machinery, owned seed, *etc.* But still in the cost analysis the value of these owned resources are considered on the basis of opportunity cost.

## Economic Costs

These are divided into explicit cost and implicit costs. Explicit costs include payments made by the entrepreneurs for purchasing and hiring of inputs and input services. They are also called paid out costs or cash costs. Entrepreneurs do not pay for the use of owned resources. The value of such resources is called as implicit costs. Costs of self-owned and self-employed resources are known as implicit costs.

## Deflated Costs

Costs if deflated by general price index are called deflated costs. By doing so the effect of inflation in an economy is taken out. *Example*: Real cost of commodities.

## Social Costs

These are also called as externalities. Firms incur both implicit and explicit costs in the production of goods and services. Their sum constitutes total cost of production. These costs we name as private costs, but from the point of view of society, these firms will give rise to some additional costs to the society in the form of environmental degradation, water, air or noise pollution *etc.*, in the areas where goods are produced by the private firms. In the absence of well-drained system, irrigation projects bring problems to the command area of the project in the form of new diseases. Such costs are called social costs.

## Separable Costs

Separable costs are the costs which can exclusively be attributed to production of output separately. Common costs are those which cannot be separated to the production of the output. So they are called joint costs. The costs are involved in the production of several products. *For example,* electricity generation, ground water use, *etc.*

## Historical Costs and Replacement Costs

Historical costs are the costs involved in the purchase of durable goods like land, buildings, machinery, equipment, *etc.* Purchase price of the asset should be considered as price of the asset and hence it is considered as historical cost in analysis. Since the costs of the assets are apportioned in computation, they are called as historical costs. Replacement costs refer to the difference between the purchase price of the asset and the current price of the same asset. Suppose a tractor is purchased 10 years ago at a price of Rs. 1,50,000, but its present price is say Rs. 2,50,000, the difference of Rs. 1,00,000 is the replacement cost.

## Establishment Costs

Construction of plant in any business activity entails some costs. Such construction costs are called establishment costs in the business analysis. They are also called first phase costs. The other costs *viz.*, licenses, site development expenditure for construction of factory, purchase of equipment, furniture, expenditure on personnel, royalties for seeking product rights, cost of raising finance, costs of maintaining raw materials *etc.*, are also included in the establishment costs.

## COST CONCEPTS

Knowledge regarding the cost functions is very much essential for optimal managerial decisions to be taken by the firm as well as the Government. In the short run, pricing and output decisions are based on short run cost curves, while in the long run, long run cost curves have crucial implications for development and growth of the firm and investment policies of the firm. Consideration of cost curves is essential and forms the basis for entry and exit of the firms in the industry. Profit maximizing rule is determined with the help of cost curves, cost functions and production functions. This rule is popularly known as marginal analysis at which MC = MR. The costs are also one of the major price determinants in all the market situations of the economy and in all the economic models which would explain the behaviour of the firms.

There are seven costs, which explain the behaviour of the firms in the production of requisite products.

### Fixed Costs

Fixed costs remain the same irrespective of level of production. These costs remain invariant in the short run but in the long run there are no fixed costs as all the inputs can be varied. Fixed costs include cost items like taxes, insurance, cess, depreciation on machinery, implements, tools, buildings, salaries of personnel working in the firm, *etc.* These are also known as indirect costs, sunk costs and overhead costs. The summation of all these costs is called total fixed costs (TFC). TFC is a horizontal straight line parallel to X-axis (Figure 4.1).

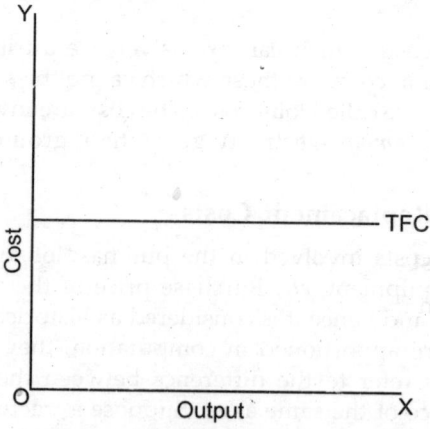

**Figure 4.1** Total fixed costs.

### Variable Costs

Variable costs as per definition vary with the level of output. These include costs of raw materials, labour, power, repairs, maintenance charges of machinery, *etc.* These are also known as working costs, operating costs, direct costs, prime costs, circulating costs and running costs. These are second phase costs. The summation of these costs refer to total variable costs (TVC). Graphically TVC has inverse 'S' shape (Figure 4.2).

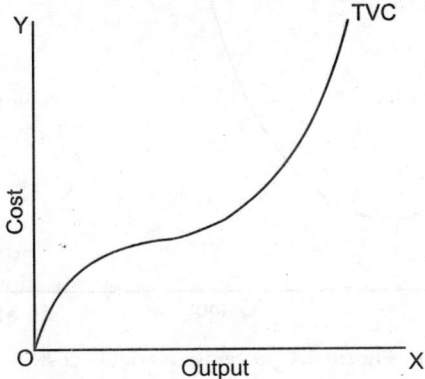

**Figure 4.2**   Total variable costs.

## Total Costs

These include total fixed costs as well as total variable costs. Its shape is similar to that of TVC (Figure 4.3).

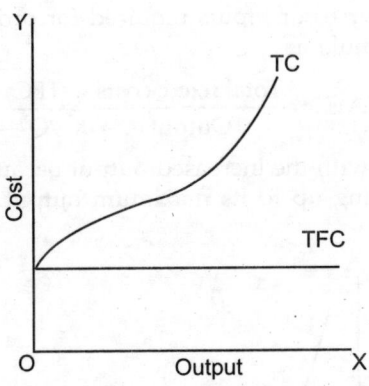

**Figure 4.3**   Total costs.

## Average Variable Cost (AVC)

It is the amount spent on the variable inputs to produce an unit of output. Algebraically it is expressed as

$$AVC = \frac{\text{Total variable costs}}{\text{Output}} = \frac{TVC}{Q}$$

When a small amount of output is produced, cost of variable input per unit of output becomes very high. This is to say in other words, that productivity of variable input increases when greater amounts are used in the production of the commodities due to economies of scale. This causes AVC to have 'U' shape when it is graphed. This is shown in the Figure 4.4. When it is 'U' shaped it becomes reciprocal of average physical product curve.

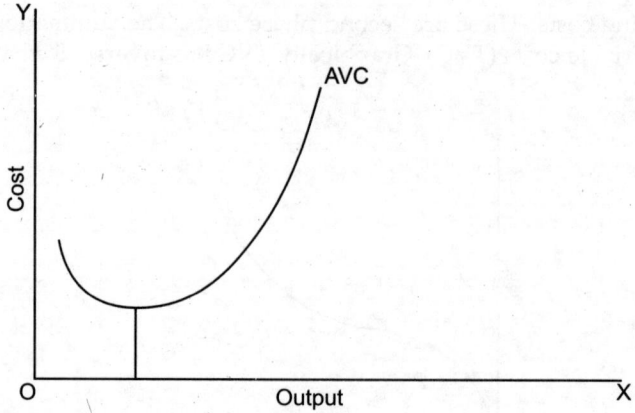

**Figure 4.4**  Average variable cost.

AVC falls to minimum level at the output level where APP is maximum. Thereafter due to production of greater amount of output, AVC rises again and becomes vertical at certain level of maximum output.

### Average Fixed Cost (AFC)

It is the cost of fixed resources or inputs required for producing one unit of output and it is given by the formula as

$$AFC = \frac{\text{Total fixed costs}}{\text{Output}} = \frac{TFC}{Q}$$

AFC curve is declining with the increased output because TFC is constant. Due to this it is continuously falling up to its maximum output. It is having the shape of hyperbola (Figure 4.5).

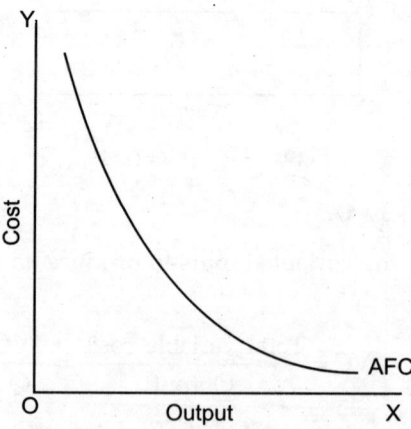

**Figure 4.5**  Average fixed cost.

### Average Total Cost or Average Cost (ATC or AC)

When the total costs are divided by output, we get ATC (Figure 4.6).

$$ATC = \frac{\text{Total costs}}{\text{Output}} = \frac{TC}{Q} = \frac{TFC + TVC}{Q}$$

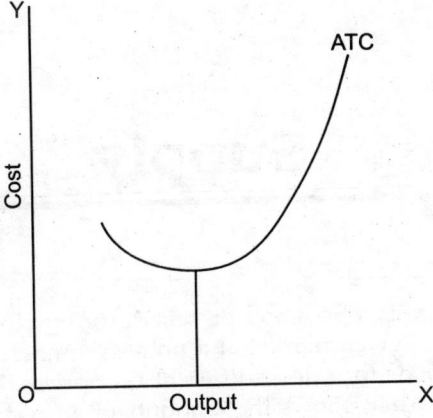

**Figure 4.6** Average total cost.

## Marginal Cost (MC)

As per the definition, it is the change in the total cost due to the change in output. Algebraically it is expressed as

$$\frac{\text{Change in total costs}}{\text{Change in output}} = \frac{\Delta TC}{\Delta Q} \text{ or } \frac{\Delta TVC}{\Delta Q}$$

Note that to compute MC, we can use TC or TVC because fixed costs cannot be changed. The only component of change in TC is TVC. The specific shape of MC curve is due to marginal product of the variable inputs. When MPP is maximum, MC is minimum. MC curve is declining when MPP curve is increasing; hence there is an inverse relationship between MPP and MC. When MPP is zero MC becomes vertical. MC curve intersects AVC and AC at their minimum points (Figure 4.7).

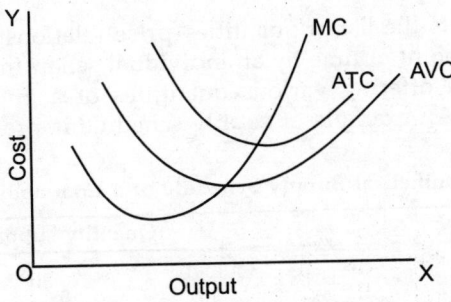

**Figure 4.7** Average and marginal costs.

# CHAPTER 5

# Supply

Supply means the quantity of a good or service offered by a producer for sale at different unit prices in a given market at a point of time. It is the willingness of the supplier to offer the goods for sale at different unit prices. More specifically, supply is defined as a schedule that shows the amounts of a product or service, sellers are willing to sell at each unit price in a set of possible prices during some specified period of time in a specified market. Meyers defined supply as *"a schedule of the amount of good that would be offered for sale at all possible prices at any one instance of time in which the condition of supply remains the same"*. Prof. Mc. Connel defined supply as *"a schedule which shows the various amounts of a product which a producer is willing to and able to produce and make available for sale in the market at each specific price in a set of possible prices during some given period"*.

## Stock

Along with the concept of supply, another concept called stock needs to be explained. Supply is drawn from the stock of the commodity. Supply is the actual quantity that a seller is willing to sell at a particular price, while stock is the amount of output that exists in a market. Depending on the demand for a commodity stock is converted into supply. For perishable commodities stock and supply are the same. For durable commodities stock and supply are different.

## Individual Supply Schedule

Supply schedule depicts the list of quantities–price relationships of a commodity in a market at a specific point of time by an individual seller. In other words, it reveals the mind of sellers in offering various quantities of a given commodity against corresponding prices. An example of supply schedule is presented in Table 5.1.

**TABLE 5.1 Hypothetical Supply Schedule of a Commodity in a Market.**

| Price (Rs./Q) | Quantity supplied (Q) |
|---|---|
| 300 | 30 |
| 325 | 40 |
| 350 | 50 |
| 375 | 60 |
| 400 | 70 |
| 425 | 80 |

It reveals that at price of Rs. 300 the quantity supplied by a seller is 30 Q at Rs. 325, 40 Q and so on. As the price per unit of the commodity rises, the quantity supplied is also increasing. As price increases, sellers are committed to increase their sales. When a supply schedule is plotted on a graph it becomes a supply curve (Figure 5.1). The supply curve will have a positive slope *i.e.*, it slopes upwards from left to right.

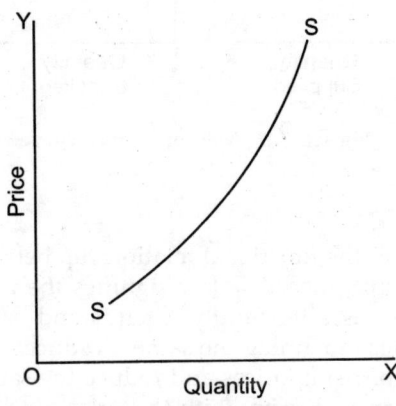

**Figure 5.1** Supply curve.

## Market Supply

It is the sum of the quantity of commodity that is brought into a market for sale by the sellers in a given market at a specific point of time. Assume that there are three sellers in a market *viz.*, A, B and C with individual supply schedules as shown in Table 5.2.

**TABLE 5.2 Market Supply Schedule.**

| Price (in Rs./Q) | Individual seller's supply/week (Q) | | | Market supply (Q) = (A + B + C) |
|---|---|---|---|---|
| | A | B | C | |
| 300 | 30 | 35 | 0 | 65 |
| 325 | 40 | 50 | 0 | 90 |
| 350 | 50 | 65 | 50 | 165 |
| 375 | 60 | 80 | 70 | 210 |
| 400 | 70 | 95 | 90 | 255 |
| 425 | 80 | 110 | 110 | 300 |

The price quantity relationship of the three sellers reveals that at Rs. 300 per quintal, seller 'A' is prepared to sell 30 Q, while seller 'B' 35 Q and seller 'C' is not prepared to sell at all at this particular price. The seller 'C' is not prepared to sell the commodity at any price less than Rs. 350/Q. Market supply is the sum total of output that is sold by the three sellers as presented in the last column of the table. Thus the market supply is 65, 90, 165 Q and so on. It is the lateral or horizontal summation of the supply of individual sellers at each unit price. The market supply curve is drawn based on the first and last columns of the table. The graphical presentation of market supply is found in Figure 5.2.

52

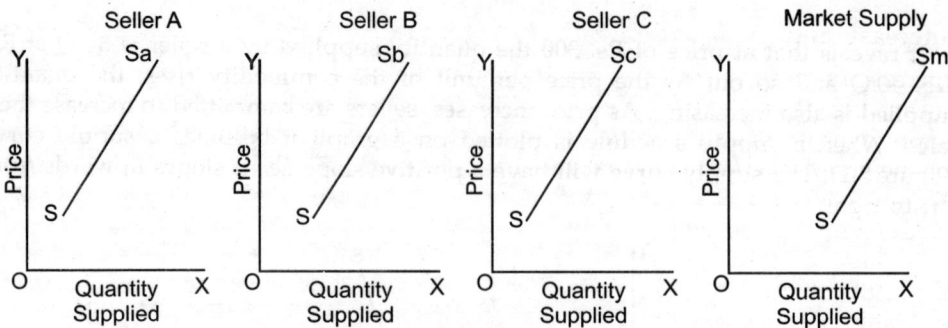

Figure 5.2  Market supply curve.

# LAW OF SUPPLY

The law of supply indicates the functional relationship between the quantity supplied of a commodity and its unit price. The law signifies the positive relationship *i.e.*, as the price of a commodity rises its supply extends and as the price falls its supply contracts, with other things remaining the same. Producers normally tend to increase the supplies in the wake of rising prices and reduce the same when the prices are on the lower side. Supply varies directly with the price, *ceteris paribus*.

### Extension and Contraction of Supply (Change in Quantity Supplied)

Extension and contraction of supply refer to the movement of product supply on the same supply curve. Extension of supply means offering more quantity for sale at a higher price, while contraction means offering less quantity at a lower price. As is seen from Table 5.2 that the quantity of commodity supplied by 'A' at Rs. 300 is 30 Q and it is 40 Q. when the price rose to Rs. 325. Here the quantity supplied has increased from 30 to 40 Q. It is the case of extension of supply, conversely if the price falls from Rs. 325 to Rs. 300, the quantity supplied too falls from 40 to 30 Q. It is the contraction of supply.

Graphically (Figure 5.3), when it is depicted, it shows that the upward movement from A to B is extension of supply and downward movement from B to A is contraction of supply.

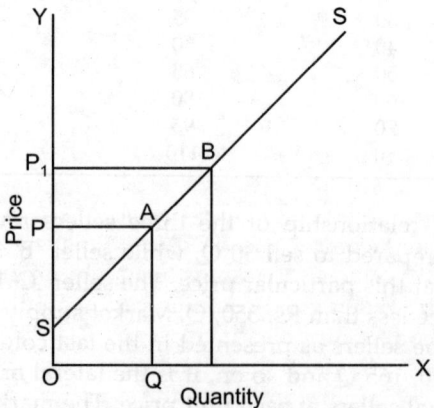

Figure 5.3  Extension and contraction of supply.

## Increase and Decrease in Supply (Shift in Supply)

Increase in supply implies more supply at the same price and decrease in supply means less supply at the same price. The change in supply (increase and decrease in supply) results in a shift of the supply curve. An increase in supply results in the shift of the supply curve towards right side of the initial supply curve SS as shown in Figure 5.4. The new supply curve is $S_1S_1$. On the other hand, a decrease in supply causes a shift of the supply curve towards the left side of the initial supply curve. The new supply curve thus formed is $S_2S_2$. Originally OQ quantity is supplied at OP price. But due to changes in supply conditions at the same price OP, $OQ_1$ quantity of commodity is supplied indicating increase in supply. On the other hand, again influenced by changing supply conditions at the same price, $OQ_2$ is supplied. This is decrease in supply.

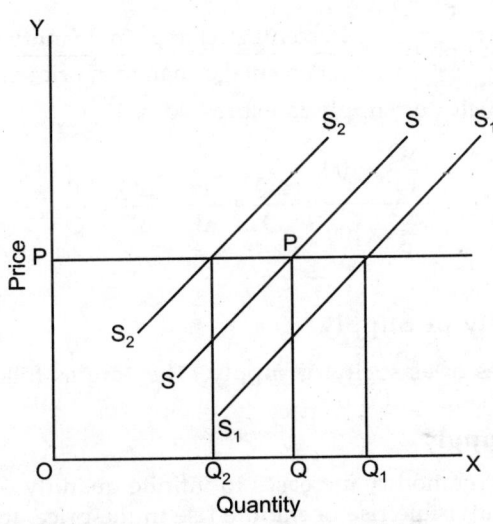

**Figure 5.4**  Increase and decrease in supply.

## Factors Causing Changes in Supply (Shift Factors)

The factors that are responsible for changes in supply are discussed below:

1. *Changes in Technology:* Technological innovations *viz.,* new varieties of crops and their consequent increased yields per unit area, help to increase the supply of the commodity.
2. *Reduction in Resource Prices:* When the prices of input factors become cheaper than before, it encourages producers to use more of them in producing more output. Supply curve shifts towards the right side.
3. *Reduction in the Relative Prices of Other Products:* A reduction in relative prices of other related products compel the producers to increase the production of that particular commodity whose prices are relatively higher.
4. *Market Infrastructure:* When good communication and transport network increase, the supply of the commodity also increases.
5. *Number of Producers:* Changes that are found regarding number of producers producing a given commodity influence the supplies. More the number of producers, greater the supply and *vice versa.*

6. *Producers' Expectations about Future Prices:* Price expectations influence the sales strategies of the producers positively.

## ELASTICITY OF SUPPLY

Elasticity of supply of a commodity is the responsiveness, or sensitiveness of supply to the changes in price. Supply is said to be elastic, if a small change in price causes considerable change in the quantity supplied. The supply is inelastic when a given change in price leads to little or less change or no change in the quantity supplied. In short, elasticity measures the adjustability of supply of a commodity to price.

Elasticity of supply (price elasticity of supply) is expressed as the ratio of percentage change in the quantity of good supplied and percentage change in price of the good *ceteris paribus.*

$$\text{Elasticity of supply} (E_s) = \frac{\text{Percentage change in quantity of good supplied}}{\text{Percentage change in price of good supplied}}$$

Algebraically elasticity of supply is expressed as

$$\frac{\dfrac{\Delta Q}{Q} \times 100}{\dfrac{\Delta P}{P} \times 100} = \frac{\Delta Q}{Q} \bullet \frac{P}{\Delta P} = \frac{\Delta Q}{\Delta P} \bullet \frac{P}{Q}$$

### Degrees of Elasticity of Supply

There are five degrees of elasticity of supply. They are as follows:

### Perfectly Elastic Supply

When the supply of commodity increases to infinite quantity or unlimited quantity, even though there is invisible rise or minute rise in the price, the elasticity of supply is said to be infinity ($E_s = \alpha$) (Figure 5.5)

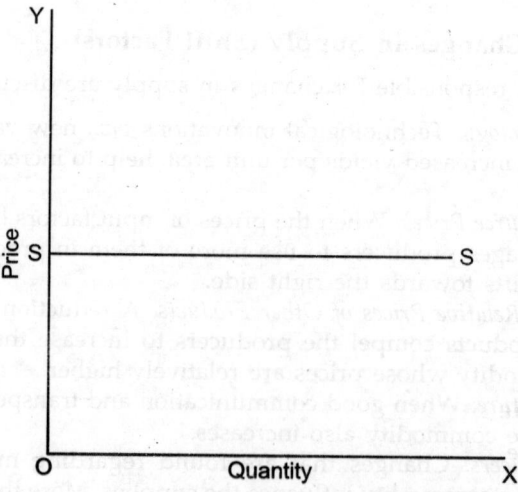

**Figure 5.5** Perfectly elastic.

## Perfectly Inelastic Supply

It means that the quantity supplied is not responsive to change in prices. Elasticity of supply in this case is zero ($E_s = 0$) (Figure 5.6).

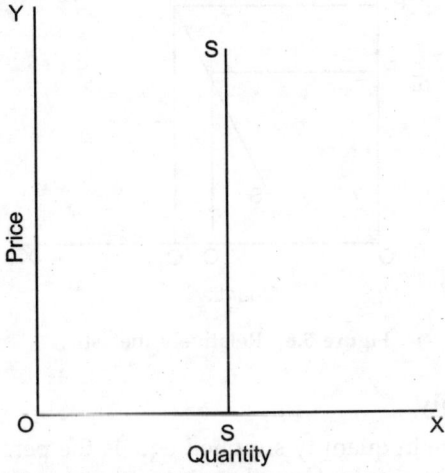

**Figure 5.6** Perfectly inelastic.

## Relatively Elastic Supply

Supply is referred as relatively elastic, when the percentage change in quantity supplied is more than the corresponding percentage change in price. It is also called elastic supply. Elasticity of supply is more than one ($E_s > 1$) (Figure 5.7).

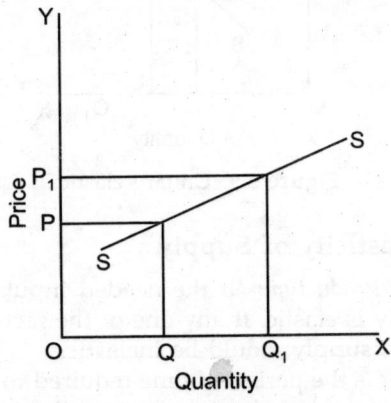

**Figure 5.7** Relatively elastic.

## Relatively Inelastic Supply

Supply is said to be relatively inelastic, when the percentage change in quantity supplied is less than the corresponding percentage change in price. In this case the elasticity of supply is less than one ($E_s < 1$) (Figure 5.8).

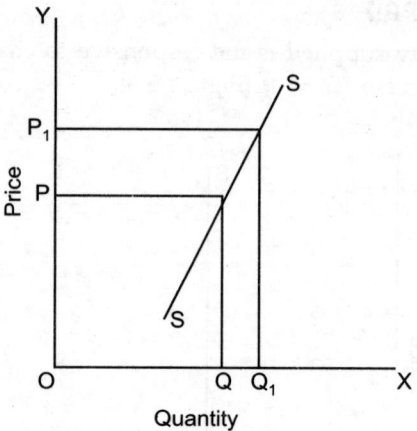

**Figure 5.8** Relatively inelastic.

## Unitary Elastic Supply

When percentage change in quantity supplied equals the percentage change in price, it is called unitary elastic supply. Here the elasticity of supply is equal to one ($E_s = 1$) (Figure 5.9).

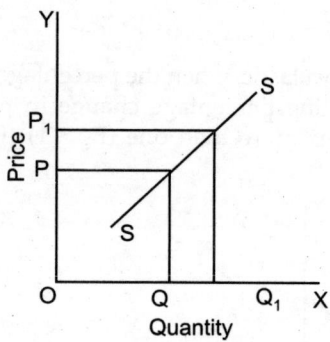

**Figure 5.9** Unitary elastic.

## Factors Influencing Elasticity of Supply

1. *Availability of Inputs of Production:* If the needed inputs are available as per the requirement, the supply is elastic. If any one of the factors is not available which is absolutely necessary, supply would be inelastic.
2. *Length of Time Period:* It is the period of time required to adjust the supplies to the changes in prices. The biological characteristics of the product dictate the changes of responsiveness.
3. *Diversification of Production Activity:* When the producer is engaged in production of a number of products and facilities exist for shifting of production from one product to the other, in such a case for each product the supply is elastic.
4. *Availability of Alternative Markets:* Suppose there exists several markets for the producer to sell the goods, a fall in price in one market would prompt him to shift his goods to other markets and a rise in price in one market induces him to shift his goods to that market. In such a case the supply is elastic.

5. *Flexibility in Starting and Winding up the Business:* If a particular production activity is quickly taken up and quickly wound up, the supply of the goods is elastic.

## PRICE DETERMINATION

Having studied the demand and supply, we know that market demand curve is a horizontal summation of individual demand curves, and similarly horizonial summation of individual supply curves become market supply curve.

Price determination can be examined arithmetically, graphically and algebraically.

### Arithmetic Approach

The information in Table 5.3 reveals that at Rs. 12, the quantities demanded and supplied are both equal *i.e.,* 80 Q. At this price, what the buyers are willing to purchase and what the sellers are willing to offer are the same. Therefore, Rs. 12 per unit is the equilibrium price and quantity amounting to 80 Q is the equilibrium output.

**TABLE 5.3 Demand, Supply and the Equilibrium Price and Output.**

| Price/ unit (Rs) | Quantity demanded (Q) | Quantity supplied (Q) |
|---|---|---|
| 14 | 60 | 120 |
| 13 | 70 | 100 |
| 12 | 80 | 80 |
| 11 | 90 | 60 |
| 10 | 100 | 40 |

### Graphic Approach

The intersection of market demand curve (DD) and the market supply curve (SS) indicates the equality of quantity demanded by the consumers and that suppli 1 by the producers Figure 5.11. This equality of quantity demanded and quantity supplied is called equilibrium quantity (OQ) and the price that occurs at this balancing point is called equilibrium price (OP). When such a condition prevails in a market, the market is said to be in equilibrium, because there are neither shortages nor surpluses of commodities (Figure 5.10).

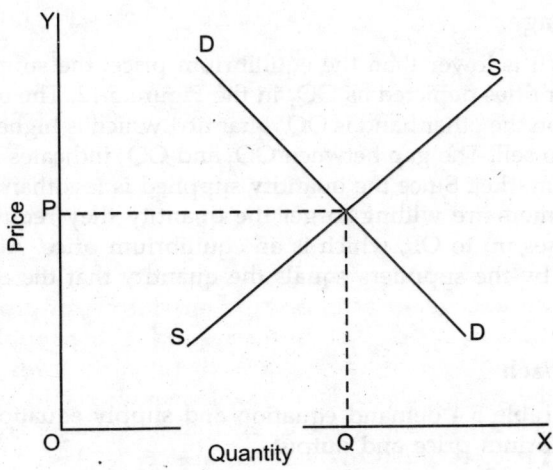

**Figure 5.10** Determination of equilibrium price and quantity.

## Market Disequilibrium

When the market equilibrium gets disturbed, it leads to market disequilibrium. This situation occurs when there is surplus or shortage of a commodity.

## A Case of Surplus

As observed in Figure 5.11 at $OP_1$ price suppliers are prepared to sell $OQ_2$ quantity, while the consumers prefer to buy $OQ_1$ quantity only. Hence $OP_1$ price is not in equilibrium as quantity to the extent of $(OQ_2 - OQ_1)$ is available as surplus. Since there is a surplus, the market is in disequilibrium. Suppose if the seller wants to sell the surplus they have to oblige the pressure of consumers to reduce the price. When the same takes place *i.e.,* when price slides down to OP the consumers are prepared to purchase what all the suppliers offer for sale.

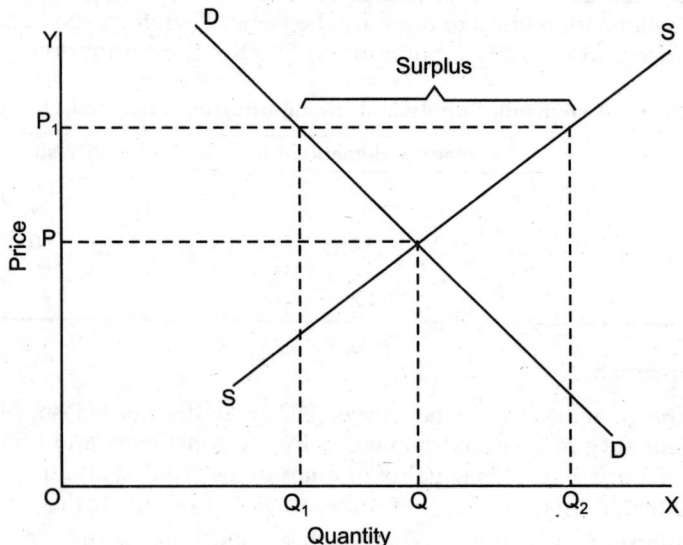

**Figure 5.11** Market disequilibrium – A case of surplus.

## A Case of Shortage

At price $OP_1$, which is lower than the equilibrium price, the suppliers are prepared to sell smaller quantities depicted as $OQ_1$ in the Figure 5.12. The quantity demanded by the consumers on the other hand is $OQ_2$ quantity, which is higher than the quantity the sellers prefer to sell. The gap between $OQ_1$ and $OQ_2$ indicates the shortage of the commodity in the market. Since the quantity supplied is less than the demand in the market, the consumers are willing to get the quantity they required consequent to which the price rises up to OP, which is an equilibrium price. At this price OP, the quantity supplied by the suppliers equals the quantity that the consumers demand *i.e.,* OQ.

## Algebraic Approach

From the data in Table 5.4 demand equation and supply equation are estimated to find out the equilibrium price and output.

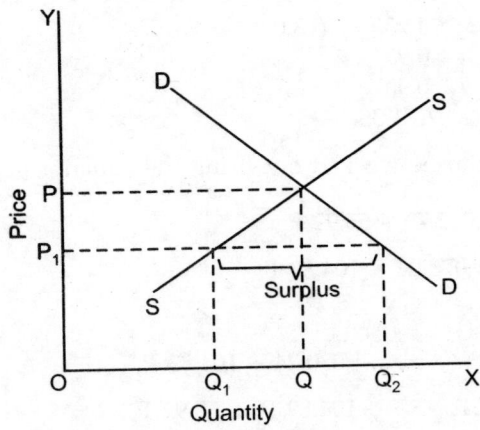

**Figure 5.12** Market disequilibrium – A case of shortage.

**TABLE 5.4 Estimation of Demand and Supply for Fish in the Market (Hypothetical Data).**

| Sl. No. | Price (Rs./kg) | Quantity demanded (kg) | Quantity supplied (kg) |
|---|---|---|---|
| 1 | 15 | 2,100 | 895 |
| 2 | 17 | 2,050 | 929 |
| 3 | 20 | 2,005 | 943 |
| 4 | 22 | 1,858 | 1,056 |
| 5 | 23 | 1,725 | 1,160 |
| 6 | 25 | 1,652 | 1,328 |
| 7 | 28 | 1,542 | 1,530 |
| 8 | 29 | 1,476 | 1,861 |
| 9 | 30 | 1,357 | 2,055 |
| 10 | 31 | 1,264 | 2,213 |
| 11 | 34 | 1,185 | 2,542 |
| 12 | 35 | 1,066 | 2,715 |
| 13 | 36 | 955 | 1,864 |
| 14 | 38 | 876 | 3,100 |
| Average | 27.36 | 1,507.93 | 1,799.36 |

Correlation coefficient: -0.9906
The demand function is estimated using the data of the Table 5.4

$$Q_t^d = 3037.75 - 55.92\ P_t$$

| | | | |
|---|---|---|---|
| SE | = | (63.07) | (2.33) |
| t | | 48.16 | (25.03) |
| $R^2$ | = | 0.98 | $R^{-2} = 0.98$ |
| F | = | 626.52 | |

The supply function is estimated using the data shown above.
Correlation (r)= 0.9667

$$Q_t^s = -1070.24 + 104.89\ P_t$$

SE                 (226.24) (8.01)

| t | | 4.73 | (13.09) |
|---|---|---|---|
| $R^2$ | = | 0.93 | |
| $R^{-2}$ | = | 0.98 | |
| F | = | 171.35 | |

Equilibrium price can be worked out using the equations

$$Q_t^d = 3{,}037.75 - 55.92\,P_t \qquad\qquad (1.6)$$

$$Q_t^s = -1{,}070.24 + 104.89\,P_t \qquad\qquad (1.7)$$

$$Q_t^d = Q_t^s \qquad\qquad (1.8)$$

$$3{,}037.75 - 55.92\,P_t = -1{,}070.24 + 104.89\,P_t$$

$$= 3{,}037.75 + 1{,}070.24 = +104.89\,P_t + 55.92\,P_t$$

$$4{,}107.99 = 160.81\,P_t$$

$$P_t = \text{Rs. } 25.54$$

$$P_t^* = \text{Rs. } 25.54$$

(equilibrium price)

Equation here P* is equilibrium price, substitute P = 25.54 either in equation (1.6) or (1.7), we can get equilibrium quantity.

Equilibrium Quantity = 1,609.24 kg

# CHAPTER 6

# Production

Production means creation of value in the goods. Production activity helps in transforming a set of inputs into goods or services. It essentially means transforming of one set of goods into another. The output which comes out of production has greater utility over the inputs combined in the production process. The inputs that are used in production of goods may be provided by the nature and/or by other industries.

## Characteristics of Modern Production

1. *Large Scale Production:* Production of goods and services on a large scale is essential in order to meet the increasing demand due to increased population, purchasing power, change in attitudes, *etc.* Besides, if production is carried out on large scale, the firm derives many benefits *viz.,* economy of buying and selling, economy of transportation, cheap credit, better utilization of by-products, *etc.* This ultimately leads to supply of goods and services at reasonable prices to the consumers.

2. *Use of Machinery:* Production of goods and services can be augmented using either labour-intensive technology or capital-intensive technology depending upon the availability of labour and capital. Labour intensive technology involves more use of labour and less of capital, whereas capital-intensive technology employs more of capital and less of labour. In general, labour-intensive technology suits well in small scale and cottage industries. Since large scale production is the order of the day, production of goods and services on a large scale with in a short period of time is possible only through the use of specialized machinery.

3. *Immense Use of Capital:* Production without capital is impossible. More so large scale production, demands huge financial requirements. Since the availability of funds with individual entrepreneur is inadequate, borrowing becomes necessary to carry out the business. In this regard a good number of institutional agencies like State Financial Corporations, Industrial Development Bank of India, and also commercial banks are readily providing the needed credit.

4. *Division of Labour:* Production of an article is divided into a number of processes and sub-processes. Each process is carried out by a set of workers and this is known as division of labour, which is associated with increased efficiency.

5. *Competition and Combination:* Large scale production results in larger amount of output. In order to have a greater share of the market for their products, there will be intense rivalry among the firms. The resultant cut-throat competition may ruin the industry. Therefore, amalgamation of the competing firms is necessary in order to reduce competition, uncertainty and entry of new firms and increasing profits.

6. *Joint Stock Companies:* The most common type of business organization is joint stock companies. This type of business organization facilitates mobilization of larger finances, through the issue of public shares to carry on the production on a large scale.

7. *Localization of Industries:* Concentration of an industry in an area is known as localization of industries. This is due to the availability of raw material, labour, transportation, communication, marketing, *etc.* Cotton mills in Bombay, leather industry in Kanpur, jute mills in Kolkata, *etc.,* are the examples.

8. *Privatization:* Government or centrally held property is tranferred to private ownership.

9. *Unemployment and Underemployment:* Extensive use of machinery replaces human labour leading to unemployment and underemployment.

10. *Price and Product Differentiation:* A product is differentiated if any significant basis exists for distinguishing the goods of one seller from those of another. The differentiation may be based on certain physical characteristics of the product, such as trademark, design, colour, *etc.*

11. *Advertisement:* It is an important marketing tool used to influence consumer preferences among different brands and products. Advertisement is an important selling cost. Advertise or perish is the modern slogan, for, the survival of the very business unit depends on its ability to popularize the product to the extent it can. Advertisement not only increases the demand for the product but also helps in the reallocation of resources.

12. *Speculation:* It is an act of buying and selling goods with the aim of making profits by correctly guessing uncertain price movements. Share market is a classical example of speculative business.

13. *Risk:* Modern production is characterized by high degree of risk and uncertainties due to changes in technology, economic environment, preferences of the consumers, *etc.*

## FACTORS OF PRODUCTION

These mean the production resources required to produce a given product. Fraser defined factor of production as *"a group or class of original productive resources"*. The factors of production have been traditionally classified as land, labour, capital and organization, the details of which are presented below:

## LAND

According to Marshall, land means *"the materials and forces which nature gives freely for man's aid, in land and water, in air and light and heat"*.

## LABOUR

The term, labour has wide and diversified meaning in economics. It can be physical work or mental work that is done by a person with an aim of earning money. It includes the work done by farmers, workers, the services of teachers, doctors, actors, *etc.* In the words of Marshall, labour is defined as *"any exertion of mind or body undergone partly or wholly with a view to earn some good other than the pleasure derived directly from the work"*. Any work that is done for pleasure does not come under labour.

## Characteristics of Labour

1. *Labour is Inseperable from Labourer:* The worker has to sell his labour in person and he has to be physically present, while delivering the work. He cannot deliver the work in absentia. It varies from labourer to labourer depending on races, climate, physical and mental alertness of labourer.
2. *Labour is Perishable:* Labour cannot be preserved which means that labourer has no reserve price. He has to sell the work without really minding the wages, for, a day's work lost is a loss forever. In other words, it is a flow resource.
3. *Labour has Very Weak Bargaining Power:* Perishability of labour is a prime factor for the labourer, which rather forces him to accept whatever the wage that is offered. The weak bargaining power of the labourer is taken as an advantage by the employer.
4. *Lack of Free Mobility:* Compared to capital, labour is less mobile. No doubt labourers move from one place to another and from one occupation to another, but it is not a common feature. Thus, labour lacks horizontal and geographical mobility. This leads to a variation in wages among the occupations as well as spatially.
5. *Supply of Labour is Independent of Demand:* Supply of labour depends on the population in a country. Population is one factor which can neither be increased nor decreased overnight. The increase or decrease is a slow process and supply of labour is independent of demand.
6. *Supply of Labour Peculiarly Changes with the Wages:* Normally the seller of a good sells more when the price per unit of commodity is higher and *vice versa.* But regarding labour a fall in wages leads to an increased supply of labour. A fall in wages leads to a reduction of their incomes. So to make good this fall in income, family members who were not working earlier also work to supplement the family income.

## Efficiency of Labour

According to economists efficiency of labour means *"the ability of labour by virture of which it is productive".* It indicates the qualitative and quantitative performance of the labourer. No doubt, it is a relative term, but when one person puts up a better performance over others, we say that he is more efficient. All individuals are not equally efficient, because of several factors which are examined below:

1. *Racial and Hereditary Characters:* Some races are known for their physical and some are for their intellectual qualities. This emerges out of climate and natural environment. Pattans are well built and physically very strong. The Punjabies and Rajputs are hard working compared to other Indians.
2. *Literacy and Training:* Education brightens the outlook of labourers and builds up character. Training imparted to the labourers make them more efficient to do specialized works.
3. *Environment:* The working environment must be congenial for the labourers to put up an impressive performance. Environment refers to conditions in the work place in which the labourers have to work and the equipment and machinery which they handle. Better physical facilities, modern machinery and equipment encourage to get best out of them.
4. *Working Hours:* Long working hours retard the performance and the efficiency of labourers. The working hours should facilitate them to relax. Short working hours keep them fresh and help to rejuvenate their energies.

5. *Fair and Prompt Payment:* Wages influence the performance of labour. Fair and prompt payment of wages encourage them to put their heart and soul into the job.
6. *Capability of Organization:* Organisation has to plan to allocate the right people at right places for the right work. Assignment of work as per the capabilities of labour leads to expected output and maximum profits.
7. *Social and Political Factors:* Social securities for the labourers and concern of State towards their conditions will be quite encouraging to improve their efficiency.
8. *Personal Qualities:* Personal qualities like intelligence, alertness, honesty, health *etc.,* improve the efficiency of labourers.
9. *Incentives:* Incentives like extra payment, and some perquisites for a good performance encourage them to be more devoted to the work.

## Division of Labour

In modern production activity, the production of a good is divided into several sub-processes, and each sub-process is entrusted to a group of workers. This is what division of labour implies.

Division of labour is meant to improve the efficiency of labourers. There are three different types of division of labour *viz.,* (1) Simple division of labour (2) Complex division of labour; and (3) Territorial division of labour

## 1. Simple Division of Labour

It is also known as occupational specialization. This means that people in a society undertake various occupations to make their livelihood. The choice of an occupation depends on the suitability of an individual in serving the society. That is how we have in a community, some are doctors, others are lawyers, some others are teachers and we have blacksmiths, goldsmiths and others craftsmen. They execute duties regularly and help the society to develop by helping themselves through their professions. Some professions like dry-cleaning, tailoring, *etc.,* have become of late very prosperous and lucrative professions. Hotelling and mechanical weaving have become giant industries, which were earlier to be taken up by only certain sections of the people as profession. From all these things we can conclude that division of labour is fast growing with full adoption of requisite technology and providing employment to millions of people.

## 2. Complex Division of Labour

It is assigning the work by task. The task here is a sub-process that is found in producing a commodity. Each group of people is given a task in which it is considered as specialists. For example, in making silk cloth, many sub-processes such as reeling, weaving, dyeing, *etc.,* are involved and for all these sub-processes we require sophisticated technology and trained people to run the silk industry. Similarly, in the manufacturing of tractors, electric motors, TVs, *etc.,* many sub-processes are involved.

## 3. Territorial Division of Labour

It refers to localization of industries. Certain areas or regions specialize in production of a commodity. The examples are textile mills in Bombay and Ahmedabad, silk sarees in Kanchi, jute mills in Kolkata, leather in Kanpur, *etc.*

## Merits of Division of Labour

1. *Increases Productivity:* In the modern industrial system, division of labour is found to augment the labour productivity to a large extent. Besides, it helps to improve the productivity of other resources such capital and management in industry.

2. *Improvement in Dexterity:* Specializing in a particular activity and performing it continuously make the workers perfect. This perfection of work helps them to devise improved methods to perform the activity. In the production of goods and services, the entrepreneurs lay more emphasis on reduction of labour costs. In this process they devise many labour saving techniques such as providing adequate skill and necessary equipment to do the maximum with minimum time and cost.

3. *Saving of Time:* Since the worker is concentrating in only one activity there is a saving of time, which otherwise would have been wasted, had he been attending to several activities in the manufacturing of a commodity. To perform such specific activities, the workers can be well trained. To save the time, the workers are to be given enough expertise to perform a work. They should be given incentives like bonus to putforth maximum concentration on the allotted work.

4. *Saving in Tools and Implements:* As a worker has to perform a specific function, he needs only a particular type of implements. In construction of farm ponds, formation of bunds, digging of wells, *etc.*, the labourers should be provided with suitable implements and machinery for turning out the work efficiently with minimum cost and time.

5. *Employment of Specialists:* In the populous country the demand for goods and services would be ever increasing. This leads to production of more goods and services. In response to this, entrepreneurs use more labour and technology intensively. The firm working for six hours earlier, would now work for more than 6 hours. Hence, 2 to 3 shifts have to be taken up by the firms to produce adequate goods for satisfying the increasing demand. Firms under such situations employ specialized labourers through intensification of work using requisite technology. Labourers are given adequate training to develop the requisite skills for executing the work efficiently.

6. *Large-scale Production:* Division of labour helps to produce goods on large scale. Further the goods can be produced at cheaper rates. For instance, we notice complex division of labour in cotton mills located at Coimbatore and Ahmedabad. The cloth making process is divided into many sub-processes and in each sub-process, the industry uses highly specialized workers for turning out the required work. Such process is basically meant for producing cloth on large scale with economies in the cost of manufacturing of required fabrics with various designs.

## Demerits of Division of Labour

1. *Monotony:* The work which is performed again and again becomes monotonous for the workers. Interest is not sustained in the work. It becomes boring for the workers.

2. *Retards Human Development:* Continuous performance of the same work, narrows his overall outlook. Since his faculties are tuned to perform a set work, his overall growth is stunted.

3. *Lack of General Responsibility:* Since many people are involved in producing a good, no body takes the general responsibility in correcting a defect, if it occurs. Everybody thinks that it is not his duty. Thus the workers become careless and irresponsible.

4. *Problem of Distribution:* Several people involve in production of a product. Based on the contribution, they should get their due share of product which is not an

easy task. This complicates the problems of distribution. This means distribution of dividend/bonus should be done scrupulously for satisfying the labour working in various divisions.

# THEORIES OF POPULATION

The theory of population is studied because the supply of labour depends upon population and its growth. Observing the abnormal growth of population and consequent fall in the standard of living, Thomas Malthus (1760-1834) an English Clergyman first studied population growth in various countries of Europe. Later he wrote a book entitled "An Essay on the Principles of Population" in 1798. His observations compelled him to foresee a gloomy future for the human race and hence emphasized the immediate need to keep the population growth under check. Now his theory is popularly known as Malthusian theory of population. To quote the theory in his own words "By nature human food increases in a slow arithmetic ratio, man himself increases in a quick geometric ratio unless want and vice stop him".

**Propositions of Malthusian Theory:** These are briefly presented below:

1. Population is necessarily limited by means of subsistence (food supply). According to Malthus in a country the size of population depends on its food production. Greater the food production, larger would be the size of population, which a country can support and smaller the food production then the country would be in a position to support smaller population only.

2. The power of population is infinitely greater than the power in the earth to produce subsistence for men. He further adds that population growth always overtakes food production, since agricultural production is constrained by the law of diminishing returns. This prevents the food production to grow as faster as that of population. As such, there is no limit to the growth of population.

3. Population increases in geometric progression, whereas food production increases in arithmetic progression. This means that population tends to grow in quick geometric progression i.e., 1, 2, 4, 8, 16, 32, etc., while, food production tends to grow in arithmetic progression i.e., 1, 2, 3, 4, 5, 6, etc. This implies that population doubles in every 25 years.

4. Population always increases, when the means of subsistence increase, unless it is prevented by some powerful checks. When the food production of a nation increases, the tendency is that it encourages people to have more children and consequently it leads to large families. After certain time, the population would become excessive relative to the food production and the per capita availability of food decreases.

5. There are two types of checks, which can keep population on a level with the means of subsistence viz., preventive checks and positive checks.

   a) **Preventive Checks:** These are man-initiated checks to keep the population under control. They aim at decreasing the birth rate. These checks are: 1) celibacy 2) late marriage 3) moral restraints, etc.

   b) **Positive Checks:** These are the checks caused by the nature when man fails to take control of the situation. These checks cause the reduction of excess population and thereby bring an equilibrium between demand and supply. We have often the famines, epidemics, wars, floods, earthquakes, *etc.* These positive checks are also called as natural checks, since they are caused by nature.

## Criticism of Malthusian Theory of Population

This theory was criticized by many economists on several grounds. Some of the important ones are presented below:

1. Presenting the mathematical precision for population growth and food production was objected because as these ratios are found to be unrealistic. There were no instances in the world, where this trend of population growth and food production recorded and the case of doubling of population in every 25 years.
2. Malthus did not foresee the scientific advances that are bound to come in agricultural production, even in the presence of operation of law of diminishing returns.
3. He did not consider total production of a nation but confined to agricultural production alone while presenting his theory. In fact it is total production of a country which gives the true picture of its economic position. Consider the example of Great Britain. It was not self-sufficient in agricultural production but it could make great progress through its industrial production. Less agricultural production cannot make the country to be called over-populous.
4. It was viewed that increase in population would cause greater demand for foodgrains, ignoring its contribution in the production of goods and services. Further it was criticised that increase in population in an over-populated country is a matter of serious concern, as it imposes heavy strain on the limited resources of the nation. In respect of an under populated country, increase in population is a welcome sign and it helps to raise economic growth and per capita income.
5. Malthus also failed to view that the education and civilization would transform the people to have small families. Hence, there would be no danger of over population.
6. Natural calamities which are infact the acts of nature are commonly found everywhere regardless of whether a particular country is over populated or under-populated. So, the statement that natural calamities would occur only in the over populated countries to reduce population is not true.

## Optimum Theory of Population

Prof. Sidgwick gave the foundation of the optimum theory in his principles of political economy. It was later developed by Edwin Cannon. It is also called as the modern theory of population.

## Definitions of Optimum Population

1. *It generally refers to that size of population which provides maximum income per head at a given amount of resources and technology.*
2. *Optimum population is that population which produces maximum economic welfare (Carr-Saunders).*
3. *Optimum population is that which gives the maximum income per head (Dalton).*
4. *Optimum population in a country ensures 1) best possible utilization of all resources; and 2) highest per capita income. More specifically, optimum population is the right size of population that is most desirable in a country consistent with its supply of resources.*

## Under Population

If the population of a country is below optimum size *i.e.,* below what it ought to be, the country is said to be under-populated. In such a case the per capita income will not be the highest, as the population is insufficient to use the available resources (both natural and capital) of the country efficiently. The resources are left unharnessed to their optimum level.

## Over Population

If the population of a country is in excess of optimum level, the country is said to be over populated. Since the resources are insufficient in relation to the population, the per capita income will not be the highest. Gainful employment is not available to people in view of the insufficiency of resources and requisite technology.

When we say optimum population, it is not fixed for all times. It keeps on changing with the development of technology and growth of capital.

In Figure 6.1 population is measured on X-axis and output per capita on Y-axis. It is observed that output per capita increases with every increase in population till OQ is reached. At this level of population (optimum population) the per capita output is the highest and is equal to QP. Any further increase of population beyond the OQ leads to reduction of output per capita. If the actual population of a country is less than optimum population (OQ), it is under populated and if it is more it is over populated.

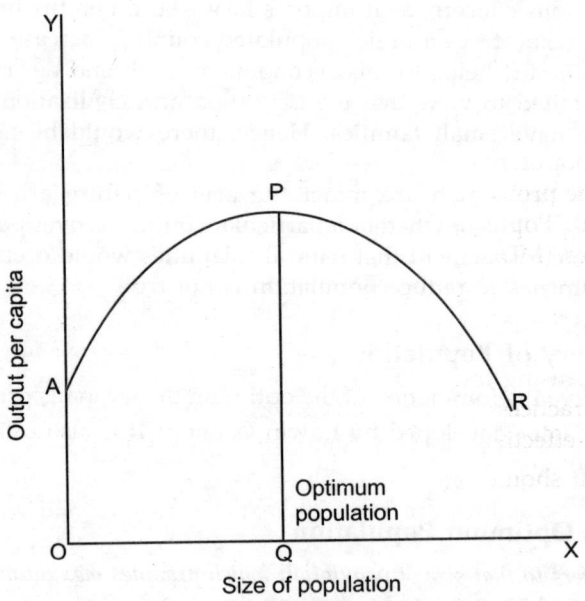

**Figure 6.1** Under, over and optimum population.

## Dalton's Formula for Maladjustment

Dalton's maladjustment means the extent of deviation of population from optimum size of population. To measure maladjustment, Dalton gave the following formula.

$$M = \frac{A - O}{O}$$

Where,

M = Maladjustment
A = Actual population
O = Optimum population

A positive 'M' indicates that the country is over-populated.

A negative 'M' indicates that the country is under-populated.

A zero value of 'M' indicates that the actual population is equivalent to optimum population.

## Malthusian Theory of Population Vs Optimum Theory of Population

1. Malthusian theory is based on the relationship between food production and the size of population, while modern theory is based on the relationship between total production of a country and the size of population.
2. Malthusian theory is more a pessimistic theory, while optimum theory is an optimistic one.
3. Malthusian theory has a limitation in its application as it is applicable to over populated countries only, whereas the optimum theory is applicable universally.

## Population Control Policy

In the countries where population is large we observe slow pace of economic development and growth. Capital and other requisite resources would be overexploited by the increasing population. As a result economic growth and development in the country retards. The problems like scarcities, hunger, malnutrition, diseases *etc.*, will become aggravated. Populous countries have to face stiff competition from more developed countries (MDCs) for resources like land, water, power, *etc.* Sometimes harsh living conditions of the population lead to very high birth rate and death rate. The problems of peace, poverty, epidemics and endemics in the country cause environmental degradation, pollution *etc.*, in the country. Urbanization would become worse sometimes causing the problems unsolvable and persistant.

The following practices should be given wide publicity for implementing population control policy effectively in the country.

1. The Government should encourage the couples to defer their marriages. Typical marriage ages for example, are 24 to 28 years for women and 26 to 30 years for men.
2. Married couples should be encouraged to follow the family planning techniques for avoiding immediate conception through incentives. They should be motivated to sign the pledges with the Government to have one child for each family as is being done at present in China. To make this policy successful, the married couples should be given incentive like bonus, additional annual increments, large sum of old-age pension, free medical care, free school education for their children, *etc.* There should also be provision to recover the incentives from the couples who violate the pledge.

3. The parents with two children should be strictly sterilized. Newly married couples should be given free and accessible contraceptives.

## Population Growth and Policy in the World

At the beginning of the Christian era *i.e.*, in the year AD1, the world population was about 250 million and the population was growing at the rate of 0.04 per cent per year. Later, the world population was increasing at the rate of 2 per cent and crossed 4 billion (400 crores). This means the population of the world was doubling in every 35 years. Now in the year 1999, the world population is nearly equal to 6 billions (600 crores), of this India constitutes 100 crores *i.e.*, 1 billion. This indicates that 1/6th of population (16.67 per cent) live in India. Another trend of the population growth to be noted cautiously is that every year, 7.8 crore people are being added to the existing population in the world. As a matter of fact, we see the higher growth rate of population in under developed countries and developing countries.

## Effects of Population Growth on Economic Development

Additional national output would be obtained as long as the marginal value product (MVP) of the persons is positive. But when MVP of the person is lower than average product of the person, further addition of persons would simply reduce the welfare of average citizen resulting in lower standard of living. We also know that the market for goods and services expands resulting in increased demand for goods and services, further, large scale farming becomes possible which eventually adds to the productivity of the persons in the economy. Due to all these obvious reasons, the per capita output increases up to certain time, as population increases. But over time if the population in continuously growing then a point would be finally reached where capital and natural resources are fully utilized.

## The Malthusian Dilemma[1]

It may be fairly pronounced, that considering the present average state of the earth the means of subsistence, under circumstances the most favourable to human industry, could not possibly made to increase faster than in arithmetical ratio[2].

In two centuries, since Malthus brought out his theory of population, food production as well as population, have followed the path of geometric pattern. However, hunger though was not eliminated totally, in the course of these two centuries it was less widespread in the past few decades. Since the year 1650, the growth of population was slow and rapid growth which is seen at present, is a recent phenomenon and it is what Malthus meant a geometric rise in population growth.

In the light of geometric expansion of population, had there been no similar growth in food production the catastrophe which Malthus predicted would have been a reality. The factors which contributed to increased food production within a century following Malthus prediction, are bringing new lands under agriculture, increased irrigation facilities, use of high yielding varieties, chemical fertilizers, shifting land to food crops from fodder crops, *etc.*

---

[1] Hunry B. Arthur and Gail L. Cramer, "Brighter forecast for the world's food supply". Harvard Business Review, May-June 1976.
[2] Thomas R. Malthus, The Principle of Population, 1878 Ed. London: Reeves and Turner, London, pp. 5-6.

In one sense *i.e.,* population growth, Malthus was right in his prediction, but the improvements in world food production averted prediction of Malthus on food front. Though the global birth rate has fallen in the past 20 years and the world's food production has overtaken the population growth, we cannot be complacent that adequate supplies will be attained in future also *i.e.,* we are not exactly sure how far into the future the situation continues. For the present, the dilemma persists.

# CAPITAL

Capital is not an original factor like land, but it is the result of man-made efforts. Man makes the capital goods to produce other goods and services. For example, machinery, raw material, transport equipment, dams, *etc.,* are considered as capital goods.

## Definitions

1. Capital is produced means of production.
2. According to Karl Marx, capital is *'crystallized labour'*. This is indicated in his book, *'Das Kapital'*.

All capital is necessarily wealth but all wealth is not necessarily capital. Money when used for the purchase of capital goods, then only it becomes capital. For instance, residential buildings are the wealth of the individuals, but these are not considered as capital.

## Characteristics of Capital

1. Capital is not a free gift of nature. It is the result of man-made efforts. Machinery, implements, *etc.,* are considered as capital goods.
2. Capital is productive, as it helps in enhancing the overall productivity of all the resources employed in the production process. Invested capital also fetches interest for its productive capacity. Farm machinery when used with skilled labourers enhances the productivity of land. Irrigation dam is considered as the capital good and with its water, we can bring out complementary effect on the productivity of other resources such as fertilizers, seeds, *etc.*
3. It is also prospective as its accumulation rewards income in future. Savings and investment in the economy leads to growth and development of the economy due to accumulation of capital over time. This leads to a rise in nation's income and consequently individuals income.
4. Capital is highly mobile as it possesses the characteristic of territorial mobility. For example, capital goods like tractor can be taken to different places of work and can be used for a variety of works.
5. Capital is supply elastic as its supply can be altered according to the need. Based on demand, supply of the capital goods can be changed.

## Classification of Capital

Capital is classified based on several criteria the details of which are presented in Figure 6.2.

72

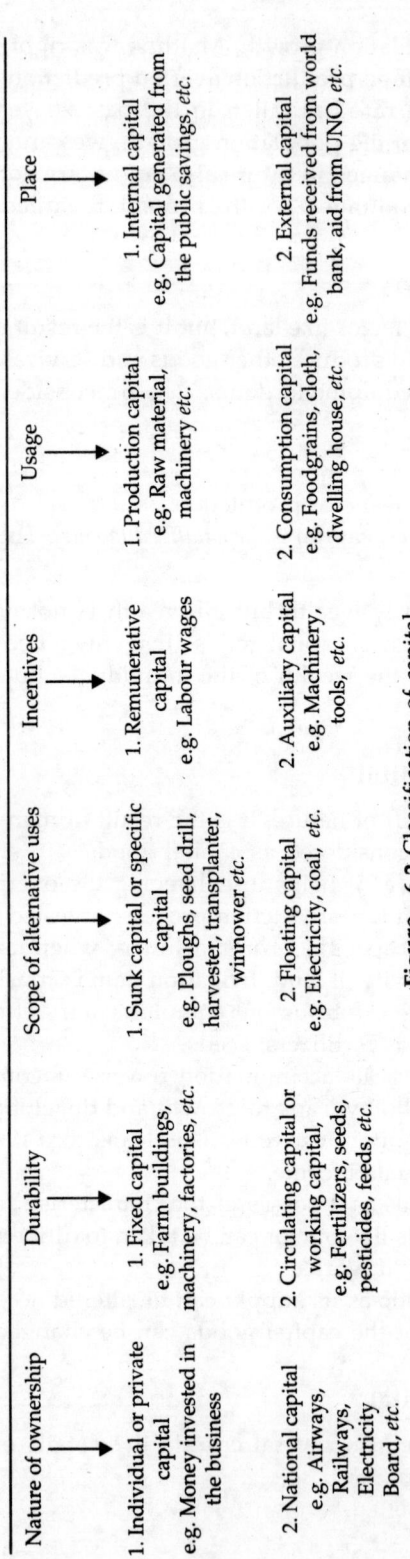

**Figure 6.2** Classification of capital.

## 1. Based on Nature of Ownership

*Individual Capital or Private Capital:* These are the assets which are owned by individuals that give service and/or income. *Example:* Money that is invested by the individuals in the business.

*National Capital:* It is the capital that is owned by the community. *Examples:* Airways, Railways, *etc.*

## 2. Based on Durability

*Fixed Capital:* It is the capital which is used time and again in the production of other goods. *Examples:* Machinery, tools, *etc.*

*Circulating Capital or Working Capital:* It is the capital that is used once and exhausts after a single use. *Examples:* Fertilizers, seeds, feeds, *etc.*

## 3. Based on the Scope of Alternative Uses

*Sunk Capital or Specific Capital:* These are the capital goods, the use of which is confined to a specific purpose. *Examples:* Plough, seed drill, harvester, transplanter, winnower, *etc.*

*Floating capital:* It is the capital good that can be used for different purposes at any time. *Examples:* Electricity, coal, *etc.*

## 4. Based on Incentives

*Remunerative Capital:* The capital when used for the payment of wages in the production process, is called remunerative capital. *Examples:* Liquid money or cash, foodgrains, *etc.*

*Auxiliary Capital:* It represents the various capital goods that help the labourers in the production process. *Examples:* Machinery, tools, *etc.*

## 5. Based on Usage

*Production Capital:* These are the capital goods which help the labourers directly in the production activity. *Examples:* Machinery, tools, *etc.*

*Consumption Capital:* These are the goods which are consumed by the labourers. These indirectly assist in the process of production. Examples: Foodgrains, cloth, money used for consumption, *etc.*

## 6. Based on Place

*Internal Capital:* The capital that is generated from the domestic savings of the public in a nation. *Examples:* Public parks, public roads, *etc.*

*External Capital:* This is the capital generated from the external source. *Examples:* Funds received from World Bank, aid from UNO, *etc.*

# ORGANIZATION

In any business activity, there is always a person who guides and controls its function. He also coordinates and regulates all the factors which are employed in the business activity. Apart from monitoring it, he takes the responsibility of the outcome. We call such a person, an entrepreneur (organizer) and the business activity which he is doing is called an enterprise or organization. The performance of an organization depends upon the capabilities of the organizer or entrepreneur. Through proper allocation of resources, the entrepreneur would be in a position to maximize productivity of the resources that are used. Hence, the success or failure of enterprise depends on the role of the entrepreneur in any business activity. Following are the prime functions of an entrepreneur.

## Functions of Entrepreneur

1. *Identification and Initiation:* Entrepreneur is the person who identifies a particular business activity and initiates the idea of commencing the business. At this stage the nature of enterprise to be started is decided by the entrepreneur. He has to venture and assume full risk for maximizing profits from his chosen business enterprise.
2. *Location of the Enterprise:* The place where the business unit is proposed to be set up is finalized by the entrepreneur. He considers both absolute and relative advantage of enterprise in choosing a particular place.
3. *Organizing:* At the outset arrangements are made to raise the required finance to commence the business. Later, the work is divided into various functions, *viz.*, production, financing, marketing, *etc*. This division of functions allows for increasing the efficiency of the individuals. Once the divisions are made, the concerned persons are given adequate authority to enable them to perform their functions efficiently. Delegation of authority to a person makes him more responsible and scrupulous. Authority and responsibility always go together.
4. *Supervision:* It is overseeing the work of the subordinates so as to ensure the desired results. Supervision includes whether the persons and their subordinates are keeping time schedules, whether the work is performed as per the requirements, and also helping them in solving their problems.
5. *Introduction of Innovation:* The entrepreneur is always on the look out for bringing innovations. It may be in terms of bringing out a variety of products, introduction of new methods of production, improving the existing method of production, making inroads into new markets, *etc*.
6. *Risk Taking:* The entrepreneur is quite prepared to accept the outcome of all his actions. In the business he may get substantial profit from organization or at times unexpectedly he may incur loss as well. He accepts them with poise, because he is responsible for all the happenings. All entrepreneurs, infact should have risk bearing ability.

# FORMS OF BUSINESS ORGANIZATION

A business activity can be organized in different ways. In practice we come across business organization owned and run by a single person or group of persons. The ownership patterns of business when looked into, reveal five distinct forms *viz.*, 1) Individual organization or single proprietorship or sole trader concern organization or the sole proprietorship; 2) Partnership; 3) Joint stock company; 4) Co-operative-organization; and 5) Sate enterprise.

The selection of each form of business organization by the entrepreneurs is influenced by several factors *viz.*, 1) type of business unit contemplated; 2) the care with which an enterprise can be run; 3) owned funds available with the entrepreneur; 4) total capital requirements; 5) possibilities of securing borrowed capital; 6) the risk and liability aspects which the entrepreneur has to assume; 7) tax aspects of different forms of business organization; 8) Organizational, managerial and controlling aspects, *etc*.

### Individual Enterprise or Single Proprietorship or Sole Trader or the Sole Proprietorship

The individual enterprise is the most common form of business organization. Many small business enterprises belong to this form. These enterprises are owned and operated by a single person, who takes all the responsibilities of outcome of the business. These enterprises are found to be small with a few exceptions here and there. This is more or less a family proprietorship with all the family members participating in the business affairs. As far as the size of the business is concerned, it is left to the desire of the entrepreneur keeping in view of the resources at his disposal.

### Merits

The owner of the business enjoys absolute freedom without the interference of anybody in the business. The firms are called by the name of the entrepreneurs and some times by the name of Gods. The owner or proprietor enjoys all the profits received from the business. Capital requirements are less. Capital is supplied from the owner's funds (equity funds) and often times there is not much distinction between personal and business assets. This type of business is more flexible allowing changes in various business decisions like investment, sales, diversifying the business activities, expanding size of the business, *etc*. In this form of business organization, there is the possibility of direct contact with the customers, so that the entrepreneur gets continuous feedback. The entrepreneur controls the entire business, unless and otherwise delegated to somebody else. This type of business is very easy to start and easy to terminate.

### Demerits

Since proprietor's funds are by far the sole source of funds, there would be limited amount of capital for the business to expand. Though funds from institutional agencies are available, the amount is restricted in view of the limited securities. Unlimited liability is the negative factor of this type of business. This means that in the event of failure of the business, creditors (lenders) are empowered to exercise every right to attach not only the assets of the business, but also the personal property of owners to make good the unpaid debts. Since the power is concentrated in the hands of a single owner of the business, there is no scope for those employees of the firm who are well trained and motivated to contribute their knowledge to the business growth. This leaves a situation of discontentment among the employees, the attitude of which will not contribute to the growth of the business. Besides, employees always will have a lurking fear that their fate depends upon the skill of one individual. The continuity of the business is also questioned, as the death of owner brings the business to a grinding halt.

## PARTNERSHIP

It is an association of two or more individuals who join together as co-owners to share profits or losses in agreed proportions. Partnership comes into existence based on the goals of the co-owners. To safeguard the business interests of the partners, normally a written partnership agreement is made covering various dimensions of business *viz.*, capital contribution, managerial responsibilities, sharing of profit and losses, withdrawal from the business, termination of the business, *etc.*

There are two kinds of partnership, *viz.*, general partnership and the limited partnership. General partnership is the most common in partnership dealings. Every partner, irrespective of the percentage of capital contributed to the business, has equal say in the management of business. Each partner has equal rights and liabilities. In limited partnership, any number of limited partners are allowed, but there should be atleast one general partner. Liability of each member is limited to the extent of investment made only. Profits are also distributed among the partners according to the contribution of capital in the business.

There are different kinds of partner, *viz.*, active partner, sleeping partner, nominal partner, secret partner, *etc.* Active partner is one who is actively involved in running the business. He performs various roles like manager, organizer and adviser of business. A sleeping partner is one who contributes capital, shares profits and takes the responsibilities of losses of the business, but he does not participate in running the business. Normally persons having capital but who do not find time in the business affairs, prefer to be sleeping partners. Nominal partner is one who joins a business but does not contribute capital. He just lends his name for the business and on his virtues the business prospers. But he is identified as a partner by the third party. In the event of loss of a business, he is liable to the third party. This is because third party thinks him as partner. Secret partner is one whose name is kept secret. His name is not disclosed to outsiders. He is liable for the losses, if any. He differs from sleeping partner in the matter that he takes part in running the business.

## Merits

The basic advantage of partnership is generation of greater financial resources coupled with diversified managerial talents. Partners pool their resources to attract larger capital to invest in the business. It can command greater amount of credit from the institutional agencies in view of its large equity capital. Management element is also strengthened as partners possess varied managerial skills. Simplicity of the business is also another feature, for it is easy to dissolve as compared to a joint stock company. It also enjoys the freedom from Government control. Risk of the business is shared by the partners and hence relatively it has less business risk.

## Demerits

Unlimited liability is the major disadvantage of partnership. All the partners are responsible for the loss arising from the partnership business. The partners stand to lose their personal assets in the event of liquidation. As against the individual enterprise, the ownership is divided in partnership business. Every partner has equal right in the business activities. Partnership has a limited size of business and uncertain life. Partnership may vertically split due to disagreement on a particular decision. The retirement, death of a partner, bankruptcy *etc.*, may bring termination of the business according to law. Besides dishonest member may spoil the business with his dishonesty activities and make other partners to be responsible of his actions. It is not that much easy to find qualified and agreeable partners for managing the business on partnership. Since the decisions are taken by the consensus of all the partners, more often it is difficult to convince all the partners on certain decisions.

# JOINT STOCK COMPANY

The drawbacks of individual enterprise and partnership business gave rise to the organization of another form called joint stock company. More specifically the discouraging aspects of unlimited liability closely followed by limitation of funds in the above forms were taken care of by joint stock company. Joint stock company has the limited liability and the involvement of large number of persons. This helps it to have adequate capital. Limited liability implies that in the event of loss for the company, shareholder is responsible to the extent of his shares only.

A joint stock company is a corporate body owned by a large number of shareholders and managed by a Board of Directors elected by the shareholders. According to Prof. L.H. Hany "A joint stock company is a voluntary association of individuals for profit, having a capital divided into transferable shares, the ownership of which is the condition of membership". In India the first Companies Act was passed in 1850 and limited liability was incorporated in 1857. Broadly there are two types of joint stock companies. These are 1) Joint stock private limited company and 2) Joint stock public limited company. To start a joint stock company, two documents, *viz., memorandum of association* and *articles of association* are to be submitted to the registrar of joint stock companies. *Memorandum of association* contains the name of the company, the location of the head office, its aims, share capital particulars, kind and value of shares and declaration of limited liability. Rules and regulations for the establishment of joint stock company are incorporated in the *articles of association*.

## Joint Stock Private Limited Company

The minimum number of members is two but the number cannot exceed 50. There is no need for the private limited company to call for a statutory meeting. Similarly, the company need not submit its annual balance sheet to the registrar of joint stock companies. The transfer of share is generally restricted by articles. It cannot issue prospectus inviting the public to subscribe to the share capital. The word 'Pvt. Ltd' must be used with the name of the company.

## Joint Stock Public Limited Company

The business can be started with seven persons and there is no maximum limit for members. The business shall commence only after getting the certificate of incorporation from the registrar of joint stock companies. The public limited company must issue a prospectus inviting the public to contribute to the share capital. A statutory meeting must be held within a prescribed period and its annual balance sheet must be submitted to the registrar of joint stock companies. The main sources of finance for the company are through shares and borrowings. The shares are freely transferable. A shareholder can sell his shares at any time he prefers. The company will not return the shares to the shareholder till it winds up the business, but shareholders can easily sell their shares through stock exchange. The common types of shares are: 1) ordinary shares 2) preference shares; and 3) deferred shares. Ordinary shares are those for which no special privilege is given. Ordinary shareholders get dividend from the net profits of the company. Preference shares are those which carry a certain fixed dividend from the net profits. This dividend is paid to the preference shareholders before dividend is paid to other kind of shareholders. At the time of liquidation of the company after paying outside creditors, preference share capital is returned. Ordinary shareholders will be paid only when preference share capital is paid in full. Deferred shareholders receive dividend, after payment of dividend to ordinary and preference shareholders.

Apart from selling shares the company raises required capital by floating debentures. A debenture is a document under the company's seal which provides for the payment of a principal sum and interest there on at regular intervals, which is usually secured by a fixed or floating charge on the company's property or undertaking and which acknowledges a loan to the company. Debentures are not shares but are just like promissory notes from which funds are raised. It is a form of loan and debenture holders (creditors) are paid interest as promised whether the company gets profits or not.

## Merits

1. *Large-scale Resource Mobilization:* It facilitates mobilization of large scale resources. Large sum of capital can be raised from large number of shareholders. There is no limit as far as the number of shareholders is considered in a public company. If more funds are required the number of shareholders can be increased.
2. *Efficient Management:* The elected board of directors and expert managers provide the needed business expertise. The efficient management of the joint stock company provides the needed impetus for business growth.
3. *Limited Liability:* Limited liability encourages many individuals to invest in shares.
4. *Less Risk for the Shareholder:* Because of limited liability even in the event of company incurring losses, the loss for the shareholder is only the face value of the share. Hence there is less risk for the shareholder.
5. *Perpetual Existence:* It is an organization with perpetual succession. The shareholders keep on changing from time to time. But the business of the company is not affected. The death or insolvency of any shareholder does not affect the existence of the company.
6. *Democratic Management:* The directors are elected by the shareholders, hence there is no scope for the continuation of undesirable directors. Moreover, every individual whomsoever wants to become a shareholder, he is welcome and shareholders come from all walks of life and places.
7. *Social Benefits:* The savings of the people which are otherwise scattered are well mobilized by companies and productively invested. Thus the society gains from the investment activities in the form of getting the goods and services they need.
8. *Economies of Scale:* Since the companies are large-scale organizations, they enjoy economies of scale and produce goods at lower costs and receive more profits.

## Demerits

1. *Concentration of Economic Powers:* The owners of the company are shareholders but management is done by different individuals. The administration is concentrated in a few hands. The shareholders, who are scattered all over, cannot influence the management. They are either powerless or not interested to act as per their desires.
2. *Fraudulent Management:* The management of the company exhibits vested interests and shows little concern for the shareholders.
3. *Delays in Decision-making:* Decisions cannot be taken quickly and they are to be taken in the meeting of board of directors or general body. It is not very easy to convene the meetings and they are time-consuming. These delays may cause further delays and result in deferred decisions.

4. *Excessive State Regulation:* The companies are governed by a number of rules of the Government. It is infact compulsory, because huge public funds are invested in the companies. Since non-compliance leads to penalties, companies have to give a lot of attention for these rules. Consequently, the main objective of the company is likely to be diverted.

5. *Evils of Factory System:* The evils of factory system like insanitation, pollution, congestion, *etc.,* are attributed to joint stock companies.

6. *Problems in Formation of Companies:* As a matter of fact a number of stages are involved in the promotion of a company by Government. Getting right persons in sufficient number for the company is a difficult proposition. A number of legal formalities are to be followed at the time of registration apart from the risks in promotion of the company.

# CO-OPERATIVE ORGANIZATION

The term, co-operation implies the self help made effective through mutual help. The philosophy behind co-operative movement embodies in a slogan called *"all for each and each for all"*. The basic objective of co-operation is protecting weaker sections of the society so that they fulfill their needs. Various types of co-operative societies are: 1) Consumers' co-operatives 2) Producers' co-operatives and 3) Credit co-operatives

## Consumers' Co-operatives

These are present in rural and urban areas. Members in an area contribute capital to form into a society. Any person, regardless of caste, creed or religion can become a member of the society. The society is run by the elected executive members. The society undertakes bulk purchases of consumer goods and sells to the members. In this process the middlemen are eliminated. Non-members are also allowed to buy the goods but they are charged extra price. The society makes small profits to cover the administrative costs. The surplus of profits are distributed among the members as dividends. A certain percentage of profits is kept aside as reserve fund for contingencies and growth of the co-operatives.

## Producers' Co-operatives

These are the associations of producers which help them in procuring inputs and in marketing their produce. These societies are formed with a sole aim of improving the economic conditions of producers. The society supplies the raw materials to these members who produce the goods. The society takes the responsibility of selling the goods. The members as workers are paid wages for their services. Part of the profits is retained as reserve fund and the balance is distributed among members. *Examples*: Weavers' societies, co-operative farming societies, *etc.*

## Credit Co-operatives

Credit co-operatives are established to protect the small farmers and other weaker sections. Here, through these co-operatives, weaker sections of the society are protected from the clutches of moneylenders, who charge exorbitant rates of interest. Credit co-operatives are categorized into two. 1) Rural Credit Co-operative Societies and 2) Urban Credit Co-operative Societies

1. Rural credit co-operatives can be formed with atleast 10 members. Individuals join as members by contributing to the share capital in the form of shares. The societies receive loans from State Co-operative Bank, and these are advanced to the members as short term loans. These societies keep up a margin while advancing loan to meet the administrative costs. The area of coverage of these societies is confined to one or two villages.
2. Urban co-operative credit societies are meant to advance loans to small traders, artisans and employees receiving small incomes. The members are provided short term loans. Liability is limited. Urban co-operatives raise their capital from Governmental agencies and members.

## Merits

1. Membership is open to every person. None can prevent any person willing to join the societies.
2. Management of the co-operatives is democratic. The members among themselves elect the board of management. Every member has equal right in electing the members irrespective of the number of shares.
3. The co-operatives purchase goods from producers directly and sell them to consumers directly. In this process the middlemen are eliminated.
4. The motto of co-operatives is service, but not profits. Co-operatives aim at spreading the virtues of discipline, integrity, honesty, mutual help, fairness in dealings, *etc*.

## Demerits

1. They suffer from timely and capital inadequacies. Societies aim at the betterment of weaker sections and the shares raised from them are of small magnitude. This limitation stands in the way of initiating a large scale enterprise.
2. Since there is no bar in entering into a society for anybody, the members are drawn from different sections of the society. This creates lack of understanding among the members. The members as a result do not take much interest and leaves everything to paid workers.
3. The transactions of the society are in cash and no credit sales are allowed. Since the members come from poorer sections of the society, they cannot always transact business with cash. Credit facilities which are found with private traders attract them to buy their requirements.
4. Societies function under the regulation of the Government. Government even nominate members to the management committee. In nominating the members of political parties take a major role and the business atmosphere suffers.

## STATE OR PUBLIC ENTERPRISE

State enterprise is an undertaking, owned and controlled by the local or State or central Government. Entire investment or major part of the investment is done by the Government. The major considerations for the States to undertake the business are heavy investment requirements, need to protect weaker sections against economically strong, and when private traders are hesitant to venture into the enterprise. State enterprises are found in manufacturing, trading and service activities. These enterprises are managed by the Government. The Government programmes are implemented through State enterprises.

## Merits

1. Industrial development is possible through State enterprises. Private sector does not show much concern for initiating projects requiring huge capital and long gestation periods.
2. Planned and balanced growth is possible through the entry of Government. Private enterprises show their preference for establishing industries in developed areas. Government is prepared to establish industries even in underdeveloped areas which ensure balanced growth in all spheres of activities.
3. Government takes over the sick units, and run them as State enterprises in the interest of the nation.
4. The profits obtained by the State concern are ploughed back into the business for further expansion and diversification and also for the welfare of the community in general.
5. Government enterprises encourage socialistic pattern of society which reduce economic disparities.
6. There is an attraction for the aspiring qualified individuals to join the Government service. It commands superior talents.
7. The employees feel greatly secured in Government service.

## Demerits

1. The proposed projects by the Government are plagued by undue delays. This is due to the complicated procedural formalities coupled with non-release of funds in time. These delays make the planned estimates go topsy-turvy, consequently the expected benefits would not be forthcoming timely.
2. Another demerit of public concern is high overhead costs. These arise out of large amounts of expenditure on unproductive items coupled with high investment on amenities for employees even before the profit is earned.
3. State enterprises when compared to private enterprises are not managed efficiently resulting in losses.
4. The security of the job of an employee in a State organization makes him not to bother too much to deliver the goods, for he gets his pay regularly.
5. Manpower planning is a lacuna in State enterprises and they employ persons disproportionate to their needs. This results in overstaffing leading to inefficiency and lethargy.
6. Redtapism is prevalent in State enterprises.
7. These are by far service oriented rather than profit oriented.

### Comparison of Co-operative Organization and Joint Stock Company.

| Particulars | Co-operative organization | Joint stock company |
| --- | --- | --- |
| 1. Membership | A co-operative society can be formed with ten persons and there is no maximum limit for the members. Members are known to each other, because the members come from a particular area. | A private limited company requires a minimum of two members and the maximum number cannot exceed fifty. A public limited company can be started with seven members and there is no maximum limit. The shareholders are not known |

| | | |
|---|---|---|
| | | to one another because they are widely scattered all over the country. |
| 2. Main objective | The main objective of a co-operative society is to meet the requirements of the members. | The main aim of a joint stock company is to earn more profits. |
| 3. Voting rights | Voting is based on one man, one vote. | Voting rights are based on one share, one vote. |
| 4. Members involvement | Members take whole-hearted interest in the affairs of the society | Shareholders are indifferent in the functioning of the joint stock company. |
| 5. Transfer of shares | A member cannot transfer his share to anybody else, all that he can do is to withdraw his share with prior intimation. | A shareholder can sell his shares. There is no provision to withdraw the shares. At the time of liquidation only the share money is returned to the shareholders. |
| 6. Distribution of profits | Profits are distributed according to the share capital | Profits are distributed among shareholders as dividends. |

# CHAPTER 7

# Market Structure

The firm's price and output decisions are made in a given market. The term market is used in different ways. Market can be viewed as the context within which voluntary exchanges among buyers and sellers take place. The most common method of classifying markets is on the basis of number of sellers and buyers and the homogeneity or degree of differentiation of the product.

The term competition always specifies the presence in a specific market of two or more sellers and two or more buyers of a definite commodity, each seller acting independently of every other seller and each buyer acting independently of buyer[*]. Perfect competition is a market in which every firm is too small to affect the market price. Alternatively, a market is said to be perfectly competitive when there are many sellers (and buyers) transacting a homogeneous product.

In real world, business people use the word competition similar to that of rivalry. In economic theory, perfect competition means no rivalry among the sellers. The perfectly competitive market is characterized by a complete absence of rivalry among the firms (sellers).

Imperfect competition is a market in which a firm can appreciably affect the market price of the product. It implies that in imperfect competition the individual sellers have some degree of control over the price of the products. Imperfect competition does not mean that a seller has absolute control over the price of its good. It is important to note that the word imperfect does not reflect upon the morals or ethics of businessmen. In imperfect competition intense rivalry exists among the firms.

**Classification of Market Structure**

| Market structure | Number of firms or producers or sellers | Degree of product differentiation | Firm's degree of control over price | Part of economy where prevalent |
|---|---|---|---|---|
| I. Perfect-competition | Many sellers | Homogeneous product | No control over price | Farm commodities |
| II. Imperfect competition | | | | |
| a. Monopolistic competition | Many sellers | Differentiated product | Some control over price | Retail trade |

---

* Clair Wilcox: The nature of competition and monopoly an American Society, TNEC, Monograph 2, 1941.

| b. Pure oligopoly | Few sellers | Homogeneous product | Some control over price | Steel, chemicals, *etc.* |
| c. Differentiated oligopoly | Few sellers | Differentiated product | Some control over price | Automobiles, computers, *etc.* |
| d. Monopoly | One seller | Product with no close substitutes | Considerable control over price | Railways, posts, electricity, *etc.* |

Monopsony is a market structure in which there is only one buyer instead of one seller.

Oligopsony is a market structure in which there are only few buyers.

Bilateral monopoly is a market structure in which a single seller faces a single buyer.

## PERFECT COMPETITION

Perfect competition is a market structure in which there are many sellers and buyers transacting a homogenous product.

### Characteristics

The following are the characteristics of perfectly competitive market.

1. *Large Number of Buyers and Sellers:* The perfectly competitive market is characterised by the presence of large number of buyers and sellers. Though a firm (seller) is large, but its supply is only a small part of the total quantity offered for sale in the market. Similarly, each buyer's demands is relatively small to the market demand. Since no seller or buyer is large to influence the market price, they take the market price as a given parameter beyond their control. The economic agents (sellers and buyers) are the price takers and quantity adjusters. There is no rivalry among buyers and sellers. The demand curve of a firm in the perfectly competitive market is infinitely elastic impling that the firm can sell any amount of output at the prevailing market price.

2. *Homogenous Product:* The commodity transacted in the perfectly competitive market is identical. There is no way to differentiate the goods produced by the different firms. The buyers have no preference of the commodity supplied by sellers and the sellers have no preference among the buyers.

3. *Free Entry or Exit of Firms:* There is no barrier on the entry or exist of firms from the industry. A firm can leave the industry if it cannot withstand losses.

4. *No Government Regulation:* Government does not place any restriction on price, output, entry of firms, *etc.* There is no Government intervention in the market.

5. *Perfect Mobility of Resources:* The factors of production can move from one firm to another. Workers can move from one job to another and from one place to another. The owners of man made and natural resources are free to use them in those economic activities where they get higher returns. There exists perfect competition in the markets of factors of production.

6. *Perfect Knowledge:* It is assumed that all economic agents (sellers and buyers) have complete knowledge of the conditions prevailing in the market. Both buyers and sellers are aware of the nature of product and prevailing market price. Therefore, no buyer will offer a price higher than the prevailing one and no seller is willing

to sell the product at the price, lower than the prevailing one. As a result, single price for the product prevails in the market.

The concept of pure competition is distinguished from that of prefect competition. The pure competition relaxes the assumptions of perfect mobility of resources and perfect knowledge. The first four characteristics are common to both perfect competition and pure competition. Markets for various farm commodities can be cited as an example for perfect competition.

## Output Decisions of the Firm–Short Run Equilibrium

The interaction of demand and supply curves determines the market equilibrium price and output. In a perfectly competitive market, given the market price, an individual firm can sell any quantity it wishes at the prevailing price. Short run is the planning period during which the rate of output can be changed by intensive use of existing plant. The number of firms in the market is fixed in the short run, since the time period is not long enough to allow any entry or exit from the market. Therefore, in the short run, the decision problems facing a firm is choosing the optimum output using the existing plant. The total curves approach and the marginal and average curves approach are used in the choice of profit maximizing level of output.

## Optimal Output Decision–Total Curves Approach

A firm is in equilibrium when it has no tendency to change its level of output *i.e.,* the firm needs neither expansion nor contraction. In equilibrium, the firm maximizes its profits. The profit, by definition is the difference between total revenue (TR) and the total cost (TC). The perfectly competitive firm is in equilibrium when it produces the output that maximizes the difference between total revenue and total costs.

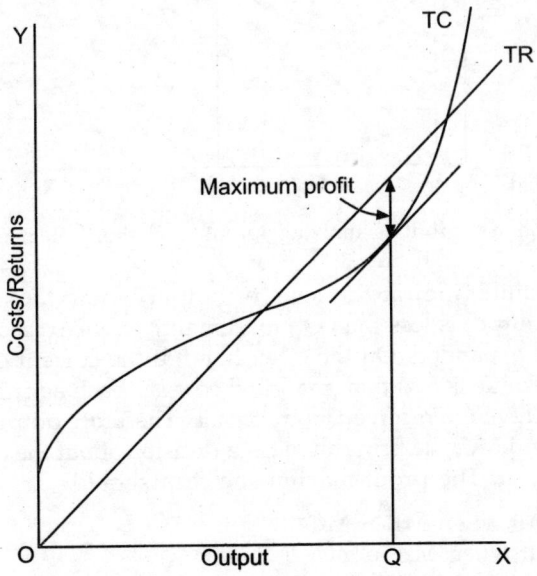

**Figure 7.1** Determination of profit maximizing output in the short run, total curves approach.

In Figure 7.1 the total revenue curve and total cost curve of a perfectly competitive firm are shown. The total revenue curve is a straight line with positive slope passing through origin indicating that the price is constant for all levels of output, since price is given for a firm in the perfect competition. If the firm produces nothing, TR will be zero. The profit maximizing or optimal output is determined where the vertical distance between TR and TC curves is greatest. The profit maximizing output is OQ units per period.

At outputs, smaller or larger than OQ, the profits earned by the firm will be reduced. Thus the firm has no intentions to change its output, and the equilibrium is a stable one.

## Marginal Curves Approach

A firm in the perfect competition can sell any amount of goods without affecting the price. Therefore, the demand or average revenue curve (AR curve) is a horizontal line at the height of the market price. Under perfect competition, any additional output can be sold at the given market price and hence the marginal revenue is also equal to the price. Accordingly in perfect competition, MR = AR = Price (Figure 7.2).

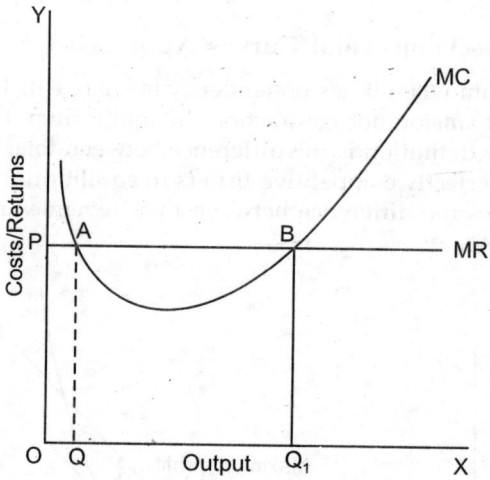

**Figure 7.2** Determination of profit maximizing output in the short run, marginal curve approach.

The firm is in equilibrium at the level of output where MC curve interests MR curve. If the firm's output is less then $Q_1$, marginal revenue exceeds the marginal cost and output must be expanded. On the other hand, if the current output is more than the $Q_1$, marginal revenue is less than marginal cost i.e., each additional unit of output costs more than the revenue received from its sale. Therefore output must be reduced. By comparing MR and MC, the firm can make a decision about the expansion, retention or reduction of output. The profit-maximizing firm should

1. Increase its output when MR > MC
2. Reduce its output when MR < MC
3. Maintain the output when MR = MC. Here profits are maximized.

Thus the first condition for the equilibrium of the firm is the equality between MR and MC. However, this condition is not sufficient, since it may be fulfilled and yet the firm may not be in equilibrium. In Figure 7.2 it is observed that the condition MR = MC is fulfilled at point A, but the firm is not in equilibrium, because profit is maximized at a greater level of output. The firm can earn more profits by producing beyond Q since MR > MC. The second condition for equilibrium requires that MC curve must cut the MR curve from below and after the point of equilibrium, it must be above MR curve. This means that the slope of the MC curve must be greater than the MR curve. At point B, the slope of MC is positive, while the slope of MR curve is zero at all levels of output. Thus at point B both the conditions for equilibrium are satisfied *i.e.*, MC = MR and slope of MC > slope of MR. Then the profit maximizing output is $Q_1$ at which MR = MC as indicated by point B.

## Output Decisions in the Short Run

Though the firm is in short run equilibrium, it does not necessarily imply that it earns excess profits. Whether the firm earns excess profits or not depends on the level of average total cost (ATC) at the short rum equilibrium. The firm's average and marginal cost curves are shown in Figure 7.3.

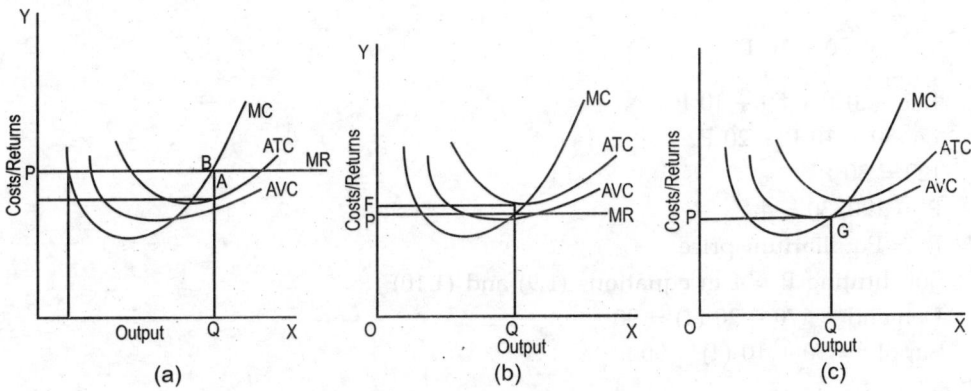

**Figure 7.3** Output decisions in the short run.

Assume that the price is Rs. P per unit. The horizontal line, demand = AR = MC shows the firm's demand, average and marginal revenue curves. ATC, AVC and MC curves show the firm's costs. Given these revenue and cost curves, the output that maximizes the profit of the firm is $Q_1$ at which MR = MC. At this level of output, the firm earns profit of AB rupees per unit and the total profit is equal to the area CPBA (Figure 7.3a). The price at equilibrium is above the ATC and therefore the firm earns excess profits. If the firm produces more than $Q_1$ level of output, MC exceeds MR and the profit will be decreased. On the other hand, if it produces less than $Q_1$ level of output, MR exceeds the MC and the total Profit will be reduced.

If the price is below the ATC the firm incurs losses (PFDE) (Figure 7.3b). In the short run, the firm will continue to produce only if it covers variable costs. If the firm fails to cover the variable costs, it will close down the operations to minimize losses. The point at which the firm covers its variable costs is called the closing down point, as indicated by the point G in (Figure 7.3c). If the price is below minimum AVC the strategy of the firm is to produce none *i.e.*, shut down the operations.

## Short Run Equilibrium Under Perfect Competition

*Numerical example*

Demand equation, $Q_d$ = f(P) $\hspace{5cm}$ (1.9)

$D = a - b\ P$

Supply equation, $Q_s$ = f (P) $\hspace{5cm}$ (1.10)

$S = C + d\ P$

Horizontal summation of all the firms gives the market supply for the commodity.

Equation (1.9) and (1.10) give demand and supply functions for a commodity. Short run market equilibrium then would be defined by market clearing condition which is specified in equation (1.11)

a-bp = C + dp $\hspace{6cm}$ (1.11)

Let the demand function be given as

$\hat{Q}_d$ = 170 – 20 P

$\hat{Q}_s$ $\hspace{0.5cm}$ = 50 + 10 P

170 – 20 P = 50 + 10 P

170-50 = 10 P + 20 P

120 = 30 P

$P^*$ = 120/30 = 4

P* = Equilibrium price

Substituting P = 4 in equations (1.9) and (1.10)

Demand = 170 – 20 (4) = 90

Supply = 50 + 10 (4) = 90

Then the equilibrium quantity is 90 units. The above estimated demand and supply equations are sensitive to the actions of buyers and sellers.

## Equilibrium of the Firm in Short Run Through Cost Function Approach

In imperfect competition the firm has 'U' shaped average cost curve and 'U' shaped marginal cost curve. These cost curves include normal returns to capital and investment and earning economic profit to the firms.

Let the estimated total cost function be

$\hat{T}C$ = 25 + 1.5 + 0.01Q² $\hspace{4cm}$ (1.12)

$\dfrac{dTC}{dQ}$ =MC=1.5 + 0.02Q

Profit of the firm will be maximized when the price per unit of output (Rs. 10) is equal to MC

10 = 1.5 + 0.02Q

$Q = \dfrac{8.5}{0.02} = 425$

Economic profit of the firm = TR-TC

$10 \times 425 = PQ - TC$

$\qquad = 25 + 1.5Q + 0.02Q^2$

$\quad 4{,}250 = 25 + 1.5(425) + 0.02(425)^2$

$TC = 5 + 1.1\ Q + 0.005Q^2$

$MC = 0 + 1.1 + 0.01Q$

$\quad P = MC$

$\quad 10 = 1.1 + 0.01Q$

$Q^* = \dfrac{10 - 1.1}{0.01} = 890$

$\pi = TR - TC$

$\quad = PQ - 5 + 1.1\ (890) + 0.005\ (890)^2$

$\quad = 8{,}900 - 5 + 979 + 3{,}960.5$

$\quad = 8{,}900 - 4{,}944.5$

$\pi = $ Rs. 3,955.5

This is the economic profit of the firm under perfect competition in short run.

## Profit Maximization in Terms of Cost Function Under Perfect Competition for the Firms

Profit of the firm is defined as the difference of the amount between total returns and total costs. Algebraically it is expressed in equation (1.13) as profit function.

$$\pi = TR - TC \qquad\qquad (1.13)$$
$$\quad = P\ (Q) - C\ (Q)$$

where,

P = Price/unit of output

Q = Quantity of output

C(Q) = Total cost function *i.e.,* cost is a function of output.

In the case of perfect competition, we get maximum profits by taking the first and second order conditions of profit function with respect to Q.

$$\frac{d\pi}{dQ} = P - \frac{dC}{dQ} = O \qquad\qquad (1.14)$$

$$\frac{d^2\pi}{dQ^2} = -\frac{dC^2}{dQ^2} < O \ \text{ or } \ \frac{dC^2}{dQ^2} > O \qquad\qquad (1.15)$$

Equation (1.14) is first order condition

Equation (1.15) is the second order condition

Equation (1.14) says that P = MC = MR

*Numerical example —1*

The revenue function of the firm R(Q) is expressed as

$$\hat{R} = 1{,}420\ Q - 3Q^2 \tag{1.16}$$

and its total cost function is given as

$$\hat{C} = 1.2\ Q^3 - 48.22\ Q^2 + 1{,}600\ Q + 1{,}000 \tag{1.17}$$

Profit function $= \pi\ (Q_1)$ is expressed as

$$\hat{\pi} = \hat{R} - \hat{C} \tag{1.18}$$

$$= (1{,}420\ Q - 3\ Q^2) - (1.2\ Q^3 - 48.22\ Q^2 + 1{,}600\ Q + 1{,}000)$$

$$= 1.2Q^3 + 45.22\ Q^2 - 180\ Q - 1{,}000 \tag{1.18}$$

Taking the first derivative of equation (1.18) with respect to Q and equating it to zero we get maximum profit

$$\frac{d\pi}{dQ} = -3.6Q^2 + 90.44Q - 180 - 0 = 0 \tag{1.19}$$

Using the quadratic equation we can find out the values of Q.

$$Q = \frac{-b \pm \sqrt{b^2 - 4ac}}{2a}$$

Where,

a = -3.6

b = 90.44

c = -180

$$Q = \frac{-90.44 \pm \sqrt{(90.44)^2 - 4(-3.6)(-180)}}{2(-3.6)}$$

$$Q = \frac{(-15.69)}{-7.2} = 2.17$$

$$Q = \frac{(-165.19)}{-7.2}$$

$$= 22.94$$

The second order condition is:

$$\frac{d^2\pi}{dQ^2} = -7.2Q + 90.44 < 0 \tag{1.20}$$

Substitute Q value of 22.94 in equation (1.20)

The value obtained is – 74.73

When Q = 22.94 units the second order conditions is satisfied. Hence the optimal level of output is 22.94 units.

Substituting Q = 22.94 in the profit equation (1.18)

We get maximum profit = Rs. 4,181.10

Substituting Q = 22.94 in equation (1.16), we get total revenue of Rs. 30,996.07

$Q^* = 22.94$ units

$$P^* = \frac{R}{Q} = \frac{30,996.07}{22.94}$$

Equilibrium price = Rs. 1,351.18

*Numerical example — 2*

The total cost function of the firm under perfect competition

is given as $\hat{C} = 0.02Q^3 - 0.6Q^2 + 8Q + 10$ $\qquad$ (1.21)

$$MC = \frac{dC}{dQ} = 0.06Q^2 - 1.2Q + 8 + 10 \qquad (1.22)$$

Putting $\qquad$ P = MC

$$P = 0.06Q^2 - 1.2Q + 8 \qquad (1.23)$$

Solving for Q in terms of P

$$Q = 0.06Q^2 - 1.2Q + (8 - P) \qquad (1.24)$$

**Using quadratic equation**

$$\frac{-b \pm \sqrt{b^2 - 4ac}}{2a}$$

**Where,**

$a = 0.06$

$b = -1.2$

$c = (8-P)$

$$= \frac{-(-1.2) \pm \sqrt{(-1.2)^2 - 4(0.06)(8 - P)}}{2(0.06)}$$

$$= \frac{+1.2 + \sqrt{1.44 - 1.92 + 0.24P}}{0.12}$$

**Supply function**

$$Q = f(P) = \frac{1.2 + \sqrt{+0.24P - 0.48}}{0.12} \qquad (1.25)$$

In the equation (1.25) negative value of $\sqrt{0.24P - 48}$ is ignored because supply curve is having positive slope.

Let the total variable cost function be given as

$$T\hat{V}C = 0.02Q^3 - 0.6Q^2 + 8Q \qquad (1.26)$$

Please note that here when constant term is dropped from the TC function, then it becomes TVC.

$$A\hat{V}C = \frac{T\hat{V}C}{Q} = \frac{0.02Q^3 - 0.6Q^2 + 8Q}{Q} \qquad (1.27)$$

Taking the first derivative of equation (1.27) and equating to zero we can solve for $Q^*$

$$\frac{d(AVC)}{dQ} = 0.04Q - 0.6 = 0$$

$$Q^* = \frac{+0.6}{0.04} = 15$$

Find the value of AVC by substituting the value of $Q^*$ in the AVC equation (1.27). We get AVC as Rs. 4.76

The supply function will be valid for all prices greater than minimum AVC i.e., Rs. 4.76.

The supply function of the firm is

$$\text{Quantity supplied } (Q_s) = \frac{1.2 + \sqrt{0.24P - 0.48}}{0.12} \quad \text{if } P \geq 4.76$$

| $Q_s$ values | P values (Rs.) |
|---|---|
| 16.78 | 4.76 |
| 17.07** | 5.00* |
| 17.64** | 5.50* |
| 19.31** | 7.20* |
| 22.25** | 11.00* |

*Assumed value;      **Derived value

Plotting the value of P on vertical axis and value of Q on horizontal axis we can draw the supply curve. Assume that, if there are 50 firms in the industry and all are having similar cost functions then the market supply of the commodity would be

$$Q_s = (50)\left( \frac{1.2 + \sqrt{0.24P - 0.48}}{0.12} \right)$$

## The Supply Curve of the Firm

Let us examine how the supply curve of the perfectly competitive firm is derived. The supply curve of the firm is derived by the points of intersection of its MC curve with successive MR curves (Figure 7.4a). If the market price is $P_2$, the firm supplies $Q_2$ units and when the price is $P_1$, it supplies $Q_1$ units. As long as the price is higher than the minimum ATC, the firm will supply the output given by MR = MC. Given the positive slope of MC curve, each higher MR curve intersects the MC curve to a point which lies to the right of the previous intersection. This means that the quantity supplied increases with the increase in price. If the price is lower than minimum AVC, the firm has to shut down its operations. Since price cannot cover AVC, continuing the production will increase the loss. A competitive firm will produce the output given by MR = MC, whenever price is equal to or greater than the minimum AVC. The profit maximizing output is zero when the price is less than minimum AVC. Therefore the portion of the MC curve on or above the AVC curve is the supply curve of the firm. The supply curve of the firm is shown in Figure 7.4b.

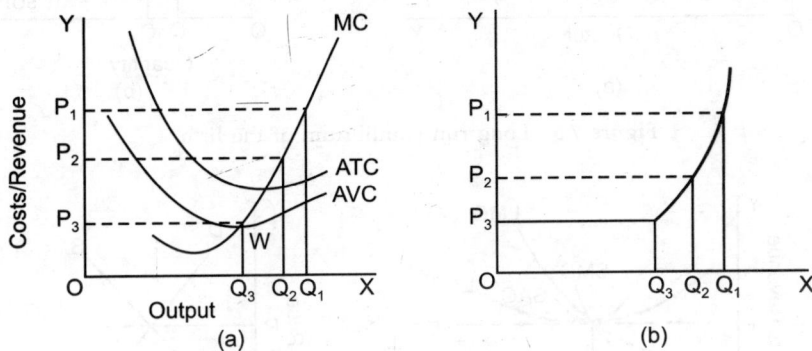

**Figure 7.4** Supply curve of the firm.

## Long Run Equilibrium of the Firm

Long run is the planning period during which a firm can make more adjustments than in the short run. In the long run the firm can adjust its plant capacity/scale of operations. Therefore, all costs are variable. Firms must earn only normal profits. If the price is more than long run average cost, the firms earn super normal profits. Attracted by these profits, new firms will enter the industry and super normal profits will be competed away. If the price is less than LAC, firms incur loss and some of the firms leave the industry so that no firm earns more than normal profits.

If the price is P, the firm makes super normal profits working with the plant whose cost is denoted by $SAC_1$ (Figure 7.5a). Excess profits encourage the firm to build new capacity and it will move along its LAC. Attracted by excess profits, new firms enter the industry. Consequently the quantity supplied increases and the supply curve shifts to the right. As a result the price falls and it reaches the level of $P_1$ (Figure 7.5b) at which the firm and industry are in long run equilibrium.

The condition for the long run equilibrium of the firm is that marginal cost equals the price and the long run average cost.

LMC = LAC = P

At equilibrium the short run marginal cost is equal to the long run marginal cost and the short run average cost is equal to the long run average cost.

SMC = LMC = LAC = SAC =P= MR

## Equilibrium of the Industry in the Long Run

The industry is long run equilibrium, when all the firms are earning normal profits. Under these conditions there is no entry or exit of firms in the industry as shown in the Figure 7.6a.

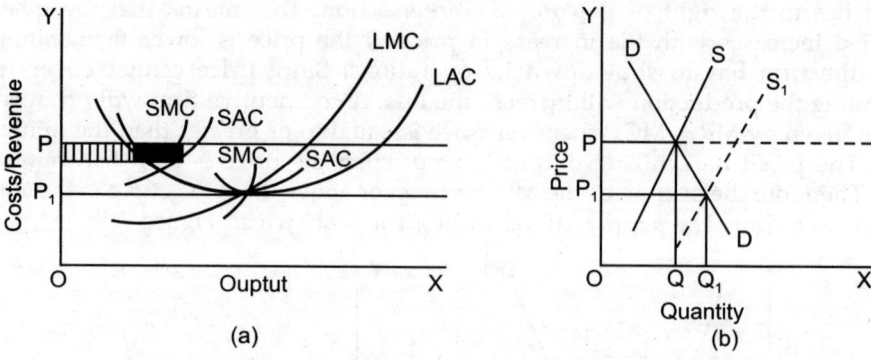

**Figure 7.5** Long run equilibrium of the firm.

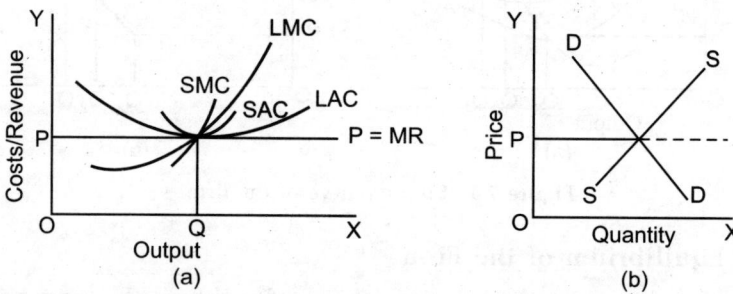

**Figure 7.6** Long run equilibrium of the industry.

When the price is P, the firms produce at their minimum cost, earning just normal profits. The firm is in equilibrium at the level of output Q.

$$LMC = SMC = P = MR$$

This equality ensures that the firm maximizes its profits at the price P. The industry is in equilibrium, because firms earn normal profits so that there is no incentive for entry or exit. It is shown by the equality of LAC = SAC= P, which is observed at the minimum point of LAC curve. As all firms in the industry are being in equilibrium, the industry supply remains stable. Given the market demand (D), the price P is a long run equilibrium price (Figure 7.6b).

## Long Run Equilibrium for a Firm Under Perfect Competition

*Numerical example*

Let the demand function for competitive industry be given as

$$Q_d = 600-6P \tag{1.28}$$

and cost function for individual firm i

$$\hat{C}_i = 100+8Q_i+1Q_i^2 \tag{1.29}$$

Where $Q_i$ = level of output of the $i^{th}$ firm. Here we assume that all firms are homogeneous with regard to production of output.

$$\hat{C}_i = 100+8Q+1Q_i^2 \tag{1.29}$$

$$AC = \frac{100}{q} + \frac{8Q}{8} + \frac{1Q^2}{Q} \qquad (1.30)$$

$$= \frac{100}{q} + 8 + 1Q$$

$$\frac{dAC}{dQ} = 1 - 100Q^2 = 0$$

$$1 = \frac{100}{Q^2} = Q^2 = 100$$

$Q = \pm 10$

$Q^* = 10$ units

The second order sufficient condition is satisfied when Q = 10, hence equilibrium output is 10.

The price of the commodity when $Q_i$ = 10

$$P = \frac{100}{10} + 8 + 1 \times 10$$

$$= 10 + 8 + 10 = 28$$

$P^*$ = 28 (equilibrium price). To get total market demand when P = 28, substitute P = 28 in equation (1.28)

$$Q_d = 600 - 6 \times 28$$

$$= 600 - 168$$

$$= 432$$

This is total market demand. Since in perfect competition all firms are identifiable and each firm produces 10 units of commodity. Hence, the number of firms in the industry are $= \dfrac{432}{10} = 43$

## Shut Down Price

In the short run managers of the firm would like to close down the operations of the firm if the price of the product is less than its AVC. If the price is greater than AVC, but less than ATC or AC, the firm would like to continue production in the short run because contribution of profits can be made to fixed costs. In the long run, all costs are variable. If the price of its product is less than AVC, the firm will close down its operations.

## Calculating Shut Down Price

*Numerical example*

Let the estimated cost function in the short run is assumed as

$$T\hat{C} = 500 + 120\,Q - 18\,Q^2 + 1.2\,Q^3 \tag{1.31}$$

Here,  Q = Quantity of output

TC = Total cost

The cost function is having a constant term $i.e.$, Rs. 500. It is called short run function. In the long run function constant term is absent.

Taking the first derivative of equation (1.31) we get marginal cost function.

$$MC = \frac{dTC}{dQ} = 0 + 120 - 36\,Q + 3.6\,Q^2 \tag{1.32}$$

The firms total variable cost function is given as

$$TVC = 120\,Q - 18\,Q^2 + 1.2\,Q^3 \tag{1.33}$$

$$AVC = \frac{TVC}{Q} = \frac{120\,Q - 18\,Q^2 + 1.2\,Q^3}{Q} \tag{1.34}$$

$$AVC = 120 - 18\,Q + 1.2\,Q^2 \tag{1.35}$$

Profit maximization condition requires that price = MC $i.e.$, MR = MC (an equality condition.)

Here, MC = minimum point of AC curve

$$MC = 120 - 36\,Q + 3.6\,Q^2$$

$$AVC = 120 - 18\,Q + 1.2\,Q^2$$

Equating (1.32) and (1.35) and rearranging we get

$$120 - 36\,Q + 3.6\,Q^2 = 120 - 18\,Q + 1.2\,Q^2 \tag{1.36}$$

$$0 - 18\,Q + 2.4\,Q^2 = 0$$

$$-18\,Q + 2.4\,Q^2 = 0 \tag{1.37}$$

$$-6\,Q\,(3 - 0.4\,Q) = 0$$

$$-6\,Q = 0;\ Q = 0$$

$$-0.4\,Q = -3$$

$$Q^* = \frac{3}{0.4} = 7.5$$

Substituting ($Q^*$ = 7.5) in MC equation $i.e.$, equation (1.32)

Here, P = MC

$$120 - 36\,(7.5) + 3.6\,(7.5)^2$$

Hence P = $120 - 36\,(7.5) + 3.6\,(7.5)^2$

= Rs. 52.5

Shut down price = Rs. 52.5

When the price of the product is below Rs. 52.5 per unit of output, the firm stops production.

# IMPERFECT COMPETITION

Any deviation from the conditions of perfect competition in a market leads to the existence of imperfect competition. More specifically the characteristic features like few sellers selling a differentiated product, absence of price information, restriction of free entry and free exit of the firms, existence of transportation costs *etc.*, represent imperfect competitive markets. Imperfect competition takes several forms, *viz.*, monopoly, monopolistic competition, duopoly, oligopoly, monopsony and oligopsony. These are briefly presented here.

# MONOPOLY

Monopoly is a market condition, wherein the entire supply of a commodity is concentrated in the hands of a single firm. There exists no close substitutes for the product produced by a monopolist. The cross elasticity of demand is very low for the product. In monopoly, the firm and the industry are the same. Though he controls both price and output policies, he cannot adopt the same simultaneously, because the monopolist has the choice of fixing his own price, but cannot sell the expected quantity at the chosen price. Normally he decides the output and fixes the price for the product. He chooses between two options. He has to produce either more output to sell at a lower price or produce less output to sell at a higher price. Monopoly is of two types, *viz.*, pure monopoly and imperfect monopoly.

The cross elasticity of demand for the product under pure monopoly is zero. The existence of pure monopoly is of theoretical interest only. It is far from reality in the real world situation. As against pure monopoly, imperfect monopoly is a condition which is found in reality with the following features.

(1) Product with a very low cross elasticity of demand.
(2) Intentions to maximise profits.
(3) No role of individual consumer influencing the price.
(4) Uniform price for all the consumers, and
(5) Absolutely no threat from the other firms.

## Average Revenue Curve and Marginal Revenue Curve Under Monopoly

The average revenue curve of the firm under monopoly slopes downwards from left to right throughout its length as shown in Figure 7.7. Marginal revenue curve slopes downwards from left to right but stationed below the average revenue curve. The numerical slope of the MR curve is twice as steep as that of AR curve. The average revenue curve is the demand curve under monopoly. The demand curve reflects the intentions of monopolist in selling large quantities of output through price reduction.

## Equilibrium in the Short Run

The intention of monopolist in maximizing the profits is achieved at the point of equality of MR and MC.

The firm is in equilibrium by producing OQ output at PQ or CO price. At this equilibrium level of output, the profit of the firm under monopoly is BCPA (Figure 7.8a).

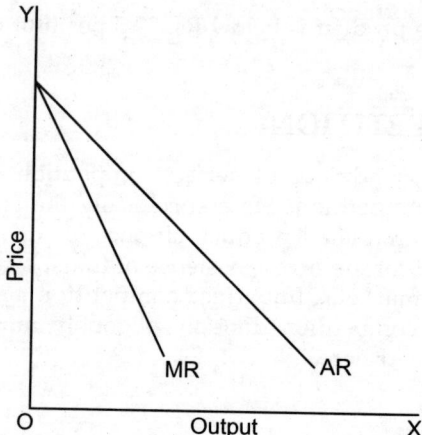

**Figure 7.7**  Average revenue curve and marginal revenue curve.

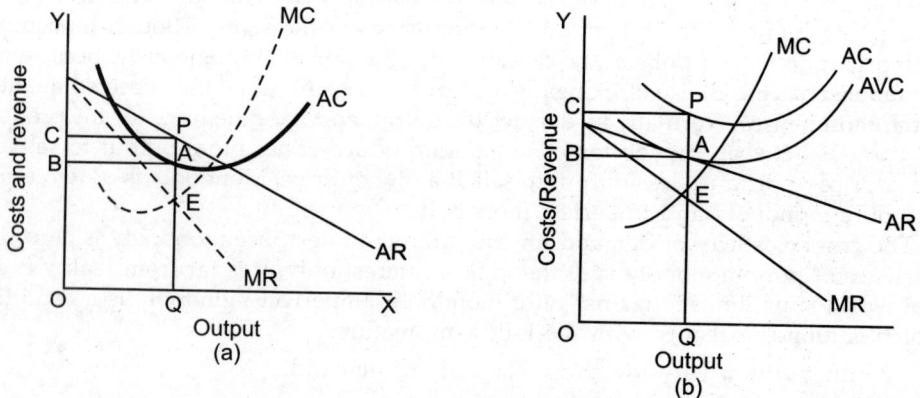

**Figure 7.8**  Equilibrium in the short run.

Another possibility for the monopoly firm in the short run is that, the price may be less than the average cost of production, covering only average variable cost of production. This situation is depicted in Figure 7.8b.

Price OB or AQ is found to be lesser than average cost but covers the average variable cost. Firms continue to operate inspite of the situation of price not covering average cost. The loss that the monopoly incurs consequently is BCPA. Even in monopoly the shut down point would be in the offing, in case the price fails to cover even the average variable cost.

## Equilibrium in the Long Run

In monopoly, though the entry of new firms is ruled out still like a **competing firm**, it makes all out efforts to adjust the supplies to the changes in demand or **even costs** that are likely to occur. Even in the long run, monopolist will be in equilibrium by equating MC and MR. On the price front, it can be equal or more than average cost in the long run, but monopolist generally opts the price to be more than long run average cost, so that it can be in a position to earn supernormal profits. This aspect is given tremendous importance in the absence of a threat from any other firm to enter the industry. As clear from **Figure 7.9** the monopolist enjoys supernormal profit of CBAP by producing OQ output at OC or PQ price. This is the firm's equilibrium position in the long run.

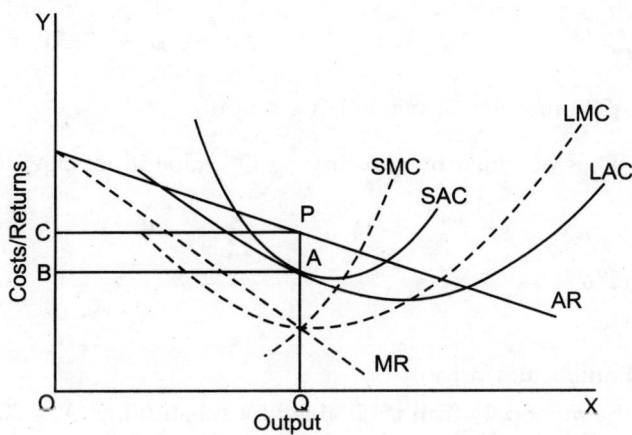

**Figure 7.9** Equilibrium in the long run.

## Equilibrium output for monopoly

*Numerical example.*
Let the estimated demand function for the monopolist be

$$\hat{Q}_d = 120 - 2P \tag{1.38}$$

Its cost function $\hat{C}$ is

$$\hat{C} = 10Q + 0.4Q^2 \tag{1.39}$$

Work out the equilibrium level of output and equilibrium price for monopoly firm. The demand function in equation (1.38) is inverted and expressed in terms of 'P' as

$$\hat{P} = \frac{120}{2} - \frac{Q}{2}$$

$$= 60 - Q/2$$

$$\hat{P} = 60 - 0.5Q \text{ (inverted demand function)} \tag{1.40}$$

Total revenue function for the monopoly firm

$$R = \hat{P}.Q$$

$$R = (60 - 0.5Q).Q$$

$$= 60Q - 0.5Q^2 \tag{1.41}$$

Profit function = $\pi$ = TR - TC

$$= (60Q - 0.5Q^2) - (10Q + 0.4Q^2)$$

$$= 60Q - 10Q - 0.5Q^2 - 0.4Q^2$$

$$\pi = 50\ Q - 0.9Q^2 \tag{1.42}$$

For profit maximization takes the first derivative of equation (1.42) and equate it to zero.

$$\frac{d\pi}{dQ} = 50 - 1.8Q = 0 \tag{1.43}$$

$$Q^* = \frac{50}{1.8} = 27.77 \text{ units (Profit maximizing output)}$$

Equilibrium P* is obtained by substituting Q* value in inverse demand function equation (1.40)

$$\pi^* = 60 - 0.5Q$$

$$= 60 - 0.5\ (27.77)$$

$$= 60 - 13.89$$

P* = 46.12 (Equilibrium price)

Total profit of the monopoly firm is obtained by substituting Q* = 27.77, in equation (1.42) *i.e.,* total profit function

$$\pi = 50Q - 0.9Q^2$$

$$= 50\ (27.77) - 0.9\ (27.77)^2$$

$$= 1,388.5 - 694.05$$

Total profits of the firm: Rs. 694.45

## Price Discrimination

Let us assume that the product supplied by monopolist is having slight differences in its qualities. For example, in a cricket match, the audience are charged different prices for different sitting arrangements.

This is similar to different rates paid by the air travellers in different classes of the aircraft. In trains, theatres, circus we will observe different price discrimination methods applied by monopolists. We should note here the identical product produced at the same cost, but sold at different prices based on consumers tastes, income levels and availability of the substitutes for the products.

When different prices are charged for the same product, then the demand curve of the product will have different elasticities. For instance, a product is sold at higher price to rich people, when it is liked by them. On the other hand, the same product is sold to a lower income group of consumers at a less price. This results in differences in the slope of the demand curve for the product. Thus, price discrimination is understood as the process of charging different prices to different consumers or buyers of the same product but with slight differences in its characteristics.

The following conditions are necessary for successful price discrimination:

1. Firm should have full control on the price.
2. It must be possible to categorize different markets based on price elasticity of demand for its products.
3. The markets must be separable *i.e.,* products cannot be purchased in one market and resold in another market.

4. It is assumed that the firm should charge **higher price** in first market and lower price in second market.

Price discrimination applies to airways, railways, **services** of a doctor, *etc.*

*Types of Price Discrimination:* There are three types of price discrimination *viz.,* (1) First degree price discrimination (2) Second degree price discrimination and (3) Third degree price discrimination.

## First-degree Price Discrimination

It is a special case of limiting price discrimination in which the consumers are charged individually the highest possible price for each corresponding unit of output. Here the demand curve and MR curve are the same as shown in Figure 7.10. As a result the monopolist receives the entire consumer's surplus (PKD) for his product.

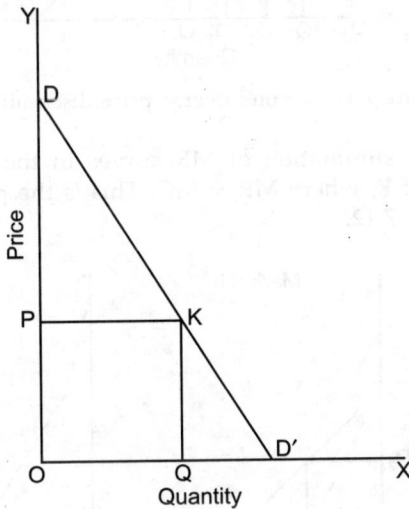

**Figure 7.10**   First-degree price discrimination.

## Second-degree Price Discrimination

In this case the monopolist will have separate negotiations with the consumers and sells his commodity at different prices as shown in Figure 7.11. For example, it sells $OQ_1$ quantity of the product at $P_1$ price to one set of consumers and for another set of consumers it sells $OQ_2$ quantity at $P_2$ price. Thus by continuing this practice, the firm receives still larger part of consumer's surplus as shown in the Figure 7.11.

## Third-degree Price Discrimination

This is the most common type of price discrimination. This involves separation of markets or consumers based on price elasticity of demand for commodities. Spatially markets are separated. The practice of selling textbooks outside the domestic market at very cheaper rate is one of the examples for this. Electricity consumption among farming and non-farming community is another example. Demand for a commodity in market I is assumed to be less elastic than that in market II. $MR_1$ and $MR_2$ are the individual marginal revenue curves. The combined marginal revenue curves is $MR_t$

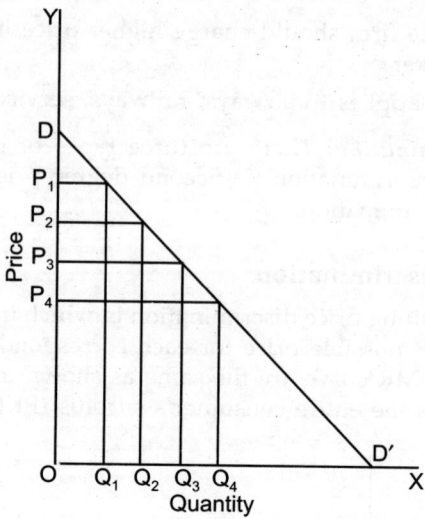

**Figure 7.11** Second-degree price discrimination.

which is the horizontal summation of MR curves in the two markets. Combined output is found at point E, where $MR_t = MC$. This is the profit maximizing output. This is shown in Figure 7.12.

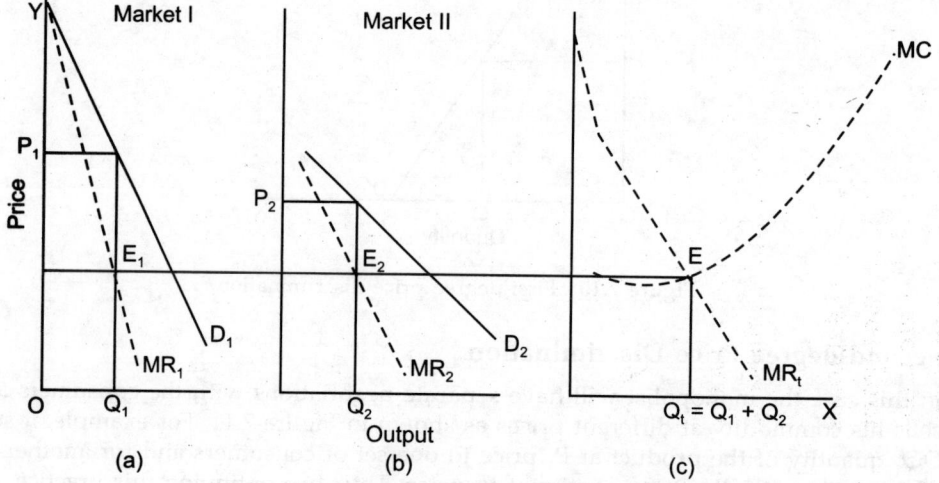

**Figure 7.12** Third-degree price discrimination.

## Second Degree Price Discrimination Under Monopoly

*Numerical example*

Let the estimated demand function for two markets where price discrimination is practiced by monopoly.

$$\hat{Q}_{1d} = 60 - P_1 \tag{1.44}$$

$$\hat{Q}_{2d} = 80 - 2P_2 \tag{1.45}$$

Inverse demand functions are

$$P_1 = 60 - Q_1 \tag{1.46}$$

$$P_2 = 80 - 0.5\,Q_2 \tag{1.47}$$

Cost function for monopolist is

$$\hat{C} = 20 + 4Q \tag{1.48}$$

Quantity sold in the two markets $Q = Q_1 + Q_2$ (1.49)

Total revenue functions for the two markets are specified as

$$
\begin{aligned}
R_1 &= P_1 \cdot Q_1 \\
&= (60 - Q_1)\,Q \\
&= 60Q - Q^2
\end{aligned}
\tag{1.50}
$$

$$
\begin{aligned}
R_2 &= P_2 \cdot Q_2 \\
&= (80 - 0.5Q_2)\,Q \\
&= 80Q - 0.5Q^2
\end{aligned}
\tag{1.51}
$$

Taking the first derivative of the total revenue function separately and making them equal to MC and solving the equation, we get optimal output

$$
MR_1 = \frac{dR_1}{dQ_1}
$$

$$
\begin{aligned}
&= 60\,Q - Q^2 \\
&= 60 - 2Q
\end{aligned}
\tag{1.52}
$$

$$
MR_2 = \frac{dR_2}{dQ_2}
$$

$$
\begin{aligned}
&= 80\,Q - 0.5Q^2 \\
&= 80 - 1Q
\end{aligned}
\tag{1.53}
$$

$$
MC = \frac{dTC}{dQ} = 4
$$

The equilibrium condition for monopolist under price discrimination is given as

$$MR_1 = MR_2 = MC$$

$60\text{-}2Q_1 = 80\text{-}1Q_2 = 4$

$60 - 2Q_1 = 4$

$80 - 1Q_2 = 4$

$-2Q_1 = 4\text{-}60$

$= Q_1 = 56/2 = 28 \text{ units}$

$-1Q_2 = -80 + 4$

$Q_2 = 76 \text{ units}$

Substituting $Q_1 = 28$ in equation (1.46)

$P_1 = 60 - 28$

$P_1 = 32$

Equilibrium price in market 1 = Rs. 32

Substituting $Q_2 = 76$ in equation (1.47)

$P_2 = 80 - 0.5\ (76)$

$= 80 - 38$

$= 42$

Equilibrium price in market 2

$\pi_1 = R_1 - C_1$

$= 32 \times 28 - 4 \times 28$

$= 896 - 112$

Profit in market 1 = 784

$\pi_2 = R_2 - C_2$

$= 42 \times 76 - 4 \times 76$

$= 3,192 - 112$

Profit in market 2 = Rs. 2,888

So the total profit of monopolist under price discrimination is

$\pi_1 + \pi_2 = $ Rs. 784 + Rs. 2,888 = Rs. 3,672

## Third Degree Price Discrimination

*Numerical example*

The demand and marginal revenue functions in the two markets are given as

| Market 1 | Market 2 |
|---|---|
| $P_1 = 50 - 2.2\ Q_1$ ... (1.54) | $P_2 = 24 - 1.2Q_2$ (1.56) |
| $MR_1 = 50 - 4.4\ Q_1$ ... (1.55) | $MR_2 = 24 - 2.4Q_2$ (1.57) |
| | $MC = 1$ |

Using third-degree price discrimination, we can find out equilibrium price and equilibrium quantity of the product. The conditions for profit maximization are:

$MR_1 = MR_2 = MC$ (1.58)

$MR_1 = MC$ *i.e.*, $50 - 4.4Q_1 = 1$

$Q_1 = -4.4Q_1 = -49$

$Q_1 = 11.14$

$MR_2 = MC$ *i.e.*, $24 - 2.4Q_2 = 1$

$-2.4Q_2 = -2.3$

$Q_2 = 9.58$

To get equilibrium price, substitute $Q_1$ and $Q_2$ values in equation (1.54) and equation (1.56).

$$P_1 = 50-2.2Q_1 \qquad (1.59)$$

Here equation (1.59) is inverse demand function in market 1.

$$= 50 - 2.2 \ (11.14)$$
$$= 50 - 24.51$$
$$= 25.49$$

$$P_2 = 24-1.2Q_2$$
$$= 24-1.2 \ (9.58)$$
$$= 24-11.50$$
$$= 12.5$$

$$P_1 = 50-2.2 \ Q_1$$

$$Q_1 = \frac{50}{2.2} = \frac{P}{2.2}$$

$$= 22.73 - 0.45 \ P$$

$$\pi_1 = TR_1 - TC \ \text{(Profits in market 1)}$$
$$= P_1 Q_1 - MC.Q_1$$
$$= P_1 \ (22.73 - 0.45 \ P - 1 \ (Q_1)$$
$$= 25.49 \ (22.73 - 0.45) \ (25.49) - 1 \ (11.14)$$
$$\pi_1 = 275.86 \qquad (1.60)$$

## Demand Function

$$P_2 = 24 - 1.2 \ Q_2 \qquad (1.61)$$

Equation (1.61) is inverse demand function is market 2.

$$Q_2 = \frac{24}{1.2} = \frac{P}{1.2}$$

$$= 20 - 0.83P$$

$$\pi_2 = TR_2 - TC_2 \ \text{(Profit in market 2)}$$
$$= P_2 \ Q_2 - MC.Q_2$$
$$= [(20 - 0.83) \ (12.5)] - 1 \ (9.58)$$
$$\pi_2 = 110.73 \qquad (1.62)$$

$$\pi_1 = Rs. \ 275.86$$
$$\pi_2 = Rs. \ 110.73 \qquad \pi = \pi_1 + \pi_2$$
$$\pi = 275.86 + 110.73$$
$$= Rs. \ 386.59$$

$\pi_1$ and $\pi_2$ are profits in market 1 and market 2 respectively. $\pi$ denotes total profits in two markets.

In the demand function, quantity demanded is a function of price but its inverse demand is expressed as

$$P = f(Q_D)$$

$$Q^* = 11.14 + 9.58$$

(Combined demand) $= 20.72$

$$P = 25.49 + 12.5$$

(Combined price) $=$ Rs. 37.99

# MONOPOLISTIC COMPETITION

It is a market situation in which the transacted products of various firms are not perfect substitutes. Firms in the industry produce heterogeneous products. Products look rather similar but possess some distinguishing features. They are not identical. Thus product differentiation is an important feature of monopolistic competition. Another feature is the production of goods under different brand names. There are large number of firms so that no single firm is in a position to influence the industry through its output and price policies.

Under competition of this nature, since every good satisfies a given want of the consumers, the quality aspects of the product of a given firm catch the attention of the consumers over the same type of product produced by other firms. Some consumers don't mind to pay premium prices for certain quality products of a given firm. Sales promotion activity *viz.*, advertisement and propaganda is an another feature of monopolistic competition.

## Short-run Equilibrium

Under monopolistic competition, each firm has a price of its own, but that danger of new firms entering into the business always lurks, if one firm earns supernormal profit. It means that super normal profits of a firm are taken away by the entry of new firms. Similarly sub-normal profits drive out a firm from the existing business. The situation of a firm earning super normal profits is presented below (Figure 7.13a). MC and MR are equal at point E, which indicates the firm's equilibrium position. The output produced at this level is OQ and the price is OP. The super normal profit is represented by BCPA. The possibility of some firms deriving super normal profit and some other firms incurring losses is not ruled out in the short run. These possibilities are expected with the established firms charging higher prices. Those firms which are relatively new to the trade may be satisfied with relatively lesser prices. The situation of a firm earning only normal profit is depicted in Figure 7.13b.

The firm is just making normal profit at an output level of OQ. At this level of output AC and AR are equal. The firm that is incurring losses is shown in Figure 7.13c.

Here the firm is not able to cover its average cost and hence incurs losses. The price fixed at OP to produce OQ covers only AVC. The loss derived is to the extent of BCPA. The above situation indicates that equilibrium of the firm is certainly a possibility, but for the group of firms as a whole the equilibrium may likely to change.

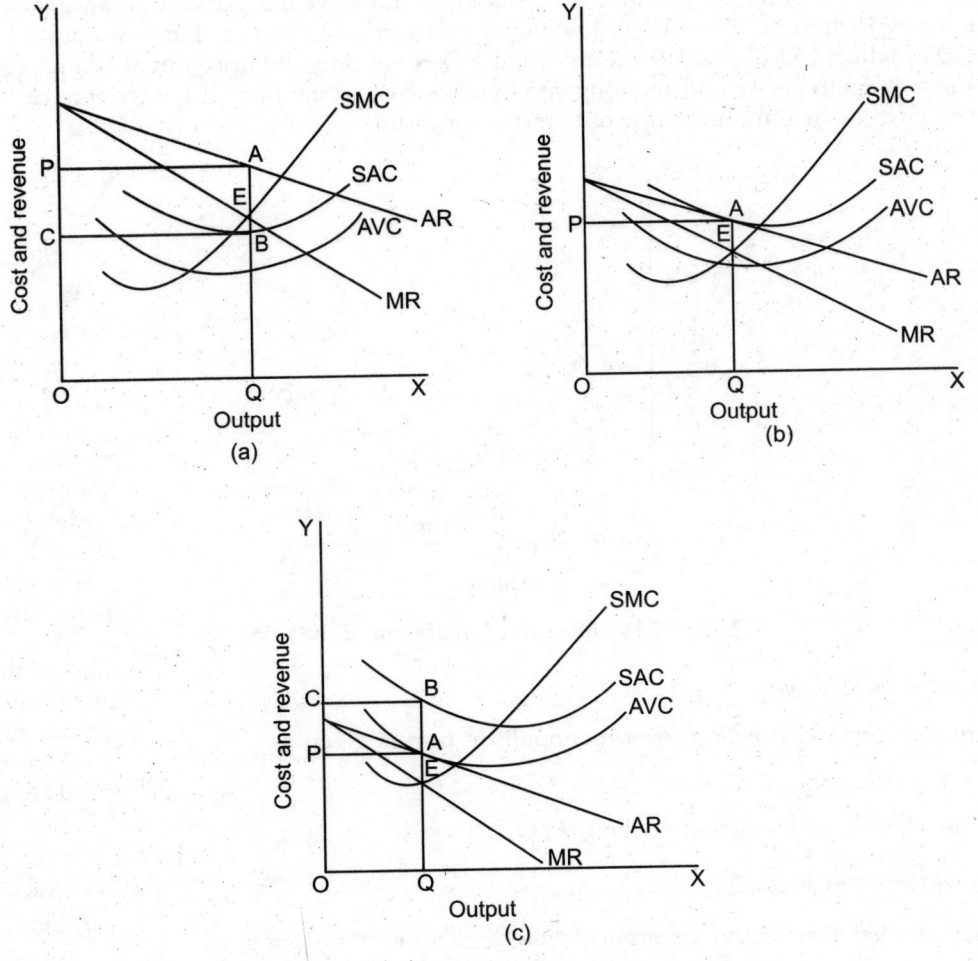

**Figure 7.13** Short run equilibrium of the firm.

## Equilibrium in the Long Run

Stability in equilibrium position of the firm as well as the group will be found in the long run. The guiding principle for price-output policy is only the equality of MC and MR. However, in the long run monopolistic competition is associated with two situations. Each firm plans to produce a given quantity of output and the aggregate production of all the firms makes the total supply in the market. For the industry to be in equilibrium, total supply should be equal to the total quantity demanded. If supply happens to be more, it will pull down the price level in which case, each firm has to make an adjustment with reference to the output and price to obtain the equality of MC and MR. The other situation in the long run is the entry of new firms attracted by the super normal profits earned by the existing firms. Their entry not only increases the supplies, but also new firms may fix prices at a lower level to have a more market share. This also forces the existing firms to lower the prices. This situation leads to the elimination of excess profits in the long run. The chances of the firm incurring losses is a remote possibility in the long run, as those firms which are likely to get losses, make an advance assessment and leave the industry because there is no restriction to do so. From the Figure 7.14 it is clear that equilibrium output is OQ at which LRMC and LRMR are equal. PQ is not only the price but also the long run average revenue and the long run average cost of the firm. It is therefore clear that firm earns only normal profits in the long run.

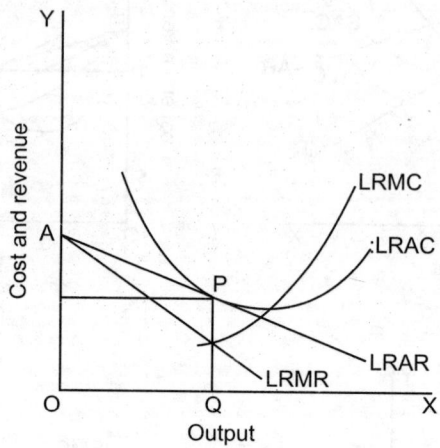

**Figure 7.14** Long run equilibrium of the firm.

*Numerical Example*

Inverse demand function for a monopolistic firm is given as

$$\hat{P} = 150 - 10Q \tag{1.63}$$

The average cost function of the firm is

$$\hat{AC} = 200 - 40Q + 2Q^2 \tag{1.64}$$

The market share curve or proportional demand curve is given as

$$\overset{*}{P} = 160 - 22Q \tag{1.65}$$

To find out the equilibrium position of the firm in the long run, the profit function of the firm is specified as

$$\pi = P . Q - C \tag{1.66}$$

$$= (150 - 10Q) Q - Q (200 - 40Q + 2Q^2)$$

$$= 150Q - 10Q^2 - 200Q + 40Q^2 - 2Q^3$$

$$= -50Q + 30Q^2 - 2Q^3 \tag{1.67}$$

Taking the first derivative of equation (1.65) with respect to Q and equate it with zero and solve for Q

$$\frac{d\pi}{dQ} = -50 + 60 - 6Q^2 = 0$$

$$= + (50 - 60Q + 6Q^2) = 0 \tag{1.68}$$

Using quadratic equation

$a = 6$

$b = -60$

$c = 50$

$$Q = \frac{-b \pm \sqrt{b^2 - 4ac}}{2a}$$

$$= \frac{+60 \pm \sqrt{(-60)^2 - 4(6)(50)}}{2(6)}$$

$$= \frac{+60 \pm \sqrt{3,600 - 1,200}}{12}$$

$$= \frac{+60 \pm \sqrt{48.98}}{12}$$

$$= 60 + 48.98 = 108.98/12 = 9.08$$

$$= \frac{60 - 48.98}{12} = \frac{1.12}{12} = 0.093$$

Optimal output = 9.08 units

Substituting Q = 9.08 in equation (1.68)

$$= 50-60 \ (9.08) + 6 \ (9.08)^2$$

$$= 0$$

To get optimal price, substitute Q = 9.08 in equation (1.63) to get equilibrium price

$P = 150 - 10Q$

$= 150 - 90.8$

$= $ Rs. 59.2

# DUOPOLY

Duopoly is a market situation in which there are only two sellers. It is very close to oligopoly in all respects barring the number of firms. Each firm keeps a close watch on the actions of the other firms as a chain reaction is imminent since the policy of one is immediately challenged by the other. This sort of rivalry goes ahead in duopoly. So as to continue in the business, one firm has to make an intelligent guess of rival's actions. Hence, the stiff competition exists between the two firms. If each seller feels that the competition in which there are lockedin is going to ruin their fortunes, a mutual agreement is arrived at for the benefit of both the firms. Thus, in duopoly, competition as well as co-operation coexist.

*Numerical Example*

The cost function of firm - I is given as

$$\hat{C}_1 = 60 + 30Q_1 \tag{1.69}$$

The cost function of firm - II is given as

$$\hat{C}_2 = 30 + 20Q_2 \tag{1.70}$$

Inverse market demand function for their products is given as

$$P = 80 - 2(Q_1 + Q_2) \tag{1.71}$$

Where,

$$Q_1 + Q_2 = Q$$

Find out equilibrium quantities, prices and profits of these two firms.

Now let us define the profit equations for the two firms as

$$\pi_1 = ((80 - 2 (Q_1 + Q_2)) \cdot Q_1 - (60 + 30 Q_1) \tag{1.72}$$
$$\pi_2 = ((80 - 2 (Q_1 + Q_2)) \cdot Q_2 - (30 + 20 Q_2) \tag{1.73}$$
$$\pi_1 = (80 Q_1 - 2Q_1^2 - 2 Q_1Q_2) - 60 - 30 Q_1 \tag{1.74}$$
$$\pi_2 = (80 Q_2 - 2Q_1Q_2 - 2Q_2^2) - 30 - 20 Q_2 \tag{1.75}$$

The first order condition for profit maximization for the two firms is given as

$$\frac{d\pi_1}{dQ_1} = 80 - 4Q_1 - 2Q_2 - 0 - 30 = 0 \tag{1.76}$$

$$= 50 - 4Q_1 - 2Q_2 = 0$$

Reaction function of firm – I

$$= -4Q_1 - 2Q_2 = -50$$
$$4Q_1 + 2Q_2 = 50 \tag{1.77}$$

$$\frac{d\pi_2}{dQ_2} = 80 - 2Q_1 - 4Q_2 - 0 - 20 = 0 \tag{1.78}$$

$$= 60 - 2Q_1 - 4Q_2 = 0$$

Reaction function of firm – II

$$= -2Q_1 - 4Q_2 = -60$$
$$= 2Q_1 + 4Q_2 = 60 \tag{1.79}$$

Solving the equations (1.77) and (1.79) simultaneously, we get $Q_1^*$ and $Q_2^*$

$$4Q_1 + 2Q_2 = 50 \tag{1.77}$$
$$2Q_1 + 4Q_2 = 60 \tag{1.79}$$

Multiply the equation (1.77) by 2 and equation (1.79) by 4 to cancel $Q_1$

$8Q_1 + 4Q_2 = 100$

$8Q_1 + 16Q_2 = 240$

$\overline{\phantom{8Q_1 + 16Q_2}}$

$- 12Q_2 = -140$

$Q_2{}^* = \dfrac{140}{12}$

= 11.66 units

Substituting $Q_2$ value in equation (1.77) we get $Q_1$ value

$4Q_1 + 23.32 = 50$

$4Q_1 = 50 - 23.32$

$Q_1{}^* = \dfrac{26.68}{4}$

= 6.67 units

Thus substituting $Q_1$ and $Q_2$ values in equations (1.72) and (1.73), we get profits of firm – I and firm – II

$\pi_1$ = ((80 –2 (6.67 + 11.66)) 6.67 – (60 + 30 × 6.67)

= (43.34) 6.67 – 260.1

= 289.07 – 260.1

= Rs. 28.97

$\pi_2$ = ((80 –2 (6.67 + 11.66)) 11.66 – 30 + 20 (11.66)

= (43.34) 11.66 – 263.2

= 505.34 – 263.2

= Rs. 242.14

To find out equilibrium price, substitute $Q_1$ and $Q_2$ values in inverse market demand function (1.71).

$P = 80 - 2 (Q_1 + Q_2)$

= 80 – 2 (6.67 + 11.66)

= 80 – 36.66

Equilibrium price = Rs. 43.34

# OLIGOPOLY

It represents the presence of a few firms in the market, producing either a homogeneous product or products which are close but not perfect substitutes to each other. Oligopoly can be divided into two forms, *viz.*, perfect oligopoly, wherein a few firms produce a homogenous product and imperfect oligopoly wherein there are a few firms producing heterogeneous products. The examples are televisions, two wheelers, four wheelers, tyres, cigarettes, textiles, *etc.*

## Characteristics of Oligopoly

1. *Presence of Few Sellers:* An important feature of oligopoly is the presence of few sellers. The product here is homogenous or heterogeneous in nature. Since the number of sellers is few, each firm commands a sizeable market share of a product.

2. *Interdependence:* Due to few number of firms in the industry, no single firm can afford to ignore the reaction of other firms to its actions. Suppose a given firm is contemplating to bring some changes in its price and output policies, it duly considers the counter actions of the other firms. Thus, the fortunes of one firm are decided upon by the policies of the other firms.

3. *Indeterminate Demand Curve:* The interdependence of the firms under oligopoly creates an uncertain atmosphere. The price and output policies resorted to by a given firm and its consequent sales of other firms can not be estimated with any amount of certainty. Suppose, if a firm would like to lower the price for increasing sales, the anticipated increase may not take place, for other firms too would have lowered the price by still a higher margin. This reaction of the other firms is a difficult proposition to be assessed. Hence, the demand or revenue curve is indeterminate.

4. *Conflicting Attitudes of Firms:* The firms in oligopoly behave as arch rivals and like to be independent to get maximum profit. By their actions and counteractions, they create an atmosphere of uncertainty. Against this action, at times they behave differently by cooperating each other to eliminate uncertainty arising out of mutual competition. By this way they join together for maximizing their profit. Certainly it is a conflicting attitude of the firms as reflected by the competition and co-operation.

5. *Competition:* There is always a constant battle among the firms in oligopoly. This rivalry continues as long as the firm exists in the business.

6. *Features of Monopoly:* Oligopoly is characterized by presence of few firms, product differentiation and a larger market share. It enjoys the superiority of a monopoly in the business as the differentiated product produced by oligopolist attracts the consumers or rather consumers are more attracted to it. The possibilities are plenty for the firms to unite and done the role of monopoly to enjoy the liberties.

7. *Price Rigidity:* Alongside competition, and monopoly element in oligopoly, price rigidity is another feature. The rivals are more passive for bringing the changes in prices, for such changes may affect the sales and consequently the revenue. The firms are well aware of the fact that any reduction or rise in price leads to price war or loss of sales. This impending danger forces them to keep quiet, consequently the price rigidity is a more common phenomenon rather than an exception in oligopoly.

8. *Lack of Uniformity in the Size of Firms:* Size of the firms in oligopoly differs considerably. Some may be very large, while others very small.

## Kinked Demand Curve and Price Rigidity

In oligopoly mutual interdependence of firms creates an environment of uncertainty for all the firms regarding price-output policy. The plans of a firm to make a given level of profits is totally influenced by the way other firms react to the activities of the one. The demand curve of the oligopolistic firm therefore cannot be definite as a given firm cannot precisely estimate the reactions of the rival firms. The demand curve which is *kinked* in this situation explains the price rigidity observed in oligopoly. It is assumed that a given group of firms produce the same product and the price OP is accepted by all the firms. At price OP, if a particular firm sells OQ quantity of output; others need not sell the same quantity (Fig 7.15). Under this situation, when we look at the demand curve (AR), 'MEN' has a *kink* at point E, because of which the demand curve has two parts *viz.*, the upper part (ME) being more elastic and lower part (EN) less elastic. The reasons for the *kink* at point E are that if this particular firm brings in a small increase in its price with other firms not reacting in the same, fearing for a fall in sales, this increase in price will cause a large reduction in sales. Hence the demand is more elastic if only one firm increases the price. Against this, if this particular firm opts to reduce the price, the competing firms also reduce the price by the same extent. Although general price cut is found here it ensures the same market share for each of the rival firms. Therefore, after point E the demand curve slopes down steeply. This leads to the point that the demand curve can be drawn given the assumption that a single price has been established in the industry with price rise by one firm not being followed by others and price decrease followed by other firms. The marginal revenue (MR) curve is discontinuous at OQ output *i.e.*, has got a gap, which is shown by dotted line OL. This is figured as a result of change from elastic nature of demand to inelastic nature of demand.

## MONOPSONY

Monopsony means the presence of a single buyer for the products produced by the firms. The example that can be cited is sugar factory. Farmers who are registered as sugarcane growers under factory's jurisdiction are supposed to sell the cane to sugar factory only. Tobacco board and coffee board can be cited as other relevant examples.

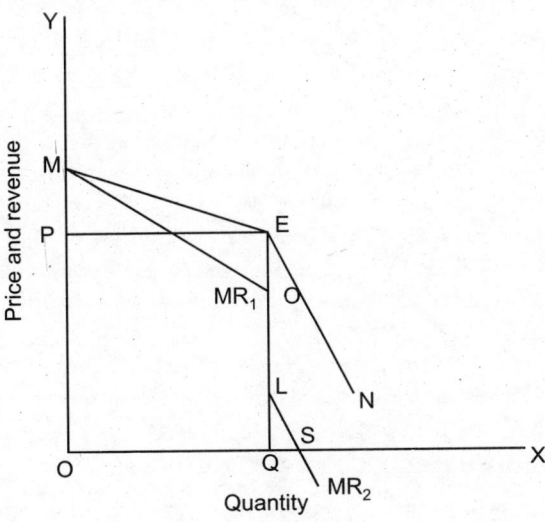

**Figure 7.15** Kinked demand curve.

## OLIGOPSONY

It is from the buyers side in a market. There are only few in number buying sizeable quantities of a product. Each individual buyer is so powerful that his buying behaviour influences the market price. The relevant example are gas and iron ores. These are purchased by a few firms in the country and final products are supplied to the market through these oligopsonies, *viz.*, Tata company for purchases iron ore, Hindustan Petroleum corporation buys the gas from Government or supplies to consumers after refinement etc.

# CHAPTER 8

# Distribution

In the process of production, the producer coordinates different factors of production *i.e.,* land, labour, capital and management. In the process of distribution the returns obtained through the production activity are apportioned to these factors that are employed in the production process. Consequently land gets rent, labour gets wages and interest is paid to capital and finally organization is rewarded with profit. Such an apportionment of returns among different factors of production is called distribution. It is also called factor pricing.

## RENT

Rent is the return for the fertility status of the land. In fact the land is defined in a broad sense. All the natural resources existing on the surface and beneath the surface of land like mines, rivers, *etc.,* are also treated as land, from which rent is received. Some resources are publicly owned, while others are privately owned. Rent is almost zero for publicly owned resource because one cannot use it for one's own purpose. These are meant for public welfare. Rent is expressed in two forms *i.e.,* one is economic rent and the other is contract rent.

### Economic Rent

Economic rent is the rent received exclusively from the use of land only. We use the term, exclusively because the rent of a building refers to the return obtained by the owner on the capital invested in the construction of building as well as the land. It is the return obtained from the combined values of both. Then it does not become rent. In farming the rent paid by a tenant to the landlord is not economic rent.

### Contract Rent

This implies the money paid by the tenant to the land lord for cultivating the land in a given year. Normally a certain rent is charged by a farmer not only for land but also for making availability of certain infrastructure on land like buildings, machinery, wells, fencing, *etc.* This means that he would like to realize some return on the investment made on the farm. Then it is not exclusively rent for land only. Thus economic rent is a part of contract rent.

## Quasi Rent

The basis for evolving this rent is the short run fixity of man made assets of production like machines, buildings, *etc*. When in the short run, the demand for these assets increases, consequently their income also increases. This results in a surplus income due to increased demand. This surplus income of assets is called quasi rent. This rise in quasi rent is a temporary phenomenon. It fact, the supply of these assets being elastic it is increased in the long run to match with the demand causing the surplus earnings to disappear. The concept of quasi rent does not apply to land because supply of land is inelastic.

## Scarcity Rent

This is the rent which arises due to scarcity of land in relation to demand. Scarcity rent is due to inelastic supply of land. This is surplus rent over the market rent for land due to increased demand for land.

## Determination of Rent for Land

We have now understand the diversified meaning of rent for land. We now go ahead to examine as to how the rent for land is determined. David Ricardo of England propounded a theory on rent, which is popularly known as Ricardian theory of rent. His theory is based on several assumptions, *viz.*, 1) The most fertile lands are cultivated first 2) There is continuous growth of population in the area 3) Rent arises due to difference in fertility of lands 4) Existence of marginal lands in every country 5) Quantitative limitations of land such as size of land 6) Land possesses certain original and indestructible power and 7) Rent is confined only to land which is a free gift of mature.

## Ricardian Theory of Rent

Ricardo applies his theory citing the example of a new country in which human habitation is yet to be found. It is being a new country people from another country slowly move into this country for settlement. It is hypothesized that there are four types of land *viz.*, A, B, C and D designated in the descending order of land fertility. It is assumed that all plots are of same size, requiring same doses of labour and capital. The people from the other countries are expected to migrate in batches to the unhabited country. The first batch thus migrating occupies more fertile land *i.e.*, type 'A' as the land is abundantly available besides being free. As the land is freely available no rent need to be paid by these migrants. Let as assume that 'A' type of lands are productive enough to produce 100 Q of wheat per unit of land in the beginning, which is sufficient for the people. Hence there is no rent for 'A' type of lands. But as the demand for wheat increases, it forces the cultivators to occupy 'B' type of lands which are relatively less fertile and uncultivated. If the productivity of this type of lands is 70 Q, 'A' type of lands begins to yield a surplus of 30 Q. This is the surplus and it is the rent of 'A' type of land. Let us assume that if still demand mounts up for wheat in view of growing population, 'C' type of lands are also occupied for cultivation, which are less fertile compared to 'B' type of lands. If the productive capacity of this type of land is assumed as 50 Q, it leaves a surplus of 20 Q for B type of lands and 50 Q for 'A' type of lands, which is claimed as rent by the respective owners of these lands. Likewise even 'D' type of land is brought under cultivation in view of the spiralling population growth.

Through his theory Ricardo described that superior or more fertile land as intra-marginal or super-marginal lands land the last type of less fertile land is called as the marginal land or no rent land. These two concepts are used depending on the situation. If still inferior type of land is brought under cultivation, it is called marginal land and the earlier marginal land becomes super-marginal land. The rent is the difference between the yield obtained from the super marginal land and that of marginal land. Rent is that extra value imparted to the type of land based on the yield surplus.

## Criticism

1. There are no original and indestructible powers of the soil.
2. It is not the fertility but scarcity which leads to rent.
3. Marginal land concept is not accepted because if the demand is higher even marginal land gets rent, and
4. Rent is not applicable for the use of land only, as it applies to all scarce inputs in economics.

## Modern Theory of Rent

Modern economists assert that rent arises for any factor of production. It is the surplus payment in excess of transfer earnings of a factor. Transfer earnings imply the amount of money which any particular unit of factor could earn in its next best alternative use.

The rent of the land is determined by the two market forces $viz.$, demand for and supply of land (Figure 8.1). Demand curve for land slopes downwards because the rent of land is influenced by the marginal productivity. On the other hand supply of land is fixed $i.e.$, supply is inelastic. Supply curve is vertical to X-axis. Equilibrium between demand and supply determines the rent of land.

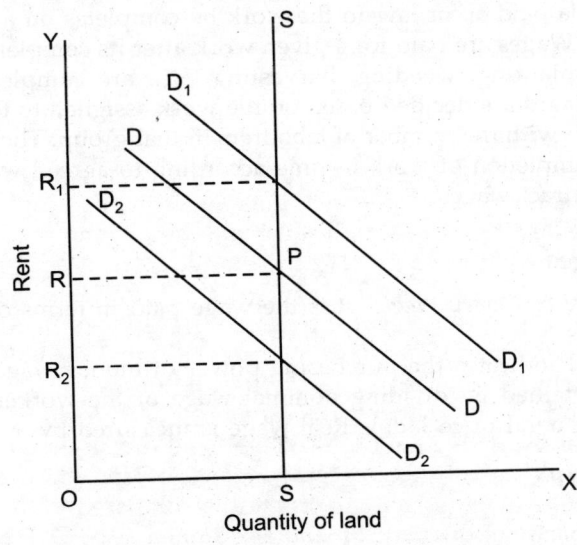

**Figure 8.1** Determination of Rent.

As seen from figure, DD is the demand curve for land and SS is the supply curve of land. The two curves intersect at 'P' and hence the rent is OR. Changes in population and consequent additional pressure on land increase the demand for land. Therefore the new demand curve is $D_1 D_1$ and the rent is $OR_1$. Contrary to this when demand for land falls the new demand curve $D_2D_2$ is formed and its corresponding rent is $OR_2$.

# WAGES

Wages are the rewards paid for the labourers for sparing their productive services. It may be paid either in cash or kind or both. *A wage may be defined as a sum of money paid under contract by an employer to a worker for services rendered (Benham).* Wages are paid for casual labourers, while salaries are paid for permanent staff and consultation fee for doctors, lawyers, *etc.*

## Methods of Wage Payment

Based on the payment, wages are classified as (1) Cash wages and kind wages (2) Time wages (3) Piece wages; and (4) Task wages.

1. *Cash Wages and Kind Wages:* Wages for the workers are paid in cash or kind or both. With the advent of currency, wages are paid in cash. However, kind payment is also in vogue along with the cash payment.
2. *Time Wages:* It is the wage per unit of time. It is the payment of wages on hourly, daily, weekly, fortnightly and monthly basis. In farming, casual labourers are paid on daily basis, workers for domestic services are paid on monthly basis and attached servants in farming are paid on half-yearly or yearly basis.
3. *Piece Wages:* It is based on the work performed by an individual in the production of goods and services. In the manufacturing process of a good, the entire production activity is divided into various sub-processes. An individual attending to a sub-process is paid according to the work he completes on a particular day.
4. *Task Wages:* Wages are paid for a given work after its completion, say in farming, paddy transplanting, weeding, harvesting, *etc.*, are completed by a group of labourers. Wages are decided based on the work assigned to the group and it has nothing to do with the number of labourers in that group. The task is given to the group for completion of work in time according to agreed wages. They are also called as contract wages.

## Types of Wages

1. *Nominal Wage or Money Wage:* It is the wage paid in terms of money at current market prices.
2. *Real Wage:* It indicates the purchasing power of money wage. Real wage of the worker is obtained by dividing nominal wage of the workers at different time periods by general price index. Real wage is measured by

$$R = W/P$$

Where,

    R = Real wage  
    W = Money wage  
    P = General price index

# THEORIES OF WAGES

## 1. Subsistence Theory of Wages

David Ricardo was the proponent of this theory. It is called as 'Iron law of wages' according to Lassalle, the German Economist. This theory emphasizes that wages prevail at such a level just to enable the workers and his family members to live at the minimum subsistence level*. In case there is a rise in wages, it attracts the workers to have large families, thereby increasing the supply of labour. The increased supply of labour brings down the wages to minimum subsistence level. On the contrary, when wages fall down, marriages are discouraged and coupled with malnutrition and consequent death, the supply of labour decreases. This results in increase of wages back to subsistence level.

## 2. Wage Fund Theory

The development of this theory is owed to J.S. Mill. According to him in any country certain fixed proportion of the capital is set aside for payment of wages to the labourers which is called wage fund. The wage at a particular point of time in a country is determined by the amount of wage fund available and the total workers. The rise and fall in wages depend on the number of workers and the amount of fund.

## 3. Residual Claimant Theory

This theory was propounded by American Economist, Walker. According to him labourers are the residual claimants in any production activity. After the rewards of the other factors, viz., rent for land, interest for capital and profit for organization have been paid, the residual amount from the production is the wages to the labourers. Wages are thus the residual after rent, interest and profits are deducted from the income obtained from a production activity.

## 4. Marginal Productivity Theory

According to this theory, under conditions of perfect competition, the wages received by a worker of same skill and ability will be equal to his marginal value productivity. Marginal productivity of labour unit is the contribution to the total output with other factors remaining the same. The marginal value productivity of a labour unit is the value of an additional unit of the output produced due to its employment.

## 5. Modern Theory

Modern theory is based on demand for and supply of labour. Demand for labour is a derived demand**. If there is a higher demand for products from consumers, there would be more demand from producers for labour for helping to bring the expected level of output.

---

\* It is the minimum amount of food, clothing and shelter which a worker in a family requires for existence.
** Derived demand for an input means the increased demand for output that would result in increased demand for the input. The term derived demand is used because the demand for input increases, not due to direct input demand, but indirectly from increased output demand.

A rise in demand for a commodity pushes up the demand for the labour involved in the production of that commodity.

The supply of labour is represented by the number of workers willing to work in a production activity at various wage rates taking into consideration the number of hours, number of days in a week, *etc*. These two forces *i.e.*, demand and supply determine the wages.

DD is the demand for labour and SS is the supply of labour which intersect at point E. The wage rate is OW (Figure 8.2).

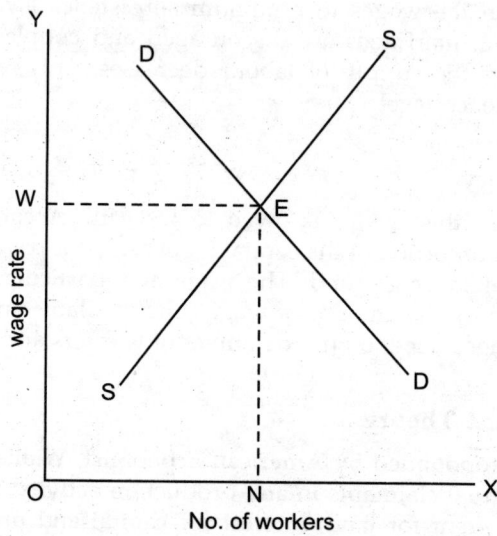

**Figure 8.2** Determination of wage rate.

## INTEREST

Interest is the amount paid by the borrower to the lender for the use of capital. According to Marshall *"the payment made by a borrower for the use of loan for, say a year, is expressed as the ratio which that payment bears to the loan"* is called interest.

*Interest is the price paid for the use of loanable funds* (Meyers).

The interest charged by a lender from the borrower is termed as gross interest. It is because lending activity is fraught with risk as well as inconvenience which are also considered. Taking these items into account the gross interest is considered under the following terms.

1. *Net or Pure Interest:* This interest is the payment exclusively made for the loan amount.
2. *Insurance Against Risk:* In the lending activity there is always the risk of not getting back the funds lent by the lender. The loan may turn out into a bad debt. This risk needs to be insured. For this purpose, some more interest is added to the net interest.
3. *Payment for Inconvenience:* Through the lending activity the lender gets interest but in the same process he is placed in inconvenience to get back his funds at the time he wants, because the funds are locked up for a certain period of time. Unless the

loan period was over he cannot get back his funds. For this inconvenience, he charges some additional interest to the net interest.

4. *Reward for Management:* The business of lending requires perfect maintenance of the records to keep a close watch on the business performance. Apart from this, the borrowers are to be pursued for prompt repayment. This calls for efficient management of the business. Hence reward for management is included in gross interest.

# THEORIES OF INTEREST

## 1. Productivity Theory

This theory is based on the fact that capital is productive. People borrow required money when they do not have enough funds. The individuals are encouraged for borrowing because they fetch additional gains in the production activity. Since the borrowed capital is helping them to generate additional income, they have to part with the money in the form of interest to the lender. The theory is based on the premise that an individual demands capital because of its productivity and naturally such a productive resource must be paid interest.

## 2. Abstinence Theory or Waiting Theory

This theory was developed by economist *viz.,* Senior, based on the supply side of money. According to him, the capital is generated from savings. Inturn savings are created from the earnings of the individual *i.e.,* individuals, by foregoing some of their present consumption requirements, keep aside some money as savings for future requirements. Thus, capital is generated from abstaining from present consumption. Interest therefore needs to be paid to the person who saves capital and offers for investment, to compensate the sacrifice he has to make in the process of generating capital*.

## 3. Austrain or Agio Theory

Bohm–Bawerk, an Austrian economist developed this theory. It is based on time preference. Naturally people tend to prefer present income and consumption to that of future, hence there is always premium on present consumption compared to some future date. Conversely, future consumption undergoes a discount compared to present consumption. Hence, interest is that discount which is paid to the lenders who postpone their present satisfaction to a future date.

## 4. Time Preference Theory

This theory was putforth by Irving Fisher. He incorporated the element of certainty for future. He extends by saying that even when the future is certain, people prefer present consumption and satisfaction to future consumption. Though people are very much interested in present consumption, but still they show willingness for saving to generate capital which needs to be given compensation in terms of interest.

---

* Here the word capital is used in the sense of money used for investment.

## Theories of Interest Rate Determination

Theories have been developed to explain as to how the interest rates are determined. They are 1) Classical or real theory 2) Loanable funds theory and 3) Keynes's liquidity preference theory.

### Classical Theory of Interest

It is based on the demand for savings to invest in capital goods and the supply of savings. Rate of interest is determined at the equilibrium point of two forces, as seen from Figure 8.3. SS is the supply curve of savings while DD is the demand curve of savings to invest in capital goods. The equilibrium position is attained at the intersection of supply and demand curves. OR is the interest rate that is determined.

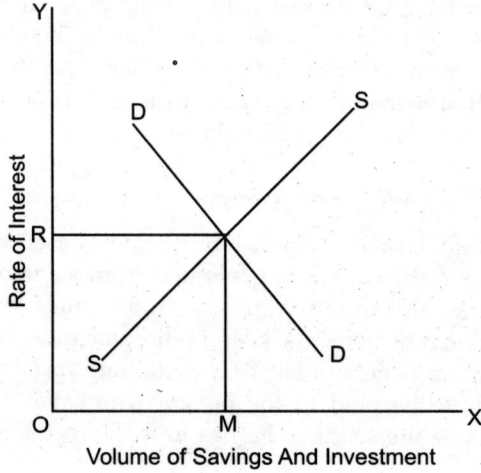

**Figure 8.3**  Determination of rate of interest.

### Loanable Funds Theory of Interest

According to Wicksell, the Swedish economist, interest rate is determined by the forces of demand for and supply of loanable funds. Loanable funds represent the sum of money supplied and demanded in money market. On the supply side of loanable funds, it comes from savings of people and money supply (credited and created through banks), *i.e.*, savings (S) and the net new money (M) (resulting from credit creation of commercial banks). The combination of these two represent the supply of loanable funds, which is represented by S + M. Here, the loanable funds (L) is nearly equal to S + M.

The demand for loanable funds results from demand for investment, in addition to demand for hoarding money. Hoarding money influences the supply of loanable funds. If the hoarded money increases, the supply of loanable funds is reduced. Conversely an increase in dishoarding results in increase in the supply of loanable funds. Demand for loanable funds ultimately is decided by the demand for investment (I) and net hoarding (H)*. I + H therefore represent the total demand for loanable funds. Interest rate is determined by the equality of (I + H) and (S + M).

---

* Difference of increase of total hoarding and increase in dishoarding.

## Liquidity Preference Theory

J.M. Keynes is the proponent of this theory. The term, liquidity means money in cash form. People have options to keep money in cash form and/or in non-liquid form like, gold, shares, fixed deposits, *etc*. According to him three motives lead to liquidity preference by the people. These are (1) Transaction motive (2) Precautionary motive and (3) Speculative motive.

1. *Transaction Motive:* People would like to hold some money in cash form to meet their daily requirements. This varies with income earners *i.e.,* those getting daily incomes, weekly incomes, fortnightly incomes, monthly incomes and annual incomes. The amount of money one wants to have in cash form for transaction motive depends on the periodicity of deriving the income.
2. *Precautionary Motive:* Apart from transaction motive, people would like to have cash balance to meet the unforeseen exigencies. To overcome the problems of raising money to meet the unforeseen expenditure, certain amount of liquid money (cash) is set aside.
3. *Speculative Motive:* Speculative motive refers to the process of taking ventures by people in high business areas where returns are more. So this motive results in keeping aside certain amount of money for obtaining higher returns.

## Determination of Interest Rate

Liquidity preference and supply of money determine the rate of interest. Liquidity preference here means the liquidity preference for speculative purpose according to Keynes. Therefore, the equilibrium between liquidity preference for speculative purpose and supply of money determine the rate of interest.

In Figure 8.4 LPS indicates liquidity preference for speculative purpose. It is the demand for money. OM represents the fixed amount of money available to meet the requirements of liquidity preference for speculative purpose. Rate of interest is determined by the equality of demand of money for speculative purpose and supply of money. The interest rate that prevails is OR, which is the equilibrium interest.

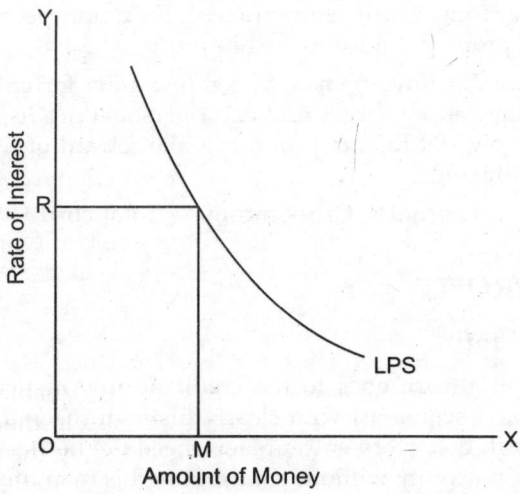

**Figure 8.4** Determination of rate of interest.

# PROFIT

Profit is the reward for entrepreneurial function of decision-making and uncertainty bearing. Profit can be either positive or negative, since it is a residual income. Profit differs from rewards of other factors of production like rent, wages and interest on the point that these are all certain, while profit is tentative. These rewards are paid even before the ultimate product is obtained, while profit happens to be the surplus of returns over total costs, and it is obtained at the end of production activity.

## Various Concepts of Profit

*Gross Profit:* It is the surplus of gross revenue over paid out costs (explicit costs) obtained in a production activity. Gross profit cannot be treated as ultimate profit, for it includes several items which are not actually profits as presented below:

*Remuneration for Entrepreneur Himself:* It is a practice for the entrepreneur to contribute his own capital in the business besides offering his services as a manager. In such a situation the opportunity cost of owned capital and the services rendered in his own business are also included in the gross profit. These imputed costs are to be deducted from the gross profit to arrive at the net profit.

*Depreciation and Maintenance Charges:* When the plant is in operation there will be wear and tear (depreciation) on equipment and machinery which needs to be duly accounted for. In addition to this, interest on fixed capital and insurance premium are other fixed cost items that are incurred in a production activity. These cost items need to be deducted from the gross profit to arrive at net profit.

*Extra Personal Profits:* These are the profits which accrue to a business unit not due to the managerial ability of the entrepreneur, but due to other factor, say the profits of monopoly. These are not due to the businessman's managerial abilities but for his position as a monopolist. The monopoly profit needs to be deducted from gross profit to arrive at net profit. Another source of getting unexpected high profit is from windfall gains or chance profits. Any sudden happening, say outbreak of war, natural calamities in certain pockets of the country *etc.*, lead to a rise in the product prices, resulting in abnormal profits for the entrepreneur. Such unexpected profit should be deducted from gross profit to find out the net profit.

*Net Profit:* The aforesaid three items *viz.*, remuneration for entrepreneur himself, depreciation and maintenance charges and extra personal profits are to be deducted from gross profit to work out the net profit. It is the reward of the entrepreneur for coordination and risk-taking.

Net profit = Gross income – Total cost.

# THEORIES OF PROFIT

## 1. Rent Theory of Profit

Development of this theory goes to the credit of an American economist, F.A. Walker. He was the first economist who clearly distinguished capitalist from entrepreneur by putting forth that every entrepreneur need not be a capitalist but still can carry on the production activity without contributing his own funds. He highlighted through his theory the similarity between rent and profit. Just like rent arising for a piece of land due to its fertility status, an entrepreneur also gets rent for his managerial ability, *i.e.*, profit. Further, as the rents for different lands are not uniform, so also the profits recovered by the entrepreneurs. Entrepreneurs with different skills are rewarded with different profit levels. Extending this similarity further between rent and profit, just as there is marginal land, there is marginal entrepreneur. As the marginal land is associated with no rent, the marginal entrepreneur too is associated with no profit. The profits of the super marginal entrepreneur are measured taking the situation of marginal entrepreneur.

The marginal entrepreneur just secures wages of management but not requisite profit. So according to Walker, profit does not enter the price, as market price is determined based on the cost of production of the product attributed to marginal entrepreneur.

## 2. Wage Theory of Profits

Prof. Taussig, an American Economist advocated this theory. In rent theory, a similar comparison was developed between rent and profit, while in this theory a similar comparison was identified between the labourer and entrepreneur. This is justified by advancing that just as labourer receives wages for his services, likewise an entrepreneur too gets profit for his services. Only the services differ in the two cases, as it is physical service in the former and organizing capacity in the latter. Through this comparison, this theory concludes that profit is a type of wage.

## 3. Dynamic Theory of Profits

Through this theory J.B. Clark, an American economist defined profit as the excess of price of goods over their costs. Further, he advocated that profits arise only due to the dynamic change that takes place in a society, and profits do not vary much in a static society. The reason putforth is that, in a static society, the parameters which influence an economy like population, technology, desire of the individuals, the available capital, *etc.*, remain the same. Everything is certain in such an economy, so economic environment in current year is the same as that of previous year. Risk and uncertainty are absolutely absent. In such a society, the prices of goods are equal to cost of production. All that what an entrepreneur gets is wage for his labour and interest on capital. If by chance, the prices of commodities are more than cost of production, competition exists, soon bringing down the prices to the level of cost of production. Hence, there is no chance of profits in a static society. Against this situation, in the dynamic economy profit arises. Dynamic economy is associated with the same parameters as indicated above of a static economy, but these are subject to change over time. The entrepreneur who takes advantage of changing demand and supply conditions of a commodity makes profit. It depends more upon the efficiency of the entrepreneur who takes advantage of it. Entrepreneurs with different efficiencies make different profit levels.

## 4. Risk Theory of Profits

Prof. Hawley, an American economist advanced this theory in 1907. His theory is based on the risk that is associated with every business activity. It explains that those who are prepared to take great risk are rewarded with greater profit. Thus greater the risk, greater would be the profits in a business activity. The profit that is obtained through risk taking exceeds the ordinary return on capital.

## 5. Uncertainty–bearing Theory of Profits

According to Prof. Knight, an American economist profit arises due to uncertainty rather than risk taking*. Further he classified risk into two categories *viz.*, foreseeable risks (insurable risks) and unforeseeable risks (non-insurable risks). Foreseeable risk is that which can be foreseen or predicted by an entrepreneur, while the unforeseeable risk cannot be foreseen or predicted. Prof. Knight calls this unforeseeable risk as uncertainty and because of this uncertainty, profits are received by the entrepreneur. Coming to risks, some risks say death, fire, theft, accident, *etc.*, in a business activity are estimable and since these are insured, it is the insurance company which undertakes these risks. As insurance is a fixed cost, the insurance premia are included in the cost of production. So the risks in a business will not help the entrepreneur to derive more profits. Moreover risk taking is not the function of an entrepreneur, but of insurance companies.

On the other hand, the unforeseen risks such as production risk, trade cycles, competition, new techniques of production, changes in Government policies, price uncertainty *etc.*, cannot be predicted and there is no insurance protection for such events. Thus, according to Prof. Knight, profits arise due to uncertainty bearing but not due to the risk bearing. Greater the uncertainty bearing, higher would be the level of profit.

## 6. Innovations Theory

This theory was postulated by Joseph A. Schumpter, an American Economist. He brought out that profits accrue to the entrepreneur as a result of innovations that are created in the production process and also in marketing of the products. Product innovations refer to the changes that are introduced in production process which will help to minimize the production costs. Innovations on marketing front mean introducing new products, making inroads into new markets, obtaining more market share, new sales promotion techniques, *etc.*

Innovations thus brought out, increase the profits. These profits are contributed by lesser production costs and/or higher product prices. However, profits which were generated through innovations do not last long, as the new firms enter into the business activity knowing the innovations and compete away with the profits. Once profits disappear from these innovations, then new innovations would be developed to derive profits which again disappear when once again new firms make their entry. So profit due to innovations would appear, disappear and reappear in regular cyclical fashion.

## 7. Monopoly Theory of Profits

Firms under monopoly or monopolistic competition will be having high market power. They will be 'price-makers' rather than 'price-takers'. They make their best efforts to safeguard their interest and keep their profits at higher level. Also in order to maintain their monopoly power, the firms will prevent the entry of new firms into the business in the form of several barriers, therefore the basic source of profits is the monopoly power of the firm.

---

* The main difference between risk and uncertainty is that risk is measured through probability concepts while uncertainty cannot be measured.

# SECTION II

# MACROECONOMICS

Macroeconomics, which is the child of the great depression of 1930s, is mainly concerned with the aggregates. It deals with aggregate output, aggregate supply, aggregate demand, aggregate income, general price level, total employment, total consumption, savings and investment and so on. Macroeconomics is the study of the behaviour of the economy as a whole. A nation's progress greatly depends upon the judicious selection of macroeconomic policies such as the level and structure of taxation, or expenditure. Macroeconomics has developed only after the great depression especially with the publication of "Keynes general theory of employment, interest and money" in 1936. Keynes theory made a genuine break from the classical and neo-classical economics and produced such fundamental and drastic changes in economic thinking that his macroeconomic analysis has earned the names "Keynesian revolution" and "New economics".

*Macroeconomics may be defined as that branch of economic analysis which studies the behaviour of not one particular unit, but of all the units combined together.*

## Macroeconomics covers the following fields

1. *Theory of income, output and employment with the two constituents viz., the theory of consumption function and the theory of investment function. The theory of business cycles is also a part and parcel of the theory of income, output and employment.*
2. The theory of prices with its constituents of inflation, deflation, reflation, etc.
3. The theory of economic growth dealing with long run growth of income, output and employment as applied to developed and developing countries.
4. Macro theory of distribution dealing with relative shares of wages and profits in national income.

# CHAPTER 9

# National Income

In an economy goods and services keep on being produced, which need to be valued. National income is nothing but the aggregate money value of all goods and services produced in a country during a year. It can also be viewed as income distributed among the factors of production in the form of rent, wages, interest and profits. It is also interchangeably used with national dividend, national output and national expenditure. According to Alfred Marshall, national income is defined as the labour and capital of a country acting on its natural resources produce annually a certain net aggregate of commodities, material and immaterial including services of all kinds. But Pigou presented the concept of national income in a different way. *According to him 'National income is that part of the objective income of the community including of course, income derived from abroad which can be measured in money.* In the words of Irwing Fischer *"the national dividend or income consists solely of services as received by ultimate consumers, whether from their material or from their human environment".*

A comparison of these definitions reveals that the one given by Marshall suffers from certain defects. In an economy in which numerous goods and services are produced, it is really a difficult proposition to estimate them correctly to arrive at the national income. Coming to farming, farmers do retain some part of production for personal consumption and the rest is only marketed. In such a case, the methodology to evaluate the amount of produce not reached the market has not been given. Another point of debate is the implicit meaning of double counting in the definition. Double counting implies the same good being counted twice; say, silkworm cocoons maybe counted in agricultural production and silk cloth in industrial production. Though the definition of Pigou is superior over that of Marshallian definition in terms of avoiding double counting and inclusion of those goods and services which can be measured in money and also inclusion of income received on account of investment in foreign countries, it too has some omissions. The stress on valuing the goods and services in money is a limitation. Only in developed countries do the commodities are exchanged for money and in under-developed countries of the world, still barter system is in practice. In these countries, we have a limitation of correct estimation of national income. Marshall and Pigou approached national income from production end. On the other hand, the approach of Fisher is from consumption end. It looks superior over Marshall's and Pigou's as he underlines the adequate concept of economic welfare, which is dependent on consumption and consumption represents our standard of living. At the same time this approach in reality poses problems.

Consumers who number in millions, consume the same good in different places, and the estimation of their total consumption is a difficult proposition. Further, consumer goods can be durable, semi-durable and non-durable. Non-durable goods may be used up in the same period they were purchased. In respect of semi-durable and durable goods which extend their benefits over a period of time, it becomes difficult to confine the period of consumption. Also the durable goods are sold from one consumer to another and with every change in the ownership the value too changes. Hence, there are problems in measurement.

## Modern View

Simon Kuznets defined national income as *"the net output of commodities and services flowing during the year from the productive system in the hands of ultimate consumers"*.

Here the income side as well as expenditure side are included to present the concept of national income.

# CONCEPTS OF NATIONAL INCOME

## Gross National Product (GNP)

It is the basic social accounting measure of the total production of goods and services in an economy. GNP is defined as the total market value of all final goods and services produced in a year. It includes the market value of such products as are produced in agriculture, mines, forests, industries *etc.*, and of services like transport, communication, banks, lawyers, doctors, teachers *etc.*, and these are added together during one year. GNP is a measure of the current output of economic activity in a country. Goods and services produced in an economy are subjected to purchase and sale several times. Therefore, to avoid the possibilities of double counting, GNP includes the market value of final goods* only, ignoring the transactions involving intermediate goods**. Those which are not included in GNP are as follows.

Services which are rendered freely are not included in GNP. Any free service that is rendered and the services that are offered by the family members to others in the family, like parents upbringing their children *etc.* is not included.

The transactions relating to the goods which are not currently produced are not taken into account while estimating GNP. Similarly, the sale and purchase of old goods and of shares, bonds, assets of the business units *etc.*, are not included, because they do not add anything to the national produce as they are simply transferred from one to another. Similarly, old age pensions, unemployment allowances *etc.*, are also not considered, as they do not provide any service. Analogously, the changes in the value of the capital assets as a result of changes in the market prices are also not taken into account, as they have nothing to do with the current production.

## Measures of GNP

Three methods *viz.*, income method, expenditure method and value added method are in vogue in estimation of GNP which are presented below:

---

* Final goods are those goods which are being purchased for final use and not for resale or further processing.
** Intermediate goods are those goods which are purchased for further processing or for resale.

## 1. Income Method

According to this method GNP is the sum total of the following items.

i. *Wages and Salaries:* The wages and salaries earned by the employees are identi-fied. These include contributions by the employees towards provident fund, in-surance, *etc.*

ii. *Rents:* Different rents earned by the individuals are taken into account. These rents represent rent of land, houses, factory, shops, *etc.,* and also estimated rates of these, if the owners use them.

iii. *Interest:* The interest earned by the individuals from various sources is found out. To this, interest on the owned funds is added.

iv. *Incomes of Non-company Business:* These are the incomes earned by individual proprietors, partners, self-employed persons, *etc.*

v. *Corporate Profits:* Corporate profits used in the calculation of GNP are equal to the sum of corporate profits, taxes plus dividends paid to the shareholders plus undistributed corporate profits.

vi. *Indirect Taxes:* Indirect taxes like sales tax and excise duty are levied on the commodities. Infact price per unit of a commodity includes these indirect taxes. But revenue from these taxes get credited to the Government but not to the factors of production. The income derived through these sources should be added to the national income.

vii. *Depreciation:* During the process of production, machines, and other capital in-vestment depreciate for which the manufacturing firms make allowance. Since this amount is not received by the factors of production, it is included in GNP.

viii. *Transfer payments:* Payments received by the individuals as pension, unemploy-ment allowance *etc.,* are called as transfer payments. These are received by the people for not doing any work, so these should be deducted from GNP.

Now GNP through income method can be expressed using the following formula.

GNP = Wages and salaries + rents + interests + profits of unincorporated firms + dividend + undistributed corporate profits + corporate taxes + indirect taxes + depreciation − transfer payments.

## 2. Expenditure Method

Through this GNP is the sum total of expenditure incurred on goods and services during a period of one year. Expenditure includes personal consumption expendi-ture, gross domestic private investment, net foreign investment and Government expenditure on goods and services.

i. *Personal Consumption Expenditure:* It includes all types of expenditure on personal consumption by the individuals of a country. It includes expenses on durable goods (scooters, watches, televisions, refrigerators, *etc.*) and expenditure on non-durable goods or consumable articles (food, clothing *etc.*) as also the expenditure incurred on services of all kinds (teachers, doctors, lawyers, transportation, *etc.*). It includes expenditure on intermediate goods. Expenditure on house is treated as investment expenditure rather than consumption expenditure.

ii. *Gross Domestic Private Investment:* It includes private investment on capital and producer goods like buildings, machinery, equipment, *etc.* These investment ac-tivities pertain to business firms. It includes capital or investment goods needed not only to replace the existing depreciated capital goods, but also the capital goods required to increase the society's production of goods and services. The term investment should reflect the real investment but not the financial investment. If an individual buys a old machinery, it may be financial investment, but not real investment, for it was included in the GNP of that year (the year of purchasing the new machinery). Hence, purchase of real investment goods such as machinery, equipment *etc.*, produced during the year is only considered. Further investment does not include, mere financial transfers such as purchase of existing old stocks and shares in the stock exchange, since it does not allow any new production.

iii. *Net Foreign Investment:* In an economy, the entire production may not be sold within the country and part of it is exported to other countries. Similarly, the country imports some finished goods from other countries during the year. The difference between the value of exports and that of imports should be worked out. If the difference is positive, it should be added to the other items of expenditure and it should be deducted from other items of expenditure, if the difference is negative.

iv. *Government Expenditure on Goods and Services*: The Central and State Governments purchase consumer goods as well as investment goods for their own enterprises. Apart from this, the Government spends on defence and police as well. Government also spends sizeable amounts on transfer payments (social security payments). However, these payments are not included in GNP, for, these payments are not the ones for currently produced goods and services.

The expression of GNP by expenditure method is

GNP =  Personal consumption expenditure (C) + Gross domestic private investment (I) + Net foreign investment (X-M) + Government expenditure on goods and services (G).

GNP = C + I + (X-M) + G

If all the items of GNP are correctly calculated, GNP worked out either by income method or expenditure method, remains the same.

## 3. Value Added Method

It is always difficult to distinguish between intermediate products (raw material, fuel etc.) and final good (machinery, equipment etc.) as raw material is a intermediate product for one industry and final product for the other industry. So to overcome this problem the value of the intermediate goods used in a manufacturing industry should be deducted from the value of the final goods. The difference that is arrived at is called value addition. If the same procedure is applied for all the industries in the economy, we can find out the GNP by value added method.

## Net National Product (NNP) or National Income at Market Prices

In the production of GNP, the capital goods employed are worn-out. The amount of decline in the value of capital goods due to wear and tear is called depreciation. To estimate the NNP, depreciation is deducted from GNP. NNP therefore is the market value of all final goods after duly accounting for the depreciation, hence the name national income at market prices. In other words, NNP is the net money value of final goods and services produced at current prices in an year in a country. NNP is found by using the formula.

NNP = GNP - depreciation      (or)

Personal consumption expenditure on goods and services (C) + Net domestic private investment (I) + Government expenditure on goods and services (G) + Net exports (X-M).

## National Income (NI) or National Income at Factor Cost or NNP at Factor Cost

*National income at factor cost implies the sum of all incomes earned by resource suppliers for their contribution of land, labour, capital and entrepreneurial ability, which go into the net production in an year.* Infact national income depicts how much it costs the society in terms of resources to produce net output. Hence, it is called national income at factor cost. The difference between national income at factor cost and NNP at market prices arises due to the indirect taxes and subsidies which pushes the NNP at market prices higher than NNP at factor cost.

National income or National income at factor cost = NNP–indirect taxes + subsidies.

## Personal Income (PI)

*It is the sum of all the incomes actually received by all individuals or households during a given year.* National income differs from personal income in the sense that national income is earned by factors of production. Social security contributions, corporate income taxes and undistributed profits are earned but not received by the individuals. Transfer payments like old age pensions, unemployment dolls, interest on public loan *etc.,* are not currently earned but received by the individuals. So to arrive at personal income, we have to subtract social security contributions, corporate income taxes and undistributed corporate profits from national income which are earned but not received and add incomes received but not currently earned.

Personal Income = National income - social security contributions – corporate income taxes – undistributed corporate profits + transfer payments.

## Disposable Income (DI)

*It is the amount of money available with the private individuals to spend.* To arrive at the disposable income, personal taxes (income and personal property taxes) are deducted from personal income. It is computed using the given formula.

Disposable income (DI) = Personal income – personal taxes (or)
Consumption expenditure + savings

# CHAPTER 10

# Money

When the human civilization was not developed, people used to exchange those goods which they produced for those which others produced. Such an act of exchanging goods for goods is called barter. But as the years rolled by and when the social organization became more complex, barter system was found to be not practicable.

The difficulties in barter system were replaced with the introduction of money. Money has been defined as the medium of exchange. According to Robertson money is defined as *"Anything which is widely acceptable in discharge of obligations"*. The different stages in the development of money are as follows.

1. *Commodity Money:* The earliest form of money consisted of goods like rice, wheat, cattle, skins, elephant tusks *etc.* These were accepted as they were all desired by all the people.
2. *Metallic Money:* As the civilization advanced, people found it difficult to carry on the exchange transaction with commodities, as they were found to be very inconvenient. Commodity money gave way for the metals to be used as money. These metals include gold, silver, copper, bronze, *etc.* From the beginning of their introduction, Government kept the right to issue coins and certify their weight and quality. Metals were converted into coins for this purpose.
3. *Paper Money:* Paper money is introduced to supplement the metallic money. When paper money was introduced, it was backed up by exactly equal amount of gold or silver kept in reserve by the issuing authority. But now paper money is not backed up by metals like gold and silver, but only proportional reserves are maintained. Their issue rests more on people's confidence on the issuing authority. Such a currency is called fiduciary issue and Indian currency largely is of fiduciary issue.
4. *Bank Money:* Paper money has been supplemented or at times replaced by bank money. It refers to the bank deposits. Deposits can be converted into money by the depositors through cheques.

## KINDS OF MONEY

The main kinds of money are (1) Metallic money and (2) Paper money. These are further divided into standard money and token money.

1. *Standard Money:* For such type of money, intrinsic value (real value) is equal to face value (the value written on the coin). It is subjected to free coinage. The coins are made of gold and/or silver. As such, no country has such a money in circulation.

2. *Token Money:* This money is made up of cheaper metal. Its face value is greater than its intrinsic value. The rupee is a standard unit of money in India, but its face value is greater than its real value and also it is not subjected to free coinage. It is a mixture of standard and token money.

## Characteristics of Money: Following are the characteristics of money.

1. *Cognisability:* As it used as medium of exchange, money should be easily recognized by one and all. Keeping this in view, the citizens of a nation give value to their respective currencies, issued by monetary authorities. In India the quantum of currency and its value are determined by RBI and Ministry of Finance.

2. *Utility:* When money is deposited by the public in different financing institutions, they will get different interest rates. This is due to the fact that money-receiving agency (borrowing agency) will have different time utilities for money. In other words, they use the money for the best alternative purpose and accordingly pay the interest to the lenders, depending on the need. Similarly money will have possession utilities and place utilities also.

3. *Portability:* It should facilitate easy carrying from one place to another without expense or inconvenience to the individual user. In other words, the bulkiness, weight and other inconveniences in the transactions are reduced with the help of currencies of different denominations, such as Rs. 5, Rs. 10, Rs. 20, Rs. 50, Rs. 100, Rs. 500 and Rs. 1,000 notes.

4. *Durability:* Coins are more durable than the paper currencies. Particularly, currencies of lower denominations *viz.,* 10 paise, 25 paise, 50 paise, one rupee and 5 rupees are being in circulation in the form of coins in order to provide durability. Similarly keeping the same purpose in view, the standard paper is used for printing currency notes.

5. *Indestructibility:* In the normal usage the coins should not get disfigured easily and the paper currency should not get torn easily in the circulation.

6. *Stability:* The value of money should not be changing. It should be more stable. For this to achieve there is need to eliminate inflation in the economy.

7. *Homogeneity:* All coins of the same metal should be as identical as possible with regard to quality and weights. Similarly notes should be printed with the same quality paper with utmost caution and use of sophisticated machinery, otherwise fake notes come into circulation and cause inflation.

8. *Universal Acceptability:* The value of the currency issued by monetary authority in the country should be recognized uniformly in all the states of the nation. Such currency of the nation will be accepted internationally also, provided the currency follows the norms of international monetary authority.

## Functions of Money: Money performs five important functions.

1. *Medium of Exchange:* Money facilitates the buying and selling of goods and services as a medium of exchange.

2. *A Unit of Account:* Money is used as the measure of value of all goods and services. Rupee is the monetary unit in India. The value of all goods and services is expressed in rupees.

3. *Standard of Deferred Payments*: Money facilitates the settlement of debts and future transactions without any risk.
4. *Store of Value*: Money is a form of holding wealth. If one feels that the available money is in excess of his requirements it is convenient to save and use it as and when the need arises.
5. *Transferable*: It facilitates the easy transfer of value. The disposal and purchase of assets can be done very easily with money.

# CHAPTER 11

# Public Expenditure

*Public expenditure is the expenditure incurred by the Government in the various sectors of economy viz., agricultural sector, industrial sector, infrastructural sector, export-import sector etc.* In the under-developed economies, the private sector is weak to invest requisite funds and induce growth and development in the economy. As a result, public sector has to take initiative to induce investment and bring about economic development. If the country lacks necessary infrastructure, its economic development is retarded. Economic overheads *viz.*, roads and buildings, railways, irrigation, power projects, educational institutions, *etc.*, are necessary for speeding up of economic development in the country. Social overheads *viz.*, hospitals, service institutions *etc.*, are also essential. In order to develop this infrastructure, expenditure by the Government is very much essential. Investment by private sector to develop overheads may be inadequate.

## NEED FOR PUBLIC EXPENDITURE

In order to bring about desired and balanced growth between backward and developed regions in the country, we require huge amounts of public expenditure.

### 1. Development of Agriculture and Industry

Economic development of the country is linked up with the simultaneous development of agricultural sector and industrial sector. Agriculture provides raw material to industry and industry gives capital goods to agriculture. Hence, both of them are essential for the country. For some countries like India, agriculture is the core sector in which nearly 70 per cent of the population thrives. Hence, Government has to incur lot of expenditure on agricultural sector particularly on irrigation, electricity, farm supply industries *etc.*, similarly, industrial sector should be developed for setting up of public enterprises *viz.*, steel plants, electricals, heavy engineering machines, tools, equipment *etc.* Investment has to be made on agro-processing industries, which will use agricultural produce and process them to serve the consumers in a better way and earn profits through export markets.

The Central Government should give grants and aids to State Governments and the local authorities to induce the private expenditure on investment.

### 2. Provision of Public Utilities

Public expenditure should be increased in order to enhance the provisions of public utility services *viz.*, water, transport, electricity, *etc.* Since private investment is low, Government assistance is essential. It is good if these utilities are supplied on competition basis as seen in capitalistic countries. According to economist Musgraves (1973)* efficient product mix between private good and social good, changes as per capita income in the country raised. This inturn raises productivity of social good. When per capita income raises it transforms the economy from agricultural – based sector to industrial based sector. This means that revenue from industry will be more due to increased output of social good, for which lot of public expenditure is essential.

## 3. Technological Changes

New innovations and inventions in the production are possible and feasible through massive investment by public sector. Due to technology, productivity of resources increases.

## 4. Requirements of Employment

Public expenditure must increase through various public works, undertakings, projects, *etc.*, in the country to provide employment opportunities. Both State Governments and Central Governments should launch various schemes and projects when resources are abundant to create employment.

Expenditure patterns by Governments should follow certain principles to get maximum benefit.

**Principles of Public Expenditure:** These principles are:

1. *Principle of Maximum Social Benefit:* It is necessary that all public expenditures should give rise to maximum social advantage, but what is required is that there should be a functional balance between social benefits and social costs. Every rupee spent by the Government should provide maximum welfare to the society as a whole, but not a particular section of the society. This means it should aim at general welfare. Every item of public expenditure should have justification and rationality.
2. *Principle of Economy:* This means, extravagance or wastage of expenditure should be avoided in spending the public money. It is also necessary to avoid duplication of expenditure and overlapping of authorities in spending the amount. Public expenditure should not affect the saving rate of the people. This means it should not lead to hyperinflation in the country as it happened in Brazil in the past.
3. *Principle of Sanction:* Every item of public expenditure should be sanctioned by competent authority with rationality. Unauthorised spending could lead to extravagancies. Therefore strict vigilance should be there for proper spending and utilizing the expenditure for which it is sanctioned.
4. *Principle of Elasticity:* The quantum of public expenditure should vary according to need and circumstances. Rigidity in the expenditure may prove a source of trouble and embarrassment. It is also necessary that scale of public expenditure should be increased gradually but not abruptly. The public expenditure should not affect production and distribution.
5. *Principle of Surplus:* The public expenditure should be kept well within the revenue levels of the state so that surplus remains at the end of the year. In other words, Government should avoid deficit budget in normal times.

---

* Musgrave, R.A. Musgrave, P.B. Public finance in theory and Practice, 1973. p. 126.

# CHAPTER 12

# Public Revenue

*This is the revenue accrued to the Government from different sources viz., direct taxes, indirect taxes, and non-tax revenue such as prices and other miscellaneous receipts.* Thus the Government would have two major sources of revenue i.e. taxes and prices. The minor sources are fee, special assessment, rates, fines, tributes and indemnities, grants, gifts and donations. Let us know about these sources of revenue.

## MAJOR SOURCES OF PUBLIC REVENUE

### 1. Taxes

Taxes are compulsory contributions levied upon persons, corporations etc. An essence of taxes is compulsory levy and it is a means of revenue for State and Central Governments. Taxes are classified as proportional, progressive, regressive and degressive. They are also classified as direct taxes and indirect taxes. A proportional tax is one in which same percentage is levied as tax irrespective of tax base or the size of the income. A progressive tax means the rate of tax increases as taxable income increases. Higher the income, greater would be tax amount. Example: Income tax. A tax is said to be regressive, when it is affecting the poor rather the rich *i.e.,* the burden of payment of tax would be more on the poor. It is just opposite to progressive tax. Example: All commodity taxes. A tax is called degressive, when the higher income groups do not make due sacrifice. This happens when a tax is mildly progressive, but not steep. A tax is progressive up to certain limit, after which a uniform tax is levied. This tax also affects the poor than the rich.

Direct taxes are the taxes directly paid by the persons. In other words, the person who pays the tax is also intended to bear it. Incidence and impact are on the same person. Example: Income Tax.

Indirect taxes are the commodity taxes. They are indirectly paid by the consumer through the dealers. Incidence is on one person and impact is on the other person. The burden of tax is passed on to other persons. Example: Sales tax.

Direct taxes are paid on income levels, while indirect taxes are paid on outlay levels or expenditure levels.

Taxes are also classified as specific taxes, advalorem taxes and value added tax (VAT).

In the case of specific tax, the amount of tax to be paid depends on the amount of commodity purchased. In respect of ad valorem tax, the amount to be paid is proportionate to the value of the commodity. Example: Stamp duty. A tax levied on the value of each of the processes carried out by a business is called value added tax.

## 2. Price

When the public authorities sell a commodity or render a service to the consumer who avails or buys the commodity on a charge is called price and this price is a source of public revenue. For example, railway fares, bus charges, electricity tariffs, water cess, *etc.*, come under the broad definition of price.

# MINOR SOURCES OF PUBLIC REVENUE

## 1. Fee

*A fee is defined as compulsory contribution of money paid by the persons, corporations, etc., under the authority of public power.* In short, a fee is a charge imposed for a specific service, which is rendered basically in public interest. Examples: License fee for vehicles, educational fee, *etc.* But we should remember that fee is not the cost of service rendered. Infact, it is much below the cost of service rendered. Tax is paid by the persons for common benefit (social benefit) whereas, prices and fee are paid for specific benefit.

## 2. Special Assessment

This is a compulsory contribution levied on persons in proportion to special benefit derived to defray the cost of specific improvement to property undertaken in the public interest. Due to the watershed development activities, value of the land would appreciate in the watershed area, and under such situation, the Government has right to appropriate a part of this unearned increment. A special benefit tax would be imposed for special purpose on the people in the watershed area. Such special benefit tax is called special assessment. Unlike a fee, there is an element of force or coercion in special assessment, which is generally imposed on the people in benefited areas.

## 3. Rates

Rates are certain kinds of taxes levied by local bodies *viz.*, municipalities, panchayats, corporations, district boards, *etc.*, on people for local purposes. These are generally levied on the immovable properties like buildings *etc.* These are not necessarily imposed for any special benefits conferred. The rates generally vary from area, locality, regions, state *etc.*, depending on the value of the immovable property.

## 4. Fines

These are the penalties imposed on persons for infringement of laws or breakdown of laws. Property rights on the land can be taken away, if the farmers do not pay land revenue for certain defined period.

## 5. Escheat

Suppose an individual dies without successors or leaves no will behind him, his property or assets will go to the State Government. This claim of the State to the deceased's assets is called escheat.

### 6. Tributes and Indemnities

Tributes are paid by the conquered countries. Indemnities are paid for any damage done to the country by way of war of aggression.

### 7. Grants, Gifts and Donations

Higher-level bodies grant funds to the lower level bodies. For example, Central Government grants funds to the State Governments for economic development or some times, for undertaking public works. ICAR sanctions 80 per cent of the funds to State Agricultural Universities under grants. Grants are basically not repayable, while loans are repayable. Gifts are granted from foreign Governments to domestic Governments for relief work at the time of natural calamities like earthquakes, droughts, cyclones *etc.* Donations are given for individuals for specific purposes or some times individuals donate funds for specific purposes *viz.*, construction of educational buildings, hospitals, relief works, *etc.*

## CANONS OF TAXATION

Adam Smith, the father of economics, made significant contributions to the economic theory and particularly in the field of taxation. His statements are considered as canons of taxation. They are simple, clear and known for clarity. Let us know about these canons of taxation.

1. *Canon of Equality:* This means equality of sacrifice. It implies the principle of equity and justice. This is the basic principle. Here equality implies that every tax payer should pay the same rate of taxation but not the same amount. It is a sort of proportional tax.

2. *Canon of Sacrifice:* This principle also means the equality of sacrifice. The amount of tax paid should be in proportion to the respective abilities of the tax payer. Here ability implies income levels. This clearly points to progressive taxation.

3. *Canon of Certainty:* Here taxation limits are fixed with certainty. The time of payment, manner of payment and amount of tax to be paid should be clear and easy to understand. Uncertainty in taxation leads to corruption and insolence. Here the element of arbitrariness is ruled out. This means that it is not left to the will of income tax department. Lot of publicity is required in the budget proposals regarding the tax component.

4. *Canon of Convenience:* This means tax ought to be levied at the time or manner in which it is most likely to be convenient for the taxpayer. The time and manner of tax payment should be clear, certain and convenient. Land revenue is collected in general when the farmer got income from farming. It can be paid either in cash or cheque according to the convenience of the taxpayer. Consumers pay commodity taxes at the time of purchasing the commodities. Tax component is indicated on the commodity along with the price.

5. *Canon of Economy:* The cost of collecting tax should be small and economical. If the incomes of the people were subjected to heavy burden of taxation, then their savings would be affected, which inturn affect capital formation in the country. The development of trade, agriculture and industry is retarded. Then certainly the tax system becomes uneconomical. Harmful drugs and intoxicants affect the health of the people; hence heavy burden of taxes on these goods is regarded as economical

and justifiable. This would result in less consumption of these commodities by the people and more income to the exchequer. But taxes on raw material are considered uneconomical because the price of the manufactured goods is increased and competitive power of the industry gets weakened.

The canons discussed so far are infact the contributions of Adam Smith. However, later some economists added some more canons to the taxation. These are (1) Fiscal adequacy and productivity, (2) Elasticity, (3) Flexibility, (4) Simplicity, (5) Diversity, (6) Achievement of social and economic objectives of the country; and (7) Neutrality.

1. *Fiscal adequacy and Productivity:* The tax revenue should adequately cover the Government expenditure and the Government should not run into deficit state. Tax system should not cripple the people in any manner and impair the productivity of the economic resources.

2. *Elasticity:* It is closely linked with fiscal adequacy. This means that tax revenue should be adequate to the Government to meet the expenditure, particularly at the time of emergency and period of stress and strain. Income tax is considered to be having good elasticity. By raising the surcharges on income levels, we can increase revenue to the Government.

3. *Flexibility:* There should not be any rigidity in the tax system. It should quickly adjust to new demand and conditions in the country.

4. *Simplicity:* It should be clear, fair, simple and easy to understand. If the tax system is complicated, taxpayer cannot understand, how much he has to pay and hence he would try to evade the tax.

5. *Diversity:* A few taxes should be imposed rather than a single largest tax. The variety of taxes should be in confirmity with the availability of taxpayers. Hence there should be wise admixture of direct and indirect taxes.

6. *Achievement of social and economic objectives:*     The revenue generated from taxes should be used for achieving the social and economic objectives of the country, without this the collection of tax becomes futile.

7. *Neutrality:*    This indicates taxation system should reduce the economic inequalities in different regions. During depression, taxation system should lead to deficit budgeting and during inflation; it should be used as a tool to fight inflation in the country. Taxation system should be used to control threats of economic instability and stagnation. During the normal years it should lead to balanced budget.

# CHAPTER 13

# Unemployment

The **goal** of public policy has been to remove unemployment and achieve full **employment.** But unemployment is the burning problem in modern societies. Unem-**ployment** results in the wastage of resources and depressed levels of income and such **a state of** economic distress affects people's emotions and family lives. Coming to the **meaning** of unemployment it is a state of inactivity suffered by a worker inspite of **his intention** to find work and inspite of his physical fitness to carry on his trade. *Unemployment is also defined as involuntary idleness of a person willing to work at the prevailing rate of pay but unable to find it. Voluntary unemployment may be defined as the unemployment which results from withdrawal of some persons from employment for diverse reasons such as (1) Absence of need to earn, when one has already made fortune or when one inherits large amount of wealth or property, (2) Social custom of certain groups discourages or forbids engagement of certain members in productive work, and (3) Lethargy.*

A **person** working 8 hours a day for 273 days of the year is regarded as employed on standard person year basis.

In sum, people with jobs are employed, people without jobs but looking for work are unemployed and people without jobs who are not looking for work are outside the labour force. The unemployment rate is the number of unemployed divided by the total labour force.

## TYPES OF UNEMPLOYMENT

Following are the types of unemployment:

### 1. Frictional Unemployment

Lack of mobility on the part of labour puts them out of work for a temporary **period,** which is called frictional unemployment. In a growing economy there are **greater** opportunities for the labourers to seek employment, as new industries come **up and** the existing ones diversified. In a situation of this nature some volume of **frictional** unemployment arises. Lack of adjustment between demand and supply of **labour** causes this type of unemployment. The reasons are that the labourers do take **some time** to get the needed skills to get into the new trade, the lack of knowledge **on the** part of the labourers about the availability of employment opportunities and **the employers** being unaware of the availability of the workers. Also breakdown of **machinery,** shortage of raw materials, *etc.*, do cause frictional unemployment. **This is** a **common** feature even for economy with full employment.

## 2. Structural Unemployment

The word 'structural' implies that the economic changes are massive, extensive and deep-seated amounting to transformation of an economic structure. Structural unemployment sets in as a result of mismatch between demand for and supply of labour. Mismatch in demand and supply can be seen as demand for one kind of labour may rise, while it may fall for another kind of labour, with supplies not readily adjusting to these changes. This is evident as we can see the imbalances accentuating in one sector and declining in another sector or region. If wages are going to be adjusted smoothly to the changing supplies and demand, then imbalances across labour markets would disappear, as the wages rise in areas of labour shortages and fall in areas of labour surpluses. But the reality is that wages do not respond that quickly to the economic shock as they take some years to adjust to shortages or surpluses of labour. If labour force increases faster than the stock of capital, all the workers cannot get employment as the instruments of production fall short of requirement. Such unemployment is termed as the structural or long term or Marxian unemployment.

## 3. Cyclical or Keynesian Unemployment

This occurs due to cyclical fluctuations of the economy and during the downswing of the business cycle. The income and output fall leading to widespread unemployment. This type of unemployment arises, as the effective demand of the community cannot absorb the entire production of goods and services that is brought into the market with the available stocks of capital. In free private enterprise economy, when the business men fail to dispose their output they tend to reduce the output. In the light of reduced production, there is a reduction in the extent of utilisation of factors of production, as a result of which some factors become unemployed. A fall in employment, which is a major source of income to the great majority of the people, brings in a fall in their incomes also.

## 4. Seasonal Unemployment

It is a case of people not getting employed throughout the year. When production activity is season bound, people remain unemployed during slack season. Agriculture is a live example of seasonal nature of production. Coming to the industries, sugar factories, rice mills, *etc.*, are the relevant examples.

## 5. Technological Unemployment

It is quite obvious that in a dynamic economy the techniques of production do undergo a change. Businessmen welcome it for these changes improve the production process. Mechanization is one such improved method in the production process, the introduction of which displaces labour, resulting in unemployment.

# MEASUREMENT OF UNEMPLOYMENT

*Chronic Unemployment or Usual Principal Status of Unemployment:* It is measured in terms of number of persons *i.e.*, persons who remained unemployed for major part of the year. This measure is more appropriate to those, who are in search of regular employment. Example: Educated and skilled persons, who may not accept casual work. It is also referred to as 'open unemployment'.

1. *Weekly Status Unemployment:* This is measured in number of persons *i.e.,* persons who did not find even an hour of working during the survey week.

2. *Daily Status Unemployment:* It is measured in person days or person years *i.e.,* persons who did not find work on a day or some days during the survey week.

3. *Under Employment:* It is a very common phenomenon in less developed economies in general and rural areas in particular. Under employment represents people employed on part time basis, seasonal basis or even as casual labourers. In jobs of this nature their productivity as well as incomes are low.

4. *Visible Under-employment:* It indicates the shorter than normal periods of work. Persons involuntarily work less than the labour time they are available for gainful employment.

5. *Invisible Under-employment:* It is the characteristic of persons whose working time is not abnormally reduced, but whose earnings are abnormally low or whose jobs do not permit full use of their capabilities or skills or who are employed in the establishment of economic unit whose productivity is abnormally low.

6. *Involuntary Unemployment:* This is defined as the unemployment due to non-availability or insufficiency of work during periods say a few weeks, a few months, or even a few years, when the worker in question wants to work.

7. *Disguised Unemployment:* This means that people are engaged in occupations, where their marginal productivity is very low (if not zero or negative). A shift to alternative occupations will improve their marginal productivity and add to the national income.

According to National Sample Survey (NSS), a person who works for 28 hours during a week on an average and is available for additional work is treated as severely under employed in rural areas.

A person who works between 29 and 42 hours a week and is available for additional work is considered as moderately under employed.

Under employment arises when a worker is not employed full time taking into account the available time of workers and the traditional or accepted norms of normal full working time of the workers.

# CHAPTER 14

# Business Cycles or Trade Cycles

A cycle is defined as a regularly occurring phenomenon. Occurrence of these cycles in aggregate employment, income, output and price level is called trade cycle. According to Keynes '*A trade cycle is composed of periods of good trade characterised by rising prices and low unemployment percentages, altering with periods of bad trade with falling prices and high unemployment percentages*'. The definition of Keynes, thus embodies two aspects *i.e.*, prices and unemployment, for fluctuations in the business cycles.

## Phases of a Typical Business Cycle

A typical business or trade cycle has five components *viz.*, depression, trough, recovery (revival), prosperity (full employment), boom (overfull employment) and recession. These are explained hereunder.

## 1. Depression

This is the initial stage of a business cycle. It is a protracted period and is marked by sharp reduction of production, mass unemployment, general fall in prices, profits, wages, interest rates, consumption expenditure, investment, credit, high rate of business failures *etc.* The prices register an all time fall, and with the costs remaining the same more or less, the industries suffer heavily. As a result, the industrial activity comes to a grinding halt with the closure of the existing industries. Construction of new factories is not possible. The fall in prices of the industrial commodities does distort the relative price structure. The farming community suffers heavily, as the fall in prices of farm products and raw material is greater than the fall in the prices of industrial commodities.

## 2. Recovery

It is a stage that sets in after depression, the lowest point in the business cycle, signalling a situation of hope from despair. To begin with, an improvement in the economy is witnessed, as the entrepreneurs are convinced that bad situation was over and they can optimistically look into the future. Business activity is improved as a result of which industrial production commences to pick up slowly but steadily. Consequently, employment shows the signs of improvement. Another development that is found is rise in prices and accrual of profits in a small measure. Wages also witness a rise alongside, but may not be in the proportion of the price rise. The profits, which are found, now start rising. The element of profits and their rising trend attracts new investors in the production of capital goods. The new investment rises the demand for bank credit therefore, the banks expand credit. All these forces contribute to the economy to reach the stage, which it had, before the depression set in. The time period for the recovery cannot be assessed accurately, for it depends upon the strength of the forces which led to recovery like technology, public and private investment, extent of exploitation of resources, *etc.*

## 3. Prosperity (Full employment)

It is a stage which is associated with increased production, high investment in basic industries, expansion of bank credit, high prices, high profits, full employment, *etc.* It is a highly encouraging stage for the businessmen. Optimism prevails in the economy. The expansionary process continues until the economy reaches a very high level of production called as boom or peak.

## 4. Boom (Overfull employment)

This stage, which indicates the continuance of investment even beyond the stage of full employment marks the end of prosperity phase as inflationary tendencies creep up. This trend prompts the businessmen to make additional investments in various sectors of the economy. These investments exert additional pressure on the factors of production, leading to an increase in their prices. These trends create a situation in which the number of jobs exceeds the number of workers available, which is called an overfull employment. The businessmen get the profits and up-trend of these profits attract new entrepreneurs into the investment. Such a spurt in the investment further rises the prices causing the occurrence of runaway inflation. The tempo of boom reaches new high, but at the same time it turns out to be the cause of self-destruction. The mounting pressure on factors of production, which are already declining cause further rise in their prices. The production costs register alarming rise. These developments compel the need to play it safe by the entrepreneurs. The businessmen apply breaks regarding expansion of the existing units, leave aside new ventures. These actions of the businessmen pave way for the emergence of a new stage called recession.

## 5. Recession

The cautious attitude which was adopted at the end of boom now makes the businessmen over pessimistic, as they are haunted by fear and suspicion. This attitude leads to a reverse in the business in the form of business failures. The event to follow is collapse in prices with which the confidence of business community is shattered. Unemployment becomes conspicuous which leads to fall in income, expenditure and profits. Once recession starts, it is accentuated and ends up as depression. Various stages of business cycle are illustrated in Figure 14.1.

148

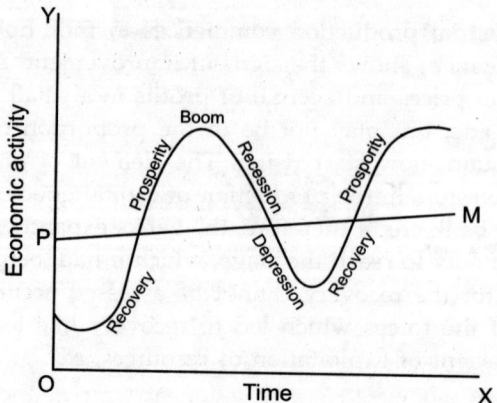

**Figure 14.1**  Stages of business cycle.

In the above diagram, PM is the full employment line. Above this we have two stages of the business cycle – a boom in the upswing and recession in the downswing. Below this line we can see recovery in the upswing and depression in downswing.

# Inflation

The term inflation generally means a situation of rising prices of goods, which lowers the purchasing power of money. It is a situation in which either the prices or the supply of money are rising because in reality both will rise together. In the Keynesian sense, true inflation commences when the supply of goods and services is unresponsive to changes in money supply. *Too much money chasing too few commodities is termed as inflation.* Inflation is a persistent and appreciable rise in the general price level. *The rate of inflation is defined as the rate of change of the price level as measured by the consumer price index (CPI)*

$$\text{Rate of inflation in } t^{\text{th}} \text{ period} = \frac{\text{Price level } (t^{\text{th}} \text{ year})(t - 1^{\text{th}} \text{ year})}{\text{Price level } (t - 1^{\text{th}} \text{ year})}.$$

## Price Indices

A price index is a weighted average of individual prices, where the weight of an each commodity's price reflects the economic importance of that commodity. The most important price indices are consumer price index, the producer price index and the GNP deflator.

## Consumer Price Index [CPI]

It is the most widely used measure of inflation, which is estimated based on prices of food items, clothing, shelter, fuel, medical care, transportation and other commodities purchased for day-to-day living.

Coming to the estimation of price index numbers, weights are given to different prices. Price indices are constructed by weighing each price by the economic importance of the commodity. In the case of CPI, each item is assigned a fixed weight proportional to its relative prices in consumer expenditure budgets.

The construction of CPI can be explained with the following example, CPI constitutes many items but for simplicity, we confine to only three commodities. Let us assume that a hypothetical budget survey in 1995 reveals that consumers spend 50 per cent of their income on food, 25 per cent on shelter and 25 per cent on health and others. The year 1995 becomes the base year and the price of each commodity is set at 100. The CPI for 1995 is therefore 100 *i.e.,* (0.50 × 100) + 0.25 × 100) + (0.25 × 100). Calculation of CPI say for the 1996 gives the rate of inflation. Let us assume that the prices of food, shelter and medical care rise by 3, 5 and 6 per cent respectively. The indices for these three items are 103, 105 and 106. Now the CPI for 1996 is calculated as follows:

$$\text{CPI for 1996} = (0.50 \times 103) + (0.25 \times 105) + (0.25 \times 106) = 104.25$$

Making use of CPI in 1995 and 1996, we can workout the rate of inflation in 1996

$$\text{Rate of inflation} = \frac{(104.25 - 100)}{100} \times 100.$$

$$= 4.25 \text{ per cent per year.}$$

## The Producer Price Index (PPI)

It measures the level of prices of food items, manufactured products, mining products, etc., at producer or wholesaler stage.

## GNP Deflator

It is the ratio of nominal GNP to real GNP which can be interpreted as a comprehensive price index.

## Inflation and Other Forms

1. *Deflation:* Opposite to inflation is deflation, which is found when the general level of prices is falling. It is a situation in which supply of money at a particular point of time is less than the demand.
   *Pigou defined deflation as the state of falling prices, which occurs at the time when the output of goods and services increases more rapidly than the volume of money income in the economy.*
2. *Reflation:* It refers to a moderate degree of controlled inflation.
3. *Disinflation:* It indicates the decline in the rate of inflation.
4. *Stagflation:* It is inflation accompanied by stagnation on the development. Stagflation is associated with high prices and high unemployment. It is a global phenomenon, as every country whether developed or developing, experiences stagflation.

## Causes of Inflation

The causes of inflation can be broadly grouped into two categories.

### 1) Factors Causing Increase in Demand

   a. Increase in money supply.
   b. Increase in disposable income.
   c. Increase in community's aggregate spending on consumption and investment goods.
   d. Excessive speculation and tendency to hoarding and profiteering on the part of the producers and traders.
   e. Increase in exports.
   f. Increase in salaries, wages or dearness allowance.
   g. Increase in population, *etc.*

## 2) *Factors Causing Decrease in Supply*

    a. Deficiency of capital equipment.
    b. Scarcity of other complementary factors of production.
    c. Increase in exports.
    d. Decrease in imports.
    e. Hoarding by traders.
    f. Natural calamities, *etc.*

**Types of Inflation:** Following are the various types of inflation:

1. *Suppressed Inflation:* Deliberate policies are adopted to prevent price rise, but the impact of these policies is only temporary, since prices rise as soon as these are relaxed. Here is a situation in which the Government does not tackle the factors causing inflation, it only imposes controls to check the price rise.

2. *Creeping Inflation:* It is the mildest type of inflation. A sustained rise of less than 3 per cent in prices per annum is called creeping inflation. It is not considered serious for the economy. The Government has sometimes resort to creeping inflation to enable the industry and trade to receive stimulus for the country to progress slowly and gradually.

3. *Walking Inflation:* When the rise in prices falls in the range of 3 per cent to 6 per cent, it is called walking inflation.

4. *Running Inflation:* When the sustained rise in prices is about 10 per cent per annum, it is called running inflation. This type of inflation is a cause of concern for the Government and warrants remedial measures.

5. *Galloping or Hyperinflation:* This is the most dangerous type of inflation and prices rise by 16 per cent or more per annum. This type of inflation should not be allowed to persist.

6. *Open Inflation:* Inflation is said to be open when the market forces are allowed to operate freely, for the prices of goods and services set without abnormal interruption by the authorities. The Government does not take any steps to check the rise in the prices.

7. *Suppressed or Repressed Inflation:* Suppressed inflation refers to a situation when, the Government actively intervenes to check the rise in the price level through the use of price control measures and rationing of scarce items in the economy.

8. *Comprehensive Inflation:* This type of inflation occurs when prices of all the commodities register a rise in the economy.

9. *Sporadic Inflation:* It is a sectoral inflation. Under this type of inflation only the prices of a few commodities show an upward trend.

10. *Demand Pull Inflation or Excess Demand Inflation:* It is often described as "too much money chasing too few goods". It occurs as a result of excessive demand for goods and services which pull prices upward.

11. *Demand-shift Inflation:* It is a special case of demand-pull inflation. This occurs when shifts take place in demand for different goods and services with total demand remaining unchanged. This arises as a result of households reducing the demand for good A, while increasing demand for B, by an equal rupee amount.

12. *Cost-push Inflation:* Enforcement of wage increase by unions and increase of profits by employers lead to cost push inflation. Here the money wages rise more rapidly than the productivity of labour, as a result of which the cost of production of commodities rises, employers, in turn increase the prices of the

commodities. Higher wages enable workers to buy as much as before, even in the light of higher prices. On the other hand unions demand higher wages in view of higher prices. In this way, the wage cost spiral continues leading to cost-push inflation to prevail.

13.  *Mixed Demand-pull and Cost-push Inflation:* This situation is termed as hybrid form of inflation. It is a situation in which some elements of demand-pull inflation and cost-push inflation are found. In reality excess demand and cost-push forces operate simultaneously and interdependently in an inflationary process. In years of poor harvest the prices of agricultural commodities rise as the demand is more than supply. The rise in the price of farm commodities causes the cost of living index to increase as farm commodities carry substantial weights. In view of rise in the cost of living index, wages rise in industries causing cost-push inflation.

On the other hand, if demand for products of an industry rises, as a result of excessive demand the price rise brings in more profits to the industry. The higher profits stimulate the industry to expand the production by employing more labour which in turn push up the wages. The result is the prevalence of a situation of wage-cost inflation.

14.  *Markup Inflation:* This type of inflation prevails when gigantic business organisations adopt a peculiar method of pricing. They work out the production costs and then add a certain mark-up to get targetted rate of profit on their capital investment. This mark-up is generally on the higher side.

15.  *Profit Induced Inflation:* Some times it happens in the economy that production cost declines as a result of which prices also tend to decline. But the Government does not allow the prices to fall down by resorting to artificial means. The Government intervention neither allows the prices to rise nor allow them to fall down. This action of the Government results in an increase in the profits of the producers.

# SECTION III
## AGRICULTURAL PRODUCTION ECONOMICS

# CHAPTER 16

# Agricultural Production Economics

## MEANING

Agricultural production economics is a field of specialization within the subject of Agricultural Economics. It is concerned with the selection of production patterns and resource use efficiency in order to optimize the objective function of farming community or the nation within a framework of limited resources. The goals of agricultural production economics are: (1) to provide guidance to individual farmers in using their resources most efficiently and (2) to facilitate the most efficient use of resources from the standpoint of economy.

### Definition

*Agricultural production economics is an applied field of science wherein the principles of choice are applied to the use of capital, labour, land and management resources in the farming industry.*

## SUBJECT MATTER

Agricultural production economics involves analysis of production relationships and principles of rational decisions in order to optimize the use of farm resources on individual farms and to rationalise the use of inputs from the nation's point of view. The primary interest is in applying economic logic to problems that occur in agriculture.

Agricultural production economics is concerned with the productivity of inputs. As a study of resource productivity, it deals with resource use efficiency, resource combination, resource allocation, resource management and resource administration.

The subject matter of production economics involves topics like factor-product-relationship, factor-factor relationship, product-product relationship, size of the farm, returns to scale, credit and risk and uncertainty, *etc.*

Any problem of farmers that falls under the scope of resource allocation and marginal productivity analysis is the subject mater of agricultural production economics. The agricultural production economist is therefore concerned with any phenomena, which have a bearing on economic efficiency in the use of resources.

## OBJECTIVES

1. To determine and outline the conditions which give the optimum use of capital, labour, land and management resources in the production of crops and livestock.
2. To determine the extent to which the existing use of resources deviates from the optimum use.
3. To analyse the forces which condition existing production pattern and resource use, and
4. To explain means and methods in getting from the existing use to optimum use of resources.

## BASIC CONCEPTS IN PRODUCTION ECONOMICS

Following are the concepts, which are used in production economics.

| | | |
|---|---|---|
| 1. | Farm | It is a piece of land, where crop and livestock enterprises are taken up under a single management and has specific boundaries. |
| 2. | Agricultural holding | The total area of land owned by an individual or joint family whether cultivated by the family or rented out. |
| 3. | Operational holding | Total land area held under a single management for the purpose of cultivation. It excludes any land leased out to another person. |
| 4. | Family holding | It refers to a farm which yields a gross income of Rs. 1600 or a net income of Rs. 1200 per annum. |
| 5. | Optimum holding | It refers to the maximum size of holding which a family should possess. Three times to the family holding is considered to be an adequate size of optimum holding. |
| 6. | Economic holding | It is defined as one which could provide a reasonable standard of living to the cultivators and give full employment for a family of normal size. |
| 7. | Marginal farmer | Farmer owning less than 2.5 acres of dry land or 1.25 acres of wet land is called a marginal farmer. |
| 8. | Small farmer | Farmer owning land holding ranging from 2.5 acres to 5.00 acres of dry land or above 1.25 acres and below 2.5 acres of wet land is called a small farmer. |
| 9. | Production | The process whereby some goods and services called inputs are transferred into other goods called products or output. |
| 10. | Product | A product is an outcome of the utilization of resources and services of resources. Examples: Paddy, sugarcane, wheat, milk, eggs, meat, *etc.* |
| 11. | Production function | A systematic and mathematical way of measuring the relationship among different quantities of inputs or input services used in the production of a commodity and corresponding quantities of output is called production function. |
| 12. | Continuous production function | This production function arises for those inputs which can be divided into smaller doses. Continuous variables can be known from measurement. Examples: |

|   |   |
|---|---|
|   | Seeds, fertilizers *etc.*, can be used from a fraction of a kg to tonnes. |
| 13. Discontinuous or discrete production function | This production function arises for those inputs or work units, which are used in whole numbers. In other words, the discrete production function is obtained for inputs, which cannot be split up into smaller doses. Discrete variables can be known from counting. Examples: Number of ploughings, weedings, harvestings, *etc.* |
| 14. Short run production function | The planning period during which one or more resources are fixed. In the short run, output can be varied only by intensive use of fixed resources. Example: $Y = f(X_1, X_2, X_3 \mid X_4, X_5.....X_n)$. The vertical bar in parenthesis separates variable inputs from fixed inputs. |
| 15. Long run production function | It is a planning period during which all resources are varied in quantity. The supply can be fully adjusted according to the demand. Example: $Y = f(X_1, X_2, X_3, X_4, X_5.....X_n)$. |
| 16. Technical coefficient | The amount of input per unit of output is called technical coefficient. |
| 17. Technical unit | It is a single convenient unit in production for which output and returns are calculated. Examples: A hectare of land, a head of livestock, *etc.* |
| 18. Plant | It means a group of technical units i.e. 5 ha. farm or 1000 birds poultry or 10 animal dairy, *etc.* |
| 19. Farm-firm | It is also known as economic unit which is run under one management. It represents an aggregation of resources for which costs and returns are worked out treating as an unit. Example: A farm holding. |
| 20. Resources | Anything that aids in production is called a resource. They physically enter the production process to transform into output. Examples: Seeds, fertilizers, feeds, veterinary medicines, *etc.* |
| 21. Resource services | The work done by a person or a machine or livestock is called resource service. Here only services are available for the production and the resources do not physically enter the production. Examples: Services of labourer, machinery, farm implements *etc.* |
| 22. Fixed resources | Resources which remain unchanged irrespective of the level of production are called fixed resources. These resources exist only in the short run. The costs associated with these resources are called fixed costs. Farmer has little control over the use of these resources. Examples: Land, buildings, machinery, implements, *etc.* |
| 23. Variable resources | Resources which change with the level of production are called variable resources. Higher the level of production, greater the use of these resources and *vice versa*. The costs which are associated with variable resources are called variable costs. These resources exist in the short run as well as in the long run. Farmer can |

exercise greater control over the use of these resources. Examples: Seeds, fertilizers, plant protection chemicals, feeds, *etc.* The distinction between fixed and variable resources ceases to exist in the long run. In the long run all resources are varied.

24. Flow resources

The resources which cannot be stored and should be used as and when they are available. For instance, if the services of a labourer available on a particular day are not used, then they are lost forever, similarly, the services of machinery, farm buildings, *etc.*

25. Stock resources

Stock resources are those which facilitate for their storage, when they are not used in one production period. Examples: Seeds, fertilizers, feed, *etc.*

Defining an input as a flow or stock, depends on the length of time period under consideration. The useful life of a tractor is assumed to be 10 years, if we take the services of a tractor for its entire useful life of 10 years, then tractor is a stock resource. But, tractor provides its services during each day of a production season. If we consider the services provided by tractor in each day, then it is a flow resource. Therefore, tractor can be regarded as flow and stock resource. The other examples are farm buildings, land, machinery, *etc.*

26. Mono-period resource

It is the resource which can be used only once in production. Examples: Seeds, fertilizers, *etc.*

27. Poly-period resource

It is the resource which is used in the production process over several periods. Examples: Machinery, implements, *etc.*

28. Production period

It is also termed as transformation period. It is the time period required for the transformation of resources or inputs into products.

29. Choice indicator

Choice indicator is an yardstick indicating which of the two or more alternatives maximizes a given end. Choice indicators are always expressed in ratios. Substitution ratio and price ratio are the examples of choice indicators.

30. Farm entrepreneur

He is the person who organizes and operates the farm business and bears responsibility of the outcome of the business.

31 Farm business manager

He is a person appointed by the entrepreneur to manage or supervise the farm business and paid for his services. He carries the instructions of the entrepreneur. He is not responsible for the outcome of the business.

32. Productivity

Productivity denotes the efficiency with which various inputs are converted into products. It signifies the relationship between output and inputs. In simple terms output per unit of input is called productivity. Example: 5 quintals of product/ha.

| | |
|---|---|
| 33. Efficiency | It means absence of wastage or using resources as effectively as possible to satisfy the farmers' needs and goals. |
| 34. Technical efficiency | It is the ratio of output to input. |
| 35. Economic efficiency | It is the expression of technical efficiency in monetary value by attaching prices. In other words, the ratio of value of output to value of input is called economic efficiency. It is the maximization of profit per unit of input. |
| 36. Allocative efficiency | It occurs when no possible reorganization of production can make any one better off without making some one else worse off. It refers to resource use efficiency |
| 37. Optimality | It is an ideal situation in which costs are minimum or profits are maximum. |
| 38. Cost of cultivation | The expenditure incurred on all inputs and input services in raising a crop on an unit area, is referred to as cost of cultivation. Example: Costs in Rs./ha |
| 39. Cost of production | The expenditure incurred in producing a unit quantity of output is called cost of production. Example: Costs in Rs./tonne or quintal. |
| 40. Variable | Any quantity which can have different values in the production process. |
| 41. Independent variable | It is a variable whose value does not depend on other variables. Such variables influence the dependent variable. Examples: Land, labour, liquid money, fertilizer, *etc.* |
| 42. Dependent variable | A variable that is governed by another variable. Example: Crop output. |
| 43. Constant | A quantity that does not change its value in a general relation between variables. |
| 44. Coefficient | When rate per unit is calculated, we use the term coefficient, a multiplying factor. For example, the regression coefficient of an input in production function denotes response of output per unit of input. Similarly, elasticity coefficient of input gives the percentage change in crop output per one per cent increase in input level. But technical coefficients refer to requirements of inputs per unit of land or per unit of crop output. |
| 45. Slope | Slope of a line represents the rate of change in one variable that occurs when another variable changes i.e., it is the rate of change in the variable on the vertical axis, per unit change in the variable on the horizontal axis. Slope is always expressed as a number. Slope varies at different points on a curve, but remains the same on all points of a given line. |

# FEATURES OF AGRICULTURAL PRODUCTION AS COMPARED TO INDUSTRIAL PRODUCTION

The conditions under which agricultural production is carried out and the nature of agricultural commodities, compared against non-agricultural sector (industrial production) necessitate a look into the distinct differences that exist between the two sectors. The differences are as follows:

1. *Farming a Way of Life:* Farmers consider farming as a way of life than a business proposition as the holdings are small and scattered with the involvement of family labour. They are contended if their family requirements are met from the farming. In industries, the entrepreneurs exhibit commercial outlook always with an eye on maximization of profits. It is purely a business approach in industries.

2. *Dependence on Weather:* Agriculture being a biological activity, is totally influenced by the nature. The vagaries of the weather like erratic rainfall, temperatures *etc.,* blunt the efforts of the farmers to derive expected results. Given the influence of vagaries of nature, the farmer has hardly any control over results of his efforts. Weather has little influence on industrial production. In industry the production activity is undertaken under the conditions, which are under the control of the entrepreneurs. The entrepreneur can plan and control the entire production process. At will he can increase or decrease the level of output according to the market situation.

3. *Seasonality of Production:* The production of the agricultural commodities and their varieties are limited by the specificity of the season. Paddy is grown in early *kharif*, late *kharif* and *rabi* seasons. Wheat is confined to *rabi* season only. Groundnut is cultivated both in *kharif* and *rabi*. Greengram is grown in *kharif* and *rabi*, while blackgram and bengalgram in *rabi*. Fruits are seasonal in nature and so also certain vegetables. Thus the production of agricultural commodities is not uniform throughout the year. Such a limitation does not exist in the production of industrial commodities.

4. *Perishable Nature of Agricultural Products:* The storage period of farm commodities ranges from few hours as in the case of flowers to few days in respect of fruits and vegetables and few years for cereals, pulses, oilseeds, *etc.* If allied activities of farming are considered few hours of storage life is found in respect of fish, prawn, milk *etc.* This characteristic of the farm products results in price variations in the same day, in the same season and among the different seasons in an year. In industrial products such price variations are not found. The industrial commodities are durable.

5. *Joint Products:* Many agricultural commodities are joint products like paddy and straw, cotton lint and cotton seed, *etc.* Since these are the joint products, the cost of production of main product and byproduct cannot be separated. In industry it is easy to estimate the cost of production of several products that are produced in the same plant.

6. *Bulkiness of Agricultural Products:* Since most of the farm products are bulky in nature they exert pressure for storage and transportation, consequent to which high costs are involved for unit of commodity in storage and transportation. Particularly high costs of transportation limit the movement of commodity from surplus areas to the dificit places. This characteristic of bulkiness is not found in respect of industrial commodities.

7. *Problems of Standardization:* The variations in the farm products regarding appearance, size, shape, colour, staple length *etc.*, are found due to the availability of a large number of varieties of crops. This poses difficulties in standardization and grading. Apart from this, quality variations bring in a wide variation in prices for the same commodity. In industrial production, there is no such problem as the products are uniform in all respects. The production practices are standardized in industries. With the help of specialized machines, it is possible to produce goods of uniform size and quality. However, in agriculture, it is not possible to recommend the uniform production practices due to variations in agro-soil-climatic conditions.

8. *Time Lag in Production of Agricultural Products:* There is a lapse of time between the decision to produce and actual realization of output in agriculture. The period is four to six months for paddy, one year for sugarcane, three months for blackgram *etc.* This time-lag in the production may upset the plans of the farmer. A farmer with a fond hope of obtaining a given net income, in view of prevailing output prices for a crop, may sow it and by the time the crop is harvested, the prices may fall down and the farmers plans may go topsy-turvy. The price fluctuations of agricultural commodities cause variations in farm incomes. Such a situation is not noticed in industrial commodities.

9. *Large Proportion of Land:* Agriculture requires larger proportion of land divided into several small and scattered holdings, which is not so in the case of industry. As in industrial production, the economies of large scale production are not a common feature in agriculture.

10. *Law of Diminishing Returns:* The law of diminishing returns is applicable to agriculture and industry as well, but the difference is that, it sets earlier in agriculture compared to industry. The obvious reasons are the dependence of agriculture on weather conditions, exhaustion of soil health in course of time, fertility variations of the land, limited scope of division of labour *etc.,*

11. *Nature of Demand:* Farm products generally being necessaries of life, the demand is relatively inelastic, while the demand for industrial goods is relatively elastic.

12. *Efficiency of Capital:* The farm business takes relatively larger time to return the investment through income, compared to industrial production. It implies that the rate of capital turnover is slow in agriculture.

13. *Producer's Share in Consumer's Rupee:* Agricultural marketing is characterized by the existence of too many middlemen and thus the share of producer in consumer's rupee is low, whereas for industrial goods there are well-defined distributing channels. Therefore, the share of producer in consumer's rupee is high.

# CHAPTER 17

# Laws of Returns

Production is the result of application of various input factors. In the process of production, the farmers combine the required input factors in various proportions. This type of usage of inputs by the farmers gives way for the operation of the laws of returns. In the production process, when a single input factor is varied keeping other required factors constant, the relationship that takes place between single variable input and the consequent output pertains to either one or a combination of the following relationships.

1. Law of increasing returns
2. Law of constant returns; and
3. Law of decreasing returns

## LAW OF INCREASING RETURNS

The addition of each successive unit of the variable factor to the fixed factors in the production processes, adds more to the total output than the previous unit i.e., each successive unit of variable factor adds more and more to the total output. The relevant data are presented in Table 17.1.

**TABLE 17.1 Law of Increasing Returns.**

| Fertilizer (kg) (X) | Total output (Q) (Y) | $\Delta X$ | $\Delta Y$ | Marginal output $\dfrac{\Delta Y}{\Delta X}$ |
|---|---|---|---|---|
| 1 | 3 | | | — |
| | | 1 | 5 | 5 |
| 2 | 8 | | | |
| | | 1 | 7 | 7 |
| 3 | 15 | | | |
| | | 1 | 8 | 8 |
| 4 | 23 | | | |
| | | 1 | 12 | 12 |
| 5 | 35 | | | |

It is clear from the table that first unit of fertilizer results in three quintals of output, second unit adds five quintals and so on. When the data is graphed the resultant curve is convex to X-axis (Figure 17.1).

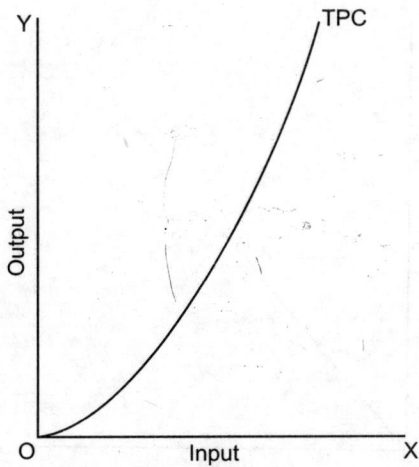

**Figure 17.1** Increasing returns.

The relationship is algebraically shown below:

$$\frac{\Delta_1 Y_1}{\Delta_1 X_1} < \frac{\Delta_2 Y_2}{\Delta_2 X_2} < \ldots\ldots < \frac{\Delta_n Y_n}{\Delta_n X_n}$$

## LAW OF CONSTANT RETURNS

The addition of each successive unit of the variable factor to the fixed factors adds the same to the output as observed for the previous unit i.e., each successive unit of variable factor results in an equal quantity of additional output. It is clear from Table 17.2 that each additional unit of fertilizer adds five quintals to the total output.

**TABLE 17.2 Law of Constant Returns.**

| Fertilizer (kg) (X) | Total output (Q) (Y) | $\Delta X$ | $\Delta Y$ | Marginal output $\dfrac{\Delta Y}{\Delta X}$ |
|:---:|:---:|:---:|:---:|:---:|
| 1 | 5 | | | |
| | | 1 | 5 | 5 |
| 2 | 10 | | | |
| | | 1 | 5 | 5 |
| 3 | 15 | | | |
| | | 1 | 5 | 5 |
| 4 | 20 | | | |
| | | 1 | 5 | 5 |
| 5 | 25 | | | |

The production function is linear (straight line) (Figure 17.2).
The algebraic form is as follows:

$$\frac{\Delta_1 Y_1}{\Delta_1 X_1} = \frac{\Delta_2 Y_2}{\Delta_2 X_2} = \ldots\ldots = \frac{\Delta_n Y_n}{\Delta_n X_n}$$

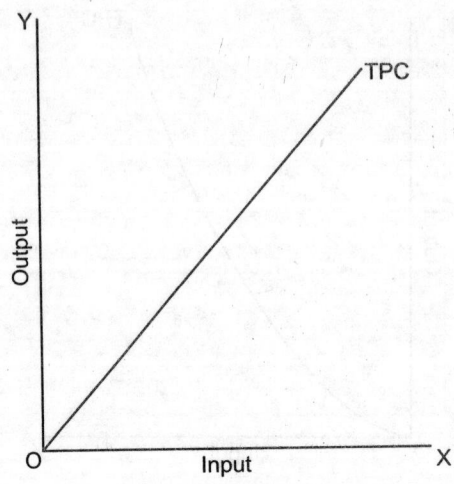

**Figure 17.2** Constant returns.

# LAW OF DECREASING RETURNS

The addition of each successive unit of the variable factor to the fixed factors in the production process, adds less to the total output than the previous unit i.e., each successive unit of variable factor adds less and less to the total output. It is evident from the table that the first unit gives 15 units, second adds 12 units, third unit adds 8 units and so on as shown in Table 17.3.

**TABLE 17.3 Law of Decreasing Returns.**

| Fertilizer (kg) (X) | Total output (Q) (Y) | $\Delta X$ | $\Delta Y$ | Marginal output $\dfrac{\Delta Y}{\Delta X}$ |
|:---:|:---:|:---:|:---:|:---:|
| 1 | 15 | | | |
| | | 1 | 12 | 12 |
| 2 | 27 | | | |
| | | 1 | 8 | 8 |
| 3 | 35 | | | |
| | | 1 | 6 | 6 |
| 4 | 41 | | | |
| | | 1 | 4 | 4 |
| 5 | 45 | | | |

The production function is concave to X-axis as shown in Figure 17.3.
This relationship is algebraically shown as follows:

$$\frac{\Delta_1 Y_1}{\Delta_1 X_1} > \frac{\Delta_2 Y_2}{\Delta_2 X_2} > \ldots\ldots > \frac{\Delta_n Y_n}{\Delta_n X_n}$$

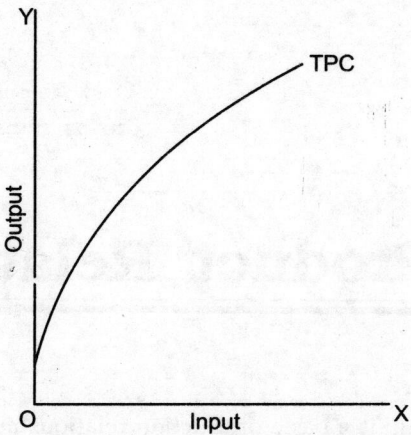

**Figure 17.3** Decreasing returns.

# CHAPTER 18

# Factor–Product Relationship

Factor–product relationship is a basic production relationship between the input and output. This is mainly concerned with resource use and its efficiency. It guides the producer in deciding as to how much to produce. The goal of this relationship is the optimization of resources. This relationship is explained by the law of diminishing returns.

The farmer as a producer has a given goal clearly cut out with the inputs/resources[*] at his disposal. These resources are put into a process called production[**]. Through the production process all the inputs get transformed into an output/product[***]. This production process may pertain to using of the resources or inputs *viz.,* seed, fertilizer, irrigation, human labour *etc.,* to produce a given quantity of output of crop enterprises like paddy, sugarcane, wheat, cotton *etc.,* and similarly employing the resources like human labour, feed and fodder, medicines *etc.,* in the production of milk, meat, eggs, fish, *etc.* Having identified the necessary inputs which facilitate the production process, the question that arises is, whether the farmer possesses the knowledge of the production activity, imbibing the physical relationship between the resources and the expected output. This knowledge is essential because the response of output to input application is at varying magnitude and hence the farmer has to make decisions of how much input to use and how much output to produce. The discussion on factor-product relationship is confined to a single variable input and output.

$$Y = f (X_1 \mid X_2, X_3, \ldots\ldots X_n)$$

Where,

$Y$ = Output from a particular enterprise

$X_1$ = Variable resource

$X_2 \ldots X_n$ = Fixed resources

$\mid$ (Vertical bar) = It separates variable resource from fixed resources

## PRODUCTION FUNCTION

*Production function is a technical and mathematical relationship describing the manner and the extent to which a particular product depends upon the quantities of inputs or services*

---

[*] An input/resource is any good or service that goes into production.

[**] Production may be defined as a process by which inputs are transformed into an output.

[***] Output/product is any good or service that comes out of production.

*of inputs, used at a given level of technology and in a given period of time.* It shows the quantity of output that can be produced using different levels of inputs.

Here, we need to understand certain concepts, which figure in the analysis of this relationship.

## Total Physical Product (TPP)

It is the total amount of output obtained by using different units of inputs, measuring in physical units like quintals, *kgs*, etc.

## Average Physical Product (APP)

It is the average amount of output produced by each corresponding unit of input. It is obtained by dividing the total output at a given level by the number of units of input applied at the corresponding level. APP reflects the efficiency of the variable input (technical efficiency).

$$APP = \frac{\text{Total physical product}}{\text{Input level}} = \frac{Y}{X}$$

## Marginal Physical Product (MPP)

MPP is the additional quantity of output, added by an additional unit of input i.e., the change in output as a result of change in the variable input. It is calculated as

$$MPP = \frac{\text{Change in total physical product}}{\text{Change in input level}} = \frac{\Delta Y}{\Delta X}$$

## Elasticity of Production (E$_p$)

It is defined as percentage change in output as a result of percentage change in input.

$$E_p = \frac{\text{Percentage change in output}}{\text{Percentage change in input}}$$

The elasticity of production can also be defined in terms of the relationship between MPP and APP as given below:

$$E_p = \frac{\left(\dfrac{\Delta Y}{Y}\right)}{\left(\dfrac{\Delta X}{X}\right)}$$

It can be written as,

$$E_p = \frac{\left(\dfrac{\Delta Y}{\Delta X}\right)}{\left(\dfrac{Y}{X}\right)}$$

We know that,

$$\frac{\Delta Y}{\Delta X} = MPP \text{ and that } \frac{Y}{X} = APP$$

Therefore,

$$E_p = \frac{MPP}{APP}$$

### Relationship between TPP, MPP and APP

As long as MPP is increasing, TPP is increasing at an increasing rate. TPP goes on increasing at an increasing rate till the point of maximum MPP. After the point of maximum MPP, TPP increases at a decreasing rate. When MPP becomes zero, TPP attains its maximum. Negative MPP results in decreasing TPP. When TPP is increasing, MPP is positive, when TPP is maximum MPP is zero and when TPP declines MPP becomes negative.

Tabular presentation of a production function is given in Table 18.1, which indicates the different levels of variable input (human labour) along with the total physical product (TPP), average physical product (APP) and marginal physical product (MPP). Land has been assumed as the fixed resource. The TPP is increasing up to 7th unit and thereafter it begins to decline. The APP up to the application of five units of human labour, increased and thereafter it tended to decline. Marginal product increased from 15 units to 30 units, with the increase in the use of human labour from first unit to fourth unit, and decreased with further increase in the labour input. Marginal product is negative when the variable input level increased to 8th unit and beyond. The production function is illustrated in Figure 18.1.

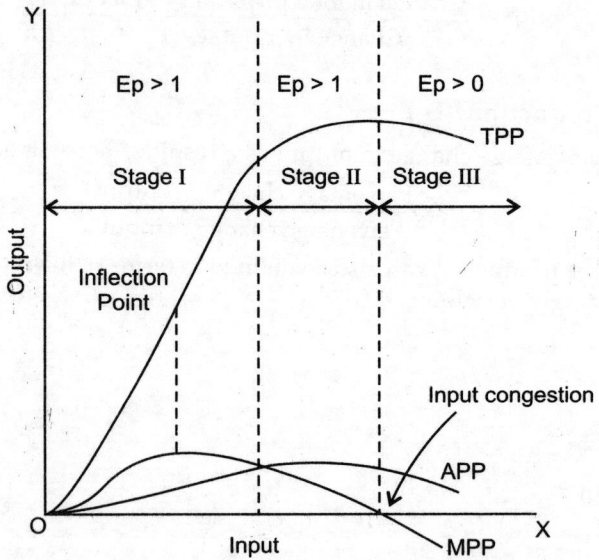

**Figure 18.1** Three stages of production function.

## THREE STAGES OF PRODUCTION FUNCTION

The classical production function can be divided into three stages or zones or regions. This is done to identify the zone in which production decisions are rational.

*Stage I:* It starts from the origin and ends at the point where, MPP = APP. In this stage, MPP > APP as a result of which APP is increasing. The MPP attains the maximum at the point of inflection[*], thereafter it begins to decline. TPP increases at an increasing rate till the inflection point and thereafter it increases at decreasing rate. $E_p$ is more than one throughout the stage I of production and $E_p$ is one at the end of this stage. In this stage fixed resources are abundant relative to variable resource. The technical efficiency of the variable resource is increasing as indicated by increasing APP. Also the technical efficiency of fixed resource is increasing as reflected by increasing TPP. This stage is regarded as irrational (sub-optimal) stage of production.

*Stage II:* It is found from the point of equality of MPP and APP and ends where MPP is zero, at which input use level, TPP is maximum. In this stage, MPP is less than APP. However, both MPP and APP exhibit declining trend. Average productivity derived from each individual unit of the variable resource is on the decline in this stage, though it is at its peak at the beginning of Stage II. TPP increases at a decreasing rate as MPP is declining. $E_p$ is less than one throughout stage II. $E_p$ is zero at the end of this stage. It is rational (optimal) stage of production. The technical efficiency of the variable resource declines as indicated by the declining APP, but the technical efficiency of fixed resource increases as indicated by increasing TPP. In this stage variable resource is abundant relative to fixed resources.

*Stage III:* The starting point of stage III is the end of stage II, at which MPP is zero. In this stage MPP becomes negative. APP continuously declines and TPP which is at its maximum at the end of stage II, begins to decline. $E_p$ is less than zero. It is an irrational (supra-optimal) stage of production. The technical efficiency of variable resource and fixed resource declines. Variable resource is in excessive quantities relative to fixed resource.

## Reasons for Increasing, Decreasing and Negative Returns

In stage I, fixed resources are in abundance and these fixed resources are not put to efficient utilization due to lack of sufficient quantity of variable resource, and infact all the fixed resources are slack i.e., they are at the disposal of the farmers. Under this situation application of more quantities of variable resource makes the hither to unutilized fixed resource efficient leading to increasing returns.

As more quantities of variable resource are applied soon we reach the point of maximum marginal physical product. Beyond this point, any further increase in the use of variable resource yields less additional output (decreasing returns). This happens as more quantity of variable resource has to accommodate with less quantity of fixed resources.

Production in general is the integrated effort of fixed and variable resources. Since the proportion of variable resource becomes too large to the very limited fixed resources, the balance which is supposed to exist between fixed resources and variable resource gets disturbed, thereby leading to the negative returns, in the third stage.

## RATIONAL AND IRRATIONAL STAGES OF PRODUCTION

Of the three stages, stage I and stage III are irrational, while stage II is rational. In stage I since the average productivity (technical efficiency) of variable resource keeps

---

[*] Inflection point is the point at which MPP is maximum. It is also the point at which TPP curve reverses its shape from convex to concave.

on increasing, it is not judicious to stop the application of the variable resource. Fixed factor which is kept as slack (idle) can be withdrawn and we can reorganize the resources. This reorganization leads to the production of more output. Since it is possible to produce more output from less quantity of resources through reorganization of fixed and variable resources, this stage is called irrational stage.

In stage III, the technical efficiency of variable resource (APP) and fixed resource (TPP) declines. The additional productivity of variable resource (MPP) becomes negative. Given the productivities of variable and fixed resources, no rational producer attempts to operate in this stage, as any attempt to continue production brings him colossal loss in the form of additional costs of variable resource and reduction in total output. At this point a look at Table 18.1 indicates that maximum production is attained with 7th unit of human labour and further increase in this resource is reducing the total output. Withdrawal of the human labour from 10 units to 7 units would increase the output. The application of 7th unit of human labour not only costs the farmer, but also the additional gains in terms of output are zero. Such a situation is called free disposability of input. Application of 8th unit of human labour, not only costs the farmer, but on the other hand, brings him loss, in the form of reduction of output; and such a situation is called weak disposability of input. Since stage III offers the opportunity of reorganization of fixed and variable resources, it is called an irrational stage.

In stage II, the point of optimality, either in the use of input or in the production of output, lies. The boundaries of this stage are the points, at which the technical efficiency of variable resource is maximum and the technical efficiency of fixed resource is maximum. Somewhere the point of optimality lies in this stage and in order to locate the point of optimality, choice indicators are needed. The choice indicators are price ratios i.e., the price per unit of input and price per unit of output. Without knowing the prices, certain generalizations can be made with regard to the use of variable and fixed resources. Assuming that fixed resource is unlimited in quantity and variable resource is scarce, the aim of the producer is to produce maximum output per unit of scarce resource i.e., human labour here. This is achieved when the average productivity (technical efficiency of variable resource) is maximum. This is seen at the beginning of stage II. When variable resources are unlimited and fixed resource is scarce; the objective again is to maximize physical production per unit of scarce resource i.e. fixed resource. This is achieved when the technical efficiency of fixed factor (TPP) is maximum i.e., the point of congestion. This point is found at the end of stage II. In stage II, since there is no possibility of recombining fixed and variable resources, it is a rational stage or optimal stage.

Production function presents the consistent relationship between different levels of variable resource applied in a production activity along with the corresponding levels of output. The producer is interested in knowing the level of input use and level of production, at which profits are maximum. To find out optimal level of input and output, we have to know the choice indicators.

# DETERMINATION OF OPTIMUM LEVEL OF INPUT

Having identified stage II as rational stage, the extent of variable resource use needs to be studied. This is done by working out and comparing marginal value product (MVP) and marginal input cost (MIC) as presented in Table 18.2.

## Marginal Value Product (MVP) of Input

It is the additional income received from using an additional unit of input. It is computed using the following formula.

$$MVP = \frac{\text{Change in the total value product}}{\text{Change in input level}} = \frac{\Delta TR}{\Delta X} = MPP \times P_Q$$

## Marginal Input Cost (MIC)

MIC is defined as the change in the total input cost by using an additional unit of input. It is expressed as

$$MIC = \frac{\text{Change in total input cost}}{\text{Change in input level}} = \frac{\Delta TC}{\Delta X} = \text{Price per unit of input}$$

In the first five rows of Table 18.2 MVP is greater than MIC. At the sixth input level, MVP and MIC are exactly equal which is the optimum level of human labour use. Beyond the sixth unit of human labour use, MVP < MIC, indicating the profit reduction as more units of input are used.

# DETERMINATION OF OPTIMUM LEVEL OF OUTPUT

Apart from identifying the profit maximizing input level, it is pertinent to find out the level of output, which maximizes profit. To carry on this analysis, it is essential to work out and compare marginal revenue (MR) and marginal cost (MC).

## Marginal Cost (MC)

MC is defined as the additional cost incurred for producing an additional unit of output. The expression is as follows:

$$MC = \frac{\text{Change in total input cost}}{\text{Change in total physical product}} = \frac{\Delta TC}{\Delta Q}$$

## Marginal Revenue (MR)

MR is defined as the additional income obtained from producing one more unit of output. It is expressed as:

$$MR = \frac{\text{Change in total income}}{\text{Change in total physical product}} = \frac{\Delta TR}{\Delta Q}$$

MR is always equal to price per unit of output. MR and MC values in Table 18.3 are compared to find out the optimal output (profit maximizing) level. In the first five rows of the table, MR is greater than MC. At the sixth input level, MR = MC, which is the optimum level of output. Using beyond sixth unit, it causes MR to fall below MC; implying profit reduction. If MR > MC, the additional unit of output enhances the profit and if MR < MC, the production of additional unit of output decreases the profit.

The analysis reveals that whether it is the equality of MVP and MIC or MR and MC, the result as well as decision-making is the same i.e., there is only one profit maximizing level of input and output for the given information of technical coefficients and prices.

**TABLE 18.1**  Total Physical Product, Average Physical Product and Marginal Physical Product.

| Input level product (MPP) | Total physical product | Average physical product (TPP) | Marginal physical product (APP) |
|---|---|---|---|
| 1 | 15 | 15 | |
| 2 | 38 | 19 | 23 |
| 3 | 66 | 22 | 28 |
| 4 | 96 | 24 | 30 |
| 5 | 120 | 24 | 24 |
| 6 | 126 | 21 | 6 |
| 7 | 126 | 18 | 0 |
| 8 | 120 | 15 | -6 |
| 9 | 90 | 10 | -30 |
| 10 | 50 | 5 | -40 |

**TABLE 18.2**  Marginal Value Product, Marginal Input Cost and the Optimum Input Level.

Labour wage = Rs. 30/- Manday
Price per kg of output = Rs. 5/-

| Input level | Total physical product (TPP) | Marginal physical product (MPP) | Total value Product (TVP) (Rs) | Marginal value product (MVP) (Rs.) | Marginal input cost (MIC) (Rs.) |
|---|---|---|---|---|---|
| 1 | 15 | | 75 | | |
| 2 | 38 | 23 | 190 | 115 | 30 |
| 3 | 66 | 28 | 330 | 140 | 30 |
| 4 | 96 | 30 | 480 | 150 | 30 |
| 5 | 120 | 24 | 600 | 120 | 30 |
| 6* | 126 | 6 | 630 | 30* | 30* |
| 7 | 126 | 0 | 630 | 0 | 30 |
| 8 | 120 | -6 | 600 | -30 | 30 |
| 9 | 90 | -30 | 450 | -150 | 30 |
| 10 | 50 | -40 | 250 | -200 | 30 |

* Optimum input level

TABLE 18.3 Marginal Revenue, Marginal Cost and the Optimum Output Level.

| Input level | Total physical product (TPP) | Marginal physical product (MPP) | Total revenue (TR) (Rs.) | Marginal revenue (MR) (Rs.) | Marginal cost (MC) (Rs.) |
|---|---|---|---|---|---|
| 1 | 15 | | 75 | | |
| | | 23 | | 5 | 1.30 |
| 2 | 38 | | 190 | | |
| | | 28 | | 5 | 1.07 |
| 3 | 66 | | 330 | | |
| | | 30 | | 5 | 1.0 |
| 4 | 96 | | 480 | | |
| | | 24 | | 5 | 1.25 |
| 5 | 120 | | 600 | | |
| | | 6 | | 5* | 5.0* |
| 6 | 126* | | 630 | | |
| | | 0 | | 5 | 0 |
| 7 | 126 | | 630 | | |
| | | −6 | | 5 | −5.0 |
| 8 | 120 | | 600 | | |
| | | −30 | | 5 | −1.0 |
| 9 | 90 | | 450 | | |
| | | −40 | | 5 | −0.75 |
| 10 | 50 | | 250 | | |

*Optimum output level

# RELATIONSHIP BETWEEN TPP, MPP AND APP OF LAND ($X_2$) AND TPP, MPP AND APP OF HUMAN LABOUR ($X_1$)

We have seen above, the three stages of production function for the variable resource, human labour ($X_1$). We can also find out the relationship between hitherto fixed factors i.e., land ($X_2$) and the output, keeping human labour ($X_1$) fixed. Now land ($X_2$) becomes variable factor and it is maintained in such a way that ratio of $X_2/X_1$, is kept as before, when the quantity of $X_2$ is increased. Table 18.4 depicts TPP, APP and MPP of variable factor i.e., human labour ($X_1$) and the land/human labour ratio. One unit of land and one unit of human labour implies a land/human labour ratio of 1/1 and one unit of land and 2 units of human labour with a ratio of ½ and so on. Using the data given in Table 18.4 we can derive TPP of $X_2$, APP of $X_2$ and MPP of $X_2$ and presented in Table 18.5. It can be seen from the table that human labour ($X_1$) is kept constant, while the land ($X_2$) is increased maintaining the ratio as before, which can be evidently seen in the Table.

A proportion of 1/9th of land ($X_2$) indicates that 1/9th unit of land ($X_2$) is combined with one unit of labour.

Estimation of TPP, MPP and APP of land ($X_2$) is presented hereunder.

$$\text{TPP of } X_2 = \frac{\text{TPP of vairable factor (Human labour)}}{\text{Land - labour ratio of fixed factor}}$$

*Example:* TPP of human labour ($X_1$) for land/human labour ratio of $1/10^{th}$ is 50, TPP of land ($X_2$) would be $50/10 = 5$

MPP and APP of $X_2$ are computed in the usual manner *i.e.*,

$$\text{MPP of } X_2 = \frac{\Delta \text{ TPP of } X_2}{\Delta X_2}; \quad \text{APP of } X_2 = \frac{\text{TPP of } X_2}{X_2}$$

Using the equations, TPP, MPP and APP are worked out and presented in Table 18.5. Stage I of $X_1$ is stage III for $X_2$ and stage III of $X_1$ is stage I for $X_2$. But stage II is common in both the cases. MPP and APP of $X_2$ are positive and declining in stage II in which economic decisions are taken. Graphical representation is found in Figure 18.2.

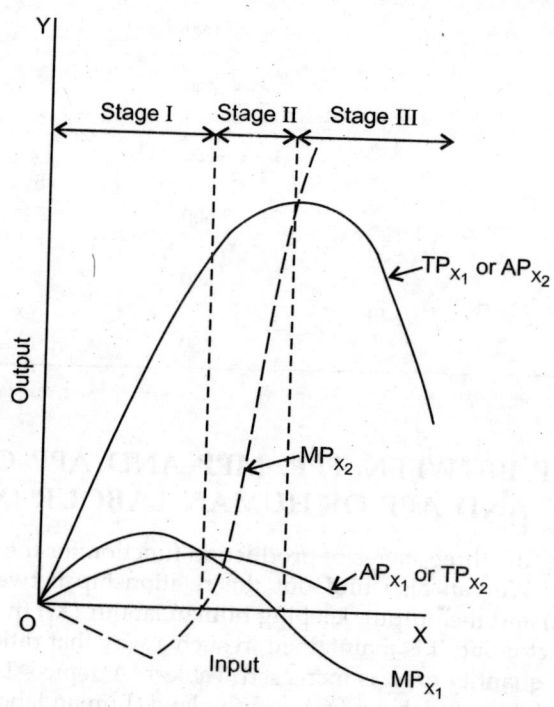

**Figure 18.2** Total, average and marginal products of fixed and variable factors.

*Please note that TPP of $X_1$ is APP of $X_2$ and APP of $X_1$ is TPP of $X_2$.*

In both the cases the characteristics of stage II are the same. In Table 18.4 we have estimated the productivity of human labour in combination with land, while in Table 18.5 the productivity of land in combination with labour is estimated. In essence the productivities of human labour and land are isolated through this analysis.

**TABLE 18.4**  Total Physical Product, Average Physical Product and Marginal Physical Product of Human Labour.

| Human labour (X) | Land ($X_2$) | Land/ human labour ratio ($X_2/X_1$) | TPP of human labour ($X_1$) | MPP of human labour ($X_1$) | APP of human labour ($X_1$) | Stages of production |
|---|---|---|---|---|---|---|
| 1 | 1 | 1/1 | 15 | | 15 | |
| | | | | 23 | | |
| 2 | 1 | 1/2 | 38 | | 19 | |
| | | | | 28 | | Stage I of $X_1$ |
| 3 | 1 | 1/3 | 66 | | 22 | |
| | | | | 30 | | |
| 4 | 1 | 1/4 | 96 | | 24 | |
| | | | | 24 | | |
| 5 | 1 | 1/5 | 120 | | 24 | |
| | | | | 6 | | |
| 6 | 1 | 1/6 | 126 | | 21 | Stage II of $X_1$ |
| 7 | 1 | 1/7 | 126 | | 18 | |
| | | | | -6 | | |
| 8 | 1 | 1/8 | 120 | | 15 | |
| | | | | -30 | | |
| 9 | 1 | 1/9 | 90 | | 10 | Stage III of $X_1$ |
| | | | | -40 | | |
| 10 | 1 | 1/10 | 50 | | 5 | |

**TABLE 18.5**  Total Physical Product, Average Physical Product and Marginal Physical Product of Land.

| Human labour ($X_1$) | Land ($X_2$) | Land/ human labour ratio ($X_2/X_1$) | TPP of $X_2$ | MPP of $X_2$ | APP of $X_2$ | Stages of production |
|---|---|---|---|---|---|---|
| 1 | 1/1 | 1/1:1 | 15 | - | 15 | |
| 2 | 1/2 | 1/2:1 | 19 | -8 | 38 | Stage III of |
| 3 | 1/3 | 1/3:1 | 22 | -18 | 66 | $X_2$ |
| 4 | 1/4 | 1/4:1 | 24 | -24 | 96 | |
| 5 | 1/5 | 1/5:1 | 24 | 0 | 120 | |
| 6 | 1/6 | 1/6:1 | 21 | 90 | 126 | Stage II of |
| 7 | 1/7 | 1/7:1 | 18 | 126 | 126 | $X_2$ |
| 8 | 1/8 | 1/8:1 | 15 | 168 | 120 | |
| 9 | 1/9 | 1/9:1 | 10 | 360 | 90 | Stage I of |
| 10 | 1/10 | 1/10:1 | 5 | 450 | 50 | $X_2$ |

## Summary of Three Stages of Production Function:

| S. No. | Stage I | Stage II | Stage III |
|---|---|---|---|
| 1. | Starts from the origin and ends where MPP = APP | Starts from where APP is maximum and ends where MPP is zero | Starts from where MPP is zero or TPP is maximum |
| 2. | TPP increases at increasing rate increasing rate | TPP increases at up to the point of inflection | TPP decreases at decreasing rate |
| 3. | APP is increasing throughout this stage | APP decreases | APP decreases |
| 4. | MPP increases up to the point | MPP decreases of inflection | MPP becomes negative |
| 5. | MPP > APP | MPP < APP | - |
| 6. | $E_p > 1$ | $E_p < 1$ | $E_p < 0$ |
| 7. | Technical efficiency of variable and fixed resources increases | Technical efficiency of variable resource decreases but of the fixed resources increases | Technical efficiency of variable and fixed resources decreases |
| 8. | Fixed resources are abundant and variable resource is scarce | Variable resource is abundant and fixed resources are scarce | Variable resource is in excess capacity |
| 9. | Sub-optimal (irrational) stage of production | Optimal (rational) stage of production | Supra-optimal (irrational) stage of production |
| 10. | Scope for the reorganization of resources | No scope for the reorganization | Scope for the reorganization |
| 11. | MVP > MIC | MVP = MIC | MVP < MIC |
| 12. | MR > MC | MR = MC | MR < MC |

# CHAPTER 19

# Factor–Factor Relationship

Any production activity requires different inputs to produce a given quantity of output. There are many ways of combining these resources or production technology in the production process. Farmers are many times confronted with the problem of making a choice of production technology. Farm production facilitates the substitution of resources. In general, a farmer producing farm products does make a choice between nutrients say, organic manures and inorganic fertilizers, human labour and machines, human labour and herbicides, *etc.* Similarly, farmers producing livestock products make an effort to decide upon the quantity of grain and hay to arrive at feed ration. The problem here is to decide up on the most appropriate resource or method or given combination that costs the farmer the least amount in producing a given level of output. The managerial problem here is to find out the least cost combination of inputs for producing a given level of output. The production function here is

$$Q^* = f(X_1, X_2)$$

Where,

$Q^*$ is fixed level of output and $X_1$ and $X_2$ are the quantities of variable inputs

The factor–factor relationship deals with two independent variables and dependent variable giving rise to three-dimensional diagram. Iso-quant is a convenient method of compressing three-dimensional diagram into two-dimensional diagram.

## ISO-QUANT

*Iso* means equal and *quant* is quantity. Iso-quant is also termed as iso-product curve or equal product curve or product indifference curve. Iso-quant in the theory of production is a counterpart of indifference curve in the theory of consumption. *The curve representing all combinations of $X_1$ and $X_2$ that produce the same level of output is called an iso-quant.* Table 19.1 shows that an output amounting to 100 units can be produced using the input combinations presented. The iso-quant (Figure 19.1) presents all combinations of $X_1$ and $X_2$ that produce 100 units of output. Iso-quants can be shown for any level of output. As such, several iso-quants can be shown for various levels of output with different levels of inputs. If a number of iso-quants are drawn in one graph, it is known as iso-quant map or iso-product contour (Figure 19.2).

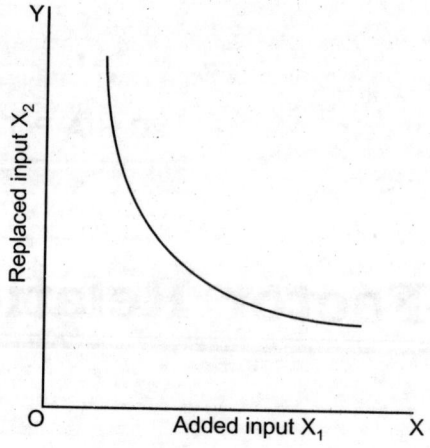

**Figure 19.1** Iso-quant

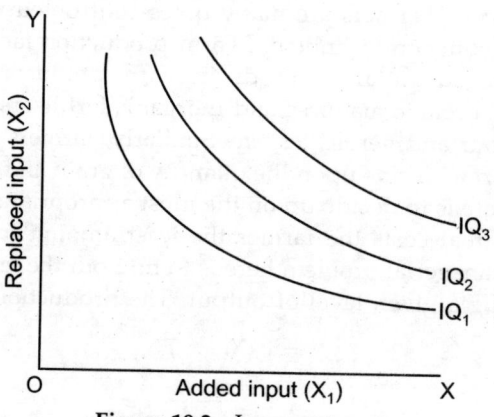

**Figure 19.2** Iso-quant map.

**TABLE 19.1  Input Combinations Producing the Same Level of Output.**

| $X_1$ | $X_2$ | Total output |
|-------|-------|--------------|
| 1 | 13 | 100 units |
| 2 | 8 | 100 units |
| 3 | 5 | 100 units |
| 4 | 3 | 100 units |
| 5 | 1 | 100 units |

## Properties of Iso-quants

1. *Iso-quants Slope Downwards from Left to Right:* If the quantity of input, say $X_1$ is increased, the quantity of other input say $X_2$ must be decreased to obtain the same level of output.
2. *Iso-quants are Convex to Origin:* The absolute slope of iso-quant decreases as we move left downwards to the right indicating diminishing rate of technical substitution. Because of diminishing marginal rate of technical substitution, each added unit of one input replaces less and less than the previous unit.

3. *Iso-quants Placed above Another Represent Higher Output*: Iso-quants placed for higher level of output normally lie above the iso-quants representing lower level of output. Alternatively, iso-quants representing higher levels of output are placed farther away from the origin.

4. *Iso-quants are Non-intersecting*: No two iso-quants intersect each other because the same combination of two input factors cannot produce two different levels of output.

## Marginal Rate of Technical Substitution (MRTS)

It is the rate of exchange between two productive resources, which are equally preferred. It indicates the absolute amount by which one productive resource is decreased to gain a unit of another productive resource. Alternatively, the quantity of one input to be sacrificed or given up in order to gain another input by one unit, in the process of substitution. MRTS of $X_1$ for $X_2$ is written as

$$MRTS_{X_1 X_2} = \frac{\Delta X_2}{\Delta X_1}$$

MRTS is computed with the following equation

$$MRTS = \frac{\text{Quantity of input scarificed}}{\text{Quantity of input gained}}$$

Algebraically,

$$MRTS_{X_1 X_2} = \frac{\Delta X_2}{\Delta X_1}$$

MRTS of $X_1$ for $X_2$ is the amount by which $X_2$ must be decreased to maintain the same level of output, when $X_1$ is increased by one unit.

$$MRTS_{X_2 X_1} = \frac{\Delta X_1}{\Delta X_2}$$

Here, $X_2$ = Added input
$X_1$ = Replaced input

The computation of marginal rate of substitution of $X_1$ for $X_2$ is presented in Table 19.2 for an output of 100 units. As the amount of $X_1$ increases each additional unit of $X_1$ replaces successively smaller quantities of $X_2$.

**TABLE 19.2** Determination of Marginal Rate of Substitution of $X_1$ for $X_2$ for an Output of 100 units.

| Units of $X_1$ | Units of $X_2$ | $\Delta X_1$ | $\Delta X_2$ | MRS of $X_1$ for $X_2$ |
|---|---|---|---|---|
| 1 | 12 | | | |
| | | 1 | −3 | −3.0 |
| 2 | 9 | | | |
| | | 1 | −2 | −2.0 |
| 3 | 7 | | | |
| | | 1 | −1 | −1.0 |
| 4 | 6 | | | |
| | | 1 | −0.5 | −0.5 |
| 5 | 5.5 | | | |
| | | 1 | −0.25 | −0.25 |
| 6 | 5.25 | | | |

The resources that are used by the farmer in the production activity are either substitutes or complements.

## Substitutes

Two resources are said to be substitutes, when change in the price of one leads to change in the demand for another. Substitutes are the range of input combinations that produce a given level of output. Decrease in the amount of one input is compensated by an increase in the amount of other input, when resources are substitutes. The marginal rate of technical substitution is negative.

## Perfect Substitutes

When the two inputs are completely interchangeable, then they are called perfect substitutes. Examples: Family labour and hired labour, owned bullock labour and hired labour, farm produced and purchased input *etc.* In this case, iso-quants are linear and negatively sloped (Figure 19.3).

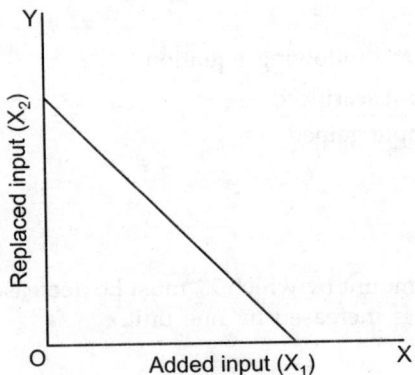

**Figure 19.3** Perfect substitutes.

## Complements

Resources, which are used together in the production process are called complements. Two resources *viz.*, $X_1$ and $X_2$ are said to be complements, when the price of $X_1$ increases the demand for $X_2$ decreases. Decrease in the amount of one cannot be compensated by increasing the amount of another. The marginal rate of substitution is zero.

## Perfect Complements

Resources which are used together in fixed proportions are called perfect complements (Figure 19.4). When inputs are perfect complements, the iso-quants are 'L' shaped. Examples: Tractor and driver, a pair of bullocks and human labour, *etc.*

## TYPES OF FACTOR SUBSTITUTION

The shape of iso-quant depends upon the manner in which resources will be combined as fairly similar and dissimilar inputs are substituted in the production process. Following are the possible types of factor substitution.

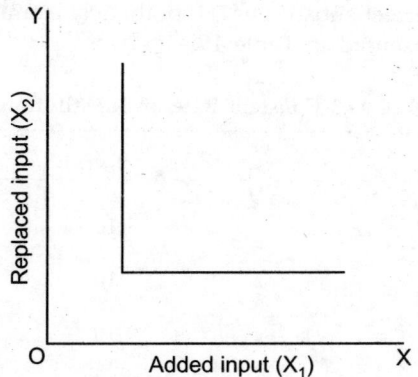

**Figure 19.4** Perfect complements: 'L' shaped iso-quant.

1. Fixed proportion combination of inputs
2. Constant rate of substitution, and
3. Decreasing rate of substitution

## Fixed Proportion Combination of Inputs

As the very nature of combination of inputs infers that a given level of output can be produced using the inputs in a fixed proportion. As such there is no problem regarding decision-making because there are no alternatives as far as the combination of input factors is concerned. There is only one way of combining the input factors. Inputs that increase the output when combined in fixed proportions are called complements. The examples that can be cited in farming are tractor and driver, fuel and grease, *etc.* The iso-quant is 'L' shaped (Figure 19.5). It is also termed as Leontief iso-quant, which was named after an economist named Leontief, who is popular for his pioneering work in input-output analysis. MRTS is zero in this situation.

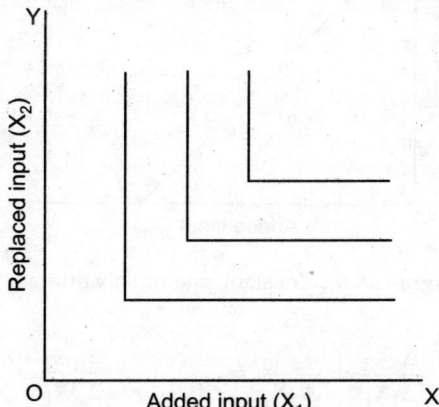

**Figure 19.5** Fixed proportion combination of inputs.

## Constant Rate of Substitution

Constant amount of one resource is sacrificed in order to gain an unit quantity of another resource throughout the process of substitution. This rate of substitution is

182

observed in the case of perfect substitutes. Hypothetical example of inputs substituting at constant rate is presented in Table 19.3.

TABLE 19.3 Constant Rate of Substitution.

| $X_1$ | $X_2$ | $\Delta X_1$ | $\Delta X_2$ | MRTS of $X_1$ for $X_2$ |
|---|---|---|---|---|
| 0 | 40 | | | |
| | | 1 | 10 | $10/1 = 10$ |
| 1 | 30 | | | |
| | | 1 | 10 | $10/1 = 10$ |
| 2 | 20 | | | |
| | | 1 | 10 | $10/1 = 10$ |
| 3 | 10 | | | |
| | | 1 | 10 | $10/1 = 10$ |
| 4 | 0 | | | |

The table shows five combinations of $X_1$ and $X_2$ that are possible to produce a given level of output. Addition of one unit of $X_1$ replaces 10 units of $X_2$ in each of the above five combinations. The MRTS of $X_1$ for $X_2$ is 10. Iso-quant is linear and negatively sloped (Figure 19.6). Family labour and hired labour, owned bullock labour and hired bullock labour, owned seed and purchased seed, different brands of urea, *etc.*, are the examples. When inputs substitute at constant rate, combination may not result in cost minimization. It is economical to use any one of the resources, and choice of use depends on their relative prices.

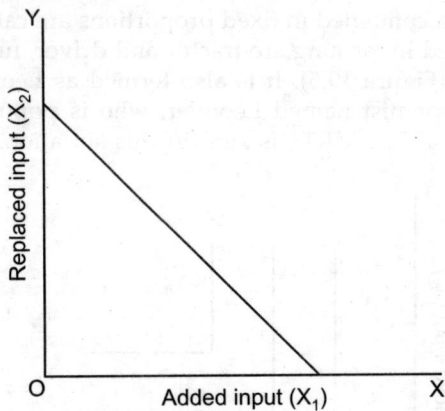

Figure 19.6  Constant rate of substitution.

$$\frac{\Delta_1 X_2}{\Delta_1 X_1} = \frac{\Delta_2 X_2}{\Delta_2 X_1} = \ldots\ldots\ldots = \frac{\Delta_n X_2}{\Delta_n X_1}$$

## Decreasing Rate of Substitution

In the process of substitution, lesser and lesser quantity of one resource is reduced so as to gain another resource by one unit which is termed as decreasing rate of substitution. When $X_1$ is increased, $X_2$ progressively decreased which leads to decreasing rate of substitution. Decreasing rate of input substitution is more common in agriculture. The examples that can be cited are labour and capital, machinery and labour, *etc.* An example of decreasing rate of substitution is presented in Table 19.4.

**TABLE 19.4 Decreasing Rate of Substitution.**

| $X_1$ | $X_2$ | $\Delta X_1$ | $\Delta X_2$ | MRTS of $X_1$ for $X_2$ ($\Delta X_2 / \Delta X_1$) |
|---|---|---|---|---|
| 1 | 12 | | | |
| | | 1 | 5 | 5/1 = 5 |
| 2 | 7 | | | |
| | | 1 | 3 | 3/1 = 3 |
| 3 | 4 | | | |
| | | 1 | 2 | 2/1 = 2 |
| 4 | 2 | | | |
| | | 1 | 1 | 1/1 = 1 |
| 5 | 1 | | | |

MRTS of $X_1$ for $X_2$ decreases numerically. Iso-quant is convex to origin (Figure 19.7).

$$\frac{\Delta_1 X_1}{\Delta_1 X_1} > \frac{\Delta_2 X_2}{\Delta_2 X_1} > \ldots \ldots > \frac{\Delta_n X_2}{\Delta_n X_1}$$

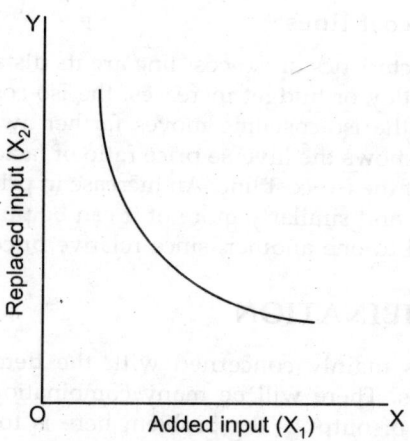

**Figure 19.7** Decreasing rate of substitution.

## ISO-COST LINE

It is known as price line or iso-outlay line or budget line. *Iso-cost lines represent various combinations of two inputs that can be purchased with the given outlay of funds.* The iso-cost line can be drawn by locating the end points of $X_1$ and $X_2$ given the total outlay and the prices per unit of $X_1$ and $X_2$. Suppose a farmer has a fund of Rs. 400 and he has to spend on two inputs *viz.*, $X_1$ and $X_2$. The price per unit of $X_1$ is Rs. 10 and that of $X_2$ is Rs. 8. Given the prices of $X_1$ and $X_2$, he can purchase 40 units of $X_1$ or 50 units of $X_2$. If the 40 units of $X_1$ and 50 units of $X_2$ are graphed, we get iso-cost line (Figure 19.8).

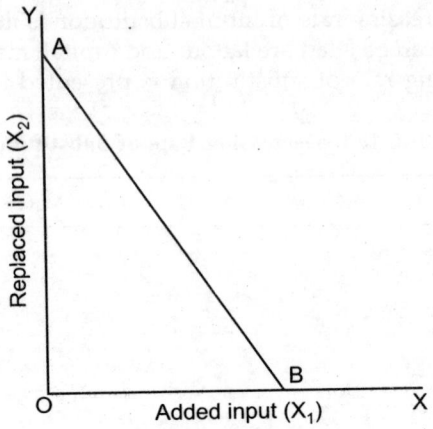

**Figure 19.8** Iso-cost line.

When the entire amount of Rs. 400 is spent on $X_1$ the quantity that can be purchased is OB (40 units) and similarly the quantity of $X_2$ that can be purchased for Rs. 400 is OA (50 units). If points A and B are joined, we get iso-cost line. For any combination of $X_1$ and $X_2$ on iso-cost line, the total cost remains the same.

## Characteristics of Iso-cost lines

The two important characteristics of iso-cost line are its distance from the origin and slope. When the total outlay or budget increases, the iso-cost line shifts upwards to the right. Alternatively, the iso-cost line moves farther away from the origin. The slope of the iso-cost line shows the inverse price ratio of factors. Changes in the input prices change the slope of the iso-cost line. An increase in price of an input means less of that can be purchased and similarly more of it can be purchased if the price falls. Iso-cost lines are parallel to one another, since relative price ratio remains constant.

## LEAST COST COMBINATION

Factor-factor analysis is mainly concerned with the determination of least cost combination of resources. There will be many combinations of two resources that produce the same level of output. The problem here is to find out that particular combination of inputs, which produces a given quantity of output with minimum cost. Following are the different methods of finding out the least cost combination.

### Tabular Method

This method can be used when there are few combinations. Given the input combinations and the prices of inputs, the total cost of each input combination can be computed. The combination which costs the least is selected (Table 19.5).

In the table the combination of 70 units of $X_1$ and 171 units of $X_2$ is the least cost combination.

### Algebraic Method

*Step* 1   Compute marginal rate of technical substitution

TABLE 19.5 Least Cost Combination of Inputs.

| $X_1$ (units) | $X_2$ (units) | $X_1$ @ Rs. 4.00 | $X_2$ @ Rs. 2.00 | Total amount (Rs.) |
|---|---|---|---|---|
| 50 | 219 | 200 | 438 | 638 |
| 55 | 206 | 220 | 412 | 632 |
| 60 | 194 | 240 | 388 | 628 |
| 65 | 182 | 260 | 364 | 624 |
| 70 | 171 | 280 | 342 | 622 |

$$\text{MRTS} = \frac{\text{Number of units of replaced resource}}{\text{Number of units of added resource}}.$$

$$\text{MRTS}_{X_1 X_2} = \frac{\Delta X_2}{\Delta X_1} \text{ (when we substitute } X_1 \text{ for } X_2)$$

$$\text{MRTS}_{X_2 X_1} = \frac{\Delta X_1}{\Delta X_2} \text{ (when we substitute } X_2 \text{ for } X_1)$$

*Step* 2    Compute the inverse price ratio (PR)

$$\text{PR} = \frac{\text{Price per unit of added resource}}{\text{Price per unit of replaced resource}}.$$

$$= \frac{P_{X_1}}{P_{X_2}} \text{ (when we substitute } X_1 \text{ for } X_2)$$

$$= \frac{P_{X_2}}{P_{X_1}} \text{ (when we substitute } X_2 \text{ for } X_1)$$

*Step* 3    Finding the least cost combination by equating marginal rate of technical substitution with inverse price ratio.

$$\frac{\Delta X_2}{\Delta X_1} = \frac{P_{X_1}}{P_{X_2}} \quad \text{or} \quad \frac{\Delta X_1}{\Delta X_2} = \frac{P_{X_2}}{P_{X_1}}$$

## Graphic Method

To find out the optimum combination of resources, through graphic method, both iso-quant and iso-cost line are drawn on the same graph. The slope of the iso-quant indicates the rate of exchangeability (MRTS) between two resources, whereas the slope of iso-cost line represents the inverse price ratio of inputs (PR). The point of tangency between iso-quant and iso-cost line indicates the least cost combination (Figure 19.9). At the point of tangency slope of iso-quant equals the slope of iso-cost line. Alternatively MRTS is equated with PR

## EXPANSION PATH

*The line or curve, connecting the points of least cost combination for different levels of output is called expansion path* (Figure 19.10). Expansion path is an isocline, on which the slope of iso-quant (MRTS) equals the slope of iso-cost line (price ratio). The expansion

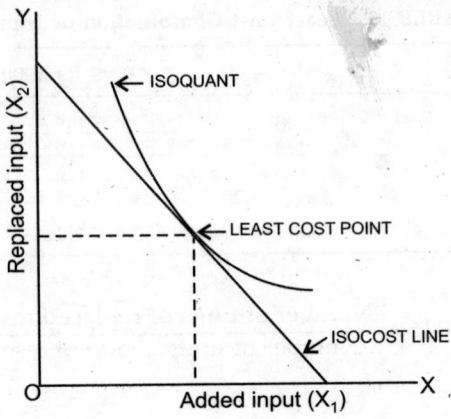

**Figure 19.9** Least cost combination.

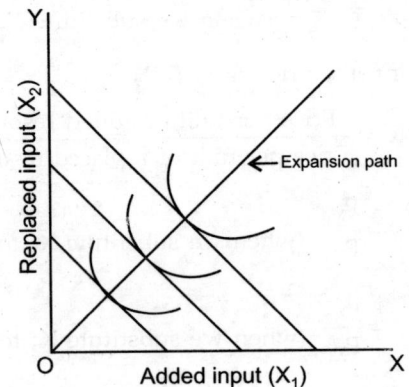

**Figure 19.10** Expansion path.

path indicates the best way of producing different levels of output, given the prices of inputs and technology. If the expansion path is a straight line starting from the origin, it implies that the inputs will be used in the same proportion at all output levels and hence it is also called scale line. On the other hand, if the expansion path is curved, it implies that the inputs will be used in various proportions.

## RIDGE LINES

Ridge lines represent the points of maximum output from each input, given a fixed quantity of other input. On the ridge lines MPP is zero. Ridge lines represent the economic relevance (Figure 19.11). Beyond the ridge lines there is no economic relevance, since the inputs are not substitutable. They indicate the limits of substitution. Within the ridge lines the iso-quants have negative slope and hence the inputs are substitutable. Within the ridge lines the MPP of both the inputs is positive but decreasing. Beyond the lower ridge line and above the upper ridge line, MPP of inputs is negative. The upper ridge line OA passes through the portions of iso-quants, where they are vertical and the MRTS is infinite. **The lower ridge** line passes through the portions of iso-quants where they are horizontal and MRTS is zero.

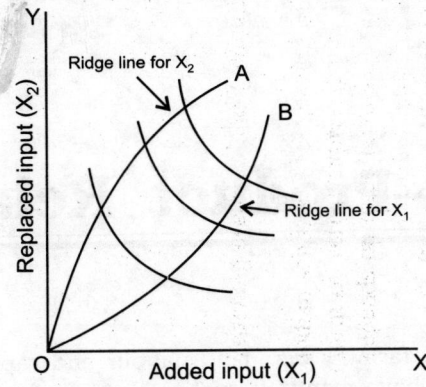

**Figure 19.11**   Ridges lines.

# CHAPTER 20

# Product–Product Relationship

The basic resources of farming *viz.*, land, labour and capital are scarce. However, these scarce resources have many alternative uses. Scarce resources can be used in producing different crops and livestock enterprises. Therefore, the farmers are faced with the management problem of what to produce. Farmers have to decide whether to produce crops alone or livestock alone or their combinations. The farmer should choose such a combination of crop and livestock enterprises that maximizes profits.

Product-product relationship deals with the allocation of resources among different crop and livestock enterprises. The objective of product-product relationship is profit maximization. In product-product relationship, resources are kept constant and the products are varied. The principle of product substitution explains the product-product relationship. Substitution and price ratios are used as choice indicators in the determination of optimum combination of enterprises. Algebraically the product-product relationship can be shown as:

$$Y_1 = f (Y_2, Y_3, Y_4 \ldots \ldots Y_n)$$

This expression reveals that a farmer is having an option of growing four or more crops in the same season in his operational holding. Then acreage proposed to be allocated under crop $Y_1$ is a function of acreage under crops $Y_2$, $Y_3$, $Y_4$ and $Y_n$.

## PRODUCTION POSSIBILITY CURVE

Production possibility curve can be drawn from production functions. Suppose a farmer has a limited input/ *i.e.* 5 acres of land. He has two alternatives i.e., the production of $Y_1$ product and $Y_2$ product. The problem here is as to how to allocate this limited input between two alternatives. The alternatives are using the entire 5 acres of land for the production of $Y_1$ alone or allotting 5 acres of land for production of $Y_2$ alone.

In between these two extreme possibilities, we have different options like allocation of 1 acre for $Y_1$ and 4 acres for $Y_2$, two acres for $Y_1$ and three acres for $Y_2$ and so on. If the entire area of 5 acres is allotted to the production of $Y_2$, 300 units of $Y_2$ would be obtained, while that of $Y_1$ is zero (Table 20.1). Analogously, if the total area of 5 acres is allotted to product $Y_1$, 300 units of $Y_1$ would be received but the production of $Y_2$ would be zero. Suppose a farmer wants to produce some quantity of $Y_1$, he has to withdraw some area of land from the production of $Y_2$. If one acre is allotted to $Y_1$, he will obtain 100 units of $Y_1$ and the remaining 4 acres in the production of $Y_2$ yield 250 units. Likewise 2 acres of land under $Y_1$ and 3 acres of land under $Y_2$ would yield an output of 150 and 190 units respectively. Other possibilities can be seen from the Table.

The different levels of land input and the corresponding levels of output of $Y_1$ and $Y_2$ represent two production functions. Production possibility curve is a convenient method of depicting two production functions on a single graph (Figure 20.1).

*Production possibility curve represents all possible combinations of two products ($Y_1$ and $Y_2$) that could be produced with given amount of resources.*

**TABLE 20.1  Possible Production Levels from the given Acreage of Land.**

| Area allotted between two products in acres | | Output | |
|---|---|---|---|
| $Y_1$ | $Y_2$ | $Y_1$ | $Y_2$ |
| 0 | 5 | 0 | 300 |
| 1 | 4 | 100 | 250 |
| 2 | 3 | 150 | 190 |
| 3 | 2 | 200 | 100 |
| 4 | 1 | 250 | 50 |
| 5 | 0 | 300 | 0 |

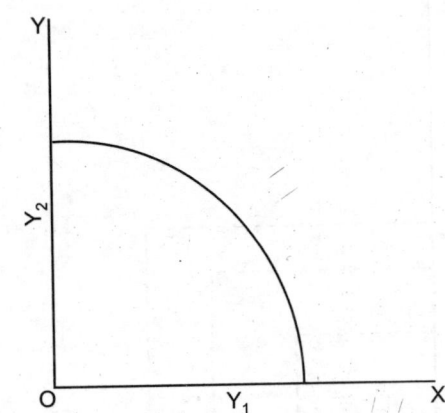

**Figure 20.1**  Production possibility curve.

Production possibility curve is also known as iso-resource curve or iso-factor curve, since all the combinations of two products require the same amount of resources. Production possibility curve presents the producer all the production opportunities available with a given amount of resources and hence it is called opportunity curve. The slope of the production possibility curve indicates the trade off between the two products. It indicates the rate at which one product is transformed into another product. Therefore it is called transformation curve. It is also a frontier because the limited resources cannot help to produce anything beyond production possibility curve. It demarcates what is possible given the available quantity of inputs. The area under the production possibility curve including the axes is called the feasible set or the attainable set of outputs.

## Characteristics of Production Possibility Curve

1. It is concave to the origin.

2. Slope of production possibility curve indicates the marginal rate of product substitution, and

3. Change in input levels, shifts the production possibility curve.

## RELATIONSHIPS AMONG THE PRODUCTS

These relationships are of different forms *viz.*, joint products, complementary products, supplementary products, competitive products and antagonistic products.

### Joint Products

Products which are produced from the same production process are called joint products. The two products derived through the production process are combined in fixed proportions. Production of one without the other is not possible. In agriculture almost all products are joint products. The proportion of the joint products can be altered or manipulated through research break-through in the long run. The examples are: paddy and straw, cotton lint and cotton seed, meat and wool, *etc.* Production possibility curve for joint products can be seen as a point for given quantity of resources. The points are as many as the levels of resources (Figure 20.2).

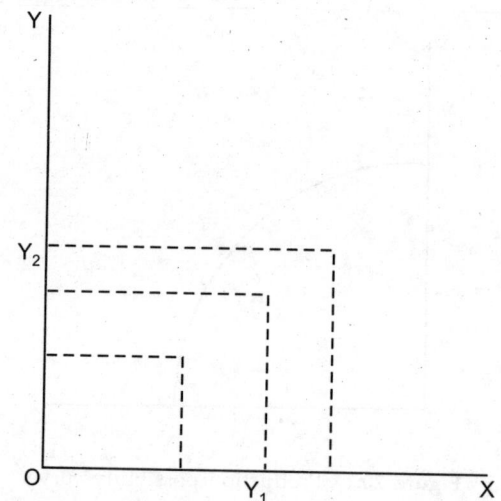

**Figure 20.2** Production possibilities for joint products.

### Complementary Products

The products are complementary, if an increase in one product causes an increase in the other product, when the total quantity of inputs used on the two products are held constant. Similarly, a decrease in the production of one product results in the decrease in the production of other product. They do not compete for the resources. One of the products contributes an element of production required by another thereby helping each other in production. An example that can be cited here is rice succeeding a legume crop. The legume fixes nitrogen thereby improving the soil fertility for the next crop. Similarly, paddy and livestock are complementary as paddy crop provides straw to livestock and livestock in turn makes the availability of farmyard manure to the paddy crop. Here these two contribute to their mutual production.

The complementary products would become competitive, when large quantities of resources are diverted to one product, affecting the production of the other. The marginal rate of product substitution is positive. The production possibility curve in Figure 20.3 presents the range of complementary of one product to the other product.

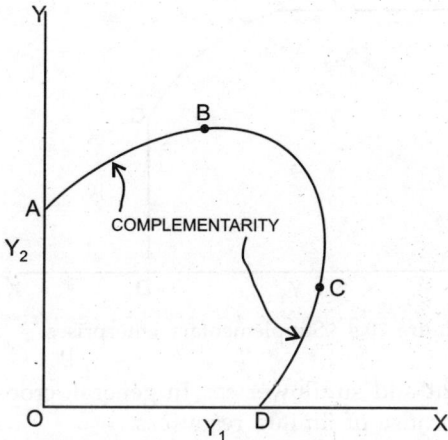

**Figure 20.3** Complementary enterprises.

As can be seen from the figure that $Y_1$ is complementary to $Y_2$ between A and B, while $Y_2$ is complementary to $Y_1$ between C and D. between the points B and C, the two products become competitive. Farmers can take the advantage of complementarity by producing both the products till they become competitive.

## Supplementary Products

Two products become supplementary, if the quantity of one product can be increased without increasing or decreasing the quantity of the other product. They do not compete for the resources. Instead they make better utilization of resources, which are being unutilized by one enterprise. They together add to the income on the farm. Crop production is seasonal in nature, and during off-season the resources are slack. They can be better utilized by adding supplementary enterprises *viz.*, a small dairy unit or poultry unit or piggery unit. A farmer should take best advantage of the products by producing both of them till they become competitive. The marginal rate of product substitution is zero. Production possibility curve for supplementary enterprises is shown in Figure 20.4. The product $Y_1$ can be increased up to AB without affecting the production of $Y_2$. If it is further increased the two become competitive. It can be seen in the diagram that the two products are competitive between the points B and C.

## Competitive Products

Two products are said to be competitive, when increase or decrease in the level of production of one results in decrease or increase in the level of production of another, given the fixed amount of resources. The marginal rate of product substitution between the products is therefore, negative. Most of the decisions regarding the selection of products involve competing products. The examples are, paddy and sugarcane, paddy

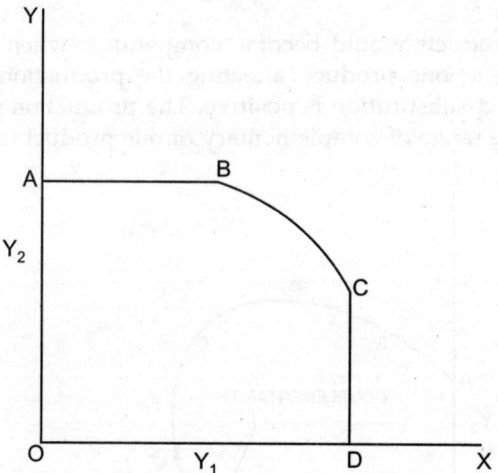

**Figure 20.4** Supplementary enterprises.

and groundnut, groundnut and sunflower *etc.* In general, crops grown in the same season are competitive because of limited resources.

### Antagonistic Products

The presence of one enterprise affects returns of another enterprises. Example: Paddy cultivation and aquaculture

## MARGINAL RATE OF PRODUCT SUBSTITUTION

Like factors, products also substitute each other. *The absolute amount, by which one product is decreased in order to gain another product by a unit is called marginal rate of product substitution.*

$$\text{Marginal rate of substitution} = \frac{\text{Number of units of replaced product}}{\text{Number of units of added product}}$$

$$MRS_{Y_1 Y_2} = \frac{\Delta Y_2}{\Delta Y_1}$$

Marginal rate of product substitution of $Y_1$ for $Y_2$ implies that the amount of $Y_2$ to be given up in order to gain $Y_1$ by one unit.

## TYPES OF PRODUCT SUBSTITUTION

The shape of the production possibility curve depends upon the type of product substitution. Competitive enterprises substitute either at constant rate or at increasing rate or at decreasing rate.

### Constant Rate of Substitution

Two products substitute at constant rate when a unit increase in the production of one replaces the same amount of another product throughout the process of substitution. In other words, a constant amount of replaced product is sacrificed in order to gain added product by one unit. Constant rate of substitution occurs when one of the production functions has an elasticity greater than one (increasing returns), the other has an elasticity of less than one or both $Y_1$ and $Y_2$ production functions have stages of both increasing and decreasing returns. The production possibility curve is linear when products substitute at constant rate. When we find two products substituting at constant rate, the production of only one product is economical based on the relative prices of the two products. This is a case of specialization. The example here is two varieties of the same farm commodity. The relevant data are presented in Table 20.2.

When we shift from A combination to B combination, $Y_1$ is increased by one unit and $Y_2$ is decreased by 6 units and the marginal rate of product substitution is 6. i.e., we need to reduce 6 units of $Y_2$ to increase $Y_1$ by one unit. Similarly, when we shift from combination B to C, C to D, D to E and E to F the amount of $Y_2$ to be given up is same. Figure 20.5 indicates the products substituting at constant rate.

**TABLE 20.2 Two Competitive Products Substituting at Constant Rate.**

| Combination | $Y_1$ | $Y_2$ | $\Delta Y_1$ | $\Delta Y_2$ | MRS of $Y_1$ for $Y_2$ |
|---|---|---|---|---|---|
| A | 0 | 60 | | | |
| | | | 1 | 6 | 6/1 = 6 |
| B | 1 | 54 | | | |
| | | | 1 | 6 | 6/1 = 6 |
| C | 2 | 48 | | | |
| | | | 1 | 6 | 6/1 = 6 |
| D | 3 | 42 | | | |
| | | | 1 | 6 | 6/1 = 6 |
| E | 4 | 36 | | | |
| | | | 1 | 6 | 6/1 = 6 |
| F | 5 | 30 | | | |

$$\frac{\Delta_1 Y_2}{\Delta_1 Y_1} = \frac{\Delta_2 Y_2}{\Delta_2 Y_1} = \text{........} = \frac{\Delta_n Y_2}{\Delta_n Y_1}$$

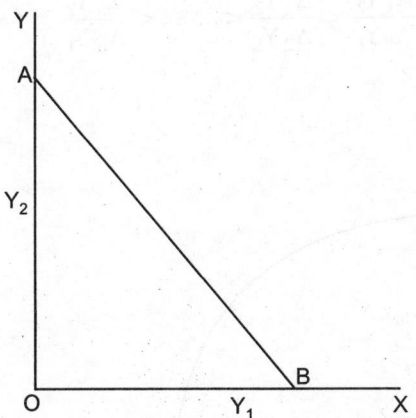

**Figure 20.5** Constant rate of product substitution.

## Increasing Rate of Substitution

Two products substitute at increasing rate when increase in one product requires larger and larger sacrifice in terms of another product. This type of substitution occurs when the production function of each independent product exhibits decreasing returns. Substitution of this *nature is more common in agricultural production* as the diminishing marginal resource productivity is a general situation in agriculture. Production possibility curve is concave to the origin when products substitute at increasing rate. The examples here are, all the crops grown in the same season *viz.*, paddy and sugarcane, groundnut and sunflower, paddy and groundnut *etc.* When products substitute at increasing rate it is economical to produce a combination of products. The general pattern of production is diversification. An hypothetical example of increasing rate of product substitution is presented in Table 20.3.

Shifting from combination A to combination B, results in increase in $Y_1$ by 8 units and decrease in $Y_2$ by 15 units. Marginal rate of substitution is 1.88. It means 1.88 units of $Y_2$ are to be sacrificed to gain $Y_1$ by one unit. When we shift from B to C, C to D, and D to E, the amount of $Y_2$ to be foregone is successively increasing as indicated by the increasing marginal rate of product substitution. The graphical representation is shown in Figure 20.6.

**TABLE 20.3 Two Competitive Products Substituting at Increasing Rate.**

| Combination | $Y_1$ | $Y_2$ | $\Delta Y_1$ | $\Delta Y_2$ | MRS of $Y_1$ for $Y_2$ |
|---|---|---|---|---|---|
| A | 0 | 75 | | | |
| | | | 8 | 15 | 1.88 |
| B | 8 | 60 | | | |
| | | | 8 | 16 | 2.0 |
| C | 16 | 44 | | | |
| | | | 8 | 18 | 2.25 |
| D | 24 | 26 | | | |
| | | | 8 | 26 | 3.25 |
| E | 32 | 0 | | | |

$$\frac{\Delta_1 Y_2}{\Delta_1 Y_1} < \frac{\Delta_2 Y_2}{\Delta_2 Y_1} < \cdots\cdots < \frac{\Delta_n Y_2}{\Delta_n Y_1}$$

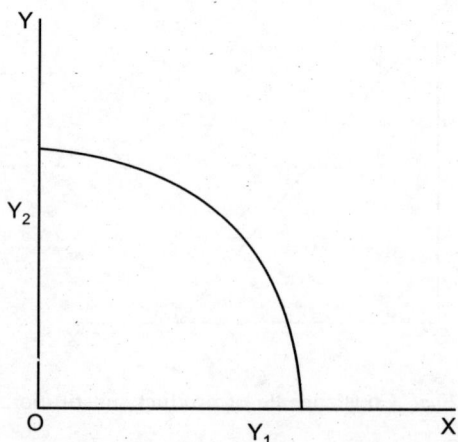

**Figure 20.6** Increasing rate of product substitution.

## Decreasing Rate of Substitution

Two products substitute at decreasing rate when increase in one product requires lesser and lesser reduction in another product. This type of substitution is observed when the production functions of both the products exhibit increasing returns. This type of substitution is very rare in production process, because increasing returns are seen in I stage of production which is irrational. It is economical to produce only one of the products depending upon relative prices of the products. The general pattern of production is specialization. The production possibility curve is convex to the origin. An hypothetical example is presented in Table 20.4.

Shifting from combination A to combination B results in increase in $Y_1$ by 2 units and decrease in $Y_2$ by 16 units. It means 8 units of $Y_2$ are to be sacrificed to gain $Y_1$ by one unit. When we shift from B to C, C to D and D to E, the amount of $Y_2$ to be forgone is successively decreasing as indicated by the decrease in marginal rate of product substitution. The graphical representation is found in Figure 20.7.

**TABLE 20.4 Two Competitive Products Substituting at Decreasing Rate.**

| Combination | $Y_1$ | $Y_2$ | $\Delta Y_1$ | $\Delta Y_2$ | MRS of $Y_1$ for $Y_2$ |
|---|---|---|---|---|---|
| A | 0 | 43 | | | |
| | | | 2 | 16 | 8 |
| B | 2 | 27 | | | |
| | | | 2 | 12 | 6 |
| C | 4 | 15 | | | |
| | | | 2 | 9 | 4.5 |
| D | 6 | 6 | | | |
| | | | 2 | 6 | 3 |
| E | 8 | 0 | | | |

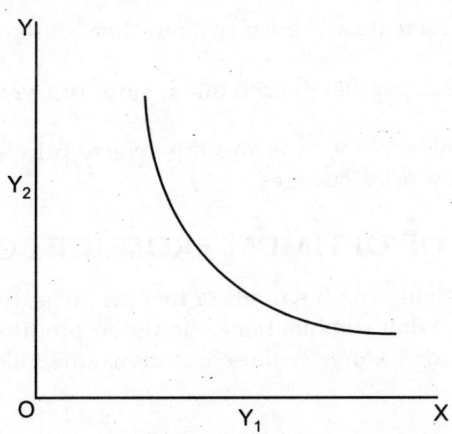

**Figure 20.7** Decreasing rate of product substitution.

$$\frac{\Delta_1 Y_2}{\Delta_1 Y_1} > \frac{\Delta_2 Y_2}{\Delta_2 Y_1} > \cdots\cdots > \frac{\Delta_n Y_2}{\Delta_n Y_1}$$

# ISO-REVENUE LINE

*It is a line, which defines all possible combinations of two products, which would yield equal revenue.* Suppose we wish to obtain total revenue of Rs. 5000, when price of $Y_1$ ($P_{Y1}$) is Rs. 10 and price of $Y_2$ ($P_{Y2}$) is Rs. 20, the expected revenue of Rs. 5000 could be earned by producing 500 units of $Y_1$ or 250 units of $Y_2$. Similarly, 300 units of $Y_1$ and 100 units of $Y_2$ or 100 units of $Y_1$ and 200 units of $Y_2$ would help to earn the same revenue. By plotting these two extreme points of 500 units of $Y_1$ and 250 units of $Y_2$ and by joining these two points, we get the iso-revenue line (Figure 20.8).

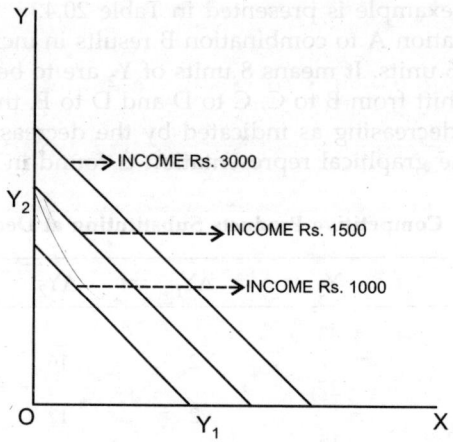

**Figure 20.8**  Iso revenue line.

## Characteristics of Iso-revenue Line

1. Iso-revenue line is a straight line, as the output prices do not change with the quantity of the output sold.
2. As the total revenue increases, the iso-revenue line shifts upwards and moves away from the origin.
3. The iso-revenue lines are parallel to each other, since price ratio remains constant, and
4. The slope of the iso-revenue line indicates the inverse price ratio of the products. The slope is affected by price changes.

# DETERMINATION OF OPTIMUM PRODUCT COMBINATION

To get the revenue maximizing combinations of two products, two relevant questions need to be answered *viz.*, what combinations should be produced and how can that combination be determined. To answer these questions, the following methods need to be examined.

## Tabular Method

Given the output combinations and prices, the total revenue of each output combination is computed as presented in Table 20.5. The output combination, which yields maximum revenue, is selected. In the given example, the combination of 16 quintals of groundnut and 44 quintals of paddy gives the maximum income of Rs. 39,600.

TABLE 20.5 Determination of Optimum Product Combination.

| Groundnut ($Y_1$) (quintals) | Paddy ($Y_2$) (quintals) | $PY_1$ @ Rs. 1100 | $PY_2$ @ Rs. 500 | Total income (Rs.) |
|---|---|---|---|---|
| 0 | 75 | 0 | 37,500 | 37,500 |
| 8 | 60 | 8,800 | 30,000 | 38,800 |
| 16 | 44 | 17,600 | 22,000 | 39,600 |
| 24 | 26 | 26,400 | 13,000 | 39,400 |
| 32 | 0 | 35,200 | 0 | 35,200 |

## Algebraic method

*Step* 1 Compute marginal rate of substitution between products

$$MRS = \frac{\text{Number of units of replaced product}}{\text{Number of units of replacing product}}$$

$$MRS_{Y_1Y_2} = \frac{\Delta Y_2}{\Delta Y_1}$$

Here, $Y_1$ is replacing product and $Y_2$ is replaced product

$$MRS\ Y_2Y_1 = \frac{\Delta Y_1}{\Delta Y_2}$$

Here, $Y_2$ is replacing product and $Y_1$ is replaced product

*Step* 2   Compute price ratio (PR)

$$PR = \frac{\text{Price per unit of replacing product}}{\text{Price per unit of replaced product}}$$

$$PR = \frac{P_{Y_1}}{P_{Y_2}} \text{ (or) } \frac{P_{Y_2}}{P_{Y_1}}$$

*Step* 3 Finding out the optimum combination of products by equating MRS with price ratio.

$$\frac{\text{Number of units of replaced product}}{\text{Number of units of replacing product}} = \frac{\text{Price per unit of replacing product}}{\text{Price per unit of replaced product}}$$

$$= \frac{\Delta Y_2}{\Delta Y_1} = \frac{P_{Y_1}}{P_{Y_2}} \text{ (or) } \frac{\Delta Y_1}{\Delta Y_2} = \frac{P_{Y_2}}{P_{Y_1}}$$

## Graphic Method

To determine the optimum combination of products through graphic method, production possibility curve and iso-revenue line are depicted on the same graph (Figure 20.9). Slope of production possibility curve indicates the marginal rate of substitution and that of iso-revenue line represents the inverse price ratio of the products. The optimum combination products are at the point where the iso-revenue line is tangent to the production possibility curve. At the point of tangency the slopes of production possibility curve and iso-revenue line are the same.

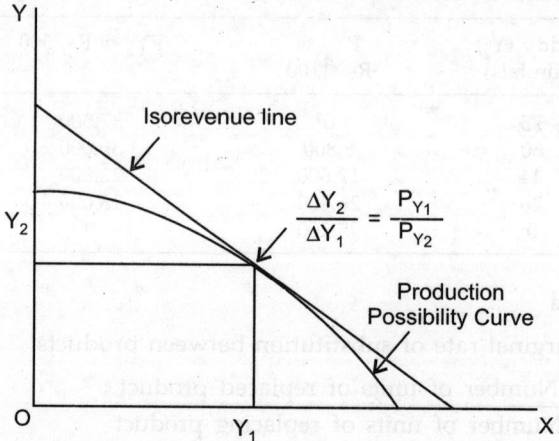

**Figure 20.9** Optimum combination of products.

## Optimal Enterprise Combination – *Numerical example*

To explain the production possibility curve (PPC) we assume two production functions in which $Q_1$ and $Q_2$ are dependent variables and $X_1$ independent variable as a fixed input, say land. These estimated production functions are specified as

$$Q_1 = f(X_1) \tag{3.1}$$

$$Q_2 = f(X_1) \tag{3.2}$$

$Q_1$ and $Q_2$ are two crop products.

Let these equations he estimated as

$$Q_1 = 30\, X_1^{0.42} \tag{3.3a}$$

$$Q_2 = 40\, X_1^{0.52} \tag{3.3b}$$

Then,

$$MP_1 = 12.6\, X_1^{-0.58} \tag{3.4}$$

$$MP_2 = 20.8\, X_1^{-0.48} \tag{3.5}$$

Further we assume that $X_1$ is the fixed input, say land, which is 2 acres in extent and it is used to produce $Q_1$ and $Q_2$ crop products.

This condition is specified in the equation as

$$X_1^0 = X_1 Q_1 + X_1 Q_2 \tag{3.6}$$

Rate of product transformation (RPT), by definition is marginal rate of substitution of one product to the other *i.e.*, $Q_1$ for $Q_2$ or $Q_2$ for $Q_1$ and this is measured as ratio of first derivative of equations (3.1) and (3.2).

So,

$$RPT = \frac{MP_1 Q_1}{MP_2 Q_2}$$

$$= \frac{12.6 X_1^{-0.58} Q_1}{20.8 X_1^{-0.48} Q_2}$$

Let us assume the price of $Q_1$ product as Rs. 500 and price of $Q_2$ product as Rs. 300. Their price ratio is $500/300 = 1.67$. Assume that $Q_1 = 20$ units and then obtain the value of $X_1$ using equation (3.3a)

$$Q_1 = 30 \, X_1^{0.42}$$

$$20 = 30 \, X_1^{0.42}$$

$$X_1^{0.42} = 20 / 30 = 0.67$$

$$X_1 = (0.67)^{2.38} = 0.39 \text{ acres}$$

Thus 0.39 acres is used in production of $Q_1$

Then $2.00 - 0.39 = 1.61$ acres are used in the production of $Q_2$.

With regard to the equality of RPT and price ratio, it is

$$= \frac{12.6(0.39)^{-0.58}}{20.8(1.61)^{-0.48}} \frac{Q_1}{Q_2} = \frac{500}{300}$$

$$300(12.6 \times 1.73)Q_1 = 500(20.8 \times 0.80)Q_2$$

$$6{,}539.4Q_1 = 8{,}320Q_2, \text{ Assumed value of } Q_1 = 20$$

$$Q_2 = \frac{6{,}539.4 \times 20}{8{,}320} = 15.62$$

$$\text{Optimal values} = Q_1^* = 20 \text{ units}$$

$$= Q_2^* = 15.72 \text{ units}$$

$$X_1 \text{ for } Q_1 = 0.39 \text{ acre}$$

$$X_1 \text{ for } Q_2 = 1.61 \text{ acre}$$

$$\text{RPT} = \text{Price ratio}$$

$$= \frac{12.6 \times 1.73 \times 20}{20.8 \times 0.8 \times 15.72} = \frac{500}{300} = 1.67$$

## OUTPUT EXPANSION PATH

Isoclines are the lines or curves that pass through the points of equal slope on a production possibility map. Output expansion path is that isocline, which connects the points of optimal product combinations on the successive production possibility curves (Figure 20.10).

## RIDGE LINES

Ridge lines are the borderlines that separate ranges of competition and product complementarity. In the Figure 20.10, OA and OB are the ridgelines. That portion of production possibility curves, falling between OA and OB have negative slope indicating the existence of competition between the products, while that portion of the production possibility curves falling outside OA and OB, have positive slopes indicating complementarity. Along the path OA, the marginal rate of substitution of $Y_1$ for $Y_2$ is zero, while along the path OB, the marginal rate of substitution of $Y_1$ for $Y_2$ is infinite.

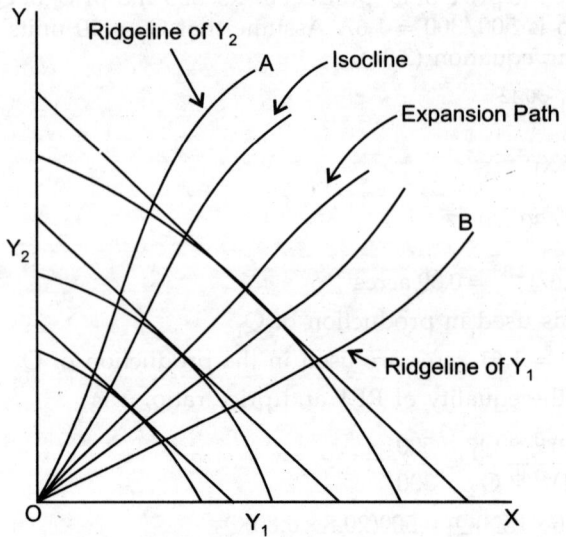

**Figure 20.10** Expansion path and ridge lines.

**Summary of Three Basic Production Relationships.**

| | Relationships | |
| --- | --- | --- |
| Factor–product | Factor–factor | Product-product |
| It is concerned with | It is concerned with resource use efficiency | resource combination and |
| Resource allocation | | |
| | Substitution | |
| Objective is optimization | Cost minimization | Profit maximization |
| of resource use | | |
| Management problem: | How to produce | What to produce |
| How much to produce | | |
| Determination of optimum input to use and optimum output to produce | Least cost combination of resources | Optimum combination of enterprises |
| **Choice indicator** | | |
| Price ratio | Substitution ratio and | Substitution ratio and price ratio          price |
| ratio | | |
| Single variable production function | Output is constant and inputs are varied | Input is constant and products are varied |
| Law of diminishing returns explains the factor–product relationship $Y = f(X_1 \mid X_2, X_3 \dots X_n)$ | Principle of factor substitution, explains factor–factor relationship $Y = f(X_1, X_2)$ | Principle of product substitution explains product-product relationship $Y_1 = f(Y_2, Y_3 \dots Y_4)$ |

# CHAPTER 21

# Returns to Scale

It refers to the change in output as a result of a given proportionate change in all the factors of production simultaneously. When all the factors or inputs involved in a production process are increased or decreased simultaneously, in a certain fixed proportion, the response of output to such an increase or decrease in the input levels, is explained through the concept of returns to scale. Returns to scale is a long run concept as all the variables are varied in quantity. Returns to scale are increasing or constant or decreasing depending on whether proportionate simultaneous increase of input factors results in an increase in output by a greater or same or small proportion. Returns to scale is illustrated with the help of hypothetical data in Table 21.1.

TABLE 21.1 Returns to Scale.

| Labour (L) | Capital (K) | Total output (Q) | Increment in output | Nature of returns to scale |
|---|---|---|---|---|
| 0 | 0 | 0 | | |
| | | | 8 | |
| 1 | 1 | 8 | | |
| | | | 10 | Increasing |
| 2 | 2 | 18 | | |
| | | | 10 | |
| 3 | 3 | 28 | | |
| | | | 10 | |
| 4 | 4 | 38 | | |
| | | | 10 | |
| 5 | 5 | 48 | | Constant |
| | | | 10 | |
| 6 | 6 | 58 | | |
| | | | 10 | |
| 7 | 7 | 68 | | |
| | | | 8 | |
| 8 | 8 | 76 | | |
| | | | 6 | Decreasing |
| 9 | 9 | 82 | | |
| | | | 4 | |
| 10 | 10 | 86 | | |

From the table it can be seen the variation in total output for changing proportion of L and K. Initially, when the input proportion is changing, output is changing by an increasing proportion. This is increasing returns to scale. This trend is seen in the use of L and K up to the ratio of 3 : 3. Constant increase in output is found till the proportion of L and K is extended up to the ratio of 7 : 7. This is constant returns to scale. The use of L and K in the proportion of 8 : 8 onwards reveals the decreasing returns to scale. This concept is graphically presented in Figure 21.1.

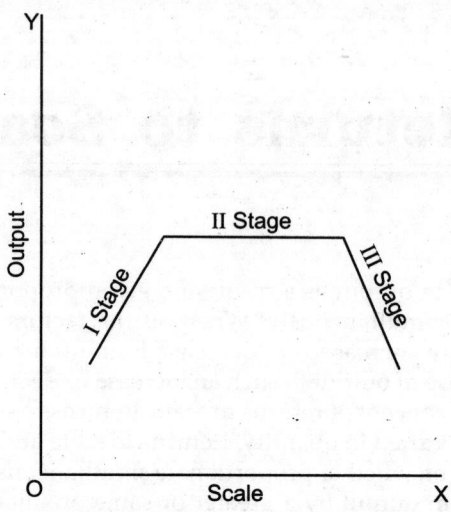

**Figure 21.1** Returns to scale.

Let the estimated Cobb-Douglas production function be represented as

$$\hat{Y} = 0.32 \, X_1^{-0.0681} \, X_2^{0.6669} \, X_3^{0.1202} \, X_4^{0.1050}$$

Y = Crop output in quintals
$X_1$ = Human labour in man days
$X_2$ = Fertilizers in kg
$X_3$ = Manures in tonnes
$X_4$ = Pesticides in litres

In Cobb-Douglas production function the returns to scale is obtained by the

summation of elasticity coefficients of the independent variables i.e., $\sum\limits_{i=1}^{n} b_i$

If $\Sigma b_i > 1$ it is increasing returns to scale
If $\Sigma b_i = 1$ it is constant returns to scale and
If $\Sigma b_i < 1$ it is decreasing returns to scale

The value of $\Sigma b_i$ from the estimated equation is 0.824, which indicates the prevalence of decreasing returns to scale.

## DIFFERENCES BETWEEN THE LAW OF VARIABLE PROPORTIONS AND RETURNS TO SCALE

The distinguishing features between law of variable proportions and returns to scale are presented in Table 21.2.

**TABLE 21.2   Differences Between the Law of Variable Proportions and Returns to Scale.**

| S. No. | Law of variable proportions | Returns to scale |
|---|---|---|
| 1 | Describes the behaviour of output when | Examines the behaviour of one input is varied output, when all inputs are varied at the same time. |
| 2 | Some factors of production are constant | All factors are varied |
| 3 | The proportion among factors varies | The proportion among the factors remains the same |
| 4 | It is a short run production function | It is a long run production function |
| 5 | Here increasing, constant and decreasing returns to a factor are observed | Here increasing, constant and decreasing returns to scale are observed |
| 6 | Increasing returns are due to the efficient utilization of fixed resources as a result of application of sufficient quantity of variable resource | Increasing returns to scale are due to scale economies of production |
| 7 | Optimum output is the result of best proportion among fixed and variable resources | The optimum output is the result of optimum size of the plant |
| 8 | The diminishing returns are due to over exploitation of fixed factor | Diminishing returns to scale are due to the operation of diseconomies of scale |
| 9 | $Y = f(X_1 \mid X_2, X_3 \ldots X_n)$ | $Y = f(X_1, X_2, X_3 \ldots X_n)$ |
| 10 | It is a reality | It is a myth |

# CHAPTER 22

# Production Functions

## Single Variable Production Functions

The production function for a single variable input and output is specified as

$$Q = f (X_1 | X_2 \ldots X_n) \tag{3.7}$$

Here, $X_1$ is variable input, $X_2 \ldots X_n$ are fixed inputs and Q is output.

Average product of $X_1$ is defined as

$$AP_1 = Q/X_1 \tag{3.8}$$

Due to the operation of the law of diminishing returns, $AP_1$ must decrease as $X_1$ increases

Marginal product of $X_1$

$$MP_1 = dQ/dX_1 \tag{3.9}$$

Elasticity of $X_1$ input

$$E_1 = \frac{dQ}{Q} \Big/ \frac{dX_1}{X_1} \tag{3.10}$$

$$= \frac{dQ}{dX_1} \cdot \frac{X_1}{Q} = \frac{dQ}{dX_1} \Big/ \frac{Q}{X_1}$$

Elasticity of production $(E_p) = MP_1/AP_1$ $\hspace{2cm}$ (3.11)

Let us assume the estimated production function with a single variable input as

$$\hat{Q}_1 = 20 + 150\, X_1 - 1.2\, X_1^2 + e \tag{3.12}$$

Equation (3.12) is called quadratic production function or second degree polynomial function. With single variable input using equations (3.8) (3.9) and (3.11) we can derive $AP_1$, and $MP_1$

$$\hat{Q} = 20 + 150\, X_1 - 1.2\, X_1^2$$

$$AP_1 = Q/X_1$$

$$= 20/X_1 + 150 - 1.2\, X_1 \tag{3.13}$$

$$MP_1 = dQ/dX_1 = 0 + 150 - 2.4\, X_1$$

$$= 150 - 2.4 \, X_1$$

$$E_p = \frac{150 - 2.4 \, X_1}{20 / X_1 + 150 - 1.2 \, X_1} = \frac{MP}{AP_1} \tag{3.14}$$

Taking derivative of $MP_1$ (second derivative)

$$\frac{dQ^2}{d^2 X_1} = -2.4$$

Since the value is negative, diminishing returns are prevailing. Assuming different values for $X_1$, we can work out $AP_1$, $MP_1$ and $E_p$ and $\hat{Q}$ and these values are given in Table 22.1.

TABLE 22.1 Calculation of $AP_1$, $MP_1$ and Q for the Given Production Function.

| Assumed values of $X_1$ (kg) | $AP_1$ (kg) | $MP_1$ (kg) | $E_p$ | $\hat{Q}$ (Quintal) |
|---|---|---|---|---|
| 1 | 168.80 | 147.6 | 0.87 | 1.69 |
| 10 | 140.00 | 126.0 | 0.90 | 14.00 |
| 20 | 127.00 | 102.0 | 0.80 | 25.40 |
| 30 | 114.67 | 78.0 | 0.68 | 34.40 |
| 40 | 102.50 | 54.0 | 0.53 | 41.00 |
| 50 | 90.40 | 30.0 | 0.33 | 45.20 |
| 60 | 78.33 | 6.0 | 0.07 | 47.0 |
| 70 | 66.28 | -18.0 | -0.27 | 46.40 |
| 80 | 54.25 | -42.0 | -0.77 | 43.40 |

Efficiency of the input use:

This occurs when $MP_1 = P_1 / P_q$

Where,

$P_1$ = Price per unit of input

$P_q$ = Price per unit of output

The details of optimal input and optimal output for corresponding assumed price ratios are furnished in Table 22.2. As the price ratio is declining optimal input levels and optimal output levels are increasing. Please note that the optimal output is almost remaining at the same level, when the price ratio is reduced from 4 to 0.25. In contrast, we notice slight increase in the optimal input level for the corresponding price ratios.

TABLE 22.2 Optimal Level of Input and Output for the Given Production Function.

| Price ratio ($P_1/P_q$) | $X_1^*$ (Optimal input) (kg) | $Q^*$ (Optimal output) (Quintal) |
|---|---|---|
| 8 | 59.17 | 46.94 |
| 7 | 59.58 | 46.97 |
| 6 | 60.00 | 47.00 |
| 5 | 60.41 | 47.02 |
| 4 | 60.83 | 47.04 |
| 3 | 61.25 | 47.06 |
| 2 | 61.67 | 47.07 |
| 1 | 62.08 | 47.07 |
| 0.5 | 62.29 | 47.07 |
| 0.25 | 62.40 | 47.07 |

# PRODUCTION FUNCTION WITH TWO VARIABLE INPUTS

The estimated production function with two variables $X_1$ and $X_2$ is given as

$$\hat{Q} = 21.27 + 10.3\,X_1 + 4.3\,X_2 - 0.32\,X_1^2 - 0.44\,X_2^2 + 0.13\,X_1X_2 \tag{3.15}$$

This is second degree polynomial response function with two variables.

$MP_1$ and $MP_2$ are estimated for the given equation (3.15)

$$MP_1 = \partial Q/\partial X_1 = 0 + 10.3 - 0.64\,X_1 + 0.13\,X_2 \tag{3.16}$$

$$MP_2 = \partial Q/\partial X_2 = 0 + 4.3 - 0.88\,X_2 + 0.13\,X_1 \tag{3.17}$$

To get maximum profit, $MP_1$ and $MP_2$ should be equated with their respective price ratios.

$$10.3 - 0.64\,X_1 + 0.13\,X_2 = P_1/P_q \tag{3.18}$$

$$4.3 + 0.13\,X_1 - 0.88\,X_2 = P_2/P_q \tag{3.19}$$

Assume the price ratios for equations (3.18) and (3.19) and solve them simultaneously to get optimal level of $X_1$ and $X_2$

$P_1/P_q$ is assumed as 0.5

$P_2/P_q$ is assumed as 1.20

To solve for the value of $X_2$, multiply equation (3.18) by 0.13 and equation (3.19) by 0.64

$$1.34 + 0.08 + 0.02\,X_2 = 0.07 \tag{3.20}$$

$$2.75 - 0.08\,X_1 - 0.56\,X_2 = 0.77 \tag{3.21}$$

$$\overline{\quad 4.09 \qquad 0.54\,X_2 = 0.84 \quad}$$

$$-0.54\,X_2 = 0 - 4.09 + 0.84$$
$$X_2 = 6.02$$

Substituting $X_2 = 6.02$ in equation (3.18) we get
$$X_1 = 17.38$$
$$X_1^* = 17.38, \quad X_2^* = 6.02$$

Here $X_1^*$ and $X_2^*$ are optimal levels of inputs

Substitute $X_1^* = 17.38$ and $X_2^* = 6.02$ in equation (3.15) to find out the optimal level of Q.

$Q^* = 127.1$ units

Here, $X_1^*$ and $X_2^*$ are the economic optimal levels of inputs and give maximum profit at that level.

The unconstrained profit function for equation (3.15) is specified as

$$= P_q\,Q - (P_1X_1 + P_2X_2) \tag{3.22}$$

$P_1$ = Rs. 6 per unit of $X_1$ input

$P_q$ = Rs. 12 per unit of output

$P_2$ = Rs. 14.4 per unit of $X_2$ input

$X_1^*$ = Rs. 17.38

$X_2^*$ = Rs. 6.02

$Q$ = 127.1 units of output

Substituting the above values in equation (3.22), we get the amount of profit.

Profit = Rs. 1,334.23

## Cobb–Douglas Production Function or Power Function or Double Log Production Function or Log-Log Production Function or Log-Linear Production Function

This is specified as:

$$\hat{Q}_i = A \prod_{j=1}^{m} X_{ij}^{bj} e^{Ui}$$

Where

$Q_i$ = Yield of crop in quintals per unit area for $i^{th}$ farm

A = Constant (it is also called efficiency parameter)

j = 1 to m inputs (n > m)

$X_{ij}$ = $j^{th}$ input applied by $i^{th}$ farm per unit of land (it is better to consider all the inputs in physical units only).

$b_j$ = Elasticity coefficient of $j^{th}$ input (it measures response of the input on the crop yield in percentage).

e = Napier base, i.e. 2.718

$\pi$ = Multiplication symbol.

$U_i$ = Stochastic (random), disturbance term, or error term.

When the estimated equation satisfies all the normal assumptions and its signs of coefficients and $R^2$ are according to theoretical expectations and significant, then we say the estimated equation is the best fitted equation to the given data.

It is a non-linear production function and becomes linear production function, when the variables are transformed into logarithms.

$$\text{Log } Y = \log A + b_1 \log X_1 + b_2 \log X_2 + \dots\dots + b_n \log X_n + \mu$$

The marginal productivity of the factor is given by

$$\left. \begin{array}{l} \dfrac{\partial Y}{\partial X_1} = b_1 \cdot \dfrac{\overline{Y}}{X_1} ; \dfrac{Y}{X_1} = \text{APP of } X_1 \\[3mm] \dfrac{\partial Y}{\partial X_2} = b_2 \cdot \dfrac{\overline{Y}}{X_2} ; \dfrac{Y}{X_2} = \text{APP of } X_2 \end{array} \right\} \begin{array}{l} \text{Here } \overline{Y}, \overline{X}_1, \overline{X}_2 \text{ are} \\ \text{their respective geometric} \\ \text{means} \end{array}$$

Elasticity of substitution ($\sigma$) is defined as percentage change in output with respect to percentage change in input.

$$\sigma = \frac{d\left(\dfrac{X_1}{X_2}\right)\left(\dfrac{X_2}{X_1}\right)}{d\left(\dfrac{X_1}{X_2}\right)\left(\dfrac{b_2}{b_1}\right)} \left(\dfrac{b_2}{b_1}\right)\left(\dfrac{X_1}{X_2}\right) = 1$$

In the Cobb-Douglas function the value of $b_i$ lies between 0 and 1. This means that 1 per cent increase in any input, holding other variables constant, would increase the output by less than 1 per cent. Returns to scale can be estimated directly from the bi values. If $\Sigma b_i > 1$, it implies increasing returns to scale. If it is exactly one, it indicates constant returns to scale and a value less than one indicates decreasing returns to scale. This should be based on test of significance of $\Sigma b_i$.

## Derivation of iso-quant from C-D function

$$Y = A \, X_1^{b1} \, X_2^{b2} \ldots \ldots X_n^{bn} \qquad (3.23)$$

Setting $Y = Y_0$ in equation (3.23) and solving equation (3.23) for $X_1$ in terms of $X_2$, the iso-quant equation is defined as

$$X_1 = \left[ \frac{Y_0}{X_2^{b2}\left(AX_3^{b3} \cdots X_n^{bn}\right)} \right]^{\frac{1}{b1}} \qquad (3.24)$$

Where,

$Y_0$ is a particular level of output defining the iso-quant.

The marginal rate of technical substitution (MRTS) of inputs is defined as the ratio of differentials of the two inputs or as the ratio of their marginal products.

$$\text{MRTS} = -\frac{\partial X_1}{\partial X_2} = \frac{\partial Y}{\partial X_2} \bigg/ \frac{\partial Y}{\partial X_2} = \frac{b_2}{b_1} \cdot \frac{X_1}{X_2} \qquad (3.25)$$

In terms of differential equation (3.25) is specified as

$$\partial Y = \frac{\partial Y}{\partial X_1} \cdot \partial X_1 + \frac{\partial Y}{\partial X_2} \cdot \partial X_2 = 0 \qquad (3.26)$$

## Advantages of Cobb-Douglas Production Function

1. It is popularly used in agricultural economics research because of its simple functional form which provides for easy computation.
2. It gives theoretically consistent and significant estimates for most of the variables used in the analysis of agricultural data.
3. Elasticities are directly measured.
4. The estimates of this function are mostly consistent with the principle of law of diminishing returns i.e. marginal productivity decreases as the input use increases
5. Returns to scale are directly estimated.
6. The inverse relationship that exists between marginal rate of substitution and factor proportions is easily computed from Cobb-Douglas function.

## Mitscherlich or Spillman Function

$Q = M - AR^{X1}$

$Q$ = Output

$M$, $A$ and $R$ = Constants

$X_1$ = Input

**Resistance function:** It is specified as

$$Q^{-1} = a_0 + \Sigma a_i \, (b_i + X_i)^{-1}$$

**Transcendental function:** It is specified as

$$Q_i = a \, \pi \, X_i^{bi} \, e^{C_i} \, i$$

# CONSTANT ELASTICITY OF SUBSTITUTION PRODUCTION FUNCTION (CESPF)

In the Cobb-Douglas production function, if elasticity of substitution ($\sigma$) is equal to 1 then, it gives constant returns to scale. Such production function is called Leontief input–output production function.

In the linear production function elasticity of substitution is infinite*. CESPE was formulated by Arrow, Chenery, Minhas and Solow. It takes the following form

$$Y = \gamma \left[ \delta K^{-\rho} + (1-\delta)L^{-\rho} \right]^{-1/\rho} \tag{3.27}$$

Where,

Y = Output

K = Capital

L = Labour

Here $\Upsilon$ is called efficiency parameter and it is equal to A in Cobb-Douglas production function.

$\delta$ is distribution parameter and its value lies between 0 and 1, $(0 \le \delta \le 1)$ and $\rho$ is substitution parameter and its limits range from –1 to $\alpha$, $(-1 \le \rho \le \alpha)$.

Elasticity of substitution $(\sigma) = 1/1+\rho$

The value of elasticity of substitution, $(\sigma) = \alpha < \alpha < 0$

If $\rho = 0$ and $\sigma = 1$, CESPF becomes Cobb-Douglas production function

If $\rho = -1$ and $\sigma = \alpha$, CESPF becomes linear production function

If $\rho = \alpha$ and $\sigma = 0$, CESPF takes the form of Leontief input-output production function.

Assuming constant returns to scale in the CES production function, we have marginal products of capital and labour as

$$\frac{\partial Y}{\partial L} = \sigma \gamma^{-\rho} \left( \frac{Y}{K} \right)^{1+\rho} \tag{3.28}$$

---

* $Y = a_1 X_1 + a_2 X_2$, There elasticity of substitution ($\sigma$) in this case is

$$\delta = \frac{d\left( \frac{X_1}{X_2} \right) / \frac{X_1}{X_2}}{d\left( \frac{a_2}{a_1} \right) / \frac{a_2}{a_1}}$$ Since $a_1$, $a_2$ are both constants, then $d\left( \frac{a_2}{a_1} \right) = 0$. This means denominator

becomes zero. Therefore, $\sigma = $ Infinite $(\infty)$.

$$\frac{\partial Y}{\partial L} = (1-\sigma)\gamma^{-\rho}\left(\frac{Y}{L}\right)^{1+\rho} \tag{3.29}$$

Dividing equation (3.28) by equation (3.29), we get the MRTS

$$\text{MRTS} = \frac{\partial L}{\partial K} = \frac{\sigma}{1-\sigma}\left(\frac{L}{K}\right)^{1+\rho} = \frac{P_K}{P_L}$$

Where,

$P_K$ = Price of capital

$P_L$ = Price of labour

CES production function can be estimated directly by using maximum likelihood technique or it can be indirectly estimated by using the relationship between average productivity of labour and its wage rate, which gives the value of elasticity of substitution as the coefficient for the wage rate.

$$\log Y = \frac{1}{1+\rho}\log P_L - \frac{1}{1+\rho}\log\left[\gamma^{-\rho}(1-\delta)\right] \tag{3.30}$$

$$\log P_L = (1+\rho)\log Y - \rho\log Y + \log(1-\delta) \tag{3.31}$$

$$C = \gamma^{-\rho}(1-\delta) \tag{3.32}$$

Where,

C = Constant

Using equations (3.30), (3.31) and (3.32) we get log-linear CESPF,

$$\log Y = \sigma \log P_L - \sigma \log C.$$

Here CES function implies a log linear relationship between average productivity of labour and wage rate.

Here $\sigma$ is the value of elasticity of substitution.

$$\sigma = \frac{d\left(\frac{K}{L}\right)}{d\left(\frac{P_L}{P_K}\right)} \cdot \frac{P_L L}{P_K K}$$

Using logarithms CESPF can be estimated as

$$\log Y = \log \gamma - \frac{1}{\rho}\log\left[\delta_1 T^{-\rho} + \delta_2 L^{-\rho} + (1-\delta_1-\delta_2)K^{-\rho}\right]$$

Where,

Y = Output in Rs.

T = Cropped area in ha.

L = Labour in Rs.

K = Interest on fixed capital in Rs.

For application and interpretation of the parameters of CES production function, please refer to Yotopoulos and Nugent (1976)[*].

# QUADRATIC PRODUCTION FUNCTION

## General Form

$$Y = a + bX - cX^2 \tag{3.33}$$

Where

Y = Yield

X = Input say, nitrogen

a = Yield due to fixed factors like phosphorous, potash, seed, irrigation, human labour *etc.*, which have been kept constant.

The simple quadratic equation with a minus sign before c denotes diminishing marginal returns. It allows both a declining and negative marginal productivity, but not both increasing and decreasing marginal products.

Sppose the estimated quadratic equation is

$$Y = 1{,}146 + 5.22\,X - 0.003X^2$$

-0.003 $X^2$ indicates that the curve is sloping downwards, which means that it shows diminishing rate of return. The optimum dose of N and Y (output) can be obtained as presented below:

$MPP_x$ of the total product curve is

$$dY/dX = 5.22 - 0.006X$$

If $P_x$ = Rs. 20/kg

and $P_y$ = Rs. 500/Qtl or Rs. 5/kg

Then,

$$5.22 - 0.006\,X = P_x/P_y = 20/5 = 4$$
$$= 5.22 - 0.006\,X = 4$$
$$= -0.006\,X = 4 - 5.22$$
$$= -0.006\,X = -1.22$$
$$X = 203.33 \text{ kg}$$

Optimum dose of nitrogen is 203.33 kg. Now substituting this value of X in the original equation

$$Y = 1{,}146 + 5.22\,(203.33) - 0.003(203.33)\,(203.33)$$
$$= 1{,}146 + 1{,}061.38 - 124.03$$
$$= 2{,}207.38 - 124.03$$
$$= 2{,}083.08 \text{ kg}$$

This is the output at the most profitable level of nitrogen application. Suppose we want to estimate the response of nitrogen, substitute the value of X in the equation

$$Y = bX - cX^2$$

---

[*] Yotopoulos P.A. and Nugent, J.B. Economics of Development. Empirical Investigation. Harper & Row, Pub. New York, 1976, pp. 47-70.

Then,

$$Y = 5.22 \ (203.33) - 0.003 \ (203.33)^2$$

$$= 1,061.38 - 124.03$$

$$= 937.35 \ \text{kg}$$

Therefore for 203.33 kg of nitrogen, the yield is 937.35 kg

# CONSTRAINED OUTPUT MAXIMIZATION–ALGEBRAIC DERIVATION

Let us assume that the Cobb-Douglas production function is in the following form

$$Q = BoL^a \ K^b$$

Bo = It is an intercept term and is called efficiency parameter, it measures the effect of technology on the output of crop, (Q)

L = Units of labour used in producing a crop

K = Units of capital used in producing a crop.

Marginal product of labour $(MP_L)$ is derived as

$$MP_L = \frac{\partial Q}{\partial L} = aBoL^{a-1} \ K^b$$

$$= a\left(BoL^a \ K^b\right)L^{-1}$$

Substitute $BoL^a \ K^b = Q$

$$= a\frac{Q}{L}$$

$$= a(AP_L)$$

Where,

$AP_L$ is average product of labour. The marginal product of capital $(MP_K)$ is defined as

$$MP_K = \frac{\partial Q}{\partial K} = bBoL^a \ K^{b-1}$$

$$= b(BoL^aK^b) \ K^{-1}$$

$$= b\frac{Q}{K}$$

$$MP_K = b \ AP_K$$

Where, $AP_K$ = average product of capital input.

$$MRTS \ L, K = \frac{\partial Q/\partial L}{\partial Q/\partial K} = \frac{a}{b} \bullet \frac{K}{L}$$

To determine the equilibrium conditions, we have to maximize the output subject to the cost constraints.

Maximize $\hat{Q} = f(L, K)$

Subject to $M° = wL + rk$

Where,

$M°$ = Money available in specified units, such as 1 unit= Rs. 100 or Rs. 1000, *etc.*

$r$ = Price (interest rate) per unit of capital

$W$ = Wage rate per unit of labour

We can solve the above problem using Lagrangean multiplier method. Here the constrained equation is rewritten as

(3.36)

$M° = wL + rk = 0$

Multiply the constrained equations (3.36) by constant term, $\lambda$, which is called Lagrangean multiplier.

(3.37)

$\lambda(M° = wL + rK) = 0$

If there is one constrained equation we should use one $\lambda$ and if there are two constraints, we should use two $\lambda$s. Here Lagrangean multiplers are undefined constants which are used for solving constrained equations.

Now let us write Lagrangean function as

(3.38)

$V = \hat{Q} + \lambda(M° = wL + rK)$

Maximization of the equation (3.38) implies maximization of output subject to cost constraint. To achieve this we have to take partial derivatives of equation (3.38) with respect to L, K and $\lambda$ and these should be made equal to zero and solve for L, K and $\lambda$.

(3.39)

$$\frac{\partial V}{\partial L} = \frac{\partial Q}{\partial L} + \lambda(-w) = 0$$

Then,

(3.40)

$$\frac{\partial V}{\partial L} = \frac{\partial Q}{\partial L} + \lambda(-r) = 0$$

Similarly,

(3.41)

$$\frac{\partial V}{\partial \lambda} = M° - wL - rK = 0$$

Solving equation (3.39) and (3.40) for $\lambda$ we get

(3.42)

$$\frac{\partial Q}{\partial L} = \lambda w$$

(3.43)

$$\lambda = \frac{\partial Q}{\partial L} \bigg/ w = \frac{MP_L}{w}$$

(or)

(3.44)

$$\lambda = \frac{\partial Q}{\partial L} \bigg/ r = \frac{MP_K}{r}$$

These two equations must be identical

$$i.e., \quad \frac{MP_L}{MP_K} = \frac{w}{r} \tag{3.45}$$

Equation (3.44) and (3.45) are first order necessary conditions. The agricultural business firm is said to be in equilibrium where it equates marginal productivities of factors to ratio of their prices.

Equation (3.44) now can be written as

$$\frac{MP_L}{w} = \frac{MP_K}{r} = \lambda \tag{3.46}$$

The second order sufficient condition should be satisfied for profit maximization under constrained cost equation. These conditions are specified as

$$\frac{\partial^2 Q}{\partial L^2} < 0 \text{ and } \frac{\partial^2 Q}{\partial K^2} < 0$$

$$\left(\frac{\partial^2 Q}{\partial L^2}\right)\left(\frac{\partial^2 Q}{\partial K^2}\right) > \left(\frac{\partial^2 Q}{\partial L \, \partial K}\right)^2$$

Then conditions are implied in the convex property of iso-quant[*].

## Constrained Output Maximization–*Numerical example*

Consider an estimated Cobb-Douglas production function, which is given as

$$\hat{Q} = 80L^{0.43} K^{0.54} \tag{3.47}$$

with $\overline{Q} = 1,800$ units of output of crop

Let us assume that wages per labour unit = Rs. 30 = w
and price per unit of capital = Rs. 40 = r

$$\frac{\partial Q}{\partial L} = 34.4 \, K^{0.54} \Big/ L^{0.57} \tag{3.48}$$

$$\frac{\partial Q}{\partial K} = 43.2 L^{0.43} \Big/ K^{0.46} \tag{3.49}$$

For output maximization, we should have

$$\frac{\partial Q / \partial L}{\partial Q / \partial K} = \frac{w}{r}$$

$$= \frac{34.4K^{0.54}/L^{0.57}}{43.2L^{0.43}/K^{0.46}} = \frac{30}{40} \tag{3.50}$$

0.54 and 0.43 are the elasticity coefficients i.e., (a and b) of capital and labour respectively.

---

[*] Henderson and Quandt, Micro Economic theory, McGraw-Hill, 1958, pp. 49-54.

$$= 40 \, (34.4 \, K^{0.54} / L^{0.57}) = 30(43.2 \, L^{0.43} / K^{0.46})$$

$$= 1,376 \, K^{0.54+0.46} \qquad = 1,296 \, L^{0.43+0.57}$$

$$= 1,367 \, K \qquad\qquad = 1,296 \, L$$

$$K = 0.94 \, L \tag{3.51}$$

$$L = 1.06 \, K \tag{3.52}$$

Equation (3.51) and (3.52) are expansion paths. To solve for optimal value of L, substitute K = 0.94 L in equation (3.47)

$$Q = 80 \, L^{0.43} \, (0.94 \, L)^{0.54}$$

$$Q = 75.2 \, L^{0.43 + 0.54}$$

$$Q = 75.2 \, L^{0.97} \tag{3.53}$$

Substitute Q = 1,800 in equation (3.53)

$$1,800 = 75.2 \, L^{0.97}$$

$$L^{0.97} = \frac{1,800}{75.2} = 23.94$$

$$L^* = 23.94^{\frac{1}{0.97}}$$

$$L^* = 26.41$$

Substitute L* = 26.41 and Q = 1,800 (assumed level) in equation (3.47)

$$1,800 = 80 \, (26.41)^{0.43} \, K^{0.54}$$

$$1,800 = 326.93 \, K^{0.54}$$

$$K^{0.54} = 5.51$$

$$K^* = (5.51)^{\frac{1}{0.54}} = 5.51^{(1.85)}$$

$$K^* = 23.58$$

To find out the value of λ, substitute value of Q = 1,800, a = 0.43 and b = 0.54 and L* =26.41 and K* = 23.58 in equations (3.54) and (3.55)

$$\frac{\partial Q}{\partial L} \Big/ w = \lambda \tag{3.54}$$

$$\frac{\partial Q}{\partial K} \Big/ r = \lambda \tag{3.55}$$

$$= \frac{0.43 \left( \dfrac{1800}{26.41} \right)}{30} = \lambda$$

$$= 0.43 \left( \frac{1800}{26.41} \right) = 30\lambda$$

$$\lambda = \frac{29.31}{30} = 0.98$$

$$\frac{\partial Q}{\partial K} \Big/ r = \lambda$$

$$\frac{\partial Q}{\partial K} = r$$

$$0.54 \left( \frac{1800}{23.58} \right) = 40\lambda$$

$$= 41.22 = 40\lambda$$

$$\lambda = \frac{41.22}{40} = 1.03$$

Difference in $\lambda$ values here is due to fractional errors. Here $\lambda$ is defined as marginal product of money. If one rupee is invested on the production of output the business firm would get one rupee and 3 paise return. So, it is the contribution to the output.

$\frac{MP_L}{w} = \frac{MP_K}{r} = \lambda$, this condition is also satisfied in the above numerical example.

# CONSTRAINED COST MINIMIZATION-ALGEBRAIC EXPRESSION

This is also called dual of output maximization. Here we minimize the costs subject to production function. This problem is in particular specified so as

Minimize $M^o = wL + rK$              (3.56)

Subject to $Q^0 = f(L, K)$              (3.57)

Equation (3.56) is a cost equation and equation (3.57) is an iso-quant equation.

The Lagrangean function $(Z)$, is now specified with Lagrangean multiplier, $\mu$.

$Z = wL + rK + \mu [(Q^0 - f(L, K)]$         (3.58)

Taking partial derivatives of equation (3.58) with respect to L, K and $\mu$ and equating them to zero we can minimize the cost.

$$\frac{\partial Z}{\partial L} = w - \mu \left( \frac{\partial Q^0}{\partial K} \right) = 0 \tag{3.59}$$

$$\frac{\partial Z}{\partial K} = r - \mu \left( \frac{\partial Q^0}{\partial K} \right) = 0 \tag{3.60}$$

$$\frac{\partial Z}{\partial \mu} = Q^0 - f(L, K) = 0 \tag{3.61}$$

Equation (3.59), (3.60) and (3.61) are first order necessary conditions for cost minimization. Second order sufficient condition is specified as

$$\frac{\partial^2 Q}{\partial L^2} > 0 \text{ and } \frac{\partial^2 Q}{\partial K^2} > 0 \tag{3.62}$$

and

$$\left(\frac{\partial^2 Q}{\partial L^2}\right)\left(\frac{\partial^2 Q}{\partial K^2}\right) < \left(\frac{\partial^2 Q}{\partial L\,\partial K}\right)^2 \qquad (3.63)$$

## MEASURING EFFICIENCY IN AGRICULTURAL PRODUCTION – APPLICATION OF FRONTIER PRODUCTION FUNCTION

Efficiency is very much required to achieve the desired growth in agricultural production. Economic efficiency by definition is the sum of technical (production) efficiency and allocative (price) efficiency under different levels of technology. The existing technology is represented by the choice of production function. For example, the particular type of production function, which is selected for the data in question represents technology i.e., technology is represented by type of production function. The concepts of efficiency measures in agricultural production have far reaching implications for policy measures viz., price policy, input, income distribution, land ceilings, etc. Yotopoulos and Lau (1971); Huang et al., (1986); Rana (1982) etc., measured the efficiency measures in agricultural production. The theory of profit function approach was developed and applied to agriculture by Lau and Yotopoulos (1973). Let us assume that profit function in logarithmic form be given as

$$\ln \pi_i = \ln f(\,L_{1i},\ Y_{1i},\ Y_{2i},\ Y_{3i},\ Y_{4i},\ Y_{5i},\ Y_{6i}) + \ln E_i$$

Where,

$\pi_i$ = Gross profit (gross income–cost of laobur i.e., family and hired labour) of $i^{th}$ farm (i=1 to n) from the production of crop.

$L_{1i}$ = Man equivalent human labour employed by $i^{th}$ farm in the production of crop.

$Y_{1i}$ = Land area in ha under the crop cultivated by $i^{th}$ farm.

$Y_{2i}$ = Costs in rupees on chemical fertilizers incurred by $i^{th}$ farm.

$Y_{3i}$ = Costs of farmyard manure in Rs. Incurred by $i^{th}$ farm.

$Y_{4i}$ = Costs of seeds incurred by $i^{th}$ farm in Rs.

$Y_{5i}$ = Costs on plant protection incurred by the $i^{th}$ farm in Rs.

$Y_{6i}$ = Miscellaneous costs incurred by the $i^{th}$ farm which include depreciation, interest on working costs, interest on fixed capital, cost of bullock labour, land revenue and machine labour incurred by the $i^{th}$ farm.

All these variables are measured in varying units by the $i^{th}$ farm at a particular point of time over space i.e., (during a particular season over an area).

Here the term $\pi$ is net of human labour cost only. The other variables viz., fertilizers, manures, seeds and plant protection chemicals are placed within the category of fixed inputs. This is because allocation decisions of these inputs thus have little bearing on the profit maximizing behaviour of the farm. The production function is specified as

$$\hat{\pi} = KL^{Bo} \prod_{j=1}^{6} Y_{ij}^{Bj}$$

Cobb-Douglas production function is thus derived as follows:
Now the profit function is expanded as

$$\pi_i = KL^{Bo} Y_1^{B1} Y_2^{B2} Y_3^{B3} Y_4^{B4} Y_5^{B5} Y_6^{B6} e^u \qquad (3.64)$$

Let the estimated profit function be written as

$$\hat{\pi}_i = 12.94 \, L^{-0.224} \, Y_1^{0.3241} \, Y_2^{0.1981} \, Y_3^{0.1624} \, Y_4^{0.1506} \, Y_5^{0.0896} \, Y_6^{0.2801}$$

$$\sum b_i = 0.981 \tag{3.65}$$

Since $\sum b_i = 0.981$, diminshing returns to sale are in operation on the farms. Let us further assume that all the variables are significant at 0.05% and 0.1% level of probability.

$$R^2 = 0.87$$

$$SSE = 0.823 = \sigma^2$$

$$\sigma = 0.907$$

Let the Cobb-Douglas production function be specified as

$$\hat{Q} = A L^{\alpha 0} \, Y_1^{\alpha 1} \, Y_2^{\alpha 2} \, Y_3^{\alpha 3} \, Y_4^{\alpha 4} \, Y_5^{\alpha 5} \, Y_6^{\alpha 6} \, e^U . \text{ Its parameters are astimated as} \tag{3.66}$$

$$A = \frac{1}{K^{(1-Bo)}}, \quad \alpha_0 = \frac{-Bo}{(1-Bo)}, \quad \alpha_1 = \frac{B_1}{(1-Bo)}, \quad \alpha_2 = \frac{B_2}{(1-Bo)} \tag{3.67}$$

$$\alpha_3 = \frac{B_3}{K^{(1-Bo)}}, \quad \alpha_4 = \frac{-B_4}{(1-Bo)}, \quad \alpha_5 = \frac{B_5}{(1-Bo)}, \quad \alpha_6 = \frac{B_6}{(1-Bo)}$$

Using the above relationships of (3.67) we can derive coefficients of equation (3.66).

$$Q_i = 0.0436 \, L^{0.183} \, Y_1^{0.2648} \, Y_2^{0.1618} \, Y_3^{0.1327} \, Y_4^{0.123} \, Y_5^{0.0732} \, Y_6^{0.2288} \tag{3.68}$$

Based on mean levels, the optimal demand for labour is worked out as 208.24 mandays. Here, the estimated profit function is in Cobb-Douglas form. The geometric means of $Y_1$ to $Y_6$ and $Q$ are considered and optimal demand for labour is worked out.

$$\text{Efficiency index of labour} = \frac{\text{Actual labour used on } i^{\text{th}} \text{ farm}}{\text{Optimal demand of labour}}$$

estimated from Cobb-Douglas function

$$= \frac{103.36}{208.24}$$

$$= 0.4964$$

Here, the labour use efficiency is nearly 50 per cent. Similarly optimal values for the remaining variables can be worked out based on elasticity coefficients of the variables and their geometric means.

# SECTION IV

## FARM MANAGEMENT

# CHAPTER 23

# Farm Management

## MEANING

A farm is a socio economic as well as a decision-making unit. It is a socio economic unit because it provides income to the farmer and also it forms a source of livelihood to the family. It is a decision-making unit as it facilitates many alternative uses for the available resources in the form of different crops and livestock enterprises. Each farm unit has the capacity to produce a given quantity of crop and livestock products. The contribution of each farm unit in the country when aggregated represents the total agricultural production of a nation. Thus, the welfare of the nation rests on the performance of the several millions of micro units. When we say the development of agriculture in the country, it is the development of all these individual farm units. Thus the prosperity of the country depends up on the prosperity of the farmers. The prosperity of the farmers in turn depends on their ability to make rational decisions in the allocation of resources and the adoption of new methods of production. In our country nearly 26 per cent of the national income is obtained from the agricultural sector. Since agricultural sector supplies raw material to industries, the prosperity of this sector depends on the farm sector.

In the context of increased accent on commercialization there is a greater need to improve the managerial abilities of the farmers. So far the managers in general have responded admirably to the technological changes that occurred in Indian agriculture. But response of some of the farmers is not in line with needed direction. We can always differentiate those farmers performing against those not performing. Hence, it is of paramount importance for the farm managers to identify the changes that take place and respond suitably, for, any lapse on his part does not help him to survive in the changing economy. This speaks of the need for the managers to sharpen the skills to tackle varying problems that crop up from time to time in the organization of farm business.

To enable the farmers to gear up to the situation, they need to have knowledge on various issues. The role of farm management, therefore, is to supply the information for the farmers for sound planning. All farm management tools are helpful to the farmers in solving their managerial problems for successful operation of the farm business. Nobody can deny the fact that it is the endeavor of the farming community only that helps realize the higher farm production. The field of management through research, teaching and training in this regard has provided the needed decision-making skills.

## SCOPE

The scope of a subject is seen in terms of the problems it solves. Farm management is considered to fall in the field of microeconomics. It treats every farm as a separate unit because of differences in the availability of resources, problems and potentialities. The main concern of farm management is the farm as a unit. Farm management deals with the allocation of inputs at the level of individual farms. The objective of farm management is to maximize returns from the farm as a whole. It is interested in the profitability along with practicability. What crops, livestock enterprises and their combination to grow, what amount of resources to be applied, how the various farm activities to be performed, *etc.*, all these fall with in the scope of farm management.

## DEFINITIONS

Various authors defined farm management in different ways as presented below:

*Farm management is defined as the science that deals with organization and operation of the farm in the context of efficiency and continuous profits* (J.N. Efferson).

*Farm management is defined as the science of organization and management of the farm enterprises for the purpose of securing greatest continuous profits* (G.F. Warren).

*Farm management is defined as the art of managing a farm successfully as measured by the test of profitableness* (Gray).

*Farm management is defined as the art of applying business and scientific principles to the organization and operation of the farm* (Andrew Boss).

*Farm management is the decision-making process whereby limited resources are allocated to a number of production alternatives to organize and operate the business in such a way to attain some objectives* (Ronald D. Kay).

*Farm management is a branch of agricultural economics, which deals with wealth earning and wealth spending activities of farmer in relation to the organization and operation of the individual farm unit for securing the maximum possible net income* (Bradford and Johnson).

*Farm management, as the sub-division of economics, which considers the allocation of limited resources within the individual farm, is a science of choice and decision-making; and thus a field requiring studied judgment* (Heady and Jensen).

## OBJECTIVES

1. To examine production pattern and resource use on the farm.
2. To identify the factors responsible for the present production pattern and resource use on the farm.
3. To determine the conditions of optimality in the resource use and production pattern on the farm.
4. To analyze the extent of sub-optimality in the resource use on the farm, and
5. To suggest ways and means in getting the present use of resources to optimality on the farm.

## FARM MANAGEMENT–ITS RELATIONSHIP WITH OTHER SCIENCES

Farm management integrates information from various disciplines *viz.*, agronomy, animal husbandry, soil science, horticulture, plant breeding, entomology, general economics, sociology, psychology, *etc.*, in its application to the farm. The relationship is examined here under.

## Biological Sciences

Biological sciences provide information on input-output relationships for various enterprises, which facilitate choices for the entrepreneur. Farm manager needs such information in the decisions related to production efficiency. Related to this, we can cite the field of agronomy, which provides the information on physical input – output relationships in crop production. Analogously, animal husbandry tells us the physical relationship between feed ingredients and body-weight of livestock. The field of agricultural engineering explains the efficiency of various implements and machinery in performing various farm operations. However, these branches of agronomy, animal husbandry, *etc.*, can only give the technical efficiency of the resources used in the production. But the farmers' interest is to measure the economic efficiency to derive the maximum net income. Farm management helps in this direction *i.e.*, identifying the point of maximum profitability.

## Economic Theory

Farm management infact is a specialized branch of wider field of economics and it draws the tools and techniques from economic theory. The law of variable proportions, law of substitution, *etc.*, which are used in farm management analyses are the tools of economic theory.

## Supporting Sciences

Statistics provide the base for the farm management specialists to provide methodology in the collection and analysis of data pertaining to the specific farm problems.

## Social Sciences

Attitudes of the farmers towards decision-making on the farm reflect their psychology. Some farmers are risk-takers in farming, while some are averse to risk. It is the mental make-up of an individual farmer to take risky decisions or not.

Certain times, philosophy and religion influence the decisions of the farm managers, particularly in the selection of livestock enterprises. Religious feelings stand in the way of choosing a particular livestock enterprise, though it is profitable to the farmers.

The cultivation of narcotics is quite a profitable proposition to the farmers. But the decision to do so is limited by sociological and ethical considerations. Looking into the harmful affects these cause, a sociologist on the ground of social justice is likely to demand for reduction in their areas. On the ethical grounds people may demand a ban on the cultivation of narcotics totally.

## Government Policies

Government policies also influence the production decisions of the farmers. A liberal export policy of the Government for a product encourages the farmer to go for it. On the other hand, restrictions in the movement of foodgrains, supply of irrigation water, *etc.*, discourage them and opt for other alternatives.

In this way farm management is related to various fields of study regarding what to produce and how much to produce, and restrictions imposed by sociology, psychology, political sciences, *etc.*

# FARM MANAGEMENT DECISIONS

Farm management implies decision-making process. Several decisions need to be made by the farmer as a manager in the organization of farm business. The management decisions are broadly classified into organizational management decisions, administrative management decisions and marketing management decisions which are discussed as below:

## 1. Organizational Management Decisions

The organizational management decisions are further sub-divided into operational management decisions and strategic management decisions.

### i) Operational Management Decisions

Those decisions, which involve less investment and are made more frequently, are called operational management decisions. The effect of these decisions is short lived. These decisions can be reversed without incurring a cost or with less cost. These decisions are what, how and how much to produce. These are also known as tactical decisions.

### a. What to produce?

Every farmer has to decide at the beginning of the every crop season about the type of farm commodities to produce with the resources available on the farm. It means whether to produce crops alone or livestock enterprises alone or a combination of crops and livestock enterprises. While selecting the enterprises and their combinations, farmers always aim at profit maximization.

### b. How to produce?

Once the decision about the enterprises and their combinations to produce is made, the next immediate operational management decision to be made is with regard to the manner in which resources are combined or the production technology to be chosen. In the selection of resources and their combinations, farmer is concerned with cost minimization.

### c. How much to produce?

After having made the above two decisions now the farmer has to decide about the amount of output to achieve in the production of farm commodities. This implies deciding upon the quantities of various inputs to be used in production as the level of production depends on amounts of inputs used.

### ii) Strategic Management Decisions

These decisions involve heavy investment and are made less frequently. The effect of these decisions is long lasting. These decisions cannot be altered. However, in the case of reversal of these decisions farmer has to incur high cost. These decisions are also known as basic decisions. Size of the farm, machinery and labour programme, construction of farm buildings, permanent improvements on the farm like development of irrigation facilities, soil conservation, reclamation, *etc.*, are some of the examples of strategic management decisions.

## a) Size of the Farm

This decision assumes greater relevance to the farmer because of slow and low rate of capital turnover, but it is very difficult to decide on the most appropriate size of the farm to be operated, as it is influenced by several factors *viz.*, availability of financial resources, state laws, managerial abilities, climate, type of farming, *etc.* There are advantages and limitations in operating the farm business on different scales. Large farms enjoy low cost of production, whereas productivity is high on small farms. The advantages and disadvantages of operating enterprises on different scales must be ascertained, while making decision on the size of the farm.

## b) Machinery and Labour Programme

One of the important management problems is to choose appropriate resources and their combinations to produce output with minimum cost. Machinery and labour are substitutes. The availability and requirement of labour, the size of the farm, the financial resources, *etc.*, are important factors in deciding the combination of labour and machinery.

## c) Construction of Farm Buildings

This decision involves huge capital requirements. Here the decisions are made on construction of farm sheds, poultry sheds, dairy sheds, storage buildings, *etc.* Once the decision is taken about the design of a farm building and implemented then it cannot be reversed, for it involves high penalty.

## d) Irrigation, Conservation and Reclamation Programmes

All these programmes help in improving soil productivity. Adoption of these programmes will have ever lasting effect on the organization of the farm business. Size of the farm, availability of funds, availability of ground water, *etc.*, influence the decision on development of irrigation facilities. Mulching, bunding, contouring, strip cropping, *etc.*, are the various alternative measures of soil conservation. Chemical and cultural practices are adapted for soil reclamation. The farmer should choose most appropriate and economical method of conservation and reclamation programmes.

## 2. Administrative Management Decisions

Besides organizational management decisions, the farmer also makes several administrative decisions like financing the farm business, supervision, accounting and adjusting his farm business according to Government policies.

a) *Financing the Farm Business:* Majority of the Indian farmers are capital starved, hence they have to depend on borrowed capital. For borrowing, he has to examine the decisions like from whom to borrow, when to borrow and how much to borrow.

b) *Supervision:* To get the desired results on the farm, farmers should keep a close watch on all the activities performed in the production of crop and livestock enterprises.

c) *Accounting:* Farmer should make a decision about the time and money to be allocated for the maintenance of farm records. Farm records provide control over the farm business.

d) *Adjusting the Farm Production Programme:* The decision of allocating farm resources in the production of farm products should be in consistent with the price policies of the Government. The Government as a welfare state exercises its control over production and marketing of farm commodities according to the situation.

## 3. Marketing Management Decisions

Marketing decisions are the most important under the changing environment of agriculture. These decisions include buying and selling.

a. *Buying:* Every farmer makes an attempt to purchase necessary inputs at the least cost source. In buying resources, a farmer has to decide the agency, the timing, and the quantity to be purchased.

b. *Selling:* Though farm product prices are not under the control of the farmer, yet by adjusting the timing of sales, farmers can obtain better prices. What to sell, where to sell, whom to sell, when to sell and how to sell are the important selling decisions that are to be made by the farmer.

# AGRICULTURAL PRODUCTION ECONOMICS VIS-À-VIS FARM MANAGEMENT

Following are the differences between agricultural production economics and farm management.

| S. No. | Agricultural Production Economics | Farm Management |
|---|---|---|
| 1. | It is a science in which the principles of choice are applied to the use of land, capital, labour and management of resources in the farming industry. | It is a science of organization and operation of farm with a view to earn continuous profits. |
| 2. | Agricultural production economics is a specialized branch of agricultural economics. | It is an integral part of agricultural production economics. |
| 3. | It is macroeconomic in its scope as it deals with the problems of farming industry. | It is microeconomic in its scope as it is concerned with the problems of individual farm. |
| 4. | It deals with allocative efficiency of the use of resources in agriculture. | It deals with economic efficiency at the farm level. |
| 5. | It is an inter-farm study. | It is an intra-farm study. |

# CHAPTER 24

# Economic Principles Applied to Farm Management

Planning is an important function in the process of management. To accomplish this function, some procedures and methods are required to guide the farmer-manager. A knowledge of economics provides a decision-maker with a set of principles for decision-making which are useful when preparing farm plans to organize a farm business. The economic principles guide the manager in setting the goals and preparing plans on the basis of optimum resource allocation, resource substitution and combination and enterprise combination. The knowledge of economic principles improves the decision-making ability and indicate the direction in which the manager should go to attain the objectives of profit maximization and maximization of family satisfaction.

Farm management is the application of economic principles in the organization and operation of farm business. The various economic principles applied to farm management are discussed below:

## LAW OF DIMINISHING RETURNS (LDR)

It explains the input-output relationship (known as factor–product relationship) and guides the producer in making one of the operational management decisions *i.e.*, how much to produce? This law helps in the determination of optimum input to use and optimum output to produce. This law is also known as *the law of variable proportions* because the proportion among factors of production varies as one of the productive resources is changed keeping the other factors of production constant. It is a fundamental economic law, applicable to any production activity and hence can be regarded as a *law of life itself*.

### Definitions

*An increase in capital and labour applied in the cultivation of land causes in general less than proportionate increase in the amount of produce raised, unless it happens to coincide with an improvement in the arts of agriculture* (Marshall).

*If the quantity of one productive service is increased by equal increments with the quantity of other resource services held constant, the increments to total product may increase, but will decrease after a 'certain point'* (Earl O. Heady).

*As additional units of variable inputs are used in combination with one or more fixed inputs, the marginal physical product will eventually begin to decline* (Kay).

These definitions explain that the application of successive units of variables factor, to fixed factors, add more and more to the total product till the point of inflection (MPP is maximum) and thereafter add less and less to the total output.

Law of diminishing returns in general applies to the field of agriculture, but its operation can be postponed under the following conditions.

1) *Improved Technology:* Technology components like high yielding varieties of seeds, irrigation, fertilizers, integrated pest management practices, *etc.*, would help in increasing the returns.
2) *New Soils:* When virgin soils are brought under plough, they give more and more yield, as their productivity is higher.
3) *Scarcity of Capital:* Scarcity of capital limits the farmers in applying sufficient quantities of variable resource leading to the prevalence of increasing returns. This situation is found in stage I.

## Reasons for the Operation of the Law of Diminishing Returns in Agriculture

1) *Excessive Dependence on Weather:* A farmer, however good he is in managing the farm may not get the expected yields as he has little control over weather. A bad weather is just enough to make his expectation go topsy-turvy.
2) *Less Scope for Division of Labour:* There is no possibility of division of labour in farming as the farmer himself performs the role of labourer, manager and capitalist, therefore the advantage of division of labour is not a possibility. Therefore, the law of diminishing returns sets in quickly in farming.
3) *Less Scope for Mechanization:* Though mechanization of the farms enhances the productivity, the small size of holdings stands against using the machines for various farm operations. Under this limitation, the farmer fails to derive the advantage of mechanization.
4) *Cultivation of Inferior Lands:* To meet the food requirements of the teeming millions of population, even inferior lands are brought under plough, the productivity of which in general is low.
5) *Continuous Cultivation:* Continuous cultivation of land drains out the fertility status of the soil, thereby leading to low productivity. It is a fact that no farmer can afford to keep the land fallow for some period, to allow the land build up its fertility status.

*For details refer the topic on factor-product relationship in Section III.*

## PRINCIPLE OF FACTOR SUBSTITUTION

Farmer as a manager, has to make an important operational management decision *viz.*, how to produce? It implies the choice of methods of production or technology or the type of combination of resources. This involves factor-factor relationship. The principle of factor substitution guides the farmer-manager in choosing the most appropriate method of production or technology that produces a given level of output with least cost. There are many alternative ways of producing farm commodities. The choice of method of production depends on the availability of resources. If the land is abundant, the producer undertakes extensive cultivation; on the other hand if the land is scarce he adopts land saving technology *i.e.*, intensive cultivation. If the labour is scarce, labour saving technology contributes to the least cost and if capital is abundant, labour saving technology or capital intensive technology is adopted.

The volume of production or the size of farm business also determines the adoption of production technology. Large scale farming in general requires capital-intensive technology. Small-scale farming is more a subsistence-oriented business requiring intensive use of labour. Having said this, we should not ignore the prices of the factors in the selection of the appropriate combination of the input factors.

A producer generally aims at choosing the most efficient method of production. A technique of production is efficient, when it produces a given level of output with minimum cost, to achieve this the producer substitutes cheaper resource for dearer resource. It means the use of more quantity of less expensive resource and less quantity of more expensive resource. Substitution is economical as long as the cost of one resource is less than the cost of another resource.

The principle of factor substitution says that it is economical to substitute one resource (added resource) to another resource (replaced resource), as long as the reduction in the cost, resulting from decreased use of replaced resource is more than the increase in the cost due to increased use of added resource. The two cost aspects are compared with the help of the principle of factor substitution, to find out least cost combination of resources.

Decrease in cost > increase in cost

*i.e.,* Quantity of replaced resource multiplied by price per unit of replaced resource  >  Quantity of added resource multiplied by price per unit of added resource

$$\frac{\text{Quantity of replaced resource}}{\text{Quantity of added resource}} > \frac{\text{Price per unit of added resource}}{\text{Price per unit of replaced resource}}$$

MRTS > PR

In the process of substitution, we shift from one input combination to another input combination, as a result of which there is an increase in the use of one resource and decrease in the use of another resource (Table 24.1). As we move on from combination A to combination B, the quantity of grain $(X_1)$ is increased by 50 units, while the quantity of hay $(X_2)$ is reduced by 125 units. Given the prices of $X_1$ at Rs. 4

**TABLE 24.1  Selecting Least Cost Feed Ration for Producing 25 kg Body Weight Price of Grain $(X_1)$ = Rs. 4/Kg; Price of Hay $(X_2)$ = Rs. 1.90/Kg.**

| Feed combinations | Grain (kg) | | Hay (kg) | | MRTS of $X_1$ for $X_2$ | Price Ratio |
|---|---|---|---|---|---|---|
| | $(X_1)$ | $\Delta X_1$ | $(X_2)$ | $\Delta X_2$ | | |
| A | 500 | 50 | 2,190 | 125 | 2.50 | 2.10 |
| B | 550 | 50 | 2,065 | 122 | 2.44 | 2.10 |
| C | 600 | 50 | 1,943 | 118 | 2.36 | 2.10 |
| D | 650 | 50 | 1,825 | 110 | 2.20 | 2.10 |
| E | 700 | 50 | 1,715 | 105 | 2.10* | 2.10* |
| F* | 750* | 50 | 1,610* | 98 | 1.96 | 2.10 |
| G | 800 | 50 | 1,512 | 87 | 1.74 | 2.10 |
| H | 850 | 50 | 1,425 | 76 | 1.52 | 2.10 |
| I | 900 | 50 | 1,349 | 60 | 1.20 | 2.10 |
| J | 950 | 50 | 1,289 | 56 | 1.12 | 2.10 |
| K | 1,000 | | 1,233 | | | |

* Least cost combination of inputs.

per unit and $X_2$ at Rs. 1.90 per unit, the amount saved from reduction of 125 units of $X_2$ is Rs. 237.50 and the increase in the cost of $X_1$ due to increase of 50 units of $X_1$ is Rs. 200. The saving thus arrived at exceeds the increased costs indicating the rationality of substituting $X_1$ for $X_2$. The process of substituting grain $(X_1)$ for hay $(X_2)$ continues till the saving in the amount from hay is equal to increase in the cost of grain. In other words, substitution of $X_1$ for $X_2$ is economical till MRTS is equal to price ratio. Least cost combination is found at the feed combination 'F' i.e., 750 units of $X_1$ and 1610 units of $X_2$.

## Profit Rules

If, MRTS > price ratio, costs can be reduced by using more of added resource.

$$\frac{\Delta X_2}{\Delta X_1} > \frac{P_{X_1}}{P_{X_2}}, \text{ use more of } X_1, \text{ if } X_1 \text{ is substituted for } X_2 \ (MRTS_{X_1X_2})$$

$$\frac{\Delta X_1}{\Delta X_2} > \frac{P_{X_2}}{P_{X_1}}, \text{ use more of } X_2, \text{ if } X_2 \text{ is substituted for } X_1 \ (MRTS_{X_2X_1})$$

If MRTS < price ratio, costs can be reduced by using more of replaced resource.

$$\frac{\Delta X_2}{\Delta X_1} < \frac{P_{X_1}}{P_{X_2}}, \text{ use more of } X_2, \text{ if } X_1 \text{ is substituted for } X_2 \ (MRTS_{X_1X_2})$$

$$\frac{\Delta X_1}{\Delta X_2} < \frac{P_{X_2}}{P_{X_1}}, \text{ use more of } X_1, \text{ if } X_2 \text{ is substituted for } X_1 \ (MRTS_{X_2X_1})$$

Least cost combination is at the point where MRTS = PR

$$\frac{\Delta X_2}{\Delta X_1} = \frac{P_{X_1}}{P_{X_2}} \text{ or } \frac{\Delta X_1}{\Delta X_2} = \frac{P_{X_2}}{P_{X_1}}$$

# PRINCIPLE OF PRODUCT SUBSTITUTION

Agricultural production is characterized by risks and uncertainties. To fight risks and uncertainties, the farmers produce several crop and livestock enterprises on their farms. This opportunity of choosing among different alternative enterprises poses an important management problem *viz.*, what to produce. This involves product-product relationship and is explained by the principle of product substitution. This principle guides the producer in the determination of optimum combination of enterprises that maximizes profits. To apply this principle there is a need to understand the following product relationships.

*Complementary Enterprises:* Two products are complementary, when increase in output of one also results in an increase in the output of the other, with resources held constant.

*Supplementary Enterprises:* Two products are supplementary, when an increase or decrease in the output of one product does not affect the output of the other product.

*Competitive Enterprises:* Two products are competitive, if output of one product can be increased only through a sacrifice in the output of the other. When products are competitive three things to be considered to determine the optimum combination *viz.,* marginal rate of substitution between products, price ratio and the cost of cultivation. If the cost of cultivation of the two products is same, the first two factors guide in the selection of the most profitable combination.

The principle of product substitution says that if the inputs are constant, it is economical to substitute one product for the other, if the returns from the first are more than that of the second. In the process of substitution, when we shift from one combination of products to another, the output of one product increases, while that of other decreases, because the input is kept constant. From the product whose level of output decreases, there is a decrease in returns. On the other hand, from the other product, whose level of output increases, there is an increase in returns. These two aspects of returns are compared using the principle of product substitution which says that we should go on increasing the level of output of a product so long as decrease in the returns from the product being replaced is less than the added returns from the product being added. Thus if $Y_1$ is being increased and $Y_2$ is being replaced, increase the production of $Y_1$ so long as

$$\text{Decrease in returns} < \text{Increase in returns}$$

*i.e.,* Quantity of output being lost      Quantity of output gained
multiplied by price per      $<$      multiplied by price per
unit of replaced product      unit of replacing product

$$\frac{\text{Quantity of output lost}}{\text{Quantity of output gained}} < \frac{\text{Price per unit of replacing product}}{\text{Price per unit of replaced product}}$$

$$\text{MRPS} < \text{PR}$$

$$\frac{\Delta Y_2}{\Delta Y_1} < \frac{P_{Y_1}}{P_{Y_2}}$$

## Determination of Optimum Product Combination

$Y_1$ and $Y_2$ are the two products, whose prices are Rs. 4.20 per unit and Rs. 6 per unit respectively. There are seven combinations of $Y_1$ and $Y_2$ products that could be produced with given amount of resources (Table 24.2). As we shift from combination A to combination B, the output of $Y_1$ is increased by 20 units, while that of $Y_2$ reduced by 4 units.

Given the prices, there is an increase in the return of $Y_1$ to the tune of Rs. 84 and decrease in the return of $Y_2$ by Rs. 24. The increase in the return of Rs. 84 due to the increase in the output of $Y_1$ is more than the decrease in the returns of Rs. 24 due to reduction in the production of $Y_2$ indicating the rationality of substituting $Y_1$ for $Y_2$. The process of substitution of $Y_1$ for $Y_2$ continues till the increase in the returns from $Y_1$ is equal to decrease in the returns of $Y_2$. In other words substitution of $Y_1$ for $Y_2$ is economical till MRS is equal to price ratio. The optimum combination of products is found at combination 'F' *i.e.,* 100 units of $Y_1$ and 16 units of $Y_2$.

## Profit Rules

1. If MRS < PR, profits can be increased by producing more of added product

TABLE 24.2 Determination of Optimum Product Combination.

$P_{Y1}$= Rs. 4.20; $P_{Y2}$ = Rs. 6.00

| | $Y_1$ units | $Y_2$ units | $\Delta Y_1$ | $\Delta Y_2$ | Increase in returns | Decrease in returns | MRS of $Y_1$ for $Y_2$ | PR |
|---|---|---|---|---|---|---|---|---|
| A | 0 | 60 | – | – | – | – | – | – |
| B | 20 | 56 | 20 | 4 | 84 | 24 | 0.20 | 0.70 |
| C | 40 | 50 | 20 | 6 | 84 | 36 | 0.30 | 0.70 |
| D | 60 | 41 | 20 | 9 | 84 | 54 | 0.45 | 0.70 |
| E | 80 | 30 | 20 | 11 | 84 | 66 | 0.55 | 0.70 |
| F* | 100 | 16 | 20 | 14 | 84 | 84 | 0.70 | 0.70 |
| G | 120 | 0 | 20 | 16 | 84 | 96 | 0.80 | 0.70 |

*Optimum combination of products.

$$\frac{\Delta Y_2}{\Delta Y_1} < \frac{P_{Y1}}{P_{Y2}}$$ increase $Y_1$, if $Y_1$ is substituted for $Y_2$ $(MRS_{Y_1Y_2})$

$$\frac{\Delta Y_1}{\Delta Y_2} < \frac{P_{Y2}}{P_{Y1}}$$ increase $Y_2$, if $Y_2$ is substituted for $Y_1$ $(MRS_{Y_2Y_1})$

2. If MRS > PR, profits can be increased by producing more of replaced product

$$\frac{\Delta Y_2}{\Delta Y_1} > \frac{P_{Y1}}{P_{Y2}}$$ increase $Y_2$, if $Y_1$ is substituted for $Y_2$ $(MRS_{Y_1Y_2})$

$$\frac{\Delta Y_1}{\Delta Y_2} > \frac{P_{Y2}}{P_{Y1}}$$ increase $Y_1$, if $Y_2$ is substituted for $Y_1$ $(MRS_{Y_2Y_1})$

3. Profit maximizing combination of enterprises will be where, MRS = PR

$$\frac{\Delta Y_2}{\Delta Y_1} = \frac{P_{Y1}}{P_{Y2}} \text{ if } MRS_{Y_1Y_2} \text{ (or) } \frac{\Delta Y_1}{\Delta Y_2} = \frac{P_{Y2}}{P_{Y1}} \text{ if } MRS_{Y_2Y_1}$$

# PRINCIPLE OF EQUIMARGINAL RETURNS

Under the conditions of unlimited resources, the law of diminishing returns helps in determining the most profitable level of resource use. In reality, most of the resources like land, irrigation, capital, *etc*, with most of the farmers are limited. When the farmers are constrained by the resources, they must prudently decide as to how the available resources should be allocated among alternative uses to gain maximum income. The principle of equimarginal returns provides guidelines and ensures that allocation of a limited input is done in such a way that profit is maximized from each unit of input. The principle is stated as follows. *The limited resources should be allocated among alternative uses in such a way that the marginal value product of the last unit of the resource is equal in all uses.* How the limited inputs should be allocated among the enterprises is shown through the following example presented in Table 24.3.

Any limited input should be allocated in that use where it brings in the greatest MVP. The limited availability of capital (here 5 units of liquid capital) must be allocated among the three crops *viz.*, sugarcane, cotton and paddy in the following manner

TABLE 24.3 Principle of Equimarginal Returns.

| Amount of liquid capital | Marginal value products per unit of '000Rs. | | |
|---|---|---|---|
| used (Unit = '000 Rs.) | Paddy | Sugarcane | Cotton |
| 1 | 2,000 (5) | 3,200 (1) | 2,200 (4) |
| 2 | 1,400 | 3,000 (2) | 1,800 |
| 3 | 1,200 | 2,500 (3) | 1,400 |
| 4 | 1,100 | 1,600 | 1,000 |
| 5 | 1,000 | 1,200 | 800 |

based on MVPs. The first dose of Rs. 1,000 has the potential of yielding MVP (added returns) of Rs. 2,000, Rs. 3,200 and Rs. 2,200 from paddy, sugarcane and cotton respectively. So, first dose of capital is invested on sugarcane, which brought in the maximum MVP among the alternatives. To apply the second dose of capital, the three opportunities are first dose to paddy (Rs. 2,000), second dose to sugarcane (Rs. 3,000) and first dose to cotton (Rs. 2,200). Among the three opportunities for the second dose of Rs. 1,000, sugarcane is yielding highest MVP. In the same manner third dose for sugarcane, fourth for cotton and fifth for paddy are allocated.

Following the principle of equimarginal returns, three units of money should be allocated to sugarcane and one unit each to cotton and paddy. Diagrammatic representation of the principle is presented in Fig 24.1. Any other allocation of capital among the crop enterprises other than the above will not help the farmer to obtain maximum returns.

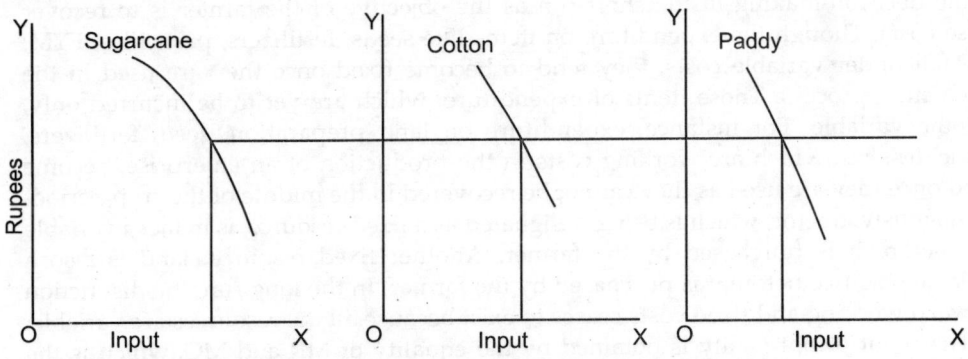

Figure 24.1 Equimarginal returns.

## OPPORTUNITY COST PRINCIPLE

If an input is used in a particular production process, it has no alternative use at that particular point of time. This means that the input will be loosing income from the alternative use and this income foregone by this input from its alternative use is called opportunity cost. *By definition, opportunity cost is the income that could have been received, if the input had been used in its most profitable alternative use.* Alternatively, it is the value of product not produced because the input was used for another purpose. The concept of opportunity cost has a bearing on the decision-making process of the farmer, particularly in decisions related to *input use*. The opportunity cost is referred to as the real cost of an input. Real cost of an input is not the purchase price of the input. It is the income earned by the input in its alternative use, which is the next best opportunity. If the returns from the current use of the input are less than its opportunity cost, then the decision is to be changed.

For example, if a farmer has Rs. 1,000 at his disposal, he has three options *i.e.*, investing on sugarcane, or cotton or paddy. As given in Table 24.3 Rs. 1,000 is spent on sugarcane which is giving a MVP of Rs. 3,200. The farmer is foregoing the other two options by spending on sugarcane. Between the two, cotton is giving greater MVP than paddy. The farmer is sacrificing Rs. 2,200 from cotton, which is the next best alternative to sugarcane, which is the opportunity cost.

## MINIMUM LOSS PRINCIPLE

In any business activity, the details of costs and returns provide an idea of profitability. Cost of production refers to the expenses incurred in producing a unit quantity of the product in a particular time period. The costs in farming in the short run can be divided into two categories *viz.*, fixed costs and variable costs. Fixed costs are those costs, which do not vary with the level of output. Fixed costs are incurred even in the absence of production. These are the costs, which arise on the investment already made in the fixed resources. Fixed costs include depreciation, interest on fixed capital, rent, land revenue, insurance premium, wages of attached (permanent) servants, family labour wages, *etc.* Variable costs on the other hand are those costs, which vary with the level of output. There are no variable costs, when there is no production. Variable costs include cost of seeds, feeds, manures, fertilizers, wages of casual labourers, electricity charges, customs hiring, *etc.* Variable costs are important in the decision-making in the short-run as the objective of the farmer is to recover these costs. Though the expenditure on items like seeds, fertilizers, pesticides, FYM, *etc.*, fall under variable costs, they tend to become fixed once they are used in the production process. Those items of expenditure, which are yet to be incurred only, become variable. For instance, expenditure on land preparation, seed, fertilizers, pesticides, *etc.*, which are working costs in the production of an enterprise, become fixed once they are used as they cannot be recovered in the middle of the crop period. Analogously, tractor, which is being designated as a fixed resource, is in fact a variable cost before it is purchased by the farmer. Another fixed resource, land is also a variable resource before it is purchased by the farmer. In the long run, the distinction between working and fixed costs ceases to exist because all the resources are variable.

The point of optimality is obtained by the equality of MR and MC, which is the profit maximizing condition. But in reality, the farmers incur loss instead of profit because MR (selling price) may not cover cost per unit of output. Then the farmers continue the farming with an objective of minimizing losses. The minimum loss principle explains, as to how, the producer minimizes losses under adverse price environment. If selling price is more than ATC, profits are expected by the producer and the objective is profit maximization. To do this the producer has to produce till MR = MC. If selling price is less than ATC, but more than AVC, loss is expected. The objective here is to minimize the loss. To accomplish the task he has to continue the production till MR = MC. In this situation loss is less than fixed costs. If selling price is less than AVC and losses are expected, loss can be minimized by stopping the production temporarily. In the long run if selling price is less than ATC, continuous losses are incurred. In this situation, the producer should stop the production permanently.

The principle is explained further with the help of hypothetical data. In sunflower the total variable costs per ha. are Rs. 4,700; total fixed costs are Rs. 2,700 and the resultant total costs stood at Rs. 7,400. The output obtained is 9.5 quintals per ha. With the price being Rs. 1,150 per quintal, gross income amounted to Rs. 10,925. Net income, which is the surplus over total costs, came to Rs. 3,525. In this situation the farmer is able to obtain profits in sunflower cultivation.

Now, assuming that the price has fallen down to Rs. 750, the gross income recorded is Rs. 7,125, which is less than the total costs of Rs. 7,400. However, the gross income covers the variable costs of Rs. 4,700 netting the farmer with an amount of Rs. 2,425. In case, the farmer decides against cultivation of sunflower, as he is not generating surplus over total costs, he would be incurring loss to the extent Rs. 2,700 representing the fixed costs. As against this if he goes ahead with the cultivation of sunflower, the loss would be to the extent of Rs. 275 only, which is the minimum loss.

*Costs and returns in sunflower cultivation per hectare*

| | | |
|---|---|---|
| TVC | = | Rs. 4,700 |
| TFC | = | Rs. 2,700 |
| TC | = | Rs. 7,400 |
| Output | = | 9.5 quintals/ha |
| Price per quintal | = | Rs. 1,150 |
| Gross income | = | Rs. 10,925 |
| Net income | = | Rs. 10,925 – 7,400 |
| | = | Rs. 3,525 |

*If price falls down to Rs. 750, then*

| | | |
|---|---|---|
| Gross income | = | Rs. 7, 125 |
| Net income | = | Rs. 7,125-7,400 |
| | = | Rs. –275 |
| AVC | = | 4,700/9.5 |
| | = | Rs. 494.74 |
| AFC | = | 2,700/9.5 |
| | = | Rs. 284.21 |
| AC | = | Rs. 778.95 |

# PRINCIPLE OF COMPARATIVE ADVANTAGE

It is true that many crop and livestock enterprises can be raised over diversified soil types and climatic conditions, but with differences in yields, costs and returns. This difference in yields, costs and returns leads to specialization in the production of farm commodities by individual farmers or regions. We observe that wheat farming is predominant in Uttar Pradesh, Punjab and Haryana, rice farming in Andhra Pradesh, West Bengal, Tamil Nadu, and Assam, cotton farming in Maharashtra and Tamil Nadu, sericulture in Karnataka, apple cultivation in Himachal Pradesh, sheep farming in Rajasthan, poultry farming in Andhra Pradesh and Tamil Nadu and fresh water prawn culture in Andhra Pradesh and Orissa. Thus, regional specialization in the production of crops and livestock enterprises is better explained by the principle of comparative advantage. The relative yields, costs and returns are to be considered as the criteria for explaining the principle.

In the production of farm commodities there are two kinds of economic advantage (1) Absolute advantage and (2) Relative advantage or comparative advantage.

The size of the margin between costs and returns from using the productive inputs represents the absolute advantage. If this margin is larger for one farm commodity in one region compared to another, we say that the first region has an absolute advantage in producing that commodity. This is illustrated with an example in Table 24.4.

**TABLE 24.4   Absolute Advantage.**

| Particulars | Region A | | Region B | |
|---|---|---|---|---|
| | Groundnut | Sunflower | Groundnut | Sunflower |
| Gross income (Rs./acre) | 5,000 | 5,010 | 7,300 | 2,500 |
| Total costs (Rs./acre) | 4,700 | 4,320 | 6,500 | 2,450 |
| Net income (Rs./acre) | 300 | 690 | 800 | 50 |
| Returns per rupee of investment | 1.06 | 1.16 | 1.12 | 1.02 |

Suppose farmers in region 'A' and region 'B' are producing two farm commodities *viz.*, groundnut and sunflower. The gross income per acre of groundnut in region 'A' is Rs. 5,000, while costs are Rs. 4,700. In region 'B', the gross income from the cultivation of groundnut per acre is Rs. 7,300 and expenses are Rs. 6,500. The net income per acre for groundnut is Rs. 300 in region 'A' and Rs. 800 in region 'B', with a return of Rs. 1.06 in region 'A' and Rs. 1.12 in region 'B' per rupee of investment. On the other hand sunflower could earn a gross income of Rs. 5,010 in region 'A' with a cost of cultivation of Rs. 4,320. The net income and returns per rupee of investment from sunflower in region 'A' are Rs. 690 and Rs. 1.16 respectively. In region 'B', the gross income from sunflower per acre is Rs. 2,500, while the cost of cultivation is Rs. 2,450. The net income per acre is Rs. 50 and the return per rupee of investment is Rs. 1.02. Region 'A' has an absolute advantage in sunflower because the size of the margin between costs and returns is greater than that for region 'B'. Region 'B' has an absolute advantage in groundnut production for the same reason.

To explain the relative or comparative advantage, let us compare region 'B' with region 'C'. In both the regions, farmers are growing redgram and groundnut.

**TABLE 24.5   Relative Advantage.**

| Particulars | Region B | | Region C | |
|---|---|---|---|---|
| | Redgram | Groundnut | Redgram | Groundnut |
| Gross income (Rs./acre) | 5,600 | 7,300 | 2,300 | 3,300 |
| Total costs (Rs./acre) | 5,200 | 6,500 | 2,000 | 3,100 |
| Net income (Rs./acre) | 400 | 800 | 300 | 200 |
| Returns per rupee of investment | 1.08 | 1.12 | 1.15 | 1.06 |

From the figures furnished in Table 24.5 it is seen that region 'B' has a greater absolute advantage in growing both redgram and groundnut than region 'C' because, the net incomes per acre are Rs. 400 and Rs. 800 respectively. In other words respective incomes are 108 per cent and 112 per cent higher than the costs. Farmers in region 'B' can earn profits by growing both the crops. But in order to earn maximum profits, farmers in region 'B' should allocate larger acreage under groundnut alone as it is related to comparative advantage. Similarly, farmers in region 'C' can make profits by growing both the crops but they have relative advantage in growing redgram. The farmers can make greatest profits by cultivating redgram as, the percentage of returns over the cost of cultivation being 115 per cent for redgram and 106 per cent for groundnut.

# TIME COMPARISON PRINCIPLE

There are two types of investments. (1) investment on operating or working inputs *viz.*, seed, fertilizers, plant protection chemicals, fuel, feeds, veterinary medicines, *etc*. and (2) investment on capital assets *viz.*, land, farm buildings, machinery and equipment. Analysis of an investment involves not only the comparison of costs and returns associated with it, but also the timings of occurrence of costs and returns.

The costs and returns from investing in operating resources occur with in a production period of a year or less. The marginal principles are used to determine the optimum level or allocation of operating resources. There is no need to bring time element into discussion, as both costs and returns were assumed to fall with in the same production cycle. The situation is different in the case of capital invested on capital assets, where the costs and returns are in different time periods. And also capital expenditure involves costs and returns over time. For example, if a farmer wants to mechanise farming, he would incur capital expenditure on tractor, fuel, repairs and maintenance, payment of wages to driver, *etc*. Establishment of an orchard involves the purchase of land, machinery, equipment, construction of buildings, digging of well, purchase of seed material, fertilizers, plant protection chemicals, payment of wages, *etc*.

While some of these expenditure are non-recurring, others are recurring expenditure. Even all non-recurring expenditure may not be incurred in a particular month or a year, but rather spread over a number of months or years. Returns may be received in future over many years. To examine the profitability of these investments it requires the recognition of time value of money. Money has time value for the following reasons.

1. *Earning Power of Money:* Whenever there are many opportunities for investment, then money possesses earning power. The earning power is represented by the opportunity cost of money (rate of interest).
2. *Inflation:* Purchasing power of money varies inversely with the price level. Thus the value of money changes with inflation and deflation. A rupee earned a year from now is less valuable than a rupee earned today.
3. *Uncertainty:* Investment deals with future and future is uncertain. Investment is concerned with the commitment of funds today with the expectation of receiving a stream of benefits in the future. As an individual since he is not certain about future cash receipts, he prefers receiving cash now. Most people have subjective preferences for present consumption over future consumption of goods and services. This may be either due to the urgency of present wants or the risk of not being in a position to enjoy future consumption that may be caused by illness, death or inflation. Also, people prefer present cash to future cash because of the available investment opportunities where their present cash earn additional cash.

Since, capital expenditure involves costs and benefits over time, it is necessary to adjust for the time value of money for conducting a meaningful investment analysis.

The adjustment for the time value of money is made through compounding and discounting.

## Compounding

Compounding is the procedure to find the future value of a present sum, given the earning power (interest rate) of money and the frequency of compounding. The future value of money is the value of an investment on a particular date in the future. This is based on the assumption that the investment will earn interest which is reinvested at the end of each year to also earn interest. The future value includes the original investment, the interest it earns and the interest on the accumulated interest. Compounding is the process used to determine costs and returns at the end of the planning horizon.

Example: Assume Rs. 100 invested in a savings account which earns 10 per cent interest is compounded annually and we would like to know the future value of this amount after 4 years. Table 24.6 illustrates the application of compounding procedure.

TABLE 24.6  Future Value of Present Sum for Lumpsum Amount.

| Year | Amount at the beginning of year (Rs.) | Interest rate (%) | Interest earned (Rs.) | Amount at the end of the year (Rs.) |
|------|------------------|-------------------|------------------|------------------|
| 1 | 100.00 | 10 | 10.00 | 110.00 |
| 2 | 110.00 | 10 | 11.00 | 121.00 |
| 3 | 121.00 | 10 | 12.10 | 133.10 |
| 4 | 133.10 | 10 | 13.31 | 146.41 |

In the example, a present sum of Rs. 100 has a future value of Rs. 146.41 invested at 10 per cent interest for four years. Compound interest is being used, when accumulated interest also earns future interest. If the interest had been withdrawn, only Rs. 100 would have earned interest each year. A total of Rs. 40 towards interest would have been withdrawn compared with Rs. 46.41 of earned interest in the example.

The procedure of finding future value of present sum become very cumbersome if the investment is for a long period of time. The calculations can be simplified by using the following equation.

$$FV = P(1 + i)^n$$

Where,

    FV = Future value
    P = Present sum
    i = Interest rate
    n = Number of years.
    $FV = 100(1+0.1)^4$
      = 100 (1.4641)
      = Rs. 146.41

The concept of compounding can also applied to an annuity. An annuity is a stream of payments (or receipts) over time. In other words, fixed payment (or receipt) each year for a specified number of years is called an annuity. Suppose Rs. 1,000 is deposited at the end of each year in savings account at the rate of 10 per cent interest.

What is the future value of this investment at the end of four years. It can be computed in the following way.

$$
\begin{array}{lllll}
\text{First Rs. 1,000} & = & 1{,}000\,(1+0.1)^3 & = & \text{Rs. 1,331.00} \\
\text{Second Rs. 1,000} & = & 1{,}000\,(1+0.1)^2 & = & \text{Rs. 1,210.00} \\
\text{Third Rs. 1,000} & = & 1{,}000\,(1+0.1)^1 & = & \text{Rs. 1,100.00} \\
\text{Fourth Rs. 1,000} & = & 1{,}000\,(1+0.1)^0 & = & \text{Rs. 1,000.00} \\
& & \text{Future value} & = & \text{Rs. 4,641.00}
\end{array}
$$

Since the money is deposited at the end of each year, the first Rs. 1,000 earns interest for only 3 years, the second Rs. 1,000 earns interest for 2 years and the third Rs. 1,000 earns interest for 1 year and the fourth Rs. 1000 earns no interest. The total interest earned is Rs. 641.00.

The following formula is used to compute future value of an annuity.

$$ FV = P\frac{(1+i)^n - 1}{i} $$

Where, $FV$ is the future value,

$P$ = Annual year-end investment

$i$ = Interest rate

$n$ = Number of years.

$$ FV = \text{Rs. } 1{,}000\frac{(1+0.1)^4 - 1}{0.1} $$

$$ = \text{Rs. } 1{,}000\frac{(1.4641) - 1}{0.1} $$

$$ = \text{Rs. } 1{,}000(4.641) = \text{Rs. } 4{,}641.00 $$

## Discounting

Discounting is the procedure to find the present value of future sum. The present value refers to the current value of a sum of money to be received in the future. A future amount is discounted because sum of money to be received in future is worth somewhat less now, because of time difference assuming positive rate of interest. The current value of future amount depends up on the rate of interest and length of time before payment is received.

The present value of single payment in the future is found out with the following formula.

$$ PV = \frac{A}{(1+i)^n} $$

Where,

$PV$ = Present value

$A$ = Amount to be received in future

$i$ = Interest rate

$n$ = Number of years

Suppose an investor wants to find out the present value of Rs. 5,000 to be received after 3 years. The interest rate is 10 per cent. By using the above formula, the present value can be found out.

$$PV = \frac{5,000}{\left(1 + \frac{10}{100}\right)^3}$$

$$= \frac{5,000}{(1 + 0.1)^3}$$

$$= \frac{5,000}{1.331} = Rs.\ 3,756.57$$

A payment of Rs. 5,000 to be received in 3 years has present value Rs. 3,756.57 at 10 per cent interest. In other words, Rs. 3,756.57 invested for 3 years at 10 per cent compound interest would have a future value of Rs. 5,000. It also implies that an investor should not pay more than Rs. 3,756.57 for an investment which can return Rs. 5,000 in 3 years, if there are other opportunities of investment which can pay 10 per cent interest or more.

An investor may receive constant amount periodically for certain number of years. The present value of an annuity cannot be found by using the formula given above. In order to find out the present value of an annuity, we have to find out the present value of the amount every year and have to aggregate all the present values to get the total present value of the stream of payments (receipts). For example, an investor, who has determined his interest rate as 8 per cent, may have an opportunity to receive an annuity of Re. 1 for five years. The present value of Re. 1 received after one year is $P = \frac{1}{1.08} = 0.9259$, after two years, $P = \frac{1}{(1.08)^2} = 0.8573$, after three years, $P = \frac{1}{(1.08)^3} = 0.7938$, after fours years, $P = \frac{1}{(1.08)^4} = 0.7350$ and after five years, $P = \frac{1}{(1.08)^5} = 0.6805$ Thus, the total present value of annuity of Re. 1 is 3.9925.

$$P = \frac{1}{(1.08)^1} + \frac{1}{(1.08)^2} + \frac{1}{(1.08)^3} + \frac{1}{(1.08)^4} + \frac{1}{(1.08)^5}$$

$$= 0.9255 + 0.8573 + 0.7938 + 0.7350 + 0.6805 = Rs.\ 3.9925$$

The following formula is used to find the present value of an annuity

$$PV = A \left[ \frac{1 - \frac{1}{(1+i)^n}}{i} \right]$$

Where,

PV = Present value

A = Constant payment (or receipt) each year

i = Rate of interest

n = Number of years

A person receives an annuity of Rs. 8,000 for five years. If the rate of interest is 8 per cent, the present value of Rs. 8,000 annuity is:

$$PV = Rs.\ 8,000 \left[ \frac{1 - \dfrac{1}{(1+0.08)^5}}{0.08} \right]$$

Rs. $8,000 \times 3.9925$ = Rs. 31, 940.

# CHAPTER 25

# Types of Farming

Types of farms refer to the nature and degree of products and their combination and various methods followed in the production of the same.

Under the types of farming, the major types that studied are (1) Specialized farming (2) Diversified farming (3) Mixed farming (4) Dry farming and (5) Ranching.

## 1. Specialized Farming

When a farm business unit derives more than 50 per cent of its income from a single enterprise it is called as a specialized farm. This means that among the possible crops or livestock enterprises taken up by a farmer, one particular crop or livestock enterprise contributes more than 50 per cent of the income. The reasons for specialized farming are; 1) assured income from the enterprise; 2) its suitability to the area; 3) its relative profitability, *etc.* The examples that can be cited are paddy farming, sugarcane farming, tobacco farming, *etc.*, among crop enterprises and poultry, sheep farming, fish farming, *etc.*, among livestock enterprises. The favourable environment though encourages specialized farming, apart from the advantages; disadvantages too are associated with it.

### Advantages

1. *Better Utilization of Land:* Land can be put to most productive use, by opting the enterprise that is best suited. A given type of land no doubt allows options for alternative crops; still there is a possibility of a particular crop capable of rewarding the farmer with better income.
2. *Better Management:* Specialization since is bestowing attention on a particular enterprise, it reduces the pressure on the farmer to care for several enterprises. Naturally it reduces the wastage of resources.
3. *Less Requirement of Equipment:* The farmer can carry on the business activity with the equipment that is required for the chosen enterprise. There are no pressing requirements to equip the farm with a variety of equipment.
4. *Increase in Skill of the Farmer:* The efficiency of the farmer increases as he can concentrate on one enterprise. His experience in the enterprise sharpens his skills in running the enterprise.
5. *Allows Better Marketing:* On marketing front, the farmer is better placed. He is saved from the pressure of finding market if he were to sell diversified products. It allows for better marketing functions *i.e.,* assembling, transport, grading, financing, *etc.*

## Disadvantages

1. *Failure of Crop:* The farmer runs the risk of loosing heavily in case, failure of crop occurs. There is no possibility of compensation. This is the biggest drawback of specialized farming.
2. *Non-utilization of Productive Resources:* Since the farmer confines to one or few enterprises, the various farm resources like land, water, labour, capital, *etc.*, may not be fully utilized. In view of the limited enterprises, some of the resources may remain untapped or under-utilized.
3. *Affect on Soil Health:* Continuous raising of one crop or few crops may be exerting greater pressure on soil health. This practice does not allow crop rotation, thereby affecting the soil health.

## 2. Diversified Farming

It is also known as general farming. Here farming is diversified *i.e.*, a number of enterprises are taken up on the farm at the same time. It also connotes production and sale of the different product at different times during a year. There is not much significance for a single enterprise under this situation. No single enterprise contributes as high as 50 per cent of the total income derived in farming. This type of farming is associated with the following advantages.

## Advantages

1. *Better Utilization of Farm Resources:* In view of the diversified cropping and crop rotations, land, labour and farm machinery and equipment are better utilized compared to specialized farming.
2. *Reduction of Farm Risks:* As a variety of crops are found, failure of one or two crops will not much affect the income from farming. Farmer can withstand the loss incurred from one or two enterprises.
3. *Flow of Income:* The farmer enjoys the advantage of deriving regular income, as different crops are grown.

## Disadvantages

1. *Ineffective Supervision:* The presence of a number of enterprises on the farm will stand in the way of the farmer in bestowing effective supervision. Effectiveness can be found when there is a limit to the number of enterprises. The diversified enterprises allow the scope for the leakages in the farm business go unnoticed. This is likely to affect the farm economy.
2. *Less Possibility for Maintaining a Variety of Implements and Machinery:* It becomes expensive to purchase and maintain the required suitable implements and machinery for the various enterprises taken up on the farm.
3. *Probable Marketing Insufficiencies:* The growing of a variety of crops is likely to bring in problems on marketing front. The farmers have to search for markets.

## 3. Mixed Farming

It represents a type of farming in which crop production and livestock production are combined to sustain and satisfy as many needs of the farmer as possible. There are limits specified regarding contribution of livestock production, poultry, fisheries, and bee keeping, *etc.*, to the gross income on the farm. These enterprises are supposed to contribute at least 10 per cent of gross income. However, this contribution should not exceed 49 per cent. Mixed farming facilitates the application of organic mannuring to soil, thus helping the maintenance of soil health. It provides employment to the farmer and his family throughout the year. Agricultural byproducts are properly used in mixed farming. It further provides a sort of stability to the farm business.

## 4. Dry Farming

Growing of crops entirely under rainfed conditions is known as dry land agriculture. Depending on the amount of rainfall received, dry land agriculture is categorized into dry farming, dry land farming and rainfed farming. Dry farming means cultivation of crops in areas where rainfall is less than 750 mm per annum. Crop failure is the most common due to prolonged dry spells during the crop period. Dry farming regions are equivalent to arid regions and moisture conservation practices are important in this region. Dry land farming is the cultivation of crops in regions with an annual rainfall of more than 750 mm. Dry spells during crop period occurs, but crop failures are relatively less frequent. Moisture conservation practices are necessary for crop production. Rainfed farming is crop production in regions with an annual rainfall of more than 1150 mm. It is practiced in humid regions where crop failures are rare and drainage is the important problem. In dry farming and dry land farming, emphasis is on soil and water conservation, sustainable crop yields and limited fertilizer use according to soil moisture availability. In rainfed agriculture the emphasis is on disposal of excess water, maximum crop yield, high levels of inputs and control of soil erosion.

## 5. Ranching

Grazing of livestock on public pastures is called ranching. These lands are not fit for cultivation.

## FACTORS INFLUENCING TYPES OF FARMING

Type of farming is influenced by several factors. These factors can be broadly grouped into two categories *viz.*, physical factors and economic factors, the details of which are presented below:

### 1) Physical Factors

1. *Climate*: It includes sunshine, rainfall, wind, length of sunlight, *etc*. These factors predominantly influence the choice of crops, thereby affecting the type of farming. Crops like paddy and sugarcane requires substantial water while others like oilseeds; millets and pulses can withstand low rainfall.

2. *Soil*: The type of soil, depth of soil and fertility status of the soil affect the selection of crops. Deep soils facilitate production of a variety of crops bringing prosperity to agriculture. If soils are fertile, manurial costs can be reduced thereby the cost of cultivation.

3. *Topography*: It means the general contour of the land, whether it is hilly or plain. Temperatures are low and growing seasons are shorter at higher elevation and therefore more suitable for establishment of plantation crops like tea, coffee, *etc*. Contour also determines the type of machinery that can be used and the rate of soil erosion, which in turn influences the choice of crop.

## 2) Economic Factors

1. *Relative Profitability:* Given the option of choosing among several crops in an area, farmers first look into the relative profitability of a given enterprise. A particular enterprise if found to be relatively profitable, farmers are inclined to go for it. In a dynamic situation the options are bound to change with changes in the relative profitability of crops. These changes occur in view of changes in yields, changes in prices, risk factors, introduction of incentives, crop restrictions, technology, *etc.*

2. *Availability of Funds:* Different enterprises require different levels of funds. For example, commercial enterprises like sugarcane, chillies, onion, prawn culture, *etc.,* require large amount of funds over enterprises like oilseeds, millets, *etc.* There-fore availability of funds determines the type of farming.

3. *Availability of the Inputs:* All the inputs like labour, seed, fertilizer, *etc.,* should be available as per the requirement. If labour shortages are common in an area during peak periods, certainly the farmers do keep this in mind in the selection of the enterprises. Timely availability of other inputs in required quantities also influences their decision with regard to selection of enterprises.

4. *Marketing Arrangements:* The marketing facilities available for the products are also given due weightage in the selection of enterprises. Mere assured output with inadequate marketing arrangements will not influence the farmers to opt for a given enterprise and

5. *Personal Choices:* Apart from aforesaid factors individual farmer's likes and dislikes have a say, in the selection of enterprises.

CHAPTER **26**

# Types of Farm Business Organizations

Types of farm business organizations are (1) Peasant farming (2) Co-operative farming (3) Capitalistic farming (4) Collective farming and (5) State farming

## 1) Peasant Farming

This is the most commonly followed form of farm business organization. In this form of farming, individual farmer is the owner, manager and organizer of the farm. He is totally independent in his decisions, as he is the owner. Compared against other farms, one drawback here is the resource limitation including the land at the disposal of the farmer. He has to judiciously use these resources to run the farm business profitably.

## 2) Co-operative Farming

It is a form of farm business organization in which persons unite voluntarily, pool their lands, resources and cultivate the lands jointly. The pooling of land, labour and capital makes it possible to fully utilize the resources and thereby increase the production capacity of the pooled land most economically. The birth of co-operative farming took place in Palestine. Jews from Palestine are the pioneers who have organized these societies.

The All India Co-operative Planning Committee (1951) classified co-operative farming into four categories *viz.*, (i) Co-operative better farming society, (ii) Co-operative joint farming society, (iii) Co-operative tenant farming society; and (iv) Co-operative collective farming society.

### (i) Co-operative Better Farming Society

Any ten or more cultivators can form into a better farming society. Here the ownership as well as the operation is individual. All members abide by a plan of cultivation adopted by the society. The society arranges for the adoption of better methods of cultivation, joint purchase of inputs, *etc*. Every member is free to follow his own plan except for the purposes agreed. The administration is done by a managing committee, elected by the general body and it governs all the activities.

### (ii) Co-operative Joint Farming Society

In this category of the society, the ownership is individual and operation is collective. Individual owners of land merge their lands and other resources in order to form into compact blocks for joint cultivation. This pooling of land permits the scientific methods of cultivation. The management of the society is in the hands of an elected body. All the members work jointly and each member gets wages for his daily work. The ownership of the land is recognized by the payment of dividend in proportion to the value of his land.

### (iii) Co-operative Tenant Farming Society

In this case landless people organize into a farming society. Here the ownership is collective but the operation is individual. Land of the society is divided into plots and leased out for cultivation to individual cultivators. Society draws a plan for the best cultivation of the land. The society arranges for the cultivation requirements and also the marketing of the produce. Every member is responsible for the payment of rent only to the society.

### (iv) Co-operative Collective Farming Society

In this society, ownership as well as operation is collective. Members do not have any right on land. Administration is the same as that of joint farming society. Members work jointly and they are paid according to the volume of work contributed. Profits are distributed according to labour and capital invested by the farmer.

### 3) Capitalistic Farming

Under capitalistic farming, the farms are large and owned by the capitalists. These farms enjoy plentiful resources. These are managed by either individuals or groups. These farms are quite efficient using the latest technology available, but this form of farming is not a common feature in India. Capitalistic farms are popular in countries like USA, Canada, UK, *etc.* Farms enjoy plenty of capital and organizational strength. The limitation of this type of farm business organization is that the actual cultivator is not the owner of the farm.

### 4) Collective Farming

Collective farm is a group of farming families who pool their resources *i.e.,* land, livestock, machinery, *etc.,* and undertakes to work together under a management committee elected by them. The management committee is responsible for the overall farm management, allocation of work and distribution of income. This type of farming is not prevalent in India. Russia and China follow this form of farm business organization. The production plan on the farms is laid down by the State and the farms have to sell a fixed portion of the output to the State. Incentives are paid to the farmers for the improvement of the farm.

### 5) State Farming

These farms are owned, managed and run by the State. State farms are equipped with latest machinery and equipment and employ scientific methods of cultivation. The workers on these State farms are wage earners. Besides producing foodgrains, the State farms produce raw products like cotton, wool, *etc.* State farming is suitable in newly reclaimed areas, where large scale mechanized cultivation is possible.

# CHAPTER 27

# Farm Planning

Farm planning is the foundation of management. Farm planning precedes all other managerial functions. Without setting the objectives and line of action to be followed, there is nothing to organize, direct or control in the organization of farm business. It is the determination of a course of action to achieve the desired results. It is deciding in advance, the production management problems *viz.*, what to produce, how to produce, when to produce; financial management problems *viz.*, how to borrow, how much to borrow, when to borrow, where to borrow, and marketing management problems *viz.*, where to buy and sell, when to buy and sell, how to buy and sell, *etc.* Farm planning bridges the gap from where we are to where we want to go. It is characterized as the process of thinking before doing. It is an intellectual process. It requires a mental predisposition to think before acting, to act in the light of facts rather than of guesses. Farm planning governs the survival progress and prosperity of farm organization in a competitive and dynamic environment. It is a continuous and unending process. Farm planning is not a new technique. It is as old as farming itself. All farmers whether progressive or backward, literate or illiterate, plan their farm business at the beginning of every crop season as to what crops and livestock enterprises to produce, what amount of resources to be used, how the various operations to be organized, what amount of credit to be borrowed, *etc.*, in their minds. This is known as informal planning. As the agriculture is becoming a more complex business, the farmers cannot remember all the necessary information required in order to plan their farm business in a systematic manner. To be more systematic in thinking, the farmers need formal or written farm plans. If the production intentions of the farmers are committed on a paper, then it is called a written plan. *Farm planning is the deliberate process of thinking, the organized foresight, and the vision based on facts and past experience that is needed for intelligent action on the farm.*

## WHY FARM PLANNING IS NECESSARY?

1. To choose different farm activities which are suited to the given farm conditions.
2. To look into the future and decide on suitable course of action.
3. To select appropriate enterprise combinations that results in the better use of resources.
4. To help the farmers in timing various jobs and operations for smooth conduct of operations without competition.
5. To avoid wastages that occur in the resource use.

6. To provide guidance and flexibility to the farmers for ensuring better use and growth of the farm.
7. To provide allocation of resources for producing the requisite products for marketing and household consumption.

### Types of Farm Plans

Farm plans are categorized into two sub-groups *viz.*, simple farm plan and complete farm plan. Simple farm plan implies planning for minor changes or for a particular enterprise. Complete farm planning envisages more number of changes in the existing organization. It is adopted for the farm as a whole.

## CHARACTERISTICS OF GOOD FARM PLAN

The following are the characteristics of a good farm plan:

1. Plans should aim at efficient utilization of all the available resources on the farm.
2. Plans should be flexible *i.e.*, they should be adaptable to changing environmental conditions.
3. Farm plans should be simple and easily understood. Complex plans consume much time and money, hence are seldom followed. They should take into account the most important suitable farm enterprises, identifying their strengths and weaknesses.
4. Farm plans should ensure balanced production programme considering the available resources on the farm. The production programme should consist of food crops, commercial crops and fodder crops.
5. The production programme included in the farm plan should aim at maintaining/improving soil fertility. This is possible through suitable crop rotation practices.
6. Farm plans should facilitate efficient marketing of farm products.
7. It should take into account up-to-date technology.
8. Farm plans should consider the goals, knowledge, training and experience of the farmers, besides their attitude towards risk.
9. Farm plans should avoid too risky enterprises.
10. Farm plans should provide for borrowing, using and repaying the credit.

## LIMITATIONS OF FARM PLANNING

Farm planning has been considered as time consuming and expensive device. Planning encourages a false sense of security against risks. Forecasting methods, statistical data supplied, *etc.*, are all inaccurate and the results of operations research cannot be applied to all cases that come under farm planning. Good farm plans should be based on the actual recorded facts, particularly giving the data on the availability and requirement of resources. The records provide adequate information for planning process, but it is unfortunate to note that relevant farm records are not given importance by the farmers in developing countries. Though some farm records are maintained by the farmers, but these suffer from inadequate and inaccurate data. The pertinent information on farms particularly in respect of climate, water supply, markets, *etc.*, is not found in the required form. The sources of data for diagnosis and planning are also lacking. As a result, farm planning is not effectively formulated and implemented. Therefore, farm standards derived from research stations and efficient farms in the locality should form the basis for scientific planning. Data from research stations should be continuously used for this purpose.

# FARM BUDGETING

Farm plan is a programme of total farm activity drawn up by the farmer in advance. The expression of farm plan in monetary terms is called farm budgeting. Farm budgets are classified into enterprise budget, partial budget and complete budget or whole farm budget. Farm budgeting is a method of examining the profitability of alternative farm plans.

# FARM ENTERPRISE BUDGET

A commodity that is being produced on the farm is called farm enterprise. Farm budgets can be developed for each potential enterprise. Enterprise budgets are prepared in terms of a common unit *i.e.*, acre, hectare, for a crop, one head of livestock, *etc*. This facilitates easy comparison between the enterprises. Enterprise budget is the estimation of expected income, costs and profit for an enterprise.

# ORGANIZATION OF ENTERPRISE BUDGET

It consists of three elements *viz.*, income, costs and profitability. Income is computed by estimating the expected output and expected price (Table 27.1). The estimated output is the average yield under normal weather conditions. Output price should be the average price expected in future. In order to estimate the variable costs we need information on quantity of inputs used and the prices at which they are purchased. Fixed costs to be included in enterprise budget are land revenue, depreciation, interest on fixed capital and rental value of owned land.

TABLE 27.1   Enterprise Budget for Groundnut per Hectare.

| | Item | Rs./ha |
|---|---|---|
| A) | Yield 9.54 quintal  @ Rs. 1332 | 12,707.28 |
| B) | i) Variable costs | |
| | Human labour | 1,843.93 |
| | Bullock labour | 920.05 |
| | Machine labour | 516.96 |
| | Seeds | 2,352.78 |
| | Manures and fertilizers | 1,724.27 |
| | Plant protection chemicals | 189.10 |
| | Repairs | 91.59 |
| | Interest on working capital | 229.16 |
| | Total variable costs | 7,867.84 |
| | ii) Fixed costs | |
| | Land revenue | 15.00 |
| | Rental value of owned land | 2,430.54 |
| | Depreciation | 137.47 |
| | Interest on fixed capital | 178.19 |
| | Total fixed costs | 2,761.20 |
| | iii) Total costs  (i + ii) | 10,629.04 |
| | Estimated profit        12,707.28 – 10,629.04      =      2,078.24 | |

# PARTIAL BUDGETING

Partial budgeting is a statement of anticipated changes in costs, returns and profitability for a minor modification.

When a farmer contemplates few modifications or minor changes in the existing organization of the farm business, partial budgeting technique is employed. It is similar to that of marginal analysis, wherein the changes in costs and returns resulting from proposed modifications are alone considered. It consists of four important elements *viz.*, added costs, added returns, reduced returns and reduced costs. Partial budgeting technique is generally used to evaluate the profitability of input substitution, enterprise substitution and scale of operation.

1. *Added Costs:* Additional costs are incurred, if the proposed modification is the introduction of a new enterprise or increase in the size of the existing enterprise.
2. *Added Returns:* Additional returns could be received when the proposed modification is the addition of a new enterprise, or increase in the size of the existing enterprise or adoption of technology that results in higher productivity.
3. *Reduced Returns:* Decrease in the returns is observed when the proposed modification involves the elimination of an existing enterprise or reduction in the size of the existing enterprise.
4. *Reduced Costs:* Decrease in the costs is found when the proposed modification involves the elimination of existing enterprise or reduction in the size of the enterprise or adoption of a technology that uses fewer amounts of resources.

## Partial Budget

*Example 1:* Proposed modification to control tikka leaf spot in JL 24 groundnut variety

| Items | | Items | |
|---|---|---|---|
| Added costs | | Added returns | |
| 200 grams Carbendazim + 500 grams Mancozeb | Rs. 628/- | Yield 187 kg @ 12.56 | Rs. 2,348.72 |
| Reduced returns | NIL | Reduced costs | NIL |
| Total of added costs and reduced returns | Rs. 628/- | Total of added returns and reduced costs | Rs. 2,348.72 |
| Net change = Rs. 2,348.72–628 | Rs. 1,720.72 | | |

The expenditure on fungicides and groundnut yield in existing and alternate situations are presented below.

| Existing situation | Alternate situation |
|---|---|
| To control tikka leaf spot of groundnut carbendazim 0.1% + mancozeb 0.25% @ 400 grams + 1000 grams = Rs. 1,257 | To control tikka leaf spot groundnut carbendazim 0.1% + mancozeb 0.25% @ 600 grams + 1500 grams = Rs. 1,885 |
| Yield = 583 kg @ Rs. 12.56 per kg. | Yield = 770 kg @ Rs. 12.56 per kg. |

The existing practice is the application of 400 grams of carbendazim and 1000 grams of mancozeb to control tikka leaf spot in JL-24 variety of groundnut. The cost of these fungicides is Rs. 1,257. The yield obtained is 583 kg of groundnut pods. If 600 grams of carbendazim and 1,500 grams of mancozeb were applied at a cost of Rs. 1,885, it would result in a yield of 770 kg. The additional yield through this practice is 187 kg. The incremental income would be Rs. 1,720.72.

*Example 2:* Substitution of sunflower for groundnut:

| S. No. | Particulars | Existing situation | Alternate situation |
|---|---|---|---|
| | | Groundnut (Rs.) | Sunflower (Rs.) |
| a) | Human labour | 1,474.00 | 1,017.00 |
| b) | Bullock labour | 872.00 | 831.00 |
| c) | Manures | 864.00 | 902.00 |
| d) | Fertilizers | 1,164.00 | 1,023.00 |
| e) | Seed | 2,200.00 | 800.00 |
| | Total operational costs/ha | 6,574.00 | 4,573.00 |
| | Gross income/ha | 9.20 Q @ 1212.82  11,158.00 | 9.86 Q @ 1144.01 Rs. 11,280.00 |

| *Added costs* | | *Added returns* | |
|---|---|---|---|
| Manures | Rs. 38.00 | | Rs. 122.00 |
| *Reduced returns* | NIL | *Reduced costs* | |
| | | Human labour | 457.00 |
| | | Bullock labour | 41.00 |
| | | Fertilizers | 141.00 |
| | | Seed | 1,400.00 |
| Added costs | Rs. 38.00 | Total reduced costs | Rs. 2,039.00 |
| Total of added costs | | Total of added returns + | |
| and reduced returns = Rs. 38 | | reduced costs = 122 + 2,039 = Rs. 2161. | |

Net change = Rs. 2161 – 38 = Rs. 2123

# COMPLETE BUDGETING

It is a method of estimating expected income, expenses and profits for the farm as a whole. Complete budgeting is employed when farmers want to overhaul the entire farm business.

# STEPS IN WHOLE FARM PLANNING AND BUDGETING

A systematic procedure is generally followed in making sound farm plans for the success of the farm business. The sound farm plan should be generally feasible, acceptable, and adaptable. To make the farm plan successful, the following steps should be adopted with relevance to a given farm and its resources.

1. Statement of objective.
2. Diagnosis of the existing organization.
3. Assessment of resource endowment on the farm.
4. Identification of enterprises to be included.
5 Preparation of enterprise budgets.
6. Identification of risks, and
7. Preparation of a plan.

## 1) Statement of Objective

The objective of the farmer may be profit maximization or cost minimization. In selecting enterprises and their combinations, the farmer aims at maximization of profits. On the other hand, while choosing resources and their combinations, he aims at cost minimization.

## 2) Diagnosis of the Existing Organization

Diagnosis and prescription are the two important components of planning. The planner has to examine the existing organization of farm business carefully and identify the weaknesses or defects or loopholes in the current plan. Once mistakes are identified, corrective steps can be taken in future. Farm plans primarily prescribe remedies for the defects of the existing plan. For example, continuous monocropping is considered as weak point in the existing plan. This should be replaced by an appropriate crop rotation practice *i.e.*, growing pulses and legumes after a cereal crop to increase soil fertility and reduce the incidence of pests and diseases. Non-adherence of the recommended package of practices is also a serious lapse on the part of the farmers resulting in heavy monetary loss. This should be appropriately weighed in the formulation of the plan. Poor drainage leads to heavy crop loss in certain farming areas. Hence, this problem should be carefully viewed in formulating the farm plans.

## 3) Assessment of Resource Endowment on the Farm

a) *Land:* Here there is a need to spell out the land holding area, type of land *i.e.*, wet land or dry land, crops grown, type of soils available, topography, texture, fertility status, drainage, soil and water development, soil and water conservation methods, *etc.* If the land is sloppy, conservation practices are very essential. If the farmer has followed any conservation measures, we have to specify the costs of such measures here. If the soil is having drainage problem then measures taken up by the farmer are indicated. If the soils are highly fertile high-income crops are grown. Thus selection of the crops particularly high yielding varieties and relevant package of practices should be based on type of the soil. Assessment regarding plant nutrients and minerals present in the soil should be taken up with the help of soil testing laboratories and recommendations should be made with regard to the use of NPK and micronutrients. If the soil is acidic or alkaline, then corrective measures should be followed consulting the soil scientists.

b) *Labour:* The extent of family labour available with the farmer *viz.*, women, men and children along with their age, household work and farm work done by them should be indicated. Permanent labourers if any engaged by the farmer, type of work done and amount of remuneration paid should be indicated. Perquisites given to the permanent labourers are also added to the remuneration. Labour supply, in the village and demand for labour for different crops in different seasons should be assessed. The actual wage rates paid for different kinds of labourers considering the peak and slack seasons should be indicated. Peak periods of labour demand and wage rates should be included. Assessment of supply and demand for cattle labour and machine labour for each type of crop in different seasons should be done realistically. The supply position with reference to livestock should be assessed correctly.

c) *Capital: (i) Working Capital:* Working capital required for raising crops should be indicated. Owned funds available and the amount of funds borrowed, from different sources, interest paid, *etc.* need to be clearly specified. Specification of repayment dates, terms and conditions, *etc.*, is also required. (ii) *Fixed Capital:* This relates to information on farm buildings, farm equipment, farm machinery, *etc.*

d) *Organization:* The farmer's knowledge in farming, his expertise, his experience in farming and confidence in adapting new potential technology should be assessed. Based on this information relevant farm plans should be devised. If the farmer is risk-averse, farm plans, which provide stable income under risk, should be generated.

e) *Irrigation Source:* Availability of different sources of irrigation, area covered under different sources, period of availability of irrigation, quantity of irrigation water available, crop demands for irrigation water, accessibility of land to the irrigation sources such as canal and tank, *etc.*, should also be indicated. In addition to this cost of irrigation needs to be mentioned.

## 4) Identification of Enterprises to be Included

List of enterprises not only grown by the farmer but also enterprises grown in that area and also crop rotations are identified. Estimate the input-output coefficients in terms of acre or hectare or head of livestock for all the enterprises, which we propose to include. Information on input and output prices should be collected so as to work out the costs and returns.

## 5) Preparation of Enterprise Budgets

Estimate the income, costs and profitability of each enterprise to be included in the plan. The preparation of enterprise budgets facilitates comparison of profitability of different enterprises.

## 6) Identification of Risks

We should list out all types of risks *viz.*, production risk, weather risk, technological risk, institutional risk, marketing risk, *etc.*, faced by the farmers. Particularly the incidence of pests, rodents and diseases, frequency of drought occurrence over time, cyclones, floods and their havoc caused to farm production should be kept in view in formulating relevant alternative farm plans. Marketing risks comprising of risks emanating from price fluctuations and failure of markets to arrest the malpractices of middlemen should be indicated.

## 7) Preparation of a Plan

Here the first step is identifying the most scarcest resources and selecting that enterprise which yields maximum returns per unit of scarcest resource. This process is repeated till all the scarce resources are put to the best use which results in optimum combination of the enterprises.

# CHAPTER 28

# Application of Operations Research Techniques to Farm Management

Of late, allocative and decision problems of agriculture are being given crucial importance in the literature and research related to agricultural economics. To such problems, operations research methods are the apt devices for taking up appropriate actions. For instance, the decision problems like what to produce, how much to produce, in what combinations and with what financial activities involved in profit maximization of the business are solved in the quantitative sense through operations research methods. Operations research methods are used for the attainment of set goals, ends or objectives under a given set of resource endowment base represented by technology, prices and resource restrictions. Operations research methods provide numerical solutions for the decision variables and hence, they have very high potential use in scientific planning.

## LINEAR PROGRAMING (LP)

In linear programming models, the objective of the typical farm *i.e.*, maximization of net profit or cost minimization is achieved through optimal plan generated from its solution. Hence, it is that method of determining the best or optimal plan to the given farm under the given linear constraints. The objective function specified, *i.e.*, profit maximization or cost minimization, is linear in form and constraints on resource restrictions are specified in linear form. In non-linear programming, objective function is specified in non-linear form, *i.e.*, power form. LP had been used in agriculture since 1950s. It provides prudent solutions to whole farm planning problems. In this context the works of Heady and Love (1952), Heady and Candler (1958), Samuel (1961) Steven (1961), *etc.*, are worth citing. As a normative tool, it always aims at combining the efficient enterprises giving weightage to constraints and profit maximization.

### Components of LP Problem

There are three quantitative components in LP model. They are: (1) an objective function; (2) resource requirements; and (3) resource availability.

Algebraically it is stated in compact form as:

Maximize $\pi = C'X$ (4.1)

Subject To

$AX \underset{\geq}{\leq} B$ (4.2)

$X \geq 0$

Where,

A is $m \times n$   matrix of technical coefficients

C is $n \times 1$   vector of prices or other weights for the objective function

X is $n \times 1$   vector of activities (crops and livestock to be produced which are unknown decision variables)

B is $m \times 1$   vector of resources or other constraints, availabilities in physical units, such as labour, credit, land, *etc.*, and the objective function.

$$C'X = \pi,$$

In expanded form it is written as:

Maximize   $\pi = c_1 X_1 + c_2 X_2 + \quad ... \quad + c_n X_n$ (4.3)

Subject To

$$a_{11} X_1 + a_{12} X_2 + \quad ... \quad + a_{1n} X_n \leq b_1$$

$$a_{21} X_1 + a_{22} X_2 + \quad ... \quad + a_{2n} X_n \leq b_2$$

$$\vdots \qquad \vdots \qquad\qquad \vdots \quad \vdots$$

$$a_{m1} X_1 + a_{m2} X_2 + \quad ...+ a_{mn} X_n \leq b_m$$ (4.4)

and $X_j$ (columns 1 to $n$) all should be specified in positive values starting from zero or any positive value ($X_j \geq 0$), $a_{ij} = i^{th}$ resource required to produce one unit of $j^{th}$ crop or livestock activity. $\pi$ is profit or gross margins from the whole farm.

With $\Sigma$ notation it is written as:

Maximize   $\pi = \sum\limits_{j=1}^{n} C'_j X_j$ (4.5)

Subject To

$$\sum\limits_{j=1}^{n} a_{ij} X_j \leq B_i$$ (4.6)

$$\sum\limits_{j=1}^{n} X_j \geq 0$$ (4.7)

$B_i = i^{th}$ resource available with farm for use in the production of crops and livestock. $i$ ranging from 1 to m denoting the number of rows (constraints) in the problem. $j$ ranging from 1 to n indicating number of columns (crop and livestock activities) in the problems.

## Assumptions of LP Problem

There are seven basic assumptions, *viz.*, (1) linearity of the objective function, (2) proportionality of activities to resources, (3) additivity of the resources and activities,

(4) divisibility of the activities as well as resources, (5) finiteness of the activities and resource restrictions, (6) single value expectation and (7) non-negativity of the decision variables:

1. *Linearity of the Objective Function:* All the decision variables in the objective function, *i.e.,* crop and livestock activities are in linear form (without power form) and the objective function is specified for example, as $\pi = 250\,X_1 + 350\,X_2 + 500\,X_3 + \ldots + 400\,X_n$. The coefficients of $X_1$ are the net returns/prices of the crops and livestock. LP problems should be specified invariably in linear form both in objective function and constraints.
2. *Proportionality of Activities to Resources:* According to this assumption, linear relationship is held between activities and resources. This means that resource requirement to produce one unit of crop or livestock activity varies directly with the level of output of crops and livestock.
3. *Additivity of the Resources and Activities:* This assumption implies that the total quantity of a resource used must be equal to the total quantity of resource used by each activity for all resources individually and collectively. If the resource is used up fully, it should equal the sum of the same resources used by all the activities appearing in the optimal solution. This condition holds good for all the resources specified in the model.
4. *Divisibility of the Activities as well as Resources:* Continuity of resources and output is implied in this assumption. This means fractional quantities such as 0.2 ha of land and 3.5 qtl of paddy, *etc.,* are allowed. But divisibility for livestock activities and labour resources appears to be unrealistic. To get integer values for such livestock activities an integer programming is being used.
5. *Finiteness of the Activities and Resource Restrictions:* With the advent of computers and availability of programmes, a large number of activities and constraints are now being specified in the model. But, there should be a limit for such number, because infinite number of activities and resource restrictions cannot be accommodated in the model. Hence, this assumption is important in the LP model. In general, it is desirable to have more number of activities than the constraints in LP model.
6. *Single Value Expectation:* This assumption connotes certainty assumption and imparts to the LP model, the name of deterministic model. According to this assumption, input-output coefficients ($a_{ij}$), resource availabilities ($B_i$) and prices of activities ($C_j$) are all specified correctly with the known quantities in the model and they all relate to a particular period of time. In the risk programming models this assumption is relaxed.
7. *Non-Negativity of the Decision Variables:* All the crops and livestock activities should have positive value in their magnitude. Negative values for such decision variables cannot make any sense. Hence, this assumption is imperative.

## Basic Concepts in LP

1. **Goals of the Programming Model:** Programming model guides the farmer to specify the farm plans which will give him maximum income under the given constraints, prices, yields and resource requirements. Cost minimisation in the cattle feeding problems, poultry feeding problems and transportation models, is considered in the objective function of LP model.
2. **Activity or Process:** The word activity is used to refer to crop and livestock enterprises being undertaken. A typical method of production with specific re-

source requirements in crops and livestock is referred to as a process or activity. Based on this concept, crops or livestock activities are delineated into separate or individual activities in the model. For example, local paddy crop requiring different levels of inputs for obtaining various output levels are treated as separate activities. Similarly, if two cows of the same breed are reared on different rations, they can be taken as separate activities in the model. A process is a method of converting a resource into a product with specified input-output relationship. This is also often referred to as technical coefficient. In the literature, we find these two terms being used synonymously.

*Types of Activities:* These are: (i) real activities, (ii) intermediate activities; (iii) purchasing, (iv) selling and (v) borrowing activities.

1. Paddy, sugarcane, poultry eggs, milch cattle, *etc.*, are real activities because they are produced on the farm for sale in the market. Real activities are also called decision variables, which are specified, in the objective function of the LP problem. The optimal solution indicates the magnitudes of real activities and hence they are called decision variables. Suppose the optimal solution obtained from the LP problem is quite different from the actual practice of the typical farm, particularly in respect of crop areas and livestock number, then the decisions of manager are to be altered according to the solution in order to get the maximum profit.

2. Fodder, though produced on the farm and if not sold in the market, it cannot become real activity, so it is intermediate activity.

3. Purchasing activities mean the inputs like fertilizers and pesticides, which are purchased from the market and used in the production process.

4. Selling activities represent the sale of products produced on the farm.

5. To supplement owned funds, depending on the need, borrowing activity is included in the LP model.

Prices for products and resources are to be ascertained with certainty. Too high or too low prices will distort the income estimates and cost and thereby profit, often leading to results of unrealistic magnitude. In general, the average prices, pooled over three to five years are considered for LP model.

3. **Restraints:** These are also called limitations or constraints. Land, labour and capital are generally considered as restraints. In the development of models for obtaining realistic results, sometimes 150 to 200 restraints are also considered by researchers in economic studies. In general, macro level studies will have more constraints than micro level studies, because of the complexities involved in macro level situation. At micro level the farmers may have restrictions regarding number of livestock animals, crop acreages, *etc.* Amount of labour availability during peak seasons of the crop growth is generally considered as the most common restriction seen in the LP model. Likewise, a farmer may have access to limited quantities of many resources. The availability and requirements in respect of machine labour, bullock labour, hired human labour, family labour, skilled labour, unskilled labour, *etc.*, in different time periods, *i.e.,* a week, a month, a season and a year may be considered in the programming model as separate restrictions or constraints. All these restrictions can be specified in the model in three types, *i.e.,* greater than equal to constraints ($\geq$) or less than equal to constraints ($\leq$) or equal to or equality constraints ($=$). The equality constraints are called balance rows. Management constraints inhibiting the desired size of the farm can be included in LP model at times.

*Procedure:* In the following sections an attempt is made to provide a better understanding of the format of the LP model and explanation of the computation procedures involved in solving the LP model. Finally, the interpretation of the optimal solution is presented.

*Statement of the Problem:* Crop production problem to maximize gross margins* from crop enterprises is shown below subject to the following restraints for a typical farmer in Chittoor district of Andhra Pradesh.

Availability of resources on the farm

| Land | 5 acres |
|---|---|
| Family labour availability in *Kharif* season | 180 days |
| Owned capital (in hundred Rs.) | Rs. 50 |

*Crop activities are:* Paddy (HYV), paddy (improved) and groundnut. Crop activities are defined in units of 1 acre. Gross margin is the return after subtracting the variables costs from gross income. These are the $X_j$ values in the objective function.

Gross margin for the given problem are

| Paddy (HYV) *i.e.*, (paddy I) | Rs. 30 |
|---|---|
| Paddy (improved) (in hundreds) (paddy II) | Rs. 28 |
| Groundnut (in hundreds) | Rs. 26 |

Resource requirements are specified in terms of crop unit areas, *i.e.*, per acre basis for land, labour and capital resources. These are specified in the matrix format of the LP problem as $a_{ij}$ coefficients, *i.e.*, $i^{th}$ resource per one acre of $j^{th}$ crop. The returns from the crop activities are given in the objective function of the problem, *i.e.*, $c_j$ values are given in the bottom row of the matrix (Table 28.1).

**TABLE 28.1 Crop Production Problem Arranged in Matrix Format (Initial Table).**

| Net income | Restriction | Restriction level (b) | Production Activities | | | Disposal activities | | |
|---|---|---|---|---|---|---|---|---|
| | | | $(X_1)$ Paddy I (1 ac) | $(X_2)$ Paddy II (1 ac) | $(X_3)$ G. nut (1 ac) | Land | Labour | Capital |
| 0 | Land | 5 | 1 | 1 | 1 | 1 | 0 | 0 |
| 0 | Labour | 180 | 90 | 80 | 70 | 0 | 1 | 0 |
| 0 | Capital | 50 | 20 | 18 | 25 | 0 | 0 | 1 |
| | Net Income | (C) | 30 | 28 | 26 | 0 | 0 | 0 |
| | Opportunity cost | (Z) | 0 | 0 | 0 | 0 | 0 | 0 |
| | Z-C (Shadow price or net evaluation row) | | -30 | -28 | -26 | 0 | 0 | 0 |

NB: 1 ac= one acre
X variables are the unknowns in the initial table of the LP problem.

---

*Gross margins are gross incomes less variable costs per acre of crop activities.

## Simplex Method

It is an algorithm adopted to solve LP problem, which allows us to choose an initial basic feasible solution with all the real activities at zero level, and disposal activities at the largest positive level to arrive at the optimal solution through iterations.

## Steps Involved in Solving the LP Problem

The LP problem is algebraically stated as

$$\text{Maximize } \pi = 30 \ X_1 + 28 \ X_2 + 26 \ X_3 \tag{4.8}$$

$$\text{Subject To} \quad 1X_1 + 1X_2 + 1X_3 \leq 5 \tag{4.9}$$

$$90 \ X_1 + 80 \ X_2 + 70 \ X_3 \leq 180 \tag{4.10}$$

$$20 \ X_1 + 18 \ X_2 + 25 \ X_3 \leq 50 \tag{4.11}$$

$$X_1 \geq 0, X_2 \geq 0 \text{ and } X_3 \geq 0 \tag{4.12}$$

*Steps 1:* All the three inequality constraints should first be reduced to equality constraints by introducing disposal activities in order to facilitate the non-use of resources. In adding disposal activities, we must consider a number of resource constraints in the problem. If there are *'m'* constraints to the given problem, then *'m'* number of disposal activities should be added. In our example, there is need to add three disposal activities and they are $X_4$ = land disposal activity, $X_5$ = labour disposal activity, $X_6$-capital disposal activity. According to this, the given problem of the equation (4.8) is restated as shown in the equation (4.13) through (4.17).

$$\text{Maximize } \pi = 30 \ X_1 + 28 \ X_2 + 26 \ X_3 + 0 \ X_4 + 0 \ X_5 + 0 \ X_6 \tag{4.13}$$

Subject To

$$1 \ X_1 + 1 \ X_2 + 1 \ X_3 + 1 \ X_4 + 0 \ X_5 + 0 \ X_6 = 5 \tag{4.14}$$

$$90 \ X_1 + 80 \ X_2 + 70 \ X_3 + 0 \ X_4 + 1 \ X_5 + 0 \ X_6 = 180 \tag{4.15}$$

$$20 \ X_1 + 18 \ X_2 + 25 \ X_3 + 0 \ X_4 + 0 \ X_5 + 1 \ X_6 = 50 \tag{4.16}$$

$$X_1 \geq 0, X_2 \geq 0, X_3 \geq 0, X_4 \geq 0, X_5 \geq 0, \text{ and } X_6 \geq 0 \tag{4.17}$$

Equation (4.14) states that land requirement of the plan including the land disposal activity ($X_4$) must be equal to 5 acres of land (*i.e.,* the available land for allocation to different crops).

Equation (4.15) reveals that labour requirement for three crops in the plan including labour disposal activity ($X_5$) must equal to 180 labour days which is the labour availability for allocating to different crops. Similar implication holds good for equation (4.17).

$\pi$, the gross margin level specified in equation (4.13) assumes zero gross margin and non-use of resources. This problem is modified in the matrix form in the following manner.

$$\text{Maximize } \pi = [30 \quad 28 \quad 26 \quad 0 \quad 0 \quad 0] \begin{pmatrix} X_1 \\ X_2 \\ X_3 \\ X_4 \\ X_5 \\ X_6 \end{pmatrix}$$

Subject To

$$\begin{pmatrix} 1 & 1 & 1 & 1 & 0 & 0 \\ 90 & 80 & 70 & 0 & 1 & 0 \\ 20 & 18 & 25 & 0 & 0 & 1 \end{pmatrix} \quad \begin{pmatrix} X_1 \\ X_2 \\ X_3 \\ X_4 \\ X_5 \\ X_6 \end{pmatrix} = \begin{pmatrix} 5 \\ 180 \\ 50 \end{pmatrix}$$

$$\qquad A \qquad\qquad\qquad Im$$

$$\begin{pmatrix} X_1 \\ X_2 \\ X_3 \\ X_4 \\ X_5 \\ X_6 \end{pmatrix} \geq 0 \qquad\qquad\qquad\qquad (4.18)$$

*Step 2:* Here we will have to define the initial feasible solution. A feasible solution is one that satisfies all the constraints including non-negativity constraints. In such a solution all real activities $(X_1, X_2, X_3)$ are at zero level. Such a solution is called an initial basic feasible solution.

Hence, $\qquad \pi = 0$

$$X_4 = 5$$

$$X_5 = 180$$

$$X_6 = 50$$

General structure of the initial table of the simplex method is shown in Section 1 of the Table 28.2. This table contains 10 columns. First column contains net income, second column name of the restriction, third column restriction level, fourth, fifth and sixth columns production activities, seventh, eight and ninth column disposal activities and finally the tenth column contains ratios. The 'C' row (gross margin) shows the net profits for each activity unit. The 'Z' row (opportunity cost row) shows for each activity the value of other activity which must be sacrificed to produce one more unit of output. All values in 'Z' row are zero because all resources are unused in the I Section.

*Step 3:* In this step, we should attempt to increase the profits from the initial basic feasible plan. Here we begin a routine substituting of real activities $(X_1, X_2$ and $X_3)$ for disposal activities $(X_4, X_5$ and $X_6)$ in the plan. This procedure leads quickly to optimal plan.

## Procedure for Computation of Iterations

(1) The column with the larger Z-C value in the shadow price row is the outgoing column from Section I and it is the incoming row in section II.

(2) To compute the ratio column, *i.e.*, column 10 in Section I, divide resource available quantities in the B column (column 3) of the I Section by the coefficients in

## TABLE 28.2 Simplex Table.

| Net price (I) | Name of restriction (II) | Resource level (b) (III) | Production Real Activities | | | Disposal Activities | | | Ratio |
|---|---|---|---|---|---|---|---|---|---|
| | | | Paddy I ($X_1$) (IV) | Paddy II ($X_2$) (V) | Groundnut ($X_3$) (VI) | Land ($X_4$) (VII) | Labour ($X_5$) (VIII) | Capital ($X_6$) (IX) | (X) |
| **SECTION I** | | | | | | | | | |
| 0 | Land ($X_4$) (acres) | 5 | 1 | 1 | 1 | 1 | 0 | 0 | 5/1 = 5 |
| 0 | Labour ($X_5$) (days) | 180 | 90* | 80 | 70 | 0 | 1 | 0 | 180/90 = 2 |
| 0 | Capital ($X_6$) (in '00 Rs.) | 50 | 20 | 18 | 25 | 0 | 0 | 1 | 50/20 =2.5 |
| | Net price (C) | | 0 | 28 | 26 | 0 | 0 | 0 | |
| | Opportunity cost (Z) | | 0 | 0 | 0 | 0 | 0 | 0 | |
| | Shadow Price (Z-C) | | -30 | -28 | -26 | 0 | 0 | 0 | |
| **SECTION II** | | | | | | | | | |
| 0 | Land ($X_4$) (acres) | 3 | 0 | 1/9 | 2/9 | 1 | -1/90 | 0 | 13½ |
| 30 | Labour ($X_1$) (days) | 2 | 1 | 8/9 | 7/9 | 0 | 1/90 | 0 | 24/7 |
| 0 | Capital ($X_6$) (in '00 Rs.) | 10 | 0 | 2/9 | 85/9* | 0 | -2/9 | 1 | 11/17 |
| | Net price (C) | | 30 | 28 | 26 | 0 | 0 | 0 | |
| | Opportunity cost (Z) | | 30 | 80/3 | 70/3 | 0 | 1/3 | 0 | |
| | Shadow Price (Z-C) | | 0 | -4/3 | -8/3 | 0 | 1/3 | 0 | |

263

## SECTION III

| C | Basis | Quantity | | | | | | | Ratio |
|---|-------|----------|---|---|---|---|---|---|-------|
| 0 | Land | $2\frac{13}{17}$ | 0 | 9/85 | 0 | 1 | -1/170 | -2/85 | $26\frac{1}{9}$ |
| 30 | Paddy | $1\frac{3}{17}$ | 1 | 74/85* | 0 | 0 | 1/34 | -7/85 | $1\frac{13}{17}$ |
| 26 | Groundnut | $1\frac{1}{17}$ | 0 | 2/85 | 1 | 0 | -2/85 | 9/85 | 45 |
| | Net price (C) | | 30 | 28 | 26 | 0 | 0 | 0 | |
| | Opportunity cost (Z) | | 30 | $26\frac{62}{85}$ | 26 | 0 | 23/85 | 24/85 | |
| | Shadow price (Z–C) | | 0 | $-1\frac{23}{85}$ | 0 | 0 | 23/85 | 24/85 | |

## SECTION IV

| C | Basis | Quantity | | | | | |
|---|-------|----------|---|---|---|---|---|
| 0 | Land ($X_4$) (acres) | $2\frac{23}{37}$ | -9/74 | 0 | -0.0135 | -0.00946 | 0 |
| 28 | Labour ($X_2$) (days) | $1\frac{13}{37}$ | 85/74 | 1 | -7/74 | 5/148 | 0 |
| 26 | Capital ($X_3$) | $1\frac{1}{37}$ | -1/37 | 0 | 9/85 | -0.0243 | 1 |
| | Net price (C) | 0 | 30 | 28 | 0 | 0 | 26 |
| | Opportunity cost (Z) | $64\frac{20}{37}$ | $31\frac{17}{37}$ | 28 | $2\frac{64}{85}$ | 0.3141 | 26 |
| | Shadow price (Z–C) | $64\frac{20}{37}$ | $1\frac{17}{37}$ | 0 | $2\frac{64}{85}$ | 0.3141 | 0 |

*Key number or Pivot number.

the outgoing column (4th). To determine the level at which the new activity, *i.e.*, incoming row the smallest positive value in the ratio column should be considered and this key row is encircled similar to outgoing column. The smallest coefficient in the ratio column indicates the level to which the incoming activity (Paddy I) could be increased in Section II.

(3) The row with the smallest ratio value in Section I is the outgoing row. This is called key row. In selecting the smallest ratio value, we should ignore the negative value.

(4) This step is concerned with the computation of rows for Section II. Identification of key number in the outgoing column and row is the first step here. Key number or pivot element lies at the intersection of the outgoing column and outgoing row. It is 90 in our problem and is designated with asterisk mark, and

(5) Compute the incoming row in Section II by dividing the entries in each column of the key row of Section I by key number, 90. This entire row thus obtained in Section II is now designated as transformed key row.

## Computation of Other Rows in Section II

To compute any new row in this section or of the subsequent sections, the following formula is to be followed:

$$N = 0 - (I \times P) \tag{4.19}$$

Where,

N = Coefficient for new row of the new Section (succeeding section).

O = Coefficient in the corresponding row of old Section (preceding section).

I = Coefficient of the transformed key row in the new Section, and

P = Coefficient of the intersection of outgoing column and outgoing row in old Section.

For the example problem, all the rows of Section II are computed using the equation (4.19).

## Computation of 'Z' Row for Section II

The coefficient in column I is multiplied by the corresponding coefficient in column IV to get the coefficient in the 'Z' row (30) under column IV.

$$0 \times 0 + 30 \times 1 + 0 \times 0 = 30$$

Similarly, the other coefficients of the 'Z' row are obtained using the above procedure. Highest negative coefficient in the Z-C row (shadow price row) is identified and its corresponding column is designated as outgoing column. The smallest positive coefficient in the ratio column helps to identify key row and key number. Outgoing column and key row are encircled and key number is star marked.

Following the same procedure adopted for computation of rows in Section II, the rows of Section III and Section IV are computed.

## Criterion for Stopping Computation of Sections

To arrive at the optimal solution, all the coefficients in Z-C row should change to positive values. This is an indication that we should stop at that section without further iterating the procedure. The values in the B column of the last section (in our example IV section) show the normative acreages under different crops. In our example, out of 5 acres of available land, 1.35 acres should be allotted to paddy (improved) and 1.03 acres to groundnut crop and remaining 2.62 acres should be kept fallow. The maximum net income derived from such plan would be Rs. 6,454.

The foregone model, which we have formulated, is much too limited in the restraints like capital and labour. As a result, the available land is not fully utilized and fallow to an extent of 2.62 acres is observed. Even with a limited number of enterprise activities and resource constraints, the requisite computations are substantial. If all the realistic constraints and the activities are introduced in the model, computation further becomes cumbersome and complicated. A realistic application of the farm planning model with 30 to 40 cropping and livestock activities and 20 to 30 resource constraints definitely requires computer facilities for solving the problem.

## Optimal Solution

The Z-C row of last section (Section–IV) gives us the shadow prices of the inputs under disposal activity. Shadow price of land is O, because land is not fully utilized, Labour shadow price is 0.31 which means that one unit of labour if it is additionally increased, it will add 0.31 to the objective function value. In this case gross margin is increased by 0.31 units. Shadow price of capital is 2.75, which means that if one unit of capital is increased, 'Z' value is increased by 2.75 units. In the shadow price row of section IV, we have the value of 1.46 for paddy I under crop activities. This means that gross margins of the farm would be decreased if one unit of paddy I activity enters the programme. This is the income penalty.

## Difficulties in Solving LP

The stated problem in the LP framework is not always solved. For some LP problems, if the data and constraints are not properly specified, we do not get optimal solution. This is due to problems of (i) infeasibility (ii) unboundedness (iii) degeneracy and (iv) scaling.

## Infeasibility

If the optimal solution for a given problem does not satisfy all the given constraints, then the LP problem is said to be infeasible. This means that there is no unique solution that satisfies all the constraints. Infeasibilities occur in solving the problem, because of mistakes in formulating the constraints. If the LP problem is large and complex with many constraints, the model builder fails to notice all theoretical and logical relationships within the model.

## Unboundedness

Sometimes feasible solution for the given LP problem exists and this has infinite values for the objective function. This would arise, if some of the activities have zero requirements. The problem of unboundedness is the most common, when there are errors in the collected data.

## Degeneracy

If the value of the objective function remains constant in one or two subsequent iterations, it is due to incoming activities entering the basis at zero level. Another reason for degeneracy is occurrence of tie for minimum ratio. Degeneracy problem is not that much serious in the sense that it does not affect the existing optimal solution, but number of iterations required would be more for obtaining the optimal solution. In some extreme case of degeneracy, we find cycling process, which means, the same sequence value in subsequent iterations occurs. When this happens, we cannot get an optimal solution for a problem. In practice degeneracy is not a serious problem. In most of the LP packages, we have built in programmes to solve the problem of tie and cycling.

## Scaling Problem

The units of measurement for activities and constraints should be the same. For example, if crop activities are specified per ha and land constraints are specified in acres we encounter scaling problem. In such contexts, we face problems in interpretation of optimal values. Similarly if inconsistencies are present in the units of $a_{ij}$ and $b_i$ values, we will have scaling problems.

## Precautions in Formulating the LP Models

### Defining the Crop Activities

In LP model, production activities *i.e.,* crops and livestock activities can be defined according to their varietal and breed characteristics and as per the methods of production *i.e.,* using technical data. Each method of production of crops and livestock requires different inputs. This is to say that we should obtain input-output data for each type of crop and livestock products and this should be incorporated in the model as a separate activity. Example: If the same paddy variety requires different levels of NPK and micronutrients in different soils, it should be treated as a separate crop activity. Technique of production refers to the requirement of inputs for unit of crop activity. If this technique differs over time and space, they should be treated as separate crop activities. In other words, crop activities are defined according to the techniques of production.

**TABLE 28.3 Quality Difference in Resources.**

|  | Irrigated crop – I | Dry land crop | Sign | $b_i$ values |
|---|---|---|---|---|
| Irrigated land (ha) | 1 | 0 | $\leq$ | $b_1$ |
| Dry land (ha) | 0 | 1 | $\leq$ | $b_2$ |

In Table 28.3 crop activity is divided based on the type of crop whether it is irrigated or not. $b_1$ value indicates the irrigated land available for raising suitable crop, $b_2$ indicates the available dry land for growing suitable varieties. Similarly, categories of labour *viz.,* family labour, hired labour, owned bullock labour, hired bullock labour, owned machine labour, hired machine labour, *etc.,* are quantified and input coefficients for different crop activities are worked out and shown under the respective crop activities in the $a_{ij}$ matrix. Each category of the labour is specified as a separate constraint ($i^{th}$ constraint). Under $b_i$ column labour supply should be indicated in appropriate units.

With regard to unit measurement requirement and availability, they should be defined with the same units. This avoids inconsistencies in the scaling of inputs.

## Seasonality

Agricultural production is seasonal in nature. In some areas where the resources are plenty, we could observe various methods of production of crops and livestock in all the seasons. On the contrary, in some areas due to scarcity of resources, for example, water availability to agricultural production is confined to *kharif* season or *rabi* season. These characteristics and aspects of farming lead to distinct seasonal pattern in the resource use, resource supplies and raising of crops and livestock. In areas where these crops are taken up in one year, it is relevant to define labour requirement in different seasons or months. Care should be taken in formulating labour constraints in different seasons/months in the sense that the requirement of labour in a particular season/month should be more than the supply. If such conditions prevail in farming situation, it is relevant to include such input constraints in the given farm model, otherwise it is redundant coefficient in the model. This is illustrated in Table 28.4

TABLE 28.4  Seasonality.

| Particulars | Crop 1 $X_1$ | Crop 2 $X_2$ | Crop 3 $X_3$ | RHS |
|---|---|---|---|---|
| Objective function (gross margin/ha) | $C_1$ | $C_2$ | $C_3$ | = Z maximize |
| *Constraints:* | | | | |
| Land (ha) | 1 | 1 | | $1 \le b_1$ |
| Labour by season | | | | |
| *Kharif:* | | | | |
| Family men (mandays) | + | + | + | $\le bk_1$ |
| Family women (mandays) | + | + | + | $\le bk_2$ |
| Hired labour (mandays) | + | + | + | $\le bk_3$ |
| *Rabi:* | | | | |
| Family men (mandays) | + | + | + | $\le bR_1$ |
| Family women (mandays) | + | + | + | $\le bR_2$ |
| Hired labour (mandays) | + | + | + | $\le bR_3$ |

Here (+ value) indicates the labour requirement in mandays per unit of crop activity (demand for labour), bi values indicate available labour supplies. $C_1$, $C_2$ and $C_3$ are the average gross margins computed over two seasons.

## Buying and Selling Activities

Leasing activities *i.e.*, leased in and leased out activity of land, buying and selling activities, *i.e.*, purchasing the necessary farm inputs and selling the output and services are the most common farm activities noticed on the farms. To get realistic and relevant optimal solutions, those activities have to be necessarily incorporated in the farm models. In doing so the negative entries in the objective function for buying, renting and hiring activities and positive entries for selling activities should be shown. Unit prices of these activities should be indicated with appropriate sign in the objective function. For example, if rent of an acre of land is Rs. 3,000, it should be entered in the farm model as a separate crop activity (Table 28.5).

## Crop Rotation Practices

These practices are observed on the farms to maintain soil fertility and obtain high yields from the crops grown. If crop 1 is alternated with crop 2 as a crop rota practice, this is shown as a constraint in Table 28.6.

## TABLE 28.5 LP Farm Model with Buying and Selling Activities.

| | Crop 1 | Crop 2 | Rent activity | Hiring of labour | Purchasing of fertilizer | Selling of water for one acre of Land | Selling of crop 1 | Selling of crop 2 | RHS |
|---|---|---|---|---|---|---|---|---|---|
| | $(X_1)$ | $(X_2)$ | $(X_3)$ | $(X_4)$ | $(X_5)$ | $(X_6)$ | $(X_7)$ | $(X_8)$ | |
| Objective function (gross margin/acre) | $C_1$ | $C_2$ | -3,000 | -WL | -PF | +WP | $+P_1$ | $+P_2$ | = Z maximize |
| Land (acre) | 1 | 1 | -1 | 0 | 0 | 0 | 0 | 0 | $\leq b_1$ |
| Labour (mandays) | + | + | 0 | -1 | 0 | 0 | 0 | 0 | $\leq bL_1$ |
| Fertilizer (kg) | + | + | 0 | 0 | -1 | 0 | 0 | 0 | $\leq bf_1$ |

$C_1$ and $C_2$ are gross margins/acre of land

RL = Rental value of land in Rupees/acre (Rs. 3,000)

WL = Wages paid to casual labour per day

PF = Price per kg of fertilizer used for production of corps

WP = Water charges per if crop

$P_1$ = Price per unit of output of crop 1

$P_2$ = Price per unit of output of crop 2

$b_1$, $bL_1$ and $bf_1$ = Available supplies of the inputs

+ Values indicate requirement of its resource for $j^{th}$ crop activity.

– 1 Indicates supply of additional unit resource for the $i^{th}$ input supply on the RHS side

## TABLE 28.6 Crop Rotation.

|  | Crop 1 | Crop 2 | RHS |
|---|---|---|---|
| Crop rotation constraint | +1 | -1 | $\leq 0$ |

## Credit Borrowing Constraints

The requirement of working capital for raising different crop and livestock activities with the given farm resources are different and vary according to the nature of the enterprise. For most of the Indian farmers, capital is a binding constraint and hence borrowing is a common activity noticed as a regular feature. This type of activity, if incorporated in the farm model, it would provide realistic farm plans. This is shown in Table 28.7.

## TABLE 28.7 Credit Borrowing Activity.

|  | Crop 1 | Crop 2 | Borrowing of credit (1 unit =Rs. 100) | RHS |
|---|---|---|---|---|
| Objective function (gross margin/acre) | $C_1$ | $C_2$ | $-i$ | = Z maximize |
| Credit row (1 unit = Rs. 100) | $+ K_1$ | $+ K_2$ | $-i$ | $\leq bK_1$ |

$K_1$ = working capital (cash component) requirement for crop 1
$K_2$ = working capital (cash component) requirement for crop 2
i = Rate of interest paid by the farmer for a loan of Rs. 100

If 11% is paid by farmer to the institution, then put –0.11 in place of – i

The farmer borrows required capital from different sources *viz.*, moneylenders, commercial banks, RRBs, co-operatives, *etc.* These sources charge different interest rates, then each source of finance should be given a separate borrowing activity with corresponding interest rates.

## Cash Flow Constraint

Cash transfer activities from one season to another are the most common features of Indian farmers. Cash earned in one season is invested in the next season. Income from livestock enterprises provide a regular source of income in all the seasons and hence cash transfer is inevitable from one season to another. Let us assume that agricultural year is divided into three seasons *viz.*, *kharif*, *rabi* and summer. Paddy is grown in all the seasons and it requires the cash input in each period and this designated as $WC_{ji}$. The yield of paddy is designated as $Q_{jt}$. Paddy production and selling activities are segregated in the farm model to provide separate picture on cash inflows and cash outflows in different seasons. Here we assume that revenue from selling paddy is collected under the selling activity and transferred to next season. Variable cost (VC coefficient) of paddy producing activity is denoted in the objective function row with minus sign. Interest rate paid by the farmer for borrowed capital in different seasons is specified with (-i) coefficient in the objective function. Selling prices of paddy in different seasons are designated as $P_1$, $P_2$ and $P_3$ with positive unit prices. Any cash surplus at the end of each period can be transferred to the next

## TABLE 28.8 Cash Flow Constraint for the Hypothetical Farm in LP Framework.

| | Season 1 | | | | Season 2 | | | Season 3 | | | RHS |
|---|---|---|---|---|---|---|---|---|---|---|---|
| | Paddy production (ac) | Borrow credit (Rs.) | Transfer cash surplus (1 unit = Rs. 100) | Sell paddy (kg) | Borrow credit (Rs.) | Transfer cash surplus (1 unit = Rs. 100) | Sell paddy (kg) | Borrow credit (Rs.) | Transfer cash surplus (1 unit = Rs. 100) | Sell paddy (kg) | |
| Objective function (Rs.) | $-VC$ | $-1$ | $0$ | $P_1$ | $-1$ | $0$ | $P_2$ | $-1$ | $0$ | $P_3$ | $= Z$ (maximize) |
| Cash balance in Rows (in '00 Rs.) | | | | | | | | | | | |
| Season 1 | $WC_1$ | $-(1+i)$ | $+1$ | $+1$ | $1$ | $0$ | $0$ | $0$ | $0$ | $0$ | $\leq b_1$ |
| Season 2 | $WC_2$ | $0$ | $0$ | $0$ | $1$ | $+1$ | $-1$ | $1$ | $0$ | $0$ | $\leq b_2$ |
| Season 3 | $WC_3$ | $0$ | $0$ | $0$ | $0$ | $0$ | $0$ | $1$ | $+1$ | $+1$ | $\leq b_3$ |
| Owned fund maintenance row (in '00 Rs.) | $0$ | $-(1+i)$ | $0$ | $0$ | $-(1+I)$ | $0$ | $0$ | $-(1+i)$ | $1$ | $0$ | $\leq b$ |
| Paddy yield constraint (kg): | | | | | | | | | | | |
| Season 1 | $-Q_1$ | $0$ | $0$ | $1$ | $0$ | $0$ | $0$ | $0$ | $0$ | $0$ | $\leq 0$ |
| Season 2 | $-Q_2$ | $0$ | $0$ | $0$ | $0$ | $0$ | $1$ | $0$ | $0$ | $0$ | $\leq 0$ |
| Season 3 | $-Q_3$ | $0$ | $0$ | $0$ | $0$ | $0$ | $0$ | $0$ | $0$ | $1$ | $\leq 0$ |

## TABLE 28.9 LP Formulation.

| 1 | Cash and food crops | | Livestock keeping activities | | | | Sales activities | | | | Purchasing activities | | | | | Borrowing activities | Consumption activities | | | Transfer activities | | RHS |
|---|---|---|---|---|---|---|---|---|---|---|---|---|---|---|---|---|---|---|---|---|---|---|
| | Crop 1 $(X_1)$ | Crop 2 $(X_2)$ | Bullock $(X_3)$ | Cow + Calf $(X_4)$ | Bull $(X_5)$ | Heifer $X_6$ | Crop 1 sale per quintal price in Rs. | Ox $(X_8)$ sale price/head | Culled $(X_9)$ price/head | Hiring out animal*/head activity $(X_{10})$ | Milk $(X_{11})$ purchases price/litre | Fertilizers $(X_{12})$ price/quintal | Seeds $(X_{13})$ price/kg | Hiring machines $(X_{14})$ Rs./hour | Labour hiring $(X_{15})$ wage/labour | $(X_{16})$ interest rate | Food crop $(X_{17})$ Qtls | Milk $(X_{18})$ (lit) | Meat (kg) $(X_{19})$ | $(X_{20})$ cash | $(X_{21})$ fodder | |
| | 2 | 3 | 4 | 5 | 6 | 7 | 8 | 9 | 10 | 11 | 12 | 13 | 14 | 15 | 16 | 17 | 18 | 19 | 20 | 21 | 22 | 23 |
| Objective function coefficients (in Rupees) | -vc | -vc | -vc | -vc | -vc | -vc | +p | +p | +p | +p | -p | -p | -p | -m | -w | -I | 0 | 0 | 0 | 0 | 0 | = Max Z |
| $i$th type of Land area cropped in ha season wise | 1 | 1 | 0 | 0 | 0 | | | | | | | | | | | | | | | | | ≤ I the type of Land area available in the season (ha) |
| Pasture area (ha) season wise | | | 0.2 | 0.3 | 0.1 | 0.01 | | | | | | | | | | | | | | | | ≤ Land under pastures (ha) |
| Labour in $i$th Period (man equivalent days) | + | + | + | + | + | + | | | | | | | | _1 | + | | | | | | | ≤ Labour availabel in the period (days) |
| Bullock labour (in pair days in $i$th period) | + | + | - | | | | | | | + | | | | | | | | | | | | ≤ Bullock pair days available in $i$th pair |
| Working capital in Rs. season wise | + | | + | + | + | + | | - | | + | + | + | + | | | | | | | | | ≤ working capital fixed costs (Rs.) |
| Bullock replacement row | | | | | + | | | 0.16 | - | | | | | | | | | | | | | ≤ 0 |
| Cow replacement row | | | | + | | | | 0.14 | | | | | | | | | | | | | | ≤ 0 |

| | | | | | | | | | |
|---|---|---|---|---|---|---|---|---|---|
| Yield balances crop yield (quintal) | −x | | | +1 | | | +1 | | ≤ Subsistence requirement of crop in quintals |
| Fodder balances row in tonnes | − | −y | + | + | + | + | + | | ≥ Minimum quantity fodder required under fodder (ha) ≤ 0 |
| Milk balances row (litres) | | | −y | | + | | | | ≥ Minimum quanitity of crops |
| Subsitence requirement crops | | | | | | | +1 | +1 | |
| Milk (litres) | | | | | | | | +1 | ≥ Minimum quantity milk requied |
| Minimum area under crops | 1 | 1 | | | | | | | ≥ Minimum area under crops |
| Crop 1 | 1 | | | | | | | | minimum area under crop 1 |
| Crop 2 | 1 | | | | | | | | minimum area under crop 2 |
| Cash transfer activity in period I | | | | | | | | 1 | ≤ Cash available |
| Cash transfer to period II | | | | | | | | −1 | ≤ 0 |
| Fodder in period I | | | | | | | | 1 | ≤ Fodder available |
| Fodder in transfer to period II | | | | | | | | −1 | ≤ 0 |

period through transfer activity. Transfer activities in the objective function would have 'O' values and + 1 in the $a_{ij}$ matrix. These details are presented in Table 28.8.

LP formulation is presented in Table 28.9

Notations shown in Table.

VC = Variable costs/acre of crop activity (paddy)

i = Rate of interest in per cent

$P_1$ = Selling price of paddy in season – I in Rs./kg

$P_2$ = Selling price of paddy in season – II in Rs./kg

$P_3$ = Selling price of paddy in season – III in Rs./kg

$WC_1$, $WC_2$ and $WC_3$ = Working capital requirements of paddy/acre in respective seasons

$Q_1$, $Q_2$ and $Q_3$ = Yields per acre of paddy in respective seasons*

In the owned fund balance row, the available borrowed amount and the owned fund available in the seasons are indicated with negative sign because these are supplies. Suppose if the farmer is obliged to pay for the clearance of debts in any season, then it should be indicated with + (1+i) sign. If the fixed costs are to be repaid in each season then these costs may be subtracted from the $b_i$ values of the respective season. Similarly, living expenses can be subtracted from the $b_i$ value.

## VC (paid out costs)

Variable costs in Rs./ha should be indicated in this objective function row with negative sign. For cows and heifers paid out costs are incurred hence they are indicated with negative sign. For bullocks and bulls VC is zero, because no paid out costs are incurred.

## Selling Activities

Selling activities are specified for food and cash crops, bullocks, culled livestock, hiring out bullocks, *etc.* The sale price per unit of these activities should be indicated in the objective function with plus sign.

## Purchasing Activities

These are specified for crops, milk, fertilizers, pesticides and borrowing of credit. The unit price of these activities is indicated with negative sign in the objective function. For borrowing activity interest rate paid by the farmer in fraction such as 0.12, 0.15, *etc.*, should be indicated with negative sign. Here 0.12 and 0.15 are the interest rates.

## Consumption Activities

These are specified for milk, food, meat, *etc.* In the objective function the coefficients under consumption activities should have zero values.

## Transfer Activities

Similarly here also zero values are indicated under transfer activities.

## Land Constraints

Land constraints under crop activities should be indicated as 1, 1, ...... here, the unit of specification is one hectare. Under right hand side (RHS) of these land constraints, the available land should be given and if land constraints are to be specified for each season and type of land such as dry land, irrigated dry land (ID land) wet land, orchard land, *etc.*, separate rows should be used to specify these land constraints.

## Pasture Area

For fodder crops the actual area under fodders is worked out and presented under livestock activities such as 0.2, 0.3, 0.7 ha, *etc.* This is true if land area under each fodder crop is less than one ha. If it is more than one ha, enter it as 1. Under pasture activity available pasture area in hectares should be specified against this constraint under RHS column.

## Replacement Constraints for Bullocks and Cows

These figures show that bullocks are to be replaced on the livestock farm once in 6 years, hence the coefficient is $1/6 = 0.16$. Cows are replaced once in 7 years, hence replacement coefficient is $1/7 = 0.14$. These should indicated with positive sign under selling or disposal activities in the $a_{ij}$ matrix. Positive values in the $a_{ij}$ matrix indicate the demand of crop and livestock activities. Negative values are the production levels of crop and livestock activities per unit. These are the supply coefficients of the respective activities.

## Subsistence Requirements

To represent subsistence requirements in the $a_{ij}$ matrix, we should give 1, 1, ... 1 *etc.* Under $b_i$ (resource columns) columns. We should enter the minimum amount of required crops for family subsistence and enter $\geq$ sign under right hand side $\geq$. Here we should note that for constraints of subsistence requirements and minimum area under crops the sign should be greater than or equal to ($\geq$).

## MOTAD *Minimisation of Total Absolute Deviation* MODEL

Hazell (1971) developed MOTAD model as a linear alternative to quadratic and semi-variance programming for farm planning under risk. This model uses linear decision criterion with expected return and mean absolute deviation. Hazell using the same data for MOTAD and QRP concluded that the optimal plans generated by these two models were nearly the same. Further, he observed that the MOTAD model could be solved with conventional linear programming packages. Risk is incorporated in the model as mean absolute deviation of farm profit. In matrix notation the MOTAD model is specified as:

$$M = S^{-1} \sum_{t=1}^{s} \left| \sum_{j=1}^{n} \left( C_{tj} - \overline{C}_j \right) X_j \right| \tag{4.20}$$

$M$ = Mean absolute deviation that can be minimized for a given level of expected profit.

$S$ = Number of years.

$C_{tj}$ = Gross margin per unit of $j^{th}$ crop or livestock activity in the $t^{th}$ year (unit here is one hectare).

$\overline{C}_j$ = Sample mean gross margin per unit of $j^{th}$ crop or livestock activity.

$X_j$ = Level of $j^{th}$ crop or live stock activity to be obtained from the solution of the model.

j = Refers to $j^{th}$ activity (j=1 to n activities).

t = Refers to $t^{th}$ year (t=1 to S years).

| | = Modulus denotes absolute value of the figures i.e., ignoring the signs within the two vertical bars.

In this model, a measure of risk of gross margins, which is given in the modulus, is incorporated into LP model of a whole farm-planning problem. The mean absolute deviation, M is minimized for a given level of expected gross margin [E (Z)], which varies parametrically over zero to some desired range.

Another simple and tidy approach is also suggested by Hazell, considering the negative deviations about the mean gross margins as mean absolute value. The objective function of the LP model is specified as:

$$D = \frac{M}{2} = S^{-1} \left| \min\left( \sum_{j=1}^{n} \left(C_{tj} - \overline{C}_j\right) X_j, 0 \right) \right| \tag{4.21}$$

min = minimize.

The negative deviations of gross margin from their mean income in the $t^{th}$ year of sample data is defined by a new variable, $Y_t$ and it is defined as:

$$Y_t = \sum_{j=1}^{n} \left(C_{tj} - \overline{C}_j\right) X_j \tag{4.22}$$

j = 1 to n crop or livestock activities,

$C_{tj}$ = Gross margin from $j^{th}$ crop or livestock activity in the $t^{th}$ year,

$\overline{C}_j$ = Mean gross margin of $j^{th}$ crop or livestock activity. The LP problem is formulated as minimization of $Y_t$ in the objective function subject to usual technical constraints and parametric constraints on expected total gross margin from crops and livestock. The MOTAD model is formulated as:

Minimise $Y_t$

Subject to

$$\sum_{j=1}^{n} a_{ij} X_j (\geq = \leq) b_i, \ i = 1 \text{ to m constraints} \tag{4.23}$$

$$\sum_{j=1}^{n} (C_{tj} - \overline{C}_j) X_j + Y_t \geq 0 \tag{4.24}$$

$$\sum_{j=1}^{n} Y_t \leq \lambda \tag{4.25}$$

$X_j \geq 0, \ Y_t \geq 0$ for j = 1 to n, t = 1 to s years $\tag{4.26}$

Equation (4.23) is technical constraint,

Equation (4.24) is deviation constraint,

Equation (4.25) is parametric constraint, and

Equation (4.26) is non-negativity constraint.

In the equation (4.24) the variable $Y_t$ is to measure the negative deviation of total gross margin from mean of crop or livestock for each year, *i.e.*, t=1 to s years. The total deviation for each year is computed in the summation term of equation (4.24). If the sum is positive, the corresponding $Y_t$ variable will be zero and this is assured by non-negativity restrictions. The total value of the objective function is limited through parametric constraint on the sum of the $Y_t$ variable in the equation (4.25). Suppose, if the sum of the gross margin deviations in any year is negative, the corresponding variable $Y_t$ will be forced to be a positive value, hence the $\lambda$ in equation (4.25) will measure the sum of the total negative deviations over s years. Changing the values for $\lambda$ in the parametric constraint in equation (4.25), the efficient E-A Plans are generated. The formulation of MOTAD model for the hypothetical farm is presented in Table 28.8.

## Illustration of MOTAD Model for a Hypothetical Farm

To formulate the model, we require information on the gross margin of different crop and livestock enterprises over time along with resource requirements and resource availabilities. The data for a hypothetical farm are assumed and furnished in Table 28.10.

TABLE 28.10 Particulars of Gross Margins from Crops and Livestock.

| Year | Gross margin/ha from crop and livestock enterprises of the farm (in '000 Rs.) | | | | | |
| --- | --- | --- | --- | --- | --- | --- |
| | Groundnut (Improved) | Groundnut (HYV) | Paddy (IR-20) | Paddy (IET -1444) | Jersi cow per head | Buffalow per head |
| | $X_1$ | $X_2$ | $X_3$ | $X_4$ | $X_5$ | $X_6$ |
| 1985 | 60.3 | 70.3 | 45.2 | 50.9 | 35.4 | 30.3 |
| 1986 | 64.2 | 68.3 | 42.8 | 48.5 | 37.1 | 32.1 |
| 1987 | 70.1 | 70.2 | 43.6 | 44.2 | 36.5 | 28.4 |
| 1988 | 66.8 | 74.5 | 45.1 | 51.3 | 38.5 | 25.6 |
| 1989 | 69.5 | 72.0 | 49.9 | 52.2 | 39.6 | 34.5 |
| 1990 | 68.3 | 76.0 | 52.8 | 57.4 | 38.3 | 33.1 |
| 1991 | 66.6 | 75.1 | 54.1 | 59.0 | 40.9 | 35.2 |
| Mean | 66.54 | 72.34 | 47.64 | 52.64 | 38.11 | 31.31 |

The MOTAD model presented in Table 28.11 is solved by using simplex algorithm by changing the $\lambda$ values. Through this different optimal crop plans are derived. The MOTAD model is very useful to provide acceptable farm plans to risk averse farmers starting from zero income levels up to infinity.

# GOAL PROGRAMMING

In goal programming several goals of the farmers are considered as against single goal, which is hitherto considered in LP model. When there are several goals to be achieved by the farmer, then we must know as to how to quantify those goals. To do so, we require introduction of auxiliary variables. These auxiliary variables do not

277

## TABLE 28.11 Initial Table of MOTAD for a Hypothetical Farm.

| Rows | Crop activities (Rs./ha) | | | | Livestock (per head) (Rs.) | | $\bar{Y}_1$ | $\bar{Y}_2$ | $\bar{Y}_3$ | $\bar{Y}_4$ | $\bar{Y}_5$ | $\bar{Y}_6$ | $\bar{Y}_7$ | RSH |
|---|---|---|---|---|---|---|---|---|---|---|---|---|---|---|
| | $X_1$ | $X_2$ | $X_3$ | $X_4$ | $X_5$ | $X_6$ | | | | | | | | |
| Objective function | | | | | | | 1 | 1 | 1 | 1 | 1 | 1 | 1 | min $=\lambda$ |
| Mean gross margin ('000 Rs.) | 66.54 | 72.34 | 47.64 | 52.64 | 38.1 | 31.31 | | | | | | | | |
| Land (ha) | 1 | 1 | 1 | 1 | | | | | | | | | | ≤5 |
| Labour (days) | 150 | 165 | 240 | 260 | 220 | 210 | | | | | | | | ≤400 |
| Capital (in '000 Rs.) | 50.23 | 48.38 | 55.42 | 60.32 | 18.54 | 19.20 | | | | | | | | ≤10,000 |
| Rotational constraint | -1 | -1 | 1 | 1 | -1 | 1 | | | | | | | | ≤0 |
| Risk rows | | | | | | | | | | | | | | |
| 1985 | -6.24 | -2.02 | -2.44 | -1.74 | -2.71 | -1.07 | 1 | | | | | | | ≥0 |
| 1986 | -2.34 | -4.04 | -4.84 | -4.84 | -1.01 | 0.79 | | 1 | | | | | | ≥0 |
| 1987 | 3.56 | -2.14 | -4.02 | -3.44 | -1.61 | -2.91 | | | 1 | | | | | ≥0 |
| 1988 | 0.26 | 2.16 | -2.54 | -1.34 | 0.39 | -5.71 | | | | 1 | | | | ≥0 |
| 1989 | 2.96 | -0.34 | 2.26 | -0.44 | 1.49 | 3.19 | | | | | 1 | | | ≥0 |
| 1990 | 1.76 | 3.66 | 5.16 | 4.76 | 0.69 | 1.79 | | | | | | 1 | | ≥0 |
| 1991 | 0.06 | 2.76 | 6.46 | 6.36 | 2.79 | 3.89 | | | | | | | 1 | ≥0 |

represent the decision problems. They are simply extra variables, which help in the formulation of model for finding the solution. Sometimes we allow these variables to be either positive or negative. When there is no bound on the negative values of the $X_j$ variable, then it can be replaced throughout the model by the difference between the positive and negative variables i.e., $X_j^+$ and $X_j^-$ so that $X_j = X_j^+ - X_j^-$, where $X_j^+$ is the positive component of $X_j$ variable and $X_j^-$ is the negative component of $X_j$ variable. For all such basic feasible solutions, we should have the property that either $X_j^+ = 0$, or $X_j^- = 0$. Here we should continue to use this type of notation with + or − superscripts for the decision variables. Suppose $X_j$ represents inventory level of particular product, then we should have $X_j^+ > 0$ and $X_j^- = 0$. Here, it means that the costs incurred on the product $X_j$ include storage expenditure and interest charges on fixed capital invested in the business. In other words if these cost values are positive, then $X_j^- > 0$ and $X_j^+ = 0$. This indicates that shortage in the inventory has occurred. If $X_j^- > 0$, there is a loss is the business. Because of this difference between +ve and −ve components, $X_j$ values and these components are simply proportional to $X_j$, hence the proportionality assumption of LP is violated.

Let us assume that unit cost of holding inventory which in positive is assumed as Rs. 18 ($C_j X_j^+$ = Rs. 18) and unit cost of holding a shortage in the inventory per unit time is given as Rs. 12. Then, to uphold proportionality assumption the objective function of LP problem should be formulated as:

$$\text{Maximize } Z = C_j (X_j^+ + X_j^-)$$
$$= 18 X_j^+ + 12 X_j^-$$

The agro-processing industry may have many goals to achieve rather than single over-riding goal. These goals, for example, are: (1) Achieving long run profit, (2) Keeping the employment of labour at desired level in the long run and (3) Achieving the expected returns for the investment. Other firms would have the goals like maintaining stable profits and stable prices in the long run, improving the workers' morale through appropriate wage incentives, maintaining family control on the agri-business, increasing the company's prestige, good will, etc. Goal programming provides solution to achieve such goals. Through goal programming, we minimize the weighted sum of these deviations of the objective function from these goals.

We have two types of goal programming techniques: (1) Non-preemptive goal programming and (2) preemptive goal programming. In non-preemptive goal programming technique, goals are roughly of comparable importance. In the preemptive goal programming, we specify the hierarchy in the priority levels for these goals. Goals of primary importance receive first priority and secondary important goals receive second priority and so on and so forth.

Let us assume that agro-industry is giving primary importance to three following goals. These are:

1. Achieving the long run profit i.e., net present worth of Rs. 12.5 million.

2. Maintaining the current employment level at 4000 employees; and

3. Holding the capital investment less than Rs. 20 million.

It is further assumed that the firm is producing three products i.e., $X_1$, $X_2$ and $X_3$. Contribution of these products to the profits and quantification of goals including the penalty weights are given in Table 28.12.

**TABLE 28.12**    Illustration of Non-Preemptive Goal Programming for Hypothetical Agro-Industry.

| S. No. | Name of the goals | Unit contribution of profits (in million Rs.) | | | Sign | Goal | Penalty weights |
|--------|-------------------|:----:|:----:|:----:|:----:|:----:|:----:|
| | | $X_1$ | $X_2$ | $X_3$ | | | |
| 1. | NPW of long run profit (in million Rs.) | 7 | 5 | 2 | $\geq$ | 12.5 | $-2$ |
| 2. | Employment goal (days in 100s) | 15 | 20 | 22 | $=$ | 40 | $+3, -4$ |
| 3. | Capital investment (in million Rs.) | 9 | 5 | 8 | $\leq$ | 20 | $+3$ |

Here the long run profit goal is a one sided goal. A minimum level of profit *i.e.*, Rs. 12.5 million is to be achieved over long run by the agro-industry, hence this constraint in designated with $\geq$ constraint. Unit contributions of $X_1$, $X_2$ and $X_3$ are formulated as $a_{ij}$ coefficients. The employment goal is a two-sided goal with $=$ sign. The capital investment goal should not exceed Rs. 20 million, hence the relevant sign is $\leq$. The coefficients of the goals, which are quantified, are entered in the $b_i$ column of LP problem. Here we have 4 penalty weights for not achieving the goals as defined in respective specific units of goals.

The overall objective function of the LP problem is specified as:

$$\text{Minimize } Z = 2(7X_1 + 5X_2 + 2X_3)^- + 3(15X_1 + 20X_2 + 22X_3)^+ + 4(15X_1 + 20X_2$$

$$+ 22X_3)^- + 3(9X_1 + 5X_2 + 8X_3)^+ \qquad \dots (4.27)$$

Since there are three goals, we should add three auxiliary variables *viz.*, $Y_1$, $Y_2$ and $Y_3$ in the following constraints.

$$Y_1 = 7X_1 + 5X_2 + 2X_3 - 12.5 \qquad (4.28)$$

$$Y_2 = 15X_1 + 20X_2 + 22X_3 - 40 \qquad (4.29)$$

$$Y_3 = 9X_1 + 5X_2 + 8X_3 - 20 \qquad (4.30)$$

We define the auxiliary variables with positive and negative components as:

$$Y_1 = Y_1^+ - Y_1^- \qquad \text{where, } Y_1^+ \geq 0, Y_1^- \geq 0 \qquad (4.31)$$

$$Y_2 = Y_2^+ - Y_2^- \qquad \text{where, } Y_2^+ \geq 0, Y_2^- \geq 0 \qquad (4.32)$$

$$Y_3 = Y_3^+ - Y_3^- \qquad \text{where, } Y_3^+ \geq 0, Y_3^- \geq 0 \qquad (4.33)$$

We introduce these auxiliary variables in the objective function of goal programming and constraints. The objective function of goal programming should consider penalty weights with respective proper signs. We do not have penalty weights for auxiliary variable *i.e.*, $Y_1^+$, *i.e.*, for exceeding the profit goal of Rs. 12.5 million. Similarly, auxiliary variable $Y_3^-$ is not having penalty weight for being under the investment goal for Rs. 20 million. Hence here in the problem, $Y_1^+ = 0$ and $Y_3^- = 0$. The overall goal programming now in the LP framework is specified. Please note that when there is $\geq$ sign in the constraint we should give $-$ penalty sign. Similarly, for $\leq$, the sign of the penalty weight coefficient should be positive. Now with auxiliary variables the LP is specified as:

Minimize $Z = 2Y_1^- + 3Y_2^+ + 4Y_2^- + 3Y_3^+$ (4.34)

Subject To,

$7X_1 + 5X_2 + 2X_3 - (Y_1^+ - Y_1^-) = 12.5$ (4.35)

$15X_1 + 20X_2 + 22X_3 - (Y_2^+ - Y_2^-) = 40$ (4.36)

$9X_1 + 5X_2 + 8X_3 - (Y_3^+ - Y_3^-) = 20$ (4.37)

$$\left. \begin{array}{l} X_1, X_2, X_3 \geq 0 \\ Y_1^+, Y_2^+, Y_3^+ \geq 0 \\ Y_1^-, Y_2^-, Y_3^- \geq 0 \end{array} \right\} \text{ these are called non-negativity constraints}$$

Now the problem is stated in the form of matrix (Table 28.13).

TABLE 28.13 Illustration of Goal Programming–Initial Table.

| | Name of the-products | | | Auxiliary variables | | | | | | RHS |
|---|---|---|---|---|---|---|---|---|---|---|
| | $X_1$ | $X_2$ | $X_3$ | $Y_1^+$ | $Y_1^-$ | $Y_2^+$ | $Y_2^-$ | $Y_3^+$ | $Y_3^-$ | |
| Objective function | 0 | 0 | 0 | 0 | 2 | 3 | 4 | 3 | 0 | = Minimize Z |
| Constraints | | | | | | | | | | |
| 1. Long run profit (in million Rs.) | 7 | 5 | 2 | -1 | +1 | 0 | 0 | 0 | 0 | = 12.5 |
| 2. Employment goals (in hundred days) | 15 | 20 | 22 | 0 | 0 | -1 | +1 | 0 | 0 | = 40 |
| 3. Capital investment (in million Rs.) | 9 | 5 | 8 | 0 | 0 | 0 | 0 | -1 | +1 | = 20 |

The above goal programming specified in LP format can be solved using simplex method. We can obtain the optimal values of $X_1$, $X_2$ and $X_3$ and these values indicate that how many units of $X_1$, $X_2$ and $X_3$ product should be produced by the agro-industry in order to achieve the goals simultaneously. Z value would indicate the penalty weight for deviating from these goals. For the given problem, suppose we get values for $Y_1$ and $Y_3$ auxiliary variables in the optimal solution, then Z value should be expressed in millions of rupees. If $Y_2$ appears in optimal solution, then Z value should be expressed in hundred man-days as penalty weights.

# GAME THEORY MODEL

When weather risk is predominant, game theory models are the apt devices to provide realistic solution. These models are developed by Swanson (1959), Walker, Heady, Teweeten and Pesek (1960), Dillon and Heady (1961) *etc.* In all these models, nature is considered as opponent of farmer and component of risk and uncertainty faced by the farmers. In this model, the general proverb that *"man proposes but God disposes"* seems to be applicable, because when the farmer selects a particular farm plan, nature forces the farmer to choose the other one. Such phenomenon is termed as two person-zero sum game. In game theory models, we use different decision criteria for helping the farmer in the selection of appropriate farm plans under a given farm risk situation. To explain the model, we require to understand utility assumptions of the model. Among the varied decision criteria, the most common ones are:

1. Wald maximin decision criterion; and
2. Savage regret criterion.

Savage regret criterion is also known as minimax criterion. These two criteria are mostly used for agricultural decision problems. Let us now understand the maximin criterion.

Let us assume that the farmer is given two farm plans, *i.e.*, $FP_1$ and $FP_2$ and each farm plan is having three states of nature, *i.e.*, $SN_1$, $SN_2$ and $SN_3$. Income from these plans are denoted as $\pi_{ij}$

where,
i refers to $i^{th}$ plan in the $j^{th}$ state of nature

i = 1 to 2 plans

j = 1 to 3 states of nature.

| Farm plans | States of nature | | | Minimum profit (in units of rupees) |
|---|---|---|---|---|
| | $SN_1$ | $SN_2$ | $SN_3$ | |
| $FP_1$ | $\pi_{11}$ | $\pi_{12}$ | $\pi_{13}$ | $\pi_{12}$ |
| $FP_2$ | $\pi_{21}$ | $\pi_{22}$ | $\pi_{23}$ | $\pi_{23}$ |

Maximin criterion is based on the pessimistic view of the farmer that whenever farmer chooses a particular plan, nature would give least profit to minimize his farm income. If the farmer selects $FP_1$ plan nature could give him $\pi_{12}$ profits. If he selects $FP_2$ plan, nature would give him $\pi_{23}$ profits. Under these circumstances the appropriate strategy for the farmer is to select the plan that has the maximum outcome under the worst state of nature. If $\pi_{12} > \pi_{23}$, then according to this criterion select $\pi_{12}$, similarly if $\pi_{23} > \pi_{12}$ then select $\pi_{23}$. McInerney (1969) used this maximin criterion in the LP farm model as follows:

Maximize M
Subject To

$$\sum_{j=j}^{n} g_{ij} X_j \geq M_i \tag{4.38}$$

$$\sum_{j=j}^{n} g_{ij} X_j \leq M_i \tag{4.39}$$

$$X_j, M \geq 0 \tag{4.40}$$

$M_i$ is income obtained from $j^{th}$ crop in the worst state of nature. Here $g_{ij}$ means gross margin of $j^{th}$ crop per unit area in $i^{th}$ state of nature.

Here state of nature refers to the data over time we have to consider the gross margin of $j^{th}$ crop per unit area of the farm over time. i refers to time period ranging from 1 to m.

$A_{ij}$ = $i^{th}$ resource required *viz.*, land, labour, capital, *etc.*, per unit of $j^{th}$ crop.

The maximin criteria give very less total gross margin, with which sometimes the farmers' needs cannot be met from the gross margin obtained.

Incorporating expected income equation Hazell (1970) and Maruyama and Kawaguchi (1971) developed LP model using K in place of equation (4.38). This model is represented as:

$$\sum_{j=j}^{n-} g_j X_j = k \tag{4.41}$$

$$\sum_{j=j}^{n} a_{ij} X_j \leq bi \tag{4.42}$$

$$X_j, M \geq 0$$

Here $\bar{g}$ refers to average gross margin computed over time of $j^{th}$ crop per unit area *i.e.*, i=1 to m years.

K=Parameterized value (it starts from minimum level of farm income to the maximum income expected).

Through the parameterization process, we obtain optimal solution for each assumed level of K. In the objective function we keep zeros under crop activities and one for worst profit activity (M) correspondingly under $a_{ij}$ matrix for each $i^{th}$ row we keep −1. This is illustrated in Table 28.14.

TABLE 28.14  Maximin Criterion.

| | Crop Activities | | | | | | |
|---|---|---|---|---|---|---|---|
| | $X_1$ | $X_2$ | $X_3$ | $X_4$ | M | Sign | RHS |
| Objective function (gross margins per acre) | 0 | 0 | 0 | 0 | 1 | = | Maximize M |
| Land (in acres) | 1 | 1 | 1 | 1 | 0 | ≤ | $b_1$ |
| Labour (mandays/acre) | + | + | + | + | 0 | ≤ | $b_2$ |
| Capital (Rs./acre) | + | + | + | + | 0 | ≤ | $b_i$ |
| Expected gross margin (Rs./acre) | — $g_1$ | — $g_2$ | — $g_3$ | — $g_4$ | 0 | = | K (expected total gross margin) |
| Year wise data of gross margin: | | | | | | | |
| 1990 | $g_{11}$ | $g_{12}$ | $g_{13}$ | $g_{14}$ | −1 | ≥ | 0 |
| 1991 | $g_{21}$ | $g_{22}$ | $g_{23}$ | $g_{24}$ | −1 | ≥ | 0 |
| 1992 | $g_{31}$ | $g_{32}$ | $g_{33}$ | $g_{34}$ | −1 | ≥ | 0 |
| 1993 | $g_{41}$ | $g_{42}$ | $g_{43}$ | $g_{44}$ | −1 | ≥ | 0 |
| 1994 | $g_{51}$ | $g_{52}$ | $g_{53}$ | $g_{54}$ | −1 | ≥ | 0 |
| — | — | — | — | — | — | — | — |
| — | — | — | — | — | — | — | — |

Here M = The worst profit activity.

K = It is value of total farm gross margin assumed ranging from minimum level to the maximum level attainable.

## Minimax Criterion or Savage Regret Criterion

Let us assume that the expected gross margins from the crop activities corresponding to the state of nature and these are given as

$g_{ij}$ = (gross margin of $j^{th}$ crop in the $i^{th}$ state of nature)

Here $i^{th}$ state of nature refers to $i^{th}$ year with $i$ ranging from 1 to m years.

The savage criterion is based on the assumption that the farmer wants to minimize the risk (remorse) that he experiences after having made decisions with perfect foresight. Let the maximum value of farm income be denoted by $\pi_i^*$ and actual realized farm income by $\pi_i$, then the difference between $\pi_I^*$ and $\pi_I$ is called regret and it is denoted as 'R'.

$$R = \pi_I^* - \pi_I \qquad (4.43)$$

Given this measure of regret (R), savage regret criterion tells us that the farmer selects the plan having the minimum value of maximum regret that is minmax criterion. Hazell (1970) developed linear programming based on minimax criterion in the following manner.

Minimize R

Subject To

$$\pi_I^* - \sum_{j=1}^{n} g_{ij} X_j \leq R \quad \text{(I ranging from 1 to m)} \qquad (4.44)$$

$$\sum_{j=1}^{n} a_{ij} X_j \leq b_i \qquad (4.45)$$

$$\sum_{j=1}^{n} g_{ij} X_j = K \qquad (4.46)$$

$$X_j, R \geq 0 \qquad (4.47)$$

Here R is the largest regret for a farm plan over time. The farm chooses the plan, which has minimum R. Equation (4.46) is called expected gross income equation. K is parameterized from minimum level to maximum attainable level. For each given level of k, we obtain one set of optimal X value $(X^*)$ along with the values of K and R. This is illustrated in the form of an initial Table 28.15.

## DISCRETE STOCHASTIC PROGRAMMING

In game theory models and risk programming models, risk in the activity gross margin $(g_{ij})$ is considered. Risk in the $a_{ij}$ and $b_i$ coefficients of LP problems is considered in discrete stochastic programming models and chance constrained LP programming models. If the resources are freely exchanged, then any random discrepancy between the requirements of input and their supplies can be captured in the objective function through buying and selling activities. Implementation of this concept requires consideration of adjustments that are to be made in the farm plan in each state of nature. To understand this more clearly let us assume that the farmer is growing 3

TABLE 28.15   Minimax Regret Criterion-an Illustration.

| | Crop Activities | | | | Largest Regret | Sign | |
|---|---|---|---|---|---|---|---|
| | $X_1$ | $X_2$ | $X_3$ | $X_4$ | R | | |
| Objective function (gross margins per acre) | 0 | 0 | 0 | 0 | 1 | = | R minimize |
| Land (in acres) | 1 | 1 | 1 | 1 | 0 | $\leq$ | $b_1$ |
| Labour (mandays/acre) | + | + | + | + | 0 | $\leq$ | $b_2$ |
| Capital (Rs./acre) | + | + | + | + | 0 | $\leq$ | $b_i$ |
| Expected gross margin (Rs./acre) | $\bar{g}_1$ | $\bar{g}_2$ | $\bar{g}_3$ | $\bar{g}_4$ | 0 | = | K (expected total gross margin) |
| Crop rotation constraint (ac) | +1 | $-1$ | +1 | $-1$ | 0 | $\leq$ | 0 |
| Year wise deta of gross margin: | | | | | | | |
| 1990 | $g_{11}$ | $g_{12}$ | $g_{13}$ | $g_{14}$ | +1 | $\geq$ | $\pi^*_1$ |
| 1991 | $g_{21}$ | $g_{22}$ | $g_{23}$ | $g_{24}$ | +1 | $\geq$ | $\pi^*_2$ |
| 1992 | $g_{31}$ | $g_{32}$ | $g_{33}$ | $g_{34}$ | +1 | $\geq$ | $\pi^*_3$ |
| 1993 | $g_{41}$ | $g_{42}$ | $g_{43}$ | $g_{44}$ | +1 | $\geq$ | $\pi^*_4$ |
| 1994 | $g_{51}$ | $g_{52}$ | $g_{53}$ | $g_{54}$ | +1 | $\geq$ | $\pi^*_5$ |
| — | — | — | — | — | — | — | — |
| — | — | — | — | — | — | — | — |

crops ($X_1$, $X_2$ and $X_3$) under 3 states of nature, ($SN_1$, $SN_2$ and $SN_3$). Gross margins ($g_{ij}$) and labour requirements ($l_{ij}$) for these three crops are subject to vary according to state of nature. Let us further assume that supply of labour depends on the state of nature. With maximin decision criterion applicable to farm risk situation, the discrete stochastic programming model is specified as given in Table 28.16.

$l_{11}$, $l_{12}$, $l_{13}$ = Labour requirements per acre for $X_1$, $X_2$ and $X_3$ crops respectively in state of nature ($SN_1$, $SN_2$ and $SN_3$).

$g_{11}$, $g_{12}$, $g_{13}$ = gross margins (gross income less paid-out costs) per acre of $X_1$, $X_2$ and $X_3$ respectively in state of nature.

$\bar{g}_1$, $\bar{g}_2$, $\bar{g}_3$ = Average gross margin computed over three states of nature.

$w_1$, $w_2$, $w_3$ = Wages of causal labour in each state of nature.

$P_1 w_1$, $P_2 w_2$, $P_3 w_3$ = propability of wages in each state of nature.

K= Parameterized value.

$b_{11}$, $b_{12}$, $b_{13}$ = supply of labour in each state of nature.

In this model, we have specified labour selling and hiring activity in each state of nature so as to enable the farmer to hire or sell different amounts of labour in each state of nature. Under selling activities it has +sign and under the hiring activity it has -ve sign. The variability in labour requirement and its supplies is transferred into gross margins rows. The maximin criterion is applied in the usual manner. Please note that the relevant entries under labour activities are weighted (multiplied) by their probabilities. If estimation of probability (P) poses a problem then equal probability weight such as 0.33 for three years and 0.2 probability for 5 years data, should be assigned. Models of this type, known as stochastic programming models were first developed by Cock (1969). The main assumption made is here that labour decisions are made in each state of nature when there is lot of risk particulary wity supply and demand of labour.

**TABLE 28.16  Distrete Stochastic Programming Model.**

| | Crop activities | | | Hire labour (mandays) | | | Sell labour (mandays) | | | Worst income (in Rs.) | RHS |
|---|---|---|---|---|---|---|---|---|---|---|---|
| | $X_1$ | $X_2$ | $X_3$ | $SN_1$ | $SN_2$ | $SN_3$ | $SN_1$ | $SN_2$ | $SN_3$ | M | |
| Objective function (Rs./ac) | 0 | 0 | 0 | 0 | 0 | 0 | 0 | 0 | 0 | 1 | = Maximize M |
| Land (ac) | 1 | 1 | 1 | 0 | 0 | 0 | 0 | 0 | 0 | 0 | $\leq b_1$ |
| Labour (mandays): | | | | | | | | | | | |
| $SN_1$ | $l_{11}$ | $l_{12}$ | $l_{13}$ | -1 | 0 | 0 | +1 | 0 | 0 | 0 | $\leq b_1$ |
| $SN_3$ | $l_{21}$ | $l_{22}$ | $l_{23}$ | 0 | -1 | 0 | 0 | +1 | 0 | 0 | $\leq b_2$ |
| $SN_3$ | $l_{31}$ | $l_{32}$ | $l_{33}$ | 0 | 0 | -1 | 0 | 0 | +1 | 0 | $\leq b3$ |
| Gross margin (in Rs./acre): | | | | | | | | | | | |
| $SN_1$ | $g_{11}$ | $g_{12}$ | $g_{13}$ | $w_1$ | 0 | 0 | $w_1$ | 0 | 0 | -1 | $\geq 0$ |
| $SN_3$ | $g_{21}$ | $g_{22}$ | $g_{23}$ | 0 | $w_2$ | 0 | 0 | $w_2$ | 0 | -1 | $\geq 0$ |
| $SN_3$ | $g_{31}$ | $g_{32}$ | $g_{33}$ | 0 | 0 | $w_3$ | 0 | 0 | $w_3$ | -1 | $\geq 0$ |
| Expected total gross margin (Rs./acre) | $\bar{g}_1$ | $\bar{g}_2$ | $\bar{g}_3$ | $P_1 w_1$ | $P_2 w_1$ | $P_3 w_3$ | $P_1 w_1$ | $P_2 w_2$ | $P_3 w_3$ | 0 | = K |

# SIMULATION

In this process, we imitate or simulate the systems behaviour of the firms particularly the production systems and marketing systems over time to evaluate alternative operating decision rules. The mathematical simulation model refers to conduct of experiments to test the underlying assumptions about the production systems' design or marketing systems' design. Then finally we evaluate alternative decision rules and policies for obtaining higher profits under their decision rules. But, most often it is dynamic in the sense that it considers data over time for evaluating alternative decisions. The demand for products of the firms which have got uncertain duration of service and time are in particular simulated and evaluated.

## Purpose of Simulation

The structure and operation of the various systems, *i.e.*, marketing and production of the firms over time is explained and analyzed in a random behaviour. Simulation models are useful to predict the behaviour of the real systems of the firms, as they evolve over time. The future decision behaviour of the firm in production and marketing is greatly influenced and controlled by the outcome of the simulation models. Monte Carlo method is one of the methods used in simulation.

## The Monte Carlo Method

This method is used when the variables of the system either in marketing or production are random and this method depends on the use of computers for the simulation or replication of systems behaviour through several runs. Even without computer, efforts were made earlier to simulate complicated situations/systems and decision problem were solved using the logistics. This method attempts to imitate the behaviour of real systems by generating random numbers for the key model variables that obey some probability laws. By conducting series of numerical examples on the decision models, it is possible to observe the systems behaviour over time and evaluate decision problems or criteria in terms of cost minimization or profit maximization. Based on that the best and efficient operating activity for the firm is suggested. Before applying the model we require an accurate account of key features and decision rules by which the firms' management exercise control over operation of the firms. Further we have to define what are the independent variables (exogeneous variables) that affect the decisions and what are the dependent variables (endogenous variables). In the simulation models, demand and rainfall are exogeneous variables and output and profit levels are endogenous variables. Costs, supplies of products, budget limits, *etc.*, are called systems parameters. Let us know the model with hypothetical example.

## Simulation Model–Numerical Example

The probabilistic demand for butter of a dairy processing unit in period 't' along with their probabilities are assumed and provided in Table 28.17. Let us represent the actual demand for butter in period 't' as $Y_t$.

TABLE 28.17   Demand for Butter in Units and their Respective Probabilities.

| | T | 1 | 2 | 3 | 4 | 5 | Total |
|---|---|---|---|---|---|---|---|
| Demand for butter in 100 grams $(Y_t)$ | 0 | 1000 | 2000 | 3000 | 4000 | 5000 | 15000 |
| Probabilities P (t) | 0 | 0.05 | 0.30 | 0.40 | 0.15 | 0.10 | 1.0 |

Please note that sum of the probabilities attached to the demand is equal to one. Let the quantity of butter supplied by the firm in $t^{th}$ period is specified in the unit of 100 grams (Qt). Systems parameters are defined as follows:

Price per unit (100 grams) of butter $(P_{bt})$ is Rs. 12.

Cost of production per unit of butter (100 grams) $(C_{bt})$ in $t^{th}$ period is Rs. 7. Salvage value $(S_{bt})$ per week is zero

Let us assume that the demand for products, supply of products and profits of the products are estimated per one week. Here 't' represents week (when t = 1, it is first week, t = 2, it is second week and so on). Profits in the respective weeks are estimated using the following formula.

$$Z_t = (P_{bt} - C_{bt})\, Y_t - (C_{bt} - S_{bt})\, (Q_t - Y_t) \qquad (4.48)$$

Equation (4.48) is a profit function for the firm in the $t^{th}$ period. Here it is assumed that there is no salvage quantity. Then equation (4.48) is reduced to (4.49)

$$Z_t = (P_{bt} - C_{bt})\, Y_t \qquad (4.49)$$

Let us assume two decision rules, which are specified below, relevant to the dairy-processing unit in question. They are:
- *First decision rule:* Demand for butter is equal to supply of the product in the $t^{th}$ period. *i.e.,* $Y_t = Q_t$.
- *Second decision rule:* Demand for butter is greater than the supply in $t^{th}$ period. *i.e.,* $Y_t > Q_t$.
1. In the first decision rule the dairy processing unit is considering its production equal to an average demand *i.e.,* 3,000 units, per week (15,000 units divided by 5 weeks which is the total number of weeks).
2. In the second decision we should assume that the firm is producing product as per the quantity demanded in the previous week *i.e.,* $Q_t = Y_{t-1}$.

*First decision rule is now expressed as* $Q_t = \overline{Y}_t = 3,000$

*Second decision rule is :* $Q_t = Y_{t-1}$

Here we can apply the simulation rule and imitate or simulate the operation of dairy processing unit for several weeks. This is done generally by generating sequence of values representing the demand and then calculate profit for each decision rule separately. Here we assume that the values generated for demand simulation occur with in the same relative frequency experienced with the past. For example, 2,000 units of demand is occurring with 0.3 probability and 3,000 units of demand with a probability of 0.4 and so on (Table 28.18).

**TABLE 28.18  Range of Random Numbers, Quantity Demanded and Probability of Demand for Butter.**

| Range of random numbers | Quantity demanded | Probability of demand | Cumulative probability |
|---|---|---|---|
| 01 to 05 | 1,000 | 0.05 | 0.05 |
| 06 to 35 | 2,000 | 0.30 | 0.35 |
| 36 to 75 | 3,000 | 0.40 | 0.75 |
| 76 to 90 | 4,000 | 0.15 | 0.90 |
| 91 to 100 | 5,000 | 0.10 | 1.00 |
| Total | 15,000 | 1.00 | |

**Note:** Range of random numbers are based on cumulative probability demand.

Quantity demanded and probability of demand are based on past data
The following steps are followed to simulate the demand for the products in different time periods.

1st step – Obtain the random number from random numbers table.
2nd step – Determine the range of random numbers in which the generated random numbers fall.
3rd step – Identify the quantity demanded ($Y_t$) corresponding to generated random number range. For example, if the random number is 18 then the corresponding demand for the product is 2,000 units. Similarly if the random number is 40 the corresponding demand for the product is 3,000 units.

**Simulation of Dairy Processing:** The relevant information is presented in Table 28.19.

**TABLE 28.19  Simulation of Dairy Processing Unit Operations (Hypothetical).**

| Week Number | Random number | Quantity demanded $Y_t$ | First Decision rule $Q_t = \overline{Y_t}$ | $Z_t$ | Second Decision rule $Q_t = Y_{t-1}$ | $Z_t$ |
|---|---|---|---|---|---|---|
| 1 | 11 | 2,000 | 2,700 | 13,500 | 0 | 0 |
| 2 | 08 | 2,000 | 2,700 | 13,500 | 2,000 | 10,000 |
| 3 | 76 | 4,000 | 2,700 | 13,500 | 2,000 | 10,000 |
| 4 | 54 | 3,000 | 2,700 | 13,500 | 4,000 | 20,000 |
| 5 | 04 | 1,000 | 2,700 | 13,500 | 3,000 | 15,000 |
| 6 | 44 | 3,000 | 2,700 | 13,500 | 1,000 | 5,000 |
| 7 | 32 | 2,000 | 2,700 | 13,500 | 3,000 | 15,000 |
| 8 | 40 | 3,000 | 2,700 | 13,500 | 2,000 | 10,000 |
| 9 | 18 | 2,000 | 2,700 | 13,500 | 3,000 | 15,000 |
| 10 | 92 | 5,000 | 2,700 | 13,500 | 2,000 | 10,000 |
|  | Total | 27,000 |  | 1,35,000 |  | 1,10,000 |

$Q_t$ is obtained by dividing the total of $Y_t$ by 10. This is equal to 2,700. The first decision rule is giving more profits, hence the diary-processing unit should produce butter based on the average of demand over time period. This ensures the long run profits for the dairy unit. Besides risk is eliminated because demand is simulated based on probabilities of demand.

# DYNAMIC PROGRAMMING

It is a method for solving optimization problem, particularly when the decision problem is having a sequence of related decisions. It is based on the Bellman's principle of optimality. This principle has the property that whatever the initial state or initial decision may be, the remaining decisions must constitute an optimal policy with respect to state resulting from the decision. We use dynamic programming method to determine the replacement policies in livestock raising, such as when to replace the cows in the dairy herd or when to replace the hens in the egg production or when to cull broilers, *etc.* Let us now see how dynamic programming helps us in the replacement decision problems in a dairy enterprise.

Let $g_i$ denote annual return of keeping a dairy cow at the age of $i^{th}$ year. The $h_i$ denotes the annual cost of maintaining a dairy cow in the $i^{th}$ year with years ranging from 0 to n years and $s_i$ indicates returns from culling cow at the age of $i^{th}$ year. Further let us assume that the dairy animal is replaced by another one. If in the year, it begins its productive life *i.e.*, then $i = 0$. Let the policy of keeping the animal is based on the keeping policy and culling is based on replacement policy. $f_i$ is the final return by following the optimal policy, if the cow is of age $i$.

Assume the following conditions to formulate dynamic problem $g_i > h_i$ and $f_i = 0$, if $i > 5$ and $f_i > 0$ for all $i_s$. Then the recursive relationship for this type of decision problem is expressed as

$$f_i = \text{Max } K,R \begin{bmatrix} K : g_i - h_i + f_{i+1} \\ R : S_i + g_i - h_0 + f_i \end{bmatrix}.$$

Let us assume the following values in rupees for $g_i$, $s_i$ and $h_i$.

| Age in years (i) | Returns in Rs./unit of livestock (in thousand Rs.) $(g_i)$ | Cost/unit of livestock (in thousand Rs.) $(h_i)$ | Returns from culling activity per unit (in thousand Rs.) $(s_i)$ |
|---|---|---|---|
| 0 | 5 | 3 | 3 |
| 1 | 8 | 4 | 4 |
| 2 | 11 | 4 | 6 |
| 3 | 10 | 5 | 8 |
| 4 | 7 | 5 | 10 |
| 5 | 6 | 5 | 7 |

Bellman's principle $(f_i) = \text{Max} \begin{bmatrix} K = g_i - h_i + f_{i+1} \\ R = S_i + g_i - h_0 + f_i \end{bmatrix}$.

Here,

K = Keeping policy

R = Replacement policy

Appling recursive principle to the assumed data in the table on dairy enterprise, we have

$$(f_5) = \text{Max} \begin{bmatrix} K = 6 - 5 + 0 \\ R = 7 + 6 - 3 + f_i \end{bmatrix} = \text{Max} \begin{bmatrix} 1 \\ 10 + f_i \end{bmatrix} = 10 + f_5 \qquad (R_5)$$

$$(f_4) = \text{Max} \begin{bmatrix} K = 7 - 5 + f_5(10) \\ R = 10 + 7 - 3 + f_i \end{bmatrix} = \text{Max} \begin{bmatrix} 12 + f_i \\ 14 + f_i \end{bmatrix} = 14 + f_4 \qquad (R_4)$$

$$(f_3^*) = \text{Max} \begin{bmatrix} K = 10 - 5 + f_4(14) \\ R = 8 + 10 - 3 + f_i \end{bmatrix} = \text{Max} \begin{bmatrix} 19 + f_i \\ 15^* + f_i \end{bmatrix} = 15 + f_3 \qquad (R_3)$$

$$(f_2) = \text{Max} \begin{bmatrix} K = 11 - 4 + f_3(19) \\ R = 6 + 11 - 3 + f_i \end{bmatrix} = \text{Max} \begin{bmatrix} 26 + f_i \\ 14 + f_i \end{bmatrix} = 14 + f_2 \qquad (R_2)$$

$$(f_1) = \text{Max} \begin{bmatrix} K = 8 - 4 + f_2(26) \\ R = 4 + 8 - 3 + f_i \end{bmatrix} = \text{Max} \begin{bmatrix} 30 + f_i \\ 9 + f_i \end{bmatrix} = 9 + f_1 \qquad (R_1)$$

In the above problem we can conclude from the Bellman's principle that the returns from replacement (R values) are highest at the end of the third year. So the dairy animal needs to be replaced at the end of the third year. Highest R value is seen in the third, in the above example. Hence the dairy animal to be replaced in the third year.

## QUADRATIC RISK PROGRAMMING

According to Stovall[*] (1966), the quadratic risk programming (QRP) is a minimization problem which is specified as follows:

---

[*] Stovall, G. John "Income variation and selection of enterprises". Journal of Farm Economics. Vol. 48(5): pp. 1975-1979, 1966.

Min, V $= (1/2)$ X'QX (4.50)

Subject To

AX $\leq$ B (4.51)

C'X $=$ K (4.52)

X $\geq$ O (4.53)

Here Q represents n × n variance and covariance matrix associated with gross margins of suitable crop activities to the given farm

A = m × n matrix of technical coefficients.

These are the $a_{ij}$ coefficients *i.e.*, $i^{th}$ resource required to produce one unit of $j^{th}$ crop activity (demand of the inputs).

B = m × 1 vector of resource supplies.

These include land, labour, working capital, crop rotational constraints, *etc.*,

C' = Transpose of $\overline{C}$ vector coefficients $\left(\overline{C}\right)$

$\overline{C}$ = Mean gross margin of $j^{th}$ crop activity (Rs./ha)

This mean is computed over the time series data

t = 1 to s years,

j = 1 to n activities

k = Some minimum fixed level of farm income required by the family. K is parameterized from minimum level of income to maximum attainable level.

X = n × 1 vector of unknown crop activity levels.

X' = transpose of X vector

## Estimation of Variance and Co-variance Matrix

To represent income risk as a proxy for all types of farm risks in the QRP model, we work out variance and covariance matrix for gross margins of suitable crops to the given farm. The gross margins of the crops are gross incomes less paid out costs. These gross margins should be computed for one production cycle period. Suppose in the given area, if length of the production cycle for majority of the crops is having 10 years time period, then gross margins for the crops should be worked out for ten years period and variance and covariance matrix should be calculated for these gross margins using the following formula.

The variances (V) of the gross margin is in general given by:

$$V = \sum_{j=1}^{n} \frac{(C_j - \overline{C})^2}{n-1} = \left[\left(\Sigma C^2 - (\Sigma C)^2 / n\right)\right] / n - 1 \text{ for } j = 1 \tag{4.54}$$

and covariances (CV) are given by

$$CV = \sum_{i=1}^{n} \sum_{j=1}^{n} C_{ij} X_i X_j \quad \text{........ for } i \neq j \tag{4.55}$$

These equations (4.54 and 4.55) are
Equivalent to X'QX of equation (4.50)

$$V = \sum_{i=1}^{n}\sum_{j=1}^{n} Q_{ij} X_i X_j \tag{4.56}$$

If $i=j$, $Q_{ij}$ is called variance of per ha of gross margin of the $j^{th}$ crop activity.

If $i{\neq}j$, then $Q_{ij}$ is covariance of gross margin of $j^{th}$ crop activity per hectare.

## Solving QRP Problem

In the QRP problem, we have coefficients in the non-linear form (power form), hence we cannot apply simplex method directly here. There is a need to convert non-linear objective function into linear objective function.

Kuhn-Tucker conditions are to be applied to QRP problem to convert it into linear form and apply the simplex method for solving the problem. Let us know the steps involved here in this regard with simple hypothetical farm problem.

$$\text{Min } Z = 6X^2_1+10X^2_2+14X^2_3-16X_1X_2+14X_1X_3-22X_2X_3-4X_1+20X_2+18X_3 \tag{4.57}$$

Subject To

$$12 X_1 + 14 X_2 + 16 X_3 \geq 16 \tag{4.58}$$

I step:

Since the constraint is in $\geq$ form, multiply the problem with – sign, then the QRP problem becomes maximization type.

$$\text{Max } Z = -6X^2_1-10X^2_2-14X^2_3+16X_1X_2-14X_1X_3+22X_2X_3+4X_1-20X_2-18X_3 \tag{4.57}$$

Subject To

$$-12X_1-14X_2-16X_3 \leq -16 \tag{4.58}$$

In the matrix form, the above problem is given

$$\text{Maximize } Z = [X_1, X_2, X_3]\begin{bmatrix} -6 & 8 & -7 \\ 8 & -10 & 11 \\ -7 & 11 & -14 \end{bmatrix} + \begin{bmatrix} X_1 \\ X_2 \\ X_3 \end{bmatrix}[4 -20 -18]$$

Subject To

$$[-12 -14 -16]\begin{bmatrix} X_1 \\ X_2 \\ X_3 \end{bmatrix} \leq (-16); C = \begin{bmatrix} -6 & 8 & -7 \\ 8 & -10 & 11 \\ -7 & 11 & -14 \end{bmatrix} : B = (-16)$$

Here

$$A = [-12 \; -14 \; -16] \; ; \; B = [-16], \; D = \begin{bmatrix} 4 \\ -20 \\ -18 \end{bmatrix}$$

Applying Kuhn-Tucker equations to the above problem.

$$\hat{A}Y = \hat{B}$$

Here

$$\hat{A} = \begin{bmatrix} A & I_1 & O_1 & O_2 \\ -2C & O_3 & -I_2 & A' \end{bmatrix} \; : \; \hat{B} = \begin{bmatrix} B \\ D \end{bmatrix}$$

A = $1 \times 3$ vector
C = $3 \times 3$ matrix i.e., variance and covariance matrix
$I_1$ = Identity matrix with $1 \times 1$ elements (m constraints=1)
$I_2$ = Identity matrix with $n \times n$ size (n variables = 3) ($3 \times 3$ size)
$O_1$ = Zero matrix with $m \times n$ size ($1 \times 3$ size)
$O_2$ = Zero matrix with $m \times m$ size ($1 \times 1$ size)
$O_3$ = Zero matrix with $n \times 1$ size ($3 \times 1$ size)

Now using the above information

$$\hat{A} = \begin{bmatrix} -12 & -14 & -16 & 1 & 0 & 0 & 0 & 0 \\ 12 & -16 & 14 & 0 & -1 & 0 & 0 & -12 \\ -16 & 20 & -22 & 0 & 0 & -1 & 0 & -14 \\ 14 & -22 & 28 & 0 & 0 & 0 & -1 & -16 \end{bmatrix} \times \begin{bmatrix} X_1 \\ X_2 \\ X_3 \\ S_1 \\ U_1 \\ U_2 \\ U_3 \\ V_1 \end{bmatrix} = \begin{bmatrix} -16 \\ 4 \\ -20 \\ -18 \end{bmatrix}$$

QRP is solved, when $P^{-T} \cdot Y = 0$ otherwise, the procedure should be continued

$$Y = \begin{bmatrix} X \\ S \\ U \\ V \end{bmatrix}; \quad \overline{Y} = [UVXS]$$

$P^{-T}$ = Coefficients of $\overline{Y}$ column vector under B column in optimal table.

X = $1 \times N$ vector $(X_1, X_2, X_3)$
S = $1 \times m$ vector $(m \times 1 = S_1)$
U = $1 \times n$ vector $(U_1 \; U_2 \; U_3)$
V = $1 \times m$ vector $(V_1)$

S is vector of slack variable, U and V are vectors of Lagrangean multipliers and $M_1$, $M_2$ are artificial variables.

Now the Q R P is transformed into LP problem as

Minimize M

Subject To

$$\hat{A}Y = \hat{B}$$

Solve the above problem as per the procedure given by Bronson[*] (1986).

## Numerical Example

We have three negative coefficients in B column vector, to make them positive, multiply through out these rows by – sign. Now we have only two variables with – 1 in the problem. This means that we should add two artificial variables and then minimize (Table 28.20).

**TABLE 28.20    Initial Table of QRP With Hypothetical Table.**

|  | $X_1$ | $X_2$ | $X_3$ | $S_1$ | $U_1$ | $U_2$ | $U_3$ | $V_1$ | $M_1$ | $M_2$ |  |
|---|---|---|---|---|---|---|---|---|---|---|---|
| Objective function | 0 | 0 | 0 | 0 | 0 | 0 | 0 | 0 | $M_1$ | $M_2$ | = Mimimize |
| Constraints | -12 | -14 | -16 | 1 | 0 | 0 | 0 | 0 | 1 | 0 | = -16 |
|  | 12 | 16 | -14 | 0 | -1 | 0 | 0 | -12 | 0 | 1 | = 4 |
|  | -16 | 20 | 22 | 0 | 0 | -1 | 0 | 14 | 0 | 0 | = -20 |
|  | 14 | 22 | 28 | 0 | 0 | 0 | -1 | -16 | 0 | 0 | = -18 |

Here $S_1$ and $U_1$ are with –1, hence we should add two artificial variables with coefficients $M_1$ and $M_2$ and solve the programme by minimizing $M_1$ and $M_2$. When $Z_j$-$C_j$ row coefficients become zero then check

$P^{-T}Y = 0$, Here Y = XSUV, and $P^{-T}$ = Coefficients of UVXS in the optimal table then the QRP is solved, otherwise follow the procedure given by Branson (1986)[**].

Here the expected gross margin (K) of the farm is parameterized from the minimum level of Rs. 1,000 to highest attainable level of Rs. 1,582. When K is Rs. 1,000 farm plan I is obtained. Similarly farm plans for given K levels of Rs. 1,250, Rs. 1,400, and Rs. 1,582 are obtained and indicated in the Table 28.23. When K is increased the variance component (risk) is also increased. This means higher level of farm income is associated with higher level of risk. In the first four plans the area of low variance and low income crops viz., bengalgram and pearl millet were predominant. This means the farmers should grow low-income crops to avoid risk. Coriander appeared in all the plans except the last plan. Similar pattern is observed for cotton crop except the first plan. Groundnut, coriander, and cotton are considered to be risky crops because of their high magnitude of variance and these crops are appearing in the farm plans at a higher farm income level (K level). At lower farm income plans, pearl millet and sorghum are having greater proportion of their areas. Similarly at low farm income levels foxtail millet any seen in the farm plan. Hence, these crops viz., foxtail millet, pearl millet and sorghum are considered to be less risky crops. These plans are consistent with the risk aversion attitude of the farmers. The farmers are

---

[*] Bronson, Richard, Theory & problems of operations research, Schanm's outline series, McGraw Hill Book Company, Singapore, 1986.
[**] See the QRP problem specified in Table 28.21 and Table 28.22 and its corresponding solution in Table 28.23.

## TABLE 28.21  Empirical Specifications of QRP Model for the Typical Farms in Gooty Taluk, A.P. Initial Table.

| Row | Crop activities (per ha) Dry crops | | | | | | | Sign | RHS |
| --- | --- | --- | --- | --- | --- | --- | --- | --- | --- |
| | Groundnut ($X_1$) | Sorghum ($X_2$) | Cotton ($X_3$) | Coriander ($X_4$) | Bengal gram ($X_5$) | Foxtail millet ($X_6$) | Peral millet ($X_7$) | | |
| Objective function | $X^1QX$ | | | | | | | = | Z minimize |
| 1. Dryland constraint (in ha) | 1 | 1 | 1 | 1 | 1 | 1 | 1 | ≤ | 1.62 |
| 2. Labour constraint (in Rs.) | 26 | 15 | 30 | 18 | 30 | 12 | 10 | ≤ | 34 |
| 3. Working capital constraint (in Rs.) | 1045 | 400 | 522 | 525 | 614 | 182 | 307 | ≤ | 1100 |
| 4. Rotational constraint (in ha) | 1 | -1 | 0 | 0 | 1 | -1 | 0 | ≤ | 0 |
| 5. Maximum crop area constraint | 1 | 1 | 0 | 0 | 0 | 0 | 0 | ≤ | 1 |
| 6. Total gross margin constraint (in Rs.) | 1347 | 760 | 1042 | 600 | 602 | 461 | 392 | = | K |

**TABLE 28.22**  Variance-Covariance Matrix of Crop Activity, Gross Margin For Typical Farm in Gotty Taluk, A.P. ($X^tQX$).

| | | | | Crop activities | | | |
|---|---|---|---|---|---|---|---|
| | | | | Dry Crops | | | |
| | Groundnut ($X_1$) | Sorghum ($X_2$) | Cotton ($X_3$) | Coriander ($X_4$) | Bengal-gram ($X_5$) | Foxtail millet ($X_6$) | Pearl millet ($X_7$) |
| $X_1$ | 8,46,078 | 2,46,819 | 4,85,410 | 3,08,900 | 13,624 | 133,125 | 2,37,566 |
| $X_2$ | | 1,00,697 | 1,66,215 | 60,512 | 9,226 | 28,799 | 73,617 |
| $X_3$ | | | 8,71,501 | 1,23,545 | 17,339 | 23,168 | 11,673 |
| $X_4$ | | | | 3,11,4,89 | 21,915 | 74,774 | 98,466 |
| $X_5$ | | | | | 97,864 | -864 | -1,290 |
| $X_6$ | | | | | | 33,007 | 1,11,446 |

Source : Subba Reddy, S. Crop planning under risk for typical dry farms in Anantapur district of Andhra Pradesh. An application of quatratic risk programming.
Unpublished Ph.D. Thesis. Dept. of Agril. Economics, Tamilnadu Agricultural University, Coimbatore, 1988.

**TABLE 28.23  Efficient E.V. A Farm Plans for the Typical Farm in Gooty Taluk, A.P. (cropped area in percentages).**

| S. No. | Particulars | Farm Plans | | | | | | | | Actual crop pattern |
|---|---|---|---|---|---|---|---|---|---|---|
| | | I | II* | III | IV* | V* | VI | VII* | VIII(LP) | |
| 1. | Dry crops | | | | | | | | | |
| | A. Groundnut | - | 8.51 | 20.31 | 22.21 | 24.14 | 26.01 | 29.81 | 43.46 | 30.38 |
| | B. Sorghum | 36.42 | 30.00 | 20.31 | 22.21 | 24.14 | 26.01 | 29.81 | 18.27 | 21.25 |
| | C. Cotton | - | 5.25 | 13.15 | 13.25 | 13.33 | 13.44 | 13.64 | 13.09 | 10.47 |
| | D. Coriander | 3.33 | 6.99 | 12.15 | 15.07 | 17.78 | 20.53 | 25.99 | - | 5.24 |
| | E. Bengalgram | 45.93 | 27.53 | - | - | - | - | - | - | - |
| | F. Foxtail millet | 9.51 | 5.68 | - | - | - | - | - | 25.18 | 32.76 |
| | G. Pearl millet | - | 13.58 | 33.88 | 27.16 | 20.37 | 13.58 | - | - | - |
| 2. | Fallow land | 4.81 | 2.90 | - | 0.10 | 0.24 | 0.43 | 0.75 | - | - |
| 3. | Total Expected gross margin (K) in Rs.1000 | | 1100 | 1250 | 1300 | 1350 | 1400 | 1500 | 1582 | - |
| 4. | Variance (V) (in million Rs.) | 0.111 | 0.222 | 0.529 | 0.567 | 0.608 | 0.651 | 0.746 | 0.829 | - |

**NB:**  *These plans are derived through interpolation.

advised to follow the crop pattern indicated in various farm plans to minimize risk and maximize farm income for the given constraints.

## DATA ENVELOPMENT ANALYSIS (DEA)

Methods of comparing relative productive efficiencies of individual economic units have attracted substantial interest from Agricultural Economists for more than half a century. Since the innovative study by Farrell (1957)[*], Linear Programming (LP) methods had been competing with regression estimates for production functions. The LP methods rationalise observed input-output combinations by providing a piece-wise linear frontier for the most efficient economic units, with other less efficient units falling within the envelope that is defined by this empirical frontier. Data envelopment analysis was initially developed by Charnes, Cooper and Rhodes (1978)[**], from the earlier work by Farrell (1957) and Farrell and Fieldhouse (1962)[***]. It is a sophisticated management tool to help the farmers in their input use decisions. DEA envelops the data and hence the name. Before presenting the model, the relevant concepts are presented below:

**Overall Technical Efficiency (OTE):** This is related to a given farm operating in constant returns to scale (CRTS). Overall technical efficient farms fall on the frontier. The overall technical efficiency can be disaggregated into two measures *viz.*, pure technical efficiency and scale efficiency.

**Pure Technical Efficiency (PTE):** This concept arises when a given farm is operating under variable returns to scale (VRTS). A decision-making unit (farm) which is identified as technically not efficient on CRTS frontier can become technical efficient, if it falls on the VRTS frontier. This unit falling on VRTS frontier is called technically efficient.

**Scale Efficiency (SE):** A decision-making unit is said to be scale efficient if it operates under constant returns to scale.

**Input Congestion:** This implies overutilisation of resources. This is found in variable returns to scale.

DEA offers a flexible approach with considerable scope for the use of diverse data (real and monetary). When Agricultural Economists are more familiar with its features, DEA based indicators may become as common as those of the production function analysis.

### LP model to estimate technical efficiency

*LP that allows only constant returns to scale*

Max $\phi_k$ (4.59)

Subject To

$X_{ij}\lambda_j \leq x_{ik}$ (4.60)

i = 1,2,3 inputs

    j = 1 to 12 farmers

[*] Farrell, M.J. 1957. The measurement of productive efficiency, Journal of the Royal Statistical Society Series A, 120: 253-281

[**] Charnes, A. Cooper, W.W. and E. Rhodes 1978. Measuring the efficiency of decision-making units. European Journal of Operations Research 2: 429-444.

[***] Farrell, M.J. and M. Fieldhouse 1962. Estimating efficiency in production under increasing returns, Journal of Royal Statistical Society Series A, 125: 252-267.

$k = k^{th}$ farmer's problem

$X_{ij} =$ $i^{th}$ input used by $j^{th}$ farmer (3 inputs are considered *viz.*, human labour in mandays, bullock labour in cattle pair days and fertilizers in kg).

$$\sum_{j=1}^{n} Y_j \lambda_j \geq \phi_k Y_k \tag{4.61}$$

Where,

$\phi_k$ and $\lambda_j$ = Unknowns

n = Number of farms

$X_{ij}$ = $i^{th}$ input used by $j^{th}$ farmer

$Y_j$ = Crop output obtained by the $j^{th}$ farmer

$\lambda_j$ = $j^{th}$ unknown parameter generated from the programme

*LP with variable returns to scale*

To the above problem, we augment

$\sum_{j=1}^{n} \lambda_j = 1$ to allow variable returns to scale.

i.e.,

$$\lambda_1 + \lambda_2 + \dots \dots \dots \dots \dots . + \lambda_n = 1$$

*LP with variable returns to scale and input congestion*

Max $\phi_k$ (4.62)

Subject To

$$Y_j \lambda_j \geq \phi_k Y$$

$$X_{ij} \lambda_j = \delta X_{ik}$$

$$\sum_{j=1}^{n} \lambda_j = 1$$

Technical efficiency of the fertilizer input is estimated using DEA analysis for a set of 12 farmers and the relevant equations are furnished below:

**Example: Technical efficiency of fertilizer input**

*LP that allows only CRTS*

$$\text{Max } \pi = 0\lambda_1 + 0\lambda_2 + \dots\dots\dots + 0\lambda_n + \overset{+}{\phi}_0 - \overset{-}{\phi}_0$$

**Subject To**

1. $176X_1+120X_2+134X_3+101X_4+113X_5+121X_6+110X_7+105X_8+110X_9+120X_{10}$
   $+135X_{11}+142X_{12}{\le}176$

2. $15X_1+11X_2+17X_3+12X_4+13X_5+14X_6+17X_7+11X_8+15X_9+15X_{10}+29X_{11}+20X_{12}\le 15$

3. $68X_1+45X_2+80X_3+14X_4+90X_5+100X_6+104X_7+27X_8+107X_9+96X_{10}+83X_{11}+135X_{12}{\le}68$

4. $-18X_1-16X_2-10X_3-14X_4-15X_5-14X_6-24X_7-15X_8-12X_9-11X_{10}-10X_{11}-15X_{12}+18X_{13}-18X_{14}\le 0$

*LP with variable returns to scale*

$$\text{Max } \pi = 0\lambda_1 + 0\lambda_2+\ldots\ldots\ldots\ldots+0\lambda_n + \overset{+}{\phi}_0 - \overset{-}{\phi}_0$$

**Subject To**

1. $176X_1+120X_2+134X_3+101X_4+113X_5+121X_6+110X_7+105X_8+110X_9+120X_{10}+135X_{11}$
   $+142X_{12}{\le}176$

2. $15\ X_1+11X_2+17X_3+12X_4+13X_5+14X_6+17X_7+11X_8+15X_9+15X_{10}+29X_{11}+20X_{12}\le 15$

3. $68X_1+45X_2+80X_3+14X_4+90X_5+100X_6+104X_7+27X_8+107X_9+96X_{10}+83X_{11}+135X_{12}\le 68$

4. $-18X_1-16X_2-10X_3-14X_4-15X_5-14X_6-24X_7-15X_8-12X_9-11X_{10}-10X_{11}-15X_{12}+18X_{13}-18X_{14}\le 0$

5. $1X_1+1X_2+1X_3+1X_4+1X_5+1X_6+1X_7+1X_8+1X_9+1X_{10}+1X_{11}+1X_{12}\le 1$

6. $-1X_1-1X_2-1X_3-1X_4-1X_5-1X_6-1X_7-1X_8-1X_9-1X_{10}-1X_{11}-1X_{12}\le -1$

*LP with variable returns to scale and input congestion*

$$\text{Max } \pi = 0\lambda_1 + 0\lambda_2+\ldots\ldots\ldots\ldots+0\lambda_n + \overset{+}{\phi}_0 - \overset{-}{\phi}_0$$

**Subject To**

1. $176X_1+120X_2+134X_3+101X_4+113X_5+121X_6+110X_7+105X_8+110X_9+120X_{10}+135X_{11}$
   $+142X_{12}-176X_{13}{\le}0$

2. $-176X_1-120X_2-134X_3-101X_4-113X_5-121X_6-110X_7-105X_8-110X_9-120X_{10}-135X_{11}-$
   $142X_{12}+176X_{13} \le 0$

3. $15X_1+11X_2+17X_3+12X_4+13X_5+14X_6+17X_7+11X_8+15X_9+15X_{10}+29X_{11}+20X_{12}-15X_{13} \le 0$

4. $-15X_1-11X_2-17X_3-12X_4-13X_5-14X_6-17X_7-11X_8-15X_9-15X_{10}-29X_{11}-20X_{12}+15X_{13} \le 0$

5. $68X_1+45X_2+80X_3+14X_4+90X_5+100X_6+104X_7+27X_8+107X_9+96X_{10}+83X_{11}+135X_{12}$
   $-68X_{13}\le 0$

6. $-68X_1-45X_2-80X_3-14X_4-90X_5-100X_6-104X_7-27X_8-107X_9-96X_{10}-83X_{11}-135X_{12}+68X_{13} \le 0$

7. $-18X_1-16X_2-10X_3-14X_4-15X_5-14X_6-24X_7-15X_8-12X_9-11X_{10}-10X_{11}-15X_{12}+18X_{14}-18X_{15}{\le}0$

8. $1X_1+1X_2+1X_3+1X_4+1X_5+1X_6+1X_7+1X_8+1X_9+1X_{10}+1X_{11}+1X_{12}{\le}1$

9. $-1X_1-1X_2-1X_3-1X_4-1X_5-1X_6-1X_7-1X_8-1X_9-1X_{10}-1X_{11}-1X_{12}\le -1$

10. $1X_{13} \le 1$

## Technical Efficiency of Fertilizer Input

From the results presented in Table 28.24 it is obvious that 4 of the 12 farmers are efficient with regard to OTE, PTE, SE and CE. The overall technical inefficiency of the eight farmers resulted more by pure technical inefficiency compared to scale inefficiency. Congestion inefficiency could be found for three of the selected farmers. On the whole, overall technical inefficiency is to an extent of 53 per cent, while the other inefficiencies *viz.*, pure technical, scale and congestion stood at 42.6 and 15 per cent respectively.

**TABLE 28.24   Different Types of Efficiency Measures.**

| Farmer No | OTE | PTE | SE | CE* |
|---|---|---|---|---|
| 1 | 1.2121 | 1.1111 | 1.0909 | 1.0000 |
| 2 | 1.0000 | 1.0000 | 1.0000 | 1.0000 |
| 3 | 2.3782 | 2.1333 | 1.1148 | 2.0015 |
| 4 | 1.0000 | 1.0000 | 1.0000 | 1.0000 |
| 5 | 1.2421 | 1.2281 | 1.0114 | 1.0000 |
| 6 | 1.4327 | 1.4285 | 1.0029 | 1.0000 |
| 7 | 1.0000 | 1.0000 | 1.0000 | 1.0000 |
| 8 | 1.0000 | 1.0000 | 1.0000 | 1.0000 |
| 9 | 1.7751 | 1.7592 | 1.0090 | 1.0000 |
| 10 | 2.5052 | 1.9393 | 1.2918 | 1.6798 |
| 11 | 1.9452 | 1.8230 | 1.067 | 1.0000 |
| 12 | 1.8904 | 1.6000 | 1.1815 | 1.0884 |
| Mean | 1.5318 | 1.4185 | 1.0641 | 1.1475 |

*CE: Congestion efficiency

## Fertilizer Wastage

Of the 12 formers, eight farmers each turned out to be overall technical inefficient, pure technical inefficient and scale inefficient (Table 28.25). Only three farmers are congestion inefficient. The inefficiency of above farmers under the respective measures resulted in an overall wastage of fertilizer touching 23.86, 23.18, 5.10 and 6.53kg/ha.

**TABLE 28.25   Fertilizer Wastage in kg/ha.**

| Farmer No. | OTE | PTE | SE | CE |
|---|---|---|---|---|
| 1 | 12.00 | 06.80 | 05.66 | 00.00 |
| 2 | 00.00 | 00.00 | 00.00 | 00.00 |
| 3 | 46.34 | 42.50 | 08.19 | 40.00 |
| 4 | 00.00 | 00.00 | 00.00 | 00.00 |
| 5 | 17.54 | 16.72 | 01.00 | 00.00 |
| 6 | 30.20 | 30.00 | 0.29 | 00.00 |
| 7 | 00.00 | 00.00 | 00.00 | 00.00 |
| 8 | 00.00 | 00.00 | 00.00 | 00.00 |
| 9 | 46.72 | 46.18 | 00.96 | 00.00 |
| 10 | 58.00 | 46.50 | 21.52 | 26.40 |
| 11 | 40.32 | 38.84 | 02.79 | 00.00 |
| 12 | 63.58 | 50.62 | 20.73 | 02.00 |
| Mean | 23.86 | 23.18 | 05.10 | 06.53 |

# CHAPTER 29

# Size of Farm

Size of the farm influences the welfare of the farmers. Operators of large sized farms enjoy better standard of living compared against their counterparts with small sized farms. Better standard of living in turn encourages their children to be in farming. This way the interest in farming makes the farmers to be committed to derive maximum production. Large farms further enjoy the scale economies as well. A country with this type of setting in agriculture regarding farm size marches ahead to produce larger quantity of farm products. In India, the law of inheritance brings down the size of farm operated by a farm family with every passing generation. Sub-division on one side reduces the size of farm but the size of holding is further reduced by fragmentation. With ever growing dependents in farming, the size of the farm had been dwindling over the years, as it is at present does not provide sufficient levels of income to ensure a reasonable standard of living. The size of farm that is operated by the farmer should doubly ensure this requirement. Apart from being small in size, these are scattered making the farming an unremunerative occupation. The various size groups are as follows:

| | | |
|---|---|---|
| 1. | Marginal | Below one hectare |
| 2. | Small | Between 1 to 2 hectares |
| 3. | Semi-medium | Between 2 to 4 hectares |
| 4. | Medium | Between 4 to 10 hectares |
| 5. | Large | 10 hectares and above |

## Size and Farm Returns

When the relevance of size of farm arises, we normally refer to large farms and small farms. Greater efficiencies in the use of resources are associated with the large farms than the small farms. During unfavourable years contributed by price and non-price factors both the categories of farms suffer, but the losses are clearly higher on large farms over small farms. But the large farms have the shock absorbing capacity to withstand, while small farms really suffer. However, for both the sizes of farms, favourable climatic conditions and favourable prices relative to costs bring in prosperity. This is what the best that can happen for a farming community. But this type of occurrence is not found frequently. Often it is found that given the favourable climatic conditions, favourable prices would not be in the offing. If favourable prices are found, climate may play havoc with the crop production.

## Small Size Farms

Small size of holdings has come as a major hurdle in the way of introduction of new and improved technology. Smallness of holdings on one hand deters the use of mechanization and on the other hand does not allow the use of new varieties of seeds, fertilizers and other modern inputs due to lack of purchasing power in the hands of small farmers. Those small holders of land who could not eke out their subsistence from their respective farm holdings either sell or lease out their holdings and join either the army of farm labourers or migrate to urban areas in search of some work.

### Advantages

1. In periods of low prices, the losses are smaller
2. The farmer can devote personal attention to every detail of farm business
3. From the viewpoint of society, the large sized farms may not be more desirable since these lead to concentration of economic power and hence anti-social. Small farms imply a wider diffusion of landed property.
4. The small farmer can withstand the period of economic depression or adverse weather conditions with relatively less loss.

### Disadvantages

1. Small size of farm makes it impossible for the farmer to use the best available tools and the best methods of production.
2. Too small a farm fails to provide full time employment to the family members all the year round.
3. The marketing economies which are well reaped by large scale farmer are absent in the case of small farming. They are rarely able to offer regular supplies of their products for sale nor are they able to buy their requisites in sufficiently large quantities to the advantages. The small farmer is exposed to the exploitation of middlemen.
4. Another difficulty encountered by the small farmer is getting required credit at reasonable rate of interest.
5. Overhead charges such as depreciation, maintenance, storage etc. per unit are high.

## Large Size Farms

### Advantages

1. Greater output and greater net income are the features of the large farms in the light of efficient use of resources, of course not forgetting the multiplier effect during the periods of unfavourable climate and unremunerative prices.
2. Greater labour efficiency is attainted on large farms because of more productive work per man and small farms don't enjoy such a privilege, as too many work on a given size of farm.
3. Large farms put the machinery and equipment to the best advantage thereby reducing the costs per unit of output.
4. The fixed costs are lower on large sized farms.
5. In the purchase of supplies as well as disposal of products, large farms enjoy an advantage over the small farms. The bulk purchase of inputs like, seeds, fertilizers,

feeds, *etc.* reduces the per unit costs. Similarly, sale of large volume of output brings down the marketing costs.

6. Diversification of the farming is a possibility on large farms compared to small farms, which normally specialize. The resource endowment base enables the large farms to enjoy such an advantage. Small farms specialize more in food crops as they are limited by the farm size and other resources.

## Disadvantages

1. Unfavourable price environment results in greater losses on large farms over small farms due to higher investment. Similarly, they are set to loose in a bigger way in the years of poor yields.

2. When the lands owned by the large farms are relatively less fertile and the conditions of which cannot be altered, they are tended to incur greater losses, of course, these are the specific cases.

3. Inefficient management of labour on large farms brings in greater disadvantage.

# FACTORS INFLUENCING SIZE OF FARM

1. *Population:* Density of the population affects the size of the farm *i.e.,* pressure of population reduces the size of farm. In India as population is increasing the size of farm is decreasing, thereby affecting the viability of the very farm business.

2. *Availability of Funds:* Farmers with greater amount of funds can not only run the farms successfully, but are in a position to acquire additional land through purchase.

3. *Climate:* Congenial nature of climate facilitates diversification of farming which increases the size of farm business.

4. *Managerial Abilities:* Farmers with good management skills can operate large farms. Size of the farm is affected by the managerial skills.

5. *Land Ceilings:* Ceilings on land impose limits on the size of holdings as indicated by the state laws.

6. *Availability of Irrigation Water:* Under the irrigated conditions generally large farms can be operated. Between tank and canal irrigated farms against well irrigated farms, the former can be run economically from the point of view of pumping costs. When these two sources of irrigation are compared, relatively large areas can be put under cultivation in the areas served by canals and tanks.

7. *Extensive and Intensive Cultivation:* In extensive cultivation, large areas can be put under plough; while in intensive cultivation the size of farm is small.

8. *Technology to be Adopted:* In land saving technology, the size of holding is small while in land augmenting technology the size of farm is larger.

9. *Perishability and Durability of Commodities:* Perishability of the commodities limits the area under cultivation as in the case of vegetables, while durability of the commodities encourage large scale farming and

10. *Law of Inheritance:* Because of the law inheritance, the size of holding declines as the given land is distributed among children.

# MEASURES OF FARM SIZE

The following measures are used to express the size of the farm.

1. *Number of Acres or Hectares:* Throughout the world, the size of farm is measured in terms of acres or hectares, which is easily understandable. But this is not a precise measure, because it does not take into account other factors like labour and capital. This method cannot be used to compare farms situated in different agro-climatic situations. For example, a five-acre farm in Rayalaseema region of Andhra Pradesh cannot be compared against a five-acre farm in delta region in Coastal Andhra.

2. *Labour:* Total labour employed (man equivalents) on farm can be a measure of size of the farm. However, it is not precise as labour employment varies with the crops and technologies adopted. Besides, it ignores land and capital inputs.

3. *Total Capital:* Total capital invested in the form of land, farm machinery, farm buildings, livestock, *etc.,* is also used as a measure of size of the farm. This is also not a precise measure, as it does not take into account labour and land.

4. *Value of Production:* This includes the gross value of all the farm products produced in a given year. This is a more accurate method of expressing the size of the farm, as it takes into account land, labour and capital resources. However, this method is also not totally free from any problem, as physical production though is same in different years, the value of production changes, because of changes in prices. Production also depends on managerial ability, which is subjective.

# CHAPTER 30

# Farm Records

Farm business has become more complex due to technological, economic and social changes. The successful organization of the farm business depends on the experience of the organizer. Farm records preserve the value of experience. No business worth its name today operates without maintenance of records. Farming is also a business and therefore there is a greater need to keep proper recording system of all the business transactions that take place on the farm. The main objectives of farm records are to control the farm business, guide future decisions and provide data required for sound farm planning.

## BENEFITS OF FARM RECORDS

The benefits of maintaining farm records are presented below:

1. *Planning:* Identification of defects in the existing organization of farm business is the first step in farm planning. Analysis of farm records helps in diagnosing the omissions in the current plan. It also provides data required for farm planning.
2. *Management:* Maintenance of farm records inculcates the business outlook as well as better insight into the working of farm business. Systematic recording of farm business transactions helps the farmer in knowing the strengths and weaknesses of the business. Once weaknesses are brought into fore, the recurrence of the same can be avoided in future and thus leading to the improvement in the management.
3. *Farm Returns:* Farm records help to check the unproductive expenditure and identify profitable enterprises. Besides, the gap between current income and potential income can be examined, as records are the sources to know the present income on the farm. Once this gap is identified, suitable corrective steps can be taken to improve the returns on the farm.
4. *Research and Government Policies:* To conduct research we need precise data on costs and returns. Similarly for formulation of various programmes for betterment of agricultural sector, the Government also needs authentic data. Well maintained records provide required information for the researcher and the Government.
5. *Credit:* The records provide information on the income generating capacity of the farm and thereby indicates the credit worthiness of the farmer. This enables the farmers to get required credit from the financial institutions.
6. *Input Management:* Records indicate the requirement of various inputs and input services in advance so as to organise the farm business smoothly.

## LIMITATIONS IN THE MAINTENANCE OF FARM RECORDS
### (difficulties in maintenance of farm records)

1. *Illiteracy:* Majority of the farmers does not have the reading and writing abilities as a result they do not show any interest in the maintenance of records.
2. *Small Size Holdings:* Small farmers do not feel it is worth to maintain the records in view of meager amount of turnover.
3. *Fear of Taxation:* Farmers are apprehensive of the taxation on their income and so they do not like to record the particulars of income.
4. *Complicated Nature:* Majority of the farmers are used to memorize the various transactions of farm business from time immemorial and they feel that it is cumbersome to maintain records, for some knowledge of accounting is required to maintain the records even the farmers are literate.
5. *Nature of Farming:* Farming is a laborious work which involves not only the physical work but also mental work. Farmer works on the farm from morning till evening, therefore he does not feel comfortable to sit and write all the business transactions that have taken place on that day.

## RECORDS MAINTAINED ON AN AVERAGE FARM

An average farmer requires the following records to be maintained for efficient management of the farm.

1. Land use records.
2. Permanent deadstock register.
3. Farm livestock records.
4. Farm labour records.
5. Input records.
6. Feed records.
7. Crop production and disposal records.
8. Livestock production and disposal records.
9. Input and feed stock register, and
10. Log book.

Formats of these records along with their utilities are presented hereunder.

### 1. Land Use Records

| Particulars | Area (in ha) |
|---|---|
| Cultivated area | |
| Irrigated | |
| Irrigated dry | |
| Dry | |
| Permanent fallows | |
| Land under buildings | |
| Problematic soils | |

The different types of land *i.e.*, irrigated, irrigated dry and unirrigated, operated by the farmer can be known through these records. Lands lying idle in operational holdings along with areas under problematic soils can also be understood.

Cropping pattern on the farm is prepared using these records. Cropping pattern which is presented below gives an idea of different crops *viz.*, cereals, pulses, commercial crops, *etc.*, that are taken up by the farmer.

## Cropping Pattern

| Crops | Area (in ha.) |
|---|---|
| *Kharif* | |
| a. | |
| b. | |
| c. | |
| d. | |
| *Rabi* | |
| a. | |
| b. | |
| c. | |
| d. | |
| *Summer* | |
| a. | |
| b. | |
| c. | |

## 2. Permanent Deadstock Register

| Particular | Year of construction/ purchase | Construction value /puchase value (in Rs.) | Amount spent on repairs per annum (in Rs.) |
|---|---|---|---|
| Farm buildings | | | |
| Cattle shed | | | |
| Wells/tube wells | | | |
| Pump house | | | |
| Tractors | | | |
| Oil engines | | | |
| Electric motors | | | |
| Threshers | | | |
| Sprayers | | | |
| Bullock carts | | | |
| Wooden ploughs | | | |
| Iron ploughs | | | |
| Cultivators | | | |
| Harrows | | | |
| Crowbars | | | |
| Spades | | | |
| Sickles | | | |
| Hand hoes, *etc.* | | | |

The particulars of year of purchase, purchase value, amount spent on repairs, *etc.*, of the deadstock on the farm are available in these records. The data facilitate the farmers the extent of wear and tear of the items so that the extent of repairs or even replacements he has to undertake can be gauged. In respect of certain items on which large amounts are incurred for repairs annually, the farmer may think in terms disposal of the same or replacing them as the case may be.

## 3. Farm Livestock Records

| Type | Number | | Breed | Age | Purchase value (Rs.) |
|------|--------|--|-------|-----|----------------------|
| | Homebred | Purchased | | | |
| Milch animals | | | | | |
| Cows | | | | | |
| She-buffaloes | | | | | |
| Draft animals | | | | | |
| Bullocks | | | | | |
| He-buffaloes | | | | | |
| Young stock | | | | | |
| Heifers | | | | | |
| Calves | | | | | |
| Others | | | | | |
| Sheep/goat | | | | | |
| Poultry birds | | | | | |
| Others | | | | | |

These records provide information on the type of animal, breed, age, value, *etc.* Planning for better breeds, purchase and disposal of animals is possible through these records. Farmers need to keep a watch on the age of animals and making use of the production level from the production records, the relevant decisions need to be taken. Similarly performance of the breeds as well can be monitored closely.

## 4. Farm Labour Records

| Season | Crop | | Variety | | Area | |
|--------|------|--|---------|--|------|--|
| Date | Operations | Owned labour | | Hired labour | | Wages (in Rs.) |
| | | TP CP M W C | | TP CP M W C | | TP CP M W C |

TP = Tractor power        CP = Cattle pair days

In these records, the name of the enterprise, the days on which human labour, bullock labour and machine labour employed, the operations performed, number of labourers employed to complete the operation, the time taken to perform the operation, wages paid in cash or kind are recorded. This record provides information on the total labour requirement for each enterprise and also for the season. Particularly the data on human labour requirement helps the farmer to plan the crops to cope up shortages of labour if any.

## 5. Input Records

| Date/month | Input | Quantity | Rate (in Rs.) | Value (in Rs.) |
| --- | --- | --- | --- | --- |
| | | | | |
| | | | | |

It gives a picture of the extent of different inputs utilized crop-wise and variety – wise. The actual use indicates whether a particular input was over-used or under-used. The imbalances if any can be corrected from the data available in these records.

## 6. Feed Records

| Date/month | Type of feed | Milch animals | | Draft animals | | Young stock | | Poultry | |
| --- | --- | --- | --- | --- | --- | --- | --- | --- | --- |
| | | Quantity (in kg or Q) | Value (in Rs.) | Quantity (in kg or Q) | Value (in Rs.) | Quantity (in kg or Q) | Value (in Rs.) | Quantity (in kg or Q) | Value (in Rs.) |
| | | | | | | | | | |
| | | | | | | | | | |

The details of daily feed expenditure figure here. The feed particulars also indicate as to whether the livestock was given required ration or not so that feed efficiency can be worked out.

## 7. Crop Production and Disposal Records

| Season | Date / month | Crop | Area | Main product (Q/tonnes) | By product (Q/tonnes) | Household consumption Quintals | Value (in Rs.) |
| --- | --- | --- | --- | --- | --- | --- | --- |
| | | | | | | | |
| | | | | | | | |

| Used as seed | | Kind payments | | Sold | | Balance | |
| --- | --- | --- | --- | --- | --- | --- | --- |
| Quantity (in Q) | Value (in Rs.) | Quantity (in Q) | Value (in Rs.) | Quantity (in Q) | Value (in Rs.) | Quantity (in Q) | Value (in Rs.) |
| | | | | | | | |
| | | | | | | | |

The details of output realized crop-wise and season-wise along with the details of consumption, kind payments, *etc.* and the quantity sold are obtained from this register. This helps to estimate the profitability or otherwise of the farm business.

## 8. Livestock Production and Disposal Records

| Date / month | Particulars | Milk | | Eggs | | Farm yard manure | |
|---|---|---|---|---|---|---|---|
| | | Quantity (in litres) | Value (in Rs.) | No. | Value (in Rs.) | Quantity (in tonnes) | Value (in Rs.) |
| | | | | | | | |

| Poultry litter | | House hold consumption | | | | Particulars of conversion | | |
|---|---|---|---|---|---|---|---|---|
| Quantity (in tonnes) | Value (in Rs.) | Milk | | Eggs | | Quantity of Milk (in litres) | Ghee obtained (in litres) | Value in Rs. |
| | | Quantity (in litres) | Value (in Rs.) | No. | Value (in Rs.) | | | |
| | | | | | | | | |

Production levels from various enterprises are made available in this register. The disposal pattern as well as consumption pattern of livestock products too are found here. The value addition through conversion of milk into ghee can also be noted here.

## 9. Input and Feedstock Register

| Date / month | Particlars of inputs / feed | Opening balance (in kg) | Receipts (in kg) | Quantity issued (in kg) | Balance (in kg) | Source of purchase | Cost (in Rs.) |
|---|---|---|---|---|---|---|---|
| | a  b  c | a b c | a b c | a b c | a b c | a b c | |
| | | | | | | | |

To assess the stock position of farm inputs and feed at a particular point of time, there arises a need to maintain the stock register.

## 10. Log Book

| Season | Crop | | | Variety | | Area | |
|---|---|---|---|---|---|---|---|
| Date/month | Name of the machinery | Operation done | Time (No. of hours) | Power Consumed | | Fuel consumed | |
| | | | | (in units) | Cost (in Rs.) | Quantity | Value (in Rs.) |
| | | | | | | | |

| Lubricants | | Repairs (in Rs.) |
|---|---|---|
| Quantity | Value (in Rs.) | |

Large farmers invest heavily on machinery and equipment. The relevant expenditure can be known by maintaining the log book.

## RECORDS MAINTAINED ON THE CORPORATE FARMS AND THE STATE FARMS

Following is the list of records maintained by the corporate farms and state farms
1. Forecast register.
2. Daily memorandum sheet (DMS).
3. Muster sheets.
4. Permanent deadstock register (PDSR).
5. Temporary deadstock register (TDSR).
6. Fertilizers and chemicals register.
7. Seed stock register.
8. FYM register.
9. Cattle feed register.
10. Tractor expenditure register.
11. Livestock register.
12. Farm produce stock register.
13. Produce stock register.
14. Indent register.
15. Sales price register.
16. Sanction register.
17. Auction register, and
18. Cash book.

### 1. Forecast Register

| Field No. | Area | Nature of work | Wage rates of labourers forecasted | | | | | |
|---|---|---|---|---|---|---|---|---|
| | | | Skilled labourers | Wage rate (in Rs.) | Semi skilled labourers | Wage rate (in Rs.) | Unskilled labourers | Wage rate (in Rs.) |
| | | | | | | | | |

| Amount | | | | | |
|---|---|---|---|---|---|
| Skilled labourers | Amount (in Rs.) | Semi skilled labourers | Amount (in Rs.) | Unskilled labourers | Amount (in Rs.) |
| | | | | | |

This is prepared a day in advance of the actual labour requirement on the farm. Keeping in view of the various operations to be performed for various crops, the requirement is forecasted. The labour units indicated in this register should not exceed the labour units given in DMS.

## 2. Daily Memorandum Sheet (DMS)

| Field no. | Purpose of work | Allotment of labourers | | | | | | Amount (in Rs.) |
|---|---|---|---|---|---|---|---|---|
| | | Skilled | Wages (in Rs.) | Semi skilled | Wages (in Rs.) | Unskilled | Wages (in Rs.) | |
| | | | | | | | | |

This deals with the distribution of work in a day along with the labour units employed and the wages paid for various operations.

## 3. Muster Sheets

| S. No. | Name of the casual labourer | Sex | Nature of work | No. of days worked in a fortnight | | | | | No. of days | Wage rate/ day (in Rs.) | Amount (in Rs.) |
|---|---|---|---|---|---|---|---|---|---|---|---|
| | | | | 1 | 2 | 3 | 4 | ...15 | | | |
| | | | | | | | | | | | |

The particulars of the labour units employed including the number of days employed and the wage bill are posted in these sheets. These sheets give an idea of fortnightly expenditure incurred on the labour wages.

## 4. Permanent Deadstock Register

| Date | Particulars | Receipts | Issues | Balance |
|---|---|---|---|---|
| | | | | |

The details of inventory of the farm along with the particulars of their issue are entered here. It also gives the balance of the inventory available.

## 5. Temporary Deadstock Register

| Date | Particulars | Receipts | Issues | Balance |
|---|---|---|---|---|
| | | | | |

The receipts of the temporary deadstock like ropes, GI wire, battery cells, crackers, *etc.*, along with their issues and balance of stock available are given in this register.

This register gives the management an idea of the stock issued and balance available so that future requirements can be assessed and undertake the purchases as required.

## 6. Fertilizers and Chemicals Register

| S. No. | Date | Particulars | Receipts | Issues | Balance |
|---|---|---|---|---|---|
| | | | | | |
| | | | | | |

The details of the different fertilizers purchased along with the purposes for which they are issued are posted here. This register presents the position of the stock of fertilizers and chemicals available at any given point of time.

## 7. Seedstock Register

| Date | Particulars | Receipts | Issues | Balance |
|---|---|---|---|---|
| | | | | |

This register gives the details of the purchases, issues and balance of the seeds of different varieties of crops grown on the farm.

## 8. FYM and Cattle Feed Register

| S. No. | Date | Particulars | Receipts | Issues | Balance |
|---|---|---|---|---|---|
| | | | | | |
| | | | | | |

This register deals with the particulars of receipts, issues and balance of FYM and cattle feed.

## 9. Tractor Expenditure Register

| Date | Particulars | Receipts | Issues | Balance |
|---|---|---|---|---|
| | | | | |

This register contains the information pertaining to the purchase of items like diesel, spare parts, *etc.*, and their use as per the requirements. The stock position of the same is also available in this register.

## 10. Livestock Register

| Date of purchase | From whom purchased / received | Description of the animal | Date of birth | Book value (in Rs.) |
|---|---|---|---|---|
| | | | | |

The description of the animal along with the source of obtaining the same and date of birth of the animal and value are entered here.

## 11. Farm Produce Stock Register

| S. No. | Dry land / wet land | Season Crop Variety | Main product (Q/tones) | Entered in page no. of stock register | Signature of the store produce | Signature of the farm superintendent keeper |
|--------|------|------|------|------|------|------|
| | | | | | | |
| | | | | | | |

The details of crop wise and variety wise main product and byproduct are entered here. This gives an account of the total produce obtained on the farm.

## 12. Produce Stock Register

| Date | Issue / sale | Receipts | Issues | Balance |
|------|------|------|------|------|
| | | | | |
| | | | | |

The information posted in the farm produce stock register is brought over here. Under the column "receipts", the quantity recorded in the farm produce stock register is entered here. The issue column indicates the details as to whether the produce was issued to the farm or sold.

## 13. Indent Register

| Indent No. | Date | Particulars | Quantity | Purpose | Receiver | Indentor | Signature of the store keeper | Signature of the farm superintendent |
|------|------|------|------|------|------|------|------|------|
| | | | | | | | | |
| | | | | | | | | |

This register presents the indents that are made. Under the column 'purpose', if the input indented is fertilizers, the crop to which it is proposed to be applied is entered here. This register holds good for all farm supplies.

## 14. Sales Price Register

| S. No. | Name of the product | Quantity proposed for sale | Rates furnished by the Secretary, Agricultural Market Committee (Rs.) | Rate per unit in the local market (Rs.) | Rate at which disposed (Rs.) |
|--------|------|------|------|------|------|
| | | | | | |
| | | | | | |

Sale particulars of the produce obtained on the farm are found in this register. The prevailing rates for the products from agricultural market committee, as well as those prevailing in the local markets are obtained and then the rates at which the produce was disposed is entered. This type of information is mostly seen with the Government farms.

## 15. Sanction Register

| S. No. | Date | Particulars cum purpose of expenditure | Quantity | Rate (unit price) (Rs.) | Amount to be sanctioned (Rs.) | Head of the account | Signature of the Farm Manager | Signature of sanctioning authority |
|--------|------|----------------------------------------|----------|-------------------------|-------------------------------|---------------------|-------------------------------|-------------------------------------|
|        |      |                                        |          |                         |                               |                     |                               |                                     |
|        |      |                                        |          |                         |                               |                     |                               |                                     |

It provides the details of the items of expenditure along with the rate per unit and amount to be sanctioned. The proposed items are purchased after due sanction from the concerned authorities

## 16. Auction Register

| S. No. | Name of the bidder | Address of the bidder | Amount (Rs.) | Signature of the bidder | Amount deposited (Rs.) | Signature of the unsuccessful bidder |
|--------|--------------------|-----------------------|--------------|-------------------------|------------------------|--------------------------------------|
|        |                    |                       |              |                         |                        |                                      |
|        |                    |                       |              |                         |                        |                                      |

The information of those items, which are auctioned, can be known from the auction register.

## 17. Cash Book

| Date | Opening balance | Sales bill No. | Amount (Rs.) | Amount remitted to the bank (Rs.) | Cash on hand (Rs.) |
|------|-----------------|----------------|--------------|-----------------------------------|--------------------|
|      |                 |                |              |                                   |                    |
|      |                 |                |              |                                   |                    |

The details of cash remittances and cash on hand are shown here.

Having studied the various records suitable to corporate farms and average farms, it is the question of their usage by the average farmers. The time and cost involved in maintenance of records need to be considered as these are the prime reasons for the failure of maintenance of records by the farmers in general. Volume of farm business i.e., small scale nature of farming, which is practiced by majority of the Indian farmers, comes in the way of maintenance of records. In respect of majority of the Indian farmers, the investment is low, income is low and even loss is low. Given this situation these farmers do not give any thought for records, which present the A to Z of the farm business. On the other hand, on the large farms the investment is high, profits are high and losses as well are high. Therefore, they need to keep a watch on the expenditure and income pattern closely on the farm. Farmers of large holdings benefit immensely from the information of farm records. The utility of farm records is higher on large farms over small farms.

## RECORDS MAINTAINED ON COMMERCIAL DAIRY FARMS

The various records maintained by a commercial dairy farm are:

1. Livestock register.
2. Milk production register.
3. Milk disposal register.
4. Calf feeding register.
5. Feeding schedule register.
6. Insemination register.
7. Calving register.
8. Gynaecological status register.
9. Health and vaccination register.
10. Mortality register.
11. History sheet, and
12. Cash book.

The formats of the registers are presented below:

### 1. Livestock Register

| S. No. | Breed | Brand No. | Tattoo No. | Date of birth | Dam No. | Sire No. | Book value (in Rs.) | Remarks |
|--------|-------|-----------|------------|---------------|---------|----------|---------------------|---------|
| 1. | | | | | | | | |
| 2. | | | | | | | | |

**Guidelines:**

Breed:          Specify the breed, say Jersey

Brand No. :    Every animal is identified by a number say J 10 (Jersey 10). The number is hot iron branded.

Tattoo No. :    For the calves born in the same dairy farm, a tattoo number is given. It is put inside the year. Say J 100.

Dam No. :     Mother's number is given, say J 1.

Sire No. :      Father's number is given, say X 2.

Book Value:   Market value of the animal is given. In commercial dairy farms, the value of the animal is evaluated in every month.

Remarks:      Special identification marks are given, for example, white pathches on the forehead, etc.

    All the animals maintained are accounted for in this register. If there is death of an animal its serial number is rounded off with red ink. It is a vital register for maintenance of dairy farm.

## 2. Milk Production Register (kg)

| Date | Animal No. (1) J 10 | | | Animal no. (2) J 18 | | | Grand total | | |
|------|------|------|------|------|------|------|------|------|------|
| | AM | PM | Total | AM | PM | Total | AM | PM | Total |
| | | | | | | | | | |
| | | | | | | | | | |

The milk production particulars of lactating animal are only recorded in this register. 12 hours gap is given between two milkings.

## 3. Milk Disposal Register

| Date | Milk production (litres) | | | Calf feeding (litres) | | | Balance (litres) | | | Disposal (litres) | | |
|------|------|------|------|------|------|------|------|------|------|------|------|------|
| | AM | PM | Total | AM | PM | Total | AM | PM | Total | AM | PM | Total |
| | | | | | | | | | | | | |

After the balance is worked out, the disposal to various individual/agencies is recorded.

## 4. Calf Feeding Register (kg)

| Date | Calf No. ..... Weight: ..... Date of birth: ..... | | | Calf No. ..... Weight: ..... Date of birth: ..... | | | Grand Total | | |
|------|------|------|------|------|------|------|------|------|------|
| | M | E | T | M | E | T | M | E | T |
| | | | | | | | | | |
| | | | | | | | | | |

M: Morning, E: Evening, T: Total.
Entries are made with regard to milk given to each calf.

## 5. Feeding Schedule Register

| S. No. | Animal No. | Date of birth | Lactation No. | Present milk yield (litres) | Pregnancy |
|------|------|------|------|------|------|
| 1 | 2 | 3 | 4 | 5 | 6 |
| | | | | | |
| | | | | | |

| Body weight | | Concentrates (kg) | | Dry fodders (kg) | |
|------|------|------|------|------|------|
| Previous month | Present month | Previous month | Present month | Previous month | Present month |
| 7 | 8 | 9 | 10 | 11 | 12 |
| | | | | | |
| | | | | | |

For each animal the ration is fixed for a month and followed. For pregnant animals extra ration is given while for those which are not giving milk, maintenance ration is given. Body weight is an important criterion for fixing ration.

## 6. Insemination Register

| S. No. | Date of AI* | Animal No. | Sire No. | No. of previous AI done | Date of base AI | Pregnancy diagnosis | Remarks |
| --- | --- | --- | --- | --- | --- | --- | --- |
| | | | | | | | |
| | | | | | | | |

AI* = Artificial insemination.
Sire No. indicate the bull used for AI.
Base AI is the date of first artificial insemination.
Pregnancy diagnosis: Whether it is positive or negative.
This register helps to know the number of calves likely to be added and the quantity of milk likely to be obtained.

## 7. Calving Register

| S. No. | Date of calving | Calf No. | Identification marks | Date of weaning | Dam No. | Sire No. | Dry period | Calving interval |
| --- | --- | --- | --- | --- | --- | --- | --- | --- |
| | | | | | | | | |
| | | | | | | | | |

Calf No. say J 100
Identification marks: Mark is fixed inside the year.
Date of weaning : It is the seperation of the young calf from the mother. In foreign breeds *i.e.,* Jersey, *etc.* calf is immediately separated. For indigenous breeds it is not so.
Dam No: Mother's number is given.
Sire No: Father's number is given.
Dry period: Dry period is to be noted. Exotic breeds continue to give milk between two calvings. So we have to stop milking forcibly for the health of the animal. In respect indigeneous cows no compulsory drying is required as they give milk for about 200 days only.
Calving interval: Record the calving interval, it may be 365 days, 390 days, 400 days, *etc.*

## 8. Gynaecological Status Register

| S. No. | Animal No. | Breed | Date of birth | Date of last calving | No. of lactation |
| --- | --- | --- | --- | --- | --- |
| 1 | 2 | 3 | 4 | 5 | 6 |
| | | | | | |
| | | | | | |

| AI particulars | Pregnancy diagnosis | Date of calving | Calf No. | Remarks |
|---|---|---|---|---|
| 7 | 8 | 9 | 10 | 11 |
|  |  |  |  |  |
|  |  |  |  |  |

| | | |
|---|---|---|
| Date of birth | : | say 1-1-94 |
| Date of last calving | : | say 1-1-97 |
| No. of lactation | : | say 1 |
| AI particulars | : | say 1-3-97 |
| Pregnancy | : | Positive |
| Date of calving | : | 10-12-97 (9 months 10 days) |

The particulars of all animals falling in the breedable age groups come into gynaecological register.

## 9. Health and Vaccination Register

| Date | Dose |
|---|---|
|  |  |
|  |  |

## 10. Mortality Register

| S. No. | Breed | Animal No. | Sex | Date of birth | Date of death | Postmortem No. | Cause of death |
|---|---|---|---|---|---|---|---|
|  |  |  |  |  |  |  |  |
|  |  |  |  |  |  |  |  |

## 11. History Sheet

Breed ....... No. .................. Date of birth ................. Birth weight ...................... Date of calving ... ..................... Dam's breed ......... No... ....... Date of birth ............ Dam's milk yield .............. Sire ............. No.

## Body weight at different ages

| Week/month | | | | Weeks | | | | | | | | Months | | | Weight at calving |
|---|---|---|---|---|---|---|---|---|---|---|---|---|---|---|---|
|  | 4 | 8 | 12 | 13 | 20 | 26 | 30 | 40 | 52 | 15 | 18 | 24 | 36 | | |
| Date |  |  |  |  |  |  |  |  |  |  |  |  |  |  |  |
| Weight |  |  |  |  |  |  |  |  |  |  |  |  |  |  |  |

| Date of service | Date of calving | Sex & weight of calf | Date of dried off | | Yield | | No. of days in | | | Calving interval |
|---|---|---|---|---|---|---|---|---|---|---|
| | | | | Total milk | 300 days milk | Peak | Milk | Dry | | |

Barring this register, all other registers represent all animals in a dairy farm. This sheet presents the single animal history.

## 12. Cash Book

The details of cashbook are furnished in farm accountancy chapter.

# CHAPTER 31

# Farm Accountancy

Farmers can make use of the accountancy to keep a close watch on the farm business. It also helps in the management of any supplementary enterprises taken up by the farmers on business lines.

## TERMS AND CONCEPTS

### Account

A summary of business transactions is an account. It is vertically divided into two parts in 'T' shape. The benefits received by that account are recorded on the left hand side and the benefits given are recorded on the right hand side.

### Accounting

Accounting is the art of recording, classifying and summarizing the business transactions. Recording refers to writing in journal. Classifying means, writing in ledger and summarizing relates to preparation of trading account, profit and loss account and balance sheet.

### Accountancy

Accountancy means the art of keeping books of accounts in a regular and systematic manner.

### Single Entry System

Under this system, the trader does not get all the information which he generally likes to have. At best he can get the information about debtors, creditors and cash but not about other matters. This system is incomplete and inaccurate.

### Double Entry System

Recording a transaction from two different angles is known as double entry system. To understand the double entry system of book keeping all that we need to do is, to remember the fundamental rule;

Debit the account which receives the benefit, and credit the account, which gives the benefit.

## Advantages of Double Entry System

1. Both personal and impersonal accounts are opened in order to keep a complete record of business transactions.
2. It provides a check on the arithmetical accuracy with the help of trial balance.
3. It reduces the chances of committing errors.
4. It helps the trader to know his debtors (customers) and creditors (suppliers) from time to time.
5. Financial position of the business unit can be known through the preparation of balance sheet, and
6. This system is useful to the tax and legal authorities.

### Rules Regarding Different Types of Accounts:

There are three types of accounts *viz.,* (1) Personal account (2) Real account and (3) Nominal account.

### 1. Personal Account

It includes accounts of (a) Natural persons (b) Artificial persons; and (c) Representative persons.

*a) Natural Persons*:  The examples of natural persons are Ramesh A/C, Suresh A/C, *etc.*

*b) Artificial Persons*: The examples are Andhra Bank A/C, Agricultural College A/C, University A/C, firm's name A/C, *etc.*

*c) Representative Persons*: Outstanding salaries A/C, pre-paid rent A/C, pre-paid commission A/C, interest due but not received A/C, rent due but not received A/C, *etc.* are the examples.

*Rule of Debit and Credit for Personal Accounts*:

The rule here is "debit the receiver and credit the giver"

### 2. Real Account

Accounts relating to assets are known as real accounts. *Examples*: Machinery A/C, furniture A/C, purchases A/C, sales A/C, building A/C, goodwill A/C, *etc.*

*Rule of Debit and Credit for Real Accounts:*

The rule that should be followed here is "debit what comes in; and credit what goes out".

### 3. Nominal Account

Nominal accounts relate to such items which exist in name only. *Examples*: Expenses, losses, incomes, gains, *etc.*

*Rule of Debit and Credit for Nominal Accounts:*

The rule is "debit all expenses and losses and credit all incomes and gains".

### Journal

All business transactions are first recorded in a book called journal. It means a daily record. The journal is also called as a "book of original entry".

## Ledger

Ledger is the chief book of accounts. A ledger is a book which contains various accounts to which the entries made in the journal are transferred (journal gives information in the form of entries, whereas ledger contains information in the form of accounts).

## Cash Book

Transactions relating to only cash are recorded in one book called cashbook. The principle here is, cash coming in is placed on debit side and cash going out is entered on credit side.

## Subsidiary Books or Subsidiary Journals

The journal is subdivided into a number of special journals called subsidiary books. They are:

1. *Purchases Journal*: It is meant for recording goods purchased on credit

2. *Purchases Returns Journal*: It is used for recording goods returned to suppliers.

3. *Sales Journal*: This is maintained for recording credit sales of goods only

4. *Sales Returns Journal*: It serves the purpose of recording goods returned by the customers

5. *Cash Book*: In this book all cash receipts and payments, including cash purchases and cash sales of goods are entered. Cash which comes in is debited and cash that goes out is credited.

## Trial Balance

The accountant prepares a list of all the ledger accounts with their closing balances, indicating the details of debit or credit, such a list of balances is known as the trial balance.

## Final Accounts

Final accounts are nothing but summaries of ledger accounts, organized in such a manner as to reveal (a) the profit made or the loss sustained during a period; and (b) the financial condition of the business concern at the end of the period. Final accounts are described as (i) Trading account (ii) Profits and loss account and (iii) Balance sheet

## Trading Account

This gives the details of purchases, sales, purchases returns, sales returns, carriage (carriage inwards), manufacturing expenses, *etc.* Trading account gives gross profit of the business.

## Profit and Loss Account

It contains the particulars of salaries, rent, rates and taxes, advertisements, carriage outwards, depreciation, commission received, interest received, *etc.*, It gives net profit or net loss of the business.

## Balance Sheet

It is a statement containing balance of fixed assets, current assets, liabilities and capital. It gives the financial position of the business. Fixed assets include buildings, machinery, furniture (tangible fixed assets) and goodwill, patents, trade marks, copy rights, *etc.*, (intangible fixed assets).

# JOURNAL

Here the format of the journal with an example and steps to make journal entries are furnished.

Journalize the following transactions

| Dec 1, | 1998 | X started a business with a capital of Rs. 10,000 |
| Dec 2, | 1998 | Purchased goods from Mahesh Rs. 4,000 |
| Dec 3, | 1998 | Sold goods to Mahesh for cash Rs. 3,000 |
| Dec 4, | 1988 | Purchased machinery from Gopal for cash Rs. 6,000 |
| Dec 5, | 1988 | Sold electric motor to Ravi Rs. 4,000 |
| Dec 6, | 1998 | Cash received from Naresh Rs. 2,000 |
| Dec 7, | 1998 | Cash paid to 'Z' Rs. 4,000 |
| Dec 8, | 1998 | Paid salary to Ravi Rs. 1,000 |
| Dec 9, | 1998 | Rent received from 'Y' Rs. 3,000 |
| Dec 10, | 1998 | Cash sales (sold goods for cash) Rs. 9,000 |
| Dec 11, | 1998 | Cash purchases Rs. 3,000 |

Given the transactions of journal, the entries are made in the journal as given in Table 31.1.

**TABLE 31.1  Journal.**

| Date | Particulars | LF | Debit (Rs.) | Credit (Rs.) |
|------|-------------|----|-------------|--------------|
| 1998 | Cash A/C                                    Dr | | 10,000 | |
| Dec. 1 | To X's capital Account | | | 10,000 |
| | (Being started a business) | | | |
| Dec. 2 | Purchases A/C                               Dr | | 4,000 | |
| | To Mahesh A/C | | | 4,000 |
| | (Being goods purchased from Mahesh) | | | |
| Dec. 3 | Cash A/C                                    Dr | | 3,000 | |
| | To sales A/C | | | 3,000 |
| | (Being goods sold for cash) | | | |
| Dec. 4 | Machinery A/C                               Dr | | | |
| | To Cash A/C | | | |
| | (Being machinery purchased for cash) | | 6,000 | 6,000 |
| Dec. 5 | Ravi A/C                                    Dr | | | |
| | To electric motor A/C | | 4,000 | |
| | (Being sold electric motor) | | | 4,000 |
| Dec. 6 | Cash A/C                                    Dr | | | |
| | To Naresh A/C | | 2,000 | |
| | (Being cash received from Naresh) | | | 2,000 |
| Dec. 7 | 'Z' A/C                                     Dr | | | |
| | To Cash A/C | | 4,000 | |
| | (Being cash paid to 'Z') | | | 4,000 |

| Dec. 8 | Salary A/C | Dr | | |
| | To Cash A/C | | 1,000 | |
| | (Being salary paid) | | | 1,000 |
| Dec. 9 | Cash A/C | Dr | | |
| | To Rent A/C | | 3,000 | |
| | (Being rent received) | | | 3,000 |
| Dec. 10 | Cash A/C | Dr | | |
| | To Sales A/C | | 9,000 | |
| | (Being goods sold for cash) | | | 9,000 |
| Dec. 11 | Purchase A/C | Dr | | |
| | To Cash A/C | | 3,000 | |
| | (Being cash purchases) | | | 3,000 |

## Steps To Make Journal Entries

1. Here 'X' is the owner and he has given Rs. 10,000 to the business, so he is giver. According to personal account principle, debit the receiver and credit the giver. So, 'X' capital account should be credited. Cash is real account, the real account principle is debit what comes in, and credit what goes out, so cash account of Rs. 10,000 should be debited.

2. Purchases are real account, purchases are coming in, so purchases account should be debited, Mahesh is giver, hence Mahesh's account should be credited.

3. Sales are real account, sales are going out, so sales account should be credited. Cash is real account, which is coming in, so cash account should be debited.

4. Machinery is a real account, which is coming in, so machinery account should be debited. Cash is a real account, it is going out, so it should be credited.

5. Electric motor is a real account, which is going out, so electric motor account should be credited, Ravi is personal account, he is receiver, so Ravi's account should be debited.

6. Cash is a real account, cash is coming in, so cash account should be debited. Naresh is personal account, he is giver, hence, Naresh's account should be credited.

7. 'Z' is a personal account, 'Z' is receiver, so 'Z' account should be debited. Cash is real account, it is going out, so cash account should be credited.

8. Salary account is a nominal account, and it is an expense. The principle of nominal account is debit all expenses and losses and credit all incomes and gains. Salary therefore should be debited. Cash is going out, so cash account should be credited.

9. Rent is a nominal account, it is an income, so rent account should be credited. Cash is coming in, hence cash account should be debited.

10. Cash is coming in, cash account should be debited. Sales are going out, hence sales account should be credited.

11. Cash is going out, cash account should be credited. Purchases are coming in, hence purchases account should be debited.

## LEDGER

After journalizing the transactions, ledger, which is the chief book of accounts is prepared. Following is the ledger containing various accounts.

## Cash A/C

| Dr. | | | Cr. |
|---|---|---|---|
| | Rs. | | |
| To X's capital A/C | 10,000 | By machinery | 6,000 |
| To sales A/C | 3,000 | By Z's A/C | 4,000 |
| To Naresh A/C | 2,000 | By salaries A/C | 1,000 |
| To Rent A/C | 3,000 | By purchases A/C | 3,000 |
| | | By balance c/d | 13,000 |
| | 27,000 | | 27,000 |

To balance brought down 13,000

## X's Capital A/C

| Dr. | | | Cr. |
|---|---|---|---|
| | Rs. | | |
| To balance c/d | 10,000 | By cash A/C | 10,000 |
| | 10,000 | | 10,000 |
| | | By balance b/d | 10,000 |

## Purchases A/C

| Dr. | | | Cr. |
|---|---|---|---|
| | Rs. | | |
| To Mahesh A/C | 4,000 | By balance c/d | 7,000 |
| To cash A/C | 3,000 | | |
| | 7,000 | | 7,000 |
| To balance b/d | 7,000 | | |

## Mahesh A/C

| Dr. | | | Cr. |
|---|---|---|---|
| | Rs. | | |
| To balance c/d | 4,000 | By purchases | 4,000 |
| | 4,000 | | 4,000 |
| | | By balance b/d | 4,000 |

## Sales A/C

| Dr. | | | Cr. |
|---|---|---|---|
| | Rs. | | |
| To balance c/d | 12,000 | By cash | 3,000 |
| | | By cash | 9,000 |
| | 12,000 | | 12,000 |
| | | By balance b/d | 12,000 |

## Machinery A/C

| Dr. | | | Cr. |
|-----|------|-----|------|
| | Rs. | | |
| To cash A/C | 6,000 | By balance c/d | 6,000 |
| | 6,000 | | 6,000 |
| To balance b/d | 6,000 | | |

## Ravi A/C

| Dr. | | | Cr. |
|-----|------|-----|------|
| | Rs. | | |
| To electric motor | 4,000 | By balance c/d | 4,000 |
| | 4,000 | | 4,000 |
| To balance b/d | 4,000 | | |

## Electric Motor A/C

| Dr. | | | Cr. |
|-----|------|-----|------|
| | Rs. | | |
| To balance c/d | 4,000 | By Ravi A/C | 4,000 |
| | 4,000 | | 4,000 |
| | | By balance b/d | 4,000 |

## Naresh A/C

| Dr. | | | Cr. |
|-----|------|-----|------|
| | Rs. | | |
| To balance c/d | 2,000 | By cash | 2,000 |
| | 2,000 | | 2,000 |
| | | By balance b/d | 2,000 |

## Z's A/C

| Dr. | | | Cr. |
|-----|------|-----|------|
| | Rs. | | |
| To cash A/C | 4,000 | By balance c/d | 4,000 |
| | 4,000 | | 4,000 |
| To balance b/d | 4,000 | | |

## Salaries A/C

| Dr. | | | Cr. |
|-----|------|-----|------|
| | Rs. | | |
| To cash A/C | 1,000 | By balance c/d | 1,000 |
| | 1,000 | | 1,000 |
| To balance b/d | 1,000 | | |

## Rent A/C

Dr.                                                             Cr.

|  | Rs. |  | Rs. |
|---|---|---|---|
| To balance c/d | 3,000 | By cash | 3,000 |
|  | 3,000 |  | 3,000 |
|  |  | By balance b/d | 3,000 |

## TRIAL BALANCE

Trial balance is prepared after the completion of ledger and presented in Table 31.2.

## FINAL ACCOUNTS

Final accounts are prepared to find out the financial position of the business at the end of the period. Given the trial balance (Table 31.3) of a business, final accounts are prepared and presented in Tables 31.4, 31.5 and 31.6.

**TABLE 31.5  Trial Balance.**

| Particulars | LF | Debit (Rs.) | Credit (Rs.) |
|---|---|---|---|
| Cash A/C |  | 13,000 |  |
| Purchases A/C |  | 7,000 |  |
| Machinery A/C |  | 6,000 |  |
| X's capital A/C |  |  | 10,000 |
| Mahesh A/C |  |  | 4,000 |
| Sales A/C |  |  | 12,000 |
| Ravi A/C |  | 4,000 |  |
| Electric Motor A/C |  |  | 4,000 |
| Naresh A/C |  |  | 2,000 |
| Z's A/C |  | 4,000 |  |
| Salaries A/C |  | 1,000 |  |
| Rent A/C |  |  | 3,000 |
|  |  | 35,000 | 35,000 |

**TABLE 31.6  Trial Balance of a Business.**

| Particulars | Debit (Rs.) | Credit (Rs.) |
|---|---|---|
| Cash at bank | 5,000 |  |
| Buildings | 10,000 |  |
| Machinery | 8,000 |  |
| Sundry debtors | 7,000 |  |
| Sundry creditors |  | 8,000 |
| Bank overdraft |  | 3,000 |
| Loan |  | 4,000 |
| Salaries | 5,000 |  |
| Rent received | 2,000 |  |
| Commission |  | 4,000 |
| Purchases | 13,000 |  |
| Sales |  | 29,000 |
| Carriage | 2,000 |  |
| Postage | 1,000 |  |
| Wages | 4,000 |  |
| Furniture | 3,000 |  |
| Capital |  | 12,000 |
|  | 60,000 | 60,000 |

Additional information: Closing stock = Rs. 30,000

The final accounts *viz.*, trading A/C, profit and loss A/C and balance sheet are prepared as given in the following Tables 31.4, 31.5 and 31.6.

### TABLE 31.4  Trading A/C.

| Dr. | | | | Cr. |
|---|---|---|---|---|
| | | Rs. | | |
| To purchases | | 13,000 | By closing stock | 15,000 |
| To carriage | | 2,000 | By sales | 29,000 |
| To wages | | 4,000 | | |
| To gross profit | | 25,000 | | |
| | | 44,000 | | 44,000 |
| Transferred to P/L Account. | | | | |

### TABLE 31.5 Profit and Loss A/C.

| Dr. | | | | Cr. |
|---|---|---|---|---|
| | | Rs. | | |
| To salaries | | 5,000 | By gross profit b/d | 25,000 |
| To commission | | 2,000 | By rent received | 4,000 |
| To postage | | 1,000 | | |
| To net profit | | 21,000 | | |
| Transferred to Capital A/C | | 29,000 | | 29,000 |

### TABLE 31.6 Balance Sheet.

| Dr. | | | | Cr. |
|---|---|---|---|---|
| Liabilities | | Rs. | Assets | Rs. |
| Sundry creditors | | 8,000 | Cash at bank | 5,000 |
| Bank overdraft | | 3,000 | Buildings | 10,000 |
| Loan | | 4,000 | Machinery | 8,000 |
| Capital | 12,000 | 33,000 | Sundry debtors | |
| Add: Net profit | 21,000 | | | 7,000 |
| | | | Furniture | 3,000 |
| | | | Closing stock | 15,000 |
| | | 48,000 | | 48,000 |

## SUBSIDIARY BOOKS OR SUBSIDIARY JOURNALS

Given the transactions subsidiary books like purchases book, sales book, purchases-returns book and sales-returns books are presented below:

### Transactions

| Dec. 1, 1998 | Purchased goods from Gopal | Rs. 3,000 |
|---|---|---|
| Dec. 2, 1998 | Sold goods to Anil | Rs. 4,000 |
| Dec. 3, 1998 | Raju purchased goods | Rs. 1,000 |
| Dec. 4, 1998 | Ravi sold goods to us | Rs. 5,000 |
| Dec. 5, 1998 | Goods returned to Naresh | Rs. 100 |
| Dec. 6, 1998 | Suresh returned goods to us | Rs. 200 |
| Dec. 7, 1998 | Sold goods to Raju for cash | Rs. 2,000 |

**Purchases Book.**

| Date | Particulars | LF | Amount (Rs.) |
|------|-------------|-----|--------------|
| Dec. 1 | Gopal | | 3,000 |
| Dec. 4 | Ravi | | 5,000 |

NB: In purchases books, goods purchased on credit basis only are entered.

**Sales Book**

| Date | Particulars | LF | Amount (Rs.) |
|------|-------------|-----|--------------|
| Dec. 2 | Anil | | 4,000 |
| Dec. 3 | Raju | | 1,000 |

NB: In sales book, goods sold on credit basis are only entered.

Item No. 7 is not entered either in purchases book or sales books, because it was not sold on credit basis.

**Purchases-Return Books**

| Date | Particulars | LF | Amount (Rs.) |
|------|-------------|-----|--------------|
| Dec. 5 | Naresh | | 100 |

**Sales-Return Book**

| Date | Particulars | LF | Amount (Rs.) |
|------|-------------|-----|--------------|
| Dec. 6 | Suresh | | 2000 |

# CASHBOOK

From the following transactions three columnar cashbook (here we have cash column, discount column and bank column) is prepared and presented in Table 31.7.

Jan. 1, 1999 Cash balance is Rs. 5,000 and cash at bank Rs. 8,000.
Jan. 2, 1999 Cash sales Rs. 9,000.
Jan. 3, 1999 Cheque received from Narendra Rs. 2,000.
Jan. 4, 1999 Paid salaries by cheque Rs. 3,000.
Jan. 5, 1999 Rent received Rs. 2,000.
Jan. 6, 1999 Cheque received from Gopal and paid to the bank Rs. 4,000.
Jan. 7, 1999 Cash received from Naresh Rs. 1960 in full settlement of claim of Rs. 2,000.
Jan. 8, 1999 Cash paid to Suresh Rs. 2,000 in full settlement of claim Rs. 21,000.
Jan. 9, 1999 Paid for postage Rs. 1,000.
Jan. 10, 1999 Withdrew from bank for personal use Rs. 500.
Jan. 11, 1999 Paid to bank Rs. 5,000.
Jan. 12, 1999 Withdrew form bank for office use Rs. 3,000.

## Rules

1. Cash balance always is put into debit side of the cash column.
2. Bank balance is put into debit side of the bank column.

**TABLE 31.7 Three Columnar Cashbook.**

| Date | Particulars | LF | Discount (Rs.) | Cash (Rs.) | Bank (Rs.) |
|---|---|---|---|---|---|
| Jan 1 '99 | To Balance b/d | | | 5,000 | 8,000 |
| Jan 2 '99 | To Sales A/C | | | 9,000 | |
| Jan 3 '99 | To Narendra A/C | | | 2,000 | |
| Jan 5 '99 | To Rent A/C | | | 2,000 | |
| Jan 6 '99 | To Gopal A/C | | | | 4,000 |
| Jan 7 '99 | To Naresh A/C | | 40 | 1960 | |
| Jan 11 '99 | To Cash A/C | 'C' | | | 5,000 |
| Jan 12 '99 | To Bank A/C | 'C' | | 3,000 | |
| | | | 40 | 22,960 | 17,000 |
| To balance b/d | | | | 14,960 | 10,500 |

| Date | Particulars | LF | Discount (Rs.) | Cash (Rs.) | Bank (Rs.) |
|---|---|---|---|---|---|
| Jan 4 '99 | By Salaries A/C | | | | 3,000 |
| Jan 8 '99 | By Suresh A/C | | 100 | 2,000 | |
| Jan 9 '99 | By Postage | | | 1,000 | |
| Jan 10 '99 | By Drawings | | | | 500 |
| Jan 11 '99 | By Bank A/C | 'C' | | 5,000 | |
| Jan 12 '99 | By Cash A/C | 'C' | | | 3,000 |
| Jan 31 '99 | By Balance c/d | | | 14,960 | 10,500 |
| | | | 100 | 22,960 | 11,500 |

3. Bank overdraft balance entered into the credit side of the bank column.
4. When cash or cheques are coming in, enter into debit side of the cash column.
5. When cash is going out, put into credit side of the cash column.
6. When cheque is going out, enter into credit side of the bank column.
7. When cheque is received and deposited in the bank on the same day, enter into debit side of the bank column
8. When money is withdrawn from bank for personal use, enter it into credit side of the bank column.
9. When deposited in bank, here bank is the receiver, so enter it into debit side of the bank column and write, to cash A/C. Since the cash is going out, the amount is entered into credit side of cash column and write, by bank A/C. Such an entry is known as contra entry which is denoted as 'C'.
10. When amount is withdrawn from bank, bank is giver, enter it into credit side of the bank column and write, by cash A/C. Since cash is coming in, the amount is entered into debit side of the cash column and write, to bank A/C, which is a contra entry.

Whenever, the amount is either deposited in the bank or withdrawn from the bank, for business purpose, the amount is entered in both sides and specified as contra entry i.e., 'C'.

# CHAPTER 32

# Farm Inventory

The list of all the physical property of a business along with their values at a specific point of time is called farm inventory. Inventory for a business is taken at two points of time in a year *i.e.*, at the beginning of the agricultural year and at the end of the year. It constitutes cash assets, depreciable assets and non-depreciable assets. The difference in the inventory at the two points of time indicates the changes in the inventory.

Farm inventory forms the basis for the preparation of income statement, balance sheet, measurers of income, *etc.* The loss in the value of the asset due to depreciation can be worked out from the farm inventory.

As per the sub-items, inventory is presented like cash assets, depreciable assets and non-depreciable assets as presented in Table 32.1.

Change in the inventory is found out by taking the differe   e of the value of assets during the two periods. As evident from the table the items that need to be included in the farm inventory are, the number of various assets along with their values. As far as recording the number of items are concerned it can be done by visual verification. The relevant weights and measurers are also noted for the corresponding items of assets.

The preparation of farm inventory involves physical verification and valuation of the assets. Physical verification of the items does not pose a problem to the farmer. Problem arises, while valuing the assets, since improper evaluation leads to errone-ous farm decisions.

## METHODS OF VALUATION

To meet this particular objective of valuing the inventory, a look at the common methods of valuation is necessary. Following are the common methods of valuation.
1. Net selling price.
2. Cost less depreciation.
3. Market price.
4. Cost.
5. Replacement cost less depreciation and
6. Income capitalization.

### 1. Net Selling Price Method

This is a common method that is followed to value the assets that are primarily meant for sale. From the market price, selling costs are deducted to arrive at the

**TABLE 32.1  Farm Inventory of a Hypothetical Farm on 1st June 1998 and May 1999.**

| S. No. | Particulars | Beginning of the year (1.6.1998) | | End of the year (31.5.1999) | |
|---|---|---|---|---|---|
| | | Quantity in Q/kg or number | Value in Rs. | Quantity in Q/kg or number | Value in Rs. |
| I. | **Cash assets** | | | | |
| | Cash on hand | xx | xx | xx | xx |
| | Savings in bank, etc. | xx | xx | xx | xx |
| | Sub-total | xx | xx | xx | Xx |
| II. | **Depreciable assets** | | | | |
| | Land (ha) | xx | xx | xx | xx |
| | Farm buildings | xx | xx | xx | xx |
| | Machinery and equipment | xx | xx | xx | xx |
| | Implements | xx | xx | xx | xx |
| | Dairy cattle | xx | xx | xx | xx |
| | Bullocks | xx | xx | xx | xx |
| | Sheep and goat | xx | xx | xx | xx |
| | Poultry birds, etc. | xx | xx | xx | xx |
| | Sub-total | xx | xx | xx | xx |
| III. | **Non-depreciable assets** | | | | |
| | Grains ready for disposal | xx | xx | xx | xx |
| | Fodder and feed | xx | xx | xx | xx |
| | Livestock products | xx | xx | xx | xx |
| | Seeds | xx | xx | xx | xx |
| | Fertilizers | xx | xx | xx | xx |
| | Pesticides and fungicides | xx | xx | xx | xx |
| | Sub-total | xx | xx | xx | xx |
| | Total of all assets | xx | xx | xx | xx |

current value of the asset. *Examples*: All farm products for sale, livestock for sale, *etc.*

## 2. Cost Less Depreciation Method

This is the method applicable in the case of working assets. Since these assets depreciate with every passing year, the depreciation amount is estimated and deducted from the purchase price. *Examples*: Machinery, buildings constructed recently, dairy cattle, carts, *etc.*

## 3. Market Price Method

This is a common method for valuing purchased farm supplies like seeds, fertilizers, pesticides, fuel, feeds, veterinary medicines, *etc.*

## 4. Cost Method

This method is used to estimate current value of the farm produced inputs like seed, FYM, *etc.* Standing crop is also valued through this method. Regarding valuing standing crop, expenses incurred in raising the crops, till the date on which inventory is taken are recorded.

## 5. Replacement Cost Less Depreciation Method

This method is confined to estimate the value of very old buildings. Replacement cost represents the cost of constructing the same type of building with the present technology at present prices. After arriving at the replacement cost, deduct the depreciation so as to arrive at the current value of the building.

*Example*:

| | | |
|---|---|---|
| Actual cost of construction of very old building | = | Rs. 10,000 |
| Useful life | = | 50 years |
| Current (Replacement) cost of construction | = | Rs. 30,000 |
| Present age of the building | = | 30 years |
| Amount of depreciation | = | Rs. 6,000 |
| Current value of the building | = | Rs. 30,000 – 6,000 = Rs. 24,000 |

## 6. Income Capitalization Method

Asset which yields income over an infinite period of time is evaluated using this method. *Example*: Land.

The formula is

V = R/r

V = Income capitalized value

R = Net income from land per annum (assumed to be constant)

r = Rate of interest

Besides this method, sale price method is also used to estimate the current value of land asset.

# CHAPTER 33

# Depreciation

Farmer uses multiplicity of resources in producing different farm commodities. Some of them are mono period resources and others are poly period resources. The expenditure incurred by the producer on mono period resources like seeds, fertilizers, plant protection chemicals, fuel, feed, *etc.,* can be taken as variable cost or operating cost or working cost for the production cycle in which these resources are committed. Therefore, there is no problem in accounting for the cost of mono period resources. The difficulty arises in accounting for the cost of poly period resources, as they provide services for a number of years to the farm business. Farmer uses machinery, equipment, farm buildings, implements and tools, livestock, *etc.,* in each production period of an agricultural year. It is not rational to account for the entire purchase price of these resources as a cost in a single production cycle, since they are used in many production cycles. Therefore, we need an accounting procedure that considers the cost of services rendered by the poly period resources in each production cycle in which they are used. Depreciation is such an accounting procedure to account for the cost of services while estimating the cost of cultivation of crops. It is also known as capital consumption allowance.

Depreciation is defined as *"the decline in the value of asset due to usage, accidental damage and time obsolescence".*

Depreciation is prorating the original cost of an asset over its useful life.

When an asset is used, it undergoes wear and tear and hence the reduction in the value of asset, which is termed as use depreciation.

Even if the asset is not used its value declines, over a period of time, because the asset may become outdated or outmodelled or go out of fashion. This is known as time depreciation. Besides, the value also declines due to accidental damage. Weather is also an important factor that causes decline in the value due to corrosion. It is important to note that depreciation is not the result of changes in the prices.

To compute depreciation, we need information on the original cost of the asset, useful life, and junk value.

Useful life is the number of years that an asset provides its services to the farm business. In other words, it is the age at which the asset completely wears out. Useful life varies with the type of asset.

Junk value is the value of the asset at the end of its useful life. It is the amount that the farmer receives when he disposes the asset after its completion of expected life. It is also known as salvage value or scrap value or terminal value.

# METHODS OF COMPUTATION

The methods of computation of depreciation work under two important assumptions. They are (1) an asset is put to constant use year after year; and (2) an asset is put to varying rates of use. Following are the methods of computation of depreciation.

1. Straight line method.
2. Diminishing balance method.
3. Sum of the years digits method; and
4. Annual revaluation method.

## 1. Straight line Method

This method is applied under the assumption of constant use of asset over the years. This is the most popular and widely used method in computing depreciation because of the simplicity in calculations. The following equation is used to compute annual amount of depreciation.

$$\text{Annual amount of depreciation} = \frac{\text{Original cost of the asset - junk value}}{\text{Useful life of the asset}}$$

*Example:*

Original cost of a plough Rs. 500
Junk value            Rs. 50
Useful life           5 years

$$\text{Annual amount of depreciation} = \frac{500 - 50}{5} = \text{Rs. 90}$$

## 2. Diminishing Balance Method

This method assumes varying rates of use of the asset, year after year. The following equation is used to compute annual amount of depreciation.

Annual amount of depreciation = (Book value of the asset) x R

R is the rate of depreciation to be charged over the useful life of the asset and is computed by dividing 100 per cent with useful life.

*Example:*

Original cost of a plough = Rs. 500
Useful life           = 5 years

$$R = \frac{100}{5} = 20\%$$

**Computation of Depreciation by Diminishing Balance Method.**

| Year | Rate of depreciation (in %) | Amount of depreciation (Rs.) | Book value (Rs.) |
|------|------|------|------|
| 1. | 20 | 500 × 20/100 = 100 | 400 |
| 2. | 20 | 400 × 20/100 = 80 | 320 |
| 3. | 20 | 320 × 20/100 = 64 | 256 |
| 4. | 20 | 256 × 20/100 = 51.20 | 204.80 |
| 5. | 20 | 204.80 × 20/100 = 40.96 | 163.84 |
| | Total | 336.16 | |

Unlike in straight-line method, where we are charging the same amount of depreciation throughout the useful life, in this method the annual amount of depreciation is declining from Rs. 100 in the first year, to Rs. 40.96 in the fifth year. We should note here that higher amount of depreciation is charged in the initial years of the asset and less amount in the subsequent years. This is because the asset is put to intensive use when it is brand new and therefore wear and tear is also rapid. When the asset becomes older, its use also gradually decreases and as a result wear and tear too is less.

## 3. Sum of the Years Digits Method

This method also assumes that an asset is put to varying rates of use year after year. Annual amount of depreciation is calculated from the following formula.

Annual amount of depreciation

$$= \text{Original cost of asset - Junk value} \times \frac{\text{Remaining years of useful life}}{\text{Sum of the digits from one through useful life}}$$

Sum of the digits can be computed quickly from the equation that is n(n+1)/2

*Example:*

| | | |
|---|---|---|
| Original cost of a plough | = Rs. 500 | |
| Junk value | = Rs. 50 | |
| Useful life | = 5 years | |
| Sum of the digits of useful life | = 5 (5 + 1)/2 = 5 × 3 = 15 | |

Sum of the digits of useful life can also be computed by adding all the digits from 1 to useful life.

5 + 4 + 3 + 2 + 1 = 15

### Computation of Depreciation by Sum of the Years Digits Method.

| Year | Rate | Amount (Rs.) | Book value (Rs.) |
|---|---|---|---|
| 1. | 5/15 | (500-50) × 5/15 = 150 | 350 |
| 2. | 4/15 | (500-50) × 4/15 = 120 | 230 |
| 3. | 3/15 | (500-50) × 3/15 = 90 | 140 |
| 4. | 2/15 | (500-50) × 2/15 = 60 | 80 |
| 5. | 1/15 | (500-50) × 1/15 = 30 | 50 |
| | Total | 450 | |

Total annual amount of depreciation + junk value = original cost

Rs. 450 + 50 = Rs. 500

The annual amount of depreciation is declining from Rs. 150 in the first year to Rs. 30 in the fifth year. The same analogy which is presented in diminishing balance method holds good here also for charging different amounts of annual depreciation with every passing year.

## 4. Annual Revaluation Method

As the very name indicates, in this method the asset is revalued every year. More specifically, the value of the asset is estimated at the beginning as well as at the end of the year. The difference thus arrived at, represents the depreciation or appreciation as the case may be for that year. In respect of livestock, as the value increases in the initial years, this method rightly catches up the amount of appreciation. However, the application of this method to all the assets poses a problem because, it is difficult to get the market prices for the assets of varying wear and tear.

# CHAPTER 34

# Farm Efficiency Measures

Farmer makes several decisions in the use of resources. Mere decision-making may not result in profitable organization of the farm business. There is need to evaluate the decisions of the farmer for which, we need some guides and standards. Here comes the role of efficiency measures, which are categorized into physical efficiency measures and financial efficiency measures. These are explained hereunder.

## PHYSICAL EFFICIENCY MEASURES

Broadly, three types of physical inputs *viz.*, land, labour and capital are used on the farms. Every farmer aims at the efficient utilization of these inputs. To measure the efficiency, we need some criteria or yardsticks or indicators. Following are the various measures to express the efficiency of these physical inputs.

### 1. Land

Generally land is measured in terms of acres or hectares owned by the farmer. Always the entire land is not put to cultivation due to the inadequacy of resources. When part of the land is slack, the efficiency of the farm is affected. Land use efficiency is measured by productivity, cropping intensity and crop yield index.

### (i) Productivity of Land

The production efficiency with respect to a particular crop, is expressed as the ratio between actual yield of a crop on the farm and average yield of the same crop in the locality. This relationship is expressed in proportionate terms.
*Example:*

| | |
|---|---|
| Paddy yield/ha. on a farm | = 40Q |
| Average yield of paddy/ha in the locality | = 35Q |
| Production efficiency | |

$$= \frac{\text{Actual crop yield on the farm}}{\text{Average yield in the locality}} \times 100$$

$$= \frac{40}{35} \times 100 = 114.28\%$$

## (ii) Cropping Intensity

It measures the extent of land use for the production of the crops during an agricultural year. Cropping intensity is worked out from the following equation.

$$\text{Cropping intensity} = \frac{\text{Gross cropped area during the agricultural year}}{\text{Net cropped area}} \times 100$$

For example, if a farmer is raising three crops in an agricultural year in his 10 hectare of irrigated land, then the cropping intensity is $\frac{30}{10} \times 100 = 300\%$

## (iii) Crop Yield Index

It is a good measure of index to compare the yields of all the crops grown on the farm with that of average yields of the same crops grown in that locality. The index is measured in terms of percentage. The computation of crop yield index is illustrated in Table 34.1.

TABLE 34.1   Estimation of Crop Yield Index.

| Crop | Average yield | | Crop acreage on selected farm (ha.) | Production efficiency | Production efficiency x crop acreage |
|---|---|---|---|---|---|
| | Given locality | Selected farm | | | |
| Paddy (Q) | 68 | 80 | 4 | 117.65 | 470.60 |
| Sugarcane (tonnes) | 90 | 110 | 6 | 112.22 | 733.32 |
| Total | — | — | 10 | — | 1,203.92 |

Crop yield index = 1,203.92/10 = 120.39.

Thus the farm in question is more efficiently organized in terms of crop yields as compared to an average farm in the area since the crop yield index is more than 100.

## 2. Labour

The profitability of the farm business is influenced by the efficiency of the labour resource. Farmers who do not introduce new methods to achieve higher labour efficiency find it difficult to continue in the agricultural profession as the wages of farm labour are on the increasing trend. Crop acreage per man equivalent year, livestock maintained per man, gross income per man and productive man work units per man equivalent year are the important measures of labour efficiency.

Farmers use different kinds of labour for executing various farm operations. These include men labour, women labour, and child labour. However, labour is measured in terms of man-days. A man-day is the productive work accomplished by a worker in eight hours of a day assuming average efficiency. As we are employing not only men labour but also women and child labour on the farm, women and children should be converted into man-days, which is a standard unit of labour measurement. Taking the efficiency into consideration, three women are considered as equivalent to two men and two children are equal to one man. All woman-days and child-days should be converted into man equivalent days by taking above ratio into consideration. The sum of all man-days and man equivalent days is divided by 365 days to get man equivalent year.

For example, paddy crop requires 100 men labour, 120 women labour and 50 child labour per hectare.

| | | |
|---|---|---|
| Mandays | = | 100 |
| Woman labour converted into man equivalent days | = | $120 \times 2/3$ |
| | = | 80 man equivalent days |
| Child labour converted into man equivalent days | = | $50 \times 1/2$ |
| | = | 25 man equivalent days |
| Sum of man days + man equivalent days | = | $100 + 80 + 25 = 205$ |
| Man equivalent year (or) person year = 205/365 | = | 0.56 |

Man equivalent year or person year can also be computed in the following manner.

| | | |
|---|---|---|
| Farmer's own contribution of labour | = | 4 months. |
| Family labour contribution | = | 8 months. |
| Hired labour contribution | = | 12 months. |
| Total contribution = 4 + 8 + 12 | = | 24 months. |

$$\text{Man equivalent year or person year} = \frac{\text{Total labour contribution}}{12 \text{ months}}$$

$$= \frac{24}{12} = 2$$

## Labour Efficiency Measures

i)   *Crop Acreage per Man Equivalent Year:* It is one of the simplest methods of measuring the productivity of labour input and is calculated by dividing the number of acres under crops by man equivalent year. Higher the ratio greater the productivity or efficiency.

ii)  *Livestock Maintained per Man:* It is arrived at by dividing the total livestock maintained on the farms divided by man equivalent year.

iii) *Gross Income per Man:* It is calculated by dividing gross income obtained on the farm by man equivalent year.

iv)  *Productive Man Work Units per Man Equivalent Year (PMWU):* Assuming average efficiency, the work done on the farm by the worker in a day of eight hours is known as productive man work unit *i.e.,* man-day.

To work out total productive man work units, we require information on the average requirement of productive man work units per unit for all the enterprises taken up on the farm, as well as the size of each enterprise. Example is presented in Table 34.2.

TABLE 34.2 Estimation PMWU Per Man Equivalent Year.

| Name of the enterprise | Area (in ha) | Average requirement of productive man work units per unit | Total productive man work units (man days × size of the enterprise) |
|---|---|---|---|
| Paddy | 5 | 250 | 1,250 |
| Groundnut | 4 | 120 | 480 |
| Dairy | 3 | 40 | 120 |
| Total | | | 1,850 |

$$\text{PMWU per man equivalent} = \frac{\text{Total productive man work units}}{\text{Man equivalent year}}$$

# FINANCIAL EFFICIENCY MEASURES

Though the physical efficiency in the use of resources is essential, but the economic efficiency ultimately decides the success or failure in the organization of the farm business. The financial efficiency measures are classified into aggregate and ratio measures.

## 1. Aggregate Measures

Aggregate measures are the measures meant to reflect the income earned by the various factors of production for the farm as a whole. Here the entire farm is considered as a unit, and efficiency is worked out for the entire farm in terms of each resource used. Following are the aggregate measures:

i) *Total Capital Invested:* Investment on fixed assets (land, building, *etc.*) and working assets (machinery, livestock, equipment, *etc.*) together represent the size of the farm.

ii) *Total Value of Production* (gross income): It is the total income derived from the sale of main product as well as by-products from the enterprises taken up by the farmer in an year. This includes value of home consumed products plus the value of products sold. This is a better measure of comparing the different farms.

It is computed as shown in Table 34.3.

TABLE 34.3 Computation of Total Value of Production.

| Crop | Area (ha.) | Main product | By-product (CL) | Price per unit (Rs.) Main product | Price per unit (Rs.) By-product | Gross income (Rs.) |
|---|---|---|---|---|---|---|
| 1 | 2 | 3 | 4 | 5 | 6 | 7 |
| Paddy | 1.0 | 35 qtls | 4 | 700 | 250 | 25,500 |
| Sugarcane | 2.0 | 90 tonnes | – | 900 | – | 81,000 |
| Total | | | | | | 1,06,500 |

iii) *Total Expenses:* These include working costs and fixed costs of the farm. In general fixed costs are worked out for the entire farm and working costs are estimated crop-wise interms of inputs, such as, labour, seeds, fertilizers, *etc.*, and totalled up for the whole farm. Fixed costs include deprecation, land revenue, interest on fixed capital, rental value of owned land, insurance, *etc.*

iv) *Net Worth:* Net worth is the surplus of total assets over the total liabilities.

Net worth or equity = Total assets – total liabilities.

The high level of net worth indicates strong financial position of the farm at a point of time. Balance sheets prepared over the years indicate financial progress of a farm business.

v) *Net Cash Income:* The surplus of cash income over cash operating expenses indicate net cash income. This is the amount available with the farmer for future investment on the farm.

Net cash income = Gross cash income–cash expenses.

vi) *Net Farm Income:* It is worked out as:

Net cash income ± change in the inventory–depreciation.

vii) *Farm Earnings:* It is computed as:

Net farm incmome + value of farm products consumed at home.

viii) *Family Labour Earnings:* It is farm earnings less interest charges on fixed capital.

ix) *Returns to Management:* It is computed using the following equation.
Returns to management = Family labour earnings – imputed value of family labour.

## 2. Ratio Measures

These measures reflect per unit returns or costs to the inputs involved in the organization of the farm business.

i) *Fertilizer Costs per Crop Acre:* It is the ratio of total expenditure on fertilizers to number of acres under crop. This varies from crop to crop and farm to farm. In reality, farmers do not have exact nutrient status of the soil but however ambitious farmers apply the fertilizers in more than required level. Under such situation, efficiency of fertilizer is brought down leading to the increase in the cost of fertilizer.

ii) *Power and Equipment Cost/Crop Acre:* This relates to running costs incurred in operating power drawn machinery per acre of cropped area. The running costs vary with make, condition and age of the machine. Similarly, some commercial crops require intensive use of machinery. Hence comparison becomes relevant only when similar farms using the same implements for the same crops are found.

iii) *Cost Ratios:* These ratios provide information to the manager to decide whether the existing costs of the business farm are higher or lower. These ratios indicate what proportion of gross income is spent for meeting different types of expenditure. These ratios are operating ratio, fixed ratio and gross ratio. The following expressions are used to compute them.

a. Operating ratio $= \dfrac{\text{Operating expeness on the farm}}{\text{Gross income on the farm}}$

b. *Fixed Ratio:* It gives the proportion of fixed expenses in gross income of the farm. This ratio is worked out as:

$$\dfrac{\text{Fixed costs on the farm}}{\text{Gross income on the farm}}$$

c. *Gross Ratio:* Gross ratio is the ratio of total costs of the farm to the gross income. The relevant formula to work out gross ratio is:

$$\frac{\text{Total costs on the farm}}{\text{Gross income on the farm}}$$

*iv) Solvency Ratios:* These ratios measure the degree of financial safety (solvency) and liquidity of comparable business farms at a point of time. These ratios include current ratio, working ratio, net capital ratio and debt-equity ratio.

a. $\quad$ Current ratio $= \dfrac{\text{Current assets}}{\text{Current liabilities}}$

b. $\quad$ Working ratio $= \dfrac{\text{Current assets} + \text{intermediate assets}}{\text{Current liabilities} + \text{intermediate liabilities}}$

c. $\quad$ Net capital ratio $= \dfrac{\text{Total assets}}{\text{Total liabilities}}$

d. $\quad$ Debt-equity ratio $= \dfrac{\text{Total liabilities}}{\text{Owner's equity}}$

*v) Income Ratios*

*Net Income per Acre:* Another important income ratio is the net income per acre. To compute this ratio we require information on (1) gross income of the farm (2) total cost on the farm; and (3) total acreage of the farm. Net income per acre is gross income less total costs. Now the net income per acre is worked out as

$$= \frac{\text{Net income}}{\text{No. of acres}}$$

Higher level of this ratio indicates greater efficiency of the farm.

*vi) Capital Ratios*

a. *Capital per Unit of Gross Income:* It is computed by dividing total capital invested with gross income.
b. *Capital per Man:* It is worked out by dividing the total capital invested by the number of man equivalent years.
c. *Rate of Capital Turnover:* It indicates the number of years it will take for the farm to return the investment through income.

$$\text{Rate of capital turnover} = \frac{\text{Gross income on the farm}}{\text{Total value of farm assets}} \times 100$$

## COST CONCEPTS AND INCOME MEASURES

Cost concepts are widely used because of their relevance in the decision-making process. This means that these costs serve as a basis to expand the size of the farm, to buy the requisite capital assets in the long run and the requisite inputs in the short run. For example, variable costs have a bearing on the level of production in the short run, on the other hand the decisions like expanding the size of the farm, buying the durable assets, *etc.,* are based on the total costs. Crop costs or costs related to livestock production are split up into various cost components, such as cost $A_1$, $A_2$, $B_1$, $B_2$, $C_1$ and $C_2$. Before knowing the cost concepts, let us know the different costs involved in the production of crop and livestock products.

*Labour Costs:* These include wages paid in cash and kind, both to casual, contract and permanent labourers. These costs come under paid out costs. Farmers do not pay for their own labour, such as family labour, owned bullock labour, owned machine labour, *etc.* The wages for these family labourers are imputed based on market wage rate. Labour cost is a variable cost.

*Machinery Costs:* These include hiring charges for machinery, implements, *etc.* Fuel and lubricant charges of the machines are considered as variable costs, whereas repairs and maintenance of machines, insurance premium, taxes, license, depreciation, and interest on capital invested are considered as fixed costs. For owned machinery, the existing market hiring charges should be used as variable costs.

*Livestock Costs:* These include cost of veterinary medicine, cost of fodders and feeds, labour services used, *etc.,* in the maintenance of the livestock. These are generally considered as variable costs. In contrast the interest rate paid on the amount spent to purchase the animals and depreciation on the animal are considered as fixed costs.

*Land Costs:* These refer to rental value of owned land, and rent of the lease-in-land. Rent of the owned land is imputed hence it is called rental value of owned land, but not the rent. Rent of the land arises only when the land is leased in. Rent of the land leased out is not land cost, but it is considered as income to the farmers. Rental value of owned land is considered as a cost though it is not incurred by the farmer, because, we use the concept of opportunity cost and impute the existing rent of similar farms in the region.

*Building Costs:* These include interest on the value of the buildings and the depreciation on the buildings. Both these costs are regarded as fixed costs.

*Input Costs:* Both farm produced and purchased inputs *viz.,* seeds, manures, plant protection chemicals, weedicides, *etc.,* are estimated for their current value at market prices.

*Interest on Working Capital and Fixed Capital:* By definition working capital vary with the level of production while fixed capital remains the same regardless of the level of production. These respective interest rates should not be clubbed into one as these are entered separately while deriving the cost concepts. Normally in farming the inputs are not used at a time but at different points of time according to the requirements of the crop. Under such circumstances, the interest calculated on the working capital should be reduced for half of the crop period. It is more an approximate method than a realistic method.

*Cost Concepts:* These include cost $A_1$, $A_2$, $B_1$, $B_2$, $C_1$ and $C_2$.

Cost $A_1$: The following items are included in cost $A_1$.

> Wages of hired human labour.
> Wages of permanent labour.

Wages of contract labour.

Wages of hired bullock labour.

Imputed value of owned bullock labour.

Charges of hired machinery.

Imputed value of owned machinery.

Market rate of manures and fertilizers.

Marker rate of seeds.

Imputed value of owned seeds.

Imputed value of manures.

Market value of pesticides, herbicides, hormones, *etc.*

Irrigation charges.

Land revenue, cess and other tax.

Depreciation on farm machinery, implements, equipment farm buildings Irrigation structures, *etc.*

Interest on working capital.

Miscellaneous expenses (value of other items which are used up in current production).

Cost $A_2$: Cost $A_1$ + Rent paid for leased in land.

Cost B: Cost $A_1$ or $A_2$ + interest on the fixed capital excluding land + rental value of owned land.

Cost C: Cost B + imputed value of family labour.

The classification of costs based on Dr. Sen's Committee report (1979) is as follows:

Cost $B_1$: Cost $A_1$ + interest on amount of owned capital invested in the business excluding the value to land.

Cost $B_2$: Cost $B_1$ + rental value of owned land less land revenue + rent paid for leased in land.

Cost $C_1$: Cost $B_1$ + imputed value of family labour.

Cost $C_2$: Cost $B_2$ + imputed value of family labour.

## Income Measures

These are the returns over different cost concepts. Different income measures are derived using the cost concepts. These measures include farm business income, family labour income, net income, farm investment income, *etc.* The following formulae are used.

1. Farm business income = Gross income – cost $A_1/A_2$.
2. Family labour income = Gross income – cost B.
3. Net income = Gross income – cost C.
4. Farm investment income = Farm business income–imputed value of family labour

(or)

Net income + imputed rental value of owned land + interest on owned fixed capital invested.

## Cost Concepts and Income Measures for a Hectare of Green Chillies.

| I. | | Operational costs | | Rs. Ps. |
|---|---|---|---|---|
| | 1) | Human labour | | 19,868.00 |
| | | Owned | | 3,346.00 |
| | | Hired | | 16,522.00 |
| | 2) | Bullock labour | | 1,479.00 |
| | | Owned | | 1,320.00 |
| | | Hired | | 159.00 |
| | 3) | Machine labour | | 141.33 |
| | | Owned | | 40.00 |
| | | Hired | | 101.33 |
| | 4) | Seeds | | 1,875.00 |
| | 5) | Manures and fertilizers | | 8,910.18 |
| | | FYM | | 2,000.00 |
| | | Fertilizers | | 6,910.18 |
| | 6) | Plant protection chemicals | | 5,543.21 |
| | 7) | Electricity charges | | 270.00 |
| | 8) | Interest on working capital | | 1,043.15 |
| | 9) | Total operational costs | | 39,129.87 |

| II. | | Fixed costs | | |
|---|---|---|---|---|
| | 1. | Depreciation | | 281.33 |
| | 2. | Land revenue | | 20.00 |
| | 3. | Rental value of owned land | | 10,000.00 |
| | 4. | Interest on fixed capital | | 730.02 |
| | 5. | Total fixed costs | | 11,031.35 |
| | | Total costs | | 50,161.22 |

## Cost Concepts

Cost $A_1$

| 1. | Hired human labour | | 16,522.00 |
|---|---|---|---|
| 2. | Bullock labour – | Owned | 1,320.00 |
| | | Hired | 159.00 |
| 3. | Machinery – | Owned | 40.00 |
| | | Hired | 101.33 |
| 4. | Seeds | | 1,875.00 |
| 5. | FYM | | 2,000.00 |
| 6. | Fertilizers | | 6,910.18 |
| 7. | Plant protection chemicals | | 5,543.21 |
| 8. | Electricity charges | | 270.00 |
| 9. | Depreciation | | 281.33 |
| 10. | Land revenue | | 20.00 |
| 11. | Interest on working capital | | 1,043.15 |
| | | Cost $A_1$ | 36,085.20 |

## Cost B

| | | |
|---|---|---|
| 1. | Cost $A_1$ | 36,085.20 |
| 2. | Rental value of owned land | 10,000.00 |
| 3. | Interest of fixed capital | 730.02 |
| | Cost B | 46,815.22 |

## Cost C

| | | |
|---|---|---|
| 1. | Cost B | 46,815.22 |
| 2. | Owned human labour | 3,346.00 |
| | Cost C | 50,161.22 |

## Income Measures

Gross returns (149.38 Q × Rs. 438) = Rs. 65,428.44

Farm business income = Gross income − Cost $A_1$
= Rs. 65,428.44 − 36,085.20
= Rs. 29,343.24

Family labour income = Gross income − Cost B
= Rs. 65,428.44 − 46,815.22
= Rs. 18,613.22

Net income = Gross income − Cost C
= Rs. 65,428.44 − 50,161.22
= Rs. 15,267.22

Farm investment income = Farm business income − imputed value of family labour
= 29,343.24 − Rs. 3,346.00
= 25,997.24
(or)
= Net income + rental value of owned land + interest on fixed capital
= Rs. 15,567.22 + Rs. 10,000 and Rs. 730.02
= Rs. 25,997.24.

# CHAPTER 35

# Management of Farm Resources

## LAND MANAGEMENT

Land is the foremost important factor of production, since the very agricultural production depends up on the land area available. Land does support the allied activities taken up by the farmer like dairy, poultry, sheep farming, *etc.* Land values generally depend on the type of soil, availability of irrigation (either surface or underground), crops grown, other infrastructural facilities like availability of motorable roads, electricity, accessibility to markets, agro-based industries, *etc.*, in addition to the demand for and supply of land. Also unirrigated lands suitable for commercial crops command good market value. It is also found that demand and supply factors for land do push up the values of land, though in general land values and productivity are not that highly correlated. An example that can be cited here is single season tank fed irrigated areas may have more or less the same market values or even more than those served by assured canal irrigation. Here the pressure for land from the farmers is the single contributing factor. In recent years, industrialization and heavy influx to urban areas are exerting tremendous pressure in urban areas leading to their horizontal expansion. Consequently, the land values even in the absence of aforesaid contributing factors, are shooting up beyond ones comprehension. In such a situation, all our calculations in terms of land values are going topsy-turvy. It may be a healthy sign from the individual's point of view, but from societal point of view, it is a tremendous setback in the long run, as the very agricultural production is going to dwindle as there is not enough compensation for the precious agricultural land lost, in terms of net area expansion under cultivation.

## Characteristics of Land

1. It is a free gift of nature.
2. Land is a permanent resource in the sense that it is indestructible unless some powerful efforts are made to destroy it.
3. Land exhibits a set of distinct characteristic features from location to location. These are inherent due to the variations in land topography and climatic factors in general.
4. Supply of land is inelastic.

*Productivity of the land can be enhanced by the following measures:*

1. Application of organic manures, crop rotation practices, cropping scheme, *etc.*
2. Preventing soil erosion through appropriate land development practices.
3. Reclaiming alkaline soils through application of *zypsum.*
4. Providing good drainage facilities and removing soil sickness through tillage operations, keeping good drainage facilities, *etc.*

## LAND USE PLANNING

Land is a fixed resource to the farmer and the main objective of the farmers is to maximize the revenue from this fixed resource using it optimally. Planning presupposes the objective of any farmer in the quest of realizing maximum revenue. Through planning he identifies the resource availability and the resource requirements so that the desired land use plan can be implemented. The plan gives a blueprint of enterprises, which he wishes to incorporate depending upon the adaptability, technical feasibility and economic viability. Enterprises and their combinations are prepared in the plan. Regarding the selection of the crops, if the field situation so permits, possible crop enterprises are listed out and selection is made based on the relative profitability of the enterprise. Relative profitability is weighed against technology, yield, prices of the products, input costs, *etc.* This plan is for a given point of time taking the entire environment into account. If the environment around the farmer undergoes a change due to technological innovations, prices, *etc.,* flexibility always exists for the farmer to adjust to the new changes by suitable adjustments in his crop and allied activities and changes in technology. Plans always envisage to make the best use of land.

In farming, one finds the presence of owner-cultivators as well as tenant-cultivators. Owner-cultivators may be those who inherited the land from their ancestors as a matter of right or those who have newly purchased the land. Tenant-cultivators are those who have taken the land on lease for a given period of time on payment of rent. Given the importance of land use planning for profitable run of the farm business, it is pertinent to examine the relative merits and demerits of owning and leasing in the land.

## OWNING THE LAND

To the owners, the very possession of land gives tremendous satisfaction. The very fact of land ownership brings in immense psychological pleasure. This is the objective of every farmer. Every farmer once gets the ownership rights would aim at acquiring more and more land through the hard-earned savings from agriculture. He makes all out efforts to keep the health of the soil intact for, it is a property, which has to last for generations.

### Merits

*Following are the specific merits of owning land*

1. *Security of Life:* Since land is a tangible asset, it provides the greatest **security** for the farmer. Let what may come, he is not perturbed for, he has an **invaluable asset,** which provides him the livelihood throughout his life.

2. *Freedom:* He enjoys the freedom in running the farm business. He is not a subordinate to any one and need not wait for any instructions to carry out any farm operation. Whatever, he feels like doing he does with total satisfaction, based on the availability of resources.

3. *Social Prestige:* Ownership brings him prestige in the society as he has some land under his possession.

4. *Encourages Development:* Owners are much concerned about land fertility. They leave no stone unturned to protect the soil health by understanding land development activities like irrigation, conservation, reclamation and other cultural practices.

5. *Incentive for Hard Work:* Whatever the income that is realized in farming is enjoyed by the owner himself. Therefore, he puts in heart and soul into the farming to maximize the income.

6. *Higher Productivity:* As a owner his main objective is to realize as higher productivity as possible without impairing soil fertility. This he achieves through judicious use of inputs, once a particular crop and variety are chosen. Particularly in the use of fertilizers, he sticks on to the balanced use, for any imbalance would have deleterious effect on the soil.

7. *Security of Tenure:* There is no worry for the farmer to be concerned about the tenure. He need not be tentative on the front of loosing tenure at any time, which normally a tenant is worried about.

8. *Provides Loan Facility:* Since land is an immovable property, it serves as an important security for obtaining loans from the institutional agencies.

9. *Property for Heirs:* Heirs enjoy the rights on the land possessed by their ancestors.

## Demerits

*Following are the demerits of owning the land.*

1. *Returns are not in Commensurate with the Worth of the Land:* Often times, the farmers feelings are that farming is not remunerative. Realizations are below the expected levels. They feel that they have a better chance of earning more money, if the land is converted into liquid form. The rate of turnover is low and slow in farming.

2. *Obligation of Standing as a Guarantor:* In the credit transactions, there is the obligation for the farmer to be as a guarantor for the other farmers (friends, relatives, *etc.*) who at times become defaulters.

3. *Chances of Unauthorized Occupation:* If for any reason the land is kept fallow for a couple of seasons, there is that lurking danger of unauthorized occupation.

4. *Small Size Holdings Prevent use of Technology:* Though technology is neutral to scale, resource non-neutrality prevents the application of technology on small-sized holdings.

## LEASING

Leasing activity is quite prevalent in farming. Normally absentee landlords would like to lease out the lands by identifying the tenants who have the reputation of being loyal tenants. Also those residing in the village, who are not interested in farming because of their pre-occupation also involve in leasing out activity. Some other farmers in view of the management problems also lease out part of their operational holdings. Coming to the tenants they may be landless labourers, small and marginal farmers and other farmers who have abilities of managing big farms. Leasing in activity when examined will throw open the following merits.

## Merits

1. *A Source of Income Augmentation:* For any one who is engaged in leasing in activity, it helps to augment their present income levels.
2. *Increase in the Size of Operational Holding:* For those who have some operational holding, leased in land add to the area of operation. Possibilities exist for better management of land resources.
3. *Crop Diversification:* Relatively larger acreage on account of leasing in activity permits crop diversification. Apart from higher income, risk in farming can be reduced.
4. *Flexibility of Rent:* Rent normally is fixed by the owner-farmer based on the prevailing rents for similar type of lands in the village. Yet flexibility does exist based on understanding between the owner and the tenant. Moreover, during the times crop failures also, the owner is not rigid in collection of rent after all he is humane and can understand the situation sympathetically.

## Demerits

1. *Insecurity of Tenure:* Normally in the leasing activities written agreements are seldom made, it is only through oral agreements. Irrespective of the likes and dislikes of the tenant, his tenure can be terminated as per the dictates of the owner-farmer.
2. *Less Scope for Land Development:* A tenant more often than not, is interested in deriving as much as he can from the leased in land in the light of the insecurity of tenure. His aim is to extract as much as he can from the available soil nutrients. Hence, no efforts are made for land development.
3. *Poor Infrastructure Development:* Farmers generally are not much concerned regarding developing the necessary farm infrastructure like farm building, fencing, *etc.,* for the lands leased out. They do not feel the need to develop these facilities, since they are not owners of the land.

## BUYING THE LAND

The purchase of cultivated land is considered as an important decision and involves large amount of money. Such practices will have long run affect on the financial structure of the business. The following factors of the land will have a bearing on the purchase of land.

1. Soil, topography and climate combine to determine the crop and livestock yield potential, thereby the expected income stream.
2. If the farm is located in the area of well-managed farms, it will have higher values and *vice-versa.*
3. Credit is an important input, hence the existence of financial institutions is considered in buying the land.
4. The farm's proximity to markets will reduce the transportation costs, increase the competition for the farm's products and hence higher price should be paid for such farms.
5. The presence of schools, hospitals, religious institutions, *etc.,* influence in buying decisions.

## LAND APPRAISAL

Appraisal is a technique to estimate the value for a given extent of land or real estate. The firms which are involved in the real estate business are well versed with the estimation of values. Two important methods are in vogue in estimating the value of land *viz.*, market data method and the income capitalization method.

## Market Data Method

Sale prices for the comparable lands which are sold very recently are taken into consideration to estimate the value for a given piece of land. If the characteristics of the land that is being appraised are more or less identical with the one that is sold recently, the estimates are quite realistic. If the characteristics are dissimilar between the one that is sold and the piece of land for which value is supposed to be appraised, the differences such as soil types, location, productivity, *etc.*, are duly adjusted before determining the value.

## Income Capitalization Method

This method uses project appraisal techniques to estimate the present value of the future income stream from the land supposed to be valued. This technique involves the estimation of annual net income, the selection of appropriate discount factor and working out the net present value.

The formula to estimate the value of land is

$$V = \frac{P}{i}$$

Where,

V = Estimated land value (Rs.)

P = Average annual net income (Rs.)

i = Discount rate

Following steps are adopted to estimate land value

1. For the given land for which the value is supposed to be estimated, find out the crops grown along with the yields and unit prices over a period of time. Yields as well as prices are based on long run averages. Using yields and prices, the gross income is estimated.
2. Estimate the total costs (working costs + fixed costs) for the crops grown.
3. Find out the average annual net income by deducting total expenses from the gross income, and
4. Finally capitalization of income is done. To do so an appropriate discount rate is selected and estimated land value is arrived at. The rate of interest charged for the farm loans by the financial institutions forms the discount rate.

## FARM LAYOUT

Farm layout deals with the determination of how, and to what extent the farm may be arranged in order to make use of the factors of production in an efficient manner. The most efficient utilization of all the inputs on the farm is the objective of farm layout. Very little attention is paid by the farmers to this important aspect and the result is that the farm is neither attractive in appearance nor convenient in operation of the farm business.

Farm layout refers to the systematic manner in which the farm is divided into fields, location and arrangement of irrigation and drainage channels, farm buildings, roads, fencing, *etc.*,

## Factors Affecting the Farm Layout

Farm layout is affected by several factors which are listed below. Hence, it is seldom possible to lay out an ideal farm.

1. *Topography:* The presence of hills and valleys makes it difficult for the farmer to get uniformity in size and shape of the fields. It not only affects the size and shape of the farm but also the location of farm buildings.

2. *Soil Types:* It is always desirable to have each field uniform in soil type in order to grow a crop on the soil best suited for it, use of machinery and adoption of better cultural practices.

3. *Other Factors:* Access to water is essential when livestock enterprises are important on the farm. Canals, high ways, railways, *etc.,* may cut the farm at such angles as to make uniformity of size or shape of field difficult.

## Effect of Farm Layout

The utilization of factors of production is effected by the shape, size, number and arrangement of the fields and the location and arrangement of the farm buildings. The farm lay out effects

1. *Land:* Small and irregularly shaped fields increase the amount of waste land or land that is not cultivable. As the fields increase in size and become more regular in shape, the percentage of total area lost to cultivation decreases. Land that is not available for cultivation also depends upon the type of implements used. The loss of the land is small when hand implements are used. The percentage of total area lost in the case of small fields is large when large power drawn machinery is employed. The amount of cultivated land is affected by the type of fence used. A smooth wire fence will permit the cultivation of more land than that of stone fence. Hedges prevent the cultivation of land to an even large extent.

2. *Labour:* Labour can be more productively used when the fields are arranged in such a way that the distance between them and the farmshed is reduced to a minimum.

3. *Capital:* The size and shape of the fields also affect the capital use. Large rectangular fields are better suited to farm machinery than small fields. Tractors and power-drawn machinery cannot be used efficiently on small and poorly arranged fields. The cost of fencing is also influenced by the shape and size of the fields. Though the cost of fencing for larger field is more, but the cost per acre decreases with the increase in the size of the field. A square field can be fenced more economically than an oblong field. Thus the shape of the field affects the cost of fencing.

## Developing a Good Farm Layout

1. *Farm Buildings and Sheds:* The farm buildings, sheds for livestock, stores, *etc.,* should be in the centre of the farm as for as possible so that the approach to different fields and supervision becomes easy. If this is not possible, the farm buildings may be located on the farm side nearer to the main road.

2. *Arrangement of the Fields:* Irregularly shaped fields waste labour, time and other resources. As far as possible, a field should be uniform in soil type. A field which is rectangular in shape is regarded to be better.

3. *Roads:* There is no need to connect each field with a road on commercial farms. Key approach roads are to be laid on the sides of the farm.

4. *Irrigation Channels:* The source of irrigation affects the layout of irrigation channels. Each field must be linked with an irrigation channel.

5. *Fencing:* The type and extent of fencing depends upon the location and type of the farm. Shrub fencing, barbed wire with wooden, concrete or iron poles, pucca brick wall, *etc.*, are the common types of fences in India.

## FARM LABOUR MANAGEMENT

Human labour is an important input factor in farm production regardless of the type of technology employed. The participation of human labour is seen right from the operations like preparatory cultivation down to sowing, fertilizer application, irrigation, weeding inter cultivation, plant protection measures, harvesting, threshing, *etc*. Several research studies have reported that human labour costs constituted about 40-50 per cent of the total cost of cultivation for various crop enterprises. Technological changes, which have brought in a sea change in Indian agriculture also created more employment opportunities for agricultural labour.

### Definition of Agricultural Labourer

The total annual income earned by labourer from agriculture or sometimes the total number of days a labourer worked in agriculture would generally serve as a basis for classifying the labourer as agricultural labourer or other kind of labourer. According to first labour enquiry of 1950-51, *the persons who were employed in agricultural operations for more than 50 per cent of their total number of days worked during the previous year are called agricultural labourers.* But the second labour enquiry of 1956-57 classified the workers as *"agricultural labourers if they derive more than 50 per cent of their annual income from agricultural sources"*. But the two enquiries felt that agricultural labourers would eke out their living from agricultural wages in India.

## CLASSIFICATION OF FARM LABOUR

The farm labour force in India is of three categories, *viz.*, (i) farmer's own labour (ii) family labour; and (iii) hired labour. The hired labourers are further split up into casual labourers, attached labourers or permanent labourers and contract labourers. The labourers could be skilled, semi-skilled and unskilled.

i. *Farmer's Own Labour:* Indian farmer performs diversified functions as a capitalist, labourer and manager. He is poor capitalist, good labourer and inefficient manager. Farmer himself contributes his services in the production of crop and livestock enterprises. This is the best type of labour because of personal interest. It is unpaid labour.

ii. *Family Labour:* Family labour consists of farmer's spouse and his children who work on the farm. It is also unpaid family labour. The main labour force for small farms is the family labour. Since the small farms cannot provide adequate employment to the available family members, they offer their services as workers on the other farms. The farmer's own labour and family labour do not receive any wages (unpaid), but there is an opportunity cost, hence family labour is imputed.

iii. *Hired Labour:* When the family labourers are not sufficient to perform farm operations, hired labourers are engaged. Hired labourers are engaged generally in operations like sowing, weeding, harvesting, *etc.* Large farmers engage more hired labourers than small farmers. Hired labour is classified into (a) casual labourer or daily labourer (b) attached labourer or permanent labourer and (c) contract labourer.

a. *Casual Labourer:* They are engaged as and when need arises. Generally casual labourers are engaged at the time of sowing, transplanting, harvesting, threshing, *etc.,*

b. *Attached Labourer or Permanent Labourer:* When a worker is hired for a production season or a year, he is called attached labourer or permanent labourer. They are paid wages in cash or kind or both. They are supposed to perform both farm and non-farm operations. Annual wages are fixed right at the time of agreement. Some times they are given a share in the crop. These workers are also provided with additional facilities like housing, clothing, *etc.*

c. *Contract Labourer:* When a group of workers are assigned a specific task to perform, on a specific payment in the production of enterprises, represent the contract labourer.

Farmers owned labour, family labour and permanent hired labour constitute fixed labour resource on the farm.

The labourers could be skilled, semi skilled and unskilled.

i. *Skilled Labourer:* When a person undergoes specialized training to perform a specific job, he is called skilled labourer. Examples: Mechanics, drivers, carpenters, *etc.* The wages of skilled workers are higher than semi-skilled and unskilled.

ii. *Semi-skilled Labourer:* He does not require any elaborate training, but needs to gain experience to perform a particular task by working under an experienced worker. Example: Rearing silkworm cocoons, *etc.*

iii. *Unskilled Labourer:* Unskilled workers are those, who engage in tending of cattle, weeding, application of FYM, *etc.,*

## CHARACTERISTICS OF FARM LABOUR

Let us now know the important characteristics of Indian farm labour.

1. The agricultural labourers in India are under employed due to the seasonal nature of agricultural production.
2. There is a predominant problem of disguised unemployment. It means that more number of persons appear to be engaged than required in the agricultural industry.
3. The productivity of labourers is low due to imbalance in the availability and use of resources on the farms.
4. Unlike industrial workers, farm workers lack organization leading to poor bargaining capacity.
5. Most of the Indian farm workers are unskilled.
6. Low wages are prevailing for farm labourers because of their excess supply.

7. Labour force is increasing in India by leaps and bounds particularly, in low-income groups due to non-adoption of family planning measures.

## Wages

Wage is the reward for the services of a worker. Agricultural wages depend on several factors *viz.,* demand for and supply of labour, efficiency of labour, nature of enterprise, nature of operation, resource endowment, economic conditions of the locality, customs, *etc.* Higher wages are paid to skilled workers because of their higher efficiency. Casual women labourers and child labourers are paid low wages compared to that of casual men labour. However, now labour wages are regulated throughout India by minimum wages act of 1948.

Following are the different methods of wage payment:

1. *Cash Wage:* When wage is paid in money, it is called cash wage.
2. *Kind Wage:* When wage is paid in kind like paddy, ragi, *etc.*, it is called kind wage.
3. *Time Wage:* When wage rate is fixed per hour or per day or per month, it is called time wage.
4. *Piece Wage:* When wage is paid after completion of work, it is called piece v age.
5. *Task Wage:* When wage is paid for completion of a specific task, it is called task wage, which is also called contract system.

# FARM LABOUR EFFICIENCY

Farm labour efficiency refers to the amount of productive work accomplished by a worker on the farm per unit of time. In other words, amount of output per unit of labour input indicates the efficiency of labour.

This can be judged by working out average and marginal productivities of labour input in hours in the production of crops and livestock products. The marginal productivity of skilled labourers is more and hence higher wages are offered. On the contrary, unskilled labourers are being paid less wages.

## Factors Affecting Farm Labour Efficiency

Higher the efficiency greater is the returns of the farm and *vice versa.* The efficiency of farm labour depends on the following factors.

1. The physical health status and interest of the agricultural labourers will have positive relation on their efficiency. If health condition of labourers is good, they will have more efficiency and similar relationship holds good if there is interest to work. Hence the manager should provide all incentives to the workers to get good amount of out turn of work.
2. All kinds of tools, implements or equipment should be provided to the labourers which are required to them so that they can do better work or increase their efficiency. In harvesting, the labourers should be provided sharp sickles. Similarly, for digging of wells, crowbars and spades are very essential implements to increase their efficiency.
3. The climatic factors play an important role on the efficiency of labourers. The exposure to hot sun during work hours leads to lot of drudgery. Therefore, labourers would become easily tired and hence low turnover of the work. Under such situations it is better to execute the farm works before 12 noon starting the work at early hours *i.e.,* at 6.00 AM.

4. The skill of the labourers is also an important factor in better performance of works. Driving the cattle in ploughing operations and sowing operations involve lot of skill. Transplanting also requires skill. Hence for these operations, skilled labourers are employed by paying higher wages.

## Measuring the Labour Efficiency

Different criteria are used to measure the labour efficiency. For example, on a dairy farm, labour efficiency is measured by considering the number of cows maintained per unit of labourer or quantity of milk produced per labourer. On a poultry farm, the number of egg laying hens maintained per labourer tells about the labour efficiency. On orchards, number of hectares of fruit crops per unit of labour indicates labour efficiency. The following formulae are used to measure the labour efficiency.

$$\text{The returns per labour per day} = \frac{\text{Gross income - cost of all inputs excluding labour}}{\text{Number of workers in days}}$$

## Labour Efficiency Index

In most of the farming areas, labour efficiency standards are being set on the basis of past experiences of labour use for various crops and livestock enterprises. The efficiency index is worked out as follows for a hypothetical farm raising crops. The average labour cost in the locality is, say Rs. 5,000 but the actual labour cost incurred by the given farm is Rs. 4,000. Then based on this cost, the labour efficiency index is

$$\frac{5000}{4000} \times 100 = \frac{500}{4} = 125\%$$

Here labourers are efficiently employed.

$$\text{Labour Efficiency Index} = \frac{\text{Requirement of labour on the farm as per the performance standar} \times \text{wage rate}}{\text{Actual labour employed on the farm} \times \text{wage rate}} \times 100$$

# CAPITAL MANAGEMENT

Capital is explained to include all the items used in farm production excluding those of land, fencing, farm buildings and labour. Therefore capital here refers to work stock, farm machinery, implements, seeds, feeds, fertilizers, manures, water, *etc*. A plan is essential to put all these items properly in the production activity as we plan land and labour use. The plan that should be put into action regarding the use of these items is shown as under.

## 1. Plan for the use of Working Capital for the Year

Planning here encompasses the package that enables the farmer in keeping ready various inputs required to be applied at various stages of crop growth. This include the plan of buying the right inputs at the right time from the authorized sources. May be that, it is not possible to plan day-wise in the crop season, still the farmer over the years of involvement in farming would be in a position to spot the period of requirement of inputs and performance of operations. Also such planning avoids the conflicts in the use of machinery. Equally sure, he is about the periods of routine work unless the farming is marred by the extreme weather conditions, which is something unpredictable. Barring these unforeseen contingencies, the farmer will have an exhaustive visualization of what is necessary and what is not in the crop period for smooth running of the farm business.

## 2. Adjustments in the Working Capital

Plan prepared by the farmer in advance for the ensuing season cannot really serve like a guide as the very nature of agricultural production stand in the way since the farmer has to alter the operations as per the weather. Some operations he needs to hasten up and some others are to be put off by some period as dictated by the weather conditions. These sudden adjustments at times increase the demand for certain machinery, implements, *etc.*, which should be dealt appropriately. Similarly, changes in prices of inputs of certain brand may pose problems to the farmer in decision-making. He has to modify the scheme of things in favour of inputs, which are cost effective, without sacrificing the quality. On the same lines, feed ratios need to be altered in livestock.

## 3. Supervision of Working Capital

Supervision of working capital in the light of avoiding wastages in the input use and services of machinery gains importance. The wastage of the inputs like seed, fertilizer, feed, *etc.*, should be strictly avoided as the wastage is part of the input cost that is accounted for a crop. Such a wastage tends to reduce the input productivity leading to inefficiency. Regarding machinery and implements, rough usage needs to be curbed so as to prolong the life period and minimizing the amount towards repairs and maintenance. Also some of the machinery and equipment are sparingly used on the farm and rest of the time they are idle and hence it is the responsibility of the supervisor to upkeep these items with minimum deterioration. Similarly work animals should be properly fed for increasing their efficiency. Overwork of the animals in the anxiety of completing the work faster should be avoided as the works will be further delayed in case the animals fall sick.

## 4. Effective Methods of Utilization of Capital

Methods of utilization of farm inputs carry a lot of significance. Right quantity of seed along with right spacing or planting at proper depth is the basic caution one has to bestow for the desired yields. Analogously, methods of application of fertilizers like broadcasting, placement, spraying, *etc.*, should be appropriately employed for the crops under cultivation. Feed rations too for livestock should be planned as per the nature of the products produced.

## 5. Soliciting the Equipment as a Means of Getting Operations on Time

Mechanization facilitates the timely operation and completion of the farm operations. The man-hour requirement for various operations can be reduced substantially using machinery; here the main benefit to the farmers is the reduction of time requirement for completion of an operation. The machinery is substituted for human labour hence managerial role picks up significance in the investment on the machinery or hiring the machinery.

CHAPTER **36**

# Risk and Uncertainty

Agricultural production is confronted with risk and uncertainty conditions. As agricultural production being biological and seasonal in nature, we don't know the nature of agricultural decisions and their possible outcomes. It is a highly difficult proposition to make right decisions when the production environment is risky and uncertain. Farmers are generally concerned with decisions on crops to be planted, seed rates, fertilizer application, application of other crucial inputs, *etc.* These decisions are subjected to change depending on the nature of weather risks and other associated risks. A livestock farmer has to take a number of decisions to expand his dairy cattle herd and he has to wait for several years to get back the investment and also income from the investment. Changes in the weather, prices and other socio-economic factors occur between time periods in which investment decisions are made and the final outcome. Due to this the farmers and ranchers have to consider various management strategies relevant to the risk and uncertainty conditions. If everything in farming and livestock rearing goes with certainty then every farmer becomes a better manager, but this is not the case with real farming situation. Only a few farmers can become efficient, particularly those who could understand risk and uncertainty situations in farming and ranching and follow the relevant risk management strategies. Let us now know about how risk is different from uncertainty.

## Risk and Uncertainty: A contrast

Earlier, *i.e.*, in the past, the economists did not make any difference between risk and uncertainty. Only about three decades ago, economists made a clear-cut distinction between these two terms. According to recent view, risk is measurable, while uncertainty is not measurable. Risk is defined as a situation when all possible outcomes are known for a given management decision and probability associated with each possible outcome is also known. Risk is measured through probability concepts. Probabilities are assigned to the events. For example, probability of rain, weather forecast, *etc.* Subjective probabilities are based on judgment and experience of individuals and these may vary from individual to individual. Subjective probabilities are measurable through certain concepts, but uncertainty situation prevails when all the possible outcomes of events are unknown, then neither the probability nor the outcomes are known. It is also difficult to estimate the associated probabilities to the possible outcomes of the events. In that case, we assign subjective probabilities to make the best estimates of true probabilities based on the information and past experience. A pure risk situation is not seen in the real world, because the true probabilities are not estimated for the events. We generally formulate a set of subjective probabilities to solve the decision problem in a situation of risk. Now let us aware of different sources of risk.

## SOURCES OF RISK

The various sources of risks are conveniently grouped into three categories viz., 1) Production risk or technical risk, 2) Price risk or marketing risk and 3) Financial risk.

### 1. Production Risk

In industrial businesses we have a technical input-output relationship with known quantity of output for a given quantity of input i.e., the production practices are standardized in industrial production. Such type of relationship does not exist in agricultural production. Output of crops and livestock is subject to change due to weather, disease, insects, weeds and inadequate technology. These factors cannot be predicted accurately and hence results in the variability of the output. In fact, the yield variations are due to many factors. Some are under control, while some others are not under the control of management. Thus, the production-risk is due to largely many factors rather than a single factor. Weather risk and technical risk are the most important components of production-risk. Yield risk may be arising due to change in production costs, institutional factors, changes in management, etc. Input-prices are subject to more variations than output prices some times. Then price trends also cause production risk. The cost of production for unit of output is also changing due to productivity levels and magnitude of costs both over time and space. When the technology is changing in a particular place, it would have greater bearing on production risk. The new technology, though it brings higher level of profit, it would also involve greater variation of the output. Because of this factor the farmers are reluctant to adopt technology as per expectation.

### 2. Price Risk or Marketing Risk

Marketing risk is also called as price risk because prices are determined due to interaction of demand and supply in the market. Production of crops and livestock is influenced by prices in the market which are beyond the control of farmers and the consumer's prices of commodities vary from year to year, season to season and exhibit seasonal variation. If there is less time lag between production and marketing activity, we could expect less price risk for agricultural commodities. But this is not the case with most of the agricultural commodities, because there is enough time for prices to fluctuate when the commodities move from production centre to consumption centre. Supply of the commodity is very much affected if there is a situation of weather risk and production risk. Demand for a commodity is mainly changing due to consumers' income, habits, tastes and preferences, export and import policies and overall, general economic measures taken up by the Government with regard to price stabilization. If there are less trade restrictions due to the increased demand in the foreign market, prices of commodities will be increasing. Similarly, in the domestic markets, if the commodity prices were very high, prices would be brought under control by import policy. Thus trade policy i.e., export import (Exim) policy would have a greater impact on price stabilization of commodities.

## 3. Financial Risk

This type of risk increases with increased amount of borrowed money in the farm business. This aspect is explained by the principle of increasing risk. According to this principle, if borrowing increases there would be greater risk of foregoing the equity capital in the event of losses and this leads to increase in the debt-equity ratio. For further details, see the principle of equity and increasing risk.

Uncertainty arises due to the changes in future interest rates and fiscal policies of RBI and finally changes in the ability of farm business to generate required cash flow for the clearance of debts. In farming and livestock rearing, all these types of risks are interrelated and mixed, for instance, the ability to repay the debt by the farmers depends on yield levels and favourable prices in the markets. Financing the production processes, and storing the commodities under scientific management require large amounts of borrowed capital. Hence a careful analysis is to be made considering all these types of risks including their sources to follow the relevant risk management strategies.

## MEASURING THE EXPECTATION AND VARIATION

In the risky farm environments, the farmers make decisions using some kind of the expectations regarding prices, costs and yields. Here what we mean is that there is no assurance or certainty regarding actual facts of these things and the outcome of the farm decisions. Let us look into as to how the expectations and variations are worked out and used in the decision-making process. Expected values are worked out as proxy for best estimates of unknown future events, so that they become one of the important components of decision-making process.

Two types of averages are used to work out an expected value. We take the actual past prices (time series data) and yield to find out the simple average over the specific time period. Here, the problem is to decide on the number of years of data that should be considered to work out average. Here the choice depends upon the subjective estimate of the decision-maker. Second type of average is called weighted average. This method considers larger weights for the more recent data and smaller weights for the remote data. This is illustrated in Table 36.1.

TABLE 36.1  Working out Weighted Averages for the Prices.

| Year | Average annual price of commodity (Rs.) | Weights | Price x Weight |
|------|------------------------------------------|---------|----------------|
| 1993 | 1,000 | 1 | 1,000 |
| 1994 | 1,200 | 2 | 2,400 |
| 1995 | 1,400 | 3 | 4,200 |
| 1996 | 1,100 | 4 | 4,400 |
| 1997 | 1,000 | 5 | 5,000 |
| 1998 | 1,300 | 6 | 7,800 |
| | 7,000 | 21 | 24,800 |
| Simple average | | | 1,166.67 |
| Weighted average | | | 1,180.95 |

For working out the simple averages we have taken sum of the annual average prices and divided by number of years *i.e.*, (6). For working out weighted average we have to multiply the individual prices by weights, obtain the total and divide by the total of weight *i.e.*, (21). Weighted average is the expected value regarding the price of the commodity. For the given data there is no price risk for the commodity, as expected value is more than simple average. On the contrary, if there is price risk its expected value would be less than the simple average. We can also work out expectations through the concept of probabilities.

## Expectation Through Probabilities

Another way of finding expectations is to choose the probability value, which is the most likely to occur for the given price or yield or cost. In this procedure, we have to elicit probabilities associated with each possible outcome (expected yield and expected price). The outcome with the highest probability would be selected as the most likely one to occur. This is shown in Table 36.2.

**TABLE 36.2 Working out Expectations Through Probabilities.**

| Crop yield/acre in quintals | Probability | Probability x Yield |
|---|---|---|
| 11 | 0.1 | 1.1 |
| 14 | 0.4 | 5.6 |
| 10 | 0.1 | 1.0 |
| 12 | 0.3 | 3.6 |
| 18 | 0.1 | 1.8 |
| Mean 13 | Total 1.0 | 13.1 |

From Table 36.2 it is clear that the most likely output based on the highest probability is 14-quintals/acre. Mathematical expectation of 13.1 quintal of commodity per acre is less than the most likely yield of 14 quintals per acre, but it is more than the average yield of crop *i.e.*, 13 quintals per acre.

## Variability

Two factors *viz.*, the variance and the expected value, in general should form the basis for selecting the best enterprise among alternative risky enterprises. Range is a measure of variability and it refers to the difference between the lowest and the highest possible outcome. Alternative enterprises should be preferred which have got smallest range value, provided their expected values are the same. Similarly alternative enterprises with lowest variance should be selected.

# MEASURES TO MANAGE FARM RISK

## 1. Diversification

Selection of suitable crop and livestock enterprises is the first step in diversification. Through diversification process, the farm entrepreneur produces several products rather than single product with the hope that when the returns from one enterprise is low, it is compensated by the higher returns from the other enterprise. Through diversification process we see that idle resources are put to use and income variability of the enterprises is reduced. This means the farmer should select enterprises in his cropping scheme with less income variability. Income variability for the enterprises would be lessened through diversification, if the following relationships among prices and yields of products were established. If the prices and/or incomes of the enterprises have correlation nearer to + 1, then such enterprises should not be included in the diversification process. If the correlation coefficient is negatively significant, then the two enterprises must be selected as the most suitable one to reduce the variability. The enterprises with zero correlation coefficient are the most suitable enterprises, for crop diversification. If their income variability and ranges are less such enterprises are selected in the cropping scheme.

Risk programming models and game theory models should be used to formulate whole farm plans under different risky situations. If enterprises are selected based on net returns over time in risky environment, the decision rule is to select the enterprises with their highest expected income. When the probabilities for the enterprises are not easily estimated, then assume equal probability for all the outcomes and the enterprise with highest expected income is selected. When the farming is subjected to high risky conditions, survival strategy should be adopted under such condition to select the enterprises with minimum net returns. Stable enterprises should be preferred rather than risky enterprises particularly in situation of farm risks.

## 2. Insurance

We have many types of insurances in farming to reduce production risk and financial risk. Crop insurance scheme reduces production risk. Livestock insurance provides safety against the fatal diseases of cattle. Farm assets are insured against theft, burglary, fire or any other damage. The decision whether to go for insurance or not is judged by the following equation.

$$\pi = F (0 - r) - P$$

Where,

$\pi$ = Profit obtained by going for insurance.
F = Financial reserve required.
O = Opportunity cost for financial resource in terms of %.
r = Interest earned on financial reserves.
P = Insurance premium paid by the farmer.

If $\pi > 0$, it is desirable for the farm to go for insurance i.e., the returns from the insurance policy are more than the cost and similarly, if $\pi < 0$, it is worthless to do insurance.

## 3. Agronomic Practices

To reduce production risk, crop rotations, suitable varieties, deep tillage, mulching, etc., should be adopted.

## 4. Market Risk Management

If commodity prices are changing to a greater degree, then the price risk arising from market is more. Following are the methods proposed to reduce the marketing risk.

(i) *Selling the Farm Products at Different Points of Time:* Due to financial obligations, farmers sell their produce immediately after the harvest. During the harvesting periods generally prices will be low for commodities in the market due to large arrivals. Such sales at harvest period are said to be distress sales. If the farmers avoid such selling practices, they can get remunerative prices for their products. The knowledge regarding supply and demand for the farm products and prices prevailing in different markets and other market information is essential. For perishable commodities, storage facilities, freezing facilities, processing facilities, *etc.*, are required to get remunerative prices. Hedging is another measure adopted by the farmers and the traders to safeguard against price-risk. Hedging practice is devised and followed by studying the future markets.

(ii) *Government Price Polices and Programmes:* Minimum support price, procurement price, levy price, issue price, *etc.*, are set for various agricultural commodities by the Government every year to bring about price stabilization. Support price provides protection for the farmers against fall in the prices of farm products.

## 5. Financial Risk Management

For reducing the financial risk we require many strategies, which are aimed at liquidity and solvency of the farm business. Solvency by definition refers to business ability to meet the long-term financial requirements of the farm. Liquidity strategies should aim at as to how to build up the farm business to meet the short-term cash requirements. At macro level fiscal policies formulated by RBI are aimed at to provide different measures to safeguard against the financial risk.

# SECTION V
## AGRICULTURAL FINANCE

# CHAPTER 37

# Agricultural Finance

## PROBLEMS OF AGRICULTURAL CREDIT IN INDIA

The mass illiteracy of Indian people in general and rural people in particular compounded the problems of agricultural credit. With around 70 to 80 per cent of the farmers being illiterate they are unaware of the various sources of farm credit, that supply at a lower rate of interest. As a result the farmers were not in a position to judge the source of credit, which supplies the same at the rate, which is lower than the one collected by the private money lenders. Another problem that adds dimension to this situation is even if the institutional sources are known to the farmers, their poor resource endowment base leaves them high and dry to offer anything as security, which is a prerequisite in credit transactions of credit institutions. Village money lenders exploit this situation by advancing liberal credit without insisting much on security but compensating the service with higher rates of interest and at times compelling the farmers to resort to forced sales of the harvested produce. The tiny and uneconomic holdings help the farmers very little in generating the marketable surplus thereby making the farmers incapable of repaying the loan, if any taken from institutional agencies. Those who have transactions with village money lenders cannot think of either marketed/marketable surplus, as it was pledged in advance.

The inaccessibility of villages in general more so in rainy season, inadequate transport and communication discourage the lending institutions to operate in those areas, for it increased the cost of loan transactions.

Illiteracy coupled with subsistence nature of farming, discourage the farmers to really look it as business and maintain records. In the absence of records the institutional agencies often find it difficult to evolve any realistic lending policies for the advantage of the farmers.

Another pressing problem for the institutional agencies is diversion of production loans for unproductive purposes. This is borne out of pressing domestic problems, as farm and family are inseparable.

In the absence of competition from any institutional sources, moneylenders acquired the status of monopolists in agricultural lendings. This led to charging exorbitant rates of interest much to the disadvantage of the farmers.

All India Rural Credit Survey Committee classified the various sources supplying rural credit in the following classes *viz.*, Government, relatives, landlords, agriculturist moneylenders, professional moneylenders, traders, cooperatives, commercial banks and others.

**Government:** The Government loans advanced to the cultivators for production of crops, land improvement and distress relief are known as taccavi loans. Taccavi loans date back to the pre-British era.

**Relatives:** Any loan that was taken free of interest by a farmer was only considered as a loan from relatives.

**Landlords:** When leasing out activity prevails in agriculture, this type of loans are often times found, when landlord advances loans to the tenants. Such loans are called loans by landlord.

**Agriculturist moneylenders:** These are those, whose main occupation is agriculture but involved in money lending activity, which is of minor importance.

**Professional moneylenders:** Those who earned substantial part of their income from money lending activity are professional moneylenders.

**Traders:** These are the individuals who are involved in trading and loans from these persons are called borrowings from traders.

**Cooperatives and commercial banks:** Borrowings from cooperatives and commercial banks are called as borrowings from institutional agencies.

## NEED OF AGRICULTURAL FINANCE

Given the requirement of finance in agricultural sector, very few farmers will have capital of their own to invest in agriculture. Therefore, a need arises to provide credit to all those farmers who require it. Even if we look into the expenditure pattern of the farm families, they have hardly any savings to fall back on. Therefore, credit enables the farmers to advantageously use seeds, fertilizers, irrigation, machinery, *etc.* Farmer has to invariably search for a source, which supplies adequate farm credit. Above all, small and marginal farmers constitute majority of the farming community. The tiny land holdings owned by these categories of farming community have the following characteristic features:

1. Allocating larger proportion of land they own for the cultivation of food crops for subsistenc e.
2. Predominance of family labour utilization in the production of farm enterprises.
3. Low marketable surplus.
4. Risk aversion.
5. More demand for consumption credit.
6. Inability to offer security due to small size of the holdings, *etc.*

Despite the above limitations, the farmers do need credit support from the institutional agencies for the development. When capital is split up, it takes two forms *viz.*, equity capital and non-equity capital. Credit (non-equity capital) plays an important role in providing the needed liquidity to the farmers, who do not have sufficient equity capital to invest in farming. Credit as such is not an input but it enables the farmer to have access to the resources, thereby removing the financial constraint. As money is not wealth, credit is not income, but it leads to income. But care should be taken while extending credit as without clear opportunities, it gets ended up as an additional consumption, instead of capital. When it is properly lent, it becomes a lever for the development. For this to attain, credit institutions should involve wholeheartedly for providing opportunities to the under developed sections of the rural areas.

A point needs to be underlined is that credit is not just one time help as the same farmer-borrower needs more and more credit in future and also demand arises from the potential borrowers. Hence, the network of credit should be expanded in the interests of the farming community and the institutional agencies.

## DEFINITIONS

Agricultural finance generally means studying, examining and analyzing the financial aspects pertaining to farm business, which is the core sector of the country. The financial aspects include money matters relating to production of agricultural products and their disposal. When we speak of the financial aspects in agriculture, issues that figure are capital required for agriculture, the way necessary funds are raised and the pattern of utilization of funds so raised. Murray (1953) has defined agricultural finance in the following words, "*it is an economic study of borrowing funds by farmers; of the organization and operation of farm lending agencies, and of society's interest in credit for agriculture.*" Tandon and Dhondyal (1962) defined agricultural finance "*as a branch of agricultural economics, which deals with the provision, and management of bank services and financial resources related to individual farm units.*"

The following are implied in the above definitions of agricultural finance:

1) All the farmers should be purveyed requisite finance,
2) Finance should stimulate and enhance the productivity of farm scarce resources, and
3) Farm finance has a vital and catalytic role for agri-economic development of the farmers.

Agricultural finance is viewed both at macro-level and micro-level. Macro finance deals with the different sources of raising funds for agriculture as a whole in the economy and it is also concerned with the lending procedures, rules, regulations, monitoring and controlling procedures of different agricultural institutions. Thus, macro finance pertains to financing agriculture at the aggregate of the individual farm business units and it is concerned with the study as to how the individual farmer considers various sources of credit to be borrowed from each source and how he allocates the same among the alternative uses within the farm. It is also concerned with future use of funds. In sum, macro finance deals with the aspects relating to total credit needs of the agricultural sector, the terms and conditions under which the credit is available and the method of using the total credit for the development of agriculture. On the contrary, micro finance refers to the financial management of the individual farm business.

## IMPORTANCE

Farm finance assumes vital importance in the agro-socio-economic development of the country both at individual/micro level and at aggregate/macro level. Its catalytic role strengthens the farming business and augments the productivity of scarce resources. For instance, new potential seeds, when combined with purchased inputs like fertilizers and plant protection chemicals in requisite proportions result in higher productivity of resources. Application of new technological inputs obtained through farm finance helps boost agricultural productivity. Accretion to farm assets and farm supporting infrastructure provided by large scale financial investment activities entail increased farm income levels, leading to the overall improvement in the living standards of rural masses. Farm finance can also contribute to reduction in regional economic imbalances and is equally good at narrowing down the inter-farm asset and wealth variations. To quote Muniraj (1987): *"Farm finance is the money extended to the farmers to stimulate the productivity of the limited farm resources. It is not a mere loan or credit of advance, it is an instrument to promote the well being of the society. Farm finance is not just a science to manage the money, but is an applied science of allocating scarce resources to derive the optimum output. It is a lever with forward and backward linkages to the economic development both at micro and macro- levels"*. Thus the role of farm finance in strengthening and development of both input and output markets in agriculture is crucial and significant. Indian agriculture is still traditional, subsistence and stagnant in nature, hence agricultural finance is needed to create the supporting infrastructure for adoption of new technology. Massive investment is needed to carryout major and minor irrigation projects, rural electrification and energisation, installation of fertilizers and chemical plants, execution of agricultural promotional programmes and poverty alleviation programmes in the country.

## REQUISITES OF GOOD CREDIT SYSTEM

The requisites of good credit system as given at regional seminar for Asia on "Agricultural Credit for Small Farmers" held in Bangkok, in October, 1974 were:

1. All the credit needs (short, medium and long-term) of the farmers (in kind or cash) should be met.
2. Credit should be made available as near to his doorsteps as possible and when needed by the farmers.
3. It should generate savings and accelerate economic growth at the socially desired rate.
4. The credit policy should reflect a compromise between the often diverse plan objectives and differing group interests *i.e.*, the farmer, the credit institution and the Government.
5. The borrower should be encouraged to adopt new technologies without which sufficient capital cannot be generated to repay loans.
6. Supply and other services too should be made available to him.
7. The lending agency had to ensure that lending machinery is matched by a recovery machinery.
8. An efficient finance system would not confine its areas of operation to a particular crop. The lending agency should be geared to finance the entire farming system, which may include crop loans, livestock loans, agro-industry loans, *etc*.
9. The credit agency should be in a position to interlink with marketing agencies to ensure the full recovery of loans.
10. It is necessary that the rate of interest charged from the farmer should be relatively low.

## CLASSIFICATION OF CREDIT OR LOANS

Loans are certain amount of money provided for certain purpose on certain conditions with some interest, which should be repaid sooner or later. It is also referred to as credit. Credit is broadly classified based on various criteria Figure 37.1.

### Based on Purpose

*1. Production Loans:* They refer to credit given to the farmers for crop production. These loans are intended to increase the production of the crops. These are also

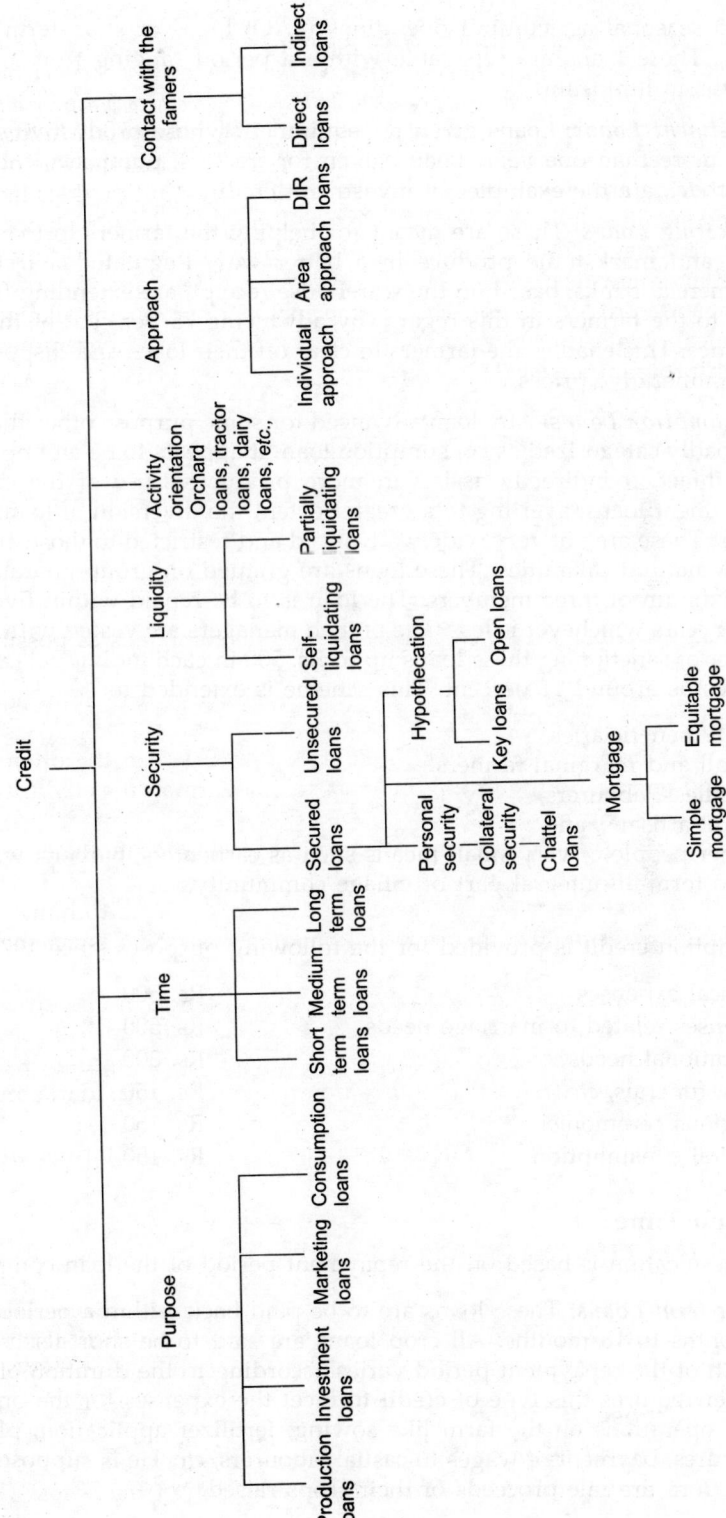

**Fig. 37.1.** Broad classification of credit.

called seasonal agricultural operations (SAO) loans or short-term loans or crop loans. These loans are repayable within a period ranging from 6 months to 18 months in lumpsum.

2. *Investment Loans:* Loans given for equipment whose productivity is distributed over more than one year. Loans given for tractors, pumpsets, tube wells, work stock, *etc.,* are the examples of investment credit.

3. *Marketing Loans:* These are meant for helping the farmers to overcome distress sales and market the produce in a better way. Regulated markets as well as commercial banks, based on the warehouse receipt, are extending financial assistance to the farmers in this regard, by advancing 75 per cent of the value of the produce. This enables the farmers to clear off their loans and dispose the produce at remunerative prices.

4. *Consumption Loans:* Any loan advanced for some purpose other than production, is broadly categorized as consumption loan. It appears to be an unproductive loan but, infact, it indirectly assists in more productive use of the crop loans and investment loans, averting to a greater extent the diversion of loans to other purposes. These are not very widely advanced and restricted to those areas, which are hit by natural calamities. These loans are granted on group guarantee basis with a maximum of three members. The loan is to be repaid within five crop seasons or 2½ years whichever is less. The branch managers are vested with the descretory power of sanctioning these loans up to Rs. 500 in each individual case. The rate of interest is around 11 percent. This scheme is extended to:

i.   IRDP beneficiaries,
ii.  Small and marginal farmers,
iii. Landless labourers,
iv.  Rural artisans, and
v.   Other people of very small means such as carpenters, barbers, washermen, *etc.,* who form an integral part of village community.

Consumption credit is provided for the following purposes since 1976:

| | |
|---|---|
| Medical expenses | Rs. 500 |
| Expenses related to marriage needs | Rs. 500 |
| Educational needs | Rs. 200 |
| Birth, funerals, *etc.* | Rs. 150 |
| Religious ceremonies | Rs. 150 |
| General consumption | Rs. 150 |

## Based on Time

This classification is based on the repayment period of the loan component.

1. *Short-term Loans:* These loans are to be paid back within a period ranging from 6 months to 18 months. All crop loans are said to be short-term loans, but the length of the repayment period varies according to the duration of the crop. The farmer requires this type of credit to meet the expenses for the ongoing agricultural operations on the farm like sowing, fertilizer application, plant protection measures, payment of wages to casual labourers, *etc.* He is supposed to repay the loan from the sale proceeds of their crops raised.

2. *Medium-term Loans:* These loans are extended for a period varying from 15 months to 5 years. These loans are required by the farmer for bringing about some improvements on his farm business by way of purchasing implements, electric motors, milch cattle, sheep and goat, *etc.* The relatively longer repayment of these loans is due to their partial-liquidating nature.

3. *Long-term Loans:* These loans fall due for repayment over a long time ranging from 5 years to more than 20 years. These loans together with medium-term loans are called investment loans or term loans. These loans are meant for bringing about permanent improvements on the land, like levelling, reclamation and conservation, construction of farm buildings, purchase of tractors, raising orchards, *etc.* Since these activities require large capital, a longer period is required for the farmers to repay the loan from additional returns obtained from these investment activities.

## Based on Security

Since the loan transactions between lender and borrower are basically governed by the confidence in the borrower, the question of security may not arise at all in advancing loans. But, this assumption is confined to private lending to certain extent, and the institutional agencies do have their own procedural formalities in credit transactions. Hence, it is imperative to classify the loans under this category into sub-categories, *viz.,* secured and unsecured.

1. *Secured Loans:* Loans advanced against some security by the borrower are termed 0as secured loans. Various forms of securities are offered in obtaining the loans, which are as follows:

1) Personal security
2) Collateral security
3) Chattel loans
4) Mortgage
   a) Simple mortgage
   b) Equitable mortgage
5) Hypothecation
   a) Key loans
   b) Open loans

i. *Personal Security:* Borrower himself stands as the guarantor. It is advanced on the farmer's promissory note. Third party guarantee may or may not be insisted upon.

ii. *Collateral Security:* It is the property that is pledged to secure a loan. The movable properties of the individuals are offered as security. Examples are: LIC bonds, fixed deposit bonds, warehouse receipts, jewellery, machinery, livestock, *etc.* These are some of the properties accepted as collateral security by the institutional lending agencies.

iii. *Chattel Loans:* These are specific type of loans with particular category of lenders. Loans obtained from pawnbrokers by pledging movable properties such as jewellery, utensils made of various metals, *etc.,* are the examples.

iv. *Mortgage:* It is the transfer of an interest in a specific immovable property for the purpose of securing the money advanced or to be advanced or an existing or a future debt or for performance of an engagement, which may give rise to a pecuniary liability. The term interest in mortgage may relate to the right of enjoying gains/income from the mortgaged property or a right to sell the property (on default) or a right to appoint a receiver (on default) or any such right. Mortgage in reality is creation of a charge on an immovable property. It is not sale of a property. Hence there is no transfer of absolute ownership or possession. The person who is creating the charge of mortgage is called as mortgagor (borrower) and the person in whose favour it is created is known as the mortgagee (banker). As against collateral security, immovable properties are presented for security purpose. For example, land, farm buildings, *etc.* There are two types of mortgages, *viz.*, simple mortgage and equitable mortgage.

(a) *Simple Mortgage:* This is done by the banking institution, when the borrower's property is inherited from the ancestors. In this process the farmer-borrower has to register his property in the name of the banking institution as security for the loan obtained. This process entails registration charges to be borne by the borrower.

(b) *Equitable Mortgage*: This applies to self-acquired property. In this case there is no such registration because the ownership rights are clearly specified in the title deeds in the name of farmer-borrower. Hence, documents will be obtained from the borrower as security by the institutional agency.

v. *Hypothecation:* It is a charge, which is preferred when the property to be taken as security is movable. It creates merely a equitable or notional charge on the property with a right to a banker to take a possession of a property and sell the goods on default or a right to sue the owner to bring the property to sale and for realization of the amount due. The person who creates the charge of hypothecation is called as hypothecator (borrower) and the person in whose favour it is created is known as hypothecate (bank) and the property, which is hypothecated, is denoted as hypothecated property. This happens in the case of tractor loans, machinery loans, *etc.* Under such loans the borrower will not have any right to sell the equipment until the loan is cleared off. The borrower is allowed to use purchased machinery or equipment so as to enable him to pay the loan installment regularly. Hypothecated loans are further categorized into two types, *viz.*, key loans and open loans.

(a) *Key Loans:* The agricultural produce of the farmer-borrower will be kept under the control of the lending institutions and the loan is advanced to the farmer. As and when the loan is repaid the produce will be handed over to the farmers. Such facility prevents the farmer from resorting to distress sales.

(b) *Open Loans:* This is another name for hypothecated loans, in which the physical possession of the purchased machinery rests with the borrower, but the legal ownership rights remain with the lending institution till the loan is repaid.

**2. Unsecured Loans:** Based on confidence between the borrower and lender the loan transactions take place. There is no mention of any type of security here.

## Based on Liquidity

Under this type, the loans are classified into self-liquidating loans and partially liquidating loans.

*Self-liquidating Loans:* The income generated through these loans helps the farmer to repay the entire loan amount in the same season or year of obtaining the loan. The productivity increase of the loan is direct in this case. Example: Short-term loans or crop loans.

*Partially-liquidating Loans:* The income generated through these borrowings will help to pay part of the loan component only. In other words, these loans are cleared over a time period by the farmer-borrower. These loans require relatively longer time for realization of benefit. Example: Term loans.

## Based on Activity Orientation

There is no other basis except the activities for which the loans are advanced by the institutional agencies. It is more a general type of classification. For example, if a loan is borrowed for sericulture, it is called sericulture loan. Similarly, tractor loans, dairy loans, orchard loans, loans for land development, *etc.*, can be cited under this category.

## Based on Approach

Under this, we have three categories:
1. Individual approach
2. Area approach, and
3. DIR loans

1. *Individual Approach:* This is advancing loans by the lending agency to any potential borrower for the purpose he needs. Examples: Crop loans, dairy loans, *etc.*

2. *Area Approach:* Here loans are advanced by selecting the contiguous area by a bank branch. 'Service area approach' followed by the banks is an apt example.

3. *DIR Loans:* Loans are advanced to the weaker sections of the community at an interest rate of 4 per cent per annum.

## Based on Contact with the Farmers

Based on contact with the farmers, the loans can be categorized into: (1) Direct loans, and (2) Indirect loans.

1. *Direct Loans:* These are advanced directly to the farmers by the institutional agencies. *Examples*: ST loans and Term loans.

2. *Indirect Loans:* Loans given to organizations for activities, which contribute to the productivity of agricultural operations, are categorized as indirect loans. Examples: Financing PACS, FSS, LAMPS, financing fertilizer manufacturing companies, financing construction of warehouses, market yards *etc.*, financing through sugar factories, financing agro-service centers and custom service units for maintaining tractors, threshers, *etc.*, financing to traders and manufacturers for distribution of agricultural inputs, financing State Electricity Boards for energisation of wells, *etc.*

# CHAPTER 38

# Institutional Agencies in Agricultural Credit

## MEANING AND DEFINITIONS OF CO-OPERATION

Co-operation generally means working together to achieve a common goal. It is an association of persons. It also implies joint effort and coordinated action of all the members of the association. The concept of co-operation is not new to India. There were various instances, where people organized into an unit to achieve common goals. The examples of co-operative life can be cited are: Hindu Sanyunkt Paarivar Pratha (Hindu joint family systems), Panchayati Pratha (Panchayati system) *etc.* According to the Co-operative Planning Committee (1946) *"Co-operation is a form of organization in which persons voluntarily associate together on the basis of equality for the promotion of their economic interests"*. M.T. Herrick defined co-operation *"as the act of persons, voluntarily united, for utilizing reciprocally their own forces, resources or both under their mutual arrangement to their common profit or loss"*. Hubert Calvert defined that *"Co-operation is a form of organization, wherein persons voluntarily associate together as human beings on the basic of equality for the promotion of common economic interest of themselves"*. According to Sir Horace Plunkett *"Co-operation is self help made effective by organization"*. Thus co-operation signifies protection of weak, provision of equal justice to all in the society and promotes societal welfare. *Each for all and all for each is the motto of co-operation.*

## MAXIMS OF CO-OPERATION

Sir Horace Plunkett the founder of Irish co-operation sums up co-operation in three famous maxims *viz.*, better farming, better business and better living.

### Better Farming

It means helping the farmers to get a better production in the farm business through the adoption of requisite technology. The farmers' objective of higher production is realized only when the resources are available in adequate quantities. If the farmer does not possess capital to acquire necessary inputs, the same should be provided by an agency at the right time. A well developed cooperative institutional set up meets this particular requirement of the farmers.

## Better Business

Farmers should get a better deal in buying the inputs as well as disposing the products. The toiling efforts of the farmer bear fruit only when an efficient marketing system is accessible to him. Farmers as a group enjoy better bargaining power. In this context co-operatives should provide inputs at reasonable rates and arrange the disposal of produce at favourable prices.

## Better Living

This means the involvement of co-operative societies in supplying the consumer goods as per the requirements at reasonable prices. This facilitates the consumers to pay less than what they pay in open markets. The successful co-operatives are instrumental in preventing private traders from taking undue advantages through exorbitant prices. All the consumer items should be provided through the co-operatives. Thus, the co-operative effort helps the producers in fulfilling their objective in production and consumers to get their domestic requirements at reasonable prices.

# ESSENTIAL PRINCIPLES OF CO-OPERATION

The essential principles of co-operation are discussed below.

1. *Principle of Open and Voluntary Association:* The admission into a co-operative society is open to everybody independent of caste, creed and religion or any social and political affiliations. It does not allow any discrimination. Not only the membership is open to whomsoever with a common interest wants to join, but also voluntary as well. There is no element of compulsion exercised on any individual to join as a member. His decision to join as a member should be borne out of his unfettered regard for the co-operative ideology. Once an individual joins as a member, there is no such compulsion to continue as such. He has every freedom to withdraw from a society, if he intends to do so. The basic idea is that when one knows the advantage of associating with a co-operative institution, he should join with a good faith. This particular principle should be taken in right perspective, in the sense that open membership does not mean that any individual as a matter of right without common objective of the society can demand for an admission with least care for success of the society.

2. *Principle of Democratic Organization:* Co-operatives are organized and managed based on the principle of democracy. Each member is given the right to vote irrespective of the magnitude of his share capital. *"One man and one vote"* is the important principle of co-operation. The working of the elected board of members is governed by the acts, rules and byelaws guiding the matters of the co-operatives.

3. *Principle of Service:* The spirit of service evokes loyalty among the members of the society, which is contrary to the profit motive of business organizations. Co-operatives' main aim is to cater to the needs of the members.

4. *Principle of Self-help and Mutual Help:* The funds of the society, are contributed by the members in the form of share capital. Members are generally weak in their financial strength. Therefore, the society needs to maintain the funds efficiently. The society can borrow requisite finance from different sources at a lower cost of credit and provide the same to the most needed members for productive pur-

poses. This may not be possible with individuals. Therefore, the principle of self-help and mutual help can work for welfare of the members.

5. *Principle of Distribution of Profits and Surpluses:* Co-operatives are discouraged from making more and more profits like business enterprises. However, they are required to run with some minimum profits through efficient working. After keeping certain percentage of profit (25%) as reserve fund, the co-operatives can distribute the remaining profit to their members based on their share capital.

6. *Principle of Political and Religious Neutrality:* The vital strength for growth of the co-operative is the unity and non-interference of political parties. The members of co-operatives should continuously work for the growth of society with harmony, integration and unbiasedness towards any religion and politics. The political and religious differences should be kept away by the members.

7. *Principle of Education:* Co-operative movement was infact started with financial provisions from Rochdale pioneers. Education to members and training to office-bearers and executives in co-operatives are imperative for promoting awareness and efficiency in the operations of co-operatives.

8. *Principle of Thrift:* The co-operatives should aim at inculcating the habit of thrift among their members. Thrift and service are part and parcel of co-operation. There should be incentive for members who save their money with co-operatives. Thrift is very basis of self-help, but it must preceed credit. This means priority should be given to the members who save, in sanctioning of credit.

9. *Principle of Publicity:* If the members are illiterate, first they should be made literate. Without education, members cannot understand what is going on in the co-operative society. Their participation will also be poor. Hence the cooperatives should make all efforts to teach their members about the society and all the dealings of the society should be made public, and

10. *Principle of Honorary Service:* Honorary personnel should simply supervise and direct the operation of co-operatives to get the desired targets. But to have efficiency in the society, trained secretaries with salaries are now helping the co-operatives. If the society is started with poor members, then it is better to have honorary office-bearers, because such societies cannot afford to bear the burden of salaries to its office bearers.

# CO-OPERATIVE MOVEMENT IN INDIA

## I). Pre-independence Era

The co-operative movement in the country in the pre-independence era can be categorized into four phases *viz.,* initiation stage (1904-1911), modification stage (1912-1918), expansion stage (1919-1929) and restructuring stage (1930-1946).

### Initiation Stage (1904-1911)

In olden days the rural credit service was replete with the dominance of non-institutional agencies particularly moneylenders who were charging usurious rates of interest from the helpless peasants. The farmers were forced to even to sell their belongings to clear the debts. This situation triggered a sort of agitation and in some parts of the country, the farmers started revolting against the moneylenders. The revolts that were found in Poona and Ahmednagar areas, even attracted the attention of the Government which having understood the precarious situation of the farmers passed three acts *viz.*, Deccan Agriculture Relief Act (1879) and later Land Improvement Loan Act (1883) and thereafter Agriculturists Loan Act (1884).

During 1892, the Madras Government appointed Frederick Nicholson to study the village banks organized on co-operative lines in Germany. On his return he submitted a report and raised a slogan *'Find Raiffeissen'*.

Indian Famine Commission in 1901 also supported the idea of Frederick Nicholson for the formation of credit societies on Raiffeissen model. Another committee in 1901 headed by Sir Edward Law also favoured the credit societies to be started on Raiffeissen model. These recommendations resulted in the enactment of co-operative credit societies Act of 1904.

## Salient Features of 1904 Act

1. Rural-urban classification of societies was made. Rural Societies are those having four-fifths of farmers, while urban societies are those with four-fifths of their members representing non-agriculturists.
2. Registrar was supposed to organise and control the societies.
3. Loans could be given to members on personal or real (immovable) security, and,
4. One-man one vote was specified in the Act.

## *Modification Stage (1912-1918)*

The shortcomings of the Act of 1904 were rectified by enacting another co-operatives societies Act of 1912. The new act provided legal protection to all types of co-operatives including central financing agencies and supervising unions. The distinction between rural and urban societies was given a new focus. The liability was limited in the case of primary societies and unlimited for central societies. Since this act gave provision for the registration of all types of co-operative societies, it led to the emergence of rural co-operatives both on credit front and non-credit front, but this growth was uneven spatially. As a consequence of this observation in 1914, the Government appointed a committee under the chairmanship of Sir Edward Mac Lagan to probe into the performance of the societies. The report of Mac Lagan committee came out in 1915. The Mac Lagan committee's recommendations and Act of 1912, introduced co-operative planning process in India. The observations of Mac Lagan committee were

1. Illiteracy among the members
2. Misappropriation of funds,
3. Rampant nepotism,
4. Delays in sanction of loans and
5. Irregular repayment of loans.

These observations prompted Mac Lagan to offer the following suggestions for effective functioning of the societies.

1. All members should be made aware of the co-operative principles,
2. Dealings should be strictly confined to the members only,
3. Honesty should be the main criterion for one to take loan,
4. Applications should be carefully scrutinized before advancing loan and there should be careful follow up for effective utilization of loan,

5. Loans should not be given for speculative purposes,
6. Ultimate authority should be with all the members but not with the office bearers
7. Thrift should be encouraged so also building up of reserve fund,
8. One member-one vote should be strictly followed,
9. Capital should be raised as far as possible from the savings of the members only, and
10. Punctual repayment should be insisted.

## Expansion Stage (1919-29)

Under the Montogue–Chelmsford Act of 1919, co-operation became a provincial subject which gave further impetus to the movement. The economic prosperity during the period between 1920-29, contributed to further growth of the movement. The same period also witnessed the birth of co-operative land mortgage banks first in Punjab and subsequently land mortgage banks were registered in Madras (1925) and Bombay (1926). The Indian Central Banking Enquiry Committee (1931) also highlighted the glaring lacunae, particularly with reference to undue delays and inadequacy of credit. Meanwhile Madras Co-operative Societies Act of 1932 and the Madras Co-operative Land Mortgage Banks Act of 1934 came into force with the former aiming at the growth of co-operative movement, while the latter for developing the long term credit.

## Restructuring Stage (1930-46)

The economic depression in early thirties and abnormal fall in prices of agricultural commodities led to the collapse of the co-operative movement. Various enquiry committees viz., Vijayaraghava Charya Committee in Madras, Rehabilitation Enquiry Committees of Travancore and Mysore, Kale Committee in Gwalior, Mehta and Bhansali Committee in Bombay and Wace Committee in Punjab etc., were appointed for examining the possibilities of restructuring and reorganization of societies. The movement picked up momentum during the period of second world war, when there was a rise in the prices of agricultural commodities. This resulted, in the recovery of overdues of the societies and betterment of finacial condition of the co-operative institutions. Prof. D.R. Gadgil, heading the Agricultural Finance Sub-Committee appointed by the Government of India, recommended in 1944, the adoption of limited liability to the co-operatives, assessing credit-worthiness based on repayment capacity of the farmer, subsidizing the cost of administration of small co-operative societies, linking of credit with marketing, etc. The Co-operative Planning Committee in 1945, under the chairmanship of R.G. Saraiya attributed the limited progress of co-operatives to the laissez-faire policy of the State, the illiteracy of the people, etc.

## II) Post Independence Period

Planning commission which was set up in March, 1950 prepared its first five year plan in 1951. The main objectives with regard to co-operatives were as follows.

1. Involvement of co-operatives in rural development programmes
2. Development of a well organized credit system
3. Extending co-operatives to the fields of industry, housing, marketing, farming, etc. and
4. Training of higher personnel engaged in co-operatives.

The *All India Rural Credit Survey Committee* (AIRCSC) appointed by the Reserve Bank of India in 1951 under the chairmanship of *Sri A.D. Gorwala* brought out that the co-operative credit was unevenly distributed, inadequate and mostly lent to the asset-oriented large cultivators. The weakest link of all, in a chain, which was weak at almost all points is the primary credit societies. It satisfies none of the requisites of either good co-operation or sound credit[1]. The report further observed that "co-operation has failed in India but must succeed". The Committee recommended an integrated scheme as a remedy to the existing situation, the salient features of which were

1. State partnership in co-operative institutions at all levels
2. Coordination between co-operative credit, marketing and processing
3. Development of warehousing and
4. Training of co-operative personnel at all levels.

During 1952, Bharat Sevak Samaj and National Advisory Committee for public co-operation were formed to review, assess, and help in encouraging people to participate in national plan process. On the recommendation of All India Rural Credit Survey Committee (1954), National Co-operative Development and Warehousing Board (NCDWB) was established by the Government in 1956. Thus, in second five year plan (1956-61) the establishment of warehousing co-operatives was stressed. Apart from these, the second five year plan also initiated the setting up of co-operative processing and producer's co-operatives. The Committee on Co-operative Credit under the chairman-ship of Sri V.L. Mehta in 1959, observed that the co-operative aspect was important as that of viability. The co-operative society cannot afford to enlarge itself into an impersonal institution. The membership should not be too large or the area too extensive. No village included in a society should be at a distance of more than 3 or 4 miles from the headquarter village[2]. Later the committee on taccavi loans and co-operative credit under the chairmanship of Sri B.P. Patel in 1961-62 felt that the co-operatives should provide loans to the farmers for the agricultural operations and land improvements and taccavi loans should be confined to the farmers only under distressed conditions. Regarding the supervision of the societies at grass-root level, *i.e.*, primary co-operative credit societies, the Committee on Co-operative Administration, under the chairmanship of Sri V.L. Mehta opined that the District Co-operative Banks should assume this responsibility. During the third five year plan (1961-66), stress was given to revitalize dormant societies apart from increased emphasis on co-operative credit and co-operative farming. National Co-operative Development Corporation (NCDC) was also established in 1963. National Federation of Co-operative Sugar Factories was also established during this plan period. *The All India Rural Credit Review Committee* (AIRCRC) which was con-stituted in *July, 1966 under the chairmanship* of Sri B. Venkatappaiah, in its final report submitted in July 1969 recommended the setting up of Small Farmers Development Agency (SFDA), the creation of Rural Electrification Corporation (REC), the reorganization of primary societies into viable units, rehabilitation of weak central co-operative banks, active administrative and policy measures to check overdues, greater flexibility in the conversion of short term loans into medium term loans, simplification of application form and disbursal of part of the loan in kind, *etc*. During the same plan,

---

[1]Reserve bank of India: All India Rural credit survey Report. Vol. II, Bombay, 1954, p. 228.
[2]Report of the committee on cooperative credit. Government of India, 1960 pp. 201-202.

a new concept of transport co-operatives was initiated. In 1967, Vaikunth Mehta National Institute of Co-operative Management (VAMNICOM) was set up in Poona. Fourth plan gave importance to the rehabilation and reorganization of district central co-operative banks for smooth flow of co-operative credit. A new concept of fertilizer co-operatives was started and the Indian Farmers Fertilizer Co-operative Ltd. (IFFCO) was established at Kandla. Success of fertilizer co-operatives in fourth-five-year-plan opened vistas for introducing new fertilizer projects during fifth-five-year-plan (1974-79). National Bank for Agriculture and Rural Development (NABARD) was formed during sixth five year plan to strengthen the credit for agriculture and other economic activities. Strengthening of dairy co-operatives was also given importance. Seventh five year plan emphasized on special recovery camps, retail sale of fertilizers by co-operatives, strengthening National State Consumer Federation (NSCF) and introduction of single window system for credit. Eighth five year plan (1992-97) emphasized the replication of Anand pattern of co-operatives regarding milk co-operatives, strengthening processing co-operatives, etc.

Soon after Independence, the Government of India following the recommendations of All India Rural Credit Survey Committee (1951) felt that co-operatives were the only alternative to promote agricultural credit and development of rural areas. Accordingly, co-operatives received substantial help in the provision of credit from the Reserve Bank of India as a part of loan policy and large-scale assistance and encouragement from the Central and State Governments for their development. Many schemes of Government with components of subsidies and concessions to the weaker sections were routed through the co-operatives. As a result, the co-operative institutions registered remarkable progress in the post-independence period. Co-operative structure was delineated into two types, i.e., three-tier structure and two-tier structure. Both co-operative credit societies and non-credit co-operative societies now have three-tier structure and two-tier structure in all the States except Bihar, Jammu and Kashmir, Maharashtra and Uttar Pradesh, where the structure is unitary. The co-operative credit structure in the country is shown in Figure 38.1.

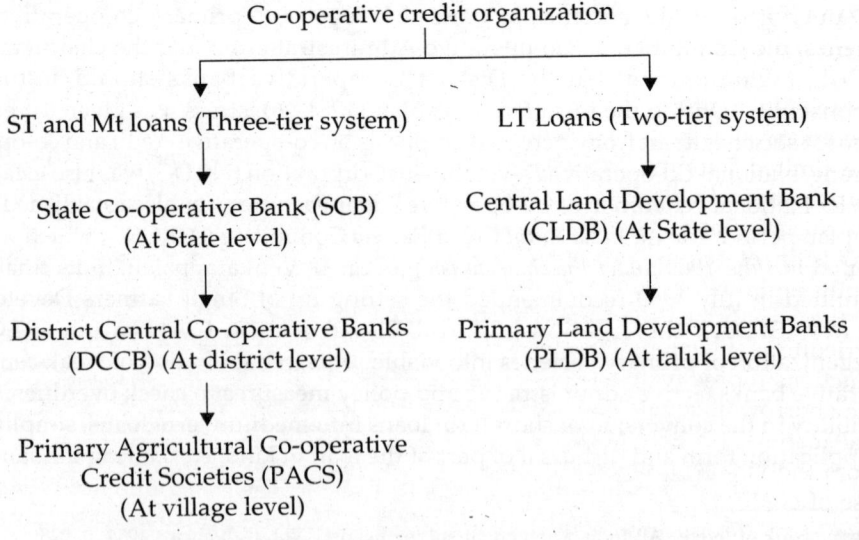

Figure 38.1   Sketch of the co-operative credit structure.

# STATE CO-OPERATIVE BANKS (SCBS)

These are the co-operative credit organizations present at the State level. DCCBs and PACS are the members of these banks. These institutions supervise the activities of the member banks and mobilize and deploy the financial resources among the member banks. They serve as a link between the RBI and the PACS. The specific functions of the State co-operative banks are: (1) they help the State Governments in formulating development plans with regard to co-operative institutions; (2) they co-ordinate the policies of the co-operatives with the Government; (3) they formulate and implement uniform credit policies regarding co-operative development in the State; (4) they act as banker's bank to DCCBs, supervise, control and guide the activities of DCCBs; (5) they grant subsidies to DCCBs for the smooth functioning of co-operatives; and (6) similar to any commercial bank, they also perform normal banking operations.

# DISTRICT CENTRAL CO-OPERATIVE BANKS (DCCBS)

These banks are, infact the link between State co-operative Banks and PACS. They are basically meant to meet the credit requirements of PACS. They also undertake bank business functions such as accepting deposits from public, collecting bills, cheque drafts, etc., and providing credit to the needy persons. The area of operation of the banks varies from the taluk to the district, but in most of the States their operations are confined to the taluk level. Membership is open to individuals and societies, working in its area of operation. Marketing societies, consumer societies, farming societies, urban banks and PACS are usually enrolled as members of the banks. The specific functions of the banks are: (1) they supervise and inspect the activities of PACS and help the credit societies run smoothly; (2) they maintain close and continuous contact and guide the primary societies and provide leadership to them; (3) they undertake non-credit activities like supply of seeds, fertilizers besides sugar, kerosene and other consumer goods; (4) they provide requisite funds to the societies under their control; and (5) they accept deposits from the member societies as well as from public.

# PRIMARY AGRICULTURAL CO-OPERATIVE CREDIT SOCIE-TIES (PACS)

Consequent to the enactment of Co-operative Societies Act of 1904, PACS came into operation following the guidelines of the Raiffeissen model. The co-operative principles like limited liability, limited area of operation, honorary management, voluntary participation of villagers, etc., were framed for the smooth functioning of the societies. The societies are at the village level and directly meant for the farmers regarding provision of requisite short-term and medium term loans. Supply of agricultural inputs and other essential commodities is also taken up by these societies. In addition to these activities, PACS are also helping in formulating and implementing the agricultural development plans. They are also undertaking advisory and welfare functions for the members. The PACS are associated with the following functions: (1) they borrow adequate and timely funds from DCCBs and help the members in financial matters; (2) they attract local savings in the form of share capital and deposits from the villagers, thereby inculcating the habit of thrift; (3) they supervise the end use of credit; (4) they distribute fertilizers, insecticides, etc., to the needy farmers; (5) they provide machinery on hire basis to the farmers; (6) they associate with the programmes and plans meant for the socio-economic development of the village; (7)

they also involve in the marketing of farm produce on behalf of the farmer-borrowers; (8) they provide storage facilities and marketing finance; and (9) they supply certain consumer goods like rice, wheat, sugar, kerosene, cloth, *etc.*, at fair prices.

## CENTRAL LAND DEVELOPMENT BANK (CLDB)

As an apex bank in the two-tier co-operative credit structure, it provides long-term finance to PLDBs and also to its affiliated branches working in the States. Branches of CLDBs, PLDBs and individual entrepreneurs are the members of the CLDB. National Bank for Agriculture and Rural Development (NABARD) and Life Insurance Corporation (LIC) subscribe for its debentures in large amounts. Infact, NABARD is the refinancing agency to the CLDBs. It acts as a link between NABARD and the Government in the long-term banking transactions. It supervises, inspects and guides the PLDBs in their banking operations. It floats debentures for raising the necessary funds. It inculcates the spirit and practice of thrift among the member banks by mobilizing savings and stimulating capital formation. The CLDB generally purveys loans to member banks for the redemption of old debts, development of land, purchase of agricultural machinery and equipment, development of minor irrigation, *etc.*

## PRIMARY LAND DEVELOPMENT BANKS (PLDBS)

The establishment of Land Mortgage Bank on co-operative lines dates back to the year 1920 in Punjab. Later during the period 1920-29, many Land Mortgage Banks were established in Punjab, Madras, Mysore, Assam and Bengal. There was not much growth in the Land Mortgage Banks till 1945, however, an alround progress of these banks was witnessed during the post-independence period, *i.e.,* 1948-53. During this period, only rich and affluent farmers derived benefit from the LMBs. Small and marginal farmers were hardly benefited. LMBs received massive support from institutional agencies like Reserve Bank of India, State Bank of India, Life Insurance Corporation and Agricultural Refinance Corporation. As a result the LMBs reoriented their lending policies towards small and marginal farmers and much emphasis was given to agricultural development. In the year 1974, LMBs were renamed as LDBs in A.P. PLDBs are generally organized to serve the farmers at taluk level. The specific functions of PLDBs are : (1) they provide long term finance to the needy farmers for the development of land, increasing agricultural production and productivity of land; (2) they provide loans for minor irrigation and for redemption of old debts and purchase of land; (3) they finance farmers in purchasing tractors, machinery and equipment; (4) the banks also provide finance for the construction of farm structure; and (5) they mobilize rural savings.

## SINGLE WINDOW SYSTEM

The farmers in Andhra Pradesh depended on PACS (in three-tier structure) for their short and medium-term credit requirements and on PLDBs (in two-tier system) for long-term credit needs till 1987, which meant that the farmers had to obtain their total loan requirements from two different types of co-operative institutions and furthermore, the performance of PLDBs was not satisfactory. Regarding marketing of the farm produce, the farmers faced hardships in getting the services of marketing co-operative societies under three-tier system. To help co-operatives serve in a more useful way, the Government of Andhra Pradesh thought that it was appropriate to bring some organizational changes in the working of co-operatives in the State. Accordingly, a Committee under the chairmanship of Sri Mohan Kanda was constituted to come out with meaningful and practicable alternatives in this regard. The Committee submitted its report in May 1985. It recommended for the establishment of 'single window system' and to this effect a bill was introduced and passed in Assembly in January 1987. The main idea of introducing this system is to supply all types of agricultural credit required by the farmers through PACS. The single window system is a three-tier structure in co-operative credit and two-tier structure in co-operative marketing. The organizational structure is sketched out in Figure 38.2.

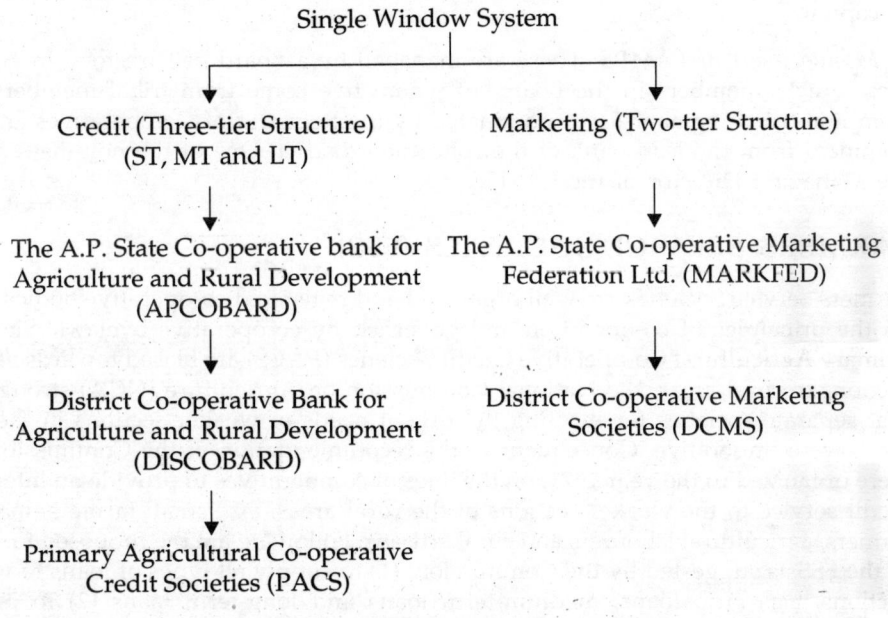

**Figure 38.2** Single window system.

Consequent to the introduction of single window system in A.P. 6,801 PACS were reduced to 4,257 and 218 Primary Co-operative Agricultural Development Banks (PCADBs) were merged with District Co-operative Central Banks and 125 Primary Co-operative Marketing Societies were amalgamated with DCMS during 1987.

The functions of PACS under single window system are: (1) to advance the ST, MT and LT loans; (2) to supply the needed farm inputs; (3) to distribute essential commodities; and (4) to arrange for marketing of the farm produce of the farmer members.

## LARGE-SIZED ADIVASI MULTIPURPOSE CO-OPERATIVE SOCIETIES (LAMPS)

Akin to the objectives of FSS, LAMPS were orgainsed for the first time in December 1971 based on the recommendations of the Bawa team appointed by the Government of India in tribal areas of the country. According to its framed objectives, these societies are expected to provide through single window, all types of credit including consumption credit. Intensification and modernization of agriculture with appropriate technical guidance and improving the marketing of agricultural and forest products in the tribal areas, are their other allied objectives.

1. *Membership and Area of Operation:* All tribes can become members of the society on voluntary basis like FSS. The area of operation of the society is larger than that of the PACS covering an area of block and some times covering an area, as big as a taluk.

2. *Sources of Capital:* Share capital of members and State Government, entrance fee, reserve fund, deposits collected from the members and non-members and loans taken from co-operative institutions and Government are the various sources of capital.

*Management of LAMPS:* These are managed by a Board of Directors. In general, there are 11 members in the Board, of whom five come from tribal members, two from non-tribal members, two nominated by the Registrar of Co-operatives and two nominees from the lead bank of the concerned district. One of the nominees acts as the Managing Director of the LAMPS.

## FARMERS SERVICE SOCIETIES (FSS)

Farmers Service Societies are well organized and registered co-operative bodies based on the principles of co-operation and governed by co-operative byelaws. Since the Primary Agricultural Co-operative Credit Societies (PACS) are biased towards affluent sections of rural areas, the National Commission on Agriculture (NCA) strongly felt that separate societies for meeting the urgent needs of weaker sections in the rural areas were imperative. Consequent to the recommendation of the Commission, FSS were organized in the year 1971, on the lines of co-operatives to provide an integrated credit service to the weaker sections of the rural areas, *viz.,* small farmers, marginal farmers, agricultural labourers and rural artisans. Following are the proposed functions of the FSS as suggested by the Commission: (1) to supply all types of loans to weaker sections, *viz.,* crop loans, medium-term loans and long-term loans (2) to provide adequate supplies of requisite inputs and technical guidance for the development of agriculture on timely and regular basis; (3) to encourage dairy, poultry, fisheries, farm forestry and other subsidiary occupations in rural areas; (4) to make arrangements for bringing about improvements in agricultural markets; and (5) to mobilize deposits and small savings from weaker sections through incentives.

*Area of Operation:* The societies have been launched in SFDA and MFAL districts. Each society has a jurisdiction of a block or a portion thereof. A district union of these societies is present at the district level to suggest ways and means of improving and organizing the societies for executing specific activities. The membership of the societies is open to those who are eligible to receive assistance under SFDA/MFAL programmes. Others may be associate members but they do not have any voting rights.

*Sponsorship:* The lead bank of the district generally sponsors the FSS in financial matters.

*Capital Structure:* The various financial sources for the society are: share capital, and funds from various sources and loans. Share capital includes share capital contributed by members, lead bank and State Government. In the case of large-sized FSS the limit of the share capital is Rs. one lakh but for a small society the same is Rs. 50,000. Funds from different sources include the funds contributed by commercial banks, co-operative societies, subsidies from SFDA and MFALs and commissions accrued to the societies through the supply of essential inputs and interest on advances.

*Management:* Depending upon the size of the society, the number of directors in the board varies from 9 to 13. One full-time Managing Director is deputed by the lead bank. Among the remaining directors, five will be elected from the members of the society, of which, three are from small and marginal categories and two from other farmers. The remaining directors are representatives of financing institutions, Block Development Office, Department of Agriculture and Co-operative societies.

# SOCIAL CONTROL AND NATIONALIZATION OF BANKS

The private sector banks being predominantly urban-oriented and controlled by a few large industrialists were not properly equipped to help the achievement of the basic socio-economic objectives. The credit needs of agriculture, small-scale industries and also weaker sections such as small traders and artisans continued to be ignored. Though agriculture is the main occupation of the country for nearly three-fourths of the population and contributed to almost half of the gross national product, the total bank credit advanced to this sector was barely one per cent as on June 1967. The bulk of the deposits that was contributed by the public was being advanced to the organized sector of the industry and trade. In the absence of financial institutional protection, the agricultural credit scene was dominated by the private money lenders, who were charging exorbitant rates of interest. All these compelled the imposition of social control over banks in 1968 with the main objectives of achieving a wider spread of bank credit, directing a larger volume of credit flow to priority sector and reducing the authority of the members of the Managing Committee, since they acted as the representatives of the industrialists, who strongly influenced the formulation of bank policies. Social control created the tempo of branch expansion as evident by the addition of 785 new offices during the first half of 1969. But this did not make any significant dent in reorienting lending activities of the banks in channelizing adequate credit requirements to the priority sector and weaker sections. In many banks, those who had been controlling their policies in the past continued to dictate terms in their functioning in one way or the other. Even the directions issued by the Government were ignored by many of the banks. This state of affairs created an impression in the Government that social control was not sufficient to make the commercial banking system a meaningful instrument of socio-economic development and hence, nationalization was considered as an alternative solution. Accordingly on 19th July 1969, the Government of India promulgated an ordinance called the Banking Companies (Acquisition and Transfer of Undertaking) Ordinance 1969, under which 14 commercial banks, which had deposits of not less than Rs. 50 crore each, were nationalized. The 14 banks together had 4,134 branches with deposits of Rs. 2,626 crore and advances of Rs. 1,813 crore. The banks were:

1. Central Bank of India
2. Bank of India
3. Punjab National Bank
4. Bank of Baroda
5. United Commercial Bank
6. Canara Bank
7. United Bank of India
8. Dena Bank
9. Union Bank of India
10. Allahabad Bank

11. Syndicate Bank
12. Indian Bank
13. Bank of Maharashtra, and
14. Indian Overseas Bank

The objectives of nationalization as set out by the then Prime Minister, Smt. Indira Gandhi were: (1) removal of control of banking business by a few industrialists; (2) elimination of the use of bank credit for speculative and unproductive purposes, particularly to the extent that it is encouraged by the association of a few leading groups; (3) expansion of credit to priority areas which were hitherto grossly neglected such as agriculture and small industry; (4) giving a professional bent to the bank management; (5) encouragement of new classes of entrepreneurs; and (6) provision of adequate training as well as reasonable terms of service to bank staff.

The broad aims of nationalization of banks as stated in the preamble to the Banking Companies Ordinance (Acquisition and Transfer of Undertakings Act 1970) were to control the peaks of the economy and better serve the needs of development of the economy in conformity with national policy and objectives. It would be the responsibility of the commercial banks to allow the free flow of credit to the hitherto neglected "priority sectors" of the economy like agriculture and allied activities, cottage industries, individual business, artisans, *etc.*

Between June 1969 and June 1975, 8,455 offices were opened by public sector banks, of which 4,337 offices were in urban areas. The number of public sector bank offices went up from 6,596 in June 1969 to 15,077 in June 1975. The average population served per bank office declined markedly to 32,000 in June 1975 from 65,000 in June 1969. Out of 8,455 offices opened between June 1969 and June 1975, the rural areas accounted for 4,092 offices.

Credit was no longer to be provided only to those who could furnish security in the form of property and projects. No time was lost by the nationalized banks to effect organizational changes. Spurred by the success of first spell of nationalization of 14 commercial banks, six more banks in the private sector, having deposits not less than Rs. 200 crore were nationalized on 15 April 1980. The six banks were:

(1) Punjab and Sind Bank
(2) Andhra Bank
(3) New Bank of India[*]
(4) Vijaya Bank
(5) Oriental Bank of Commerce, and
(6) Corporation Bank

Branch expansion received a further spurt between January 1979 and March 1984. Over this period the public sector banks opened 7,612 new offices, of which 5,384 offices were in unbanked centers. Priority sector advances of public sector banks at the end of June 1992 stood at Rs. 44,995 crore, of which the share of agricultural advances was 16.2 per cent. The percentage of advances towards priority sector was 39.3 of net bank credit. Priority sector advances of public sector banks at the end of March, 2012 stood at Rs. 11,24,148 crore, of which the share of agricultural advances was 42.64 per cent.

---

[*] In September 1993, the New Bank of India was merged with Punjab National Bank consequent to which the nationalized banks as on today stand at 19.

# STATE BANK OF INDIA (SBI)

Imperial Bank of India was formed in 1921 by amalgamating Bank of Bengal (founded in 1809), Bank of Bombay (founded in 1840) and Bank of Madras (founded in 1843) as there was no Central Bank in the country at that time. Imperial Bank besides, issuing notes performed other functions of a Central Bank until the emergence of RBI as a Central Bank in 1935. Subsequently, in 1955 Imperial Bank of India was nationalized with other State Banks (subsidiary banks) and was converted into SBI. It is the largest commercial bank in India. The particulars of SBI and its associate banks are presented in Table 38.1.

TABLE 38.1 SBI and its Associate Banks.

| Name of the Bank | Headquarters |
| --- | --- |
| State Bank of India | Mumbai |
| State Bank of Bikaner and Jaipur | Jaipur |
| State Bank of Hyderabad | Hyderabad |
| State Bank of Patiala | Patiala |
| State Bank of Travancore | Trivandrum |
| State Bank of Mysore | Bangalore |
| State Bank of Saurashtra | Bhavnagar |
| State Bank of Indore | Indore |

# SPECIALIZED BRANCHES

The commercial banks opened special bank offices to cater to the needs of weaker sections in the rural areas. In this context, Agricultural Development Branches (ADBs), Agricultural Banking Divisions (ABDs) of SBI and its associate banks, Grama Vikas Kendras (GVKs) of Bank of Baroda, Rural Service Centers (RSCs) of Dena bank, Farm Clinics (FCs) of Syndicate Bank, and Rural Credit and Development Division (RCDD) of Indian Overseas Bank need to be mentioned. These branches were established to overcome practical difficulties in lending activities, reducing the high cost of operations, strengthening follow-up measures, supervision, etc. The branch managers of these specialized branches have discretionary powers in respect of the sanctioning higher amounts of credit to farmer-borrowers, compared to the branch managers of commercial banks. Similar to RRBs, they have concessions in maintaining Statutory Liquidity Ratios (SLRs) and Cash Reserve Ratios (CRRs). They give half percent more interest on the deposits over commercial banks. Analogous to FSSs, these branches are equipped with technical staff to provide technical guidance with regard to farming.

The Bank of Baroda has set up GVKs on the lines of ADBs of SBI group. The GVKs not only provide advances to agriculture, but also extend credit facilities to business, retail sale, transport operations, etc. Multi-Service Agency (MSA) Cells were set up in those rural branches where full-fledged GVKs did not operate. At some places MSAs operate as independent rural branches.

Satellite offices are micro features of State Bank of Bikaner and Jaipur. They do not have independent existence and branches, but function as integrated parts of full-fledged branches. The officials of parent branches visit these centres through well-intimated schedules.

# MULTI-AGENCY APPROACH

This concept was originated after the first spell of nationalization of 14 commercial banks in 1969. Considering the magnitude of credit requirements to agriculture, no single financial agency can cater to the credit needs, so we have institutions like co-operatives, commercial banks, Regional Rural Banks and Farmers Service Societies purveying credit to agriculture. The All India Rural Credit Review Committee (1969) opined that though the co-operatives were the pioneering institutions in agricultural credit and increased their coverage (in terms of area, number of farmers and quantum of loans) since 1950s, in view of the growing demand for institutional credit in the field of agriculture, it is imperative to have other financial institutions in the form of commercial banks to act as supporting agencies. The report also mentioned that there should not be any room for conflicts among these institutions. Green revolution, which flashed Indian agriculture during the late sixties, also forced the need of providing adequate credit for the farmers to reap the benefits of technology. Though multi-agency approach was introduced for a good cause, some problems cropped up in practice. The most common problems were: (1) double financing, multiple financing, over-financing and under-financing: (2) unproductive use of credit; (3) failure of the financial institutions to formulate meaningful credit programmes; (4) problems in recovering loans; (5) greater diversity in the procedural formalities followed by the different financial institutions; (6) varying interest rates; and (7) unnecessary supervision charges.

Having noticed the problems that percolated with the introduction of multi-agency approach, RBI had appointed a Working Group under the Chairmanship of C.E. Kamath in August 1976 to probe into the problems that surfaced and suggest meaningful recommendations. The Committee submitted its report in 1978 with the following recommendations:

1.  *Geographical Demarcation of Functions:* The area of operation of each institutional agency needs to be demarcated.

2.  *Prime Role of Co-operatives:* Co-operatives were given the status of ideal institutions in view of their spread and accessibility. The other institutional agencies have to play a supplementary role to support and strengthen the co-operative sector.

3.  *Stipulated Role of Lending Institutions:* The members of co-operative societies should not be financed by the commercial banks and other institutional agencies. These banks can advance loans to the needy but without giving room to the element of competition.

4.  *Uniformity in Rates of Interest:* The Working Group expressed the view that there was a strong case for the adoption of uniform pattern of interest rates by all the institutional agencies to avoid a feeling of dissatisfaction and discrimination among different types of borrowers in homogenous areas.

5.  *Streamlining of Inspection Procedure:* It should aim at meaningful and productive follow-up of advances.

6.  *Agricultural Pass Book:* Suggestion was made for the introduction of agricultural passbook, so that the present practice of insisting upon production of non-encumbrance certificate can be dispensed with. The improvement will successfully prevent the farmers from going for double finance. The other alternative use of passbook is that it avoids periodical verification of records like 10-1 and adangal.

7.  *Inspection of the End Use of Loan:* Here emphasis should be laid on quality of inspection rather than periodicity of inspection. In its absence there is a possibility of the loans being unproductive and in this process, repayment capacity of the farmer gets affected, leading to the poor recovery performance of the bank.

## BRANCH EXPANSION SCHEME

Simple earmarking of funds to the agriculturists will not serve the purpose of solving the credit needs of farmers, unless the distribution pattern of financial institutions is also examined. This emphasized that it is not enough, if we increased the number of branches of any institutional agency without considering the distribution aspect. The two important institutional agencies in multi-agency approach, *viz.,* commercial banks and regional rural banks, which are only the supporting agencies to the co-operatives institutions, should operate only where the co-operatives are ineffective owing to financial or managerial incompetence.

At the end of March 1992, there were 60,528 branches of commercial banks (including RRBs) in the country, of which 35,275 (58.3 per cent) were in rural areas. As a result of rapid branch expansion witnessed since 1969, the average population per bank office, which was 65,000 at the time of nationalization, came down to 14,000 (as per the 1981 census) at the end of March 1992. Of the 60,528 branches at the end of March 1992, 8,563 (14 per cent) belonged to the State Bank of India, 3,752 (6 per cent) to associate banks of SBI, 29,681 (49 per cent) to nationalised banks, 4,007 (7 per cent) to other commercial banks and 14,525 (24 per cent) to RRBs. Of the total number of 60,528 braches at the end of March 1992, 58.3 per cent was in rural areas as against 22.3 per cent at the time of bank nationalization in July 1969. At the end of March 2001, there were 65,908 branches of commercial banks (including RRBs) in the country of which 32,533 were in rural areas. The average population per bank office was 15,000 (March, 2001). Advances to priority sector by the commercial banks stood at Rs. 1,26,309 crore at the end of March, 1999. Outstanding credit of scheduled commercial banks from agricultural sector was Rs. 45,638.27crore as on March, 2000.

## VILLAGE ADOPTION SCHEME

As per the guidelines of RBI, SBI had first conceived 'village adoption scheme' with an intention to do intensive financing in the rural areas, where there is lot of potential for agricultural development. The scheme aims at achieving full advantages from concentrated and coordinated efforts of banking activities. The scheme is not meant to serve the interests of few residents of a village but instead it has to cover all farmers without exception in a phased manner through extensive financing. This did not preclude other banks to finance the villages adopted by a particular bank. It is for the adopted bank to take special interest in overall agricultural development of the village.

## LEAD BANK SCHEME

The National Credit Council (NCC) had appointed a Study Group in 1969 under the Chairmanship of Prof. **D.R. Gadgil** to suggest an appropriate organizational framework for effective implementation of social objectives. The Study Group recommended the 'Area Approach' for the development of financial structure through intensive efforts.

In the same year, RBI appointed Sri F.K.F Nariman Committee to study this recommendation. The Committee endorsed the views of the Study Group on 'Area Approach' and recommended the formulation of 'Lead Bank Scheme'. The RBI accepted the recommendation and 'Lead Bank Scheme' came into force from 1969. As per the scheme, specific districts are allotted to each bank, which would take the lead role in identifying the potential areas for banking and banking development and in expanding credit facilities.

Lead bank is the leading bank among the commercial banks in a district thus usually has maximum number of branches in the district. It is a consortium leader for coordinating the efforts of all credit institutions in each of the allotted district for the expansion of branch banking facilities and for meeting the credit needs of the rural economy. The lead bank scheme has two important phases: phase I was concerned with the survey of the districts by the respective lead banks and phase II was marked by the preparation of the district credit plans.

*Phase I: Survey of the lead district.*

The RBI, while implementing the lead bank scheme, mentioned the following functions of lead bank.

1. Surveying the potential areas for development of banking in the district.
2. Identifying the business establishments which were hitherto dependent up on non-institutional agencies and financing them so as to enable them to raise their resources and surpluses from the advances made by the bank.
3. Examining the marketing facilities available for disposal of agricultural and industrial commodities and linking credit with marketing.
4. Invoking the cooperation of other banks operating in the district for opening the branches.
5. Estimating the credit gaps in various sectors of the district economy.
6. Developing contacts and maintaining liaison with the Government and other agencies.

*Phase II: Preparation of the district credit plans.*

While formulating the district credit plans, the RBI emphasised that the lead bank should:

1. Formulate bankable loaning schemes having labour intensive orientation so as to generate employment.
2. Disburse loans to increase the productivity of land and allied activities so as to reduce under-employment and thus increase income level.
3. Give maximum credit to weaker sections of the society mainly for productive purposes.

## Monitoring of Lead Bank Scheme

As part of the implementation of the lead bank scheme, some important fora like the State Level Bankers' Committee, the District Consultative Committee, District Level Review Meetings and Block Level Bankers' Committee were created. As indicated earlier, every district has a lead bank which is the leader for preparation, consolidation and implementation of credit plans of a district. The lead bank has a lead bank officer who does the coordination of all the activities/schemes proposed to be implemented and acts as a liaison between banks and the Government departments. Besides, a lead bank has a district coordinator in the non-district who may be one of the branch managers of the main branch where there is no controlling office. The progress made in implementation of the lead bank scheme particularly with reference to district credit plans, is evaluated by the standing committee of the District Consultative Committee.

Lead bank scheme expects the banker to become an important participant in the developmental process of the area of his operation in rural areas, while the "service area approach" put the banker in the position of implementing the development plans/schemes.

## REGIONAL RURAL BANKS (RRBS)

All India Rural Credit Review Committee (1969) pointed out that over large parts of the country the small farmers have been handicapped in having access to the co-operative credit both for current inputs and investment. Therefore, a need arose for the establishment of institutional agencies. This led to the first spell of nationalisation of the banks with greater expectations. Though they did add to the institutional structure, they simultaneously created some problems too.

The subject was examined by the Government of India and appointed a Working Group in 1975 under the Chairmanship of Sri M. Narasimham to go into the financial assistance rendered to the weaker sections in the rural areas. The Working Group came up with the recommendation of setting up of rural-based institutional agencies called 'Regional Rural Banks' after having identified shortcomings in the t inctioning of commercial banks and co-operatives. The Government of India accepted the recommendation and the RRBs came into existence through Regional Rural Banks Ordinance on 26 September, 1975 and initially five rural banks, sponsored by commercial banks were set up on a pilot basis in the country on 2 October, 1975. This ordinance of 1975 was immediately replaced by the Regional Rural Bank Act 1976. The major premises on which the establishment of RRBs recommended were that the existing credit institutions even after necessary restructuring and modifications cannot be expected to meet the varied and growing needs for rural credit. The purpose of introducing the RRBs is to have an institutional agency with clear understanding of rural problems and local familiarity which the co-operatives possessed and the business outlook which the commercial banks were known for, to serve the rural community with much more dedication. These banks were conceived as low-cost ones to uplift the lot of rural economy by financing agriculture, trade and industry in general and small and marginal farmers, agricultural labourers, artisans and small entrepreneurs, in particular. RRBs were expected to play a vital role in mobilizing the savings of the small and marginal farmers, artisans and agricultural labourers and initiate banking habit among the rural people. These institutions were also expected to plug the gap created in extending credit to rural areas by the largely urban-oriented commercial banks and the rural co-operatives, which have the close contact with rural areas, but fall short in terms of funds.

### Lists of RRBs

The list of RRBs first opened in the country is shown in Table 38.2.

**TABLE 38.2   List of RRBs.**

| Sl. No. | Sponsor Bank | RRB | Headquarters |
|---------|--------------|-----|--------------|
| 1. | Syndicate Bank | Prathma Bank | Moradabad (UP) |
| 2. | State Bank of India | Gorakhpur | Gorakhpur (UP) |
| 3. | United Bank of India | Gaur Grameena Bank | Malda (WB) |
| 4. | Punjab National Bank | Haryana Kshetriya Grameena Bank | Bhiwani (Haryana) |
| 5. | United Commercial Bank | Jaipur Nagalur Anchalik Grameena Bank | Jaipur (Rajasthan) |

The objectives assigned to RRBs were: (1) to develop rural economy; (2) to provide credit for agricultural and allied activities; (3) to encourage village industries, artisans, carpenters, craftsmen, *etc.*; (4) to reduce dependence of weaker sections on money-lenders; (5) to fill up the gap created by morotorium on borrowings from money-lenders; (6) to help the poor financially for their consumption needs; and (7) to make backward and tribal areas economically better off by opening new branches.

## Characteristic Features of Regional Rural Banks

1. *Sponsorship*: RRBs are sponsored by scheduled commercial banks.

2. *Jurisdiction*: The operational area to be covered by each RRB varies from one to two districts for efficient functioning. The number of branches in the area covered by each RRB may range from 50 to 60, keeping in view the operational and financial efficiency. Coming to the population to be served by each branch, it has been kept at 20,000 roughly. However, these are subject to changes as per the direction of the Central Bank of the country.

3. *Management*: The management of the bank is in the hands of a Board of directors numbering eight, headed by a Chairman, who is an officer of the sponsoring bank. Of the eight directors, three are nominees of the sponsoring bank, two from the State Government dealing with district developmental programmes and three from the Central Government. The Regional Rural Banks are sponsored by commercial banks, generally the lead bank in the district. In some areas State Co-operative Banks and private commercial banks are allowed to sponsor RRBs. The sponsoring bank provides assistance to RRBs for the first five years.

4. *Share Capital*: The authorized share capital of RRB has been fixed at Rs. one crore and issued capital at Rs. 25 lakh. This is contributed by the *Central Government, State Government and the sponsoring bank* in the ratio of 50:15:35 respectively.

5. *Functions*: The main function is, to grant loans and advances particularly to small and marginal farmers, agricultural labourers, co-operative societies, co-operative farming societies for agricultural purposes, artisans, small entrepreneurs, *etc.*, within the operational area of the RRB. They have been asked to extend other banking facilities like issue of drafts, collection of cheques, *etc.* They act as vital instruments in schemes, like IRDP, 20-point economic programme, *etc.*

6. *Rate of Interest*: The rate of interest on the loans charged is the same as collected by PACS. They have been allowed to offer 0.50 per cent interest more on deposits than that of commercial banks

7.  *Special Concessions to RRBs*: (1) Statutory Liquidity Ratio (SLR) to be maintained is fixed at 25 per cent as against 38 per cent by commercial banks; (2) Regional Rural Banks are allowed to pay half per cent higher interest to its depositors over the interest rate offered by commercial banks; (3) Cash-reserve requirements of 3 per cent to be maintained with RBI as against 10 per cent by commercial banks; (4) They are allowed to draw refinance from NABARD to the extent of 50 per cent or more depending upon the type of advance of the eligible outstanding loans at a concessional interest rate of 7 per cent per annum; and (5) The RRBs are registered as insured banks with Deposit Insurance and Credit Guarantee Corporation of India (DICGC). All deposits up to Rs. 30,000 in each bank are accordingly insured with the DICGC thus providing protection to the depositors. As on 30 June 1996, there were 196 RRBs covering 405 districts in 23 States of the country with a network of 14,547 branches. A survey conducted by RBI has shown that the deposits mobilized by these RRBs had moved on from Rs. 252.85 crore at the end of June 1981 to over Rs. 14,443.51 crore by the end of June, 1986. At the end of September 1989, 196 RRBs were functioning through 14,279 branches covering 370 districts in 23 States. Till 1991, while the number of RRBs remained unchanged at 196, the total number of districts covered by RRBs increased to 385. At the end of September 1991, RRBs had 14,531 branches. Among the States, U.P. had the highest number of RRB branches having 3,055. Aggregate deposits at the end of September, 1991 stood at Rs. 5,141 crore. Aggregate advances at the same period were Rs. 3,804 crore. The number of offices was 14,639 as on March, 2000. Aggregate deposits at the end of March, 2000 amounted to Rs. 32,034.68 crore. Amount outstanding at the end of March, 2000 was Rs. 13,126.10 crore. Aggregate deposits and advances of all RRBs as on 31-03-2014 stood at Rs. 2,39,511 crore and Rs. 2,18,110 crore respectively.

Based on the recommendation of Baldev Singh Working Group, the RRBs have simplified the procedural formalities in respect of agricultural finance. All RRBs use local languages in their dealings and financial operations. Their cost of operation is low as compared to commercial banks.

Some of the causes identified by the various committees that are plaguing the viability of RRBs are as follows.

1.  Since the RRBs are mainly expected to meet the credit requirements of weaker sections in the rural areas, the banks had to cater to the needs of large number of small accountees, which increased the cost of services.
2.  The small amounts advanced to large number of borrowers reduced the productive capacity of RRBs.
3.  RRBs are located in unbanked, interior rural villages. In these areas the business potential remains on a lower profile. Coupled to this limitation, the RRBs are directed to restrict their business to targeted groups as result of which the income generating capacity of the funds lent is normally gets reduced.
4.  Enhancement of pay structure of the staff on par with the staff of commercial banks and
5.  High rates of interest offered on the deposits against declining rates of interest on lendings over the years.

In the light of the above observations which are hindering the performance of RRBs, Kelkar Committee (1986) came out with the following suggestions for improving the viability of the RRBs.

1. Raising the issued capital of RRBs from Rs. 25 lakh to Rs. 100 lakh and the authorized capital from Rs. one crore to Rs. 5 crore.
2. Charging a lower rate of interest, *i.e.*, 7 per cent against 8.5 per cent by sponsor bank on the refinance to RRBs.
3. Investing the deposits of RRBs in the long dated Government securities, which are presently kept in non-interest earning current account.
4. Setting up of internal audit cells.
5. Adequate training for both officers and clerks.
6. Continuation of deputed staff from sponsor bank beyond five years and as long as felt necessary.

## Progress of RRBs

There were 196 RRBs in 23 States with 14,500 branches in the year 1996. They raised the aggregate deposits to the tune of Rs. 14,200 crore and advanced Rs. 7,500 crore by way of crop and term loans for agricultural activities, rural artisans, village and cottage industries, retail trade, self-employment and consumption loans during the year 1996. More than 95 per cent of the loans was provided to the weaker sections by RRBs. In deed, it is noteworthy performance of RRBs. RRBs are primarily intended to replace moneylenders in rural areas to supplement the cooperative credit system. These are also implementing differential interest rates schemes for weaker sections and physically handicapped persons for their gainful employment. An amount of Rs. 2,500 was given per borrower. Of late RRBs are financially becoming weak like cooperatives. To strengthen them, the RBI relaxed some of the conditions related to interest rates policy and adopting suitable measures for making RRBs viable and grow over time. RRBs are following instructions given by RBI and Government of India regarding procedures, loan policies, interest rates polices, *etc.* Due to these changing policies, RRBs are growing in number and strengthening their activities and provide for all the activities of development of agriculture, trade, commerce, industries, *etc.*, in the rural areas. In addition to the weaker sections, other sections of rural people are now being financed through RRBs. Bangladesh has made significant headway in the performance of RRBs. World Bank recently appreciated its progress in RRBs. There were 64 RRBs and 17,856 branches in the country in the year 2013.

Narasimham Committee was appointed to evaluate the progress of RRBs. According to the committee, there were three basic problems to the growth of RRBs. On account of many restrictions imposed on the businesses of RRBs, they did have low earning capacity as compared to that of commercial banks. Wages and pay scales are increasing steadily and scales are approximately equal to that of commercial banks. This means the important rationale for setting up of RRBs is lost. The sponsoring banks of RRBs were having their own rural branches in the area of operation of RRBs. This gave rise to certain anomalies and unavoidable expenditure on controls and administration.

According to the Narasimham Committee's report the major problem confronting RRBs is activating the basic objectives of RRBs along with their viability. In this direction, the Committee indicated two factors. (1) There should be enough competition in rural areas for provision of business credit and (2) The credit gap between rural India and urban India should be bridged. For this, the establishment of viable banking structure is the need of the hour. Keeping these factors in view, the Committee has recommended that commercial banks should separate the operation of their rural branches through the formation of one or more subsidiaries. Each rural subsidiary should have a compact area of operation, recruitment and deployment of manpower needed for the banking activity. There should be effective improvement in the control, supervision and information of the bank. Much emphasis should be given to the business credit operations. Banks should be given freedom to extend its rural branches. Those rural subsidiaries should be treated on par with RRBs with regard to cash reserve ratio (CRR), statutory liquidity ratio (SLR) and refinance facilities from NABARD. Adequate measures should be taken to reduce the cost of administration. This committee leaves the option to RRBs and their sponsoring banks, as to whether the former should retain their identity or they should be merged with sponsoring bank. Such merging should be based on commercial considerations. In case RRBs retain their identify, they should engage in all types of business activities in rural area. They should finance not only the target groups but also other groups, who are involved in rural economic development.

## Khusro Committee on RRBs

Agricultural Review Committee under the chairmanship of Dr. A.M. Khusro observed that the problems of RRBs are endemic and their non-viability is due to their structure. Most RRBs are incurring huge losses and eroding part of their deposits. Besides, RRBs were not serving interests of large group in a manner expected of them. Hence Khurso committee suggested that they should be merged with the sponsoring bank.

## Bhandari Committee

This committee was appointed by Reserve Bank of India under the chairmanship of M.C. Bhandari to suggest suitable measures for restructuring the RRBs. All the recommendations except few of them are in force. Issued share capital of RRBs was raised from Rs. 75 lakh to Rs. 1 crore. Based on the recommendations, NABARD has devised and implemented a package of short-term measures for restructuring the RRBs.

1. RRBs are allowed to increase their non-target group financing from 40 per cent to 60 per cent.
2. They are permitted to relocate some of their loss-incurring branches and make them viable.
3. They are given freedom to open extension counters in the high demand areas for credit.
4. They are allowed to upgrade and deepen wide range of activities to cover non-credit activities.
5. As many as 49 RRBs in the country were selected for restructuring.

## Agricultural Review Committee–Recommendations for Rural Credit

RBI appointed Agricultural Review Committee in 1989 under the chairmanship of A.M. Khusro to make in depth review of major problems and issues confronting rural credit in India. The major observations and recommendations of this committee are as follows.

1. The Committee recommended that there should be two categories of lending rates for agriculture, for small and marginal farmers, interest rate concession should be 1.5 per cent on the loans and on their deposits 1.5 per cent higher interest should

be given. Higher rate of interest should be charged for other farmers with ceiling interest rate of 15.5 per cent.

2. National Co-operative Bank of India should be created for acting as balancing centre for co-operative credit at the apex level in the country.

3. The Committee also recommended for establishment of Agricultural and Rural Infrastructure Development Corporation in the eastern regions of the country comprising the States of Bihar, Orissa, West Bengal and North Eastern States. This corporation evolves a strategy for increasing agricultural lending through building up of necessary forward and backward linkages and formulating location-specific projects/schemes for accelerating transformation of agriculture.

4. The present crop insurance scheme is deficient in its operation and its work should be done by statutory crop insurance corporation. It also recommended for setting up of expert committee of insurance experts, agricultural economists and actuaries.

5. Common legal framework should be devised for recovery of dues. For this purpose State level tribunals for adjudication, i.e., determining the case legally should be established.

## DIFFERENTIAL RATE OF INTEREST SCHEME (DIR SCHEME)

The Ministry of Finance instructed all the public sector banks to introduce DIR scheme on the recommendations of RBI Committee under the Chairmanship of Dr. B.K. Hazare.

It is implemented since 1975 by all the commercial banks under the public sector. However, private banks too volunteered to participate in the scheme from 1977 onwards. Under this scheme loans are being extended to the weaker sections of the society, who do not possess tangible security, at a concessional rate of 4 per cent interest per annum. The scheme was originally applicable at selected branches located in 265 backward districts of the country. Later in 1971, the scheme was extended to cover all parts of the country. DIR Scheme is also expected to cover all backward districts including SFDA and MFAL areas. Eligible members for DIR scheme are marginal farmers and agricultural labourers, scheduled castes and scheduled tribes engaged in agriculture, people having rural industries and cottage industries, persons engaged in small business and service units such as tailors, road-side hoteliers, rickshaw pullers, cobblers, basket-makers, carpenters, physically handicapped persons, orphans and indigent students having higher education. The DIR loans are covered by Small Loans Guarantee Scheme of the Deposit Insurance and Credit Guarantee Corporation of India. Since 1981, the banks were allowed to route their advances under DIR scheme through RRBs in their area of operation on refinance basis which was taken into account by the sponsoring bank towards their lendings under the scheme. The amounts of loan advanced were up to Rs. 3,000 and Rs. 2,000 in urban or semi-urban and rural areas respectively. However, the restrictions on the loan amounts are relaxed for persons belonging to scheduled castes and scheduled tribes. The commercial banks are required to advance one half to one per cent of their aggregate lendings towards this scheme, and 40 per cent of the total amount available under the scheme should be made available to SC and ST borrowers. Family income of the borrowers from all the sources should not exceed Rs. 7,200 per annum in urban and semi-urban areas and Rs. 6,400 in rural areas. The size of borrower's holding should not exceed one acre in the irrigated land and 2.5 acres in the case of dry land. According to social justice the interests of the weaker sections should be safeguarded.

DIR scheme is one such measure. This is based on the principle of negative taxation, which means low-income groups must be subsidized in their borrowings.

At the end of June 1990, the outstanding advance by public sector banks under DIR scheme amounted to Rs. 708 crore in 42.87 lakh borrower accounts and at the end of March 1992, outstanding advance of public sector banks amounted to Rs. 727 crore. The Government of India had set up in April 1983, a Task Force to examine various provisions of the DIR scheme and make modifications, if any, considered necessary. After considering the recommendations of the Task Force, it has been decided that the DIR scheme, IRDP and Self Employment Programme for the Urban Poor (SEPUP) would be mutually exclusive. In other words, if a person is assisted under IRDP or SEPUP, he will not be eligible for the benefit under the DIR scheme. Furthermore, the benefit of DIR scheme will be available only to those borrowers within the prescribed eligibility criterion who are not assisted under any of the subsidy-linked schemes of Central/State Governments and State-owned corporations. Besides, the ceiling of family income of the borrowers under DIR scheme is revised from Rs. 2,000 per annum in rural areas and Rs. 3,000 per annum in urban or semi-urban areas to Rs. 6,400 and Rs. 7,200 respectively. Family income under DIR scheme was raised to Rs. 18,000 in rural areas and Rs.24,000 in urban and semi-urban areas from 2009. At the end of 2012, the outstanding advances by public sector banks under DIR scheme amounted to RS. 8,020 crore.

## FARM GRADUATE SCHEME

In order to enlarge the extent and scope of its lending activities and enable the agricultural graduate who have necessary technical expertise but lack financial sources to stand on their own feet, Farm Graduate Scheme was introduced by SBI.

*Eligibility Criterion*: Any one who has (1) proven integrity; (2) degree in Agriculture/Veterinary Science/Dairy Science/Agricultural Engineering; (3) the land either in his own name or jointly with others; (4) worthwhile project of agriculture/ and allied activities; and (5) requisite technical ability to run the farm; (6) an enterprising spirit; determination and urge to put knowledge and ideas to profitable use and (7) proper orientation and aptitude for rural life.

In exceptional circumstances, advances are also permitted to agricultural diploma holders and to those who hold a degree in a subject other than agriculture, but have at least one year training in agriculture.

The scheme is applicable to those who do not own land but want to join those who own land in partnership, subject to it being ensured that only *bona fide* ventures are financed under this scheme.

The scheme is intended to be made operative only at selected centres in areas which are most responsive and where extension of bank's activities results in substantial benefit.

1. *Activities Covered*: Production of foodgrains and commercial crops including hybrid and high-yielding varieties of seeds and special farming activities such as poultry, dairy, piggery, fisheries, horticulture, *etc.*, are being covered.

2. *Nature of Assistance*: Assistance under the scheme is intended to cover the purposes like development of land, provision of irrigation facilities, construction of farmhouses, purchase of tractors and other agricultural implements, *etc.* In exceptional circumstances, the loan is granted for the purchase of land, depending up on the merits of the case. Advances, for the purchase of land, if granted, should

be restricted to Rs. 25,000. Working capital loans are also advanced for the purchase of seeds, fertilizers and pesticides.

3. *Quantum of Assistance*: The financial assistance under the scheme for the projects considered is restricted to Rs. 2 lakh. The actual quantum of assistance depends up on the requirements of each project, the governing factors being the total cost of the project and resources, if any, already available with the borrower. The branches are permitted to provide credit for reasonable consumption expenses during the gestation period. The requirements may be calculated at Rs. 250 per month and upper limit for such loan could be fixed at Rs. 3,000 per year subject to an overall ceiling of Rs. 2 lakh.

4. *Margin*: The applicant is required to offer as much as possible from his own resources, and in case he had no resources of his own, no margin is insisted up on.

5. *Technical Scrutiny*: Although the scheme is intended to assist technically qualified persons, as a safeguard, initial scrutiny and certification should be arranged through an agency of the Government or bank's own technical staff, with a view to ensuring the technical feasibility and economic viability of the scheme.

6. *Disbursement*: The advances are disbursed in such a manner as to avoid any misuse of funds. As far as possible, direct payment to the suppliers of inputs is arranged.

7. *Security*: Mortgage of land with structures and hypothecation of tractor, machinery, and implements are followed.

8. *Repayment Period*: The repayment period is fixed at a maximum of 10 years.

# DEPOSIT INSURANCE AND CREDIT GUARANTEE CORPORATION OF INDIA (DICGC)

The failure of two scheduled banks, *viz.,* Palai Central Bank Ltd. (Kerala) and Laxmi Bank Ltd. (Maharashtra) in 1960 gave a rude shock to the stability of the banking system in the country. This forced RBI to frame legislative measures so as to arrest bank mortality and create confidence in depositors. In 1961, RBI formulated proposals for the establishment of Deposit Insurance Corporation (DIC) on the model of the Federal Deposit Insurance Corporation in USA. The Deposit Insurance Corporation bill was passed and the Corporation came into existence on January 1, 1962. In order to provide safety to the banking system from risks involved in lending to priority sectors, the Government of India established Credit Guarantee Corporation of India Limited (CGCI) in 1971. The CGCI is associated with Credit Guarantee Organisation (CGO), set up in 1960 to provide guarantee in respect of lending to small-scale industries. Subsequently, in 1978 CGCI and CGO were merged with Deposit Insurance Corporation of India (DIC) and new institution by the name of Deposit Insurance and Credit Guarantee Corporation of India (DICGC) was established.

## Role of the Corporation

1. The Corporation gives protection to the depositors particularly the small depositors, form the risk of loss of their savings in the event of a bank's failure; such protection increases the confidence of the depositors in the individual banks and reduces the occurrence of panicky withdrawals of deposits;
2. The Corporation contributes to the stability and orderly growth of the individual banks as well as collectively of the banking system; and

3. It plays an active role in developing the banking habits of the people and ensures a larger mobilization of their savings.

DICGC is a wholly owned subsidiary of RBI with a paid up share capital of Rs. 50 crore, contributed entirely by RBI. The Corporation maintains three funds *viz.,* general fund, deposit insurance fund and credit guarantee fund. The share capital of the Corporation is held in the general fund and the administrative expenses are met from the interest from this fund. The deposit insurance fund is built up by the premia received from the insured banks and augmented by the interest earned on the fund investment. The deposit insurance fund is utilized exclusively for meeting the deposit insurance claims. The Corporation has formulated schemes *viz.,* (1) Small loan guarentee scheme 1971, (2) Guarantee scheme for SSI and (3) Small loans (co-operative credit societies) guarantee scheme.

The credit guarantee fund is used for meeting the credit guarantee claims, among the guarantee schemes introduced by the Corporation. The small loan guarantee scheme, 1971, covers among others, the credit facilities granted by commercial banks and RRBs to farmers. The scheme covers advances granted to borrowers, undertaking various types of agricultural and allied activities such as raising of crops, poultry farming, pisciculture dairy farming and animal husbandry. The Corporation does not guarantee losses to the full extent of defaults. The guarantee cover has been varying from time to time. Currently the guarantee is provided at 60 per cent of the amount in default.

In 1982, the Small Loans (co-operative credit societies) Guarantee Scheme was formulated with a view to providing guarantee cover for advances granted by co-operative credit institutions at the primary level for agriculture and allied activities. However, this scheme had not evoked sufficient interest and has failed to take off in the case of co-operatives.

A credit institution has to comply with the following three conditions for invoking the guarantee cover.

(a) The advance under the guarantee has not been repaid within one month from the date on which the notice of demand for repayment of the entire dues has been served to the borrower.

(b) The advance is treated by the credit institutions as bad or doubtful of recovery and

(c) It is provided or accounted for as such in the books of the credit institution.

The Corporation's guarantee is against bad advances and not against mere defaults. An advance is considered as bad if it is doubtful of recovery after credit agencies have exhausted all possible avenues for recovery including legal actions for acquisition of the assets created out of the loan. Since it is impossible for the Corporations to verify whether sources open to the credit agency for recovery have been exhausted, the corporation has further laid down that the period of three years should have been elapsed between the due date for loan/installment of loans and the invocation of guarantee.

From January 4, 1997 several banks have withdrawn from the coverage of DICGC for the purpose of credit guarantee. However, insurance coverage continues.

## SERVICE AREA APPROACH

After the first spell of nationalization of banks in 1969, to reach the agriculturists and those involved in allied activities, village adoption scheme was introduced by many of the banks. But, there was no bar for entry by the other banks in an area operated by a particular bank in extending credit facilities. This liberal entry of several banks created innumerable problems in a planned approach. To emphasise the planned approach to the rural areas the 'Service Area Approach' was proposed in 1988, which is not an altogether a new concept from 'Village Adoption Scheme', but it has certainly some merits. It is aimed at improving the productivity of bank credit in the rural areas. Scattered lendings over wide areas diluted the quality of lending and also post-disbursement supervision was inadequate. Hence, increased attention is being paid by each rural branch in lending activities within a compact area.

At the instance of RBI, the chief executives of the nationalized banks evaluated the rural credit delivery system in 88 districts of 21 States in November 1987. The reports submitted by the executives were discussed in a seminar held on 9th and 10th of January 1988, which was addressed by the then Union Minister of Finance. One of the recommendations that emerged out of this was to assign a specific service area to each bank branch. Thereafter, a committee was set up under the chairmanship of Dr. P.D. Ojha to examine the operational aspects in implementation of 'Service Area Approach'. After getting convinced about the recommendation, the RBI implemented the 'Service Area Approach' for bank branches in 1989.

*Need for Service Area Approach*: This is based on the following aspects: (1) to ensure planned development of villages by qualitative lending, *i.e.*, to make credit much more productive, (2) to make branch managers responsible for developing villages under their jurisdiction, and (3) emphasizing the grass-root level planning involving branch managers, and block/village level officials of the Government.

The following stages are involved in the 'Service Area Approach' (1) identification and allocation of service areas for each branch of commercial banks in rural and urban areas, (2) a survey to identify the scope of lending, (3) preparation of branch credit plans, (4) coordination between different agencies for development, and (5) a continued system of monitoring the performance under credit planning, branch wise or bank wise.

*Objectives of Service Area Approach*: The objectives are: (1) to make the area of operation of the bank compact and accessible; (2) to improve the quality of lending in rural areas; (3) to envisage systematized credit planning and supervision; and (4) to remove regional imbalances in credit lending.

The special feature of the 'Service Area Approach' is that the branch manager is made responsible for the preparation of credit plans of the villages attached to his bank. The distinguishing features of 'Service Area Approach' from that of 'Village Adoption Scheme' are presented in Table 38.3.

## CROP LOAN SYSTEM

Though All India Rural Credit Survey Committee 1954 and the V.L. Mehra Committee on Co-operative Credit (1960) recommended the adoption of crop loan system in all the States, for one reason or the other, it was not implemented in several States of the country immediately. After a lapse of five years this was introduced in the country during 1965 and in Andhra Pradesh from *kharif* 1966. The scheme has been implemented with the twin objectives of treating crop as security instead of landed property and fixing the scale of finance depending upon the actual farm expenditure.

**TABLE 38.3** Distinguishing Features between 'Village Adoption Scheme' and 'Service Area Approach'.

| Sl. No. | Particulars | Village Adoption Scheme | Service Area Approach |
|---|---|---|---|
| 1. | Identification of villages. | According to the will of the branch manager based on potentiality of business the villages were identified. | According to the criteria such as proximity, contiguity, dominant share, IRDP allocation, etc., the villages are identified. |
| 2. | Operation | Operation was not strictly within the adopted villages. | Operation is strictly within the assigned villages only. |
| 3. | Basis for development | Detailed surveys were not conducted for the implementation of all relevant schemes. | It entails a detailed village survey for deciding the potentiality for various branch schemes by the manager. |
| 4. | Coverage of families | There was no firm commitment about provision of credit to all the needy families. | All the families needing credit facilities would be covered in a phased manner. |
| 5. | Accountability | The branch manager was not accountable for full development of the villages | The branch manager is fully accountable for providing need-based credit to the villages. |
| 6. | Integration | There was no integration between, credit and non-credit activities. | Perfect integration is expected between credit and non-credit activities. |
| 7. | Involvement of agencies | There was no involvement/ commitment of other agencies. | There is involvement of other agencies. |
| 8. | Forum for coordination | There was no forum for coordination. | The forum for coordination is Block Level Bankers' Committee (BLBC). |

## Salient Features of the System

(1) The credit requirements of the farmers are to be based on the cost of cultivation (variable costs) of the crop.

(2) The eligibility to receive the loan is not gauged by the ownership of the land but by the factor that he is a *bona fide* farmer who needs credit for cultivation.

(3) The crop loan should be given based on the hypothecation of the crop.

(4) The disbursement and recovery of loan are to be made in accordance with the crop production schedule.

(5) The loan should be given both in cash and kind, and kind component is related to quantum of actual total inputs required for production of a particular crop.

(6) The quantum of loan should be fixed according to the variety and the season in which it is grown and type of crop, *i.e.*, irrigated or rainfed.

(7) Crop loan should be recovered with tie-up arrangement, *i.e.*, linking credit with marketing, and

(8) Crop loan is fixed by the district consultative committee which consists of experts from the fields of agriculture, animal husbandry, banking, *etc.*

# DISTRICT CREDIT PLANS (DCPs)

Scattered lendings over a large area do not create any desired impact of institutional finance on agricultural development. However, the impact of bank finance can be achieved if the lendings are advanced in a compact area. The district credit plans introduced through the 'Lead Bank Scheme' have become instruments for the Government to implement the developmental activities, identify the priority sectors and reduce the imbalances in the operational area. The 'District Credit Plans' are prepared keeping in view the credit requirements for agriculture and allied activities and small industries. Regulating the flow of funds as per the plan priorities is the main objective of 'District Credit Plans'. It is a blueprint of bankers containing the technically viable and economically feasible schemes, which will be implemented by the collective efforts of all institutional agencies. According to the original scheme, these plans used to be prepared by Lead Bank Officers covering a period of 3 to 5 years based on the development plans of various participating agencies and conveniently phased into 'Annual Action Plans' for smooth implementation. However, with the implementation of 'Service Area Approach', the bank branches are required to prepare their 'Branch Credit Plans' for each financial year, which are aggregated at district level into 'Annual Credit Plan' by the lead bank officer. With the introduction of 'Service Area Approach' the method for preparation of 'Annual Action Plans' at district level has undergone a sea change. According to the new method, the branch managers are required to prepare branch credit plans, which after finalization by 'Mandal Level Bankers' Committee/Taluk Level Bankers' Committee', are aggregated at joint mandal level/taluk level and district level. Till now the exercise used to be top down while hereafter it will be bottom-up.

# DISTRICT CONSULTATIVE COMMITTEE (DCC)

It is formed at the district level. Lead bank officer is the convener and the district collector/district magistrate the chairman. The participants are Chief Executive Officer of Zilla Parished, Project Director of DRDA, Executive Officers of SC/BC Corporation, General Manager of District Industries Centre, District level functionaries of agriculture/animal husbandry/sericulture, Regional Managers of the participating banks, representatives of NABARD, RBI and representation of co-operative institutions.

# FUNCTIONS OF DCC

1. Identifying potential areas for development and formulation of bankable schemes for inclusion in DCP/AAP;
2. Discussing and finalizing DCP/AAP/IRDP plans;
3. Reviewing the implementation of the plans and identifying unbanked areas for branch expansion;
4. Reviewing advances made by all institutional agencies to various sectors;
5. Reviewing the recovery performance and rendering necessary help for the recovery of overdues;
6. Reviewing the progress achieved in implementation of various Governmental sponsored programmes;
7. Reviewing the problems faced by the banks and Governmental agencies (officials) in implementing various Governmental programmes; and;

8. Considering the security arrangements and other infrastructural facilities for rural branches.

## NEW PRIVATE SECTOR BANKS

As part of the reforms process in the financial services sector, the Reserve Bank of India had issued guidelines for licensing of new banks in the private sector in January 1993. Since then, on the basis of these guidelines, the RBI granted licenses to 10 applicants. After eight years, the RBI has revised the guidelines in January 2001 and issued fresh guidelines for entry of new private sector bank in the Indian banking industry. The licenses granted include Centurion bank, HDFC bank, ICICI bank, IDBI bank, Indus Ind Bank, Bank of Punjab and UTI bank. One bank, viz. Times bank has been taken over by HDFC bank. The number of new private sector banks was 23 as on September, 2013 with offices standing at 9,905. The total deposits stood at Rs. 95,98,485 million, while advances were Rs. 81,14,551 million at the same point of time.

Some of the important conditions of the revised guidelines issued by the RBI for entry of new banks in the private sector are:

1. *Paid-up Capital:* The initial minimum paid-up capital for new bank would be Rs. 200 crore. The initial capital should be raised to Rs. 300 crore within three years of commencement of the business. The overall capital structure of the proposed bank would have to be approved by the RBI.

2. *Promoters' Contribution:* The promoters' contribution will be a minimum of 40 per cent of the paid-up capital of the bank at any point of time. The initial capital, other than the promoters', contribution, can be raised through public issue. In case the promoters' contribution to the initial capital is in excess of the minimum proportion of 40 per cent, their excess stake will have to be diluted after one year of the bank's operations. Otherwise specific approval of the RBI will have to be obtained. The promoters' contribution of 40 per cent of the initial capital will be locked in for five years from the date of licensing the bank. Similar conditions would apply while augmenting capital to Rs. 300 crore within three years of commencement of the business.

3. *Foreign Investment:* Non-resident Indian participation in the primary equity of a new bank will be to the maximum extent of 40 per cent. In the case of a foreign banking company or finance company (including multilateral institutions) acting as a technical collaborator or a co-promoter, equity participation will be restricted to 20 per cent within the above ceiling of 40 per cent. In the case of shortfall in foreign equity contributions by NRIs, designated multilateral institutions would be allowed to contribute foreign equity to the extent of the shortfall in NRI contribution to the equity. The necessary approvals of the Foreign Investment Promotion Board will also have to be obtained.

4. *Banks Promoted by Large Industrial Houses:* A large industrial house will not be allowed to promote a new bank. However, individual companies, directly or indirectly connected with large industrial houses, may be permitted to participate in the equity of a new private sector bank up to a maximum of 10 per cent of the equity but will not have controlling interest in the bank. The 10 per cent limit would apply to all inter-connected companies belonging to the concerned large industrial houses. Further, the proposed banks will maintain an arms length relationship with business entities in the promoter group and the individual company/companies investing up to 10 per cent of the equity as stipulated above.

It will not extend any credit facilities to the promoter and the company/companies investing up to 10 per cent of the equity. The relationship between the business entities in the promoter group and the proposed bank will be of a similar nature as between two independent and unconnected entities.

5. *Capital Adequacy Requirements:* The bank will be required to maintain a minimum capital adequacy ratio of 10 per cent on a continuous basis from the commencement of its operations.

6. *Other Requirements:*

1. The new bank will have to observe priority sector lending target of 40 per cent of net bank credit as applicable to other domestic banks.
2. The new bank will be required to open 25 per cent of its branches in rural and semi-urban areas to avoid over concentration of their branches in metropolitan areas and cities on the same lines as new private sector banks established under guidelines laid down by RBI in January 1993.
3. The new bank will not be allowed to set up a subsidiary or mutual fund for at least three years from the date of commencement of business, and
4. The headquarters of the proposed new bank could be in any location in India as decided by the promoters.

### Approach to Public Sector Banks

The decision of RBI is also to be viewed in the backdrop of the privatization policy of public sector enterprises (PSEs) being pursued by the Government at present. The Government is proposing to dilute its stake in public sector banks (PSBs) to 33 per cent of the equity. If the RBI is not allowing large industrial houses to enter commercial banking by granting new licenses, it will also prevent large industrial houses from taking a controlling stake in PSBs. The RBI is not allowing industrial houses to enter commercial banking on account of the moral hazards and risks associated with lending to companies within the same group. The fresh guidelines for new private banks clearly indicate that there will be very few new entrants. In addition to the banks granted licenses since 1993, we will have less than ten banks which will offer substantial competition to the PSBs and foreign banks. Foreign banks will be allowed by the RBI to expand their branch network only on a selective basis. As the PSBs control 80 per cent of the banking business in the country, these banks will continue to be the critical element to this industry.

## REPORT OF THE HIGH-LEVEL COMMITTEE ON AGRICULTURAL CREDIT THROUGH COMMERCIAL BANKS

A high-level committee on agricultural credit through commercial banks was constituted by RBI under the chairmanship of Shri R.V. Gupta, Ex-Deputy Governor of RBI with the following terms of reference.

A. To study the working of credit delivery system for agriculture through field level interaction with the farmers, borrowers of commercial banks and bank staff and analyze the constraints in the flow of such credit to the farm sector.
B. To make suggestions for the simplification and improvements in systems and procedures including reduction in paper work for delivery of farm credit.
C. To make proposals on other initiatives that may be taken up by commercial banks to facilitate credit delivery to the farmers. The committee in its report submitted

in April, 1998, examined in detail, the issues related to flow of credit to agricultural sector and made extensive recommendations, which are as follows:

1. *Simplification of Procedures Regarding Loan Agreements*: Indian Banks Association is being requested to work out simplified loan application forms, documents, *etc.*, for agricultural loans.

2. *Pre-sanction Appraisal*: Banks may ensure that the pre-sanction appraisal of the borrower focusses on income stream of the borrower, his credibility, his capability for taking up the activities proposed, integrity, *etc.*, and technical viability of the project.

3. *Delegation of Powers to Branch Managers*: Banks should delegate adequate powers to the branch managers so as to enable the disposal of 90 per cent of loan applications at the branch level. The powers at the branch level may have to be modified if the pattern of lending so warrants. Bank management's may therefore review the position branch-wise and take suitable action.

4. *Composite Cash Credit*: Banks may introduce annual cash credit limits to all agricultural borrowing families. This limit should cover all aspects relating directly and indirectly to production, including the need for seasonal as well as post-harvest/household requirements of the farmers after taking into account the total family income and the estimated repaying capacity.

5. *New Loan Product*: Farmers cultivating cash crops have a tendency, during cash rich periods to invest in gold, land, *etc.*, and as a result such farmers are vulnerable during times of price fluctuations and natural calamities. Banks may therefore consider designing a fully liquid savings module with an appropriate return and in-build the same in the loan product which will take care of farmers' interests in the event of adverse price fluctuations or natural calamities.

6. *Cash Disbursement of Loans*: With a view to providing farmers wider choice as also eliminating undesirable practice, banks may disburse all loans for agricultural purposes in cash which will facilitate dealer choice to borrowers and foster an environment of trust. However, bank may continue the practice of obtaining receipts from borrowers.

7. *No Dues Certificate*: In order to avoid delay in sanction and disbursement of loans, production of 'No dues certificate' from banks/institutions by borrowers as a compulsory requirement is dispensed with. Obtaining 'No dues certificate' is now left to the discretion of lending banker. However, the banks have to satisfy themselves regarding the status of the borrower before granting any loan.

8. *Recovery Climate*: Banks may evolve appropriate mechanism to monitor recovery performance in respect of agricultural loans particularly with regard to old sticky loans.

9. *Additional Collateral Security*: Banks may ensure that the value of security taken is in commensurate with the size of the loan and desist from asking additional collateral by way of guarantors where the land mortgaged, is considered adequate.

10. *Margin/Security Requirements*: Banks may use their discretion on matters relating to margin/security requirements for agricultural loans above Rs. 10,000. However, in respect of Government sponsored schemes existing instructions will continue.

11. *Statement of Credit Facilities with Details of Levies Charged*: In order to introduce an element of transparency, banks may furnish the borrowers with a clear statement of the credit facilities availed by them separately indicating various fees and charges levied.

12. *Visits of Internal Inspection Officials to a few Service Area Villages*: Banks may ensure that the officials of the internal inspection team visit a few service area villages covered by each branch and meet the farmers to discuss their problems during the course of their inspection so as to obtain a feedback on the quality of interaction, which branch officials have at the ground level.

13. *Chairman and Managing Director's Visit to Rural Branches*: Banks may ensure that CMDs of banks during their tours to States also pay surprise visits to their rural branches which will send appropriate signals to the field level staff. CMDs may consider paying at least one such visit every month.

14. *Rationalisation of Internal Returns*: IBA is being requested to undertake a detailed exercise to rationalise the number of internal returns to be submitted by bank branches to reduce avoidable paper work including elimination of adhoc returns for monitoring agricultural and other advances.

15. *Performance Appraisal System, Staff Incentives, Training Facilities, etc.*: Performance appraisal system in vogue in commercial banks gives substantial weightage to deposit mobilization. Instead banks may consider increase in the outreach measured by number of new clients and incremental increase in flow of credit and loan recoveries. Banks may with a view to motivating the staff also consider evolving package of incentives, which may include foreign training, posting people to centre of choice, meeting educational expenses of children, *etc.* In order to improve, the skills and knowledge of bank managers, the training systems at the Bank's training institutions may be strengthened by introducing programmes which emphasise on borrower appraisal.

16. *New Loan Product for Borrowers in Non-farm Sector*: Banks may design specific loan projects to the borrowers in non-farm sector to enable them to pursue a number of non-agricultural activities alongside their agricultural activities for supplementing their income.

17. *Model Set of Documents for Self-help Groups*: Banks may ensure operationalisation of set of documents prescribed by the working group on non-Governmental organizations and self-help groups relating to agreements among members, loan formats, application forms, *etc.*

18. *Dissemination of Information to Farmers Regarding Type and Kind of Inputs to be Used, etc.*: With a view to ensuring that farmers are given right advice regarding cultivation of corps, inputs to be used *etc.*, banks may work out a system. This system should ensure that the package of practices for the cultivation of crops specific to their areas is disseminated to the farmers on a regular basis.

19. *Link up of Rural Branches with Farmers' Clubs*: In order to facilitate technology transfer regarding agricultural operations, wherever possible, rural branches of banks may link up their activities with '*farmers clubs*' established with initiative of NABARD.

20. *Hi-tech Agricultural Branches*: Banks may quickly undertake a review of working of their hi-tech agricultural branches which are not working well due to inad-

equate expertise and/or lack of climate. Banks may also ensure that these hi-tech branches are used to disseminate information relating to agricultural operations specifically for providing technical information regarding the cultivation of high value crops.

These recommendations are being implemented by the commercial banks as per the directive of the Reserve Bank of India.

# HIGHER FINANCING AGENCIES

## World Bank

The International Bank for Reconstruction and Development (IBRD) is an international organization established in 1945 and is owned by its member countries. IBRD's main goal is reducing poverty by promoting sustainable economic development. It pursues this goal primarily by providing loans, guarantees and related technical assistance for projects and programmes in its developing member countries. IBRD's ability to intermediate funds from international capital markets for lending to its developing member countries is an important element in achieving its development goals. IBRD's objective is not to maximize profit, but to earn adequate net income to ensure its financial strength and to sustain its development activities.

The financial strength of IBRD is based on the support it receives from its shareholders and on its array of financial policies and practices. Shareholder support for IBRD is reflected in the capital backing it has received from its members and in the record of its borrowing members in meeting their debt service obligations to it. IBRD's principal assets are its loans to the member countries.

### Development Activities

IBRD offers loans, related to hedging products, and guarantees to its borrowing member countries to help meet their development needs. It also provides technical assistance and other services to support poverty reduction in these countries.

### Loans

From its establishment through June 30, 2000, IBRD had approved loans, net of cancellations, totalling $309,839 million to borrowers in 129 countries. Cumulative loan repayments at June 30, 2000, based on U.S. dollar equivalents at the time of receipt, were $141,265 million.

Generally, the appraisal of projects is carried out by IBRD's operational staff (engineers, financial analysts, economists and other sector and country specialists). With certain exceptions, each loan must be approved by IBRD's Executive Directors. Loan disbursements are subject to the fulfillment of conditions set out in the loan agreement. During the project implementation, IRBD staff with experience in the sector or the country involved periodically visit project sites to review progress, monitor compliance with IBRD policies and assist in resolving any problems that may arise. After completion, projects are evaluated by an independent unit and the findings reported directly to the Executive Directors to determine the extent to which the project's major objectives were met.

## Other Activities

### Consultation

In addition to its financial operations, IBRD provides technical assistance to its member countries, both in connection with, and independently of, loan operations. There is a growing demand from borrowers for strategic advice, knowledge transfer, and capacity building. Such assistance includes assigning qualified professionals to survey developmental opportunities in member countries, analyzing their fiscal, economic and developmental environment, assisting member countries in devising coordinated development programmes, appraising projects suitable for investment and assisting member countries in improving their asset and liability management techniques.

### Research and Training

To assist its developing member countries, IBRD–through the World Bank Institute–provides courses and other training activities related to economic policy development and administration for Governments and organizations that work closely with IBRD. The World Bank Institute also makes contributions for research and other developmental activities.

### Trust Fund Administration

IBRD, alone or jointly with IDA, administers on behalf of donors, funds restricted for specific uses. These funds are held in trust and are not included in the assets of IBRD.

### Investment Management

IBRD has leveraged its treasury management capacity and infrastructure to provide investment management services to an external institution for a fee. These funds are not included in the assets of IBRD.

### Purpose and Affiliated Organizations

The International Bank for Reconstruction and Development (IBRD) is an international organization, which commenced operations in 1946. The principal purpose of IBRD is to reduce poverty through promoting sustainable economic development in its member countries, primarily by providing loans and related technical assistance for specific projects and for programmes of economic reforms in developing member countries. The activities of IBRD are complemented by those of three affiliated organizations viz., the International Development Association (IDA), the International Finance Corporation (IFC), and the Multilateral Investment Guarantee Agency (MIGA). IBRD, IDA, IFC, and MIGA are collectively known as the World Bank Group. Each of these other organizations in the World Bank Group is legally and financially independent from IBRD, with separate assets and liabilities, and IBRD is not liable for their respective obligations. IDA's main goal is to reduce poverty through promoting economic development in the less developed areas of the world included in IDA's membership by providing financing on concessionary terms. IFC's purpose is to encourage the growth of productive private enterprises in its member countries through loans and equity investments in such enterprises without a member's guarantee. MIGA was established to encourage the flow of investments for productive purposes among member countries and, in particular, to developing member countries by providing guarantees against noncommercial risks for foreign investment in its developing member countries.

# RESERVE BANK OF INDIA (RBI)

The Reserve Bank of India was established in 1935 under the Reserve Bank of India Act, 1934. The bank was set up to regulate the issue of bank notes and keeping up resources with a view to securing monetary stability in the country and operate the currency and credit system to its advantage. When examined, the role of RBI in the sphere of agricultural credit, the creation of Agricultural Credit Department (ACD) comes to light. The primary functions of ACD were to coordinate the functions of RBI with regard to agricultural credit with other banks and State Co-operative Banks, to maintain expert staff to study all the questions of agricultural credit and be available for consultation by Central Government, State Governments, scheduled commercial banks and State Co-operative Banks and to provide legislations to check private money-lending and checking malpractices. All India Rural Credit Survey Committee (AIRCSC) in 1954 suggested several recommendations with regard to the activities of RBI in the sphere of rural credit. Subsequently two funds were established after amending the RBI Act, *viz.*, the National Agricultural Credit (Long Term Operations) Fund and the National Agricultural Credit (Stabilization) Fund. The National Agricultural Credit (Long-term operations) fund was set up in 1955 with an initial contribution of Rs. 10 crore and a sum of Rs. 5 crore was to be further added every year from the profits of RBI in the first five years. In May 1960, as per the recommendations of the committee on co-operative credit, the RBI was authorized by the Government of India to credit Rs. 15 crore to the fund annually as against Rs. 5 crore contemplated to be added earlier. This fund was meant to provide long-term loans to various State Governments with a view to enabling them to contribute to the share capital of different types of co-operative societies including Land Mortgage Banks. Loans and advances out of this fund are made to State Governments for periods not exceeding 20 years, for purchasing debentures of State Land Development Banks and for providing funds to the Agricultural Refinance and Development Corporation for periods not exceeding 20 years. Since the creation of this fund, medium term loans are provided by the Reserve Bank of India only out of this fund.

The second fund, *viz.*, National Agricultural Credit (stabilization) Fund was established in June 1956 with an initial contribution of Rs. 1 crore and subsequent annual contributions of Rs. 1 crore. The fund is utilized for the purpose of granting medium term loans to state co-operative banks, specially during the times of famine, drought or other calamities, when they are not able to repay their short-term loans to RBI. This fund is also used for converting short-term loans which are due to RBI from the central co-operative banks in areas affected by natural calamities. The State and central co-operative banks, and Primary Agricultural Co-operative Credit Societies in turn provide a similar facility to the farmer borrowers regarding short-term production loans taken for the crops affected by the natural calamities. This makes them eligible for further agricultural finance at the same time reducing their burden of repaying the loans.

The role of RBI in the sphere of rural credit can be seen under three aspects *viz.*, provision of finance, promotional activities, and regulatory functions.

## ı. Provision of Finance

Reserve Bank of India provides necessary finance needed by the agriculturists through the commercial banks, co-operatives and regional rural banks. It advances long-term loans to State Governments for their contribution to the share capital of the co-operative credit institutions, *i.e.*, apex and district banks. Refinance facility is extended to RRBs only, to an extent of 50 per cent of their outstanding advances.

## 2. Promotional Activities

The RBI's efforts on this front can be seen in the appointment of study teams in organizing and running the co-operative credit institutions in the country. It conducts a number of studies and surveys pertaining to rural credit aspects in the country. The All India Rural Credit Survey, the All India Rural Debt and Investment Surveys, *etc.*, can be cited as the most comprehensive ones. The Committees which need special mention are: The Committee on Co-operative Land Development Banks (1974), the Committee on Integration of Co-operative Credit Institutions (1976), and the Committee to study the Interest Rates Spreads in Agricultural Lending Sector. The Reserve Bank of India felt that co-operatives are the major force in the sphere of agricultural credit and accordingly the following policies were made for strengthening the co-operatives.

a) Reorganization of the State and Central Co-operative banks on the principle of one apex bank for each State and one Central Bank for each district.
b) Rehabilitation of those Central Co-operative banks, which are financially and administratively weak for reasons such as mounting overdues, inadequacy of internal resources, untrained staff, poor management, *etc.*
c) Strengthening of PACS to ensure their financial and operational viability and
d) Arranging suitable training programmes for the personnel of co-operative institutions.

## 3. Regulatory Functions of RBI

Apart from lending aspects, RBI is concerned with efficiency of channels through which credit is purveyed to rural sector. Banking Regulation Act 1966 of RBI enables it to exercise effective supervision over co-operative banks and commercial banks. The co-operative banks should get prior authorization from RBI for providing finance beyond a certain limit as per the Credit Authorization Scheme of 1976. Based on institutional demand for credit, credit limits are fixed. The Cash Liquidity Ratio (CLR) and Cash Reserve Ratio (CRR) are fixed by RBI for co-operatives, FSSs, RRBs and ADBs at lower level than those fixed for commercial banks. For these banks the bank rate is 3 per cent less than that of commercial banks. They are permitted by RBI to pay ½ per cent higher rate of interest on their deposits.

Historically, the role of RBI in rural credit has passed through the following five phases.

| | |
|---|---|
| *Phase – I* : 1935-1943: | The period of initiation and preparation. |
| *Phase – II*: 1944-1954: | Beginning of the active role in rural credit. |
| *Phase – III*: 1955-1968: | Building up of rural credit institutions based on co-operative pattern. |
| *Phase – IV*: 1969-1982: | Period of the multi-agency system for rural credit. More aggressive promotional role. |
| *Phase – V*: 1982 onwards: | Establishment of NABARD and the confinement of RBI to only the policy issues in rural credit. |

## Credit Control or Credit Squeeze

The term refers to the regulation by monetary authority (RBI) of the volume and direction of credit (loans and advances) of the banking system, particularly the commercial banks. In times of inflation, credit control operations aim at contraction of credit, while during deflation, they aim at expansion of credit. Credit control is meant for price and exchange stability, avoiding business fluctuations, halting the gold drain, ensuring full employment and to accelerate development of the economy. Different devices or levers used for this process are:

a) The RBI can raise or reduce bank rate.
b) The RBI can engage in the open market operations in the Government Securities market to absorb or increase the supply of funds.
c) It can raise or lower the level of reserves that commercial banks must maintain; and.
d) The RBI can set limits on the credit terms offered in loans on securities, mortgages and consumer credit.

The first three levers or methods augment or limit the volume of money supply to commercial banks for credit expansion. They also exert an indirect impact on the price, interest rate charged for new credit, and influence indirectly the size and quantum of new loans that can be made by the banks. Only the specific controls on securities, mortgages and consumer credit can be called credit rationing. Fiscal policies are also a sort of credit control.

## Fiscal Policy

This refers to the use of Government's spending and revenue producing activities in order to achieve specific objectives of full employment with price stability. When Government taxes to a greater extent than it spends, it causes a net reduction in the flow of income, thereby reducing the aggregate demand. When it spends a greater amount of money than it receives, it raises national income and aggregate demand. Budget deficits or surpluses are the tools of fiscal policy meant for regulating the economic stability and growth. The instruments of credit control are:

1. Discount rate or bank rate,
2. Open market operations,
3. Rationing of credit,
4. Direct action,
5. Variation of cash reserves,
6. Regulation of consumer credit,
7. Regulation of margin money,
8. Minimum secondary reserves, and
9. Moral suasion or publicity.

## Credit Rationing

This refers to the art of rationing loans by non-price means in situations of excess demand for credit by financial intermediaries. It may assume two forms.

1. *Variable Portfolio Ceiling*: This is the system under which RBI fixes a ceiling or maximum amounts of loans and advances for every commercial bank.

2. *Variable Capital Asset Ratio*: This refers to the system by which RBI fixes the ratio, which the capital of commercial banks should have to the total assets of the banks.

If the loan ceiling to commercial banks is fixed with reference to total amount, it is quantitative control, but if it is done with reference to specific type of credit, it assumes a qualitative character.

Credit rationing also refers to the power of RBI to allow only a fixed amount of accommodation to member banks by means of rediscount.

# AGRICULTURAL REFINANCE AND DEVELOPMENT CORPORATION (ARDC)

Prior to independence, the long-term credit requirements for agricultural development were by far met by the moneylenders and to a small extent by the state Government. Considering this, All India Rural Credit Survey Committee (1951) and Committee on Co-operative Credit in 1960 stressed the inadequacy of term finance for investment in agriculture and suggested the establishment of an institution at the apex level. The Standing Advisory Committee of RBI on agricultural credit had also supported the recommendations. Consequent to their recommendations, Parliament through an Act of 1963 provided for the establishment of Agricultural Refinance Corporation (ARC) from 1st July 1963. It was basically a refinancing agency, meant for promotion and development of agriculture through long-term financial assistance. Considering its developmental and promotional role, it was renamed as Agricultural Refinance and Development Corporation (ARDC). Eversince its inception it is providing term finance, which includes medium-term and long-term loans for major agricultural development projects, which were hitherto not financed by existing credit agencies. The Corporation was primarily meant to refinance, assist and guide the State Co-operative Land Development Banks. But later, it extended its financial assistance to scheduled commercial banks and State Co-operative Banks. On 30th June 1980, a provision to the Act of ARDC was promulgated to provide short-term refinance facility. The broad functions of ARDC were as follows:

1. To help commercial banks for their participation in investments in agricultural development in a big way;
2. To extend needed assistance to the Governmental agencies in the formulation of technically feasible and economically viable projects;
3. To offer needed strength to the member banks in the aspects pertaining to operation and finance;
4. To provide greater assistance to the small farmers;
5. To diversify its lending activities to achieve overall growth in the nation's economy;
6. To play a role as an effective development bank; and
7. To strive hard for reducing regional inequalities in growth and development by concentrating assistance in backward and under developed areas.

The range of refinance facilities in respect of agricultural development projects provided by the Corporation was very wide and the following were the specific investment projects.

1. Afforestation programmes.
2. Horticultural development and plantation crops like cardamom, pepper, cloves, coffee, tea, rubber, *etc.*

3. Minor irrigation in the form of tube-wells, dug wells, installation of filter point pumpsets, energisation of wells, *etc*. Minor irrigation claimed a lion's share in corporations refinance.
4. Soil conservation, reclamation and dry farming programmes.
5. Infrastructural development projects like development of market yards, warehouses, cold storage facilities, *etc*.
6. Dairy, sheep breeding and rearing, fisheries, poultry, sericulture, shrimps, *etc*.
7. Selective farm mechanization, agro-service centres, *etc*.

Of all the above, 80 to 90 per cent of the refinance of the corporation goes in minor irrigation, land development and farm mechanization activities.

As a development banker, the Corporation made earnest efforts to remove regional imbalances by instructing commercial banks and other financial institutions to formulate bankable projects in the areas where co-operatives were weak. All community development blocks in the country were covered by one scheme or the other in a phased manner under the guidance of the Corporation. The Corporation provided special concessions to small and marginal farmers in the SFDA, MFAL and IRDP areas with relatively lower rate of interest at 9½ per cent. A longer repayment period was allowed with concessions in repayment. Increased quantum of refinance facility at the rate of 90 per cent of the total disbursement of the loans was made to commercial banks functioning in these areas. It rendered specific assistance to all the banks in these areas in drawing up of bankable projects in collaboration with State Governments and other authorities concerned with the development of weaker sections. Generally, it provided 75 per cent of the bankable projects through refinance in developed regions, but in the backward regions, to eliminate the regional imbalances in development, it extended refinance to an extent of 90 per cent of the bankable projects. In the year 1982, ARDC was merged with NABARD.

### Sources of Funds

The authorized share capital of the Corporation was Rs. 25 crore with a paid-up share capital of Rs. 5 crore. The share of the ARDC was guaranteed by the Government of India. The sources of funds were share capital, issue and sale of bonds guaranteed by the Government of India and loans from RBI.

### Management

It was managed by a Board consisting of 9 members with Deputy Governor of RBI as the Chairman. Managing Director was appointed by RBI and one more Director was also nominated by RBI. Three directors were nominated by the Government of India. Three more directors were elected from among share-holders of State Co-operative Land Development Banks, State Co-operative Banks, scheduled commercial banks, Life Insurance Corporation of India and other insurance companies.

## NATIONAL BANK FOR AGRICULTURE AND RURAL DEVELOPMENT (NABARD)

### Genesis

ARDC has not made much headway in the field of direct financing and delivery of rural credit against the massive credit demand for rural development. Its role to meet the challenges of integrated rural credit through institutional buildings, training, research, policy-making, planning and providing expertise in the diverse disciplines of finance was inadequate and insufficient. As a result, many Committees and Commissions, *viz.*, Banking Commission (1972), National Commission on Agriculture (1976) and Committee to Review Arrangements for Institutional Credit for Agriculture and Rural Development (CRAFICARD) in 1979 were constituded CRAFICARD under the chairmanship of B. Sivaraman, former member of Planning Commission were constituted, recommended the setting up of a national level institution called NABARD for providing all types of production and investment credit for agriculture and rural development. In pursuance of its recommendations, NABARD came into existence on 12 July 1982. The then existing national level institutions such as Agricultural Refinance and Development Corporation (ARDC), Agricultural Credit Department (ACD) and Rural Planning and Credit Cell (RPCC) of RBI were merged with NABARD with a paid up capital of Rs. 500 crore equally contributed by Government of India and Reserve Bank of India. It operates through its head office at Mumbai, 17 regional offices, one each in major States, 10 sub offices in smaller States/union territories and 213 district offices.

## Board of Management

All the directors in the Management Board are appointed by the Central Government in consultation with RBI. The Board is envisaged with the role of providing direction, management and supervision of various financial affairs of NABARD. The Chairman and the Managing Director are the two major top level executives. The Managing Director is the chief executive of NABARD and he is primarily responsible for the various operations and performance of the bank.

In addition to Chairman and Managing Director, the Board consists of 13 other directors and these directors form the Advisory Council of NABARD. Of the 13 directors two are experts in rural economics and rural development. Three directors are representatives of co-operatives and three from commercial banks. Three directors are officials of the Government of India and two belong to State Governments.

## Sources of Funds

| | |
|---|---|
| Authorized Share Capital of NABARD | Rs. 500 crore |
| Issued and paid up capital | Rs. 100 crore |

*Other sources are*

1. Borrowings from the Government of India and any institution approved by the Government of India.
2. Issue and sale of bonds by the Government of India
3. Borrowings from RBI
4. Deposits from State Governments and local authorities and
5. Gifts and grants received.

## Objectives

As an apex refinancing institution, NABARD purveys all types of credit needed for the farm sector and rural development. It is also vested with the responsibility of promoting and integrating rural development activities through refinance. The bank is also providing direct credit to any institution or organization or an individual, subject to the approval of the Central Government. It has close links with RBI for guidance and assistance in financial matters. As an effective catalytic agent for rural development and in formulating appropriate rural development plans and policies, its role is remarkable.

## Functions

The activities of NABARD are presented under three broad categories.

A. Credit activities
B. Developmental activities and
C. Regulatory activities

## A. Credit Activities

1. It prepares for each district annually a potential linked credit plan, which form the basis for district credit plan
2. It participates in finalisation of annual action plan at block, district and State level
3. It monitors implementation of credit plans
4. It lays down the terms and conditions to be followed by credit institutions in financing production, marketing and investment activities of rural farm and non-farm sectors and
5. It provides refinance facilities as detailed below:

## Short-term Refinance

The purposes for which short-term refinance extended are

1. Agricultural production operations and marketing of crops by the farmers and farmers' cooperatives, *etc.*
2. Marketing and distribution of inputs like fertilizers, seeds, pesticides, *etc.* and
3. Production and marketing activities of village and cottage industries, handicrafts, handlooms, powerlooms, artisans and other rural non-farm enterprises.

The eligible institutions are State Co-operative Banks (SCBs), Regional Rural Banks (RRBs) and commercial banks and other financial institutions approved by RBI. The time period is up to 12 months.

## Medium and Long-term Refinance

The purposes for which medium and long-term refinance extended are

1. Investment in agriculture and allied activities such as minor irrigation, farm mechanization, land development, soil conservation, dairy, sheep rearing, poultry, piggery, plantation/horticulture, forestry, fishery, setting up of storage and market yards, sericulture, apiculture, animals and animal driven carts, agro-processing, agro-service centres, compost plants, installation of pumpsets, *etc.*
2. Investment activities of artisans, small, scale industries, village and cottage industries, handicrafts, hand looms, power looms, *etc.*
3. Activities of voluntary agencies/self-help groups of rural poor and
4. Investment in share capital/securities of institutions concerned with agriculture and rural development. The period is up to a maximum of 15 days.

## Conversion and Rescheduling of Refinance

Under conditions of drought, famines or other natural calamities, military operations, *etc.*, short-term refinance facilities granted for production activities of farm and non-farm sectors are converted into medium term refinance. Similarly, installments falling due for repayment under medium and long term are rescheduled for repayment in future years.

## Criteria and Extent of Refinance

The criteria of refinance to any scheme submitted by financing institutions are

1. Technical feasibility of the project and adequate response from prospective beneficiaries
2. Financial viability and adequate incremental income to the ultimate borrower to repay the loan within a reasonable period and
3. Organizational arrangements to ensure close supervision by financing banks.

While all the funds are routed through institutional agencies, the ultimate beneficiaries of investment finance may be individuals, partnership concerns, State owned corporations, co-operative societies, *etc.* The ultimate beneficiaries of production credit are generally individuals.

NABARD's refinance ranges from 50 per cent to 10 per cent, the balance being met by banks or the concerned State Governments/Government of India in the case of State land development banks.

The extent of refinance under various activities is as under.

1. Pilot rainfed farming projects (100%)
2. Waste land development schemes where individuals are beneficiaries (100%)
3. Non-farm sector schemes (outside the purview of IRDP) (100%)
4. Agro-processing units (75%)
5. Bio-gas schemes (75%)
6. All other schemes including IRDP (70%)
7. Farm mechanization (50%)
8. Rural Electrification Corporation (REC) (50%).

The rate of interest on refinance and the ultimate lending rates by the banks are fixed from time to time. The banks get a margin of 3 per cent to 5 per cent. NABARD stipulates beneficiary's contribution to the project cost in order to ensure his stake in the investment. Such margin money varies from 5 to 25 per cent according to the type of investment and the class of borrowers. Corporate borrowers such as Irrigation Corporations, Forest Development Corporations, *etc.*, provide higher contribution up to 25 per cent of the investment cost.

Apart from refinance facilities to credit institutions, NABARD also provides direct finance to

i) State Governments and
ii) State sponsored corporations directly for development of infrastructure projects in priority areas like minor and medium irrigation, land development, soil conservation, *etc.* NABARD also finances State Governments for contributing to the share capital of co-operative credit institutions and thereby improving their financial strength.

NABARD pays special attention to monitoring the assisted projects in order to ensure their proper implementation. It also undertakes consultancy work for projects even in cases where refinance is not availed from NABARD.

## B. Development Activities

The following are the developmental activities undertaken by NABARD for the productive use of credit.

## I. Institutional Development

1. It helps co-operative banks and RRBs to undertake a critical review of their own operations and come up with development action plans (DAPs) for themselves.
2. It enters into memoranda of understanding (MOUs) with State Governments and co-operative banks, specifying the obligations of each to improve the banks affairs within a stipulated time
3. It helps RRBs and their sponsor banks to enter into MOUs specifying their commitments and relative obligations to improve affairs of RRBs in stipulated time
4. It provides financial assistant to co-operative and RRBs for the establishment of technical and monitoring cells with them to improve the quality of their project formulation, implementation and project monitoring
5. It provides organization development intervention (ODI) through training institute (Bankers Institute for Rural Development (BIRD), Lucknow, National Bank staff college, Lucknow, College of Agricultural Banking (CAB, Pune) to staff of selected RRBs and co-operative banks
6. It provides financial assistance for the training institute of co-operative banks
7. It provides training for senior and middle level executives of commercial banks, RRBs and co-operative banks at training institutes of NABARD (BIRD, Lucknow and CAB, Pune)
8. It provides borrowers education on ethics of repayment in selected areas through Vikas Volunteer Vahini, *etc.*

## II. Research and Development Fund

NABARD maintains a research and development fund for
1) Supporting operational research projects aimed at upgradation and transfer of technology from lab to land
2) Conducting research studies on subjects of topical interest in economics and banking
3) Organising national and international seminars, conferences, symposia, *etc.*, on subjects related to rural development and banking
4) Conducting programmes for upgrading skills of prospective borrowers
5) Providing grants to select Krishi Vikas Kendras

## III. Agricultural and Rural Enterprises Incubation Fund (AREIF)

Under this fund, assistance is provided on flexible terms to new and innovative rural small and micro enterprises undertaking innovative and potentially viable ventures carrying technology and market risks.

## IV. Rural Promotion Corpus Fund (RPCF)

It is meant to provide assistance for enterprises promotional concepts like training cum production centres, rural entrepreneurship development programmes, district rural industries project, women and environment programmes, consultations, technical monitoring and evaluation cells, Vikas Volunteer Vahini, *etc.*

## V. Credit and Financial Services Fund (CFSF)

It is aimed at providing the assistance for innovations in rural banking and credit system, supporting institutions for the research activities, surveys, meets, *etc.*

## VI. LINKING SHGS TO CREDIT INSTITUTIONS

NABARD has launched a pilot project for linking self-help groups with credit institutions. Under the linkage programme, it provides 100 per cent refinance to banks for loans extended to SHGs.

## C. Regulatory Activities

As the apex development bank, NABARD shares with RBI, some of the regulatory and supervisory functions in respect of co-operative banks and RRBs.

1. Under banking regulation Act, 1949, NABARD undertakes inspection of RRBs and co-operative banks (other than primary co-operative societies) and assesses if the banks meet the statutory requirements of the real value of their capital and reserves and continue to be managed in ways not detrimental to the interest of depositors
2. Any RRB or co-operative bank seeking permission of RBI, for opening branches, *etc.*, needs recommendation of NABARD for the purpose and
3. The State and district central co-operative Banks also need authorization from NABARD for extending assistance to units outside the co-operative sector and non-credit co-operatives for certain purposes beyond a cut-off limit.

The refinance support to co-operative banks, commercial banks and RRBs and loans to State Governments, NGOs and other agencies during 2000-01 aggregated Rs. 16,461 crore compared to Rs. 14,178 crore during 1999-2000, registering a growth of 16.1 per cent.

## AGRICULTURAL FINANCE CORPORATION (AFC)

In view of the inexperience of the commercial banks in financing agriculture, a need was felt to set up an institutional agency at the national level to take care of this aspect. Accordingly, the Agricultural Finance Corporation was promoted by the Indian Banks' Association. It was incorporated on 10th April 1968 under the Indian Companies Act 1956, with an authorized share capital of Rs. 100 crore and issued share capital of Rs. 10 crore. Basically AFC is a consortium of commercial banks and consultancy agency of member banks in the formulation of projects for agriculture and rural development. Scheduled commercial banks numbering 37, notified under the RBI Act of 1934 had subscribed to the share capital of the corporation.

The corporation has two distinct roles, *viz.*, financing the individuals/institutions/organizations involving agricultural development and promoting commercial bank advances for agricultural development.

### (i) Financing Role

In order to gain experience in financing, only certain projects and areas are selected and farmers are financed for these projects. If the projects are successful, its experience in financing will be passed on to the commercial banks. In this regard the corporation formulates projects, works out the economics and invites the commercial banks to join with it in financing the projects. Top priority is being given by the Corporation to the following projects.

a. Sinking and deepening irrigation wells and energiging the same;
b. Production, distribution and marketing of agricultural inputs such as seeds, fertilizers, insecticides, implements and machinery;
c. Construction of storage structures for foodgrains and fertilizers, and
d. Establishment of agricultural service units, *etc.*

### (ii) Promotional Role

This is indeed a challenging task for AFC. It provides expertise in the formulation of appropriate projects to all commercial banks working under its guidance and advancing loans. To increase the credit absorbing capacity in agriculture for modernizing agriculture, it suggests the following steps to be taken by the commercial banks. (1) commercialisation and industrialization of agriculture; (2) development of requisite infrastructure for rapid agricultural development; (3) formulation of potential projects financed by commercial banks; (4) removal of various difficulties and handicaps experienced by the commercial banks and the farmer borrowers; (5) simplification and streamlining the procedures in sanctioning the loans, and (6) development of co-operation, coordination and consortium arrangement among different lending agencies and co-operatives involved in agricultural financing.

A national level consultative committee for bringing out coordination among the different lending agencies has already been set up by the Corporation. Besides being a financing agency the Corporation has emerged as a consultancy organization to the State Governments in the block level planning.

It organizes seminars, workshops and training programmes to bank staff. Its association with FAO and World Bank is also significant.

## OUTLINES OF RECOMMENDATIONS OF KHUSRO COMMITTEE AND NARASIMHAM COMMITTEE

Two committees *viz.*, Khusro Committee and Narasimhan Committee examined the formal agricultural credit system in India and recommended certain important changes regarding modifying rural banking structure, interest rates policy and co-operative credit structure. The Agricultural Credit Review Committee (ACRC) under the Chairmanship of A.M. Khusro was appointed by RBI in August 1986 and it submitted its report in August, 1989. Later the Committee on financial systems which was appointed by Government of India in August, 1991 submitted its report in November 1991. The outlines of the recommendations of the two committees are briefed hereunder.

### Khusro Committee's Report

Its recommendations pertained to the co-operative credit system, RRBs and commercial banks. It estimated the gross interest margins and suitably recommended lending rates to these institutions. The ceiling for the lending rate for commercial banks towards agricultural lendings was fixed at 15.5 per cent, whereas in the case of PLDBs it was 5 per cent and 8.65 per cent for RRBs. The Committee recommended that any shortfall in the interest earnings might be made good to the credit institutions by the Government. The Committee did not thoroughly examine the margins required for improved measures in lending rates. It made two broad recommendations on the structure of agricultural credit institutions in India. Firstly, it recommended the merger of RRBs into sponsoring commercial banks. This is very essential in areas where the performance of RRBs is not economically viable. Secondly, it recommended the creation of National Co-operative Bank to function as national apex bank for all co-operative institutions in the country.

## Narasimham Committee's Report

The first set of recommendations related to the extent of directed investment and directed credit programmes. These are:

1. To cut the Statutory Liquidity Ratio (SLR) from 38.5 per cent to 25 per cent over five years;
2. To reduce Cash Reserve Ratio (CRR) progressively;
3. To bring down the priority sector credit target to 10 per cent of the total bank credit (here priority sector is defined in terms of small and marginal farmers, small business and transport operators, village and cottage industries, rural artisans and other weaker sections); and
4. Left out priority sector should be made eligible for preferential refinance.

Second set of recommendations related to Bank of International Settlement (BIS) standards of capital adequacy in a planned manner.

Third set of recommendations pertained to:

1. Ruling out further nationalization of banks, abolition of branch licensing and increasing the operational autonomy of the banks and financial institutions,
2. Creating four tier structure comprising three to four large banks with international character, eight to ten national banks with country-wide universal banking, and
3. Increasing competition between commercial banks and other financial institutions, etc.

## PROGRAMMES OF RURAL DEVELOPMENT

## SMALL FARMERS DEVELOPMENT AGENCY (SFDA) AND MARGINAL FARMERS AND AGRICULTURAL LABOURERS DEVELOPMENT AGENCY (MFAL)

Many of the small farmers did not reap the benefits from the nationalization of commercial banks due to cumbersome loaning procedures and their inadequacy of tangible securities in obtaining loan, undue delays in the disbursement of loans and high cost of credit. As a result, the small farmers depended mostly on the moneylenders for their credit needs paying usurious rates of interest. Hence, there was a need to develop and uplift the small and marginal farmers in the rural areas. All India Rural Credit Review Committee (1969), keeping this in view, recommended the establishment of SFDA and MFAL in areas where there was scope for development. SFDA and MFAL came into operation in 1971. In general, marginal farmers are not potentially viable; hence depend upon agricultural wages and subsidiary occupations. To help them come out from the abject poverty, the Government formulated a pilot project for marginal farmers and agricultural labourers and this project was implemented by an agency called Marginal Farmers and Agricultural labourers Development Agency (MFAL). These projects were sponsored by the Government of India and executed in selected districts of different States with substantial financial assistance. These agencies were created primarily to provide employment avenues in rural areas and foster rapid increased agricultural production by helping the small and marginal farmers. These agencies were expected to coordinate with the credit institutions in the matter of disbursal of loans to weaker sections. It was pointed out that these agencies were resorting to undue delays in releasing the subsidy amount to deserving farmer borrowers. As a result, the credit institutions were in irksome position with regard to sanctioning timely credit to these farmers. Several schemes like minor irrigation, dairy, poultry, sheep rearing, etc., which were being implemented by these agencies were meant to raise the standard of living of the small and marginal farmers. But many of the schemes could not be implemented in time due to lack of supporting infrastructure, consequently granting of loans for small farmers was delayed. Most of the loaned funds by the agencies to small farmers were misutilised due to lack of supervisory mechanism and inadequacy of the staff.

## Functions of SFDA

1. To identify eligible small farmers and their problems.
2. To arrange for the services and supplies of various inputs.
3. To arrange for developing irrigation sources in the area through financing schemes like digging of wells, deepening of wells, installation of motors, etc.,
4. To promote production activities in the given area; and,
5. To help small farmers in securing facilities of storage, transportation, processing, marketing, etc.

## Functions of MFAL

1. To identify the eligible marginal farmers and agricultural labourers;
2. To investigate their problems and offer relevant solutions to such problems;
3. To formulate economic programmes for providing gainful employment;
4. To promote rural industries in the given area;
5. To evolve adequate institutional, financial and administrative arrangements for implementing various programmes;
6. To organize labour contracts required for farming;
7. To construct and develop minor irrigation structures; and
8. To establish facilities for storage, etc.

These agencies gave subsidies at the following rates to the individual beneficiary.

Small farmers were eligible to receive 25 per cent of the total investment, while

marginal farmers and agricultural labourers* $33\frac{1}{3}$ per cent.

---

*Small farmer is one with 2 ha dryland or 1 ha. of wetland. Marginal farmer is one whose holding is less than 1 ha of dryland or 0.5 ha of wetland. A person without any land holding but having a homestead and deriving more than 50 per cent of his wage income from agriculture is called agricultural labourer.

# INTEGRATED RURAL DEVELOPMENT PROGRAMME (IRDP)

A number of programmes and agencies have been started in the country during the past two decades for the improvement of economic conditions of the rural poor. But these programmes did not create the expected impact in the rural areas. The reasons behind this state of affairs were that none of these programmes covered the entire country, frequent overlapping of the schemes in the same area, lack of coordination and enthusiasm among these agencies, lack of earnest efforts and motivation from the officials of these agencies, *etc.* In essence these programmes and agencies, by far remained as subsidy-giving programmes rather than development oriented ones. Hence, it was decided to replace all the programmes and agencies by one single integrated programme, which aims at poverty alleviation and rural development. This programme was named as Integrated Rural Development Programme (IRDP) and launched during 1978-79 by the Ministry of Rural Reconstruction with the twin objectives of eliminating poverty and unemployment in rural areas. Initially IRDP covered 2,000 blocks in the country. With effect from $2^{nd}$ October 1980, the programme was extended to all the 5,011 development blocks in the country. It is the largest poverty alleviation programme. The main aim of IRDP is to improve the economic conditions of the families (in rural areas) living below the poverty line through the creation of new assets or the generation of new employment. Basically, it is an action-oriented and time bound programme. Under this programme, hitherto existing programmes like SFDA, MFAL, Drought Prone Area Programme (DPAP), Command Area Development Authority (CADA) National Rural Employment Programme (NREP) and Training of Rural Youth for self employment (TRYSEM), *etc.*, have been merged.

IRDP is popularly known as an anti-poverty programme. Under this programme, in addition to small and marginal farmers, agricultural labourers, landless workers, sharecroppers, artisans, scheduled castes and scheduled tribes and others living below the poverty line[*] are being covered.

## Specific Objectives

IRDP aims at achieving the following specific objectives:
1. Increasing the productivity of land by providing the needed inputs in required quantities at right time, thereby raising the productivity and production in agriculture.
2. Creating tangible assets for the rural poor to improve their economic conditions.
3. Augmenting the resources and income levels of weaker sections.
4. Diversifying agriculture through poultry, dairy, fishery, sericulture, *etc.*, and,
5. Providing infrastructural facilities like processing, storage, organized marketing, milk chilling and collecting centres, artificial insemination centres, *etc.*

## Identification of Beneficiaries

Those people living below the poverty line are eligible to be covered under this programme. Income is the criterion to decide the poverty line. The different categories of people falling under the purview of the scheme besides those falling below the poverty line are small and marginal farmers, agricultural and non-agricultural labourers, rural artisans and scheduled castes and scheduled tribe families.

---

[*] A family of 5 members is said to be below the poverty line, if it earns an annual income of less than Rs. 11,600 in rural areas and Rs. 12,800 in urban areas.

The IRDP scheme should be financed in the villages in service area allotted to the branch. In regard to provision of finance for infrastructure under IRDP, viable activities proposed by IRDP, provision of veterinary facilities, purchases of IRDP assets, supply of animal fodder, which could be in the nature of backward and forward linkages are financed by the bank.

## Purpose of Advance

Three major kinds of activities capable of income generation on continuing basis have been contemplated for the target families.

*Primary Sector:* The activities are agricultural, animal husbandry, fisheries, farm forestry, *etc.*

*Secondary Sector:* Activities like khadi and village industries, handlooms, handicrafts, blacksmithy, pottery, carpentry, *etc.*, are included here.

*Tertiary Sector:* Activities like transport, small business and other service activities like tailoring, workshops, repair shops, *etc.* are included. IRDP is funded by central and State Governments in the ratio of 50:50

## Implementation Agencies

At the National Level IRDP is administered by the Ministry of Rural Areas and Employment (earlier known as Ministry of Rural Development). The States and union territories have set up bodies known as DRDAs (District Rural Development Agencies). It is DRDA, which identifies beneficiaries, draws up income generating projects for them and brings them into contact with banks. DRDA provides capital subsidy to the identified families and supply the list of such families along with the suggested economic activities to the financing institutions for extending loan assistance. DRDA also ensures backward linkage (inputs, technical advice, *etc.*) and forward linkage (processing facility, marketing arrangements, *etc.*) in respect of the proposed economic activities.

## Subsidies

The subsidy in the case of small farmers is 25 per cent while for marginal farmers, agricultural labourers, and non-agricultural labourers it is $33\frac{1}{3}$ per cent, and for scheduled tribes 50 per cent.

## Progress of IRDP

The total number of beneficiaries covered during the period from 1980 to 1985 was 16.56 millions as against the target of 15 millions. Of the beneficiaries covered, 6.46 millions belonged to SC/ST categories. During the period from 1985 to 1990 the target to be covered was put at 20 million households and for the period 1985-86 to Dec 1988 the number of families assisted was 4 millions. During the year 1991-92 (April-March), under IRDP, banks assisted 25.17 lakh beneficiaries and an amount of Rs. 1,133.27 crore was distributed as loan and Rs. 800.99 crore as subsidy. Out of these beneficiaries, 12.78 lakhs belonged to SCs and STs and 8.33 lakhs were women. Loans and subsidy disbursed during the Seventh Plan amounted to Rs. 5,373 crore and Rs. 3,316 crore respectively. The proportion of recovery to demand of public sector banks under IRDP loans was 41.4 per cent during 1990-91.

## Merits of IRDP over Earlier Programmes

Most of the earlier programmes, *viz.*, SFDA, DPAP, CADA, *etc.*, were implemented by the Government with specific objectives and time framework. Since no extra staff was given to implement these programmes they have become part and parcel of the work of the officials involved in the development programmes. Contrary to these programmes IRDP is an integrated and permanent development programme with accepted objectives and action-oriented programme with direct involvement of the officials. Hence the staff of the programme takes keen interest in implementation of the programme. IRDP has efficiently combined and strengthened both area and beneficiary approach. Compared to the earlier programmes IRDP distinguishes itself in identifying target group and variations in the amount of subsidy under various development programmes. Overall it is a poverty alleviation programme meant for improving the standard of living of families falling below the poverty line.

## IRDP–Mehta Committee Recommendations

For improvement of IRDP, the following recommendations were made by Mehta Committee in 1994-95.

## Identification of Beneficiaries

The below poverty line lists (BPL) drawn by block authorities should be approved by the panchayats in the meetings convened for this purpose, which are attended by bank officials also.

## Project Profiles

District level technical groups consisting of lead district officers of RBI, district development officers of NABARD (wherever posted), lead bank manger, technical officials of State Governments and non-Governmental consultants are being set up for preparation of project profiles.

## Animal Husbandry

Before disbursing loans to beneficiaries for animal husbandry, banks should ensure that adequate number of good quality animals and also the required linkages (infrastructual arrangements for supply of feed, and marketing) are available.

## Investment

Branches should ensure that the loan components are enhanced suitably to cover the project cost.

## Security Norms

For loans up to Rs. 25,000 under IRDP for all activities, branches should not obtain mortgage of land. However, for loans above Rs. 25,000 but up to Rs. 50,000 branches may obtain mortgage, without asking for any collateral security.

## Infrastructure

The controllers and lead district manager attending the DCC meetings and the branch managers attending, the BLBC meetings should list the infrastructure, which is required in the service area villages for pursuing a number of activities. These requirements should be included in the perspective plans prepared by the DRDA.

## Training Facilities

The lead district managers may organize orientation training programmes for IRDP beneficiaries.

## Institutional Support

The minimum repayment period for IRDP loans has been increased (from the earlier 3 years) to 5 years. However, the repayment schedule should be fixed in a realistic manner after taking into account the level of income generation from the activity and the economic life of the assets. Moratorium may also be allowed, wherever necessary.

## Role of SHGs

The possibility of routing the assistance to the families falling below poverty line through SHGs (self-help groups) on a large scale should be explored.

## Improving Recoveries

For improving the recoveries of loan (1) group loans may be considered, wherever necessary (2) rescheduling of loans may be considered, wherever necessary (3) more attention should be devoted to appraisal of loans (4) wherever required, adequate gestation or moratorium should be allowed in such a way that the commencement of recovery coincides with accrual of incremental income from the activity. In the case of projects where accrual of income is low in the beginning but goes up over a period of time, such of loan installments in the initial period should be suitably reduced.

# SELF HELP GROUPS (SHGs)

Despite the vast expansion of institutional credit in India, the dependence on the moneylenders continues particularly for meeting urgent credit needs. Such dependence is more pronounced in respect of small and marginal farmers, agricultural lobourers, petty traders, artisans, *etc.*, generally and more so in areas endowed with poor resource base.

The crux of the problem in agricultural finance is the high transaction cost to the banks in financing a large number of small and marginal farmers. These farmers also incur high transaction cost, while visiting the banks due to the distance, other procedural formalities and finally small amount of funds. Moreover, the less privileged classes have a belief that, the banks function to serve the needs of others. Moreover the small and marginal farmers need some sort of financial assistance for consumption purpose, the provision for which is not there in traditional banking generally. This placed these categories to rely on informal credit channels (moneylenders, market vendors, friends, relatives, *etc.*). The advantage with the informal credit channels is that, credit is available as and when required without any documentation formalities. However, interest rates will be on higher side compared to institutional agencies. Also, the observation is that the interest rates are not very high, but condition like pre-harvest sale of the crop is insisted upon which is detrimental to the interests of the farmers. Availability of alternative financial services would do a world of good to these categories of the farmers.

A number of poverty alleviation programmes have been tried in the past in the country to uplift the lot of poorer sections. The main focal point of many of the poverty alleviation programmes has been income generation with support of micro credit. The success stories of Bangladesh Grameen Bank and experiences of Indonesia, Korea and Nepal in involving participatory approach to the micro-credit programmes for poverty alleviation provided momentum in India also. With a view to developing a supplementary credit delivery mechanism to reach the poor in a cost effective and sustainable manner, the National Bank for Agriculture and Rural Development (NABARD) introduced a pilot project for linking 500 Self-Help Groups (SHGs) with banks in 1992 after thorough discussions with the RBI, the commercial banks and non-Governmental organisations (NGOs). The pilot phase was followed by setting up of working group on non-Governmental organisations (NGOs) and SHGs by the Reserve Bank of India in 1994 which came out with wide ranging recommendations on internalisation of the SHG concept as potential intervention tool in the area of banking with the poor. Group finance is a informal finance existing in the rural areas in which a group of individuals carryout pooling their savings and lending to each other or to persons outside the group. Self-help group is also a type of group finance wherein savings and lendings are mutual. A self-help group (SHG) is a small economically homogenous and affinity group of rural poor voluntarily coming together (1) to save small amounts regularly (2) to mutually agree to contribute to a common fund (3) to meet their emergency needs (4) to have collective decision-making (5) to solve conflicts through collective leadership and mutual discussion and (6) to provide collateral free loans with terms decided by the group at market rates. Self-help groups are voluntary associations of people formed to attain a collective goal. People who are homogenous with respect to social background, heritage or traditional occupations come together for a common cause to raise and manage resources for the benefit of group members. SHGs may be organised in a village or a cluster of villages either by reputed voluntary agencies (VAs), non-Government Organisations (NGOs) or at the initiative of branch managers of banks. Normally the promoting VAs/NGOs provided training, extension and support facilities to the group and its members. The core problem of rural finance is the high transaction cost to the banks in financing a large number of small borrowers who require credit frequently and in small quantity. The same holds true of the costs involved in providing saving facilities to the small, scattered savers in the rural areas. The rural savers and borrowers also face high transaction costs while dealing with banks due to distances, small value of the financial transactions, *etc.* In a recent unpublished study, the transaction costs to a small rural borrower raising a loan from the commercial bank under a poverty alleviation programme, was placed at 24.6 per cent. Further, the transaction cost of operating a savings account with a bank was placed as high as 10 per cent of the savings on the assumption of only one transaction per month. Involvement of SHGs with banks could help in overcoming the problem of high transaction costs in providing credit to the poor, by passing on some banking responsibilities regarding loan appraisal, follow-up, recovery, *etc.*, to the poor themselves. In addition, the character of the voluntary association and their relation with members offered ways of overcoming the problem of collateral, excessive documentation and physical access which reduced the capacity of the formal institution to serve the poor.

## Criteria for Selection of SHGs for the Linkage with Banks

For selection of SHGs for the linkage programme with banks, NABARD has set out the following for the use of bankers and non-Governmental Organisations (NGOs).

1. The group should be in active existence for atleast a period of six months.
2. Have successfully undertaken savings and credit operation from its own resources.
3. Democratic working wherein all members feel that they have a say.
4. The banker should be convinced that a group has not come into existence only for the sake of participation in the programme and availing benefits (credit) thereunder. There should be a genuine need to help each other and work together among the members.
5. Members preferably have homogeneous background and interest.

In SHG linkage programme, RBI and NABARD have tried to promote "relationship banking, *i.e.,* improving the existing relationship between the poor and bankers with social intermediation by NGOs".

## NABARD and the SHG Linkage[1] Programme

1. Conceptualization and introduction of pilot phase of the SHG linkage programme in February 1992 for linking 500 SHGs with the banks after consultations with RBI, banks and NGOs.
2. Developing a conducive policy framework through provision of opening savings banks accounts in the names of SHGs, relaxation of collateral norms, documentation, delegation of all credit decisions and terms to SHGs.
3. Introduction of bulk lending scheme in 1993 for encouraging the NGOs which were keen to try group approach and other financial services in the rural areas.
4. Training and sensitisation of bank officials and standardisation of training modules.
5. Experimenting with RRBs for promotion of SHGs.
6. Selective support to NGOs for promotion of SHGs.
7. Close monitoring through State level forums.
8. Dissemination through seminars, workshops, occasional papers, *etc.* .

A comprehensive strategy has been thought of for substantially up-scaling the linkage programme. NABARD envisages to cover one million groups by the year 2008. NABARD has set up a new department known as Micro Credit Innovations Department (MCID) at the head office and MCID cell at RO level. To strengthen the skills of the staff at the NABARD level, crash training programmes for the staff posted in these cells, district development managers, district development officers and nominee directors of the boards of RRBs are being organised.

The SHG and the linkage programme have the following unique features.

1. Decisions are made by members collectively.
2. SHGs provide the needed financial service at the door steps, *i.e.,* they provide consumption credit and production credit.
3. SHGs are helping in mobilisation of thrift, low transaction costs and near full repayments.
4. SHGs have exclusive focus on individuals of lower stratum.
5. There is no subsidy dependence syndrome.

---

[1]Trainers Training Programme on self-help groups 10-5-99 to 14-5-99. Bankers Institute of Rural Development, Lucknow.

6. Its special feature is the empowerment process that it initiates among the poor.
7. Through the linkage programme, the policy-makers, banks and NGOs have become involved in the SHG movement.

## Linkage Models[1]

The linkage between the self-help groups and the banks is expected to be of symbiotic in nature. The linkage concept is based on savings linked credit. There are two basic models, with number of modifications to depict the relationship between SHGs and banks are at present found in the country.

### 1. Direct Linkage Model

In the case of direct linkage model, the bank identifies the group and deals with the SHG directly for both mobilising the savings and for making available credit facilities to the group as a whole or individual members. Group members act as collateral security. In this model the credit is generally made available to the group and individuals to be financed are identified by the group itself which takes the responsibility of loan repayment.

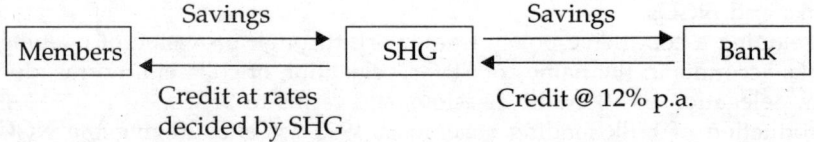

### 2. Modified Direct Linkage Model I

Here the activity and member to whom loan is given are identified by the group. The group is morally responsible for repayment but credit is given as in any individual loan. The advantage of this model is if a member raises above the normal capability status of the group and needs more credit for further development, the need can be met by individually financing him/her.

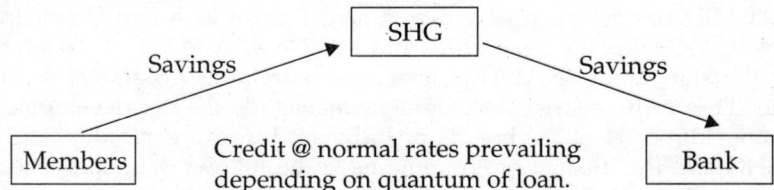

### 3. Modified Direct Linkage Model II

In this model NGO is not the financial intermediary. The NGO's role is only in group evaluation and stabilisation, whereas the financial linkage is directly with the group. This is the most popular model in India, as it can promote sustainable linkage between SHG and bank even after withdrawal of NGO.

---

[1]Sheokand S.M. Re-orienting banking with the poor — The SHG Bank—Linkage way.

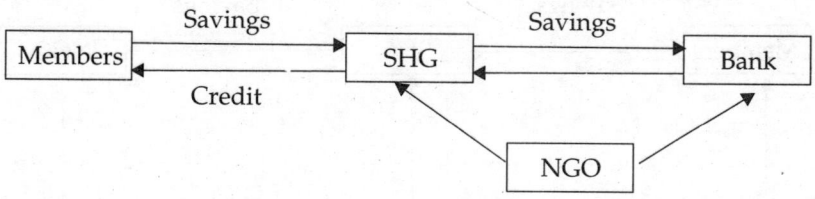

## 4. IFAD Model

In this model, the NGO is involved as in 'Modified direct linkage Model II', but the line departments of Government like women development corporations, sericulture, and rural development are also involved in identification of activity, beneficiaries, *etc.* The model is in existence in areas where IFAD projects are being implemented like in Tamil Nadu, Maharashtra and Uttar Pradesh.

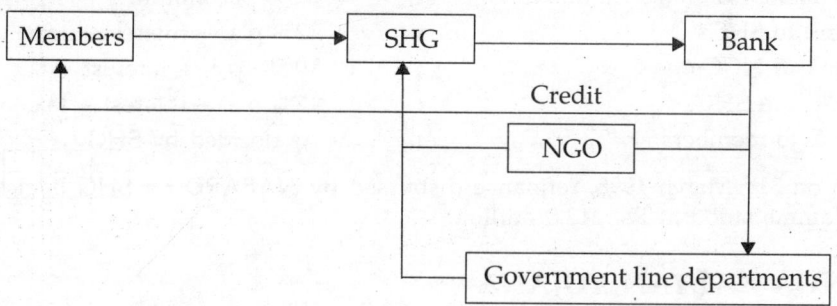

## 5. Indirect Linkage Model

In this model, the funds flow through the NGO, *i.e.*, NGO is financial intermediary. Therefore it is called indirect linkage model. Although in the initial stages of linkage, bank may follow this model to gain self-confidence for lending to SHGs, this is not a very healthy model. It distorts the role of NGO, and also created dependency of groups on NGOs for assessing credit.

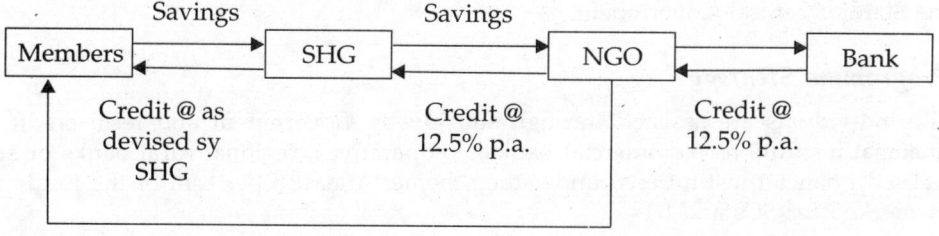

## 6. Modified Indirect Model

This model exists in cases where the groups are artisans/handicraft groups and NGO support for marketing is also available like SEWA, Lucknow. This model is limited to those areas where NGOs are directly helping the SHG members in income generating activity, marketing, *etc.* This model has very limited scope.

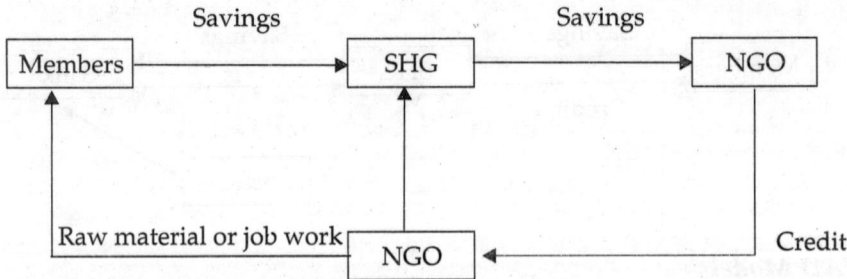

NABARD gives 100 per cent refinance under SHG linkage programme at an interest rate of 6.5 per cent to banks.

The interest rate structure stipulated by NABARD under linkage programme is as under:

| | | |
|---|---|---|
| NABARD to banks (Refinance) | : | 6.5% per annum |
| Bank to SHGs | : | 12% p.a. + interest + tax |
| Bank to NGOs | : | 10.5% p.a. + interest + tax |
| NGOs to SHGs | : | 12% p.a. + interest + tax |
| SHG to members | : | As decided by SHG |

As on 31st March 1998, refinance disbursed by NABARD for SHG bank linkage programme stood at Rs. 213.8 million.

# GANGA KALYAN YOJANA

A centrally sponsored scheme *viz.*, Ganga Kalyan Yojana (GKY) has been launched with effect from 1-2-1997.

## Objective

The objective of the scheme is to provide irrigation through exploitation of groundwater through tube wells and bore wells to individuals and groups of beneficiaries belonging to the target groups, *i.e.*, small and marginal farmers falling below poverty line other than those who have already been assisted under any minor irrigation programme of the State or central Government.

## Programme Strategy

The individuals are assisted through subsidy by Government and term credit by financial institutions (commercial banks, co-operatives, regional rural banks or specialised financial institutions) under the scheme. Atleast 5 per cent of the funds are earmarked for SCs and STs.

## Selection of Beneficiaries

The list of those falling below the poverty line is prepared after it is approved by the Gram Sabha in an open general body meetings, in the presence of officers/staff of the block, revenue staff, bank officials and representatives of NGOs, where such NGOs are involved in the formation of groups.

### Funding Pattern

The funding pattern is 80 per cent by the central Government and 20 per cent by the State Government.

### Area of Operation

The scheme is in operation in the entire country except the following:

(a) In the areas where the level of ground water development for irrigation has already exceeded 65 per cent as per the Central Ground Water Board (CGWB), *i.e.*, areas declared as over-exploited (level of exploitation exceeds 100%), dark (between 85 and 100%) or gray (between 65 and 85%).

(b) In the coastal areas 15-20 kms. from the sea-shore.

(c) The State Government identifies and excludes pockets where significant localised fluctuation in ground water table is observed during the dry season.

*Implementing Agency:* DRDA/Zilla Parishads.

## Subsidy

1. *Group Schemes*: Subsidy is at the rate of 75 per cent for SC/ST and physically disabled groups and 50 per cent for others. Monetary ceiling on subsidy per group is Rs. 40,000.

2. *Individual Beneficiaries*: Subsidy is @ Rs. 5,000 per acre of land under the scheme subject to a ceiling of Rs. 12,500 per beneficiary.

## Users' Association and Facilitators

Group schemes are available to beneficiaries who form themselves into users' associations. The assets provided under the scheme vest in the users' association.

## Monitoring and Supervision

The district collector along with the project director DRDA is the incharge of monitoring and supervision of the scheme in the district.

## IRDP and MWS

With the introduction of GKY, bore-well/tube-well schemes under IRDP and million wells scheme (MWS) have been subsumed under the GKY and all assistance for bore-well/tube-well is given under GKY only.

# KISAN CREDIT CARD (KCC)

The union finance minister in his budget speech for the year 1998-99 had stated about the introduction of Kisan Credit Card (KCC) by the banks. The scheme has been designed as per the model scheme prepared by NABARD.

## Salient Features

1. *Objectives*: Kisan card scheme aims at adequate and timely support from the banking system to the farmers for their short term production credit needs for cultivation of crops, purchase of inputs, *etc.*, in a flexible and cost effective manner.

2. *Eligibility*: Existing borrower farmers having good track record for the last two years and who require production credit limit of Rs. 5,000 and above are eligible. However, if the branch manager is satisfied about the credit worthiness of the new borrower the facility may be extended to him also.

3. *Assessment of Credit Limit*: The credit limit under the card may be fixed on the basis of operational land holding, cropping pattern and scale of finance as recommended by the District Level Technical Committee (DLTC)/State Level Technical Committee (SLTC). Where the DLTC/SLTC have not recommended the scale of finance for any crops, branch may fix appropriate scale of finance after the consultation with the controlling authorities. While fixing the limit, the branch may take into account the entire production credit requirements of the farmer for the full year including the credit requirement of the farmer for the auxiliary activities related to crop production such as maintenance of agricultural machinery/implements. Branches will fix appropriate limit, taking into account the seasonally in credit requirements. A projected cash flow can be worked out on the basis of expenses required for cultivation purposes and repayment plan of the farmer and the peak requirements of funds can be assessed there from. Where the land is in the joint names of two or more persons, the account under the scheme can be operated by one person provided the joint account holders request the bank in writing.

4. *Contingent Credit Needs*: As per the R.V. Gupta Committee recommendations on "the flow of credit to agricultural sector" apart from production needs of the farmer, the contingent credit needs of the farmers also are required to be financed. In view of this, financing the contingent needs can be suitably built into *kisan credit* facility. The contingent needs pertaining to medical expenses, expenses for education of children and expenses related for marriages, funerals, births and certain religious ceremonies can be financed under the arrangement. The drawing power for the contingent needs should be added to the total limit under the card scheme and the farmer has to avail the facility as and when required to an extent of 15 per cent of the peak credit limit granted (subject to a maximum of Rs. 10,000) for production purposes.

Farmers would be issued a credit card — cum pass book incorporating the name, address, particulars of land holding, borrowing limit, validity period, *etc.*, which will serve both as an identity card as well as facilitate recording of the transactions on on-going basis. Regarding security it is similar to agricultural advances.

5. *Renewal of Limit*: The credit card should normally be valid for 3 years subject to an annual review. The repayment should coincide with the harvest and sale of crops which has to be built into the projected loan disbursal and repayment statement while fixing the drawing power.

When the branch has granted extension and/or reschedulement of the period of repayment on account of natural calamities affecting the farmer or for reasons beyond the control of the borrower, the period for reckoning the status of the operations as satisfactory or otherwise would get extended together with the extended amount of limit.

As a measure of incentive for cardholders with good performances the branch extends the credit limit suitably to take care of increase in cost of inputs/labour, change in cropping pattern, *etc.*

The credit card account would be deemed to be a non-performing asset (NPA) if it remains inoperative for a period of two crop seasons. The introduction of KCC scheme is expected to facilitate the flow of credit to agricultural sector, thereby enhancing disbursement of crop loans to the farmers.

## COMPREHENSIVE CROP INSURANCE SCHEME (CCIS)

Agricultural production is fraught with risk and uncertainty conditions. Risks are very many and generally measurable, whereas uncertainty situations are not amenable to measurement. This general distinction is lost in the modern usage and hence, both terms are used interchangeably. Weather uncertainty, price risk and production risks are the major types confronting Indian agriculture. However, risks also emanate from the institutional changes along with natural hazards like flood, drought, hail, cyclones, hurricanes, earthquakes, *etc.* Natural hazards cause widespread devastation and grave loss to the properties of humans and their lives. Under such situations, disaster management programmes will be implemented by the Government with large-scale help coming both from the people and the Government.

Market risks arise from the fluctuations in demand and supply situations of goods and over-indulgence of middlemen in the market system. Vegetable products, fruits, commercial agricultural products, *etc.,* do have substantial market risk, sometimes forcing the producer into helpless distress sales. Strategies like storage, processing, hedging, *etc.* are being followed to overcome such situations.

Risks that are associated with the weather aberrations such as changes in the rainfall distribution, quantum of rainfall, floods, droughts, *etc.,* result in heavy losses of crop and livestock products. Inefficient management of the farms due to illiteracy, ill-health and ignorance along with institutional risk also causes losses in production of crops and allied products. The production risks in dryland agriculture due to these factors are not only higher in magnitude but also more frequent. Hence, the farmers in these regions developed risk averse attitudes. Many management strategies have been evolved over time to combat production risks. They are practices of improvement of soil moisture holding capacity through bunding, levelling, mulching, strip cropping, mixed cropping, coupled with other agronomic practices and tenancy practices. Crop insurance, of late is being given prime importance by the Government of India primarily with the idea of safeguarding the farmers against crop losses during drought and flood years.

### Origin of Crop Insurance Scheme

The desire to introduce two pilot schemes, *viz.,* crop insurance scheme and cattle insurance scheme with the objective of protecting the farmers from the heavy losses of crop and livestock by the Government of India, dates back to 1948 soon after the independence. But, none of the State Governments agreed to implement the schemes due to paucity of funds. In the year 1970, an Expert Committee on crop insurance under the Chairmanship of Dharamnarain was appointed by the Government of India to examine and analyse administrative and financial implications of the scheme with a view to introducing it. In his report he ruled out the possibility of implementing the scheme in India. Contrary to this report, Prof. Dandekar strongly defended the implementation of the scheme. In 1973, the Government of India had set up General Insurance Company (GIC) to carry out all types of insurance business throughout the nation with four subsidiary insurance companies.

At the instance of the Government of India, GIC first introduced the crop insurance scheme in 1973 on experimental basis as a pilot scheme in selected centres of Gujarat. Only H4 cotton was considered for implementation of the scheme. Later, the same was extended to West Bengal, Tamil Nadu and Andhra Pradesh for the cotton crop and this scheme was in operation till 1979 except the year 1977.

Area based crop insurance scheme was subsequently introduced from 1979 by GIC on a pilot basis in selected areas. If the actual average yield of the crop in the selected area was less than the guaranteed yield of the crop, then the indemnity would become payable to all the insured farmer-borrowers. Sum assured was 100 per cent of the crop loan but a ceiling was imposed with regard to payment of indemnity, *i.e.*, Rs. 5000 per farmer borrower in the case of dry land and Rs. 10,000 per farmer-borrower in the case of irrigated areas. This scheme was implemented by 12 States in India up to 1984.

In the year 1985 Comprehensive Crop Insurance Scheme (CCIS) was introduced by GIC in all the States. The scheme covers all farmers availing crop loans and it is limited to cereals such as rice and wheat, millets, oilseeds and pulses. The loans granted from 1st April to 30th September qualify for *kharif* insurance business. The loans granted from 1st October to 31st March of next year qualify for *rabi* insurance business. All farmers availing crop loans for crops specified are covered under insurance. Insurance cover is built-in-aspect of crop loan. Crop insurance risk is shared by the GIC and the State Government in the ratio of 2 : 1. The sum insured is 100 per cent of the crop loans for the crops disbursed to a farmer during the season. Presently the sum is restricted to Rs. 10,000 per farmer (for all insurable crops) irrespective of the quantum of loan taken by the farmer. Only that part of crop loan is insurable which is actually utilized for the purpose of covering insured crops. The insurance premium is fixed at 2 per cent of the sum insured for rice, wheat, and millets, whereas for oilseeds and pulses it is one per cent. The premium is sanctioned as an additional loan to the farmers and should not be deducted from the loan amount. In the case of small and marginal farmers, 50 per cent of the insurance premium is subsidized by the central and State Governments in equal proportion. Indemnity payable under the scheme is ascertained on the basis of "Threshold yield" which is equal to 80 per cent of the average yield per hectare of the defined area for the previous years. More specifically 80 per cent of the average annual yield of the given crop in a given area (block level) over the last preceding five years is considered as 'threshold yield' in that area. Shortfall in the yield of the crop is the difference between threshold yield and actual yield of the crop in a particular area in the year under reference. Indemnity is calculated based on the following formula.

$$\text{Indemnity} = \frac{\text{Shortfall in the yield of the crop}}{\text{Threshold yield of the crop}} \times \text{Sum insured}$$

The yield data for this purpose is the data collected through crop cutting experiments conducted by the State Government in accordance with the prescribed procedure approved by the national sample survey organization, ministry of planning, Government of India.

## Advantages of Crop Insurance

The following are the specific advantages of the present scheme, which is now in operation in all the States since 1985.

1. It stabilizes the farm business during the period of crop failure
2. The farmer can act much more confidently in farm business as there is protection against hazards of farming.

3. It prevents the farmers to approach non-institutional agencies during the periods of crop failure.
4. It enhances the use of modern inputs to boost the productivity in agriculture, and
5. In high-risk areas crop insurance serves as a catalyst in bringing areas under cultivation which otherwise would have remained uncultivated.

The drawbacks identified in the existing comprehensive crop insurance scheme are:

1. It provides coverage only to a limited number of crops–wheat, paddy, oilseeds, millets and pulses excluding important crops such as sugarcane, potato, cotton, *etc.*
2. The coverage was restricted to rainfed crops only and because of this, the scheme was not effective in States such as Punjab, Haryana and Western Uttar Pradesh and
3. The scheme covered only those farmers who had taken loans from financial institutions and the sum insured was limited to a maximum of Rs. 10,000 only.

The eminent economists regarding satisfactory functioning and improvement of the crop insurance scheme made certain suggestions. The important suggestions among them are:

1. All crops and all farmers should be brought under the purview of the scheme.
2. The premium rates should vary with the nature and index of crop production in different areas;
3. The defined unit area for paying indemnity should be a village or group of villages as against block, as is being considered at present
4. Threshold yield should be worked out by considering indices of crop production over a 10 year period as against five year period, *etc.*

## National Agricultural Insurance Scheme (NAIS)

National Agricultural Insurance Scheme (NAIS) a vastly improved insurance package over the existing comprehensive crop insurance scheme (CCIS) was launched by the Primer Minster on 23.6.99. The new package is expected to take the concept of crop insurance closer to the farmers. It was from 1999-2000 *rabi* season, the NAIS would provide insurance facilities to all farmers, irrespective of the size of their holdings. The scheme would cover all crops, including coarse crops, all pulses and oilseeds. In addition, three cash crops *viz.*, sugarcane, potato, and cotton were also brought under the purview of the scheme in the first year. All the other crops including horticulture and other commercial crops would the placed under insurance cover within 2002. There was no restriction on the total sum insured.

The premium rates would be 3.5 per cent of the sum insured for bajra and oilseeds and 2.5 per cent for other *kharif* crops. It would be 1.5 per cent of the sum assured for wheat and two per cent for other *rabi* crops. The scheme also had provision of 50 per cent subsidy on the premium amount for small and marginal farmers. However, the subsidy would be phased out over a period of five years.

The scheme would be operated on the basis of area approach and all the farmers of a defined area, affected by a calamity would be entitled to payment of insurance claim according to the indemnity rates prescribed for the area. However, in the event of a localized calamity, individual claims of affected farmers would be entertained separately. Localized calamities would include hailstorm, landslip, cyclones, floods, *etc.*

Initially, the scheme would be implemented by the General Insurance Corporation but a separate organization would be created subsequently for the purpose. The new organization would be called 'Indian Agricultural Insurance Corporation'.

The progressive farmers who usually manage to reap better harvests, would have the option to insure for higher compensation by paying more premium. The settlement of the claims would be the responsibility of the insurance agency, which would be required to clear all cases before the beginning of the new cropping season.

As the name suggests, the *Rashtriya Krishi Bima Yojana'* would be applicable all over the country. The States opting to offer this facility to its farmers would be free to join the scheme but with two conditions. First, the scheme would have to be continued for a period of at least three years, second, all the crops would have to be covered.

The features of the new scheme are so conceived as to bring down the ratio of premium to claims to a more manageable 1:1.4 or even lower. Moreover, the Government proposes to give freedom to the GIC's subsidiary company which would operate the scheme to alter or modify insurance charges depending upon the prevailing circumstances. The aim is to make the venture internally viable to be able to survive and serve the intended objective.

According to the preliminary estimates by the Government, the implementation of the new scheme would cost about Rs. 413.18 crore to the centre and an equal amount to the States in the first year of its operation. The expenditure would decline to Rs. 239.47 crore for the centre and an additional equal amount to the States in the second year and further down to Rs. 216.17 crore to each of them in the third year. The likely total cost for the first three years would thus be around Rs. 868.82 crore on the central account and the same amount on the State account. The expenditure would go on diminishing every year till the scheme became fully self-sustaining in five years from the year of commencement.

## Modified National Agricultural Insurance Scheme (MNAIS)

To improve further and make the scheme easier and more farmer-friendly, a Joint Group was constituted by GOI to study the existing crop insurance schemes. Based on the recommendations of the Joint Group and views/comments of various stakeholders, a proposal on Modified National Agricultural Insurance Scheme (MNAIS) was prepared which has been approved for implementation on pilot basis in 50 districts during the remaining period of 11th plan from *rabi* 2010-11. The salient improvements made in MNAIS are as under:

- Actuarial premium with subsidy in premium ranging up to 75 per cent to all farmers;
- Only upfront premium subsidy is shared by the Central and State Governments on 50:50 basis and all claims liability is on the insurance company.
- Unit area of insurance was reduced to village / village panchayat level, for major crops.
- Indemnity for prevented sowing / planting risk and for post harvest losses due to cyclone (coastal areas);
- On account payment up to 25 per cent of likely claims as immediate relief to farmers;
- Uniform seasonality discipline for loanee and non-loanee farmers;
- More proficient basis for calculation of threshold yield; and minimum indemnity level increased to 70 per cent instead of earlier 60 per cent;

- The scheme is compulsory for loanee farmers and voluntary for non - loanee farmers;
- Participation of private sector insurers for creation of competitive environment for crop insurance.
- Setting up a catastrophic fund at the national level contributed by the central and the state government on 50:50 basis to provide protection to the insurance companies in the event of premium to claim ratio exceeds 1:5 at national level and failure to procure appropriate reinsurance cover at competitive rates;
- NAIS is withdrawn from those area(s) / crop(s) where MNAIS is implemented.

The scheme has been implemented in all the 50 districts during *rabi* 2011 - 12 season and in 44 districts during *kharif* 2012 and is being implemented in 35 districts during *rabi* 2012-13. Since inception of the pilot scheme, 33.26 lakh farmers have been covered over an area of 36.27 lakh hectares insuring a sum amounting to Rs.8,063.73 crore. The claims amounting to about Rs.234.27 crore have become payable against the premium of about Rs.824.38 crore benefitting about 2.29 lakh farmers (upto *kharif* 2012 season).

## Pilot Weather Based Crop Insurance Scheme (WBCIS)

With the objective to bring more farmers under the fold of crop insurance, a Pilot Weather Based Crop Insurance Scheme (WBCIS) was launched in 20 States (as announced in the Union Budget 2007). WBCIS aims to provide insurance protection to the farmers against adverse weather incidence, such as deficit and excess rainfall, high or low temperature, humidity etc. which are deemed to impact adversely the crop production. It has the advantage to settle the claims within shortest possible time. The WBCIS is based on actuarial rates of premium but to make the scheme attractive, premium actually charged from farmers is restricted at par with NAIS. In addition to Agricultural Insurance Company of India Ltd. (AIC) private General Insurance Companies i.e. ICICI - Lombard, IFFCO - TOKIO, HDFC - ERGO and Cholamandalam MS Ltd. have also been allowed for implementation of the scheme. From Current *kharif* 2013 season, five more private insurance companies have also been allowed.

Since inception of the scheme in *kharif* 2007, 323.74 lakh farmers have been covered over an area of 450.31 lakh hectares insuring a sum amounting to Rs.55,813.40 crore. The claims amounting to about Rs. 2,764.35 crore have become payable against the premium of about RS. 5,113.25 crore benefiting about 181.26 lakh farmers (upto *kharif* 2012 season).

# CHAPTER 39

# Tests of Farm Credit Proposals

The technological break-through, which revolutionized Indian agriculture made it capital-intensive. In our country, most of the farmers are capital-starved necessitating the institutional agencies to provide the needed capital base through credit. The farmers need the credit at the right time from the right agency to derive maximum productivity out of it. This is from the farmers' point of view. In the view of the bankers, on the other hand, when a farmer approaches an institutional agency with a proposal for loan, the banker should be convinced about the economic viability of the proposed investment. In this connection, some guidelines are essential for the banker to ponder over, for, each investment activity is different from the other in terms of productivity. Keeping this in view, various farm credit proposals including the details of repayment plans are furnished in this Chapter.

## ECONOMIC FEASIBILITY TESTS OF CREDIT

When we take up this issue, three basic financial aspects are assessed by the banker.

(1) If the loan is advanced, will it generate returns more than the costs;
(2) Will the returns have surplus, to repay the loan when it falls due, and
(3) Will the farmer stand up to risk and uncertainty in farming.

These three aspects are popularly known as 'Three Rs.' of credit, which are as follows.

(1) Returns from the investment;
(2) Repayment capacity the investment generates, and
(3) Risk-bearing ability of the farmer-borrower.

## RETURNS FROM THE INVESTMENT

This is an important measure in the credit analysis. The banker needs to have an idea about the extent of returns likely to be obtained from the proposed investment. The farmer's demand for credit can be accepted only when he will be able to generate returns that will enable him to tide over the costs. Returns depend upon the decisions like what to grow, how to grow, how much to grow, when to sell, where to sell, *etc.*, which the farmers take in their production activities. The main concern here is that the farmers should be able to generate incremental income when they go for the additional returns from the borrowed funds. To estimate the additional costs to be made good by the borrowed funds, we can apply partial budgeting technique as presented in Table 39.1.

TABLE 39.1   Partial Budgeting Technique.

| Season/Crop | Existing plan | | | | Season/Crop | Alternative farm plan | | | |
|---|---|---|---|---|---|---|---|---|---|
| | Area (ha) | Gross returns (Rs.) | Costs (Rs.) | Net Income (Rs.) | | Area (ha) | Gross returns (Rs.) | Costs (Rs.) | Net income (Rs.) |
| *Kharif* | | | | | *Kharif* | | | | |
| Paddy 1.0 (Improved) | 1.0 | 7,000 | 3,500 | 3,500 | Paddy (H Y V) | 1.0 | 11,500 | 5,200 | 6,300 |
| *Rabi* | | | | | *Rabi* | | | | |
| Paddy 1.0 (Improved) | 1.0 | 7,400 | 3,900 | 3,500 | Paddy (H Y V) | 1.0 | 12,000 | 5,600 | 6,400 |

A. i) Added costs = 5,200 + 5,600 = Rs. 10,800

    ii) Reduced returns = 7,000 + 7,400 = Rs. 14,400

      Total of A = Rs. 25,200

B. i) Added returns = 11,500 + 12,000 = Rs. 23,500

    ii) Reduced costs = 3,500 + 3,900 = Rs. 7,400

      Total of B = Rs. 30,900

Incremental income = 30,900 − 25,200 = Rs. 5,700

By getting a loan amount of Rs. 4,500 and 4,600 in *kharif* and *rabi* respectively, the farmer can switch over from improved varieties of paddy to high yielding varieties in both the seasons. The borrowed fund is quite productive in generating an incremental amount of Rs. 5,700 per hectare of land. It is an important positive factor in favour of the farmer to present his claims for the loan amount from the institutional agency.

## REPAYMENT CAPACITY

This simply means the ability of the farmer to clear off the loan obtained for production purposes within the time stipulated by the bank. The loan amount may be productive enough to generate additional income to the borrower, but it may not be productive enough to repay the loan. Hence, the necessary condition here is that the loan should not only be profitable but also have potential for effecting repayment. Then only the farmer has a favourable point on his side. The above condition emerges out of the fact that repayment capacity not only depends upon the returns, but also on several other factors as given below:

$$Y = \frac{f(X_1, X_2, X_3, X_4 \quad X_5, X_6)}{\text{Quantitative} \quad \text{Qualitative}}$$
variables    variables

Where,

Y = Repayment capacity (in Rs.)

$X_1$ (+) = Gross returns from the enterprise for which loan was taken during a season/year (in Rs.)

$X_2$ (-) = Working expenses (in Rs.)

$X_3$ (-) = Family consumption expenditure (in Rs.)

$X_4$ (-) = other loans due (in Rs.)

$X_5$ (+) = Literacy

$X_6$ (+) = Managerial skill

The signs in parentheses are *apriori* expected signs based on the past studies.

Though the returns are encouraging, other factors may offset the returns reducing the farmer to a helpless condition with regard to repayment capacity. The estimation of repayment capacity varies from crop loans (self-liquidating loans) to term-loans (partially-liquidating loans). In the case of self-liquidating loans the repayment capacity is as follows:

For simplicity, only quantitative variables are taken into consideration.

Repayment Capacity = Gross income – (working expenses excluding crop loan + family living expenses + other loans due + miscellaneous expenditure + crop loan).

A hypothetical example is presented in Table 39.2, which is generally practiced by the bankers.

Another hypothetical example for partially-liquidating loans is given in Table 39.3.

**TABLE 39.2  Estimation of Repayment Capacity for Self-liquidating Loans.**

| Particulars | Amount (in Rs.) | |
|---|---|---|
| | Without loan | With loan |
| Gross returns | 28,000 | 40,500 |
| Working expenses excluding crop loan | 8,800 | 8,800 |
| Family living expenses | 10,000 | 10,000 |
| Other loans due | 4,000 | 4,000 |
| Miscellaneous expenditure | 600 | 600 |
| Crop loan | — | 5,000 |
| Repayment capacity | | 12,100 |

It is evident from Table 39.2 that the farmer has generated gross income of Rs. 40,500 with a loan amount of Rs. 5,000. His repayment capacity stood at Rs. 12,100 after clearing the loan which indicates his credit-worthiness.

In respect of partially liquidating loans, the repayment capacity is estimated in the following manner.

Repayment capacity = Gross income – (Working expenses including short term loans + family living expenses + other loans due + miscellaneous expenditure + annual installment due for term loan).

TABLE 39.3   Estimation of Repayment Capacity for Partially Liquidating Loans.

| Particulars | Amount (in Rs.) | |
| --- | --- | --- |
| | Without loan | With loan |
| Gross returns | 38,000 | 54,000 |
| Working expenses including short term loan | 18,600 | 23,600 |
| Family living expenses | 10,000 | 10,000 |
| Other loans due | 4,000 | 4,000 |
| Miscellaneous expenditure | 600 | 600 |
| Annual installment due for term loan | — | 5,617 |
| Repayment capacity | | 10,183 |

The particulars in Table 39.3 reveal that the farmer has taken an investment loan of Rs. 20,000 which is payable in 5 equal annual installments of Rs. 5,617 each. In this case also the term loan is productive enough to augment gross income to clear off annual installment quite comfortably. The repayment capacity stood at Rs. 10,183 after deducting the annual installment.

**Causes for Poor Repayment Capacity:** Following are the causes for poor repayment capacity of the Indian farmers.
1. Small size of land holdings;
2. Low productivity and production;
3. Low prices and fluctuations of prices for agricultural commodities;
4. High family expenditure;
5. Using farm credit for unproductive purposes;
6. Low farmer's equity;
7. Lack of adoption of improved technology; and
8. Poor management of farm resources.

**Measures to Strengthen Repayment Capacity:** Following are the measures to be adopted to strengthen the repayment capacity of the farmers.
1. Increasing net income by proper organisation and operation of the farm business;
2. Adopting the potential technology for increasing production and reducing the farm expenses;
3. Removing the imbalances in the resource availability;
4. Scheduling the loan repayment plans according to the flow of income;
5. Strengthening networth of the farm households;
6. Diversifying the farm enterprises;
7. Adopting the risk management strategies like crop insurance/cattle insurance/machinery insurance, hedging to control price variations, *etc.*

## RISK BEARING ABILITY

It is the ability of the farmer to withstand the risks that arise due to financial loss. Risk can be quantified through statistical techniques like coefficient of variation, standard deviation, programming models, *etc.*

Probabilities can be estimated and ascribed to the measurement of uncertainty phenomenon. Some of the types or sources of risk in farming are:

1. Production risk;
2. Technological risk;

3. Risk caused by illiteracy and ignorance;
4. Inefficiency by sickness of the farmer (personal risk);
5. Institutional risk;
6. Weather uncertainty;
7. Price uncertainty; *etc.*

The farmer may satisfy the banker with regard to returns and repayment capacity, but yet another factor to be fulfilled is risk-bearing ability. This is vital because at times our estimates go awry and the expected output may not be forthcoming because of the risks enumerated above may stand in the way. Consequently our plans turn topsy-turvy. Here what we wish to know is whether the farmer has got shock-absorbing capacity to withstand the onslaught of the unforeseen events or not. How is the risk-bearing ability estimated under such situations? The productivity of any enterprise or investment activity is gauged by its past performance. Similarly, in estimating the risk-bearing ability, we need to find out what has been the variation in the yields or returns for the given enterprise over the past 5 or 10 years. This variation can be computed using coefficient of variation technique (we have confined ourselves to a simple statistical tool here). The gross returns are deflated to the extent of variability of income CV. For example, if the CV of paddy yields in a given area is 15 per cent, the expected yield or gross returns are deflated by 15 per cent to arrive at the corrected yield or income. From this income so arrived at, the repayment capacity is estimated. In this exercise if the farmer comes out successfully, his credit worthiness is cent per cent endorsed.

Repayment Capacity under risk = Deflated gross returns − (working expenses
(Risk bearing ability) excluding proposed loan + family living expenses + other loans due + miscellaneous expenditure + crop loan).

An hypothetical example pertaining to this can be seen in Table 39.4. Suppose the gross income expected is Rs. 48,000 and the variability in gross income is 15 per cent, then the deflated gross income is Rs. 40,800, *i.e.*, [48,000 − (48,000 × 0.15)].

TABLE 39.4 Estimation of Risk Bearing Ability or Repayment Capacity Under Risk.

| Particulars | Amount (in Rs.) | |
|---|---|---|
| | Without loan | With loan |
| Deflated gross returns | 28,000 | 40,800 |
| Working expenses excluding crop loan | 8,800 | 8,800 |
| Family living expenses | 10,000 | 10,000 |
| Other loans due | 4,000 | 4,000 |
| Miscellaneous expenditure | 600 | 600 |
| Crop loan | — | 5,000 |
| Repayment capacity under risk | | 12,400 |

After allowing the possible reduction in gross income the repayment capacity is also increased with the loan amount. It infers that the farmer has the risk bearing ability in using the borrowed funds. His is a very sound case for consideration of extending loan by the banker.

## Measures to Strengthen Risk Bearing Ability

These are: (1) developing owner's equity which is the backbone of risk bearing ability, (2) developing moral character, *i.e.*, honesty, integrity, dependability, feeling responsibility, *etc.*, which are also called good credit rating, (3) reducing farm and family expenditure, (4) taking up stable and reliable enterprises, (5) providing ability to borrow in both good and bad periods, particularly during the bad periods, the farmer should get the funds, (6) creating ability to earn money and save money, *i.e.*, an individual farmer may be very good in farming but part of the earnings should be saved to meet uncertainty, and (7) taking up crop and other insurance, *etc.*

## 'FIVE Cs' OF CREDIT

Next to the 'Three Rs.' of credit the other tests that can be applied to study the economic viability of a scheme or investment activity are the 'Five Cs' of credit, *viz.*,

1.   Character
2.   Capacity
3.   Capital
4.   Condition, and
5.   Commonsense

### 1. Character

The basis for credit transactions is the trust, the trust that the banker has on his borrowers. No doubt the bank insists upon security for any loan, even then, the element of trust has greater say in the mind of the banker, before he takes a decision in considering the proposal of a prospective borrower. The confidence which the institutional agency keeps is influenced by the moral qualities like honesty, integrity, commitment, hard work, promptness, *etc.*, which the borrower exhibits. In essence it means the mental as well as moral characters of the borrower. Generally, people with good mental and moral character will have a good credit character.

### 2. Capacity

This is related to the capacity of an individual borrower to repay loans when they fall due. It is synonymous with repayment capacity. It largely depends upon the income obtained in the farm business, *i.e.*, $C = f(Y)$, where, $C$ = Capacity and $Y$ = Income.

### 3. Capital

Capital implies the availability of money with the farmer-borrower, when character and capacity proved to be inadequate. It represents the networth of the individual. It is related to repayment capacity and risk-bearing ability.

### 4. Condition

This refers to the conditions needed for obtaining a loan from the financial institutions (presented in detail under the topic, procedural formalities followed in obtaining a loan).

### 5. Commonsense

This relates to perfect understanding between the lender and the borrower in credit transactions. This is in fact a *prima facie* requirement for obtaining credit for the borrower.

## 'SEVEN Ps' OF CREDIT

The role of financial institutions in the light of the technological changes that have been brought in, on the agricultural front, lies in evolving principles of farm finance which are expected to bring not only commercial gains to the bankers but also social benefits. The principles thus evolved by the institutional agencies are supposed to have universal validity. These are popularly known as 'Seven Ps' of credit which are listed and explained hereunder:

1. Principle of productive purpose;
2. Principle of personality;
3. Principle of productivity;
4. Principle of phased disbursement;
5. Principle of proper utilization;
6. Principle of payment; and
7. Principle of protection.

### 1. Principle of Productive Purpose

When owned capital is a limiting factor on the farms, the credit needs of the farmers are many and varied. The requirements of credit commence right from short-term loans to term loans. This capital limitation is visible on all the farms but more pronounced on small and marginal farms. The farmers of these tiny holdings require another type of credit, which the large farmers do not need, *i.e.*, the consumption loan. In the absence of consumption loans for the small and marginal farmers, the crop loans advanced may not be as productive as they are expected to be, because of their diversion for other purposes. But inspite of this known fact, the consumption credit is relegated to the backseat by the institutional agencies. When the loan is diverted for other purposes, the productivity of the loan receives a setback and the desired results will be a far cry. But the principle of productive purpose says that the loan distributed to any borrower should be capable of generating incremental income. If one wants the principle of productive purpose to hold good, the short-term loans of small and marginal farmers can be made productive, if they are provided with other income augmenting assets through term loans. The income generated from these productive assets will add to the income obtained from farming. In this process, the term loans not only turn out to be productive but also help in enhancing the productivity of crop loans taken by these categories of farmers. To cite some of the assets for which term loans required are dairy animals, sheep and goat (grazing or stall-feeding), poultry, installation of pumpsets on group action, *etc.*

### 2. Principle of Personality

The 'Three Rs.' which were explained earlier are the sound indictors of credit-worthiness of the farmers. Credit-worthiness of the farmer makes him eligible for the loan he desires from the institutional agencies. Over the years of experience in lending, the bankers have identified an important factor in credit transactions, *i.e.*, the trust-worthiness of the borrower. It has relevance to personality of the individual. When the farmer-borrower fails to repay the loan in the event of natural calamities, his is a case of non-wilful default. He has to be bracketed in the category of defaulters, not by his own fault, but by the natural forces that influence farming, which are beyond the control of human beings. But a large farmer who profitably uses the loan, and still falls in the category of defaulters means, his is a case of sheer wilful default. This character is born out of the dishonesty of the individual. When this habit becomes perpetual with large farmers, who borrow substantial funds, the very functioning of the institutional business gets crippled. Thus, the safety element of the loan is not totally dependent up on the security of the loan alone, but also on the personality (character) of the borrower. The growth and progress of the lending institutions have dependence on this major influencing factor, i.e., personality. The personality of the individual and growth of the financial institutions, thus are inter-linked.

## 3. Principle of Productivity

This principle emphasizes that the credit, which is advanced, is not just meant for increasing production from that enterprise alone, but should be able to increase the productivity of other factors employed in the enterprise. For example, for taking up any enterprise we need resources (factors of production), but the resource productivity (marginal value productivity) of the factors employed exhibit a varying trend among the enterprises choosen. To cite a few more examples in crop enterprises, preferring HYV to improved variety among the competing crops, choosing the one, which gives relatively higher returns, and in livestock, selecting the breed which is superior among alternatives, etc. Here what we understand is that by our above decisions of varietal preference in crops, better competing crops and superior breeds, not only increase the returns by themselves, but also augment the productivity of other complementary factors employed in the respective production activities. The main concern here is that since we are using scarce borrowed capital resources, no stone should be left unturned in realizing as much productivity as possible from each resource employed. Thus, this principle is based on the point of making the resources as productive as possible by choosing the most appropriate enterprises.

## 4. Principle of Phased Disbursement

Ensuring the end-use of the funds is the most vital aspect of institutional lending. No enterprise or investment activity needs all the required funds at a time and the requirements of funds is spread over a period of time. In paddy crop enterprise, the need for capital is felt over 4 or 5 months for different operations, for sugarcane over an year and investment activities like digging a well or installation of pumpsets require an altogether different time schedule. Relevant to this situation, the principle of phased disbursement underlines that the loan amount needs to be distributed in phases or spells to make it productive and the banker can also make himself sure about the end use of the borrowed funds. This procedure holds good in perennial crops and investment activities, where the phased disbursal of the loan helps to overcome the misuse or diversion of funds, but the demerit of this system is that it will make the cost of credit higher.

## 5. Principle of Proper Utilization

Proper utilization implies using the borrowed funds for the purpose for which they are advanced. It sounds pretty good because every banker by heart and soul wishes this particular aspect for the mutual benefit. This, to certain extent, depends upon the situation prevailing in the rural areas. Explaining, this a bit further, this means whether the farmers are getting the type of resources they need at the right time and in right quantities. Are the resources like seeds, fertilizers, pesticides, *etc.*, free from adulteration to guarantee the farmer to take full advantage of their use? Whether the technical advise is available with regard to production problems that crop up from time to time? Whether infrastructure facilities like storage, transportation, marketing, *etc.*, are available? Is price stability in existence to help the farmer plan the cropping pattern for effective use of funds? Proper utilization of funds is possible, when the suitable conditions for investment of funds exist.

## 6. Principle of Payment

This principle deals with the fixing of repayment schedules of the loans advanced by the institutional agencies. As far as the investment credit is concerned, say for irrigation structures, tractors, *etc.*, the annual repayments are fixed over a given number of years depending upon the incremental returns that are supposed to be obtained after duly accounting for consumption needs of the farmers. With reference to crop loans (barring perennial crops) the loan is to be repaid in lumpsum because he gets the output only once. Two to three months are allowed after the harvest of the crop to enable the farmer to get a reasonable price for his produce; otherwise, he will resort to distress sales. Whenever the crop fails due to the unfavourable weather conditions, the repayment is not insisted upon immediately, and the repayment period is extended besides assisting the farmer with another fresh loan to enable him carry on the farm business.

## 7. Principle of Protection

In view of the unforeseen calamities striking farming more often than not, banks cannot abstain themselves from extending loans to the farmers. Instead, what they do is that they demand the security for the advances they make, otherwise, the overdues resulting due to non-payment of loans by the farmers owing to the natural calamities, affect the recycling of bank funds adversely. To tide over the situation of this nature, the institutional agencies resort to safety measures, *viz.*, (i) insurance coverage, (ii) linking credit with marketing or tie-up arrangement, (iii) provision of finance on production of warehouse receipt, (iv) covering credit under small loan guarantee scheme of Deposit Insurance and Credit Guarantee Corporation of India, and (v) taking securities.

1. *Insurance Coverage:* The loans for certain crops and investment activities like poultry, dairy, piggery, irrigation structures, *etc.*, are insured. Suppose any eventuality breaks out and brings colossal loss to the farmers, it is beyond their capacity to repay the loan, more so if the affected happens to belong to small and marginal categories. Under such situations, the insurance agencies estimate the losses and indemnity is paid to the farmer, from which banks recover their dues.

2. *Linking Credit with Marketing or Tie-up Arrangement:* By linking credit with marketing, the banker is quite safe in recovering the loan. Let us take the case of a sugarcane grower-borrower who supplies cane to the factory as per the agreement. The loan particulars of the sugarcane farmer are let known to the sugar factory. As soon as the crop is harvested, it is supplied to the factory. The fac-

tory will not pay the proceeds of the entire cane received, but deducts the loan component and the balance is paid to the grower. The loan amount so deducted will be credited to the bank against the loan amount taken by the farmer.

3. *Provision of Finance Against the Warehouse Receipt:* When the prevailing product prices are not acceptable to the farmers, they need not submit to the situation. They can store the produce in the warehouse and based on the warehouse receipt, the financial institution advances loans to the extent of 75 per cent of the value of the produce. It is a symbiotic process wherein the bank can recover the loans and the farmers can derive the price benefits when they sell after the glut period was over.

4. *Credit Guarantee:* When the banks fail to recover the loans advanced to the weaker sections, Deposit Insurance and Credit Guarantee Corporation of India (DICGC) reimburses the loans to them.

5. *Taking Sureties:* The banks advance loans either by hypothecation or mortgage of assests

## Acquisition of Capital

By all possible means, the farm financial manager has to raise the needed capital for running the farm business. The sources through which he can raise the capital are savings that have been generated in the farm business in the previous year or years and borrowings. For many farming households in our country, credit (non-equity capital) is a sine-quo-non in running the farm business. As far as possible, credit transactions are maintained with the institutional agencies. In the utilization of non-equity capital, risk is an associate factor. This can be clearly explained with the principle of equity and increasing risk.

# PRINCIPLE OF OWNER'S EQUITY AND INCREASING RISK

The principle hinges upon the fact that the risk in farming tends to increase at an increasing rate, as the owner's equity decreases. When owned capital (equity capital) is limited with the farmers, borrowings will become necessary to raise the needed capital for production. We have to mention here the concept of leverage, which is nothing but the ratio of debt to equity. The leverage will be higher on the farms using more and more of non-equity capital. As the leverage stands higher and the farmers get expected returns, it contributes to the prosperity of the farm business. The non-equity is productive, when every thing goes right on farms, but it is equally destructive when the farmer's expectations go topsy-turvy. The borrowed capital, which earlier brought prosperity to the farmers, now spells doom.

## Statement of the Principle

As the debt-equity ratio or leverage increases, the borrower runs a greater risk of loosing owned capital. This principle is clearly explained with an example presented in Table 39.5. Before that, we should know the formula of percentage of equity.

$$\text{Percentage of equity} = \frac{\text{Owned capital}}{\text{Owned capital + borrowed capital}} \times 100$$

We have considered five farms with an identical owned capital of Rs. 5,000, each. Barring farm A, the other four farms have borrowed an amount of Rs. 2,500, Rs. 5,000,

Rs. 7,500 and Rs. 10,000 respectively. Consequently, the total capital available with A, B, C, D and E farms exhibited an increasing trend. Assuming that a gain of 15 per cent uniformly on all the five farms, the resultant profit would be Rs. 750, Rs. 1,125, Rs. 1,500 Rs. 1,875 and Rs. 2,250 on the corresponding farms.

After paying the interest to the lending agency, owned capital increased in ascending order for the respective farms in question. The percentage gain on owned capital too, exhibited a similar trend. Now let us assume that there is a 15 per cent loss uniformly on the five farms. Analogous to the profit trend, the loss calculated too reveals an increasing trend on farms A, B, C, D and E receptively. The position of owned capital after the loss resulted in progressive decline leading to increased percentage loss on owned capital on the above farms. An overview of the particulars in Table 39.5 refers that, as the borrowed funds increase the percentage of equity gets reduced under the condition of losses. No doubt borrowed capital is productive as long as there is no setback, but in the context of any eventuality, it is equally destructive to the farmer. This amply demonstrates that credit is a double edged knife and hence the requisite amount only, needs to be borrowed by the farmers.

TABLE 39.5 Principle of Equity and Increasing Risk.

| S. No. | Items | Farms | | | | |
|---|---|---|---|---|---|---|
| | | A | B | C | D | E |
| 1. | Owned capital (Rs.) (equity capital) | 5,000 | 5,000 | 5,000 | 5,000 | 5,000 |
| 2. | Borrowed capital (Rs.) (non-equity capital) | — | 2,500 | 5,000 | 7,500 | 10,000 |
| 3. | Owned + borrowed capital (Rs.) (Total capital) | 5,000 | 7,500 | 10,000 | 12,500 | 15,000 |
| 4. | Leverage ratio | 0 | 0.5 | 1.0 | 1.5 | 2.0 |
| 5. | Gain @ 15% on total capital | 750 | 1,125 | 1,500 | 1,875 | 2,250 |
| 6. | Interest to be paid to the institutional agency @ 12% | — | 300 | 600 | 900 | 1,200 |
| 7. | Owned capital at the end of year (Rs.) | 5,750 | 5,825 | 5,900 | 5,975 | 6,050 |
| 8. | Gain on owned capital in % | 15.0 | 16.5 | 18.0 | 19.5 | 21.0 |
| 9. | Loss @ 15% on total capital | 750 | 1,125 | 1,500 | 1,875 | 2,250 |
| 10. | Interest to be paid to the institutional agency @ 12% | — | 300 | 600 | 900 | 1,200 |
| 11. | Owned capital at the end of the year (Rs.) | 4,250 | 3,575 | 2,900 | 2,225 | 1,550 |
| 12. | Loss on owned capital in % | 15.0 | 28.5 | 42.0 | 55.5 | 69.0 |

# PROCEDURAL FORMALITIES IN SANCTION OF FARM LOANS

The financing bank is vested with the powers either to accept or reject the farmer's loan application. This is sequel to an objective appraisal of farm credit proposals and procedures and formalities followed in the processing of loans. Here an attempt is made to explain the set of procedures and formalities required in processing of a farm loan application. The processing procedure is detailed under the following sub-heads.

1. Interview with the farmer;
2. Submission of loan application by the farmer;
3. Scrutiny of records;
4. Visit to the farmer's field before sanction of loan;
5. Criteria for loan eligibility;
6. Sanction of loan;
7. Submission of requisite documents;
8. Disbursement of loan;
9. Post-credit follow-up measures; and
10. Recovery of loan.

## 1) Interview with the Farmer

The banker studies the farmer-borrower in the interview regarding his credit characteristics such as honesty, integrity, frankness, progressive thinking, indebtedness, repayment capacity, *etc.* The banker explains to the farmer the terms and conditions under which the loan is going to be sanctioned. Interview helps the banker to understand the genuine credit needs of the farmer. So interview is more than a mere formality, as it facilitates the banker to study the farmer in detail and assess his credit requirements.

## 2) Submission of Loan Application by the Farmer

After getting satisfied with the credentials of the farmer, the banker gives a loan application form to him. Details regarding the location of the farm, purpose of the loan, cost of the scheme, credit requirements, farm budgets, financial statements, *etc.,* as required in the form are filled in by the farmer. Certificates such as 10·1 (indicates ownership of the land or title deeds) and Adangal (statement showing cropping pattern adopted by the farmer-borrower), farm map, no objection certificate from the co-operatives, non-encumbrance certificate from Sub-Registrar of Land Assurances, affidavit from the borrower regarding his non-mortgage of land elsewhere are appended to the loan application. A passport size photograph is affixed to the loan application form.

## 3) Scrutiny of Records

The ownership and extent of land as indicated in the relevant certificates are verified by the bank officials with village karanams or village revenue officials.

## 4) Visit to the Farmer's Fields Before Sanction of Loan

After verifying the records, the field officer of the bank pays a visit to the farm to verify the particulars given by the farmer. The pre-sanction visit is expected to help the banker to identify the farmer and guarantor, locate the boundaries of land as per the map and assess the managerial capacity of the farmer in farming and allied enterprises and the farmer's attitude towards latest technology. Details on economics of crop and livestock enterprises, feasibilities for implementing proposed plans, and farmer's loan position with the non-institutional sources are ascertained in the pre-sanction visit. Thus, pre-sanction visit of the bank officials is very important to verify credit-worthiness and trust-worthiness of the farmer-borrower. While appraising different types of loans, different aspects should be verified. For example, to advance loan for well digging, the location of proposed well, ground water availability, distance from the nearby well, rainfall, command area of the well, *etc.*, are verified in the pre-sanction visit. Similarly, for other loans, the pertinent aspects are verified. All these aspects are included in the report submitted to the branch manager for taking up final decision in the sanction of the loan.

## 5) Criteria for Loan Eligibility

The following aspects are considered in judging the eligibility of a farmer-borrower to receive the loan.
(1) He should have sound character and financial integrity,
(2) His dealings with friends, neighbours, financial institutions, *etc.*, must be proper (He should not be a defaulter in the past),
(3) He must have progressive outlook and be receptive to modern technology,
(4) He should sincerely implement the proposed scheme and ensure proper use of credit,
(5) The security provided by the farmer must be free from any sort of encumbrance and litigation.

## 6) Sanction of Loan

After examining all the aspects presented in the pre-sanction farm inspection report, the branch manager takes a decision as to whether to sanction the loan or not. Before sanctioning, the branch manager considers the technical feasibility, economic viability and bankability of proposed projects including the repayment capacity, risk-bearing ability and sureties offered by the farmer-borrower. If the loan amount is beyond the sanctioning power of the branch manager, it is forwarded to the Regional Manager or head office of the bank, incorporating his recommendations. The authorities at the respective offices take the final decision on the proposed projects, and communicate their decision to the branch manager for further action.

## 7) Submission of Requisite Documents

Before sanctioning the stipulated amount to the farmer-borrower, the following documents are obtained by the banking institution.
(1) Demand promissory note;
(2) Deed of hypothecation;
(3) Guarantee letter;
(4) Installment letter;
(5) Authorization letter regarding the payments of loan from the marketing agencies or intermediaries on behalf of the farmer; and
(6) Mortgage deeds.
Title deeds are examined by the legal officer of the bank and his opinion with regard to clear, marketable and unlitigated title is sought.

Simple mortgage is followed in the case of ancestral property and equitable mortgage in respect of acquired property. However, the opinion of the bank's legal officer is obtained in this regard. Mortgage of land is done prior to obtaining non-encumbrance certificate and sanction of loan.

## 8) Disbursement of Loan

As soon as the execution of documents is completed, the loan amount is credited to the borrower's account. The loan amount is disbursed in a phased manner, that too after ensuring that the loan is used by the farmer-borrower properly. A realistic repayment plan is framed and given to the farmer keeping in view the income flow of the proposed project.

## 9) Post Credit Follow-up Measures

The branch manager or agricultural officer pays a visit to the farmer to ascertain the proper use of the credit. This also benefits the farmer, for they can get the technical advice if any needed from the agricultural officer in the implementation of the scheme. These visits are also meant for developing a close rapport between the farmer and the banker. These visits are more informal than formal which are supposed to inculcate the feeling of friendliness and underlying the obligation of the farmer to repay the loan when it falls due. Such visits also facilitate in assessing any further requirement of supplementary credit to complete the scheme.

## 10) Recovery of Loan

The bank reminds the farmer-borrower in advance about the repayment of loan in time. If needed, recovery camps, special drives, village meetings, *etc.*, are organized at an appropriate time. All appropriate measures are taken to persuade the farmer-borrower to repay the loan in time. In the case of default, the reasons for the same are ascertained to find out whether the borrower is a deliberate defaulter or not. If the reason is genuine, the borrower is further helped by extending finance to accelerate farm production. In such situations, a closer supervision is necessary. If the bank officials find that the borrowers are wilful defaulters, stringent measures are initiated to recover the loans through court of law.

In all possible cases the bank officers make tie-up arrangements, *i.e.*, the recovery of the loan is linked with marketing. Rephasing of repayment plan is allowed in the case of justifiable cases.

# REPAYMENT PLANS

For term loans, which are characterized by partially liquidating nature, the loan repayment plan is not as similar as that of short term loans. These loans are recovered through a given number of installments depending upon the nature of the asset and the amount advanced for the asset in question. Various repayment plans in vogue are listed and briefly explained here.

1. Straight-end payment plan or single repayment plan or lumpsum repayment plan;
2. Partial repayment plan;
3. Amortised repayment plan;
a. Amortised decreasing repayment plan;
b. Amortised even repayment plan;

4. Variable repayment plan, and
5. Reserve repayment plan.

## 1. Straight-End Payment Plan or Single Repayment Plan or Lumpsum Repayment Plan

The entire loan amount is to be cleared off after the expiry of loan period stipulated. More clearly in this method, the principal component is repaid by the farmer at a time in lumpsum when the loan matures, while the interest component is paid each year.

## 2. Partial Repayment Plan or Balloon Repayment Plan

The farmer is expected to settle the entire loan amount in quarterly, half-yearly or annual instalments (principal + interest). It implies that repayment of loan will be done partially over the years. Usually, the instalment amount will be decreasing as the years pass by except in the maturity year (final year) during which the investment generates sufficient revenue for liquidation. Table 39.6 illustrates this.

*Example*:   Loan amount Rs. 10,000
Time period 6 years
Rate of interest      12%

This is also known as balloon repayment plan, as the large final payment is made at the end of the loan period following a series of smaller partial payments.

**TABLE 39.6   Partial Repayment Plan.**

| Year | Principal (in Rs.) | Interest (in Rs.) | Installment (in Rs.) | Balance amount (in Rs.) |
|------|------|------|------|------|
| 1 | 1,000 | 1,200 | 2,200 | 9,000 |
| 2 | 1,000 | 1,080 | 2,080 | 8,000 |
| 3 | 1,000 | 960 | 1,960 | 7,000 |
| 4 | 1,000 | 840 | 1,840 | 6,000 |
| 5 | 1,000 | 720 | 1,720 | 5,000 |
| 6 | 5,000 | 600 | 5,600 | — |
| Total | 10,000 | 5,400 | 15,400 | — |

## 3. Amortised Repayment Plan

It is an extended version of partial repayment plan. Amortisation means the repayment of the entire loan amount in a series of installments. Here we have two types of amortisation plans, *viz.*, amortised decreasing repayment plan and amortised even repayment plan.

1.   *Amortised Decreasing Repayment Plan*: In this repayment plan, the principal component remains constant over the entire repayment period, while the interest part decreases continuously. With the principal amount remaining fixed and interest amount decreasing, the annual installment amount decreases over the years. The advance made for the purchase of machinery is one of the suitable examples under this category, for the machinery does not demand much repairs in the initial years of loan payments enabling the farmer to repay large amounts of

installments. The diagrammatic representation of the repayment schedule is shown in Figure 39.1.

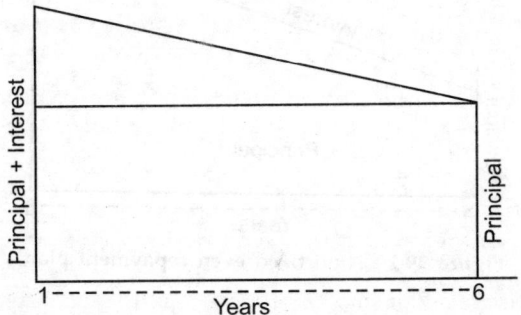

**Figure 39.1** Amortised decreasing repayment plan.

The arithmetic calculation of the plan is embodied in Table 39.7.

*Example:*      Loan amount           Rs. 10,000
            Time period           6 Years
            Rate of interest      12%

**TABLE 39.7  Amortised Decreasing Repayment Plan.**

| Year | Principal (in Rs.) | Interest (in Rs.) | Installment (in Rs.) | Balance amount (in Rs.) |
|------|------|------|------|------|
| 1 | 1,666.67 | 1,200.00 | 2,866.67 | 8,333.33 |
| 2 | 1,666.67 | 999.99 | 2,666.66 | 6,666.67 |
| 3 | 1,666.67 | 799.99 | 2,466.66 | 5,000.00 |
| 4 | 1,666.67 | 600.00 | 2,266.67 | 3,333.33 |
| 5 | 1,666.67 | 399.99 | 2,066.66 | 1,666.67 |
| 6 | 1,666.67 | 199.99 | 1,866.67 | |
| Total | 10,000.00 | 4,199.96 | 14,199.96 | |

2. *Amortised Even Repayment Plan*: This is called equated annual installment method. The annual installment over the entire loan period remains the same in this method. The principal portion of the instalment increases continuously, while the interest part declines gradually (Table 39.8). This method is mostly adopted for term loans. Loans granted for farm development, digging of wells, deepening of old wells, construction of godowns, dairy, poultry, *etc.*, are the examples. This is depicted diagrammatically in Figure 39.2.

The annual installment is arrived at through the formula given below:

$$I = B\frac{i}{1-(1+i)^{-n}}$$

Where,

I = Annual installment in Rs.
B = Principal amount borrowed in Rs.
n = Loan period in years
i = Annual interest rate in fraction

458

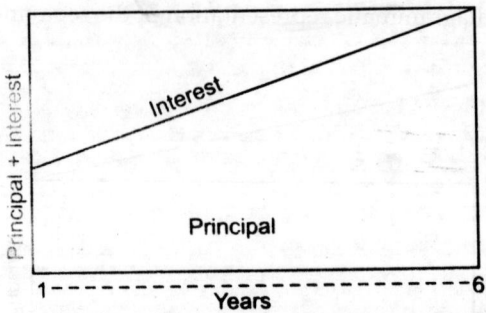

**Figure 39.2** Amortized even repayment plan.

or $I = B \dfrac{1}{a_{\overline{n}|i}}$

Where,

a = Annuity in Rs.
n = Loan period in years
i = Annual interest rate in fraction

*Example:* Loan amount     Rs. 10,000
           Time period      6 years
           Rate of interest    12%

$$I = B \dfrac{i}{1 - (1 + i)^{-n}}$$

$$= 10,000 \times \dfrac{0.12}{1 - (1 + 0.12)^{-6}}$$

$$= 10,000 \times \dfrac{0.12}{1 - \left( \dfrac{1}{(1.12)^6} \right)}$$

$$= 10,000 \times 0.243225$$
$$= \text{Rs. } 2,432.25$$

**TABLE 39.8 Amortised Even Repayment Plan.**

| Year | Instalment (in Rs.) | Principal (in Rs.) | Interest (in Rs.) | Balance amount (in Rs.) |
|------|---------------------|--------------------|-------------------|-------------------------|
| 1 | 2,432.25 | 1,232.25 | 1,200.00 | 8,767.75 |
| 2 | 2,432.25 | 1,380.12 | 1,052.13 | 7,387.63 |
| 3 | 2,432.25 | 1,545.73 | 886.52 | 5,841.90 |
| 4 | 2,432.25 | 1,731.22 | 701.03 | 4,110.68 |
| 5 | 2,432.25 | 1,938.97 | 493.28 | 2,171.71 |
| 6 | 2,432.25 | 2,171.64 | 260.61 | — |
| Total | 14,593.50 | 9,999.93 | 4,593.57 | — |

## 4. Variable Repayment Plan

As the very name indicates, various levels of installments are paid by the borrower over the loan period. In times of good harvest a higher instalment is paid, while in periods of low yields lesser amount is credited towards instalment to the lender. According to the convenience, the borrower effects the repayment. This method is not found with institutional borrowings.

## 5. Reserve Repayment Plan or Future Payments

This type of repayment is made by the borrowers in areas which are subject to high income variability of farms. The impending problem here is that the farmers are haunted by the fear that they may not be able to keep up their promise of repaying loans at scheduled time. To overcome such situations, the farmers make advance payments of the loan realized from the savings of the previous year. The farmer is not a looser in this transaction by any means since he is paid interest at the rate charged on the loans for the advance amount credited. This type of repayment is advantageous to the banker as the institutional agency need not worry regarding loan collection during the periods of crop failure. The farmer too gains here as he can keep up his integrity in credit transactions.

# CHAPTER 40

# Tools of Farm Financial Analysis

Every farm financial manager has to assess the performance of his business in order to act suitably. Various tools of financial analysis, *viz.*, balance sheet, income statement, cash flow statement, break-even analysis, *etc.*, are available to him in this regard, which are presented below:

## BALANCE SHEET OR NETWORTH STATEMENT

Any farmer, whether small, medium or large, measures financial performance of the farm business during an agricultural year or over a period of time. There is a possibility in the variation of degree of keenness that is shown by the different categories of farmers. In other words, as the size of the farm gets increased, the capital requirement too gets enlarged forcing the farmer to be more vigilant in running the farm business, since the risk element is much higher in the event of any unforeseen eventuality. Management component plays a pivotal role in managing higher financial outlays. Nevertheless, management of finance is equally important even for a small farmer, if not, at the magnitude that is viewed at by a large farmer in the farm business. The balance sheet indicates an account of total assets and total liabilities of the farm business revealing the financial solvency of the business. More specifically it is a statement of the financial position of a farm business at a particular time, showing its assets, liabilities and equity. If the assets are more than the liabilities it is called networth or equity and its converse is known as net deficit. The typical balance sheet shows assets on the left hand side and liabilities and equity on the right hand side. Both sides are always in balance hence the name balance sheet. Networth is placed on the right hand side, along with liabilities, in order to indicate that like any other creditor the farmer has a claim against the farm business equal to the equity amount. The balance sheet can be easily prepared by the farmer in the presence of farm records. It can be prepared at any point of time to know the financial position of the farm business. It can also be prepared to study the performance of a business over years by preparing the same number of balance sheets. If the networth increases over the different periods, it indicates efficient performance of the business. To prepare a balance sheet the prime requisites are total assets and total liabilities of the farm.

*Assets:* Assets are those, which are owned by the farmer.

*Liabilities:* These refer to all things, which are owed to others by the farmer.

Assets are the three types, *viz.,* current, intermediate or working and long-term or fixed. So also the liabilities. This classification of assets facilitates the analysis of liquidity of the farm business.

*Current Assets:* They are very liquid or short-term assets. They can be converted into cash, within a short time, usually one year. For example, cash on hand, agricultural produce ready for disposal, *i.e.,* stocks of paddy, blackgram, jowar, wheat, *etc.*

*Intermediate or Working Assets:* Intermediate assets are less liquid than the current assests. Examples: Machinery, equipment, livestock, tractors, trucks, *etc.*

*Long-term Assets or Fixed Assets:* An asset that is permanent or will be used continuously for several years is called a long-term asset. It takes longer time to convert into cash due to verification of records, legal transactions, *etc. Examples*: Land, farm buildings, *etc.*

*Current Liabilities:* Debts that must be paid in the short term or in very near future. Examples: Crop loans, accounts payable, hand loans, *etc.*

*Intermediate Liabilities:* These loans are due for the repayment within a period of two to five years. *Examples*: Livestock loans, machinery loans, *etc.*

*Long-term Liabilities:* The duration of loan repayment is five or more years. *Examples*: Tractor loan, orchard loan, land development loan, *etc.*

**Test Ratios:** The balance sheet is analysed with the help of ratio measures so as to know the exact financial position and stability of the farm business.

The test ratios, *viz.,* current ratio, intermediate ratio, net capital ratio, quick ratio, current liability ratio, debt-equity ratio and equity-value ratio are used to analyse the balance sheet (Table 40.1).

**TABLE 40.1 Balance Sheet of a Hypothetical Farm.**

| Assets | Amount (in Rs.) | Liabilities | Amount (in Rs.) |
|---|---|---|---|
| **Current assets** | | **Current liabilities** | |
| Cash on hand | 10,000 | Crop loans to be repaid to institutional agencies | 8,000 |
| Savings in bank | 8,000 | Accounts payable | 11,000 |
| Value of grains ready for disposal | 38,500 | | |
| Livestock products (eggs, birds, *etc.*) | 60,000 | Hand loans | 5,000 |
| Fruits, vegetables, fodder and feed ready for sale | 8,000 | Money owed to input suppliers | 25,000 |
| Value of bonds and shares to be realized in the same year | 2,000 | Annual installments of MT and LT loans | 19,000 |
| Sub-total | 1,26,500 | Sub-total | 68,000 |

| Intermediate assets | | Intermediate liabilities | |
|---|---|---|---|
| Dairy cattle | 10,000 | Livestock loan (outstanding amount) | 8,000 |
| Bullocks | 9,000 | Machinery loan (outstanding amount) | 15,000 |
| Poultry birds | 15,000 | Unsecured loans (outstanding amount) | 10,000 |
| Machinery and equipment | 15,000 | | |
| Tractor | 1,75,000 | | |
| Sub-total | 2,24,000 | Sub-total | 33,000 |
| **Long-term assets** | | **Long-term liabilities** | |
| Land | 6,00,000 | Tractor loan (outstanding amount) | 1,20,000 |
| Farm buildings | 25,000 | Orchard loan (outstanding amount) | 25,000 |
| | | Unsecured loans (land development) | 10,000 |
| Sub-total | 6,25,000 | Sub-total | 1,55,000 |
| | | **Total of liabilities** | **2,56,000** |
| **Total of assets** | **9,75,500** | **Networth or equity** | **7,19,500** |
| | | **Total of liabilities + networth** | **9,75,500** |

$$1.\ \text{Current Ratio} = \frac{\text{Total current assets}}{\text{Total current liabilities}}$$

$$= \frac{1,26,500}{68,000} = 1.86$$

This ratio indicates the capacity of the farmer to meet immediate financial obligations (liquidity). If current assets are more than current liabilities and if the borrower fails to repay the loan, this is a case of wilful-default in spite of his position being solvent. This type of wilful-default is more common in respect of large farmer-borrowers of financial institutions. If by chance the ratio falls less than one due to certain unforeseen contingencies, his case for further lendings cannot be ruled out by the institutional agencies, as it is a temporary setback and he may be given a chance to prove his credit-worthiness. A ratio of more than one indicates a favourable run of the farm business. Current ratio reflects liquidity within one year's time.

$$2.\ \begin{array}{c}\text{Intermediate Ratio}\\\text{or Working Ratio}\end{array} = \frac{\text{Total current assets} + \text{Total intermediate assets}}{\text{Total current liabilities} + \text{Total intermediate liabilities}}$$

$$= \frac{1,26,500 + 2,24,000}{68,000 + 33,000} = \frac{3,50,500}{1,01,000} = 3.47$$

This indicates the liquidity position of the farm business over an intermediate period of time, ranging from 2 to 5 years. Here certain time is allowed for the farmer to build up the farm business to improve his liquidity position. This ratio should also be more than one to indicate sound running of the farm business. The progressive intermediate ratio observed for a given farm business over time implies, the increase in the value of current and intermediate assets due to minimal physical loss and price decline. The steady growth of this ratio over a period is a healthy sign of the business.

$$\text{3. Net Capital Ratio} = \frac{\text{Total assets}}{\text{Total liabilities}}$$

$$= \frac{9,75,500}{2,56,000} = 3.81$$

It indicates the solvency position of the farmers. If the net capital ratio is more than one, the funds of the institutional agencies are safe. A consistently increasing ratio over the years reveals the sound financial growth of farm business. The farmer with this record should be a very prompt repayer of all types of credit obligations. This ratio is also the most important measure of overall solvency position of the farmer-borrowers.

$$\text{4. Acid Test Ratio or Quick Ratio} = \frac{\text{Cash receipts + accounts receivable + marketable securities (bonds, shares, etc.) available in more than one year}}{\text{Total current liabilities}}$$

This reflects adequacy of cash and income surpluses to cover all current liabilities during the period of one to two years. If there is no difference in income position of a farmer within that period, current ratio and acid test ratio reflect the same position.

$$\text{5. Current Liability Ratio} = \frac{\text{Current liabilities}}{\text{Owner's equity}}$$

$$= \frac{68,000}{7,19,500} = 0.09$$

This ratio indicates the farmer's immediate financial obligations against the net worth. A ratio of less than one indicates a healthy performance of the farm business and over the years the ratio should become smaller and smaller to reflect a consistently good performance.

$$\text{6. Debt-equity Ratio (Leverage Ratio)} = \frac{\text{Total debts}}{\text{Owner's equity}}$$

$$= \frac{2,56,000}{7,19,500} = 0.36$$

This presents the capacity of the farmer to meet the long-term commitments. Also it throws light on the extent of indebtedness in the farm business or conversely the amount of capital raised by the farmer in running the farm business. A consistently falling ratio indicates a very heartening performance of farming and the ability of the farmer to reduce dependence on borrowings.

$$\text{7. Equity-value Ratio} = \frac{\text{Owner's equity}}{\text{Value of assets}}$$

$$= \frac{7,19,500}{9,75,500} = 0.74$$

This ratio highlights the productivity gained by the farmer in relation to the assets he has. The improvement in the ratio over the years makes it crystal clear regarding the increased strength in the financial structure of the farm business. This ratio has a direct bearing on the type of assets one has. Managerial ability of the farmer is an essential element in raising the productivity of the assets.

## INCOME STATEMENT OR PROFIT AND LOSS STATEMENT

This is entirely different from a balance sheet in the sense that in a balance sheet, we considered assets and liabilities and did not consider operational efficiency in terms of receipts and expenses. In income statement, the items included are receipts, expenses, gains and losses. It could be defined as a summary of receipts and gains minus expenses and losses during a specified period. It is prepared for the entire farm for one agricultural year. In income statement monetary values are assigned to inputs and output. It is also prepared over time. The advantages of this statement are that it indicates the trend in various cost items and whether there has been any over expenditure on the farm. Thus, it helps to know the success or failure of a farm business over time. Income statement basically constitutes three items, *viz.*, receipts, expenses and net income. Income statement of a hypothetical farm is presented in Table 40.2.

*Receipts:* They mean the returns obtained from the sale of crop produce and other supplementary products like milk and eggs, wages, gifts, *etc.* Gain in the form of appreciation in the value of assets is also included in the receipts. However, returns from the sale of capital assets, such as livestock, machinery, farm buildings, *etc.* are not included because such returns/income are not really obtained during the period.

*Expenses:* Operating and fixed costs are recorded here. Losses in the form of depreciation on the asset value fall under the expenditure item. However, the amounts incurred on the purchase of capital assets are not considered.

Income statement prepared for a given farm for a given year may present a very bright picture of the farm. The same position cannot be taken for granted as the actual position of the farm, since the said year might have been a good agricultural year with respect to weather, yields, prices, *etc.* A realistic position on the performance of a farm can be gauged by preparing income statements over years to show the actual situation, as the parameters influencing farm business are subject to fluctuations.

**TABLE 40.2 Income Statement of a Hypothetical Farm.**

| Particulars | Amount (in Rs.) |
| --- | --- |
| I. *Receipts* | |
| A. Returns from the sale of crop output (Paddy + pulse) | 52,000 |
| B. i. Revenue from milk and milk products | 5,000 |
| ii. Returns from poultry enterprise | 12,000 |
| Returns from supplementary enterprises | 17,000 |
| C. Gifts | 2,000 |
| D. Gross cash income | 71,000 |
| E. Appreciation on the value of assets | 3,000 |
| F. Gross income | 74,000 |

II. *Expenses*

| | |
|---|---:|
| *Operating expenses or costs* | |
| A. Hired human labour | 10,500 |
| B. Bullock labour | 900 |
| C. Machine labour | 1,500 |
| D. Seeds | 1,100 |
| E. Feeds | 5,000 |
| F. Manures & fertilizers | 3,000 |
| G. Plant protection measures | 1,550 |
| H. Veterinary aid | 500 |
| I. Irrigation | 250 |
| J. Miscellaneous | 2,000 |
| K. Interest on working capital | 2,100 |
| Total operating expenses | 28,400 |
| *Fixed expenses or costs* | |
| A. Depreciation | 3,000 |
| B. Land revenue | 200 |
| C. Interest on fixed capital (includes interest of Rs. 1500 paid towards term loan) | 3,200 |
| D. Rental value of owned land | 10,000 |
| E. Total Fixed costs | 16,400 |

III. *Net income*  $\qquad$ 74,000 − 44,800 = $\qquad$ 29,200

# CASH FLOW STATEMENT

This is also known as cash flow summary or cash flow budget or flow of funds statement. Earlier, we have discussed about the balance sheet and income statement. These two financial management tools have inherent weaknesses in presenting certain valuable information, hence another tool in the form of cash flow statement bridges these deficiencies. Cash flow statement is a summary of cash inflows and cash outflows of a business organization in a particular period, say a season or a year. It is usually prepared for the future, hence the name cash flow budget. The merit of this particular statement is that, it helps to assess the time at which the funds are required for farming and other allied enterprises, sources from which these can be raised, the purpose for which the loan is required, the need of sale and purchase of capital assets, the time and quantum of repayment, *etc*. Now, let us see why a farmer borrows funds from a particular source or sources; why he resorts to transactions like selling of farm products and livestock products and selling and buying of capital assets. The answers to these questions are that the small and marginal farmers have poor resource base, and therefore, borrowings aid them in continuing the farm business. Large farmers too borrow for farm operations depending up on the need and time during which they cannot properly recycle the funds. Farmers resort to sale of farm assets like milch cattle, machinery, *etc*., because they might have worn out, for which replacements are to be made through purchases.

Cash flow statement is prepared at the beginning of the agricultural year and checked every quarterly. For convenience, quarterly checks are made. The statement prepared over the years serves the purpose of studying the pattern of expenditure and cash receipts and cash balance that have been raised. A close scrutiny of the statement throws light on the performance of the business.

The example provided in Table 40.3 is briefed hereunder.

TABLE 40.3   Cash Flow Statement of a Hypothetical Farm.

| S. No. | Particulars | I quarter (June–Aug) | II quarter (Sept. – Nov.) | III quarter (Dec. – Feb.) | IV quarter (Mar. – May) | Total |
|---|---|---|---|---|---|---|
| I | Cash receipts (in Rs.) | | | | | |
| 1. | Cash balance (brought forward from previous year) | 3,000 | - | - | - | 3,000 |
| 2. | Total operating sales (farm and livestock products) | 1,350 | 1,400 | 30,200 | 7,800 | 40,750 |
| 3. | Total capital sales (milch cattle) | - | 5,000 | - | - | 5,000 |
| 4. | Non-farm income (family members working elsewhere) | 2,000 | 1,500 | 2,000 | 3,200 | 8,700 |
| 5. | Borrowings (ST, MT and LT loans from institutional agencies) | 7,500 | - | - | - | 7,500 |
| 6. | Total | 13,850 | 7,900 | 32,200 | 11,000 | 64,950 |
| II | Cash expenses (in Rs.) | | | | | |
| 1. | Operating expenses | 8,500 | 6,750 | 6,200 | 5,300 | 26,750 |
| 2. | Capital investment (purchase of milch cattle) | - | - | 6,000 | - | 6,000 |
| 3. | Family living expenses | 2,400 | 2,800 | 3,200 | 3,000 | 11,400 |
| 4. | Payment of previous years debt | 500 | - | - | - | 500 |
| 5. | Payment of ST loans and instalments on investment loans | - | - | 7,968 | - | 7,968 |
| | Total | 11,400 | 9,550 | 23,368 | 8,300 | 52,618 |
| III | Cash balance (in Rs.) | 2,450 | -1,650 | 8,832 | 2,700 | 12,332 |

## I. Cash Receipts

1. *Cash Balance:* This is the surplus amount of previous year with the farmer which stood at Rs. 3,000.

2. *Total Operating Sales:* These are the returns obtained from the sale of farm products and livestock products. Lesser amounts are discernible in the first and second quarters, while the returns to be obtained in the third and fourth quarter are on the higher side. The farmer is sure of getting returns from milk for about 250 days in a year, which is more or less uniform in the first three quarters. The returns from crop production will be received in the third quarter for *kharif* and the returns from *rabi* crops obtained in the last quarter. The total operating sales amount to Rs. 40,750 at the end of the year.

3. *Total Capital Sales:* The farmer is contemplating to sell the she-buffalo, which he possesses, in the second quarter and the amount to be received will be Rs. 5,000.

4. *Non-farm income:* It is the income which will be added by the family members by their earnings elsewhere.

5. *Borrowings:* The farmer wishes to borrow an amount of Rs. 7,500 for *kharif* crop operations.

6. *Total:* It is the summation of particulars of 1 to 5 rows, which presents the total cash receipts to be obtained in the year.

## II. Cash Expenses

1. *Operating Expenses:* These include the expenditure to be incurred on the *kharif* as well as *rabi* crops and the dairy cattle.

2. *Capital Investment:* Since the farmer proposes to dispose the dairy cattle in second quarter, he wants to buy a new one in lactation in the third quarter.

3. *Family Living Expenses:* These include the expenditure towards food, shelter, medical, education and other items.

4. *Payment of Previous Year's Debts:* A handloan of Rs. 500 is due to be paid in the first quarter.

5. *Payment of ST Loans and Installments on Investment Loans:* Since the farmer is proposing to take crop loan, the repayment of same falls due in the third quarter. Along with the interest and installments, the amount due to be paid would be Rs. 7,968.

6. *Total:* It is the total expenditure to be incurred.

## III. Cash Balance

It is the sum of amount to be realized after deducting expenditure from cash receipts. Barring the second quarter, the farmer is expected to have a surplus in the remaining three quarters. The deficit of Rs. 1,650 in the second quarter can easily be cleared off from the savings of previous quarter, *i.e.,* first quarter. Overall, the net surplus would be Rs. 12,332.

### Advantages of Cash Flow Budget

It is a summary of all the financial matters of the farmer in a comprehensive report. This helps (i) to estimate the total credit needs (ST, MT and LT) of the farmer along with time and quantum; (ii) to plan the repayment schedule, (iii) in making purchases and sales at the appropriate time thereby helping to minimize the credit dependence, so that the farmers can keep limits to avoid wastages (iv) to keep ready input requirements well in advance so that the last minute rush can be avoided (v) to know the farm household's expenditure pattern (vi) the farmer to exercise a check on farm costs, (vii) the farmer in preparing the farm business plans for the ensuing years, (viii) the banker for revising the scales of finance, rescheduling loans, *etc.,* and (ix) finally, as a tool of financial control to the farmer.

## BREAK-EVEN ANALYSIS

Break-even analysis indicates costs-volume-profit relations in the short run. This analysis relies on the assumption of constant factor prices, constant technology and constant selling prices. Despite, this limitation, break-even analysis is very important because in the short run, the cost and revenue structure is reasonably stable. The point at which the two curves i.e., total cost curve and total revenue curve intersect is called the break-even point (BEP), which indicates the level of production at which the producer neither looses money nor makes profit. It is a point of no-profit and no-loss. Apart from helping to spot the break-even point, this analysis helps to develop an understanding of the relationship of costs, price and volume within a farm's range of operations. A given farm is said to be at break-even point, when its costs are equal to revenue i.e., when the contribution margin is exactly equal to the fixed costs. Contribution margin is estimated by deducting variable costs per unit, from price per unit of output. At quantities smaller than break-even point, there is a loss and at larger quantities there is a profit.

## Estimation of Break-even Point

### Graphic Approach

Break-even point is estimated for small farms of onion crop. The total cost curve is represented by SL and the total revenue curve by OK. The point at which the two curves intersect is the break-even point (B) as indicated in Figure 40.1.

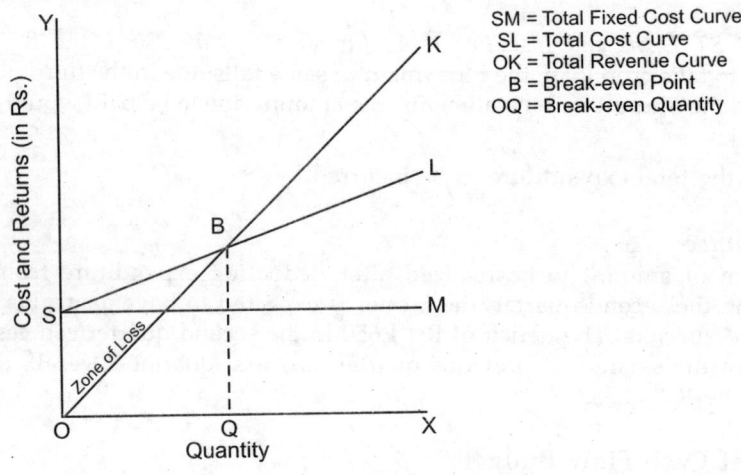

SM = Total Fixed Cost Curve
SL = Total Cost Curve
OK = Total Revenue Curve
B = Break-even Point
OQ = Break-even Quantity

**Figure 40.1** Break-even output.

### Algebraic Approach

Algebraically, break-even point can be estimated by using the formula.

$$BEP = F/P-V$$

Where,
BEP = Break-even point
F = Fixed costs in Rs. per hectare of onion
P = Price per quintal of onion in Rs.
V = Variable costs per quintal of onion in Rs.

$$= \frac{12175.07}{305.45 - 139.16}$$

$$= 73.22 \text{ Quintals}$$

Break-even output on small farms is 73.22 quintals (Table 40.4).

**TABLE 40.4  Break-even Analysis in Onion Cultivation.**

| Size of farm | Fixed costs/ ha (in Rs.) | Variable costs/ ha (in Rs.) | Total costs/ ha (in Rs.) | Price per quintal (in Rs.) | Output/ha (in Q) | Total revenue/ ha (in Rs.) | Variable costs/ unit (in Rs.) | Break-even output (in Q) | Margin of safety (in Q) |
|---|---|---|---|---|---|---|---|---|---|
| Small | 12,175.07 | 20,874.48 | 33,049.55 | 305.45 | 150 | 45,817.5 | 139.16 | 73.22 | 76.78 |
| Large | 13,362.26 | 22,920.58 | 36,282.84 | 305.45 | 175 | 53,453.75 | 130.97 | 76.58 | 98.42 |

It is said earlier that a farm is said to be at break-even point, when the contribution margin is exactly equal to fixed costs.

Contribution margin = Price per unit of output–variable costs per unit of output.

## Margin of Safety

The exess of production over the break-even point is called the margin of safety. It indicates the strength of the enterprise. A high margin of safety indicates that the enterprise will make profits even if there is a fall in the output. In other words, a high margin of safety reveals the shock absorbing capacity of the enterprise in the event of fluctuations in returns against anticipation owing to an unforeseen eventuality.

Margin of safety (in physical units) = Total output – output at BEP
= 150 - 73.22
= 76.78 quintals

Analogously, BEP can also be estimated for large farms.

## Angle of Incidence

It is the angle that is formed between total revenue curve and total cost curve at the point of intersection. The wider the angle, greater is the margin of safety.

# CHAPTER 41

# Agricultural Projects

## MEANING

Projects are the cutting edges of development. They are meant for increasing the output from the given resources. Evaluation of projects needs projecting the future trend of output, sales, costs, returns, flow of funds, *etc*. The World Bank has recognized six important aspects in the project preparation. They are technical, administrative, organizational, commercial, financial and economic aspects. All the technological aspects in the project must be thoroughly studied under technical analysis. Goods and services required for the project execution need a detailed assessment. The awareness of the lending agency regarding the technology to be used, *i.e.*, capital-intensive technology or labour-intensive technology, is to be assessed. Another important factor is the technical feasibility, which determines the size of the project based on capital requirement, future and present demand for product, cost-benefit aspects, *etc*. The selected area for the project must be adequate in the resource endowment base and infrastructural facilities. The need for a particular project picks up the priority in the project implementation. Under administrative coverage, managerial aspects, project staff, extension personnel, credit agencies and farmers (beneficiaries) will be studied. Organisational aspects deal with relationship of project administration and the Government, training arrangements, disbursement of wages, *etc*. Regarding commercial aspects of the projects, arrangements for the supply of input materials, services needed for the project, marketing of output, *etc.*, are to be assessed. Regarding the financial aspects the items which fall under this category are source of funds, cost of funds, repayment, *etc*. The estimated costs based on technical aspects, estimated sales based on commercial analysis and probable profits from the operation of the project are to be properly evaluated. Financial gains accrued as well as incentives offered to the farmers (participants) in the project should also be viewed. On the other hand economic analysis concentrates, in determining project's contribution to the development of the economy as a whole and justifying the use of scarce resources. Proper identification of costs and benefits is an important aspect in the economic analysis of projects.

## DEFINITION

Project is an investment activity wherein we spend the capital resources to create a productive asset for realizing benefits over time. Generally "Project is an activity on which we spend money in expectation of returns, which lends itself to planning, financing and implementation as a unit. It also refers to specific activity, with specific starting point and specific end point to achieve a specific objective. It should be measurable in costs and returns. It must have priorities for area development and reach specific clientele group."

In sum, the meaning of project may be comprehended as follows. "It is an investment activity meant for providing the returns for specific clientele group for specific activity, specific objective and specific area development. It should facilitate analysis in planning, financing, implementation, monitoring, controlling and evaluation."

## PHASES IN PROJECT CYCLE

The important phases in project cycle are:

1. Conception or identification,
2. Formulation or preparation of the project,
3. Appraisal or analysis,
4. Implementation,
5. Monitoring, and
6. Evaluation.

Project is considered as a cycle (Figure 41.1) because, each phase not only grows out of the preceding one, but leads into the subsequent phase and it is a self-renewing cycle, so that new projects come out of the old ones in a continuous manner.

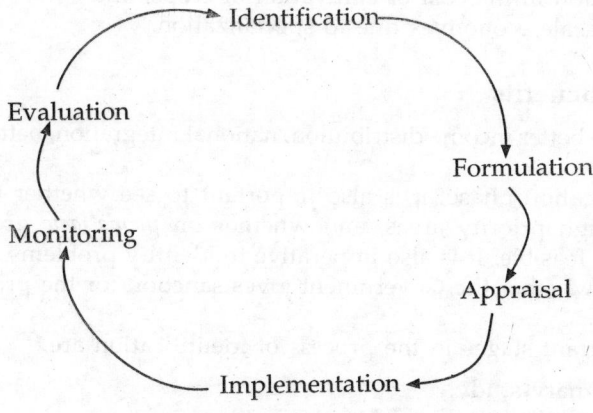

**Figure 41.1** Project cycle.

### Conception or Identification of the Project

In agricultural project, costs are easier to identify than benefits because the expenditure pattern is easily visualized. The various types of costs involved in the project are:

1. *Project Costs*: These include the value of the resources in maintaining and operating the projects.

2. *Associate Costs*: These include producing immediate products and services of the projects for use or sale.

3. *Primary Costs or Direct Costs*: These represent the costs incurred in construction, maintenance and execution of the project.

4.  *Indirect Costs or Secondary Costs*: The value of goods and services incurred in providing indirect benefits from the projects such as houses, schools, hospitals, *etc.*

5.  *Real Costs and Nominal Costs*: Costs at current market prices are nominal costs, whereas if costs are deflated by general price index, these are termed as real costs.

6.  *Social Costs*: These are technological externalities and technological spill-over accrued to the society due to the presence of projects, *i.e.*, pollution problems, health hazards, salinity conditions, *etc.*

Next to identifying the costs, the estimation of benefits is imperative to ascertain the impact of the project. This is generally done by taking into account two situations, *i.e.*, 'with' and 'without' the projects. The difference is the net additional benefit arising out of the project. Benefits are split into two: tangible and intangible benefits.

## Tangible Benefits

Incremental income due to the existence of projects is obtained from the following changes.

(i) Improvement in cropping pattern involving high value crops.
(ii) An increase in the productivity of crops.
(iii) Adoption of recommended package of practices.
(iv) Increase in the intensity of cropping.
(v) Reduction in the cost of cultivation of crops, and
(vi) Large scale economies due to specialization.

## Intangible Benefits

These include better income distribution, national integration, better standard of living, *etc.*

In identification phase, it is also important to see whether the project is implemented in high priority areas, and whether on *prima-facie* grounds the project is economically feasible. It is also imperative to identify problems and objectives of the projects and whether the Government gives sanction for the project implementation or not.

The important stages in the process of identification are:

(i) Preliminary study,
(ii) Pre-feasibility study, and
(iii) Project report.

In these stages we assess whether the project proposed on the grounds of *prima-facie* is feasible and the objectives of the project are achieved. On this ground, the preliminary study should embody the investment proposals, benefits extended from the projects and method of implementation. Assessment of the demand for the project's products, technical feasibility of the project, import and export requirements, marketing aspects, investment prospects, *etc.*, should be exhaustively covered by the feasibility studies, including the analysis of sensitivity.

## Some of the Sources Through which the Projects Identified are

1.  Agricultural and allied programmes proposed in the plans of the country as well as States;

2. Areas identified as potential of further development through Governmental surveys;
3. Special developmental programmes like IRDP;
4. Irrigation projects which offer scope for development through forward and backward linkages; and
5. New projects emerging out of existing projects, *etc.*

## Formulation or Preparation

The following points are considered while formulating the projects. The location of the project and project site must be based on technical analysis and technical feasibility of the project. The location of the project depends upon available physical resources, market conditions, marketing facilities, alternative investment prospects, administrative experience, farmers' objectives, technical skill, motivations, demand for products, *etc.* Technical analysis must take into consideration all aspects of technology to be used in the project, and account for all inputs of goods and services. Assessment of suitability and adequacy of natural resources in advance based on the scientific investigations is also essential. Alternatives to the resource use are to be considered in formulation of the project. Due consideration is to be given to all the aspects such as technical, financial, commercial, managerial, organizational, social, economic, *etc.*, in the formulation of the projects. Identification of the missing links in the infrastructure system, particularly in relation to adequacy of communication systems, markets and storage facilities is important.

## The Above Aspects of the Project are Briefly Outlined Here Under

**1) Technical Aspects**: The issues, which need technical examination, are thoroughly analyzed here. For example, in the case of agricultural projects, the issues include the aspects relating to the pre-production, production and post-production aspects. To begin with, we have to examine the soil types, problems associated with different types of soils, potentiality which the soils offer for development, irrigation supply and availability, crops to be grown, availability of desired variety of seeds in required quantities, availability of other complementary inputs as per choice, credit facilities, pests and disease problems, possibility of mechanization, expected yields, storage, processing, marketing facilities, *etc.* for confirming technical feasibility.

**2) Financial Aspects**: Here, we have to find out the sources of raising financial assistance and terms and conditions of obtaining finance from the credit agencies. The implementing agency should be in a position to estimate the financial requirements and anticipated returns, through farm planning and budgeting. Once the incremental income is arrived at, the repayment capacity duly giving allowance for risk and uncertainty can be worked out. Cash flow chart can be used here.

**3) Commercial Aspects**: These aspects focus on the estimation of effective demand, availability of input supplies and arrangements for the output marketing. Market potentiality for the products needs a careful scrutiny.

**4) Managerial Aspects**: If we want successful implementation of the project, effective managerial issues are very crucial. The need to identify the beneficiary is the foremost item. Once identification process is over, we have to find out the managerial abilities of the beneficiaries. The managerial skills can be sharpened, if necessary;

technical skills are imparted by the extension agencies who have the onerous responsibility in the transfer of technology. Ability of the implementing agency needs a closer scrutiny.

5) **Organisational Aspects**: Organisation refers to the process of putting the priorities in an orderly form. Prepare the organizational hierarchy of the implementing agency. The availability of staff at various cadres, demarcation of authority and linking of authority and responsibility, *etc.*, are expected to be dealt with, under this aspect. For proper administration of the projects, efficient personnel and other requirements are indispensable.

6) **Social Aspects**: Here customs, culture, traditions, habits, *etc.*, of the beneficiaries are considered. The other relevant implications like the probable changes in the living standards, material welfare, consumption habits, income distribution effects, *etc.*, fall under this heading.

7) **Economic Aspects**: Here we have to examine the benefits, which the project is going to contribute in terms of the utilization of scarce resources of the nation. The point of merit is to whom the project is going to benefit, *i.e.*, to one section of the society or the entire area of the project. The indirect effect like, the income distribution, needs to be assessed. Under income distribution, we are interested to know whether income imbalances are going to be narrowed down or accentuated as a result of proposed projects. Overall, what one expects of a project is that it should bring in the greatest contribution to the national economy.

## Appraisal or Analysis

Appraisal should take place before the implementation of the project. It is done independently by specialists. In the appraisal stage, it is important to know whether the project is technically feasible according to the data available. The technical data for assessing the feasibility of the project should be consistent with the information available in the office of the sanctioning authority or elsewhere. Managerial aspects play a key role in the project appraisal. Projects become abortive due to the failure to consider managerial aspects, *i.e.*, such as new skills and information gained by the farmers in the project area, including adoption of new technology. The managerial capabilities and capacity of administrative personnel must also be assessed in project appraisal.

## Implementation

This is the most crucial phase of the project cycle. The secret of successful implementation depends up on the extent of realism put into the plans drawn before hand. It is often not uncommon to notice our plans getting deviated form the reality. The role of wise decisions by the personnel incharge of implementation to tackle the situation comes into play. Project implementation can be divided into three different periods, *viz.*, investment period, development period, and full-production period. Investment period may range from few months to few years depending up on the nature of assets to be acquired. Assets proposed should be of superior quality. Development period too consumes time. Implementing agency should make all efforts to reduce the gestation period as per the plan envisaged in the beginning. Full production period is the time during which the beneficiaries start reaping the benefits of the project. Implementing agency should ensure that the beneficiaries do continue to receive benefits during the entire life span of the project.

## Monitoring

Monitoring is the timely collection and analysis of data on the progress of a project, with the objective of identifying constraints, which impede successful implementation. This is highly desirable, particularly when projects fail to be completed as per time schedule or in the process of attaining the set goals. It is imperative to get the feedback on the problems faced so that the effective measures can be taken up to plug the deficiencies, which hamper the speedy implementation. Monitoring has to be done continuously to offset various shortcomings that crop up from time to time with regard to various aspects of implementation.

## Evaluation

This is the last phase of the project cycle. It is not confined to the completed project. Evaluation can be done several times during the life of a project. In the evaluation process, it is important to see, how far the objectives set out in the project are achieved. Deficiencies, snags or failures to achieve the objectives may be analyzed and appropriate solutions to such failures answered. Evaluation process is to be completed in three phases. They are midcourse evaluation, concurrent evaluation and ex-post evaluation. In the first phase, evaluation is attempted before any change occurs in the existing situation. This is primarily meant to assess economic feasibility of the projects, since it is done at the very beginning. This type of analysis is otherwise called pre-project evaluation. Sometimes it is also important to take up evaluation when the project is in execution, and such an evaluation is called concurrent evaluation. This type of evaluation is basically meant for identifying and analyzing the pitfalls in the execution of the project. Evaluation is also resorted to particularly when the project is completed in all its phases, in order to assess the achievement of ends or objectives set out by the projects. Such evaluation is called ex-post evaluation or end-evaluation. Evaluation is done by the agency other than the implementing one, like financing bank or sponsoring agency or Government.

# INVESTMENT ANALYSIS (CAPITAL BUDGETING)

Investment in agriculture is of two types. The first type involves operating investment such as seed, feed, fertilizers, *etc.*, and the second one is concerned with capital assets such as land, machines, livestock, *etc.* Analysis of investment is different for these categories of investment, owing to the differences in timing of expenses and their associated returns.

In the case of operating investment, expenses and returns fall within a production cycle or year. The time element is not brought into consideration in analyzing the profitability of operating inputs because both expenses and returns are assumed to fall in the same production cycle. But, investments in agricultural projects are made in different time periods and the returns are also spread over time. In order to assess the returns from investments, available alternatives must be weighed for different lengths of time in respect of costs and returns, *i.e.*, recognition of time value of money.

## Time Value of Money

Future Value of Present Money. A rupee today is worth more than a rupee in future. This is primarily due to its opportunity cost, *i.e.*, interest. Interest will be added to the principal over time and hence its value increases. Future value of present sum is an important concept in financial analysis and this is called compounding. In the compounding process, the interest is added to the principal at the end of each time period, which, in turn, earns interest. The future value of present investment is calculated by using the well-known formula of compound interest:

$$A = P (1 + i)^t \qquad (5.1)$$

Where,

A = **Future** value of the present sum invested,
P = **Principal** amount invested,
i = **Interest** rate in per cent, and
t = **Number** of years.

**Annuity**: By definition annuity means a stream of payments or returns over time. The **future** value of annuity can be estimated using the following equation.

$$A = P \; \frac{(1+i)^t - 1}{i} \qquad (5.2)$$

Where,
A = Future value,
P = Annual investment,
t = Time period, and
i = Rate of interest.

## Present Value of Future Money

The present value of future sum is the current value of investment to be received in the future at a specified date. This present value is worked out through discounting process in which the future sum is discounted back to the present time to find out its current or present value. The rationale behind this process is that a sum to be received in future is somewhat less now, because of time difference assuming a positive interest rate. Discounting is the inverse procedure of compounding. A present sum is compounded to know the future value and future sum is discounted to know the present value.

$$PW = \frac{P}{(1+i)^t} \text{ or } P \; \frac{1}{(1+i)^t} \qquad (5.3)$$

Where,
PW = Present value or worth of future money,
P = Money value in future
i = Rate of interest,
t = Project life period in years.

The present value of annuity or stream of constant annual payment is found out using the following formula

$$PW = P \; \frac{1 - (1+i)^{-t}}{i} \qquad (5.4)$$

Where,
PW = Present worth of future money,
P = Money value in future

i     = Rate of interest, and

t     = Project life period in years.

Investment analysis is also called capital budgeting. The profitability of two or more alternative investment projects is determined through capital-budgeting technique. Four components are required for the analysis of investment. They are: (1) Cash revenues from different projects, (2) Costs, (3) Terminal or salvage value of investment, and (4) Interest or discount rate to be used.

Cash receipts less cash expenses give net cash revenue resulting from the alternative proposed projects. The cost of investment is the actual total expenditure for its implementation. The terminal value of the project will also be estimated and it is set equal to the junk value for depreciable assets of the project. For simplicity junk value is assumed to be zero. The land values of the projects should be estimated at the market rate, at the time at which the project is terminated.

Another problem in financial analysis is with the estimation of discount rate. This discount rate is the opportunity cost of capital, which represents the minimum rate of return for justifying the investment. If the proposed investment in the project fails to earn this minimum rate of interest, then the capital should not be invested in the said project and alternative projects must be chosen as worthy of investment.

If capital is to be borrowed for investment on the project, then the discount rate chosen should be higher than the cost of borrowed capital. Under risk situations, the discount rate is to be equalled to the expected rate of return from the alternative projects of equal risk. Many problems are involved in deciding upon the actual rate of discount in project evaluation particularly, when the discount rate is to be adjusted to the risk.

Broadly there are two methods of project appraisal, *viz.*, undiscounted measures and discounted measures. In the undiscounted measures, payback period, ranking by inspection, proceeds per rupee of outlay, average annual proceeds of rupee outlay, *etc.*, are important. Under discounted measures, Net Present Worth (NPW), Benefit-Cost Ratio (B-C ratio), Internal Rate of Return (IRR), and Profitability Index are prominent.

# UNDISCOUNTED MEASURES

The undiscounted measures are the naive methods of choosing among the alternative projects. The methods listed under these measures often mislead in ranking of the projects and hence, choices go wrong.

## Ranking by Inspection

It is based on the size of costs and length of cash-flow stream. Suppose if the two projects are with the same investment and the same net value of production, but with difference in the length of the period, then the project with longer duration is preferred to the one with shorter time period. This leads to bias in the choice obviously due to the absence of more elaborate and appropriate analysis.

## Payback Period

Another simple method of ranking a project is the length of time required to get back the investment on the project.

The pay back period of the project is estimated by using the straight forward formula:

$$P = \frac{I}{E}$$

(5.5)

Where,

P = Payback period of the project in years,

I = Investment of the project in Rs. and

E = Annual net cash revenue in Rs.

The preference of a particular project is based on the shorter payback period. This is shown in Table 41.1.

**TABLE 41.1  Estimation of Payback Period.**

Initial investment = Rs. 20,000

| Year | Cash Flow (in Rs.) | |
|------|--------------------|--------------------|
|      | Project 'A'        | Project 'B'        |
| 0    | - 20,000           | - 20,000           |
| 1    | 5,000              | 4,000              |
| 2    | 5,000              | 4,000              |
| 3    | 5,000              | 4,000              |
| 4    | 5,000              | 4,000              |
| 5    | 5,000              | 4,000              |
| 6    | 5,000              | 4,000              |

$$\text{Project 'A'} = \frac{\text{Rs. } 20,000}{\text{Rs. } 5,000} = 4 \text{ years}$$

$$\text{Project 'B'} = \frac{\text{Rs. } 20,000}{\text{Rs. } 4,000} = 5 \text{ years}$$

It is inadequate to exercise the option among the alternatives, because it fails to consider very important points like, consistency of running, timing of the proceeds, returns after the payback period and whether the cash-flows would be positive or negative in future.

## Proceeds Per Rupee of Outlay

This is worked out by dividing the total proceeds with the total amount of investment, and a given project is ranked based on the highest magnitude of the parameter.

## Average Annual Proceeds of Rupee Outlay

This is another simple choice criterion and in this procedure, total receipts are first divided by the project life span and the average proceeds obtained per year are divided by the initial investment on the project. Here too, ranking is given to the projects, based on the highest magnitude of the estimate.

The major drawback with undiscounted measures is that for the same data of the project, we get different rankings, hence, choice process becomes useless. Rankings by these methods are inconsistent and incompatible.

# DISCOUNTED MEASURES

Cash flows are the yearly net benefits accrued from the project. If they are weighed by discount rate, they become discounted cash flows. These discounted cash flows are the best estimates to decide on the worth of the project. This approach will give the net present worth of the project. The present worth of the costs is subtracted from the present worth of the benefits in order to arrive at the net present worth of the project every year.

## Measurement of the Cash Flow of the Project

From the annual stream of gross benefits of the project, the capital invested and the other input costs like labour, machinery, fertilizers, pesticides, management, *etc.,* are deducted. From the residual, the return of capital, and return on capital or return to capital, *i.e.,* investment made in the project (depreciation) and compensation for the use of money (interest) are computed. This residual is called cash flow of the project. In financial analysis the cash flow is the net incremental benefits of the project. But, in accounting, the term implies the sum of cash flows of projects plus depreciation allowance. The concept of cash flow in the financial analysis includes, both return of capital and return to capital. We generally do not resort to deduction of depreciation, *i.e.,* allowance of return of capital or interest in the economic analysis, because our analytical technique automatically takes care of return of capital in determining the worth of the project. In economic analysis, income taxes, sales taxes, custom duties, *etc.,* are only the transfer payments, but not payments used in the production process. Hence, from the gross returns these are not deducted. But in the financial analysis taxes are the costs which individuals must pay for the use of capital.

By far, financial analysis aims at the estimation of return to all resources employed in the project. Hence, borrowed capital is considered as benefits received, while, its interest is considered as cost and it is deducted from the gross returns. In economic analysis, this consideration is ruled out because of the assumption, that all the resources employed in the project belong to someone or the other within the society. In the economic analysis, it is important that the price of some of the inputs must be the shadow prices. In financial analysis all prices are market prices and they must include taxes and subsidies. For clear distinction between cash inflows the economic analysis vis-à-vis financial analysis (Gittinger, 1976) may be referred.

## Net Present Worth (NPW)

This is simply the present worth of the cash flow stream. Sometimes, it is referred to as Net Present Value (NPV). The selection criterion of the project depends on positive value of NPW when discounted at the opportunity cost of the capital. This could be satisfactorily done, provided there is a correct estimate of opportunity cost of capital. NPW is an absolute measure, but not relative.

NPW of the project is estimated using the following equation:

$$\text{NPW} = \frac{P_1}{(1+i)^{t_1}} + \frac{P_2}{(1+i)^{t_2}} + \cdots + \frac{P_n}{(1+i)^{t_n}} - C \tag{5.6}$$

Where,

$P_1$ = Net cash flow in first year,

i  = Discount rate,

t  = Time period, and

C = Initial cost of the investment.

Project with positive NPWs are given weightage in the selection compared to those with negative present values, while zero NPW makes the investor indifferent. Table 41.2 presents the particulars of NPV calculations for two projects.

## Benefit-Cost Ratio (B-C RATIO)

Here, we compare the present worth of costs with present worth of benefits. Absolute value of the benefit-cost ratio will change based on the interest rate choosen. While ranking the projects depending upon the B-C ratio, the most common procedure of selecting a project is, to choose the project, having B-C ratio of more than one, when discounted at opportunity cost of capital. Finally, the given project is opted for implementation, among alternatives based on the highest B-C ratio. Following formula depicts the estimation of B-C ratio.

$$\text{B} - \text{C Ratio} = \frac{\sum\limits_{t=1}^{n} \dfrac{B_t}{(1+r)^t}}{\sum\limits_{t=1}^{n} \dfrac{C_t}{(1+r)^t}} \tag{5.7}$$

The estimation procedure of B-C ratio is embodied in Table 41.3

## Internal Rate of Return (IRR)

In the computation of Internal Rate of Return (IRR), the time value of money is accounted. The method of working IRR provides the knowledge of actual rate of return from the different projects. Thus IRR is known as 'marginal efficiency' of capital or yield on the investment. It is the discount rate at which the present values of the net cash flows are just equal to zero, *i.e.,* NPW = zero. The IRR must be found out by trial and error with some approximation. The procedure is elucidated for the projects on sericulture and mango in Tables 41.4 and 41.5 respectively.

In the working procedure, an arbitrary discount rate is assumed and its corresponding NPW is arrived at. The positive NPW value of the project indicates that IRR is still higher and the next assumed arbitrary IRR value must be comparatively higher than the initial level. This process is continued until NPW becomes negative. Then by interpolation method the exact IRR is found out using the following equation.

$$\begin{bmatrix} \text{Internal} \\ \text{rate of} \\ \text{return} \end{bmatrix} = \begin{bmatrix} \text{Lower} \\ \text{discount} \\ \text{rate} \end{bmatrix} + \begin{bmatrix} \text{Difference} \\ \text{between the two} \\ \text{discount rates} \end{bmatrix} \times \begin{bmatrix} \dfrac{\text{Present worth of the}}{\text{cash flow at the lower discount rate}} \\ \text{Absolute difference between the} \\ \text{present worths of the cash flow at} \\ \text{the two discount rates} \end{bmatrix}$$

**TABLE 41.2  Estimation of NPW for Two Projects (Hypothetical).**

| | Sericulture (one ha) | | | | |
|---|---|---|---|---|---|
| Year | Costs (in Rs.) | Returns (in Rs.) | Net income (in Rs.) | Discount factor at 12% | NPW (in Rs.) |
| 1. | 38,900 | — | -38,900 | 0.8929 | -34,733.81 |
| 2. | 9,239 | 28,475 | 19,236 | 0.7972 | 15,334.94 |
| 3. | 10,575 | 32,550 | 21,975 | 0.7118 | 15,641.81 |
| 4. | 11,952 | 35,610 | 23,658 | 0.6355 | 15,034.66 |
| 5. | 12,858 | 39,802 | 26,944 | 0.5674 | 15,288.03 |
| | | | | NPW | 26,565.63 |

| | Mango Orchard (one ha) | | | | |
|---|---|---|---|---|---|
| Year | Costs (in Rs.) | Returns (in Rs.) | Net income (in Rs.) | Discount factor at 12% | NPW (in Rs.) |
| At the end of 6th year | 25,000 | — | -25,000 | 0.507 | -12,675 |
| „ 7th year | 4,250 | 10,260 | 6,010 | 0.452 | 2,716.52 |
| „ 8th year | 4,792 | 12,550 | 7,758 | 0.404 | 3,134.23 |
| „ 9th year | 5,368 | 14,530 | 9,162 | 0.361 | 3,307.48 |
| „ 10th year | 5,975 | 16,275 | 10,300 | 0.322 | 3,316.60 |
| „ 11th year | 6,456 | 19,396 | 12,940 | 0.287 | 3,713.78 |
| „ 12th year | 7,187 | 21,470 | 14,283 | 0.257 | 3,670.73 |
| | | | | NPW | 7,184.34 |

**TABLE 41.3** Estimation of Benefit-cost Ratio (BCR) for 2 Projects (Hypothetical).

Sericulture (one ha)

| Year | Costs (in Rs.) | Gross returns (in Rs.) | Discount factor at 12% | Present worth of costs (in Rs.) | Present worth of gross returns (in Rs.) |
|---|---|---|---|---|---|
| 1. | 38,900 | — | 0.8929 | 34,733.81 | — |
| 2. | 9,239 | 28,475 | 0.7972 | 7,365.33 | 22,700.27 |
| 3. | 10,575 | 32,550 | 0.7118 | 7,527.29 | 23,169.09 |
| 4. | 11,952 | 35,610 | 0.6355 | 7,595.50 | 22,630.16 |
| 5. | 12,858 | 39,802 | 0.5674 | 7,295.63 | 22,583.65 |
| | | | | 64,517.56 | 91,083.17 |

Mango Orchard (one ha)

| Year | Costs (in Rs.) | Returns (in Rs.) | Discount factor at 12% | Present worth of costs (in Rs.) | Present worth of gross returns (in Rs.) |
|---|---|---|---|---|---|
| At the end of 6th year | 25,000 | — | 0.507 | 12,675.00 | — |
| „ 7th year | 4,250 | 10,260 | 0.452 | 1,921.00 | 4,637.52 |
| „ 8th year | 4,792 | 12,550 | 0.404 | 1,935.97 | 5,070.20 |
| „ 9th year | 5,368 | 14,530 | 0.361 | 1,937.85 | 5,245.33 |
| „ 10th year | 5,975 | 16,275 | 0.322 | 1,923.95 | 5,240.55 |
| „ 11th year | 6,456 | 19,396 | 0.287 | 1,852.87 | 5,566.55 |
| „ 12th year | 7,187 | 21,470 | 0.257 | 1,847.06 | 5,517.79 |
| | | | | 24,093.70 | 31,278.04 |

$$\text{Benefit-cost-ratio} = \frac{\text{Present worth of gross returns}}{\text{Present worth of costs}} = \frac{91,083.18}{64,517.56}$$

$$= 1.41$$

$$\text{Benefit-cost-ratio} = \frac{31,278.04}{24,093.70}$$

$$= 1.30$$

**TABLE 41.4 Estimation of IRR for Sericulture (One Hectare) (Hypothetical).**

| Year | Costs (in Rs.) | Gross income (in Rs.) | Net income (in Rs.) | Discount factor (40%) | Net present worth (in Rs.) | Discount factor (43%) | Net present worth (in Rs.) |
|------|------|------|------|------|------|------|------|
| 1. | 38,900 | – | -38,900 | 0.7143 | -27,786.27 | 0.6993 | -27,202.77 |
| 2. | 9,239 | 28,475 | 19,236 | 0.5102 | 9,814.21 | 0.48902 | 9,406.4 |
| 3. | 10,575 | 32,550 | 21,975 | 0.3644 | 8,007.69 | 0.3419 | 7,513.25 |
| 4. | 11,952 | 35,610 | 23,658 | 0.2603 | 6,158,17 | 0.2391 | 5,656.62 |
| 5. | 12,858 | 39,802 | 26,944 | 0.1859 | 5,008.89 | 0.1672 | 4,505.04 |
| | | | 52,913 | | 1,202.69 | | –121.46 |

**N.B:** The entire lifespan of mango orchard should be considered for working out IRR. For want of data we considered here only for seven years for illustration purpose.

$$IRR = 40 + 3\left[\frac{1202.69}{1,202.69+121.46}\right]$$

$$= 40 + 3(0.9083)$$

$$= 40 + 2.7249$$

$$= 42.7249\%$$

$$= 42.7\%$$

**TABLE 41.5 Estimation of IRR for Mango Orchard (One Hectare) (Hypothetical).**

| Year (in Rs.) | Costs (in Rs.) | Gross income (in Rs.) | Net income (in Rs.) | Discount factor (25%) | Net present worth (in Rs.) | Discount factor (30%) | Net present worth (in Rs.) |
|------|------|------|------|------|------|------|------|
| End of 6th year | 25,000 | - | -25,000 | 0.262 | -6,550 | 0.207 | -5,175.00 |
| End of 7th year | 4,250 | 10,260 | 6,010 | 0.21 | 1,262.01 | 0.159 | 955.59 |
| End of 8th year | 4,792 | 12,550 | 7,758 | 0.168 | 1,303.30 | 0.123 | 954.23 |
| End of 9th year | 5,368 | 14,530 | 9,162 | 0.134 | 1,227.71 | 0.094 | 861.23 |
| End of 10th year | 5,975 | 16,275 | 10,300 | 0.107 | 1,102.10 | 0.073 | 751.90 |
| End of 11th year | 6,456 | 19,396 | 12,940 | 0.086 | 1,112.84 | 0.056 | 724.64 |
| End of 12th year | 7,187 | 21,470 | 14,283 | 0.069 | 985.53 | 0.043 | 614.17 |
| | | | 35,453 | | 443.49 | | -313.24 |

$$IRR = 25 + 5\left[\frac{443.49}{443.49 + 313.24}\right]$$

$$= 25 + 5(0.586)$$
$$= 25 + 2.93$$
$$= 27.93\%$$

## Profitability Index

Here we relate the NPV of the cash flows of the project to the total capital required (Cr) for a project through "profitability index". It is defined as the ratio of net present values of the cash flows to the initial capital expenditure (Co). Assuming that all the capital expenditure is incurred in year zero, the profitability index (PI) is as follows (Table 41.6):

$$PI = \frac{NPV}{Co} = \frac{1}{Co} \sum_{1=0}^{n} \frac{Cr}{(1+i)^n} \tag{5.9}$$

**TABLE 41.6   Estimation of Profitability Index.**

Original amount invested in a project = Rs. 60,000

| Year | Cash flow (in Rs.) | Discounting factor (15%) | Net Present Worth (in Rs.) |
|------|--------------------|--------------------------|----------------------------|
| 1.   | 14,500             | 0.8929                   | 12,947                     |
| 2.   | 14,900             | 0.7972                   | 11,878                     |
| 3.   | 16,600             | 0.7118                   | 11,816                     |
| 4.   | 18,700             | 0.6355                   | 11,884                     |
| 5.   | 19,000             | 0.5674                   | 10,781                     |
| 6.   | 20,000             | 0.5066                   | 10,132                     |
|      | 1,03,700           |                          | 69,438                     |

$$PI = \frac{\text{Net present value of cash flows}}{\text{Original amount invested}}$$

$$= \frac{69,438}{60,000} = 1.16$$

## SENSITIVITY ANALYSIS

Project appraisal techniques above, provide us certain measure of project's worth and this is related to a certain period of time and we will be forming this measure of the project under the assumption that the data used in the project evaluation remain unchanged over a length of time. But, in reality this is not a valid assumption because our estimates of costs and returns goes awry over time, as prices of agricultural produce as well as the costs of inputs are subject to change. Under these conditions our estimates of financial analysis will be misleading. Hence, there is a need for considering the probable changes in the data required for the project appraisal. It is also sometimes necessary to know as to how far our estimates of project appraisal remain constant under the changing situation of costs, prices and yields. If any analysis is able to provide clues to all these questions, we call such an analysis as 'sensitivity analysis'. If a thought is given to the forecasting behaviour of costs and prices in the sensitivity analysis, indeed, it becomes very much useful to the policy makers and planners of development. Since forecasting process is a difficult proposition, our project appraisal tends to go wrong. But, a simplified procedure of sensitivity analysis, which is not subject to criticism, is always welcome. The sensitivity analysis of the project appraisal includes the following points.

1. Consideration of the length of the period over the existing one;
2. Changes (increase or decrease) in the prices of goods and services by certain proportions of the project, say, by 10 per cent, 20 per cent, 30 per cent, 40 per cent, 50 per cent, *etc.*,
3. Changes (increase or decrease) in the levels of costs, say, by 10 per cent, 20 per cent, 30 per cent, 40 per cent, *etc.*,
4. Changes (increase or decrease) in the yield of crops and livestock; and
5. Delays in the implementation, *i.e.*, varying gestation periods.

Assuming the changed values for the above parameters, by a definite proportion, the project worth is calculated time and again. This procedure deals with the question of risk and uncertainty to a certain extent in the project analysis. But, in fact, this is not adequate. Elaborate risk analysis using probability analysis and simulation models exploring the randomization, are the most appropriate tools to indicate the real worth of the project under the conditions of risk and uncertainty.

# SECTION VI
## AGRICULTURAL MARKETING

# CHAPTER 42

# Agricultural Marketing

## MEANING

The term, agricultural marketing implies selling of goods and services by the farmers and ranchers. It includes various functions *viz.*, assembling, transportation, storing, buying, selling, standardization, grading, processing, sales promotion, *etc.* In general, marketing is the performance or operation of various business activities, which direct the goods and services from the producers to the ultimate consumers. In agricultural marketing the starting point is the farm or ranch and it is the basic source of market supply. The marketing process begins at this point and continues up to the point of final consumption. We have basically two markets *i.e.*, input market and output market. The input market supplies various inputs from the manufacturing firms to the farms and ranches. Output market deals with various activities of the crops and livestock products in their marketing channels. In each marketing channel of the product, utilities are created by the marketing efforts causing the transfer of ownership of the commodities among the middlemen over time and space. In these marketing processes, goods will have form, time, place and possession utilities. Form utility is imparted to commodity through processing activity of marketing. Time utility is added to commodity by the storage function of the marketing. Transportation service, which is found in marketing activity, creates place utility. When commodities are transferred from one person to another person in marketing channel through buying and selling activities, possession utility is imparted. Thus, through various functions of marketing, utilities are created to the commodities at each level of the marketing, involving marketing costs and margins to the middlemen.

## ROLE OF MARKETING

From the welfare point of view of farmers, consumers and middlemen, agricultural marketing plays a crucial role. Producer's share in the consumer's rupee for a commodity is based on the development of marketing system in the economy. Higher their share, greater would be the welfare to the producers. This depends up on the marketing system, market information and marketing facilities available in the country over time and space. Consumers get maximum welfare through marketing activity. The consumer's needs and desires are satisfied, if the supply of the commodities is ensured to the consumer with normal prices without much price risk. This is possible only when the markets are efficient in supplying the needed commodities to the consumers. For majority of the middlemen marketing is a source of livelihood. Greater the development of marketing system, more would be the involvement of middlemen in the marketing of goods and services. Middlemen wish to obtain higher margin for their marketing services. Hence, an efficient marketing system should always direct towards bringing the welfare to all these categories. It is from the point of view that marketing has become an important and integral part of economic activity and depends mostly on the magnitude of modern production technology. Hence, market policy as an integral part of agricultural development policy should be revised from time to time, keeping in view of the needs of the economy, which are desired by the producers, consumers and middlemen.

## DEFINITIONS

*Market: Market is a place where goods and services are exchanged. Market consists of buyers and sellers with facilities to communicate with each other for transactions of goods and services.*

*Marketing: Marketing is the economic process by which goods and services are exchanged between the producers and the consumers and their values determined in terms of money prices.*

The term agricultural marketing is to be understood from the two words, *viz.*, agriculture and marketing. Agriculture in its broadest sense comprises of all the farm activities, which use the natural resources for human welfare. Generally, it means growing or raising of crops and livestock products with the given level of resources and technology. Marketing comprises of all the economic activities involved in facilitating the flow of goods and services from production centre to consumption centre. According to Thomsen, *"agricultural marketing comprises all the operations, and the agencies conducting them involved in the movement of farm produced foods, raw materials and their derivatives, such as textiles, from the farms to the final consumer and effects of such operations on farmers, middlemen and consumers."*

According to Acharya and Agarwal, agricultural marketing is *"the study of all activities, agencies and policies involved in the procurement of farm inputs by the farmers and the movement of agricultural products from the farmers to the consumers."* Thus, marketing provides a link between farm and non-farm sectors. Various aspects of organizations supplying raw materials to the processing industries, analyzing, assessing and revamping the marketing policies for farm products and farm inputs are studied under agricultural marketing.

National Commission on Agriculture (NCA) in its XIII report observed that agricultural marketing involves all aspects of agricultural market structure and system, *i.e.*, both functional and institutional aspects based on technical and economic considerations and include post-harvest operations, assembling, storage, transportation, distribution, *etc.*

According to Richard Kohls agricultural marketing is *"the performance of all business activities involved in the flow of goods and services from the point of initial agricultural production until they are in the hands of the ultimate consumer."*

## SIGNIFICANCE OF AGRICULTURAL MARKETING

With the advent of new farm technologies in the form of green revolution, white revolution, blue revolution, yellow revolution, *etc.*, we have achieved self-sufficiency in agriculture and allied sectors. Now the economy is involved in international trade to siphon off excess supplies of commodities to the needy foreign countries. Farmers now are producing products for international marketing. Input marketing is growing at a rapid rate. The consumption of fertilizers and pesticides is fast increasing. There is a growing importance for improved seeds, machinery, irrigation, farm finance, *etc.* The modern agricultural production is continuously searching for new technology for tapping the input potential over time and space. All these trends in modern agriculture are increasing the scope of agricultural marketing, from the point of view of product marketing and input marketing. Greater thrust in the policy areas is laid on the problems of marketing and finding solution to such problems through relevant marketing facilities. Prices are determined based on the interaction of demand and supply for the commodities in the market. These prices give signals to producers with regard to what products to be produced and in what quantities. Thus, marketing gives signals to increase production and thereby ensures the availability of goods, and services. If the marketing activity is developed, demand for goods increases as a result, production of goods also increases. Due to increased production, the demand for inputs increases *i.e.*, the demand for input is derived from the increase in demand for the output. To distribute the required inputs to the farm sector, the input marketing has to be strengthened.

## SCOPE OF AGRICULTURAL MARKETING

The scope of the field of marketing can be examined from five angles *viz.*, producers' interest, consumers' interest, societal interest, traders' interest and Government role.

### Farmers' Interest

Farmers as the scare resource users are always on the look out for the most rewarding benefits in farming. An assured market environment for the products enlivens the spirit of the farmers to use the resources most judiciously. It is well known that technology is scale neutral but it is resource non-neutral, yet the farmers of all categories do not let the opportunity to go to tap as much productivity as possible from the resources they employ. Thus a healthy marketing system acts as an incentive for the farmers to use the resources prudently. Thus efficient input marketing and output marketing systems are indispensable to bring desired level of welfare to the farmers. Farmer infact may turn out to be a major beneficiary if the market system properly functions.

### Consumers' Interest

Marketing is a system that facilitates the movement of farm commodities from production centres to consumption centres. Thus it provides scope to the consumers to choose farm commodities of their choice to satisfy their needs. Consumers' welfare is brought about through increased marketing output by following efficient methods of marketing.

### Society's Interest

It is an extension of individual consumer's interest. When the consumption require-ments are met by an effective marketing system, society at large gets benefit in this process. It enhances the standard of living of the people. Society's resources are distributed efficiently among population in the desired direction. That is to say people's welfare is directly influenced by the efficient marketing system. Marketing is a source of livelihood to several people. An efficient marketing system also brings in price stabilization.

### Traders' Interest

Market intermediaries or middlemen, are those functionaries apart from the Government who facilitate the movement of the products from producers. They may be wholesalers or commission agents or retailers, *etc.* Through the process of marketing they not only fulfill the needs of the producers and consumers, but also in the process make out their living.

### Government Role

Government as a custodian of people's welfare has to perform certain functions. These include the procurement of foodgrains for the maintenance of buffer stocks as well as to meet the public distribution system through purchases by FCI. Government activities are also evident from the marketing assistance rendered by Cotton Corporation of India, Jute Corporation of India, commodity boards, *etc.* Government regulates marketing activities through various legislations and marketing policies. Regulated markets are created to check the malpractices of middlemen and bring about efficiency in various marketing operations. Government facilitates any marketing function to be performed efficiently keeping in view of the welfare of producers, consumers and stabilization of prices.

## CHARACTERISTICS OF AGRICULTURAL COMMODITIES

Agricultural commodities have special characteristics that are different from industrial goods and due to these special characteristics the subject of agricultural marketing is treated in isolation. These special characteristics of agricultural commodities bring changes in the supply and demand pattern. Let us now know how these characteristics are affecting the agricultural marketing operations.

### 1) Seasonality of Agricultural Production

Agricultural year is divided into two to three seasons depending on the agro-climatic conditions. In areas with scanty rainfall, agricultural production is confined to *kharif* season only. In areas of adequate and well distributed rainfall farmers are raising crops in three seasons *viz., kharif, rabi* and *summer* season. Thus, over the length and breadth of the country agricultural production is subjected to change based on rainfall distribution and availability of irrigation facilities. This means we have production risk. This finally leads to marketing risk. It means in the years of good production, prices are falling and in the years of low production, prices are increasing. This trend results in wide fluctuations in the prices of agricultural commodities. On the contrary, the supply of manufactured products is continuous and regular and their product prices are remaining nearly at the same level in a given period with less price risk.

### 2) Perishability of the Product

Most of the agricultural products are perishable in nature. The degree of perishability is very high for fruits and vegetables. This calls for immediate marketing soon after harvest. If marketing does not take place for perishable products, it immediately ruins the farmers' economy. The prices and arrivals of perishable products are subjected to fluctuations. If the arrivals of perishable products to the markets were less, prices would be very high and *vice versa.* As a result, consumers and producers are the sufferers. Marketing facilities should be strengthened particularly processing and storage, to overcome the problems emanating from perishable character of agricultural products. This characteristic is almost absent for industrial products.

## 3) Bulkiness of Agricultural Products

Most of the farm products are bulky in nature. This characteristic of bulkiness adds to the transportation, storage and labour costs. The price spread which is the difference between consumer's price and producer's price is very high when the products are bulky in nature.

## 4) Quality of the Products

We have large variations in the quality aspects of agricultural products such as size, colour, freshness, maturity, appearance, smell, *etc.* These quality differences form the basis for standardization and grading. But this is not a simple task as standardization and grading of products have greater effect on the prices of the commodities. This calls for scrupulous methods of standardization and grading of agricultural products, which are acceptable to consumers. Such problems are not there for manufactured goods.

## 5) Irregular Supply of Agricultural Commodities

Due to seasonality of agricultural products, supply is subject to fluctuations over space and time. Variations in supply, with demand remaining the same, bring about changes in the prices of the commodities causing marketing risk. But the supply of industrial goods is smooth and uniform.

## 6) Small Size of Holdings and Scattered Production

Seventy per cent of the holdings in India are small in size (below 2 ha) and these small holdings are scattered and fragmented over space. Under such circumstances it is difficult, to estimate the supply as a result of which market policy formulation regarding marketing development becomes difficult.

## 7) Inadequate Processing Facilities

Processing of agricultural commodities require adequate capital investment. Establishment of sugar factory meant for processing sugarcane into sugar will have special advantages of being one seller and one buyer *i.e.*, the advantages of monopoly and monopsony. Similarly processing facilities help to avoid problems of perishability and ensure smooth supply of commodities to the consumer. Further, storage problems, problems arising out of bulkiness, *etc.*, are reduced partly by the processing activity.

# CLASSIFICATION OF MARKETS

Markets are classified based on various criteria as presented here under:

## 1) On the Basis of Number of Commodities

Following this criterion we have a) general markets b) specialized markets.

*a. General Markets:* In general markets all types of commodities are sold. These commodities for example range from foodgrains to textile goods and so on.

*b. Specialized Markets:* Specialized market deals with a specific commodity. The markets are named after such commodities. For example, if vegetables are sold in the market, we call such markets as vegetable markets. Similarly, markets transacting goods like wool, cotton, jute and fish are called wool market, cotton market, jute market and fish market respectively.

## 2) On the Basis of Market Area

Here, we classify the markets based on the area covered by the markets into various categories *viz.,* local markets, regional markets, national markets and international markets.

*a. Local Markets:* These markets are also called village markets or primary markets or hats. The area covered by the market is limited to some group of villages, which are nearby or close to each other. In these village markets, perishable commodities like vegetables, fruits, fish, milk, *etc.,* are being transacted. Shandies, fairs *etc.,* which are held occasionally on special important days would also come under local markets or primary markets. In some areas, local markets are being conducted daily, while in other areas, these are conducted once in a week or twice in a week. Tribal markets in Madhya Pradesh are a sort of primary markets. Cattle markets, sheep markets *etc.,* come under primary markets.

*b. Regional Markets:* Here the area of operation of the market is relatively larger than that of local market. This market covers four to five districts. Sometimes regional markets cover a State. Food grain markets are the examples to be cited under regional markets. Fruit markets operated in the State are called regional markets. They are regular in conducting business transactions in notified commodities.

*c. National Markets:* These markets cover the entire country in their operation. National markets are found for commodities having demand over the entire country. Textile markets, jute markets, tea markets *etc.,* are the relevant examples.

*d. International Markets:* Here the commodities are sold in all the nations of the world. The market area of operation is extended over the entire globe. These markets exist for commodities like cashew, coffee, tea, spices and condiments, gold, silver, diamonds, machinery, *etc.* Recently, even textiles, rice, wheat, sugar, cut flowers, fruits, processed products, *etc.,* have international market.

## 3) On the Basis of Location

The location of the markets serves as a basis for classifying markets into village markets, primary wholesale markets, secondary wholesale markets, terminal markets and seaboard markets.

*a. Village Markets:* The area of operation of these markets is confined to a small village or a group of villages. Major transactions of goods and services take place among the buyers and sellers of these villages. Such markets may be regular or occasional in their operation.

*b. Primary Wholesale Markets:* These are located in big towns or taluks or mandal headquarters. All types of agricultural commodities from the village markets are pooled here. Transactions take place between the producers and the traders.

*c. Secondary Wholesale Markets:* These markets are found in the district headquarters dealing with major agricultural commodities like rice, pulses, oilseeds, chillies *etc.* Wholesalers and village traders are the main participants in these markets. Bulk of the arrivals comes from primary wholesale markets or village markets. Transactions of the commodities take place in large quantities. We find many commission agents, brokers, hamalies, weighmen, *etc.,* working in these markets for facilitating the marketing operations.

*d. Terminal Markets:* These are located in big cities/State capitals/seaports. These are well-organized markets and controlled by the Government to see that all modern methods of marketing operations take place. Processing and storage activities are predominant in these markets. Consumers, wholesalers and marketing agents are seen in these markets with rigorous transaction activities. Future marketing or forward marketing takes place in these markets. These are situated in big cities like Chennai, Bangalore, Mumbai, *etc.*

*e. Seaboard Markets:* These are primarily meant for export and import of commodities. Scientifically standardized and graded commodities are transacted. These are located in Mumbai, Chennai, Kolkata, Visakhapatnam, *etc.*

## 4) On the Basis of Time

Based on time markets are generally classified into three categories *viz.,* short period markets, long period markets; and secular markets.

*a. Short Period Markets:* These are held for a brief period in a day. Here the supply of commodity is fixed. Fish market in a particular place *i.e.,* village or town or city, vegetables markets, flower markets, *etc.,* are to be cited here. Since supply is fixed here, we can notice price variation based on demand in a day in the transaction of these commodities. Here the supply is zero elastic.

*b. Long Period Markets:* Durable commodities, which can be stored for some time, are transacted in these markets. The prices for the products are governed by supply and demand. The examples are foodgrains, oil seeds, *etc.*

*c. Secular Markets:* These are permanent markets. Manufactured goods, machinery, *etc.,* are transacted in these markets. Godown facilities and processing facilities are highly developed in these markets. These are well-organized markets. They deal with the export and import transactions.

## 5) On the Basis of Volume of Business

According to this criterion markets are classified into two types *viz.,* wholesale markets and retail markets.

*a. Wholesale Markets:* When large quantities of a commodity are bought and sold in the market among the traders, such markets are called wholesale markets.

*b. Retail Markets:* These are the markets in which retailers sell commodities to the consumers in very small quantities as per their requirements. Producers, retailers and consumers are seen in these markets.

## 6) On the Basis of Nature of Transaction

Markets are classified into cash markets and forward markets based on the type of transaction.

*a. Cash Markets:* If there are cash transactions in buying and selling of the goods, such markets are called cash markets or spot markets.

*b. Forward Markets:* These are the markets in which future sales and purchase of commodities take place at the current time. This process is called hedging.

## 7) On the Basis of Competition

On the basis of competition, the markets are categorized into perfectly competitive markets and imperfectly competitive markets (for details see Chapter 7 in Section I).

## 8) On the Basis of Government Intervention and Regulation

Based on the regulation by the Government, markets are classified into two categories *viz.,* regulated markets and unregulated markets.

*a. Regulated Markets:* The statutory market committees govern regulated markets and the Government from time to time makes marketing acts. The marketing costs, margins, fee, *etc.,* are standardized. Marketing practices are regulated and facilities created for the smooth conduct of marketing. Prices prevailing in different markets are displayed through various mass media.

*b. Unregulated Markets:* In unregulated markets, business is conducted without any supervision. There is absence of rules and regulations. The middlemen exploit the farmers and the consumers to the maximum extent. Sometimes producers are put to loss as middlemen exploit them in weighment, measurement, payment, *etc.*

## 9) On the Basis of Nature of Commodity

Based on the nature of commodity, markets are classified into commodity markets and capital markets.

*a. Commodity Markets:* Commodity markets deal with the buying and selling of the commodities. Examples are cotton market, wheat market, chillies market, cattle market, bullion market, *etc.*

*b. Capital Markets:* Capital markets are those markets in which shares, securities, bonds, *etc.,* are being purchased and sold. Share market and money market are the specific examples.

## 10) On the Basis of Vision

Based on vision, they are classified into black markets and open markets.

*a. Black Market or Invisible Markets:* In these markets the goods are not placed in the shops, but they are kept in the godown of the market. Hence, the name black market. The goods cannot be seen by the naked eye. On demand, the goods are delivered to the buyer on cash transaction. Any good, which is in short of supply and anything which is having high effective demand will be sold in the black market. During wars, drought, floods, catastrophes, *etc.,* there is a rigorous black marketing activity for scarce goods. Black markets are closed markets.

*b. Open Markets or Visible Markets:* These are visible markets and transactions take place between buyers and sellers and the price is determined by demand and supply.

## PRODUCER'S SURPLUS

The marketing activity is in reality looks into the ways and means of moving the surpluses available with the farmers, not exactly the actual production of the commodities. This is the producer's surplus. More specifically it is the quantity of produce, which is actually made available to the non-farm population of the country. The producer's surplus can be split up into two categories *viz.,* marketable surplus and marketed surplus.

### Marketable Surplus

This is the actual quantity of a commodity that is available with the farmers after meeting his seed requirements, family requirements, kind payments as wages, and payments to others to whom he pays for their services.

### Marketed Surplus

It is the quantity of a commodity, which a farmer actually sells in the market. This need not necessarily have a bearing on the other requirements of the farmers. This can be examined by building a relationship between marketable surplus and marketed surplus. The possible relationship between these two is that marketed surplus may remain more or less or equal to the marketable surplus. The situation of marketed surplus being more than marketable surplus arises, when the farmers cannot really afford to keep aside genuine quantity of the commodity for various purposes before disposing the produce. This is termed as distress sale, which is resorted to by small and marginal farmers, as their pressing requirements of making cash payments force them to do so.

### Factors Influencing Marketable Surplus

Marketable surplus is influenced by the factors presented below:

1. *Size of Holding*: Marketable surplus is directly influenced by the size of hidings. Larger the size of holding, greater is the quantum of marketable surplus. There is ample empirical evidence to support this observation.
2. *Level of Production*: Higher levels of production help to generate larger marketable surplus. Higher productivity results in higher production. Productivity is influenced by the efficient use of various resources employed on the farms.
3. *Family Consumption Requirements*: Higher consumption requirements for the family members reduce the quantum of marketable surplus.
4. *Other Requirements*: The requirements of products for seed purpose, kind payments, payments for religious and social purposes, *etc.,* influence the marketable surplus. Higher the requirement for these purposes, lower the level of marketable surplus, and
5. *Prices of Products*: Prices exhibit a direct relationship with the marketable surplus. Higher prices always encourage the farmers to produce and sell more and converse is the case when the prices are low.

# CHAPTER 43

# Process of Agricultural Marketing

The marketing system is not complete unless the commodities are made available in the form they are needed, at the time they are needed and at a place they are needed by the consumers. In making availability of the commodities from the producers to the ultimate consumers, three processes are involved *viz.,* 1. Assembling (concentration), 2. Equivalisation and 3. Dispersion (distribution).

## 1) Assembling

Assembling begins with the collection of surpluses of individual farmers. This becomes essential in moving the produce to the consuming markets, in sufficient quantities to permit efficient processing, transportation, storage, *etc*. This process of concentration is called as assembling. Assembling takes two forms *i.e.,* one is primary assembling and the other is secondary assembling.

*a. Primary Assembling:* The produce is assembled in the villages and primary markets as the farmers feel it convenient in view of small quantities of produce, pressing demand for cash, lack of transport facilities, paucity of information on the prices, *etc.,* prevailing in the other markets. The prominent functionaries are village merchants and itinerant traders or merchants[*].

*b. Secondary Assembling:* It succeeds primary assembling. It is associated with greater concentration. These are initially found in the producing areas and then in the consuming areas. New functionaries like commission agents join in the secondary assembling. These are the wholesale markets. The percentage of marketable surplus that arrives stands at around 70 to 75 per cent at this stage, before it is dispersed to different consuming areas. In the secondary assembling, the role of commission agents is of paramount importance.

## 2) Equivalisation

It is the adjustment of supply and demand as per the requirements based on the time, place and quantity. Equivalisation though tends to be confined to the wholesale markets, but to some degree or the other, it is also found in marketing channels as

---

[*] These are the petty traders generally without shops, who move from village to village with their carts, buy from the farmers and sell in the near by markets. It is a continuous activity for them.

well. With demand being spread throughout the year, the supplies should be made available as per the consumption requirements of the people. Storage has an important function here, to hold back the stock for timely release. Also certain agricultural commodities are area-specific and there are certain areas with abundant production and some other areas with deficit production. Through the process of equivalisation the products are moved from surplus areas to deficit areas. Transportation should be well developed in this regard. Adjustment of quantities as per the requirements is another important aspect of equivalisation. The requirements of one market to other market vary. Keeping this in view, the required quantity should be moved from assembling centers to the consuming markets.

## 3) Dispersion

The produce that is collected from producing areas should be made available to the millions of consumers through the consuming markets. From the major wholesale markets the process of dispersion starts. Dispersion is seen through various marketing channels. In respect of those products, which need processing before they are suitable for consumption, are moved to processors. The processed products then are channeled to the ultimate consumers. Commodities like paddy, cotton, sugarcane, *etc.*, need to be processed before consumption in the form of rice, yarn and cloth, sugar or jaggery, *etc.*, respectively.

## MARKETING FUNCTIONS

Marketing facilitates the transfer of ownership of the products from the producers to the consumers. In the transfer of ownership from the producer to the ultimate user, several activities are performed. *Now marketing function can be identified as a single activity, which facilitates the movement of the product from the point of its production till it reaches the ultimate consumer.* Marketing function encompasses the broad functions of concentration and dispersion. Marketing functions vary for different commodities. The marketing functions of fruits and vegetables vary from that of cereals and pulses. For the same commodity, the farmer may sell part of the commodity directly to the consumer and the rest of the product may be bagged and transported to the nearby primary markets. With this background, marketing functions are presented as per the following classification.

(1) Function of Transfer of Ownership
    (a) Selling
    (b) Buying
    (c) Demand creation and
    (d) Price determination.

(2) Function of Physical Movement
    (a) Transportation and
    (b) Storage

(3) Function of changing the form of the product
    (a) Standardization and grading and
    (b) Packaging

(4) Facilitating Functions
    (a) Market financing

    (b) Risk bearing and

    (c) Market information

## 1. Function of Transfer of Ownership

Selling and buying are the two important functions involved in the transfer of ownership. Selling is the process of finding the buyers and convincing them to buy the product at a price that is satisfactory to both sellers and buyers. Buying includes identifying ones needs, finding the source of supply of the goods and procuring them at the right price.

*a) Selling:* Selling function consists of the following sub-functions:

1. *Product Planning and Development:* The needs of the consumers should be taken into consideration in selling. We identify that product which is required by the buyer and sell the same.

2. *Contractual Function:* It involves identifying the potential consumers for the product and initiating and maintaining contacts with them for selling the commodity.

3. *Demand Creation:* Once the potential consumers identified we introduce various sales techniques to stimulate them so that the desired sales target can be achieved.

4. *The Function of Negotiation:* Some important factors to be considered at the time of selling are quality and quantity of the commodity proposed to be transacted, time of transfer, particulars of packing, mode of payment, *etc*. These should be well negotiated to avert any future conflicts between the buyers and the sellers.

*b) Buying:* Buying includes identifying ones needs, finding the source of supply of products and procuring them at the right price. Following are the sub-functions of buying.

1. *The Function of Planning the Purchases:* The buyers must plan their needs and undertake the purchases. Also the buyers should survey their own markets to identify the quality and quantity of the goods that are required.

2. *Contractual Function:* It is the identification of the sources of supply to confirm the suppliers of a commodity so that the flow of supplies can be made continuous to the market.

3. *The Function of Negotiation:* The terms and conditions of purchase along with the prices are negotiated with the sellers. Once the negotiations are completed, the goods are transferred to the buyers. The prevailing methods of buying and selling in the markets are as follows:

## a) Sale Under Cover of a Cloth

Under this method of sale, the prices of the products are settled through a transaction, which is done under the cover of a cloth. The transaction under the cover is that a given price is fixed between the buyers and the commission agents of the sellers by pressing the fingers of each other, for which code symbols are arranged. These traders are well versed with these codes. This practice continues between the commission agents and the prospective buyers, till an acceptable price is settled. The highest bidder is privileged to buy the produce, whose bidding price is conveyed to the buyer by the commission agent. This method suffers from the severe drawback that actual price offered by each buyer to the commission agent of the seller is not let known as this system does not provide any provision for such an open announcement. Sometimes the price that is finally settled may not be the highest one though theoretically it is claimed as the highest one. Although this method of sale is legally not permitted, still found in some markets.

## b) Sale Through Negotiations

This is a method, which facilitates direct contact between the buyers and sellers. The seller approaches the buyer with a sample of the produce proposed to be sold. The buyer quotes the price, if he is satisfied with the quality and if the seller accepts the price, the deal is completed. The seller gets advantage here, as he is in a direct contact with the buyer.

## c) Sales Based on Samples

In this method, commission agents approach the buyers with the sample of the lots proposed to be disposed. The commission agents go round the shops of the buyers and produce is offered to that buyer, who offered the highest price per unit of the produce.

## d) Morghum Method

A common method of sale found in villages when the farmers borrow funds from local moneylenders. The transactions are effected based on the oral agreement that is made between the buyers and sellers. In these transactions harvest price is the selling price. It is a common method of sale found in villages when the farmers borrow funds from the local moneylenders.

## e) Open Auction Sale

This is the method, which prevails in most of the regulated markets. The prospective buyers examine the lots of the produce kept for sale and offer their bids openly. Highest bidder has the right to take the possession of the commodity, provided the price offered is acceptable to the farmer (seller). The amount is settled immediately after the sale.

## f) Closed Tender System

This is, more or less, the same as that of the open auction system except that the bids are offered in the form of closed tenders. This method is followed in some regulated markets. The lots of produce that are kept for sale at display units are given lot numbers. The prospective buyers inspect the lots and offer their prices on a slip of paper and put in a sealed box kept for that purpose. When all the buyers complete their turn the slips deposited in the sealed box are taken out and arranged according to the lot number. Highest bidder has the right to take the possession of the produce in the lot. How well this type of sales is conducted depends upon the efficiency with which the officials of the markets conduct this method of sale.

*c) Demand Creation:* When commodities are basic requirements of the consuming section, the demand is automatically created. The need of demand creation arises for those products with which the consumers are not familiar and they are likely to consume by an act of persuasion highlighting the merits of the products through personal approach of the salesmen, advertisement through various mass media like newspapers, posters, pamphlets, radio, television, *etc*. Besides sellers resort to sales promotion activities like distribution of free samples, price discounts, exhibitions, sales by installments, *etc*.

*d) Price Determination:* In the marketing system all the participants who help the movement of the products to the ultimate consumers should be rewarded. This is possible by pricing the products at various stages, as they move from the farmers to the consumers. At this stage, the concepts of price determination and price discovery need to be specified. Prices are determined by the aggregate forces of demand and supply in a market, while prices are discovered at each stage in the marketing channel *i.e.*, market intermediaries discover the prices based on the availability of the commodities and the demand for the commodities from the buyers at each stage. The prices that are discovered facilitate to clear the supplies at each stage so that the distribution of the commodities flows smoothly among the persons who need them.

## 2. Function of Physical Movement

Transportation and storage facilitate the physical movement of products from the producer to the consumer.

*a) Transportation:* Transportation creates, place utility. The products rights from the farmer's fields to the ultimate consumers are moved through transportation. The spatial variations in agricultural production need the movement of the commodities from the places of surplus production to the places of deficit production. Besides, the output produced in an area cannot be consumed there only. It needs transportation. Transportation helps in development of markets, reduction in spatial price differences, provision of employment, *etc*. The common modes of transportation are bullock carts, tractors, trucks, rail, *etc*.

*b) Storage:* Seasonality of production is a specific characteristic of agricultural products. On the other hand, consumption of the products is regular and continuous. This unmatching situation requires the need of storing the commodities after harvest to make them available throughout the year. Storage is involved at various stages in the marketing. Producers, who can afford to sell the produce at a later date, store the produce. The middlemen in the process of buying and selling activity also store the produce to take advantage of the market situation. The consumers who can purchase the commodities in bulk particularly those of foodgrains, tamarind, *etc*., when the prices are on the lower side also go for storage. Though the commodities are stored to take the price advantages by all those concerned, storage is associated with risks as well. The sources of risks are moisture loss, loss caused by storage pests, rodents, *etc*. At times faulty storage too contributes to storage loss. In spite of all these losses, if the per unit price at the time of sale covers these risks, the concerned are not too unhappy about storing the farm commodities.

## 3. Function of Changing the Form of the Product

It facilitates making the products available to the consumers with various specifications in the usable form and also as per the choice of the consumers.

*a) Standardization and Grading:* Standardization preceeds grading. The characteristics based on which the standards are determined are: freshness, ripeness, size, weight, colour, foreign material, moisture content, *etc*. Pyle has defined standardization as *"the determination of the basic limits on grades or the establishment of model processes and methods of producing, handling and selling goods and services."* Grading means the sorting of produce into different lots having the same characteristics with respect to quality specifications. It is the process of dividing the lots of commodities with already set standards. Grading is of two types. Fixed grading/mandatory grading and permissive/variable grading. In fixed grading, the standards set out are fixed and it is mandatory for the individuals to follow the set standards if they are going to sell the graded commodities. The Agricultural Marketing Advisor, Government of India, fixes these standards for majority of the agricultural commodities. Fixed grading is a must for commodities which are exported. Permissive grading is subject to variations over time. Individual choice for grading is permitted under these methods. Grading of produce done by farmer is *Kutcha* or *crude* grading. Merchants also resort to grading and this grading is called *pucca* grading.

Any graded product fetches a better price to the farmer. Grading acts as an incentive to produce a better quality product. Hence there is no extra effort needed to ensure the quality of the product. Grading ensures a wide market for the product, as the quality of the product need not be tested in the distant markets. Consumers will be greatly benefited, as price paid is in commensurate with the quality of the product. Also, they have the choice of picking the goods from various alternative grades available in line with the prices fixed.

**b) Packaging:** To make the products move from farm gate to merchants and finally to different users, some kind of packaging is essential. Gunnies are the commonly used packing means for cereals, pulses, oilseeds, limes, vegetables like potato, ladyfingers, *etc.* Grapes and apples are packed in boxes, pots, *etc.,* while we pack fish in baskets, oil and milk in sachets, tins, *etc.* It facilitates easy handling, reduces spoilage, ensures cleanliness, reduces marketing costs, prolongs storage life, *etc.*

**4) Facilitating Functions:** These include market financing, risk bearing and market information.

## a) Market Financing

The need of financial assistance applies to both the farmers as well as traders. Farmers find it difficult to continue the farm business with their owned funds and they need a helping hand from external sources. Apart from production credit, farmers need market finance too before they dispose the produce at a favourable price. The traders who are found at various stages of marketing too find their owned funds short of requirement to purchase the stock and carry on other functions like packaging, processing, storage, *etc.* Distress sales are averted if adequate marketing finance is given to the farmers.

## b) Risk Bearing

There is always a time lag between the harvesting and final consumption. Risk is imminent under such a situation and this is borne by the producers, traders and others involved in the marketing process. Following are the kinds of risks associated in the marketing process.

1. *Physical Risks:* Physical risk is caused during weighment, bagging, transportation, storage, *etc.* Physical risk consists of loss of quantity as well as loss of quality of the product

2.  *Price Risk:* Because of the special characteristics of the agricultural products, price risk is a common phenomenon. Price fluctuates during the same day, from week to week, from month to month and from year to year. Price rise can help the farmer and trader and equally they are at loss, if the price falls.

3.  *Institutional Risk:* Government policies like movement restrictions of foodgrains, imposition of levies, *etc.* bring losses to the marketers.

## Measures to Reduce Risks

1.  *Improving Storage Structures:* Physical losses can be reduced by bringing suitable changes in storage structures at the farmers' level to prevent the possible losses. An example here is *'puri'* which is a very common storage structure made with twisted paddy straw has been converted into *'cement based puri'* for protecting the produce particularly from rodents.

2.  *Insurance:* Insuring the produce against the risks, which the insurance companies cover is an important safety measure during the periods of quantity losses in storage. Insurance dose not eliminate all risks, but the farmers' risk is transferred to the general population through insurance companies.

3.  *Market Information:* Market information is broadly defined as communication or reception of knowledge or intelligence. It includes all the facts, estimates, opinions and other information which affect the marketing of goods and services (Tousely *et al.,* 1968)[*]. This information is of great importance to the farmers, merchants and Government as well. Price information is very vital for farmers in their decision of timing the sales. Merchants require market information to carry on their routine transactions like buying, storing and selling. This information facilitates them in planning their strategies like quantities to be purchased, quantities to be sold immediately and quantities to be stored in the market, where they should plan their sales (local markets or distant markets), *etc.* Government too needs this information to keep an eye on the price trends and for market intervention, maintenance of buffer stock, *etc.*

Market information constitutes market news and market intelligence. Market news present information on prices of the commodities, market arrivals, stock, directions of outflows, *etc.* The availability of market information helps the farmers to plan their sales. Timely availability of this information is of utmost importance. Market intelligence is the historical record of market situation. It is a continuous study of market behaviour with reference to price trends, arrivals over time, stocks over time, outflows over time, *etc.* Based on the available information one can peep into the future to know the price trends, supply and demand situation, *etc.*

The sources from which one can get the market information are: news papers, Bulletin of Agricultural Prices (weekly), Agricultural Situation in India (monthly), Agricultural Prices in India (annual), *etc.,* and the reports of the Bureau of Economics and Statistics, the Directorate of Marketing and Inspection (DMI), the Directorate of Economics and Statistics, regulated markets, *etc.*

---

[*] Tousley R.D. and Others - Principles of marketing. MacMillan Publishing Company, New York, 1968: pp. 496.

# MARKETING EFFICIENCY

Marketing activities and the intensity or the degree of market performance reveal about the marketing efficiency. But, the concepts of marketing efficiency are very broad, complex and dynamic. Infact, we consider many theoretical concepts and many practical aspects of marketing to know about marketing efficiency. In this context the following aspects are to be studied: (1) Objectives of marketing in terms of cost minimization or maximizing the goals of marketing and (2) Expectations of the participants in the marketing system. For example, farmers expect higher prices (remunerative price) for their produce. Here, what is expected is that the marketing system should provide facilities to the farmers to dispose their produce at reasonable prices with less marketing costs per unit of marketed output. Consumers should receive the goods and services in the form and quality and quantity of the produce they desire at lower and reasonable prices. (3) Marketing functionaries *i.e.*, market-middlemen expect stable and increasing revenue for the services they render. (4) Government expects the marketing system to protect the interests of all the people *viz.*, farmer-producers, middlemen and consumers and bring about long run welfare of the society. From the above points we know that marketing efficiency is broad, complex and dynamic. Hence in its broad implication lies all its theoretical and practical implications. Let us now know some of the definitions on marketing efficiency.

Kohls and Uhl (1980)[*] defined *"marketing efficiency as the ratio of market output (satisfaction) to marketing input (cost of the resources used in the marketing)"*. A higher value of this ratio indicates improved marketing efficiency and lower value denotes reduced efficiency. Improvement in the marketing efficiency is either due to reduction in the costs for the same level of satisfaction or increase in the satisfaction of marketing services for the given marketing costs.

According to Clark (1954)[**] marketing efficiency; should include the three following components: (1) Effectiveness of the marketing system with which marketing service is performed (2) The cost at which the service is performed and (3) The effect of this marketing cost and the method of performing the service. Of the three components, the last two are the most important because consumers get maximum satisfaction at the lowest possible cost and it should comply with the need of striking a balance between production of commodities and consumption needs of consumers.

# EFFICIENT MARKETING

As per the demands of the consumers, the goods move from producers to ultimate consumers at the lowest possible cost, and such a movement with lowest cost is known as efficient marketing. Sometimes a change in the marketing functions takes place with the reduction in the cost, but without reducing consumer's satisfaction. This is due to the improvement in the efficiency of marketing. In a nutshell efficient marketing system for farm products ensures the following:

1.  Increased farm production leads to proportionate increase in the level of real income in the economy. This acts as an incentive to have additional output and input surpluses.

---

[*] Kohis R.L. and J.N. Uhl Marketing of Agricultural Products. MacMillan Pub. Co. Inc. New York, 1980, pp. 260.
[**] Clark F.E Principles of Marketing, 1954.

2. In the years of bumper production, producers would have higher revenues through effective marketing functions *viz.*, storage, transportation, distribution, *etc.*, based on the demand of their commodities.

3. Consumers get the highest satisfaction at the lowest cost, if the marketing system is efficient. Thus, an efficient marketing system is an important means for rising the income levels of people in the country and thereby the standard of living of the people.

# APPROACHES TO MARKETING EFFICIENCY ESTIMATION

We in general, study the efficiency of marketing system from the two angles *viz.*, technical efficiency or physical efficiency or operational efficiency and pricing efficiency or allocative efficiency.

## Technical Efficiency

Any change that reduces the marketing costs per tonne or quintal of output is desirable and directly adds to marketing efficiency. The technological development that is embodied in the various marketing functions *viz.*, assembling, transportation, storage, grading and standardization, processing, packing, *etc.*, reduces marketing costs and increases marketing efficiency. Such process is called technical efficiency. All marketing activities cannot be brought about with less marketing costs. Sometimes changes in the marketing activities take place. These changes bring down the marketing costs as well as consumers' satisfaction. For instance if certain commodities like betel leaves, fish, fruits, vegetables, *etc.*, are transported to distant markets without taking proper care in grading, packing and transportation, the marketing costs may come down partially, but the consumers' satisfaction is very much reduced and finally marketing efficiency is greatly affected. From this we could know that reduction in the marketing costs need not necessarily cause efficiency in the marketing as it reduces the consumers' satisfaction. Ultimately economic efficiency would also fall.

## Pricing Efficiency

Prices give signals to the producers and the traders with regard to allocation of farm products over time and space and the consumers receive signals from prices and create demand for farm products over time and space, so that marketing system is finally developed for the welfare of the producers, consumers and middlemen. Thus, the economy would be on the wheels of growth, if pricing system for the commodities in the marketing is efficient over time and space. In simple terms, pricing efficiency in the marketing system occurs under the following conditions:

1. Transportation costs add the price differential of the product between two distant markets.
2. Price differential is also due to storage of the commodity over time, and
3. Processing costs lead to price differences at a particular point of time and space. For example, price of wheat and bread, price of paddy and rice, price of oil seeds and oil, *etc.*

Whenever marketing functions are performed for a commodity at a particular point of time and place, cost is increased and hence the commodity would have higher value or price at every point of marketing functions. This means, utilities are imparted to commodities due to various marketing functions and in this process the efficiency of marketing is concerned with the extent to which prices of the commodities deviate from the cost of performing these marketing functions. Thus, if the price deviation for the commodity in this process is more, to that extent the marketing efficiency is also affected by the extent of competition in the markets, market information, nature, attitude and objectives of the functionaries. In this context, pricing efficiency and the marketing efficiency are similar. But, in general one is at the cost of the other. Pricing efficiency is concerned with how accurately prices reflect consumers' demand through marketing channels. Thus, pricing efficiency is affected by the rigidity of the marketing costs and nature and degree of competition in the marketing. If the marketing functionaries are ambitious and exploit the market situations, it leads to pricing inefficiency. High marketing costs for commodities lead to pricing inefficiency. Pricing efficiency is improved through improvement in disseminating the relevant market information, market news, market research, *etc.*

## APPROACHES TO THE STUDY OF MARKETING

We can study the marketing aspects through various approaches. Among these the most common are: (I) Functional approach (II) Institutional approach and (III) Market structure approach.

### I. Functional Approach

The various marketing activities involved in getting the farm products from the farmers to the consumers are called marketing functions. These are done by the private agencies, co-operative marketing societies and the Government. In this functional approach, we estimate the costs of these functions and compare these against the efficient standards of doing the same functions. The functions are classified based on three purposes *viz.*, (i) exchange functions (buying and selling activities) (ii) physical functions (processing, storage and transportation) and (iii) facilitating functions (grading, standardization, financing, risk bearing, market information, market research, *etc.*).

Studying these functions over time and space for different commodities would enable us to know about the marketing costs and its efficiency. We have marketing channels for every type of commodity. Greater the length of the marketing channel, greater would be the magnitude of exchange function. Exchange function includes buyers bidding for supply of commodities and sellers offering the commodities at a specified price. The buying function locates the supplies of the commodity and helps in the assembling process for shipment. The selling function differs based on the stage of marketing channel, and involves various functions like packaging, labelling, advertisement, promotion and other business activities.

### 1. Physical Functions

These functions add form, time and place utilities to commodities, for instance processing activity adds form utility to the commodity. In agricultural marketing, the processing activity comprises all the manufacturing activities in bringing the commodity nearer to the point of consumption. Wheat into bread, fruits into jams and jellies, tomato into sauce, maize into corn flakes, milk into cheese, *etc.*, are the examples. Storage function adds time utility to the product by holding the commodity from the time of harvest and distributing the same to the different markets over time, as and when it is required by the marketing system. For example, wholesalers maintain large stock of grain for future trading through storage function. This adds storage cost to the price of the commodity and creates time utility at all levels in the marketing channels. Transportation process imparts place utility to the product.

## 2. Facilitating Functions

These functions would help improve the performance of marketing system by enhancing pricing efficiency and operational efficiency. Standardization and grading help the buyers to know about the exact qualities of the commodities, which they are buying without seeing the same. When the commodities are standardized, they are made uniform regarding quality and durability. In the market we have 'AGMARK' ghee products, milk products, jams, squashes, jellies, *etc*. Through the process of standardization, the cost of exchange activities are greatly reduced because the buyers don't have to incur travelling costs to purchase the standardized item. When standardization is employed to the commodity it ensures the consumers the accepted quality characteristics of the products. Weights and measures were already standardized in 1956. AGMARK seal for export commodities and BIS for industrial products are necessary as per the instruction of Directorate of Marketing and Inspection (DMI).

Financing activity is indispensable for all the marketing activities. There is a time lag between the time some one buys the commodity and the lapsed time some one sells the commodity. Money is held up in these exchange activities; as a result the interest on money is foregone. The interest foregone is the real cost of financing in the purchasing activity and storing activity of the commodity. If the middlemen borrow the funds to purchase and hold the commodities, the interest amount paid on the borrowed funds is the cost of financing.

## 3. Risk-bearing Functions

Marketing risks or losses fall on the person who buys the commodity. These losses may be physical or price related. Physical risks refer to the losses of the commodities due to quality deterioration. These are generally caused by the excessive moisture, wind, fire, cyclone, storage pests, rodents, *etc*. These physical losses can be reduced or lessened through insurance programme and scientific storage practices.

Price risks occur during holding the food grain inventories or due to abnormal price movements. These price risks can be minimized or reduced by hedging process or future trading.

## 4. Market Information

Market information is infact the data collected on prices of the commodities, and on marketing costs and margins of the commodity, analyzing and disseminating the required information through various mass media such as news papers, bulletins, radio, TV *etc*. Market information is necessary for smooth operation of the marketing system. The buyers and sellers coming to the market should be well informed about the demand, supply and prices of commodities in different markets.

## II. Institutional Approach

This approach examines various marketing activities and assesses the marketing costs, marketing margins and marketing efficiency. According to this approach, the middlemen are classified into five categories *viz.*,

1. Merchant middlemen (wholesalers and retailers)
2. Agent middlemen (commission agents, brokers, *etc.*)
3. Speculative middlemen
4. Facilitative organizations and
5. Processors.

## 1. Merchant Middlemen

Wholesalers buy the commodities from producers and producer merchants and sell to industrial users (processors) and retailers. Different types of wholesalers exist in the different marketing activities. Their primary function is directed towards holding food grain inventories in bulk. The products in lots are moved according to the consumers' requirements to different areas. They also provide credit to the retailers and offer assistance and help in merchandising or business activity. In doing so, they assume different forms of risk and reap the benefits.

Retail merchants purchase the required lots of commodities from wholesalers and handle the products to serve the needs of consumers. They generally purchase the commodities for resale to consumers on retail basis.

## 2. Agent Middlemen

They are distinguished from merchant middlemen in the possession of the commodities. In general the agent middlemen do not take the title to goods. This means that they neither buy nor sell the commodity through money transactions. They simply help in the negotiation activity, so that selling takes place from seller to buyer. This task is accomplished successfully through special knowledge of the product and marketing services which are endowed with the agent middlemen.

*Commission Agents:* These are the agent middlemen in the marketing of the agricultural commodities. They help in buying and selling of the commodities but do not possess the title of the commodity. They take over the physical handling of the produce, bargain with the buyers and settle the price. They charge commission and expenses incurred in the transaction and remit the balance amount to the seller.

*Brokers:* Brokers perform the duty of bringing buyers and sellers together and charge brokerage.[*] Brokers are very much required in the marketing activity of foodgrains, fruits, vegetables, livestock products, *etc*. Commission agents are generally seen in foodgrains, fruits, vegetable and livestock marketing. These are the licensed brokers meant to help the selling activity. Any product consigned to the commission agent is sold at the best price available in the market because they collect relevant information in the market on the sales of the commodity, deduct their commission[**] for their services and remit the balance to the seller.

## 3. Speculative Middlemen

They actually take the ownership of the commodities through purchasing activity and attempt to make profit from the future price movements. The process of making profits from the future price movements is called speculation. The speculative mid-

---

[*] Brokerage is a fixed sum of amount for helping to perform a particular marketing activity.

[**] Commission refers to certain fixed proportion or percentage on the value of commodity for helping to perform a particular marketing activity.

dlemen are seen in share markets related to agriculture and industry. They are technically called scalpers or day-traders or floor-brokers in commodity markets.

### 4. Facilitative Organization

It includes various trade associations involved in the marketing functions. These institutions provide general data on prices, supply, demand of the commodity and also provide physical facilities for marketing activities. They also guide the rules of trading among competitive firms or middlemen.

### 5. Processors

Processors transform raw agricultural products into different and required consumable forms. Almost all the agricultural commodities are subjected to some degree of processing or the other, between the points of production and consumption. In this context, the bakery units, sugar factories, rice mills, *etc.*, are worth to mention.

## III. Market Structure Approach

This analysis gives importance to the nature of market competition existing in the different marketing systems. It attempts to relate the variables of market performance to the various types of market structure and market conduct. By market structure generally we mean the description of number and nature of marketing participants in the marketing activities. This tells us about the number and size distribution of buyers and sellers in the marketing activities, the degree of product differentiation and the various barriers or restrictions to the potential marketing entrants.

The behaviour of the marketing firms is studied under the heading *i.e.*, market conduct. Profit seekers in the process of searching for relevant and remunerative prices determine their selling price and determine the quality of output they want to sell. In that process they use their market powers to weaken and eliminate their competitors. If the market structure in the industry resembles monopoly rather than pure competition, then the market performance and efficiency would be very poor.

## MARKET RISKS

Risk is apparent in all marketing activities. Loss of produce occurs in the marketing activity due to fire, rodents, storage pests, moisture, *etc.* Price fluctuations for commodities are seen due to changes in supply and demand, which in turn are due to time lag between production and consumption. The longer is the time lag, the greater would be the magnitude of marketing risks. Whenever there are marketing risks, the margins and costs greatly vary and accrue to the market middlemen. Institutional risks arise due to changes in Government policies, programmes, changes in taxes and tariffs, changes in the movement restrictions of foodgrains between the states and regions, *etc.* Price control measures, imposition of levies, taxes, *etc.*, are also changing over time and create market risks. Such risks are broadly categorized under institutional risks. Let us now know about risk management strategies.

## MARKETING RISK-MANAGEMENT STRATEGIES

Middlemen involved in the marketing adopt different practices to minimize market risks. Marketing risks are associated with not only profits to middlemen but also losses. Persons who don't take risks in marketing activities, cannot get higher profit, *i.e.*, higher profits in the marketing necessarily involve higher marketing risk. Let us examine some of the risk management practices. These are broadly grouped under the management of physical losses and the marketing practices to minimize price risks.

## I. Management of Physical Loss

The physical losses or quantity losses of the product are reduced by the adoption of the following measures.

1. Using the improved storage structures and following the preventive measures of avoiding fire and quantity loss due to various factors in the storage process.
2. Using appropriate chemicals for avoiding storage losses.
3. Using better and quicker transport methods and better handling practices during transportation.
4. Using proper package practices and storage practices, and
5. Insuring the produce with the concerned insurance agencies, *etc.*

## II. Marketing Practices to Minimize Price Risk

The following practices are adopted to minimize the price risk.
1. Fixation of minimum and maximum prices for major commodities by the Government and allowing the prices to move within the defined price ranges.
2. Strengthening the marketing facilities for dissemination of information on prices, demand and supply of commodities over time and space to all the concerned with the marketing.
3. Strengthening of advertising system and strengthening of forward marketing through hedging and speculation.

# SPECULATION

It is a process of transacting the goods or financial obligation that assumes risks in the hope of favourable price movements. Speculation is done in the marketing to keep the marketing activities competitive and to keep the average cost of future transaction or future trading as low as possible. It also creates the volume of trading to allow producers to get into contract or to go out of contracts. Speculation generally refers to the process of gambling but to the economists it is not so. A gambler unnecessarily takes risks and the gambler deliberately creates risks and he does not serve the society with fixed purpose. On the other hand, speculator assumes risks in marketing. Speculation means doing the business with the sole objective of making profits by correctly guessing the price movements. Speculation leads to hoarding which creates artificial scarcity. The physical delivery of the produce is neither taken nor given. Speculation is not a regular marketing practice. It is of two types *viz.*, speculation proper (legitimate speculation) and speculation improper (illegitimate speculation).

## Speculation Proper

In the process of speculation proper, speculators devote their whole time in the marketing activity. They collect relevant information with regard to future price movements to make intelligent forecasts and do speculation. These actions are beneficial to the economy.

## Speculation Improper

Illegitimate speculation is a sort of gamble. In this method of speculation, the speculators create artificial scarcity leading to rise in the prices of commodities in the future for making very large profits. Their actions are not rational and hence the economy surffers due to their actions.

# HEDGING

It is a trading technique of transferring the future price risk. According to Shepherd (1965)[*] *"hedging is executing opposite sales or purchases in the future markets to effect the purchases or sales of physical products made in the cash market"*. Through hedging practice, the future buying and selling the produce at a specified date is done to offset an equal and opposite price trend in the cash markets. By doing so the traders avoid the risk of uncertain fluctuations in the prices of commodities. In this approach each unit sale of the commodity transaction is entered with an equivalent unit purchase of the commodity in two different markets. Hence it is assumed that the prices in two markets move exactly parallel to strike a balance. This means losses arising in one market are offset by the profits in the other markets.

## Advantages of Hedging

1. It avoids price risk and enables the trader to earn normal trade profits.
2. It keeps the margins at a very low level and ensures no risk in the future.
3. It facilitates financing of inventories.

In the developed countries hedging is a common practice in livestock marketing. In India hedging is done in the food grain marketing.

**TABLE 43.1 Hedging.**

| Cash market (Buying) | Future market (Selling) |
|---|---|
| May 15th 1998: Buys 100 Q of rice @ Rs. 1,200 per Q | Sept 15th 1998: Sells 100 Q of rice @ Rs. 1,200 per Q |
| Sept 15th 1998: Sells 100 Q of rice @ Rs. 1,140 per Q <br> Rs. – 60 (loss) | Sept 15th 1998: Buys 100 Q of rice @ Rs. 1,140 per Q <br> Rs. + 60 (profit) |

## How Hedging Helps in Price Risk Management?

The hedging practice adopted by wholesale trader in rice is assumed and provided in Table 43.1. On 15th May, 1998 he buys 100 Q of rice @ Rs. 1,200 per quintal from the cash market and makes agreement to sell the same quantity at the same price *i.e.,* Rs. 1,200 per quintal in future market on 15th Sept. 1998. Similarly, the wholesalers are

[*] Shepherd G.S. (1965) Marketing farm product-Economic analysis. IOWA State University Press, Ames IOWA, 1965. pp. 153-154.

also fearful regarding price fall for rice in future date. To safeguard his trading practice against price fall, he would also resort to future trading in future market on September 15, 1998. He would make an agreement to buy 100 quintals at a lower price of Rs. 1,140 in future market and also to sell the same quantity at the same price of Rs. 1,140 in the cash market.

This practice is adopted by the wholesaler to overcome the price risk (price fall). In the above illustration the loss of Rs. 60 per quintal incurred in the cash market is offset by an equivalent gain from future trading (hedging). Thus through hedging process the traders safeguard themselves against price fall in future. The gain obtained in future market covers the cost of holding the inventory plus normal rate of return on investment *i.e.,* normal profit for undertaking the business.

# CHAPTER 44

# Problems in Agricultural Marketing

Over the years, major changes came into affect to improve the agricultural marketing system. Many institutions *viz.*, the regulated markets, marketing boards, co-operative marketing institutions, State Warehousing Corporations, Food Corporation of India, dairy corporations, *etc.*, have been established primarily to help the farmers. However, various studies indicated that modernization in agricultural marketing could not keep pace with the technological adoptions in agriculture. The various marketing functions *viz.*, grading, standardization, storage, market intelligence, *etc.*, need to be improved to meet the present day requirements of the farmers. Let us know about the various marketing problems. Indeed such knowledge of marketing problems would give information as to why markets are not developing and what measures are required to develop the markets.

### 1. Large Number of Middlemen

The field of agricultural marketing is viewed as a complex process and it involves a large number of intermediaries handling a variety of agricultural commodities, which are characterized by seasonality, bulkiness, perishability, *etc.* The prevalence of these intermediaries varies with the commodities and the marketing channels of the products. Because of the intervention of many middlemen, the producer's share in consumer's rupee is reduced.

### 2. Small and Scattered Holdings

The agricultural holdings are very small and scattered, as a result of which the marketable surplus generated is very meager. Moreover the farmers have no business awareness.

### 3. Forced Sales

The financial obligations committed during production period force them to dispose the commodity immediately after the harvest though the prices are very low. Such forced sales or distress sales will keep the farmers in vicious cycle of poverty. The National Planning Committee on rural marketing and finance remarked that the farmer, in general, sells his produce at an unfavourable place and at an unfavourable time and usually he gets unfavourable terms.

## 4. Technological Development Problems in Farm Production

Evidence is that technological changes in performing certain farm operations brought in new problems in agricultural marketing. For example, paddy harvesters are identified to increase the moisture problem in paddy, mechanical picking of cotton is associated with the problem of mixing trash with cotton, potato diggers are found to cause cuts to the potato, sugarcane harvesters effect the problems of trash mix with the cane, *etc.* These problems lead to the reduction of price for the farm products. Unless corrective measures are effected, the production technologies accentuate the marketing problems.

## 5. Poor Handling and Packing

For efficient and orderly marketing of agricultural products, careful handling and suitable packing are required. Present packing and handling are inadequate. For instance, many times we see rough and careless treatment in the packing and initial handling of fruits and vegetables. Green vegetables are packed in heavy sacks or piled on a lorry in bulk, which will be heated up quickly at the centre, wilt and rot soon. Workers or passengers are allowed to ride on the top of a load of vegetables, which results in physical damage. Careless handling of fruits and insanitary handling of the produce are other problems. Poor handling and packing expose the products to substantial physical damage and quality deterioration. Not only do these losses cut down the supply of products reaching the consumers, but also raise the price of the remaining portion, which must bear all the costs.

## 6. Lack of Standardization and Grading

Inadequacies exist in scientific grading of the produce in the country. In the absence of standardization and grading, adulteration is the consequence. Each middleman may adulterate the produce to his short run advantage. This poses a problem in assigning prices to the commodities as per the quality specifications. It is alleged that no proper relation exists between the price and quality of the agricultural commodities and this situation thwarts the farmers in getting a remunerative price in consistent with quality of the product. The transaction of such products hardly encourages the farmers and the consumers who are also denied the privilege of buying a good in relation to the price he pays. The true beneficiaries in a situation of such inadequacies of standardization and grading are the market intermediaries as they are at liberty to quote any commodity as inferior and offer a low price.

## 7. Inadequate Storage Capacity

Inadequate storage facilities are the causes of heavy losses to farmers in many parts of the world; and resulted in serious wastage of foodstuffs, and increased costs to producers. There are no scientific storage facilities for perishable products (fruits, vegetables, milk, meat, fish, *etc.*). In developing countries, the losses from inadequate storage are estimated to be 15 to 20 per cent of the production levels of farm products. These storage losses of foodgrains occur at all stages between the farm level and the final level of consumption. Such losses occur from physical damage due to infestation of pests, rodents and quality deterioration, discolouration and unpleasant odour which would make the products unfit for human consumption. The final result of such inadequacy is substantial loss in food, which indirectly means increase in the cost of living of consumers. Hence, the produce needs to be stored in the safe storage structures, if the farmer wants to wait for some time for the prevalence of a better price. The National Commission on Agriculture (NCA) observed that the farm produce is usually stored in dwelling houses, tins, drums, specially improvised big earthenware, underground mud-walled cellars and RCC bins. There are also storage structures like *kottu, puri, bamboo structures, cement bins, cement based puri, tin bins, etc.* 'Kottu' is a storage structure in which the produce can be stored for a longer period with minimum damage. It is rodent proof and well ventilated to avoid excessive moisture loss. But in spite of the many positive factors, many farmers do not prefer, as it involves a very high initial cost. *'Puri'* is the commonly used storage structure. *Puri* is installed at the time the produce is harvested and it lasts for one season only. The incidence of storage losses due to storage pests, rats, moisture loss, *etc.*, is more in *'puri'* compared to bamboo storage structures. Mostly small farmers use bamboo structures. Though losses are relatively less, due to their less storage capacity, they do not find the favour of the other farmers. Cement bin though a very safe storage structure, its initial cost is the limiting factor for its wide adoption by the farmers. *Cement based puri is the puri* made on a cement construction of about 2 meters height. This structure is very safe against rodent damage. The inadequacies of storage structures are estimated to cause storage losses to the extent of 6.2 per cent as per the National Commission on Agriculture. The network of Central Warehousing Corporation (CWC) and the State Warehousing Corporation (SWC) is mostly confined to major market areas. In rural areas the network is conspicuous by its absence. Most of the rural godowns are run by co-operatives with inadequate capacity and the services are extended to only to their members.

## 8. Malpractices in Markets

Manipulation of weights and measures is still prevalent in the unregulated markets in spite of the introduction of uniform system of weights. Arbitrary deductions in the name of sampling are a common feature. The farmers are burdened in the form of forced deductions for religious and charitable purposes. Farmers' voice against these practices goes unheeded. Sale under cover is also another feature of these unregulated markets.

## 9. Multiplicity of Market Charges

The farmers are forced to part with their income for various purposes. The expenses commence at the entry point of the market itself. The farmers have to contact commission agents or brokers to get in touch with the prospective buyer. Commission should be paid for their services. Other charges down the line are unloading charges, weighment charges, amounts towards charity fund, generous amounts to other workers, like sweepers, watchmen, watermen, *etc.* By the time the farmer sells his produce he may have to shell down some unwilling amount.

## 10. Lack of Market Information

Market information is essential for producers, traders, consumers as well as the Government, if market mechanism has to work efficiently. The relevant market information deals with character and volume of supply of commodities, the present and expected level of consumers' demand, current price quotations and future price trends for different farm products and their probable impact on prices, *etc.* Market information is of two types *viz.*, Market intelligence and Market News. Market intelligence indicates a record of past information in relation to prices, market arrivals, *etc*. It essentially helps to take decisions in future based on the past information. Market news on the other hand, provides current information on prices, arrivals, *etc*. But in reality the farmers more often than not, are in total dark as far as this information is concerned. The farmers do not know the information on the existing prices of the products in the important markets. By and large, the farmers rely on the price information furnished by the traders. The price information provided generally is quite advantageous to the traders, rather than to the farmers.

## 11. Lack of Farmers' Organization

The farmers are scattered over a wide area without any common organization. In the absence of such an organization farmers do not get anybody to guide them and protect their interests. On the other hand, traders are an organized body. Thus, the marketing system, therefore, constitutes unorganized farming community on one side and organized and powerful traders on the other side. Under such situations farmers will be generally exploited and do not get remunerative prices for their produce.

## 12. Inadequate Means of Transport

One of the pressing problems of agricultural marketing is the lack of adequate transport services at reasonable cost. Lack of transport services refer to absence of the transport service in important agricultural marketing areas, seasonality of transport service, high freight charges due to inadequacy, lack of all weather roads and transport vehicles, unsuitability of the existing transport facilities for the transportation of some products like, fruits, vegetables, eggs, *etc*., from rural areas. About 50 per cent of the 5,76,000 villages in the country is not connected by motorable roads. *Katcha* roads figure in many parts in rural India. Roads not at all connect inferior rural areas. The most common means of transport in the rural areas is still bullock drawn cart. Hence, these inaccessible problems with the movement of the products remain a problem of great concern. Although mechanized means of transport have also developed, they are not available to all the areas and all the farmers. Railways though are providing transport, farmers and merchants have a cause of concern regarding the availability of parcel vans for transport of perishable products such as fruits, vegetables, betel leaves, *etc*.

## 13. Communication Problem

One of the key elements of efficient marketing system is the availability of proper communication infrastructure. Rural areas are inadequately placed with reference to posts, telegraphs and telephones. The literacy rate being low among the farmers, it poses difficulty of the communication tasks.

# CHAPTER 45

# Suggestions to Improve Agricultural Marketing

Improving the marketing system of agricultural products would help the farmer to better his economy. Following are the measures that reflect an improved marketing system.

## 1. Establishment of Regulated Markets

A regulated market is one, which aims at the elimination of the unhealthy and unscrupulous practices, reducing marketing charges and providing facilities to producers. In 1886, under the Hyderabad Residency Order, the first regulated market (Karanjia Cotton Market) was established with a view to arranging supplies of pure cotton at reasonable prices to the textile mills in Manchester UK. The first legislation was, however, the Berar Cotton and Grain Market Act of 1897, which was enacted to induce Indian farmers to raise cotton, which empowered the British Resident to declare any place in the assigned district a market for the sale and purchase of agricultural produce and to constitute a committee to supervise the regulated market. Later in 1927, the then Government of Bombay province enacted the Bombay Cotton Market Act, which attempted to regulate the agricultural product markets with a view to evolving their market practices (Acharya and Agarwal, 1994)[*].

The subsequent Acts, were virtually based on the general principles embodied in the Berar Cotton and Grain Market Act of 1897, the salient features of which are:

1. All the markets, which existed on the date of the enforcement of the law, came under its fold.
2. The resident could declare any additional market for the sale of agricultural produce, and
3. The committee was authorized to appoint a sub-committee or joint committee.

An important landmark in the agricultural marketing scene in the country had been the recommendation of the Royal Commission on Agriculture in 1928, for regulation of marketing practices and establishment of regulated markets. The idea of regulated market was imported into India from Great Britain in the pre-independence period. By 1920, there were 32 such Acts in the UK. They were mainly concerned with the sale and slaughter of cattle and with transactions in wheat markets run by local

---

[*] Acharya S.S. and N.L. Agarwal (1992). Agricultural Marketing in India. Oxford & IBH Pub. Co., New Delhi.

sheriffs. In 1922, a Committee was set up to investigate and comment on the prevalent legal chaos and on British Agricultural Markets. It was presided over by Lord Linlithgow and resulted in the formation of market branch of Ministry of Agriculture in the year 1924. He also suggested an autonomous body to manage the markets. Since Lord Linlithgow subsequently became Viceroy of India and headed the Royal Commission on Agriculture in 1928, he wanted perhaps to reenact the model in India. Subsequently Central Banking Enquiry Committee (1931), National Planning Committee (1947) and other adhoc Committees, like Shirname Committee, Banse Committee, Krishnaswamy Naidu committee including Ford Foundation supported the expunsion of regulated market system. Based on these recommendations, a Central Marketing Department, later renamed as the Directorate of Marketing and Inspection (DMI) was established in 1935, for advising the Government on matters relating to the improvement in the conditions of primary agricultural produce markets. Though several provinces and States enacted legislation for the regulation of agricultural produce markets, not much headway was made till independence. The number of regulated markets even up to mid-sixties was 1012 (Acharya, 1994)[*], which was only about 15 per cent of the total wholesale assembling markets in the country. Since then owing to the concerted efforts made by DMI and the State Governments, the regulation programme got momentum and the number of regulated markets had gone up to 6,836 by March, 1995. With the expansion in the regulation programme, State Agricultural Marketing Boards were established and separate agricultural marketing departments were created in several States and for overseeing the functioning of the programme as also of the work of the market committees. Almost all the States, barring a few like Jammu and Kashmir and Mizoram have by now enacted the Agricultural Produce Market Acts.

## Management of Regulated Markets

The management of the regulated markets is vested with market committees in which the members are growers, traders, officials of the marketing societies, officials of Agricultural or Animal Husbandry department, *etc*. Market committee is responsible for the enforcement of the rules and regulations.

## Functions of Market Committee

To help the farmers in the fair disposal of the produce, the market committee is empowered to perform the following functions.

1. The market committee is empowered to implement the provisions of Farm Produce Markets Act, rules and bylaws made there under.
2. To provide marketing facilities for notified crops in the notified area.
3. To provide regulation on the marketing of notified crops.
4. To establish market yards with the required infra structural facilities.
5. To grant or deny licenses to market functionaries and renew, suspend or cancel such licenses or prescribe the conditions of licenses.
6. To determine market practices and market charges.
7. To provide the services of personnel to settle disputes.

---

[*] Acharya S.S. (1994) Marketing environment for farm products-Emerging issues and challenges. Presidential address-VIII National Conference of Agricultural Marketing. Indian Journal of Agricultural Marketing, Vol. 8: July-December.

8. To collect, maintain and disseminate market information.
9. To prevent adulteration and promote grading.
10. To levy market fee and license fee as entitled.
11. To prosecute the offenders for violating the provisions of the Act, rules and by laws.
12. To employ necessary manpower for the effective use implementation of the Act, rules and bylaws.
13. To ensure correct weighment of the produce.
14. To implement the directions given by the Government from time to time in the establishment and development of markets.
15. To contribute to the State agricultural marketing boards.

The Director of Agricultural Marketing is the authority to supervise the overall functions of the Market Committees to see that it implements the provisions of the Act.

## Sources of Revenue

Market committees by the provisions of the Act are the independent self-financing statutory autonomous institutions. Every market committee raises its funds through license fee and market fee on the notified agricultural produce transacted in the premises of the market yard.

## Benefits of Regulated Markets

1. The farmers are encouraged to bring the produce directly to the markets.
2. Farmers are protected from the exploitation of market functionaries.
3. Market charges are specified and enforced for an orderly marketing system.
4. Farmers are ensured better prices for their products.
5. Availability of the staff for the settlement of disputes between producers and traders.
6. Facilities like platforms for sales transactions, storage facilities, cart parking facilities, rest houses, *etc.*, are made available.
7. Farmers have access to up-to-date market information.
8. Open auction method of sales transaction is common in regulated markets.
9. The marketable surplus of the farmers will be increased.
10. The village sales will be reduced.
11. Marketing costs are lowered and producer's share in consumer's rupee will be increased eventually.

## *Suggestions*

Regulated markets were developed as an efficient marketing system to benefit the farmers. These regulated markets were established in different parts of the country with the same objective. The number of regulated markets in the country rose from 286 to 7,001 during the period from 1950 to 1998. These developments, however have not created expected dent on the creation of infrastructural facilities in the regulated markets in proportion to the arrivals. No doubt regulated markets are providing services like standard weights, competitive bidding, competitive price, immediate payment, direct contact with the farmers, *etc*, if further services as suggested by several researchers like provision of transport system by the market committees to help the farmers from the high private transport costs, procurement at the production centre, provision of market finance as a mandatory function, market information etc., are made available, it would encourage the farmers to market at regulated markets.

Eighth plan document relating to agricultural marketing emphasized this aspect and it is said "the augmentation and streamlining of the marketing infrastructure becomes necessary if diversification envisaged under the new agricultural policy is to succeed and the primary producers are to realize a fair share of the price paid by the consumers. Recent changes in the economic and industrial policies envisage a market led growth, and the farm sector cannot remain aloof from this trend." Taking note of this observation, the Government of India set up a High Power Committee on agricultural marketing to review the State enactments and working of various marketing bodies and recommended measures to strengthen them.

## Major Recommendations are

1. At least 15,000 rural primary markets out of over 30,000 markets in the country must be brought under the ambit of market regulation and taken for development.
2. The term 'agricultural marketing' in the State Acts should be defined as to include all activities.
3. All commodities should be brought under regulation except the ones for which Commodity Boards exist.
4. Agricultural produce markets act should be enacted by all the States covering all regions.
5. A separate ministry should be set up.
6. Establishing a separate bank to provide finance like pledge loan, *etc.*

## 2. Standardization and Grading

Standard specifications and grading should be designed to be useful to as many producers, traders and consumers as possible *i.e.*, standards should reflect market needs and wants. One grade should have the same implications to producers, traders and consumers in the quality of the product. It must have mutually acceptable description. They should reflect commodity characteristics that all types of buyers recognize. The grading should be simple, clear and easily understood.

The AGMARK scheme initiated by the Agricultural produce (Grading and Marking) Act, 1937 to ensure the quality of agricultural produce is enforced by the Directorate of Marketing and Inspection. The mandatory label contains the logo including map of India with the word 'AGMARK' inscribed vertically and horizontally and a rising sun on one side. The labels are affixed to the packages containing the graded agricultural produce. The act regulates not the grade standards prescribed but also the specifications that are to be followed for packing, sealing, *etc.* Only those traders who are willing to follow the regulations are given a 'Certificate of Authorization'. Random inspection of the products is done both at the packer's level and also in the market. 'AGMARK' label ensures the quality of the produce and consumers are rest assured of it. Commodities presently graded under 'AGMARK' include rice, wheat, atta, spices and condiments, vegetable oils, butter, ghee, *etc.* The quality of the product is determined with reference to characteristics like size, shape, variety, weight, colour, moisture, *etc.* Grading under 'AGMARK' for internal consumption is a purely voluntary programme. But the Export (quality control and inspection). Act was enacted in 1963 and the Export Inspection Council was set up to enforce compulsory quality control on commodities meant for export. The DMI is presently acting as an export inspection agency on behalf of the Export Inspection Council.

To make the consumers aware of the 'AGMARK' products, advertisements are issued in the newspapers, along with the effective use of electronic media, besides AGMARK exhibitions in important cities. The AGMARK awareness campaign is being undertaken by the DMI.

## 3. Standard Weights and Measures

Lack of uniform standards in weights and measures was a cause of concern as the traders took it to their advantage at the cost of the farmers. Government made an effort in this direction by passing the Standard Weights Act in 1939, which however could not make significant impact. Subsequently the Standard Weights and Measures Act was passed in 1958 giving the provision of using Government weights and measures. In 1963, the metric system of weights was introduced, which is in operation all over the country uniformly in transactions.

## 4. Improvement in Handling and Packing

This refers to the adoption of new techniques for the physical handling of commodities throughout the various phases of marketing, for instance, the use of cold storage (mechanical refrigeration) in handling perishables, new methods of packing, etc. The most appropriate handling and suitable containers among the available ones are meant to use against dust, heat, rain, flies, etc., to prevent considerable physical losses and quality deterioration.

## 5. Provision of Storage Facilities

Reduction of physical damage and quality deterioration in the products can be brought about through the application of the scientific techniques and provision of appropriate storage facilities depending on the nature and characteristics of products and the climatic conditions of an area. In this regard the role of Warehousing Corporation in providing scientific storage facilities is discussed here.

## History of Warehousing Corporation

The need for establishing licensed warehouse in India was early realized in the year 1928 by the Royal Commission on Agriculture. In 1945 the Agricultural Finance Sub-committee (Gadgil Committee) made specific recommendations for establishing licensed warehouses. It also emphasized the need to establish scientifically managed warehouses with an objective to provide warehouse receipt through which the commercial banks could provide finance to both the producers and dealers. The All India Rural Credit Survey Committee had fully endorsed the views of Gadgil Committee and made specific recommendations as early as in 1954. The parliament had passed the Agricultural Produce (Development and Warehousing) Corporation Act 1956. This act resulted in the initiation of three organizations viz., (1) National Co-operative Development and Warehousing Board (2) State Warehousing Corporation (SWC) and (3) Central Warehousing Corporation (CWC). The National Co-operative Development and Warehousing Board had been changed into National Co-operative Development Corporation by retaining co-operative warehousing with it. Warehousing has a three tier programme viz., Central Warehousing, State Warehousing and Co-operative Warehousing. Central Warehousing Corporation had established warehouses at markets of all India importance. The number of Central Warehouses as on 31.3.1999 was 447 with a storage capacity of 73.48 lakh tonnes. The percentage of average utilization of warehousing capacity is 74 per cent. State Warehouses are set up at a taluk or district level of with a storage capacity ranging from 500 Mts to 5000 Mts. Central Warehousing Corporation holds 50 per cent of shares in the equity of all the State Warehousing Corporations through out the country. As on 31.3.1999, 16 SWCs were operating with 1405 warehouses with an aggregate capacity of 113.89 lakh tonnes. Nearly 52,600 PACS are now having godown facilities with requisite funds provided by National Co-operative Development Corporation (NCDC). These facilities are further supplemented by Rural Godown Scheme (RGS), which was sponsored by market committees and SWCs. Besides warehouses, there are cold storage units which help the farmers to avoid distress sales of perishable commodities *viz.*, onions, fruits, fish, meat, dairy products, *etc.* At present the country is having cold storage capacity of 8 million tonnes. Co-operative warehousing covers a big village or a group of small villages covered by a storage capacity of 500 Mts.

## Objectives of CWC

1. To reduce the wastage in storage of various commodities by providing scientific storage facilities.
2. To assist the Government in orderly marketing of agricultural commodities by introducing standard grade specifications.
3. To issue warehouse receipt, a negotiable instrument in which commercial banks advance finance to the producers and dealers.
4. To assist the Government in the scheme of price support operations, and
5. To provide training to the personnel to run warehouses.

## Functions of CWC

1. To acquire and build godowns at suitable places in the country.
2. To run warehouses for storage of agricultural commodities, seeds, manures, fertilizers, farm implements and other notified commodities offered by producers, co-operatives and other institutions.
3. To act as an agent for the Government for the purpose of purchase, sale, storage and the distribution of commodities.
4. To provide disinfestation extension service to all needed.
5. To provide storage facilities at airports for imports/exports.
6. To provide warehousing facilities to the export oriented industries in the free trade zone.
7. To provide cold storage for perishables like fruits, vegetables, dairy products, drugs, pharmaceuticals, *etc.*
8. To subscribe to the share capital of State Warehousing Corporations (SWCs).

The functions of CWCs and SWCs are more or less similar. The CWC will subscribe 50 per cent of the shares of State Warehousing corporations (SWCs). It also provides technical guidance and finance to the SWC for training their personnel and construction of godowns. In the constitution of warehousing corporation, central Government subscribes 40 per cent of its shares and 60 per cent of shares by scheduled banks, investment trusts and other financial institutions.

## Procedure for Storing the Produce

The depositor intending to store the produce in the warehouse has to present a written requisition in the application prescribed by the warehouse. The commodity meant for storage is properly packed and delivered at the warehouse. The depositor should disclose all the details of the commodities including market value in the application form. The commodities brought for storage are graded and weighed by the trained technical personnel. Those requirements are to be fulfilled before the commodity is stored.

## Issue of Warehouse Receipts

For each lot stored by the producer or dealer, a warehouse receipt will be issued in which the particulars of commodity, insured amount and the market value are furnished. Based on the grade and the market value given in the warehouse receipt, commercial banks advance loans to the depositor to an extent of 75 per cent of the value of the commodity. The rate of interest on such type of loan ranges from 10 to 18 per cent. As soon as the warehouse receipt is pledged and the depositor financed the warehouse is informed. This completes the formalities of pledging the warehouse receipt with the bank by the depositor. The depositor may repay the loan either in lumpsum or installments and take delivery of the stocks as directed by the banks. Normally the minimum storage period allowed for each commodity is three months initially which may be extended from time to time subject to the satisfactory condition of the goods stored.

## Storage Charges

Different rates are charged for different commodities. The standard rate for foodgrains for bag of one quintal weight is Rs. 1/month. The storage charges are normally collected on weekly/monthly basis.

## Insurance

The stocks offered for storage are insured against possible risks of fire, theft and floods. Insurance coverage is also taken against risk of strikes and civil commotion. For the purpose of insurance, the goods are categorized into three classes *viz.*, 1) extra hazardous (cotton, jute, *etc.*) 2) hazardous (empty gunnies, pesticides, and Ammonia fertilizers, *etc.*) and 3) non-hazardous (all agricultural commodities).

## Scientific Preservation

Scientific storage facilities are provided for the lots brought for storage. A systematic stacking plan is prepared for proper preservation. Facilities are extended to prevent moisture seepage into the grain from the floor. The stocks, which are received in infested condition, are subjected to quarantine fumigation before storage. The bags are sprayed periodically with chemicals like Malathion and DDVP (Nuvon), *etc.* Fumigation is also resorted to with Aluminium Phosphide, EDB (Ethylene dibromide), MBR (Methyl bromide reagent) if the infestation rate is high. However, whenever fumigation is done stacks are aerated thoroughly after brushing. The corporation assures that the stocks, which are deposited, are delivered to the depositors in the same condition as it was during time of deposit. In order to maintain the quality and health of the stocks, storage practices have been evolved for more than 200 commodities, which include various agricultural commodities, industrial goods, chemicals and other notified commodities. Due to the adoption of modern techniques in storage, the corporation has been able to keep the storage losses well within the control and during the year 1999 the overall storage losses in foodgrains was 0.45 per cent.

Apart from the scientific facilities extended for proper storage of the goods, warehousing corporation also extends education to the farmers on the need of scientific storage and the facilities available in the warehouses. This particular objective is to help the farmers to minimize the wastage of the commodities, which they produce after the grueling toil on the farms, as a grain saved is a grain produced.

The Central Warehousing Corporation had launched the 'Farmers Extension Service Scheme' (FESS) in 1978-79 to educate the farmers in post-harvesting preservation techniques. The technical personnel of the corporation organize demonstrations about the use of pesticides and preservation techniques in the villages. The Central Warehousing Corporation is conducting training programmes every year to the selected farmers to impart knowledge of preservation and at times incentives in the form of bins for storage are given. The 'Disinfestations Extension Service Scheme' (DESS) is extended at the doorsteps of the customers including the farmers. Under this scheme, Pest Control Services are provided at nominal rates for control of insects, rodents, *etc.* In addition to the above, fumigation of exportable foodgrains on vessel is also undertaken. Under this scheme the corporation is also disinfesting aircrafts, railway coaches, various office building premises, *etc.*

## 6. Improving Transport Facilities

The National Commission on Agriculture pointed out the "Link-up and associated road development is *sine qua non* for the success of market structure". The availability of efficient transportation encourages the farmers to go to the markets of their option to derive the price benefits. Rural roads particularly are in a bad state during all seasons and more so during rainy season. Investment on roads should be given top priority. Coming to rail transport, small exporters normally depend on it as road transport is costlier for them. The problem is that parcel vans coming from the starting point were closed and sealed to be opened at the destination point. This is causing tremendous inconvenience to exporters and others who wish to book luggage in the stations enroute. The railways should provide minimum quota on all trains to promote small exporters. And also another problem is that perishable items cannot be transported in closed wagons and hence there is a need to provide necessary ventilation.

## 7. Market Information

As such we have newspapers, price bulletins, reports of the Government agencies, *etc.*, which provide market information. This information would be much more useful if an educational programme is made available to analyze and interpret the information at the markets. The raw data no doubt provides valuable information but skilful interpretation makes it still useful to the farmers.

## 8. Market Research

Hanson[*] defined market research as '*the study of consumer demand by a firm so that it may expand its output and market its product*'. It centers around consumers needs,

---

[*] Hanson, J.L. A. Dictionary of Economics and commerece. The English language Book Society, London, 1974: p. 318.

preferences, impressions of a product, accessibility of markets, efficiency of marketing, *etc*. Marketing research needs to be given top priority to improve up on the marketing system. Directorate of Marketing and Inspection (DMI) has been conducting commodity surveys and publishing reports on the marketing aspects. Apart from DMI, State Agricultural Marketing Boards, National Council Applied Economic Research, State Agricultural Universities and Centre for Agricultural Marketing Research are involved in marketing research.

## 9. Market Extension

Directorate of Marketing and Inspection has an extension cell, which is involved in dissemination of needed information on marketing to producers. The farmers are advised on consumer preference, grading, packaging, transport, *etc.*, in order to help them to secure better returns. Apart from DMI, regulated markets are also helping in the orderly marketing.

## 10. Provision of Agricultural Marketing Training to the Farmers

Though training of marketing personnel was realized in India as early as **1945**, very little attention has been paid in providing training to the farmers in agricultural marketing. No doubt in 1962, an extension cell was established in DMI to supply material on agricultural marketing to the States, but the impression is that on the whole the provision of training to the farmers in agricultural marketing is not up to the desired levels. Provision of training is of utmost importance in view of the malpractices resorted to by various market functionaries. The farmers need to be trained in product planning *i.e.*, crops and varieties to be grown, preparation of produce for marketing, malpractices and rules and regulations, market information, promotion of group marketing, *etc*[*]. In this regard the efforts of Indian Society of Agricultural Marketing in organizing a national level training to farmers in 1990 at Bangalore is highly commendable. The department of marketing or Directorate of Marketing in States should undertake training programmes to the farmers on the field and off the field developing a suitable curriculum.

## 11. Co-operative Marketing

The agricultural marketing system basically should ensure that the producer is encouraged to increase production, besides assuring the farmer remunerative prices for his produce and supplying the commodities to the consumers at reasonable prices. In view of the importance of marketing, co-operating marketing societies were established for meeting the requirements of the farmer. They have a history of more than eight decades. The scenario of co-operative marketing societies is presented in Table 45.1.

Co-operative marketing is the organized sale of farm products on a non-profit basis in the interests of the individual producer. Co-operative marketing societies are organized by the farmers themselves and the profits are distributed among the farmer-members based on the quantity of the produce marketed by them.

---

[*]Sidhu D.S. 1992. Training for farmers in Agricultural Marketing in India. Indian Journal of Agricultural Marketing Vol. 6(1): p. 6.

TABLE 45.1   Scenario of Co-operative Marketing Societies (As on 31st March 1996).

| Category | No. of Societies | Membership | Business operations (Rs. in crores) |
|---|---|---|---|
| **A) National, State, Central Societies** | | | |
| 1. All marketing societies | 501 | 13,61,151 | 10,590 |
|  | (100) | (100) | (100) |
| 2. General purpose societies | 387 | 11,66,436 | 8,382 |
|  | (77.25) | (85.69) | (79.15) |
| 3. Specialized societies | 114 | 1,94,715 | 2,208 |
|  | (22.75) | (14.31) | (20.85) |
| **B) Primary Societies** | | | |
| 1. All marketing societies | 8,049 | 40,02,849 | 5,479 |
|  | (100) | (100) | (100) |
| 2. General purpose societies | 2,692 | 29,15,041 | 4,881 |
|  | (33.45) | (72.82) | (89.09) |
| 3. Specialized societies | 5,357 | 10,87,808 | 598 |
|  | (66.55) | (27.18) | (10.91) |

Note: Figures in parentheses are percentages to total.

Source:   Important items of data: credit and non-credit co-operative societies, 1995-96. National Bank for Agriculture and Rural development, Mumbai, July 1997.

## Need for Co-operative Marketing

To make the produce available to the consumers from the farmers the services of intermediaries are pressed into action. For the services they render, middlemen collect service charges, which are disproportionate to the services rendered, due to which the producer does not receive his due share in the consumer's rupee. Also various malpractices exist in agricultural marketing system, to the disadvantage of the farmer. Hence an effective action is needed to make the producer to market the goods himself so that he can receive his due share in the marketing. Since it is not a possibility, group activity warrants itself, i.e., there is a need for co-operative action in the field of agricultural marketing. Institutional approach is the best course, and establishing a producer-owned marketing system run on co-operative lines would help the farmers to receive benefits due to them.

## Objectives

1. To make arrangements for the sale of produce of the members.
2. To provide credit facilities to the members on the security of agricultural produce.
3. To provide grading facilities, which would result in better price.
4. To make arrangements, for scientific storage of the member's produce.
5. To arrange for the supply of inputs required by the farmers.
6. To undertake the system of pooling the produce of the members to enhance the bargaining power through unity of action.

7. To arrange for the export of the produce to enable the farmers to get better returns.

8. To act as an agent of the Government in the procurement of foodgrains, *etc.*

Co-operative marketing structure constitutes National, State, Central and primary co-operative marketing societies. Primary co-operative marketing societies are the ones with which farmers have direct contacts. The societies are broadly categorized into general purpose societies and specialized societies. These specialized societies deal with specific commodities only. The details pertaining to the co-operative marketing societies are also presented in Table 45.1. Business turnover was distinctly higher in the higher level societies over primary societies to the tune of Rs. 10,590 crore. But an interesting observation is that at the higher level, general purpose societies dominated the scene with 387 societies against specialized societies numbering 114. The general-purpose societies accounted for 77.25 per cent in number, 85.69 per cent in membership and 79.15 per cent in business turnover. On the other hand, in respect of primary societies specialized societies dominated the co-operative marketing scene in terms of number only as they formed 66.55 per cent. The general purpose societies though they were only 33.45 per cent in terms of their number; they registered an overwhelming membership and business. It reveals that they are quite a good number of specialized primary societies, but as far as business operations are concerned their performance is meager. Marketing societies are established primarily to provide marketing facilities to the farmer-members.

## Suggestions for Improving the Co-operative Marketing

In the light of the higher quantities of agricultural produce available for handling in future, the marketing societies need to gear up themselves exploiting the opportunities. Following suggestions need to be examined in this regard.

1. Sticking on to the principle of open membership in right earnest, societies should target to enroll as many number of farmers as possible in the area of operation. This helps them to serve the common interest of the farmers for whom these have been established. Once the societies make their functions transparent they can muster the support of the farmers, which automatically make them stronger.

2. Once the institutional foundation is strengthened necessary marketing infrastructure needs to be developed. Also a point to be kept in view is that as co-operatives have to compete with the private traders a greater urgency needs to be exhibited by the societies with regard to services. They need to provide better services than those provided by the private traders.

3. Arrangements regarding assembling the produce of the farmers under the area of jurisdiction of the society need to be planned. Options are less regarding bringing the produce from the village. To lessen the burden of the farmers, assembling centres can be opened in villages where from, society arranges to transport the produce to the headquarters. The society then undertakes all the activities involved in market preparation and arranges for sale. For sales it has to identity the outlets. One is arranging for sale in market yards nearby. By increasing the arrivals of the marketing co-operatives in the yards, they turn out to be the strong competitors to private traders in pricing mechanism. For certain commodities in case they need to route it to private traders, trust worthy private agencies need to be identified and collaboration needs to be established.

4. The experience over the functioning of the co-operative marketing societies it warrants a change in their functioning. In the sense that marketing of agricultural produce should be given top priority *i.e.*, it should be the principal activity. Other activities of the societies like agricultural input sales, sale of consumer goods, *etc.*, should be their secondary function.

5. Coming to the capital base it should be allowed to raise funds in the capital market on its own credibility so that societies will be cautious in using their funds judiciously.

## 12. Provision of Cold Storage Facilities

Unlike foodgrains, for perishable commodities like fruits and vegetables, quality losses are enormous and hence it would be necessary to take measures and devise methods to control and minimize losses. Preservation is, thus, a necessary adjunct of production and a vital link between production and consumption. Cold storage is the most important for the proper marketing of horticultural produce, because it has a definite season of production and the quality of the produce deteriorates quickly after harvest. Most fruits and vegetables loose moisture to the surrounding air almost any time if the humidity of the air is less than saturated. It is possible to maintain high humidity of 80 to 95 per cent in proper cold storages and hence refrigeration is also beneficial in reducing moisture losses as moisture-holding capacity of air at low temperatures is much less than it is on warm temperatures.

It is reported that the first cold storage was established as early as in 1892 in Kolkata. But not much progress in the expansion of cold storage industry was made until 1947. According to the survey conducted by the Directorate of Marketing and Inspection, Ministry of Agriculture, Government of India in 1955, it was revealed that the total available storage capacity in the country was only 77,145 tones. However, after that, rapid strides have taken place in this respect and cold storage capacity increased manifold. Potato is the main product stored in cold storage. Approximately 90 per cent of the cold storage capacity is used for potatoes.

In India cold storage facilities are available only in the terminal markets and served little purpose by the time the fruits and vegetables are brought, as they are not in a state for storing. It is therefore necessary that the marketing agencies and growers' co-operatives should be encouraged to set up cold storages in the producing areas. The available cold storage capacity in the country was 310.10 lakh tones.

## 13. Reduction in Post-harvest Losses

Not withstanding the breakthrough in production, inadequate and unscientific post-harvest management of fruits and vegetables have led to an estimated post-harvest loss (20-40%) of about Rs. 30 billion per annum. All out efforts need to be made to prevent these losses. The expertise and technologies available with AFMA member countries should be shared effectively.

## 14. Development of Physical Market

Physical markets handling fruit and vegetables suffer from operational and management inadequacies. A country level plan to identify markets of national importance for fruits and vegetables and provisions of need-based infrastructure from export point of view in all these markets is imperative.

## 15. Refrigerated Transport

Refrigerated transport for perishables needs to be provided during their movement in marketing channels. Besides road transport, railway wagons should also be suitably modified for transport of perishables.

# CHAPTER 46

# Marketing Channels

The agricultural commodities move from the farmers to consumers over time and space. This movement is made possible through various market intermediaries that operate in the marketing system. The chain of intermediaries through which the various farm commodities pass between producers and consumers is called a marketing channel. Marketing channels differ from commodity to commodity. The marketing channels of foodgrains are different from those of fruits and vegetables and those of processed products. Also the selling behaviour of small farmers do vary against the large farmers. The development of market for the products brings in the changes in the length of the marketing channels. Following are some of the channels for selected agricultural and livestock commodities presented in Figure 46.1, 46.2, 46.3, respectively.

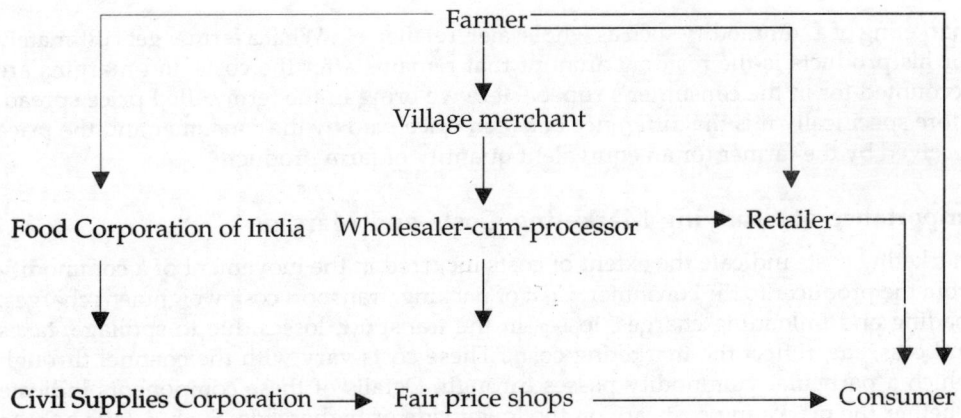

**Figure 46.1** Marketing Channels for Foodgrains.

## MARKETING COSTS AND MARGINS

The movement of the products from the producers to the ultimate consumers involve costs, taxes and cess which are called marketing costs. These costs vary with the channels through which a particular commodity passes through. Marketing margin represents the difference between price paid and received by a given market intermediary in the

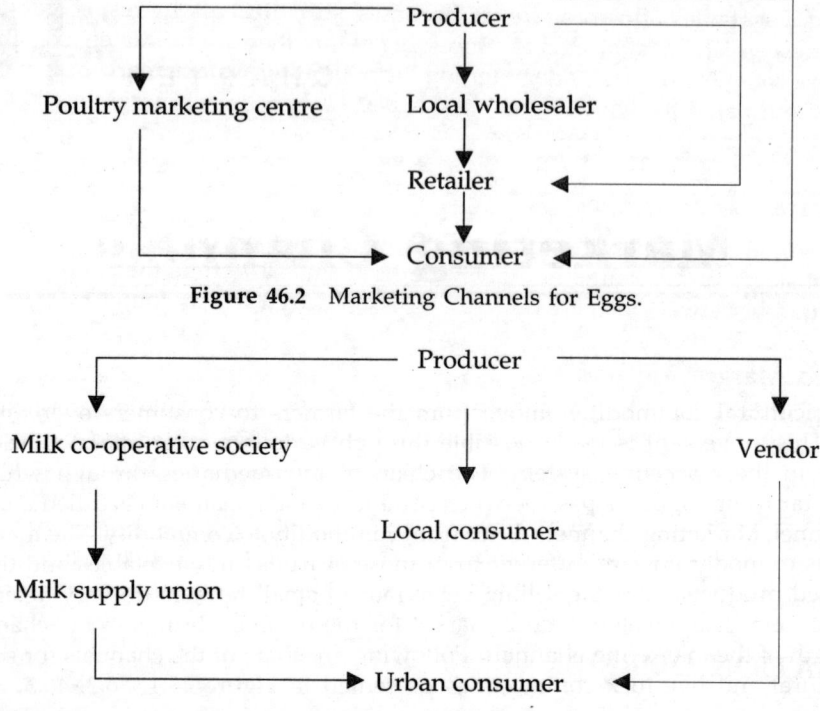

Figure 46.2 Marketing Channels for Eggs.

Figure 46.3 Marketing channels for milk.

marketing of a commodity such as wholesaler, retailer, *etc.* What a farmer gets ultimately for his products is the residual amount that remains after the costs and margins are accounted for in the consumer's rupee. Here we bring in the term called price spread. More specifically it is the difference between price paid by the consumer and the price received by the farmer for an equivalent quantity of farm product.

## Importance of Studying Marketing Costs and Margins

Marketing costs indicate the extent of costs incurred in the movement of a commodity from the producer to the consumer. Cost of packing, transport cost, weighment charges, loading and unloading charges, losses in the transport, losses due to spoilage, taxes and cess, *etc.*, reflect the marketing costs. These costs vary with the channel through which a particular commodity passes through. Details of these components indicate whether the marketing costs are on the lower side or higher side. Such details help to think in terms of reducing marketing costs if they are on the higher side.

Marketing margins reveal the remuneration that the intermediaries receive for their services in moving the commodity in the marketing channels. Estimation of marketing margins helps us to estimate the efficiency of the marketing system. If the analysis indicates, the exploitation by the intermediaries in terms of excessive margins, it helps the policy-maker to bring in some changes in the marketing system.

## Estimation of Marketing Margins

Estimation of margins in reality poses a problem for the farm products as it is difficult to follow the path of the channel for a given quantity of commodity, say foodgrains, fruits and vegetables, flowers, *etc*. It becomes still difficult in respect of those commodities, which are subjected to processing before they are ultimately consumed. Given these complexities, for understanding the marketing system, marketing margins need to be estimated for which two methods *viz.*, (1) concurrent margin method and (2) lagged margin method, are used.

## 1. Concurrent Margin Method

This method stresses on the difference in prices that prevail for a commodity at successive stages of marketing at a given point of time. The time that elapses in real world situation between buying and selling is not taken into account.

## 2. Lagged Margin Method

This method takes into account the time that elapses between buying and selling of a commodity by the intermediaries and also between the farmer and the ultimate consumer. Lagged margin indicates the difference of price received by an agency and the one paid by the same agency in purchasing an equivalent quantity of commodity. It is a superior method as it takes into account the time that normally elapses in buying and selling in realistic situation. However, this method poses problems in the collection of data taking time lag into account.

# ESTIMATION OF PRICE SPREAD

## Mango

The procedure of estimating price spread of one tonne of mango is presented below:

### Marketing Channel

Producer → pre-harvest contractor → local wholesaler → distant wholesaler → retailer → consumer.

Price received by the producer/price paid by pre-harvest contractor = Rs. 3,542.09
Costs incurred by the pre-harvest contractor:

> Labour charges = Rs. 55.88
> Transpotation = Rs. 142.94

Pre-harvester's sale price/local wholesaler's purchase price = Rs. 4,329.41
Costs borne by local wholesaler:

> Unloading       = Rs. 27.05
> Grading         = Rs. 17.05
> Storage charges = Rs. 5.58
> Spoilage losses = Rs. 88.0

Telephone, electricity, charges and tax = Rs. 16.34
Local wholesaler's sale price/distant wholesaler's purchase price = Rs. 5,017.54
Costs incurred by distant wholesaler:

> Loading               = Rs. 30.00
> Unloading             = Rs. 25.00
> Cleaning and grading  = Rs. 21.00
> Transportation        = Rs. 1,450.00
> Packing material      = Rs. 5.00

| | | |
|---|---|---|
| Weighing charges | = | Rs. 1.25 |
| Storage losses | = | Rs. 29.0 |
| Spoilage losses | = | Rs. 300.00 |
| Telephone and electricity charges | = | Rs. 18.00 |
| Material used for ripening | = | Rs. 5.00 |

Distant wholesaler's sale price/Retailer's purchase price = Rs. 8,911.66

Costs incurred by retailer:

| | | |
|---|---|---|
| Loading | = | Rs. 33.00 |
| Unloading | = | Rs. 32.06 |
| Transportation | = | Rs. 69.66 |
| Storage charges | = | Rs. 23.06 |
| Spoilage losses | = | Rs. 710.00 |
| Tax | = | Rs. 42.66 |
| Rent for push cart | = | Rs. 36.66 |

Retailer's sale price/consumer's purchase price= Rs. 12,000.

Price spread is presented in Table 46.1.

**TABLE 46.1  Price Spread for a Tonne of Mango.**

| Particulars | Amount (in Rs.) | Percentage share in consumer's price |
|---|---|---|
| Price received by the producer | 3,542.09 | 29.52 |
| Costs incurred by pre-harvest contractor | 198.82 | 1.66 |
| Pre-harvest contractor's margin | 588.50 | 4.91 |
| Costs incurred by local wholesaler | 1,54.04 | 1.28 |
| Local wholesaler's margin | 534.09 | 4.45 |
| Costs incurred by distant wholesaler | 1,884.25 | 15.70 |
| Distant wholesaler's margin | 2,009.87 | 16.75 |
| Costs incurred by retailer | 947.1 | 7.89 |
| Retailer's margin | 2,141.24 | 17.84 |
| Gross marketing margin | 8,457.91 | 70.48 |
| Consumer's price | 12,000.00 | 100.00 |

## Marketing Margins

**Pre-harvest contractor's margin** = Pre-harvester's sale price - (pre-harvest contractor's price + costs incurred)
= Rs. 4,329.41 - (3,542-09 + 198.82)
= Rs. 588-50

**Local wholesaler's margin** = Local wholesaler's sale price - (local wholesaler's purchase price + costs incurred)
= Rs. 5,017.54 - (4,329.41 + 154.04)
= Rs. 534.09

**Distant wholesaler's margin** = Distant wholesaler's sale price-(distant whole saler's purchase price + costs incurred)
= Rs. 8,911.66-(5,017.54 + 1,884.25)
= Rs. 2,009.87

**Retailer's margin** = Retailer's sale price-(Retailer's purchase price + costs incurred)
= Rs. 12, 000-(8,911.66 + 947.1)
= Rs. 2,141.24

$$\text{Producer's share in consumer's rupee} = \frac{\text{Price received by the producer}}{\text{Price paid by the consumer}} \times 100$$

$$= \frac{3,542.09}{12,000} \times 100 \qquad = 29.52\%$$

## Groundnut

### Marketing channel

Producer → Wholesaler → Decorating unit → Oil miller → Retailer → Consumer.

| | | |
|---|---|---|
| Producer's sale price per quintal | = | Rs. 1,119.48 |
| Marketing costs incurred by the wholesaler | | |
| Cost of gunny bags | = | Rs. 3.75 |
| Labour (bagging, loading and unloading) | = | Rs. 6.50 |
| Transportation charges | = | Rs. 12.50 |
| Commission charges | = | Rs. 6.25 |
| Taxes | = | Rs. 2.50 |
| Rent | = | Rs. 2.25 |
| Miscellaneous charges | = | Rs. 1.50 |
| Wholesaler's sale price/purchasing price of decorticating unit | = | Rs. 1,225.73 |
| Costs borne by decorticating unit | | |
| Market fee | = | Rs. 4.35 |
| Costs of gunny bags | = | Rs. 3.75 |
| Labour charges | = | Rs. 6.00 |
| Transportation costs | = | Rs. 5.00 |
| Rent | = | Rs. 4.80 |
| Processing charges | | |
| (a) Electricity charges | = | Rs. 1.75 |
| (b) Labour | = | Rs. 3.00 |
| (c) Depreciation | = | Rs. 0.59 |
| (d) Repairs and maintenance | = | Rs. 0.81 |
| (e) Interest on fixed capital | = | Rs. 0.13 |
| (f) Sub-total | = | Rs. 6.28 |
| Tax | = | Rs. 0.44 |
| Miscellaneous charges | = | Rs. 1.00 |
| Sale price of decorticating unit/purchase Price of oil miller | = | Rs. 1,297.35 |
| Costs borne by oil miller | | |
| Cost of bags | = | Rs. 3.75 |
| Labour charges | = | Rs. 7.50 |
| Transportation charges | = | Rs. 2.00 |
| Rent | = | Rs. 6.90 |
| Crushing charges | | |
| Electricity | = | Rs. 10.41 |
| Labour | = | Rs. 3.50 |
| Depreciation | = | Rs. 4.86 |
| Repairs and maintenance | = | Rs. 2.60 |
| Interest on fixed capital | = | Rs. 0.67 |
| Sub-total | = | Rs. 22.04 |
| Tax | = | Rs. 1.36 |

| | | | |
|---|---|---|---|
| Miscellaneous charges | | = | Rs. 1.00 |
| Oil miller's sale price/retailer's purchase price | | = | Rs. 1,374.40 |

Costs borne by retailer

| | | | |
|---|---|---|---|
| Transportation charges | | = | Rs. 5.00 |
| Labour | | = | Rs. 2.00 |
| Container costs | | = | Rs. 6.00 |
| Retailer's sale price/consumer's purchase price | | = | Rs. 1,404.20 |

Price spread is presented in Table 46.2

**TABLE 46.2   Price Spread in Groundnut Marketing.**

| S. No | Particulars | Amount (in Rs.) | Percentage to consumer's purchase price |
|---|---|---|---|
| I | Producer's sale price | 1,119.48 | 79.78 |
| II | Wholesaler | | |
| | 1. Purchase price | 1,119.48 | - |
| | 2. Marketing costs | 35.25 | 2.51 |
| | 3. Margin | 71.00 | 5.06 |
| III | Decorticating unit | | |
| | 1. Purchase price | 1,225.73 | - |
| | 2. Marketing costs | 31.62 | 2.25 |
| | 3. Margin | 40.00 | 2.85 |
| IV | Oil miller | | |
| | 1. Purchase price | 1,297.35 | - |
| | 2. Marketing costs | 44.55 | 3.17 |
| | 3. Margin | 32.50 | 2.32 |
| V | Retailer | | |
| | 1. Purchase price | 1,374.40 | - |
| | 2. Marketing costs | 13.00 | 0.86 |
| | 3. Margin | 16.80 | 1.20 |
| VI | Consumer's purchase price | 1,404.20 | 100.00 |

## Chilies

Green chillies

Producer → Commission agent → Retailer → Consumer

Dry chillies

Producer → Commission agent → Wholesaler → Retailer → Consumer

Marketing costs of green chilies are presented in Table 46.3

**TABLE 46.3   Marketing Costs for a Quintal of Green Chilies.**

| S. No | Particulars | Amount (Rs.) | % to Total costs |
|---|---|---|---|
| I | Producer | | |
| | 1) Transportation | 10.00 | 16.13 |
| | 2) Loading and unloading | 4.00 | 6.45 |
| | 3) Weighing | 2.00 | 3.22 |
| | 4) Cost of bags | 16.00 | 25.81 |
| | 5) Commission (6%) | 30.00 | 48.39 |
| | Total costs | 62.00 | 100.00 |
| II | Commission agents | | |
| | 1) Rent for shop | 1.66 | 100.00 |
| | 2) Total costs | 1.66 | 100.00 |

| III | Retailer | | |
|-----|----------|---|---|
| | 1) Loading and unloading | 4.00 | 26.54 |
| | 2) Spoilage | 10.00 | 66.36 |
| | 3) Municipal fee | 1.07 | 7.10 |
| | Total costs | 15.07 | 100.00 |

Price spread is presented in Table 46.4.

### TABLE 46.4 Price Spread for a Quintal of Green Chilies.

| S. No | Particulars | Amount (Rs.) | % to consumer's purchase price |
|-------|-------------|--------------|-------------------------------|
| I | Price received by producer | 500.00 | 83.33 |
| | 1) Market costs | 62.00 | 10.33 |
| | 2) Net price | 438.00 | 73.00 |
| II | Commission agents | | |
| | 1) Commission charge | 30.00 | 5.00 |
| | 2) Market costs | 1.66 | 0.28 |
| | 3) Margin | 28.34 | 4.72 |
| III | Retailer | | |
| | 1) Purchase price | 500.00 | 83.33 |
| | 2) Market cost | 15.07 | 2.51 |
| | 3) Margin | 84.93 | 14.15 |
| IV | Consumer's purchase | | |
| | Price/retailer's sale price | 600.00 | 100.00 |

Marketing costs of dry chilies are found in Table 46.5.

### TABLE 46.5 Marketing Costs for a Quintal of Dry Chilies.

| S. No | Particulars | Amount (Rs.) | % to Total costs |
|-------|-------------|--------------|------------------|
| I | Producer | | |
| | 1) Transportation | 165.00 | 41.77 |
| | 2) Loading and unloading | 6.66 | 1.69 |
| | 3) Weighing | 3.33 | 0.84 |
| | 4) Cost of bags | 40.00 | 10.13 |
| | 5) Commission (6%) | 180.00 | 45.57 |
| | Total costs | 394.99 | 100.00 |
| II | Commission agent | | |
| | 1) Rent for shop | 1.66 | 100.00 |
| | 2) Total costs | 1.66 | 100.00 |
| III | Wholesaler | | |
| | 1. Transportation | 165.00 | 56.58 |
| | 2. Loading and unloading | 6.66 | 2.28 |
| | 3. Market fee | 120.00 | 41.15 |
| | Total costs | 291.66 | 100.00 |
| IV | Retailer | | |
| | 1) Loading and unloading | 6.66 | 58.83 |
| | 2) Spoilage | 3.00 | 26.50 |
| | 3) Municipal fee | 1.66 | 14.67 |
| | Total costs | 11.32 | 100.00 |

Price spread is shown in Table 46.6.

TABLE 46.6   Price Spread for a Quintal of Dry Chilies.

| Sl. No | Particulars | Amount (Rs.) | % share in consumer's rupee |
|---|---|---|---|
| I | Price received by producer | 3,000.00 | 82.19 |
| | 1. Market costs | 394.99 | 10.82 |
| | 2. Net price | 2,605.01 | 71.37 |
| II | Commission agent | | |
| | 1. Commission charge | 180.00 | 4.93 |
| | 2. Market costs | 1.66 | 0.05 |
| | 3. Margin | 178.34 | 4.87 |
| III | Wholesaler | | |
| | 1. Purchase price | 3,000.00 | 82.19 |
| | 2. Market costs | 291.60 | 7.99 |
| | 3. Margin | 158.40 | 4.34 |
| IV | Retailer | | |
| | 1. Purchase price | 3,450.00 | 94.52 |
| | 2. Market costs | 11.32 | 0.31 |
| | 3. Margin | 188.68 | 5.17 |
| V | Consumer's purchase price/ retailer's sale price | 3,650.00 | 100.00 |

# REASONS FOR HIGHER MARKETING COSTS OF FARM COMMODITIES

Various empirical studies have indicated that the marketing costs are higher for farm commodities, the reasons of which are as follows:

1. *Bulkiness*: The character of bulkiness of farm commodities requires more space in storage, transport, *etc.* Transportation costs particularly are on the higher side for farm commodities.
2. *Perishability of the Products*: Perishability of the agricultural products lead to losses in transport. Because of perishability, products need special transport facilities, consequent to which transportation costs are pushed up.
3. *Storage Requirements*: Seasonality of agricultural production leads to the need of storage, which adds to the costs marketing. Losses in storage due to storage pests, rodents, moisture loss, *etc.*, also lead to additional costs in marketing.
4. *Specialized Packing*: In the case of highly perishable products like fruits, special packing is required before they are finally consumed. This is an additional marketing cost.
5. *Large Number of Intermediaries*: The length of the marketing channel also affects the marketing costs. Longer marketing channels *i.e.*, more number of intermediaries between producer and ultimate consumer lead to higher marketing costs and
6. *Seasonality of Production*: Seasonality of production influences the supply. When the supplies are on the higher side, it is obvious that prices fall down but the marketing costs stand as before. This leads to an increase in the marketing costs per unit basis.

# CHAPTER 47

# Agricultural Prices

## MEANING

Consumers go to the market in order to purchase the required commodities. In the process of purchasing, consumers pay for the value of the commodity in terms of currency units of the nation. Price of the commodity refers to the value of the goods in terms of money units. Prior to evolution of money, goods were exchanged, one for the other depending on the needs of the people. This process of marketing is known as barter system. Barter system was replaced with the advent of money. However, we find the barter system in vogue among tribals. Prices give signals to the producers regarding what are all the commodities to be produced in the economy and how to earn money and sustain in the process of production. Similarly, prices also give signals to the consumers to sustain the demand or restrain the demand. Changes in supply and demand of the commodity change the price level and consequently the welfare of producers and consumers in *Laissez-Faire* economy (free economy). In the countries, where consumers and producers enjoy more economic freedom, prices play a greater role compared to the centrally planned economies, where less freedom is given to the producers and middlemen. Agricultural prices are volatile in the sense that they are subject to changes over time and space. The irregular fluctuations in prices of commodities affect the fortunes of individuals in the economy.

In agricultural-based countries like India, prices of farm products undergo wide variations than the prices of industrial goods. They have profound affect on the growth, equity and stability of the economy. The incomes and living standards of the farmers, labourers and non-farming population *i.e.*, consumers are very much affected by price fluctuation. High and rising prices in general provide incentives to the producers to increase the output. The reverse also holds good if the prices are allowed to fall over time. The functions of prices in any economic system are as follows:

## FUNCTIONS OF PRICES

### 1. What and How Much to Produce

In the competitive economic system prices of the commodities give signals to the producers regarding the type and quantity of commodity to be produced in particular place at a particular point of time. Consumers are guided by the prices of the commodities and plan their purchases from the given income, so as to maintain their desired levels of living. Thus, the directions given by the prices vary according to various groups of consumers and producers. For example, in a situation of inflation, producers would get higher incentives to produce the required quantities at higher magnitude. On the contrary, consumers with fixed income have to cut down their consumption levels because of high prices. The reverse situation would result in the periods of deflation *i.e.,* falling prices.

## 2. How to Produce

Prices play an important role in deciding methods of production. Every producer aims at producing a commodity, with efficient methods of production. A process of production is efficient when a given amount of output is produced with minimum cost. The choice of resources depends on their relative prices. Producer uses more expensive resources in smaller quantities and less expensive resources in larger amounts. If labour is cheap, adoption of labour-intensive technology contributes to the least cost production. On the other hand, adoption of capital-intensive technology results in cost minimization, when it is available in abundant quantities. The producers who fail to adopt least cost methods of production find it difficult to survive in the business.

## 3. Prices Serve Basis for Allocation of Resources in the Production Process

Employment of scarce resources *viz.,* land, labour, capital, water, *etc.,* are to be directed for producing the commodities, which are having higher price. If this is not followed it results in losses in the production, particularly when the economy is in inflation. This is to say that high value commodities should be produced so as to sustain in the process of production.

## 4. Prices Help to Strike a Balance between Demand and Supply

This means when the demand is more than the supply, prices would give incentives to the producers. Similarly when the supply is more than the demand prices are brought down to lower level.

## 5. Prices Help the Consumers to Allocate their Income in the Purchase of Goods and Services

In purchasing the commodities consumers seek to maximize satisfaction from the given level of income. The low income consumers spend higher proportion of their income on necessaries *i.e.,* food, clothing, *etc.,* and they spend relatively less income on luxuries or manufactured goods like transistors, TVs, *etc.*

But at higher level of their income, the proportion of income spent on necessaries decreases. This is the Engel's law of family consumption. Similarly, when relative prices of commodities are changing, low-income groups prefer low priced goods. Only high-income groups do have a preference to luxuries. Thus, the relative prices of the commodity help to allocate the consumers' income on different goods, which are having varying relative prices. Prices affect the transfer of income from the farm sector to non-farm sector through terms of trade.*

---

* Terms of trade = $\dfrac{\text{Indices of prices received by the farmer}}{\text{Indices of prices paid by the farmer}} \times 100$

## 6. Distribution of Income Among Different Groups of Farmers

Marketable surplus would be high with the large farmers relative to small farmers. If there is a price rise for agricultural commodities, large farmers with substantial marketable surplus would be more benefited by the price rise of the commodity compared to small farmers with low marketable surplus.

## 7. Movement of Commodities Over Time and Place

Transportation of commodities from one area to another area *i.e.*, from surplus area to deficit area takes place due to changes in the price. As commodities are produced based on agro climatic characteristics and seasons, there is a great need for marketing of the commodities over space and time due to price differential and consumption needs. This is to say that commodities are transported to deficit areas and the movement and magnitude of goods is regulated by the price differentials and transport costs.

## 8. Capital Formation in Agriculture

Prices affect capital formation in agriculture, industries and allied sectors. They also affect the production potential of these goods in the long run. Higher prices for farm products lead to increased value of farm output. Consequently farm investment increases and with the increased investment, there will be further increase in the value of farm output and this further leads to higher consumption and standard of living. Thus, capital formation takes place through accelerator and multiplier effect. Farm workers demand higher wages when prices are increased. This inturn squeezes farm profit and lowers the incentives to the farmers. On the other hand, the increased value of farm output would lead to investment in the non-farm sector partly. Thus capital formation would also take place in the non-farm sector.

## 9. High Prices and Inflation

If the prices are persistently increasing, it leads to inflationary trends in the economy. As a result, a large number of economic variables like per capita income, per capita consumption, employment, real wages, interest rate policies, *etc.*, are very much affected by the price rise.

## 10. Monetary and Fiscal Policies

When prices are changing we require appropriate monetary and fiscal policies to be formulated in the economy. The policies related to money supply, interest rates, deficit financing, direct and indirect taxes, subsidies, levies, quotas, *etc.*, are to be regulated when prices are fluctuating widely in the economy.

# INSTABILITY IN PRICES

Price instability means lack of stability in prices. Instability in prices is defined as the state in which the prices continue to change over time and space. Fluctuations in the prices are associated with varying lengths of time due to various factors. We have long-term price variations (cyclical variations), short term price variations (seasonal variations or intra-year variations), inter-year variations (trend variations) and finally irregular fluctuations (random variations).

Let us clearly understand as to how the instability is related to measurement concept. Suppose prices in a country over the years rise at a constant rate of 6 per cent, this is certainly a situation of unstable prices, but the direction and extent of price movement is measured and known with certainty. This means that there is no uncertainty element in the movement of prices.

## TYPES OF PRICE INSTABILITY

Following categories of price instabilities are to be understood in order to have a comprehensive picture of the prices.

(1) Short term instability
(2) Intra-year instability
(3) Inter-year instability and
(4) Long term instability

Short term instability refers to the fluctuation in prices which last for comparatively short period of time. Prices may change from week to week, from day to day and within a day. Intra year-instability is due to the behaviour of prices seasonally in a year. Inter-year instability means the price movements from year to year. Long-term instability is related to price behaviour over a period of time.

## MEASUREMENT OF INSTABILITY

Instability is measured through two approaches *viz.*, graphical approach and quantitative approach.

*Graphical Approach*: In the graphical presentation we draw scatter diagrams by taking prices on Y-axis and time on X-axis and plot the points and draw the curve by joining the plotted points on X, Y plane. Such curves show upward and downward movement in the price of that commodity. If there are more irregular movements on the curves for the commodity, we say that price of that commodity is instable. This is illustrated for commodities A and B in Figure 47.1.

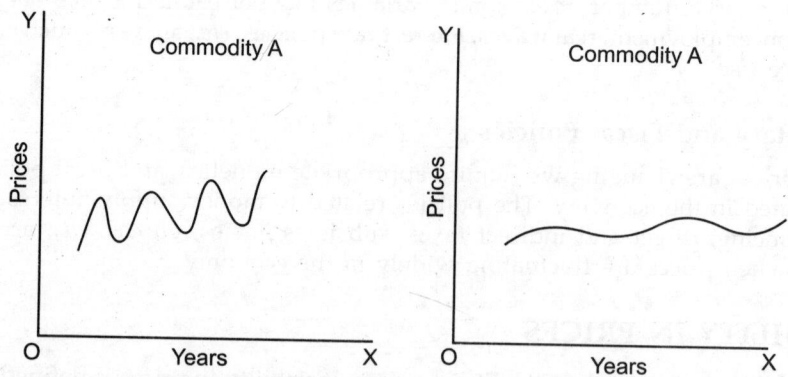

**Figure 47.1** Instability in prices.

The graphs indicate that commodity A is having more fluctuations over commodity B.

## Quantitative Approach

Here we use statistical tools like range, standard deviation and coefficient of variation. The difference between the highest price and the lowest price is called range.

$$\text{Coefficient of variation (CV)} = \frac{\text{SD}}{\text{Mean of price variable}} \times 100$$

SD = Standard deviation

$$\text{Variance} = \frac{\sum_{i=1}^{n}\left(P_i - \overline{P}\right)^2}{n-1}$$

$$\text{SD} = \sqrt{\text{Variance}}$$

Greater the CV, more would be the price instability. The CV of price beyond 5 per cent indicates instability and calls for price stabilization measures.

These three measures are applicable under static conditions. These are not applicable to measure the instability in the situation of rising trends. We have to compute here the long term instability index. This index measures the proportion of variation in the price of the commodity not explained by the price trend line over long term period. This index can be obtained using the following formula.

$$I = (1 - R^2)\,100$$

I = Index of dynamic instability in prices
$R^2$ = Coefficient of multiple determination

For example, $R^2$ for the estimated trend equation is 0.8, and then the price instability is

(1-0.8) × 100
= 0.2 × 100
= 20%

Price instability is 20 per cent. This instability is estimated over time and it is called dynamic instability index.

## TIME SERIES ANALYSIS

We separately study the nature and behaviour of various components of prices and other time series variables in the time-series analysis. To analyze the nature of inter temporal behaviour of prices, we require time series data on prices for various agricultural commodities over time and space. The time series data may belong to a particular week, month, season, year and decade, *etc.* The time series data on prices in general will have 4 components *viz.*, trend (T), cyclical (C), seasonal (S) and irregular factor (I). In the case of time-series data obtained for year-wise prices, we do not have seasonal component. There are two hypotheses regarding the estimation of these time series components. They are multiplicative model and additive model. Multiplicative model is specified as follows:

*Monthly data:*

$P_t$ = T.C.S.T.

$P_t$ = Price of commodity in $t^{th}$ period *i.e.,* month of the year.

*Yearly data:*

$P_t$ = T.C.I

$P_t$ = Price of commodity in $t^{th}$ period.

Similarly we have additive model in the following form.

*Monthly data*

$P_t$ = T + C + S + I

$P_t$ = Monthly data on the prices of the commodity in $t^{th}$ period

*Yearly data*

$P_t$ = T + C + I

$P_t$ = Price of commodity in the $t^{th}$ Year

Here, T = Trend factor

C = Cyclical factor

S = Seasonal factor

I = Irregular factor

*Trend factors*

Trend factors are those, which reflect movement in the economic variables over time. If the time series data is collected over different years, then using this data we can analyze the trend factor and measure the growth or recession of the variable over several years. The prices of the commodities, real GNP of nation, production of goods and services, population, *etc.*, can be analyzed and measured, if time series data are available over several years, year-wise.

## Cyclical Factors

These are the long-term fluctuations. We require very large time series data to analyze the cyclical component. The cyclical factor is found in the economic variables *viz.*, GNP, unemployment, prices, rainfall, *etc.* We require data on the time series variable for a minimum of period of 30 to 40 years to analyze the cyclical factors. The length of the time period of business cycle is large and it possesses six components. For example, one cycle is identified from peak to peak position for time series variable of employment in the country. Peak stage is followed by recession, depression and trough in the economy. On reaching the trough stage, the economy moves on the path of recovery, prosperity and reaches peak or boom stage.

## Cobweb Model

The prices of agricultural commodities follow the pattern of cycles indicated in the cobweb theories. According to this theory, the effect of price change on the production of commodity is felt with one time lag. However, the length of the lag varies from commodity to commodity, depending on the technology underlying the production. In agriculture the existence of time lag is due to inherent characteristics of crops.

Production of the crops can be affected even before the commencement of sowing season because of the changes in the area under crops due to changes in the price. For dry land agriculture, the time lag is one year, while for irrigation crops it may be a season, for livestock and orchards the length of time lag is still longer. The supply function in the cobweb model is specified as $S_t = f(P_{t-1})$.

Here $S_t$ refers supply or production of the crop in the $t'^{th}$ year. $P_{t-1}$ is the price in the previous year. Here the time lag is one year. If the time lag is two years, the supply function is represented as

$$S_t = f(P_{t-2})$$

In the cobweb model the price-demand relationship involves no time lag. This means the demand for the crop in the current year depends on the price in the current year.

$$D_t = f(P_t)$$

The supply and demand equations are jointly used to explain the phenomenon of price cycles for agricultural commodities.

In these cobweb models the quantities demanded, quantities supplied and the price in the subsequent periods moves towards equilibrium, or away from equilibrium depending upon the relative slopes of demand and supply curve and the relative price elasticities of demand and supply. Thus the magnitude of price elasticities of demand and supply determine the following three forms of cobweb cycles in the prices of the agricultural commodities.

| Forms of cobweb cycle | Price elasticity of demand and supply $(E_d, E_s)$ |
|---|---|
| Convergent | $E_d > E_s$ |
| Divergent or explosive | $E_d < E_s$ |
| Regular or uniform | $E_d = E_s$ |

In cobweb cycle, if price elasticity of demand is less than that of supply we get the divergent or explosive type of cobweb cycle. If elasticity of demand is greater than elasticity of supply, we get convergent type of cobweb. In case the elasticity of demand and supply are equal, there are regular oscillations between quantities and price levels.

## Convergent Cobweb Model

Let us assume that the price prevailed in the initial period ($t_1$) is $P_1$ Figure 47.2. This price makes the farmers to produce $Q_2$ level of output in period $t_2$. Given the demand DD and quantity supplied $Q_2$ in the market, the market clearing price is $P_2$ in period $t_2$. This low price of $P_2$, which is lesser than $P_1$ compels the farmers to reduce the output to $Q_3$ in period $t_3$. The fall in the quantity produced, pushes up the price to $P_3$ and so on for the following time periods. It is noticed that price ($P_1$, $P_2$, $P_3$ and $P_4$) oscillates above and below the equilibrium price and gets closer to the equilibrium price ($P_e$) as the adjustment process continues. This is evident when the movement of the price is observed. Such a model is called convergent cobweb model or model of damped oscillation. A cobweb model converges if the absolute slope of the demand curve is less than the absolute slope of the supply curve.

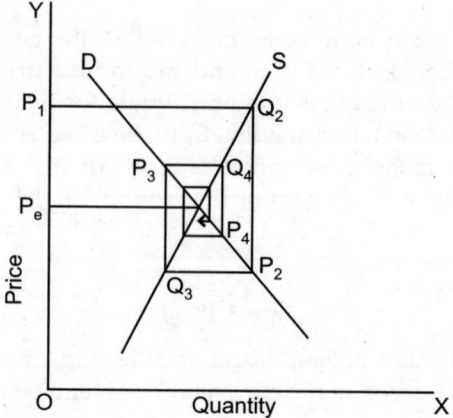

**Figure 47.2** Convergent cobweb model.

## Divergent Cobweb Model

Again it is assumed that the price in the initial period is $P_1$ (Figure 47.3). This price encourages the farmer to produce $Q_2$ quantity in $t_2$ period. The price that prevails in $t_2$ is $P_2$, which is lower than $P_1$ price. Discouraged by the lower price of $P_2$, the farmers produce $Q_3$ quantity of output in period $t_3$, which fetches them a price of $P_3$. This price encourages the farmers to produce $Q_4$ quantity of output in period $t_4$ only to realize that the price has fallen down to $P_4$. This phenomenon continues so on. It is seen that the price oscillates around the equilibrium price ($P_e$) and the diversion from equilibrium price gets larger and larger continuously. Such a model is called divergent model or model of explosive oscillation. A cobweb model diverges if, the absolute slope of demand curve is greater than the absolute slope of the supply curve.

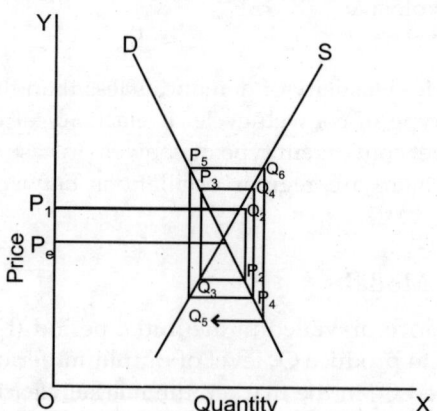

**Figure 47.3** Divergent cobweb model.

## Perpetual Oscillating Cobweb Model

The initial price of $P_1$ results in $Q_2$ output in period $t_2$ leading to the prevalence of a lower price $P_2$ (Figure 47.4). This lower price forces the farmers to reduce the output to $Q_3$ level in period $t_3$. This lower quantity helps the price to move up to $P_1$. This situation encourages the farmers to produce $Q_2$ quantity of output in period $t_4$ which only help them to realize $P_2$ price, the cycle of which goes on. The price oscillates around the equilibrium price $(P_e)$ perpetually. A Cobweb model converges or oscillates perpetually if the absolute slope of demand curve is equal to the absolute slope of the supply curve.

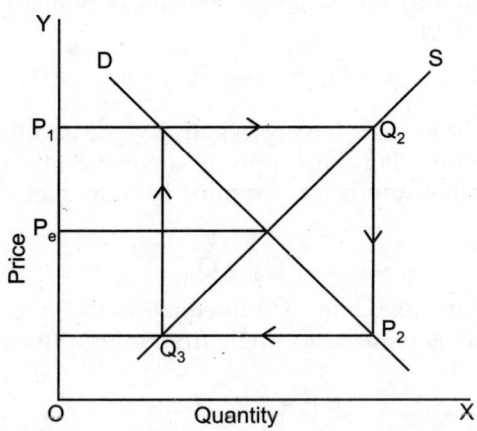

**Figure 47.4**  Perpetual oscillating cobweb model.

## Cobweb Model—An Illustration

The supply function in general is specified as

$$Q_{st} = f(P_t) \tag{6.1}$$

Here price in $t^{th}$ period determines the quantity supplied to the market in the $t^{th}$ period. But in the case of agricultural commodity the supply function of the commodity is specified as

$$Q_{s\,t+1} = f(P_t) \tag{6.2}$$

Here equation (6.2) assumes that price of commodity in $t^{th}$ period determines the supply of commodity in $t+1^{th}$ period. Shifting the subscript t by one period, equation (6.2) is restated as

$$Q_{st} = f(P_{t-1}) \tag{6.3}$$

The supply function of equation (6.3) is interacted with demand function of the same commodity in the market and dynamic price patterns of the commodity in different periods are set in the market. The relevant demand function of the commodity is specified as

$$Q_{dt} = f(P_t) \tag{6.4}$$

In equation (6.4) we assume that price of the commodity in $t^{th}$ period determines the demand for the commodity without any time lag. In estimation of equation (6.3) and (6.4), we receive data on all other variables which affect demand and supply of the commodity in question. After estimating the demand and supply functions considering all the relevant factors, the final demand and supply functions are expressed in terms of price only by adding the coefficient value of the other factors (coefficients multiplied by their mean levels) to their respective constant terms. The final estimated demand function for same commodity over time is specified as

$$\hat{Q}_{dt} = \hat{a} - \hat{b}\,P_t + e_t \tag{6.5}$$

Here the intercept term represents the effect of all other variables affecting the demand. The $\hat{b}$ coefficient indicates effect of price of the same commodity in question. Here b is having negative value because of the inverse relationship between price per unit of commodity and quantity demanded. Similarly the estimated supply function is represented as

$$\hat{Q}_{st} = -\hat{c} + \hat{d}\, P_{t-1} + e_t \tag{6.6}$$

Here the intercept term, $\hat{C}$ is having negative value and this represents the effect of all other factors influencing the supply of a commodity in the market in the $t^{th}$ period. To get the equilibrium price, we must have market clearing conditions and this is specified as

$$\hat{Q}_{dt} = \hat{Q}_{st} \tag{6.7}$$

Substituting equations (6.5) and (6.6) in equation (6.7) we get the cobweb model and this cobweb model is reduced to single first order differential equation which is stated as follows:

$$b\, P_t - d\, P_{t-1} = a + c \tag{6.8}$$

In order to solve the equation it is desirable to normalize the equation and shift the time subscript by one period ahead. This causes the 't' to be replaced by $t + 1$ and the resultant equation is now specified as

$$b\, P_{t+1} - d\, P_t = a + c \tag{6.9}$$

Throughout divide the equation (6.8) by b

$$P_{t+1} - \frac{d}{b} P_t = \frac{a+c}{b} \tag{6.10}$$

Now let us consider the time path equation which is given as

$$Y_t \left( Y_o - \frac{c}{1+a} \right)(-a)^t + \frac{c}{1+a} \tag{6.11}$$

Now substitute $Y_t = P_t$; $Y_o = P_o$, $a = d/b$, $c = \frac{a+c}{b}$ in equation (6.11)

$$P_t = \left( P_o - \frac{a+c}{b+d} - \left( \frac{d}{b} \right)^t + \frac{a+c}{b+d} \right) \tag{6.12}$$

Where $P_o$ is initial price of the commodity.

$\frac{a+c}{b+d} = \bar{P}$, this is equilibrium price. Substitute this equation (6.12) to get time path equation.

$$P_t = (P_o - \bar{P}) - \left( \frac{d}{b} \right)^t + \bar{P} \tag{6.13}$$

Given the values of $P_o$ when $t = 1$, we get the value of $P_t$ later, when $t = 2, 3, 4$ and so on, then we can get the corresponding $P_t$ value

If $d > b$ — We get diverging or explosive type of cobweb model.
If $d < b$ — We get converging or damping type of cobweb model.
If $d = b$ — It is perpetual type of cobweb

*Numerical Example*

If d > b, let the estimated supply function be specified as

$$\hat{Q}_{st} = -1,200 + 10 \ P_{t-1}$$

$$\hat{Q}_{dt} = 1,600 - 9 \ P_t$$

Equilibrium price = Rs. 400 = $\overline{P}$
Let the initial price be Rs. 410/unit
Where t = 1, using time path equation (6.13), the corresponding $P_t$ is Rs. 421.11.

Let $\hat{P}_t$ is considered as $P_o$ (initial price) for the subsequent time period

*i.e.,* $P_{t+1}$ = Rs. 432.22

Similarly, for $P_{t+2}$ = Rs. 443.33, $P_{t+3}$ = Rs. 454.44, $P_{t+4}$ = Rs. 465.55, $P_{t+5}$ = Rs. 476.66

Hence, it is observed that the price is increasing in subsequent time periods. Here the cobweb cycle is divergent or explosive type. This is due to supply elasticity being more than demand elasticity.

If d < b let us assume the following estimated equations of supply and demand

$$\hat{Q}_{st} = -7,869 + 0.1 \ P_{t-2}$$

$$Q_{dt} = 1,068 - 0.8 \ P_t$$

Equilibrium price = Rs. 9,687.14 = $\overline{P}$
Let the initial price be = Rs. 9,500

When t = 1 using time path equation (6.13), the corresponding $P_t$ is Rs. 9,476.61 when t = 2 the corresponding $P_t$ is Rs. 9,453.21. Similarly for 3rd, 4th and 5th periods, the corresponding values of $P_t$ are Rs. 9,429.82, Rs. 9,406.43 and Rs. 9,383.04. Based on the $P_t$ values in subsequent years, it is evident that the cobweb model is of convergent type.

## Seasonal Factors

These relate to specific season of the year or month or some times week. If the time series data are available over seasons of the years or months of the year, we can separate seasonal factor and measure the seasonal component of time-series data for making appropriate decisions regarding seasonal variations. Time series data on prices of the agricultural commodities recorded over different seasons or months can be used for analyzing the seasonal effects. The consumption of meat, chicken, eggs, *etc.,* undergoes variations over different seasons or months. Rainfall data recorded over months or seasons would have wide variations. Using such data we can analyze the seasonal factors and draw relevant solutions to the decision problems.

## Irregular Factor or Random Factor

This factor includes all other omitted factors, which influence the value and magnitude of economic variable. Changes in the tastes and preferences, which are not specifically related to passage of time or trend are the factors to be cited as random factors. Similarly, selling costs *i.e.,* advertisement costs would enhance the value of random factor in the time series analysis.

## Forecasting Methods

Analyzing the time series data and splitting the same into different components *viz.*, trend, cyclical, seasonal and random (irregular) would give us many uses. Forecasting techniques, which are very much useful in decision-making process, are developed from the time series analysis. These techniques at micro and macro level would give us implications for policy making and management of the farms and use of the inputs efficiently by the farms. In prediction of prices of commodities, GNP, investment spending, consumption expenditure, spending by the Government, export and imports of the commodities, supply and demand of the commodities, *etc.*, we use the forecasting tools. A business firm is interested to forecast its product supply and demand, prices and sales of their commodities weekly, monthly, and yearly.

We have many types of forecasting methods. Some of the important among these are (1) trend analysis (2) seasonal analysis (3) cyclical analysis (4) barometric forecasting (5) input-output analysis, *etc.*, Among these forecasting methods trend analysis is the most popular method, which is explained below:

## Trend Analysis

Trend component of the economic variable is generally calculated using the least squares method. Free hand method and moving average method were the earlier methods to estimate the trend. These are not sophisticated because of large magnitude of forecasting errors. Trend is estimated by solving the following normal equations.

$$\Sigma Y = Na + b \Sigma T$$

$$\Sigma XY = a \Sigma T + b \Sigma T^2$$

Where t = years

Estimation of trend for hypothetical demand of the product is presented in Table 47.1

**TABLE 47.1  Calculation of Trend for the Hypothetical Demand of the Product.**

| Year | Actual quantity demanded (units) Y | T | TY | $T^2$ | Forecast |
|------|------|------|------|------|------|
| 1984 | 210 | -6 | -1,260 | 36 | 177.96 |
| 1985 | 225 | -5 | -1,125 | 25 | 204.39 |
| 1986 | 236 | -4 | -944 | 16 | 230.82 |
| 1987 | 242 | -3 | -726 | 9 | 257.25 |
| 1988 | 267 | -2 | -534 | 4 | 283.68 |
| 1989 | 292 | -1 | -292 | 1 | 310.11 |
| 1990 | 310 | 0 | 0 | 0 | 336.54 |
| 1991 | 345 | 1 | 345 | 1 | 362.97 |
| 1992 | 372 | 2 | 744 | 4 | 389.4 |
| 1993 | 410 | 3 | 1,230 | 9 | 415.83 |
| 1994 | 467 | 4 | 1,868 | 16 | 442.26 |
| 1995 | 489 | 5 | 2,445 | 25 | 468.69 |
| 1996 | 510 | 6 | 3,060 | 36 | 495.12 |
| | $\Sigma Y = 4,375$ | $\Sigma T = 0$ | $\Sigma TY = 4,811$ | $\Sigma T^2 = 182$ | |

$$\Sigma Y = Na + b \ \Sigma T$$

$$\Sigma TY = a \ \Sigma T + b \ \Sigma T^2$$

$$4,375= 13 \ (a) + b \ (O)$$

$$4,811= a \ (O) + b \ (182)$$

$$a = \frac{4375}{13} = 336.54$$

$$b = \frac{4811}{182} = 26.43$$

Y = 336.54 + 26.43 T i.e., Y = a + bT

While estimating the equation we assume that

T = 0 for the year 1990. The demand for future period *i.e.*, 1999 is estimated by substituting t= 9 in estimated equation (1), we get the predicted demand for the product in 1999.

$\quad$ Y′ (Predicted demand) $\quad$ = 336.54 + 26.43 T

$$= 574.41 \text{ units}$$

Here we will have to assume that the future demand would follow the normal distribution with the value 574.41 as its mean.

The particulars of testing the accuracy of demand forecast is presented in Table 47.2.

### TABLE 47.2 Accuracy of the Demand Forecast.

| Year | Actual quantity demanded (Y) | Forecast demand (X) | Deviation (d) d = 2-3 | $d^2$ | Cumulative deviation |
|---|---|---|---|---|---|
| 1 | 2 | 3 | 4 | 5 | 6 |
| 1984 | 210 | 177.96 | 32.04 | 1026.56 | 32.04 |
| 1985 | 225 | 204.39 | 20.61 | 424.77 | 52.65 |
| 1986 | 236 | 230.82 | 5.18 | 26.83 | 57.83 |
| 1987 | 242 | 257.25 | -15.25 | 232.56 | 42.58 |
| 1988 | 267 | 283.68 | -16.68 | 278.22 | 25.90 |
| 1989 | 292 | 310.11 | -18.11 | 327.97 | 7.79 |
| 1990 | 310 | 336.54 | -26.54 | 704.37 | -18.75 |
| 1991 | 345 | 362.97 | -17.97 | 322.92 | -36.72 |
| 1992 | 372 | 389.40 | -17.4 | 302.76 | -54.12 |
| 1993 | 410 | 415.83 | -5.83 | 33.99 | -59.95 |
| 1994 | 467 | 442.26 | 24.74 | 612.07 | -35.21 |
| 1995 | 489 | 468.69 | 20.31 | 412.50 | -14.9 |
| 1996 | 510 | 495.12 | 14.88 | 221.41 | -0.02 |

$\Sigma$ X = 4,375.02 $\Sigma$d = -0.02 $\Sigma$ $d^2$ = 4,926.93

Mean $(\overline{X})$= 336.65

Mean absolute deviation (MAD) is worked out by taking the difference between actual demand and forecast demand for the product of over 13 years *i.e.*, from 1984 to 1996. Here negative values are ignored for summing.

Total deviation = 235.54

Here, sum of the negative deviations = 117.78
Sum of the positive deviations = 117.76
Total deviations = 235.54

$$\text{MAD} = \frac{\text{Total deviation}}{\text{No. of years}}$$

$$= \frac{235.54}{13} = 18.12$$

$$\text{Tracking the signal} = \frac{\text{Cumulative deviation}}{\text{MAD}}$$

$$= \frac{-0.02}{18.12} = -0.001$$

Absolute value of the tracking signal = 0.001

Here the tracking signal is less than 4.00. Hence the forecast value is accurate, but we cannot infer at this time that the forecast demand is accurate. We will have to work out sum of deviation squares and work out the control limits (confidence limits) for the forecast demand.

$$\text{Sum of deviation square } (S_f) = \sqrt{\frac{\sum d^2}{n-1}}$$

$$= \sqrt{\frac{4926.93}{12}}$$

$$= \sqrt{410.57}$$

$$= 20.26$$

Here N is less than 30 so it follows 't' distribution 't' value for n-1 = 12 d.f. at 10% probability is 1.782

Mean forecast value = 336.55

Control limits for 90% probability level with t values 1.782 are worked out as follows. Mean forecast value ± 't' value ($S_f$ factor)

$$336.55 + 1.782 (20.26)$$

$$= 372.65 \text{ units}$$

$$= 336.55 - 1.782 (20.26)$$

$$= 300.45 \text{ units}$$

The forecast demand ranges from minimum level of 300.45 units to 372.65 units.

Difference = 372.65 – 300.45 = 72.2

The variability between the lower and upper limit is 72.2, which is high. This implies that uncertainty is associated in the process of forecasting. A large sample of more than 100 observations would give us better forecast estimates with closer confidence limits or control limits.

## Seasonal Analysis

## Moving Average Method

This method is employed for time series data collected month-wise and over years. The time series variables *viz.*, price of the product, rainfall, *etc.*, may be used for illustration. We generally use the moving average method to remove the cyclical component as well as seasonal component for price series data or rainfall data. The following steps are involved in the calculation of moving averages.

If, $P_t = T + C + I$

Then moving averages is $T + I$ for additive model

$$(MA_t) = T + I$$

For multiplicative model moving average is $T \times I$

$$(MA_t) = T \times I$$

$C_t$ (Cyclical component) $= P_t - MA_t$ for additive models

$$C_t = \frac{P_t}{MA_t} \text{ for multiplicative model}$$

Where, $P_t$ = Price index in the $t^{th}$ period
  T = Trend component
  C = Cyclical component
  I = Irregular component in the time series data for the $t^{th}$ year.

Method of working out moving average for price indices is shown in Table 47.3.

**TABLE 47.3  Computation of Moving Averages for Indices of Egg Prices (Hypothetical).**

| Year | Index No. of egg prices | 3 year moving average | 5 year moving average |
|------|------|------|------|
| 1978 | 102.44 | - | - |
| 1979 | 110.42 | 103.84 | - |
| 1980 | 98.67 | 109.74 | 112.36 |
| 1981 | 120.12 | 116.31 | 114.00 |
| 1982 | 130.14 | 120.31 | 115.08 |
| 1983 | 110.66 | 118.87 | 122.29 |
| 1984 | 115.82 | 120.38 | 122.82 |
| 1985 | 134.67 | 124.45 | 129.13 |
| 1986 | 122.82 | 139.72 | 137.54 |
| 1987 | 161.68 | 145.73 | 142.87 |
| 1988 | 152.69 | 152.27 | 142.36 |
| 1989 | 142.45 | 142.43 | 147.94 |
| 1990 | 132.16 | 141.76 | 149.79 |
| 1991 | 150.68 | 151.27 | 157.12 |
| 1992 | 170.98 | 170.33 | 168.96 |
| 1993 | 189.34 | 187.32 | 183.69 |
| 1994 | 201.64 | 198.93 | 194.04 |
| 1995 | 205.81 | 203.30 | - |
| 1996 | 202.44 | - | - |

**NB:** Among the computed moving averages, the best method of moving average would have least variability.

## Forecasting With Seasonal Demand

In the seasonal analysis, we generally take the time series data, which are recorded month-wise or quarter-wise. Let us consider the chicken demand for five years recorded season-wise. Here a year is divided into 4 quarters *viz.*, Quarter I (June – Aug); Quarter II (Sept – Nov); Quarter III (Dec – Feb); and Quarter IV (March – May). The hypothetical data is presented in Table 47.4.

**TABLE 47.4 Hypothetical Demand for Chicken (quarter–wise).**

| Year | Quarter | Actual demand (Y=TCSR) | Four quarter moving average (MA) | Centred moving average (CMA=TC) | Specific seasonal index (SSI) |
|------|---------|------------------------|----------------------------------|---------------------------------|-------------------------------|
| 1992 | $Q_1$ | 386 | - | - | - |
|      | $Q_2$ | 401 | 400.25 | - | - |
|      | $Q_3$ | 502 | 394.25 | 397.25 | 1.26 |
|      | $Q_4$ | 312 | 399.25 | 396.75 | 0.79 |
| 1993 | $Q_1$ | 362 | 394.75 | 397.0 | 0.91 |
|      | $Q_2$ | 421 | 393.25 | 394.0 | 1.07 |
|      | $Q_3$ | 484 | 408.25 | 400.75 | 1.21 |
|      | $Q_4$ | 306 | 444.25 | 426.25 | 0.72 |
| 1994 | $Q_1$ | 422 | 453.25 | 448.75 | 0.94 |
|      | $Q_2$ | 565 | 474.25 | 463.75 | 1.22 |
|      | $Q_3$ | 520 | 492.5 | 483.25 | 1.08 |
|      | $Q_4$ | 390 | 475.0 | 485.25 | 0.81 |
| 1995 | $Q_1$ | 495 | 497.5 | 486.25 | 1.02 |
|      | $Q_2$ | 495 | 515.27 | 506.38 | 0.98 |
|      | $Q_3$ | 610 | 533.25 | 524.26 | 1.16 |
|      | $Q_4$ | 461 | 563.0 | 548.13 | 0.84 |
| 1996 | $Q_1$ | 567 | 584.5 | 573.75 | 0.99 |
|      | $Q_2$ | 614 | 599.25 | 591.88 | 1.05 |
|      | $Q_3$ | 696 | 587.5 | 593.38 | - |
|      | $Q_4$ | 520 | - | - | - |
| 1997 | $Q_1$ | 520 | - | 585.75 | - |

**TABLE 47.5 Computation of Typical Seasonal Index Values from Specific Seasonal Value.**

| Year | $Q_1$ | $Q_2$ | $Q_3$ | $Q_4$ |
|------|-------|-------|-------|-------|
| 1992 | - | - | 1.26 | 0.79 |
| 1993 | 0.91 | 1.07 | 1.21 | 0.72 |
| 1994 | 0.94 | 1.22 | 1.08 | 0.81 |
| 1995 | 1.02 | 0.98 | 1.16 | 0.82 |
| 1996 | 0.99 | 1.05 | - | - |
| Total | 3.86 | 4.32 | 4.71 | 3.14 |
| Average | 0.97 | 1.08 | 1.17 | 0.78=4.00 |

Estimate the trend equation

a    = 478.52
b    = 11.10
Y    = 478.52 + 11.10 T

| | |
|---|---|
| Predicted demand for 2nd quarter of 1997 due to trend | = 619.20 |
| Predicted demand for 2nd quarter of 1997 adjusted by seasonal index | = 668.74 |
| Predicted demand for 3rd quarter of 1997 due to trend | = 630.64 |
| Predicted demand for 3rd quarter of 1997 due to seasonal index | = 737.85 |
| Predicted demand for 4th quarter of 1997 due to trend | = 642.08 |
| Predicted demand for 4th quarter of 1997 due to seasonal index | = 500.82 |
| Predicted demand for 1st quarter of 1998 due to trend | = 653.53 |
| Predicted demand for 1st quarter of 1998 due to seasonal index | = 633.92 |

## Procedure

Time series data on demand for chicken are considered for working out season-wise demand forecast. The procedure consists of the following steps.

1) We have worked out moving averages by pooling the data of four subsequent quarters. For example $MA_1$ is worked out as follows:

$$MA_1 = \frac{386 + 401 + 502 + 312}{4} = 400.25$$

$$MA_2 = \frac{401 + 502 + 312 + 362}{4} = 394.25$$

Similarly, the other moving averages are worked out and shown in column No. 4 of the Table 47.4. Here the yearly data are split into 4 quarters, hence four quarter moving averages seem to be appropriate for removing seasonal influences on demand. If the data are divided into months or weeks then it is better to take 12-month moving average or 52 week moving average. By doing so we get deseasonalized measure of demand for chicken and then account for trend and cyclical effect. Thus centred moving averages are worked out by summing up these successive moving averages of column 4 and dividing by two. For example, $CMA_1$.

$$\frac{400.25 + 25 + 394.25}{2} = 397.25$$

Similarly $CMA_2 = \frac{394.25 + 399.25}{2} = 396.75$

These figures are furnished in column 5 of the table. With regard to computation of specific seasonal index, the actual quantity of demand for each quarter is divided by the respective centred moving average figure. For example, for quarter 3 of the year 1992, the specific seasonal index $(SSI) = \frac{502}{397.25} = 1.26$

For quarter 4 of 1992 $= \frac{312}{396.75} = 0.79$

Taking the specific seasonal indices of the column 6 in the Table 47.4 they are separated quarter-wise for all the years under consideration and these are presented in Table 47.5. The summation for each quarter is worked out and their averages are arrived at. Since these average values when summed up are equal to 4 there is need for adjustment. In some cases the summation would not be equal to 4 then we have to adjust these averages to four. The typical seasonal index value for the demand of chicken computed over the five years data are as follows:

S $Q_1$ (First quarter seasonal index)     = 0.97
S $Q_2$ (Second quarter seasonal index)     = 1.08
S $Q_3$ (Third quarter seasonal index)     = 1.17
S $Q_4$ (Fourth quarter seasonal index)     = 0.78

## Forecasting the Demand for Quarters

We have to work out trend for the demand data recorded quarter-wise. For the example problem the trend equation is Y = 478.52 + 11.10 T, here X = 0 for $3^{rd}$ quarter of 1994. Now we can forecast the demand for $Q_2$ in 1997. Put the value of X = 11, then Y = 619.20. This is due to trend component. Forecast due to season is worked out by multiplying the trend forecast values with respective seasonal index $i.e.$, 619.20 x 1.08 = 668.74. Similarly the forecast of the demand for the subsequent quarters is computed.

The predicted demand for chicken is highest in III quarter (Dec–Feb) followed by $2^{nd}$ quarter (Sept–Nov), first quarter (June–Aug) and fourth quarter (March–May). Poultry farmers should produce adequate chicken to the markets, keeping in view the demand in different quarters. Based on this information marketing personnel should probe into the reasons for varying levels of demand in the various quarters of the year and evolve relevant strategies for raising the demand in the lean months and stabilize the prices for chicken in the interest of the consumers as well as producers.

## Price Forecasting

Taking the monthly modal prices of maize in Nizamabad market of Andhra Pradesh for the period from April, 2000 to September, 2013, the prices have been forecasted for the months commencing from October, 2013 to January, 2014, using Auto Regressive Integrated Moving Average (ARIMA) model. The procedure has been presented hereunder.

## Monthly modal prices of maize in Nizamabad market (Andhra Pradesh)

| Year | Month | Price Rs./Qtl | Year | Month | Price Rs./Qtl |
|------|-------|------|------|-------|------|
| 2000 | Apri | 503 | | May | 560 |
| | May | 485 | | June | 579 |
| | June | 504 | | July | 565 |
| | July | 447 | | Aug | 433 |
| | Aug | 470 | | Sep | 509 |
| | Sep | 370 | | Oct | 453 |
| | Oct | 354 | | Nov | 490 |
| | Nov | 386 | | Dec | 482 |
| | Dec | 383 | 2004 | Jan | 494 |
| 2001 | Jan | 370 | | Feb | 492 |
| | Feb | 428 | | Mar | 496 |
| | Mar | 440 | | Apri | 528 |

| Year | Month | Price Rs./Qtl | Year | Month | Price Rs./Qtl |
|------|-------|---------------|------|-------|---------------|
|      | Apri  | 364 |      | May  | 558 |
|      | May   | 373 |      | June | 552 |
|      | June  | 400 |      | July | 573 |
|      | July  | 433 |      | Aug  | 547 |
|      | Aug   | 404 |      | Sep  | 550 |
|      | Sep   | 385 |      | Oct  | 525 |
|      | Oct   | 393 |      | Nov  | 515 |
|      | Nov   | 383 |      | Dec  | 520 |
|      | Dec   | 375 | 2005 | Jan  | 525 |
| 2002 | Jan   | 433 |      | Feb  | 527 |
|      | Feb   | 450 |      | Mar  | 530 |
|      | Mar   | 517 |      | Apri | 525 |
|      | Apri  | 519 |      | May  | 535 |
|      | May   | 482 |      | June | 530 |
|      | June  | 515 |      | July | 540 |
|      | July  | 623 |      | Aug  | 537 |
|      | Aug   | 540 |      | Sep  | 545 |
|      | Sep   | 492 |      | Oct  | 530 |
|      | Oct   | 473 |      | Nov  | 540 |
|      | Nov   | 513 |      | Dec  | 535 |
|      | Dec   | 510 | 2006 | Jan  | 545 |
| 2003 | Jan   | 512 |      | Feb  | 525 |
|      | Feb   | 587 |      | Mar  | 540 |
|      | Mar   | 550 |      | Apri | 510 |
|      | Apri  | 537 |      | May  | 515 |
|      | June  | 525 | 2010 | Jan  | 838 |
|      | July  | 580 |      | Feb  | 805 |
|      | Aug   | 620 |      | Mar  | 800 |
|      | Sep   | 570 |      | Apri | 900 |
|      | Oct   | 585 |      | May  | 860 |
|      | Nov   | 705 |      | June | 971 |
|      | Dec   | 680 |      | July | 925 |
| 2007 | Jan   | 650 |      | Aug  | 950 |
|      | Feb   | 790 |      | Sep  | 1101 |
|      | Mar   | 715 |      | Oct  | 910 |
|      | Apri  | 735 |      | Nov  | 876 |
|      | May   | 670 |      | Dec  | 915 |
|      | June  | 675 | 2011 | Jan  | 940 |
|      | July  | 700 |      | Feb  | 960 |
|      | Aug   | 653 |      | Mar  | 1140 |
|      | Sep   | 625 |      | Apri | 1205 |
|      | Oct   | 610 |      | May  | 1200 |
|      | Nov   | 660 |      | June | 1125 |
|      | Dec   | 650 |      | July | 1015 |
| 2008 | Jan   | 665 |      | Aug  | 910 |
|      | Feb   | 650 |      | Sep  | 960 |
|      | Mar   | 735 |      | Oct  | 960 |
|      | Apri  | 700 |      | Nov  | 980 |
|      | May   | 730 |      | Dec  | 1000 |
|      | June  | 800 | 2012 | Jan  | 1025 |
|      | July  | 890 |      | Feb  | 1078 |
|      | Aug   | 900 |      | Mar  | 1100 |
|      | Sep   | 970 |      | Apri | 1140 |
|      | Oct   | 830 |      | May  | 1020 |

| Year | Month | Price Rs./Qtl | Year | Month | Price Rs./Qtl |
|------|-------|---------------|------|-------|---------------|
|      | Nov   | 800           |      | June  | 1100          |
|      | Dec   | 810           |      | July  | 1251          |
| 2009 | Jan   | 770           |      | Aug   | 1235          |
|      | Feb   | 760           |      | Sep   | 1301          |
|      | Mar   | 760           |      | Oct   | 1275          |
|      | Apri  | 800           |      | Nov   | 1281          |
|      | May   | 800           |      | Dec   | 1270          |
|      | June  | 860           | 2013 | Jan   | 1250          |
|      | July  | 815           |      | Feb   | 1270          |
|      | Aug   | 870           |      | mar   | 1280          |
|      | Sep   | 840           |      | Apri  | 1175          |
|      | Oct   | 845           |      | May   | 1175          |
|      | Nov   | 860           |      | June  | 1400          |
|      | Dec   | 900           |      | July  | 1425          |
|      |       |               |      | Aug   | 1380          |
|      |       |               |      | Sep   | 1450          |

Data written to the working file.
3 variables and 162 cases written.

| Variable: YEAR | Type: Number | Format: F11.2 |
|---|---|---|
| Variable: MONTH | Type: String | Format: A8 |
| Variable: PRICE | Type: Number | Format: F11.2 |

The following new variables are being created:

| Name | Label |
|---|---|
| YEAR_ | YEAR, not periodic |
| MONTH_ | MONTH, period 12 |
| DATE_ | DATE. FORMAT: "MMM YYYY" |

## Arima

MODEL: MOD_1
Model Description:
Variable: PRICE
Regressors: NONE
Non-seasonal differencing: 0
Seasonal differencing: 0
Length of Seasonal Cycle: 12

**Parameters:**
AR1 _____ < value originating from estimation >
SAR1 _____ < value originating from estimation >
CONSTANT _____ < value originating from estimation >
95.00 percent confidence intervals will be generated.
Split group number: 1 Series length: 162
No missing data.
Melard's algorithm will be used for estimation.
Termination criteria:
Parameter epsilon: .001
Maximum Marquardt constant: 1.00E+09
SSQ Percentage: .001
Maximum number of iterations: 10

**Initial values:**

| | |
|---|---|
| AR1 | .95543 |
| SAR1 | .87031 |
| CONSTANT | 928.9426 |

Marquardt constant = .001
Adjusted sum of squares = 837768.61

### Iteration History:

| Iteration | Adj. Sum of Squares | Marquardt Constant |
|---|---|---|
| 1 | 531579.64 | .00100000 |
| 2 | 525434.29 | .00010000 |
| 3 | 524915.43 | .00001000 |
| 4 | 524643.13 | .00000100 |
| 5 | 524565.22 | .00000010 |
| 6 | 524530.84 | .00000001 |
| 7 | 524516.92 | .00000000 |

Conclusion of estimation phase.
Estimation terminated at iteration number 8 because:
Sum of squares decreased by less than .001 percent.

**FINAL PARAMETERS:**
Number of residuals 162

| | |
|---|---|
| Standard error | 56.747393 |
| Log likelihood | −884.56207 |
| AIC* | 1775.1241 |
| SBC** | 1784.3869 |

### Analysis of Variance:

| | DF | Adj. Sum of Squares | Residual Variance |
|---|---|---|---|
| Residuals | 159 | 524512.25 | 3220.2667 |

### Variables in the Model:

| | B | SEB | T-RATIO | APPROX. PROB. |
|---|---|---|---|---|
| AR1 | .98647 | .01651 | 59.739007 | .00000000 |
| SAR1 | .09754 | .08533 | 1.143001 | .25475626 |
| CONSTANT | 850.97218 | 263.81339 | 3.225659 | .00152587 |

### Covariance Matrix:

| | AR1 | SAR1 |
|---|---|---|
| AR1 | .00027268 | −.00024156 |
| SAR1 | −.00024156 | .00728200 |

### Correlation Matrix:

| | AR1 | SAR1 |
|---|---|---|
| AR1 | 1.0000000 | −.1714247 |
| SAR1 | −.1714247 | 1.0000000 |

---

\* Akike Information Coefficient
\*\* Schwarz's Bayesian Information Criterion

**Regressor Covariance Matrix:**

CONSTANT
CONSTANT        69597.507

**Regressor Correlation Matrix:**

CONSTANT
CONSTANT        1.0000000

The following new variables are being created:

| Name | Label |
|------|-------|
| FIT_1 | Fit for PRICE from ARIMA, MOD_1 CON |
| ERR_1 | Error for PRICE from ARIMA, MOD_1 CON |
| LCL_1 | 95% LCL for PRICE from ARIMA, MOD_1 CON |
| UCL_1 | 95% UCL for PRICE from ARIMA, MOD_1 CON |
| SEP_1 | SE of fit for PRICE from ARIMA, MOD_1 CON |

6 new cases have been added.

## ACF*

MODEL: MOD_2.
Variable: ERR_1      Missing cases: 6      Valid cases: 162
Autocorrelations:   ERR_1   Error for PRICE from ARIMA, MOD_1 CON

| Lag | Auto- Corr. | Stand. Err. | −1 −.75 −.5 −.25 0 .25 .5 .75 1 | Box-Ljung | Prob. |
|-----|-------------|-------------|-------------------------------------|-----------|-------|
| | | | +----+----+----+----+----+----+----+----+ | | |
| 1 | -.072 | .078 | . *I . | .866 | .352 |
| 2 | -.069 | .078 | . *I . | 1.658 | .436 |
| 3 | .061 | .077 | . I* . | 2.280 | .516 |
| 4 | -.051 | .077 | . *I . | 2.717 | .606 |
| 5 | -.061 | .077 | . *I . | 3.337 | .648 |
| 6 | -.011 | .077 | . * . | 3.357 | .763 |
| 7 | -.020 | .076 | . * . | 3.424 | .843 |
| 8 | .031 | .076 | . I* . | 3.593 | .892 |
| 9 | .069 | .076 | . I* . | 4.409 | .882 |
| 10 | -.038 | .076 | . *I . | 4.663 | .913 |
| 11 | .068 | .075 | . I* . | 5.474 | .906 |
| 12 | .036 | .075 | . I* . | 5.703 | .930 |
| 13 | -.064 | .075 | . *I . | 6.443 | .929 |
| 14 | -.101 | .075 | .**I . | 8.285 | .874 |
| 15 | .056 | .074 | . I* . | 8.859 | .885 |
| 16 | .060 | .074 | . I* . | 9.521 | .890 |

Plot Symbols:     Autocorrelations     * Two Standard Error Limits .
Total cases:     168     Computable first lags:     161

* Auto correlation function

**Partial Autocorrelations: ERR_1 Error for PRICE from ARIMA, MOD_1 CON**

| Lag | Pr-Aut- Corr. | Stand. Err. | −1 | −.75 | −.5 | −.25 | 0 | .25 | .5 | .75 | 1 |
|-----|------|------|----|------|-----|------|---|-----|----|-----|---|
| 1 | −.072 | .079 | | | | | . *I . | | | | |
| 2 | -.075 | .079 | | | | | . *I . | | | | |
| 3 | .051 | .079 | | | | | . I* . | | | | |
| 4 | -.048 | .079 | | | | | . *I . | | | | |
| 5 | -.061 | .079 | | | | | . *I . | | | | |
| 6 | -.030 | .079 | | | | | . *I . | | | | |
| 7 | -.027 | .079 | | | | | . *I . | | | | |
| 8 | .029 | .079 | | | | | . I* . | | | | |
| 9 | .067 | .079 | | | | | . I* . | | | | |
| 10 | -.028 | .079 | | | | | . *I . | | | | |
| 11 | .066 | .079 | | | | | . I* . | | | | |
| 12 | .036 | .079 | | | | | . I* . | | | | |
| 13 | -.038 | .079 | | | | | . *I . | | | | |
| 14 | -.108 | .079 | | | | | .**I . | | | | |
| 15 | .038 | .079 | | | | | . I* . | | | | |
| 16 | .073 | .079 | | | | | . I* . | | | | |

Plot Symbols:     Autocorrelations *       Two Standard Error Limits .
Total cases:   168     Computable first lags:   161

Error for PRICE from ARIMA, MOD_1 CON

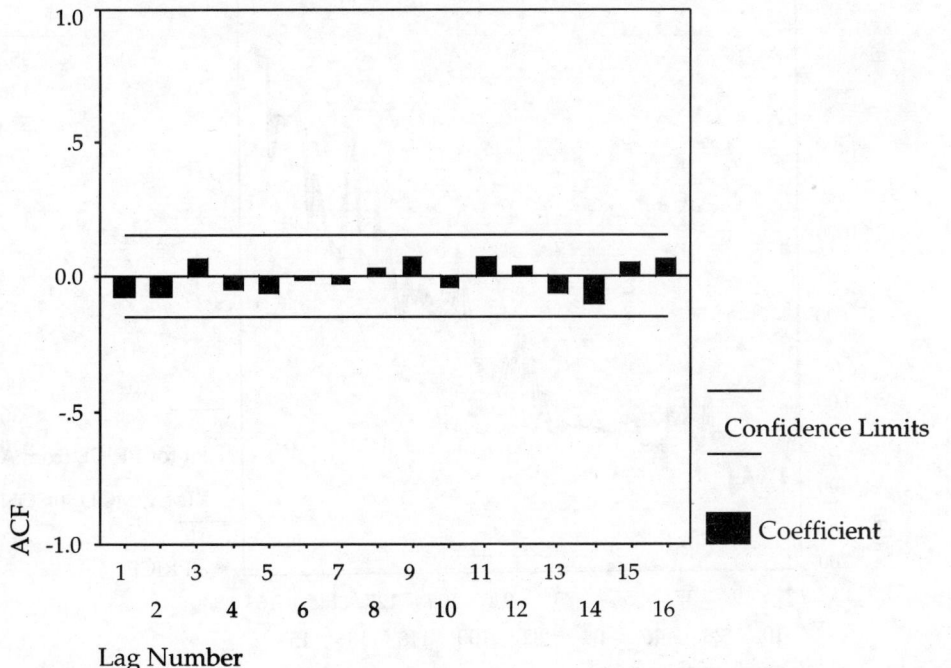

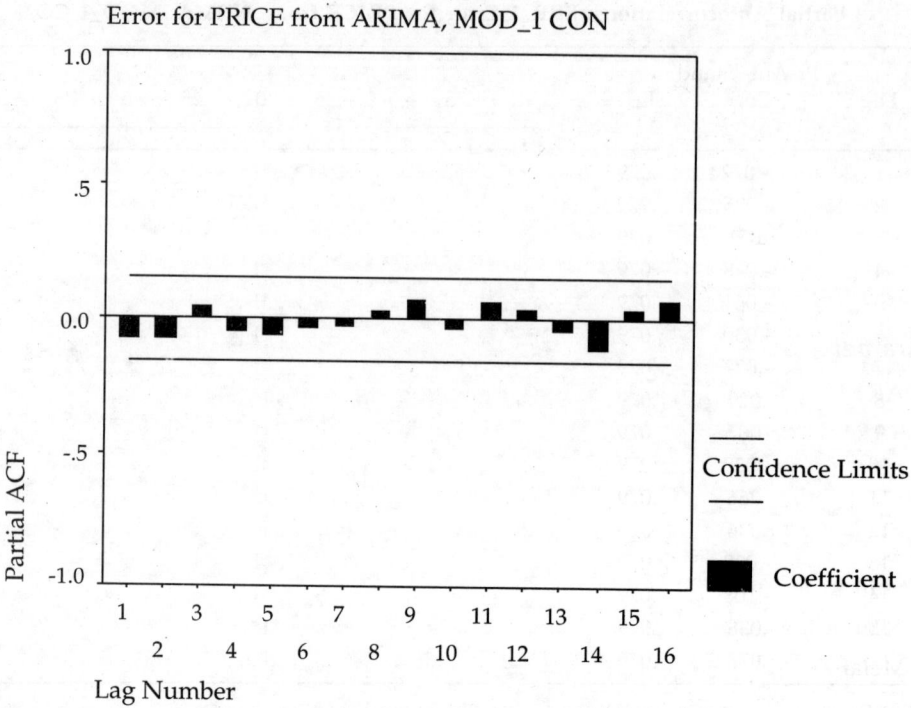

Error for PRICE from ARIMA, MOD_1 CON

**Graph**

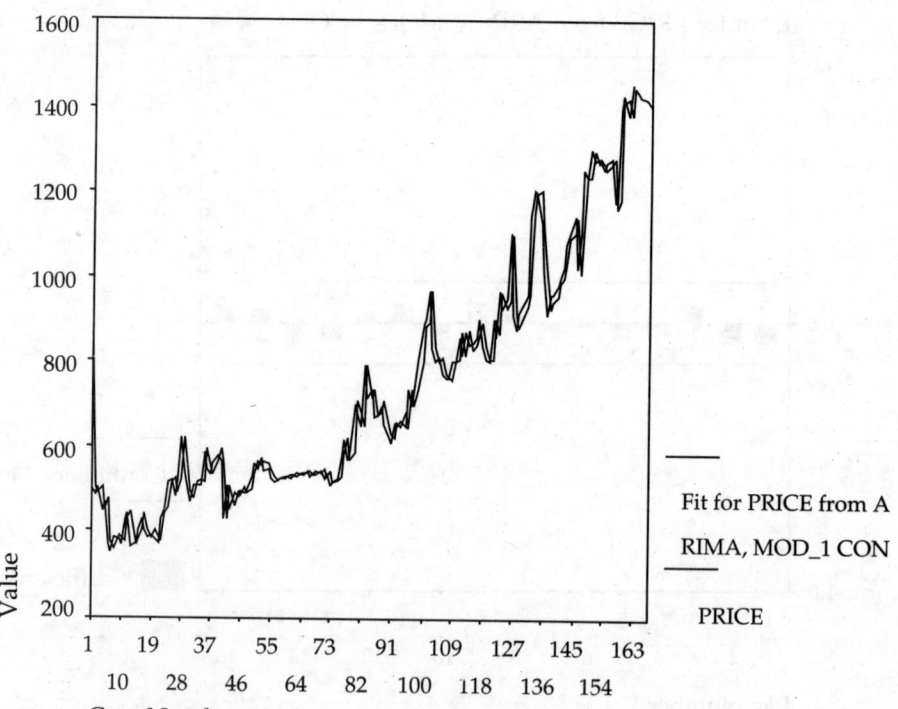

## Arima

MODEL: MOD_3
Model Description:
Variable: PRICE
Regressors: NONE
  Non-seasonal differencing: 1
     Seasonal differencing: 1
  Length of Seasonal Cycle: 12

## Parameters:

AR1 _____ < value originating from estimation >
SAR1 _____ < value originating from estimation >
CONSTANT _____ < value originating from estimation >

95.00 percent confidence intervals will be generated.

Split group number: 1 Series length: 162
Number of cases skipped at end because of missing values: 6
Melard's algorithm will be used for estimation.

## Termination criteria:

Parameter epsilon: .001
Maximum Marquardt constant: 1.00E+09
SSQ Percentage: .001
Maximum number of iterations: 10
Initial values:
AR1          −.11006
SAR1         −.45284
CONSTANT   1.80454
Marquardt constant = .001
Adjusted sum of squares = 643938.1

### Iteration History:

| Iteration | Adj. Sum of Squares | Marquardt Constant |
|-----------|---------------------|--------------------|
| 1 | 641112.06 | .00100000 |
| 2 | 641075.58 | .00010000 |

Conclusion of estimation phase.
Estimation terminated at iteration number 3 because:
Sum of squares decreased by less than .001 percent.

## FINAL PARAMETERS:

| | |
|---|---|
| Number of residuals | 149 |
| Standard error | 65.493707 |
| Log likelihood | −834.76037 |
| AIC | 1675.5207 |
| SBC | 1684.5326 |

### Analysis of Variance:

| | DF | Adj. Sum of Squares | Residual Variance |
|---|---|---|---|
| Residuals | 146 | 641075.06 | 4289.4256 |

### Variables in the Model:

| | B | SEB | T-RATIO | APPROX. PROB. |
|---|---|---|---|---|
| AR1 | −.0694456 | .0824290 | −.8424899 | .40089177 |
| SAR1 | −.5017237 | .0741052 | −6.7704279 | .00000000 |
| CONSTANT | 1.7978641 | 3.4361308 | .5232234 | .60161199 |

### Covariance Matrix:

| | AR1 | SAR1 |
|---|---|---|
| AR1 | .00679454 | −.00078842 |
| SAR1 | −.00078842 | .00549158 |

### Correlation Matrix:

| | AR1 | SAR1 |
|---|---|---|
| AR1 | 1.0000000 | −.1290715 |
| SAR1 | −.1290715 | 1.0000000 |

### Regressor Covariance Matrix:

| | CONSTANT |
|---|---|
| CONSTANT | 11.806995 |

### Regressor Correlation Matrix:

| | CONSTANT |
|---|---|
| CONSTANT | 1.0000000 |

The following new variables are being created:

| Name | Label |
|---|---|
| FIT_2 | Fit for PRICE from ARIMA, MOD_3 CON |
| ERR_2 | Error for PRICE from ARIMA, MOD_3 CON |
| LCL_2 | 95% LCL for PRICE from ARIMA, MOD_3 CON |
| UCL_2 | 95% UCL for PRICE from ARIMA, MOD_3 CON |
| SEP_2 | SE of fit for PRICE from ARIMA, MOD_3 CON |

# ACF

MODEL: MOD_4.
Variable: ERR_2          Missing cases: 19          Valid cases: 149
Autocorrelations:  ERR_2  Error for PRICE from ARIMA, MOD_3 CON

| Lag | Auto- Corr. | Stand. Err. | −1 −.75 −.5 −.25  0  .25  .5  .75  1 | Box-Ljung | Prob. |
|-----|-------------|-------------|-------------------------------------|-----------|-------|
| 1 | −.008 | .081 | . * . | .010 | .921 |
| 2 | −.072 | .081 | . *I . | .797 | .671 |
| 3 | −.050 | .081 | . *I . | 1.180 | .758 |
| 4 | −.055 | .080 | . *I . | 1.642 | .801 |
| 5 | −.202 | .080 | *.**I . | 8.030 | .155 |
| 6 | −.072 | .080 | . *I . | 8.853 | .182 |
| 7 | −.060 | .079 | . *I . | 9.417 | .224 |
| 8 | .038 | .079 | . I* . | 9.649 | .291 |
| 9 | .001 | .079 | . * . | 9.649 | .380 |
| 10 | .024 | .079 | . * . | 9.746 | .463 |
| 11 | .141 | .078 | . I*** | 12.967 | .296 |
| 12 | −.114 | .078 | .**I . | 15.085 | .237 |
| 13 | −.058 | .078 | . *I . | 15.647 | .269 |
| 14 | −.170 | .077 | ***I . | 20.447 | .117 |
| 15 | .042 | .077 | . I* . | 20.745 | .145 |
| 16 | .091 | .077 | . I**. | 22.159 | .138 |

**Plot Symbols:**     Autocorrelations *       Two Standard Error Limits .
**Total cases:**  168     Computable first lags:  148
artial Autocorrelations: ERR_2  Error for PRICE from ARIMA, MOD_3 CON

| Lag | Pr-Aut- Corr. | Stand. Err. | −1 −.75 −.5 −.25  0  .25  .5  .75  1 |
|-----|---------------|-------------|-------------------------------------|
| 1 | −.008 | .082 | . * . |
| 2 | −.072 | .082 | . *I . |
| 3 | −.051 | .082 | . *I . |
| 4 | −.061 | .082 | . *I . |
| 5 | −.214 | .082 | *.**I . |
| 6 | −.098 | .082 | .**I . |
| 7 | −.114 | 082 | .**I . |
| 8 | −.016 | .082 | . * . |
| 9 | −.058 | .082 | . *I . |
| 10 | −.048 | .082 | . *I . |
| 11 | .094 | .082 | . I**. |
| 12 | −.163 | .082 | ***I . |
| 13 | −.067 | .082 | . *I . |
| 14 | −.233 | .082 | **.**I . |
| 15 | −.008 | .082 | . * . |
| 16 | .068 | .082 | . I* . |

**Plot Symbols:**     Autocorrelations *       Two Standard Error Limits .
**Total cases:**  168     Computable first lags:  148

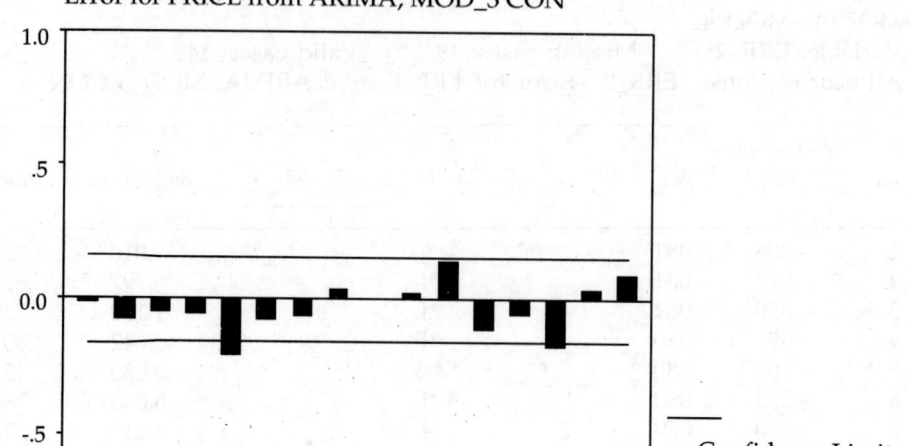

Error for PRICE from ARIMA, MOD_3 CON

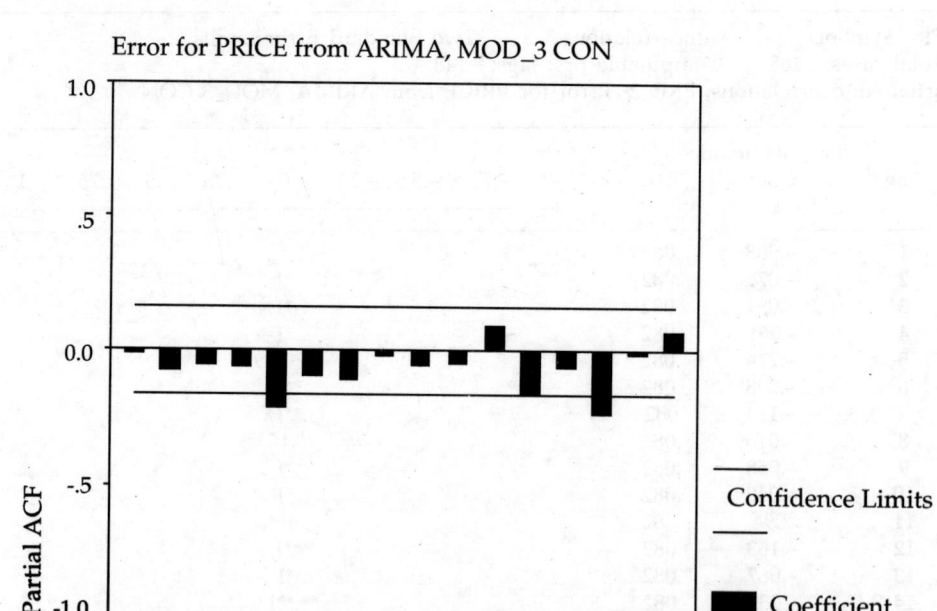

Error for PRICE from ARIMA, MOD_3 CON

# Graph

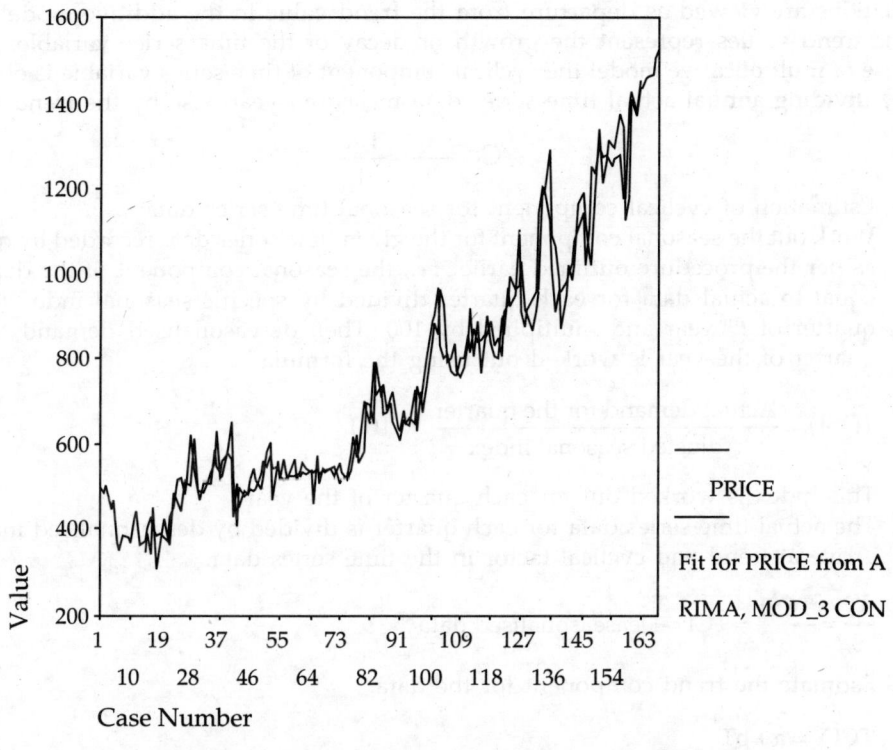

Case Number

| PDQ Values | AIC Values | SBC Values |
|---|---|---|
| 100 | 1775.12 | 1784.38 |
| 110 | 1675.52 | 1684.53 |

PDQ 100 - with white noise
PDQ 110- without white noise

### Forecasted Prices (Rs/Qtl)

| 2013 | Fit 1 | Fit 2 |
|---|---|---|
| Oct | 1440 | 1439 |
| Nov | 1433 | 1455 |
| Dec | 1425 | 1462 |
| 2014 Jan | 1416 | 1467 |

Fit 1 - fulfilled the white noise condition
Fit 2 - not fulfilled white noise condition

## Estimation of Cyclical Demand for Annual Time Series Data

When the time series data on the demand or prices of the commodities are given annually, such data include two major components *viz.*, trend and cyclical factor ($Y_t$

= C.T). **Here** we assume that random influences on the variable are negligible compared to cyclical fluctuations. When this assumption is maintained the cyclical fluctuations are viewed as departure from the trend value in the additive model. Here the trend values represent the growth or decay of the time series variable. In the case of multiplicative model the cyclical component of time series variable is obtained by dividing annual actual time series data measured year-wise by the trend value

$$C_t = \frac{Y_t}{T} = \frac{T.C}{T}$$

Estimation of cyclical component for seasonal time series data:

1. Work out the seasonal component for the given time series data recorded by quarter as per the procedure outlined earlier *i.e.*, the seasonal component of the data $S_t$ is equal to actual data for each quarter divided by specific seasonal index for the quarter of $t^{th}$ year and multiplied by 100. Then deseasonalised demand for the quarter of the year is worked out using the formula.

$$(DSI) = \frac{\text{Actual demand for the quarter}}{\text{Adjusted seasonal index}} \times 100$$

This index is worked out for each quarter of the year.

2. The actual time series data for each quarter is divided by deseasonalised index to arrive at trend and cyclical factor in the time series data.

$$\frac{Y_t}{S} = \frac{TCSI}{S} = TCI = \text{Deaseasonalised data}$$

3. **Estimate** the trend component for the data.

$$TCI \hat{Y} = a + bT$$

4. **Then** estimate the cyclical component as follows.

$$CI = \frac{TCI}{\hat{Y}} \quad \text{where, } \hat{Y} = \text{Trend factor}$$

Here 'I' refers to irregular fluctuation in the quarter of the $t^{th}$ year. These we assume negligible for given data and estimate cyclical component.

With this type of analysis we cannot get cycles having uniform length of time period. To get uniform length of time period, we use the technique of Fourier analysis.

## Fourier Analysis

Steps in Fourier analysis

1. In the first step we calculate the trend value for the time series data, say for example, rainfall data collected year wise.
2. Obtain stationary time series data on rainfall ($X_t$) through the following formula

$$X_t = Yt - \hat{Y}_t$$

Where, $Y_t$ = Actual data of rainfall

$\hat{Y}_t$ = Estimated trend of rainfall data

$X_t$ = Stationary variable on rainfall data which is devoid of trend component.

At first trend value is estimated based on the highest $R^2$ and significant level of the polynomial regression. The stationary time series data $(X_t)$ is conveniently grouped for investigating a given period 'P' (length of cycle period).

Where mp is equal to T (total number of years)

3. Grouping of the stationary data based on the 'P' value is shown below:

**Grouping the stationary data for Fourier analysis.**

| $X_1$ | $X_2$............................ | $X_P$ |
|---|---|---|
| $X_{p+1}$ | $X_{p+2}$...................... | $X_p$ |
| . | . | . |
| . | . | . |
| . | . | . |
| . | . | . |
| . | . | . |
| $X_{(m-1)p+1}$ | $X_{(m-1)p+2}$..................... | $X_{mp}$ |
| Sum: $U_1$ | $U_2$............................... | $U_P$ |

Here 'm' represents 'm' groups of time series variable based on 'P' *i.e.*, length of cycle assumed.

$U_1$ = Sum of first group.
$U_2$ = Sum of second group.

4. Fourier constants *viz.*, $A_p$ and $B_p$ are worked out using the following formula:

$$A_p = \frac{2\left(\sum_{j=1}^{P} U_j \cos\left(\frac{360j}{P}\right)\right)}{mp}$$

$$B_p = \frac{2\left(\sum_{j=1}^{P} U_j \sin\left(\frac{360j}{P}\right)\right)}{mp}$$

5. **The squared amplitude of the cycle** $\left(R_p^2\right)$ is obtained using the formula.

$$R_p^2 = A_p^2 + B_p^2$$

6. **We can test the significance of the cycles through 'K' values.**

$$K = \frac{R_p^2}{R_1^2}$$

$K > P_s$, then the assumed cycle is said to be significant

If $K < P_s$, then the assumed cycle is non-significant

Here $P_s = \exp^{-K}$

7. $\quad R_P^2 = \dfrac{4\sigma^2}{T}$

$\sigma^2$ = sum of squares of error term computed for T number of observations. For further details refer to Tintner (1952)[*].

Here 'P' is the length of the cycle and it is assumed from the values 2, 3, ...... 20 for 40 years data.

If it is two years cycle there will be 20 variables (m), then we find out sums for these 20 variables ($U_1$, $U_2$ ......... $U_{20}$).

Subba Reddy (1988)[**] identified discernible cycles for the time series variables of Anantapur district in Andhra Pradesh. The selected variables are total area, production, productivity and wholesale prices of major agricultural commodities and monthly rainfall in mm. The data relate to the period from 1956 to 1983. Trend values were estimated for the selected variables using second-degree polynomial functions.

Cycle ranging in the length of time period from 3 years to 12 years was hypothesized and for each period of the cycle Fourier coefficients are worked out viz., $A_p$, $B_p$, $R_P^2$, $R_T^2$ and K and presented in tables. The hypothesized cycles are tested for their

significance using the value of $K\left(\hat{K} = \dfrac{R_P^2}{R_T^2}\right)$. All the values of Fourier coefficients of

hypothesized cycles, though worked out they are not presented in Tables for want of space. But instead, the highest K value and its related Fourier coefficients for hypothesized cycles are presented in the tables. The significance of these cycles are judged based on the calculated K value ($\hat{K}$). If $\hat{K}$ value is greater than table K value (ps) i.e., the cycle is said to be significant (Table 47.6, 47.7 and 47.8).

**TABLE 47.6** Fourier Coefficients, Amplitude Square, Mean Square Amplitude and K Values for Production Cycles of Major Crops in Anantapur District in Andhra Pradesh.

(Production unit = '000 tonnes)

| Sl. No. | Crop | Fourier $A_p$ | Coefficients $B_p$ | Amplitude square $R_p{}^2$ | Mean square amplitude $R_t{}^2$ | K value | Time period of significant cycle |
|---|---|---|---|---|---|---|---|
| 1. | Horsegram | 1.461 | -3.142 | 12.005 | 2.369 | 5.068 | 10 year cycle |
| 2. | Rice | 16.005 | 19.34 | 630.20 | 135.006 | 4.668 | " |
| 3. | Ragi | -2.326 | -5.691 | 37.799 | 10.436 | 3.662 | " |
| 4. | Sugarcane | -2.536 | 5.926 | 41.554 | 12.215 | 3.401 | 4 year cycle |

**N.B:** For crops namely sorghum, pearl millet, cotton, red gram and groundnut all the hypothesized production cycles were non-significant, hence the particulars are not presented.

[*] Tintner, G.—Econometric Theory. Ames. IOWA: IOWA University Press, 1952.
[**] Subba Reddy S. 1988. Crop Planning under Risk for Typical Dry Land Farms in Anantapur district A.P. Ph.D. thesis, TNAU, Coimbatore.

**TABLE 47.7** Fourier Coefficients, Amplitude Square, Mean Square Amplitude and K Values for Productivity and Total Cycles of Major Crops in Anantapur District in Andhra Pradesh.

(Productivity in Qtl/ha and area in '000 ha)

| Sl. No. | Crop | Fourier Coefficients $A_p$ | $B_p$ | Amplitude square $R_p^2$ | Mean square amplitude $R_t^2$ | K value | Time period of significance cycle |
|---------|------|------|------|------|------|------|------|
| | | | **Productivity** | | | | |
| 1. | Horsegram | 0.212 | -0.348 | 0.166 | 0.04 | 4.18 | 10 year cycle |
| 2. | Sugarcane | -2.42 | 9.799 | 101.872 | 27.749 | 3.67 | 5 year cycle |
| | | **Area** | | | | | |
| 1. | Sorghum | -23.34 | 3.736 | 558.912 | 165.249 | 3.382 | 11 year cycle |
| 2. | Rice | 11.872 | -10.614 | 253.61 | 39.589 | 6.406 | 10 year cycle |

**N.B:** For rest of the crops the cycles were non-significant and hence they are not presented.

**TABLE 47.8** Fourier Coefficients, Amplitude Square, Mean Square Amplitude and K Values for Cycles of Rainfall (mm) in Anantapur District in Andhra Pradesh.

| Sl. No. | Crop | Fourier $A_p$ | Coefficients $B_p$ | Amplitude square $R_p^2$ | Mean square amplitude $R_t^2$ | K value | Time period of significance cycle |
|---------|------|------|------|------|------|------|------|
| 1. | July rainfall | 46.459 | 9.551 | 2248.884 | 310.864 | 7.234 | 10 year cycle |
| 2. | August rainfall | -35.14 | 8.658 | 1309.771 | 276.652 | 4.734 | 6 year cycle |
| 3. | Sept. rainfall | 3.845 | -43.693 | 1923.864 | 631.826 | 3.045 | 5 year cycle |
| 4. | Oct. rainfall | 11.444 | 37.989 | 1574.12 | 462.02 | 3.407 | 6 year cycle |
| 5. | South–west monsoon (June – Oct) | 57.327 | 20.607 | 3710.975 | 1017.862 | 3.646 | 10 year cycle |
| 6. | May rainfall | -18.694 | 14.524 | 560.429 | 150.398 | 3.726 | 10 year cycle |
| 7. | North–east monsoon (Nov - Feb) | 23.53 | 37.477 | 1958.199 | 477.557 | 4.1 | 6 year cycle |
| 8. | Summer rainfall (Mar-May) | -18.107 | 13.557 | 511.613 | 170.268 | 3.005 | 10 year cycle |
| | Total annual rainfall | 73.982 | 24.941 | 6093.445 | 1369.366 | 4.45 | 10 year cycle |

**Note:** The hypothesized price cycles for major agricultural commodities were non-significant, hence they are not presented.

Estimated quadratic quations are presented in Table 47.9.

---

Nerlove Marc "**Distributed lags and Estimation** of long run Supply and Demand Elasticities. Theoretical Considerations". **Journal of Farm Economics** Vol. 40, 1958. pp. 302-311.

**TABLE 47.9  Estimated Quadratic Equations for the Time Series Data (1956-1983) in Anantapur District in Andhra Pradesh.**

| Variable used | Crop | Estimated quadratic equation |
|---|---|---|
| Production ('000 tonnes) | Horsegram | $\hat{Y} = 15.2692 - 0.6057t + 0.0166t^2$ |
| | Rice | $\hat{Y} = 64.9511 + 5.0514t - 0.1064t^2$ |
| | Ragi | $\hat{Y} = 37.1489 + 0.2536t - 0.0004t^2$ |
| | Sugarcane | $\hat{Y} = 16.1275 + 2.8123t - 0.0949t^2$ |
| Productivity (Qtl/ha) | Horsegram | $\hat{Y} = 1.9998 - 0.0799t + 0.0033t^2$ |
| | Sugarcane | $\hat{Y} = 77.6960 - 0.8254t + 0.0353t^2$ |
| Area ('000 ha) | Sorghum | $\hat{Y} = 167.7541 + 0.3779t - 0.1162t^2$ |
| | Rice | $Y = 54.8965 + 2.0857t - 0.0590t^2$ |
| Rainfall (in mm): | | |
| July rainfall | | $\hat{Y} = 73.6411 - 0.227t + 0.0329t^2$ |
| August rainfall | | $\hat{Y} = 53.9909 + 3.8032t - 0.133t^2$ |
| September rainfall | | $\hat{Y} = 143.8613 - 3.6637t + .1854t^2$ |
| October rainfall | | $\hat{Y} = 78.6984 + 6.7943t - 0.2711t^2$ |
| South-west monsoon (June-October) | | $\hat{Y} = 426.1993 - 17.1988t - 0.5324t^2$ |
| May rainfall | | $\hat{Y} = 37.0546 + 3.3786t - 0.114t^2$ |
| North-east monsoon (November-February) | | $\hat{Y} = 138.8809 + 3.2159t - 0.1302t^2$ |
| Summer rainfall (March-May) | | $\hat{Y} = 71.882 - 0.2019t + 0.0464t^2$ |
| Total rainfall | | $\hat{Y} = 636.0210 - 13.8402t + 0.4349t^2$ |

## Uses of Fourier Analysis

Farmers should device suitable crop plans based on the length of the cycles for rainfall variable in the individual months. As area, production and productivity of major crops exhibited 10 year cycles, 10 years time period of the cycles, should be considered in forming relevant farm plans. If the production cycles of crops reveal 10 year length of time period, then it is relevant to use 10 years time series data in the risk programming model for obtaining reliable and accurate farm plans under different types of risk. The time period of the cycles for these variables should be considered for formulating relevant programmes and minimizing the risk.

## PARTIAL ADJUSTMENT MODEL

Marc Nerlove* model which is popularly known as partial adjustment model is another form of adoptive expectation model developed by Koyck. This model is popularly used by Agricultural Economists to estimate effect of prices on the acreage of crops. In this model it is assumed that farmers would react to the expected price and also to the last year's price and then decide the acreage under the crops. Hence this model is also known as price expectation model or acreage adjustment model. This model takes the following form:

$$A_t = a + b\overset{*}{P_t} + U_t \tag{6.14}$$

$$\left(\overset{*}{P_t} - \overset{*}{P_{t-1}}\right) = \beta\left(P_{t-1} - \overset{*}{P_{t-1}}\right) \tag{6.15}$$

$O < \beta < 1$

**Where,**

$A_t$ = Actual acreage in year, t

$\overset{*}{P_t}$ = Expected price of the crop in year, t

$\overset{*}{P_{t-1}}$ = Expected price of the crop in year the preceding year *i.e.*, t - 1

$P_{t-1}$ = Actual price of the crop in year, t - 1

$U_t$ = Error term

a = Intercept

b = Parameter to be estimated

ß = Coefficient of expectation

Here in the Nerlovian model, price variability is considered as explanatory variable to explain variation in the acreage of the crops. Later, other economists included variables like rainfall, relative crop yields and total irrigated area to explain the price variability. Such types of models are called price expectation models.

In the acreage adjustment model (supply response model), the desired acreage under the crop $A_t$ depends on the last year's price and the coefficient of adjustment in the crop acreage. This coefficient of adjustment depends on difference between the desired acreage of the crop during the current year ($\overset{*}{A_t}$) and the last year's actual crop acreage *i.e.*, $A_{t-1}$ This model is specified as follows:

$$\overset{*}{A_t} = a + bP_{t-1} + U_t \tag{6.16}$$

$$A_t - A_{t-1} = \lambda\left(\overset{*}{A_t} - A_{t-1}\right). \tag{6.17}$$

$\overset{*}{A_t}$ = Desired acreage in the current year. This is not seen in the time series data.

$\overset{*}{A_t} - A_{t-1}$ = Desired change in the acreage

$A_t - A_{t-1}$ = Actual change in the acreage

$\lambda$ = Coefficient of adjustment in the acreage

Actual change in the acreage at any given period is some function of $\lambda$ of desired change in the acreage. If $\lambda = 1$, it means that actual acreage is equal to desired change in acreage. If $\lambda = O$, this means that last year's acreage and current year acreage are the same. But in reality, the crop acreage adjustment lies between O and 1, so $O < \lambda < 1$. Hence this model is called partial acreage adjustment model.

$$A_t = \lambda\overset{*}{A_t} - \lambda A_{t-1} + A_{t-1} \tag{6.18}$$

$$= \lambda \overset{*}{A_t} + (1 - \lambda) A_{t-1} \tag{6.19}$$

Substitute $\overset{*}{A_t}$ value of (equation 6.16) in equation (6.19)

$$= \lambda(a + bP_{t-1} + U_t) + (1 - \lambda)A_{t-1}$$

$$A_t = a\lambda + bP_{t-1} + (1 - )A_{t-1} + U_t \tag{6.20}$$

Equation (6.16) is long run acreage equation, whereas equation (6.20) is called short run acreage equation.

Let us assume $\hat{\alpha} = \lambda a$, $\hat{\beta} = \lambda b$, $\hat{r} = 1 - \lambda$, $W_t = \lambda U_t$.

$$A_t = \hat{\alpha} + \hat{\beta} P_{t-1} + r A_{t-1} + W_t. \tag{6.21}$$

Now estimate the equation (6.21) using time series data on At, $P_{t-1}$ and $A_{t-1}$ of the crop under study.

$\hat{r} = 1 - \lambda$

$\lambda = 1 - \hat{r}$ = short run adjustment coefficients.

$\beta = \lambda b$, then $b = \dfrac{\beta}{\lambda}$

Here 'b' represents long run adjustment coefficient. Numerical example is presented in Table 47.10.

**TABLE 47.10 Acreage Adjustment of Groundnut–Ananthapur District, AP.**

| Year | Area (in ha) | Price (Rs./qtl) | Index of area | Index of price | $\hat{A}_t$ |
|------|------|------|------|------|------|
| 1971-72 | 2,64,881 | 129 | 79.68 | 51.64 | 83.57 |
| 1972-73 | 2,25,026 | 234.24 | 67.69 | 93.76 | 78.37 |
| 1973-74 | 2,55,129 | 173.51 | 76.75 | 69.45 | 97.78 |
| 1974-75 | 3,00,275 | 255.56 | 90.33 | 102.30 | 86.58 |
| 1975-76 | 3,24,406 | 146.35 | 97.59 | 58.58 | 101.72 |
| 1976-77 | 2,73,111 | 294.59 | 82.16 | 117.92 | 81.57 |
| 1977-78 | 2,94,306 | 244.18 | 88.54 | 97.74 | 108.92 |
| 1978-79 | 3,38,048 | 210.78 | 101.69 | 94.37 | 99.26 |
| 1979-80 | 3,64,889 | 294.50 | 109.77 | 117.88 | 98.07 |
| 1980-81 | 3,32,414 | 249.82 | 100.00 | 100.00 | 108.90 |
| 1981-82 | 3,91,173 | 417.17 | 117.68 | 166.99 | 100.66 |
| 1982-83 | 3,97,106 | 404.12 | 119.46 | 161.76 | 131.54 |
| 1983-84 | 4,51,232 | 437.57 | 135.74 | 175.15 | 129.13 |
| 1984-85 | 4,73,361 | 485.62 | 142.40 | 194.39 | 135.30 |
| 1985-86 | 4,86,529 | 462.31 | 146.36 | 185.06 | 144.16 |
| 1986-87 | 4,41,720 | 608.41 | 132.88 | 243.54 | 139.86 |
| 1987-88 | 5,45,322 | 618.97 | 164.05 | 247.77 | 166.82 |
| 1988-89 | 4,91,190 | 563.23 | 147.76 | 225.45 | 168.77 |
| 1989-90 | 7,30,102 | 759.42 | 219.64 | 303.99 | 158.48 |
| 1990-91 | 7,41,459 | 964.46 | 223.05 | 386.06 | 194.68 |
| 1991-92 | 7,20,859 | 1052.0 | 216.86 | 421.10 | 232.51 |
| 1992-93 | 7,42,749 | 831.06 | 223.44 | 332.66 | 248.65 |
| 1993-94 | 7,31,105 | 977.38 | 219.94 | 391.23 | 207.89 |
| Mean |  |  | 134.93 | 174.37 |  |

| Sl. No | $A_t$ Index of area | $P_{t-1}$ | $\hat{A}_{t-1}$ |
|--------|---------------------|-----------|------------------|
| (1) | (2) | (3) | (4) |
| 1 | 79.68 | 62.96 | 72.32 |
| 2 | 67.69 | 51.64 | 83.57 |
| 3 | 76.75 | 93.76 | 78.37 |
| 4 | 90.33 | 69.45 | 97.78 |
| 5 | 97.59 | 102.3 | 86.58 |
| 6 | 82.16 | 58.58 | 101.72 |
| 7 | 88.54 | 117.92 | 81.57 |
| 8 | 101.69 | 97.74 | 108.92 |
| 9 | 109.77 | 94.37 | 99.26 |
| 10 | 100 | 117.88 | 98.07 |
| 11 | 117.68 | 100 | 108.9 |
| 12 | 119.46 | 166.99 | 100.66 |
| 13 | 135.74 | 161.76 | 131.54 |
| 14 | 143.40 | 175.15 | 129.13 |
| 15 | 146.36 | 194.39 | 135.30 |
| 16 | 132.88 | 185.06 | 144.16 |
| 17 | 164.05 | 243.54 | 139.86 |
| 18 | 147.76 | 247.77 | 166.82 |
| 19 | 219.64 | 225.45 | 168.77 |
| 20 | 223.05 | 303.99 | 158.48 |
| 21 | 216.86 | 386.06 | 194.68 |
| 22 | 223.44 | 421.10 | 232.51 |
| 23 | 219.94 | 332.66 | 248.65 |
| Mean | 134.93 | 174.37 | 127.68 |

Zero order Correlation matrix

| | | $P_{t-1}$ | $A_{t-1}$ | $A_t$ |
|---|---|-----------|-----------|-------|
| | $P_{t-1}$ | 1.0000 | 0.9114 | 0.9332 |
| | $A_{t-1}$ | 0.9114 | 1.0000 | 0.9034 |
| | $A_{t-}$ | 0.9332 | 0.9034 | 1.0000 |

$$A_t = 37.2058 + 0.3229\, P_{t-1} + 0.3244\, \hat{A}_{t-1} \qquad (6.22)$$

| SE = | 12.81 | 0.09 | 0.19 |
|------|-------|------|------|
| t = | 2.90 | 3.58 | 1.68 |

$$R^2 = 0.89$$

$$\text{Error } \sigma = 343.83$$

$$R^{-2} = 0.88$$

$$F = 78.26$$

Elasticity coefficient of $P_{t-1} = \dfrac{134.93}{174.3} \times 0.3229$

$$= 0.2499$$

It is the long run price elasticity with reference to price. One percent increase in lagged price would result in an increase of 0.25 per cent increase in the acreage in the long run.

Elasticity coefficient of $\quad \hat{A}_{t-1} = \dfrac{134.93}{127.68} \times 0.3244$

$$= 0.3428$$

Short run acreage adjustment $\quad = 1 - 0.3428$

$$= 0.6572$$

About 66 per cent of the previous year's groundnut acreage is planted in the current year.

Short run price elasticity = Long run price elasticity × coefficient of adjustment in the short run

$$= 0.2499 \times 0.6572$$

$$= 0.1642$$

One percent increase in the lagged price would result in 0.1642 per cent increase in area in the short run.

## DYNAMIC MODELS OF MARKETING

Let us write the following regression equation as:

$$Q_t = a + bx_t + CX_{t-1} + dX_{t-2} + U_1 \tag{6.23}$$

$$Q_t = a + \beta x_t + \gamma Q_{t-1} + e_1 \tag{6.24}$$

In equation (6.23) $Q_t$ is current dependent variable, $X_t$ is the current explanatory variable whereas $X_{t-1}$ and $X_{t-2}$ are the lagged explanatory variables. $U_1$ is error term.

The parameters c and d are the coefficients of lagged explanatory variables. This model is called distributed lag model. In equation (6.24) we have lagged dependent variable $(Q_{t-1})$ as explanatory variable.

Such model is called dynamic model because this model gives the time path of the dependent variable in relation to its past values. This model is also known as auto-regressive model. These models have applications in economics to measure consumption activity, in the estimation of demand and supply for agricultural commodities, in estimation of demand for money, *etc.* Let us consider the following consumption function for the country.

$$\hat{C} = b_1 + b_2 I_t^* + u_t \tag{6.25}$$

In the equation (6.25) we assume consumption, C is linearly related to permanent income in cash in $t^{\text{th}}$ period. Estimation of this equation is not possible, because $I_t$ is not directly quantifiable. Hence, we propose the following adoptive expectation hypothesis as

$$I_t^* - I_{t-1}^* = d(I_t - I_{t-1}^*) \tag{6.26}$$

Here the value of 'd' lies between 0 and 1 and it is called coefficient of expectation. The hypothesis of equation (6.26) is known as adoptive expectation and this implies that consumers would adopt their expectations in the light of past experience. This means that their expectations about the permanent income are received by a fraction of 'd' value leaving from their mistakes in the past. This expectation is provided in the following equation.

$$\overset{\bullet}{I_t} = d\,\overset{\bullet}{I_t} + (1-d)\,\overset{\bullet}{I}_{t-1} \tag{6.27}$$

This shows the expected value of permanent income of the time, 't' is the actual value of permanent income at time 't' plus its expected value in the previous period.

If $d = 1$, then $\overset{\bullet}{I_t} = I_t$ $\tag{6.28}$

This equation (6.28) implies that the expectations about permanent income are fully realized and immediately in the same period.

If $d = 0$, $\overset{\bullet}{I_t} = \overset{\bullet}{I}_{t-1}$ $\tag{6.29}$

This means conditions prevailing today hold good in the future period. Substituting equation (6.27) in equation (6.25) we get

$$\hat{C} = b_1 + b_2(d\overset{\bullet}{I_t} + (1-d)\overset{\bullet}{I}_{t-1}) + U_t = b_1 + b_2(1-d)\overset{\bullet}{I}_{t-1} + U_t \tag{6.30}$$

Now take the lag of equation (6.25) by one period, and multiply the same by 1-d. Subtract this product from equation (6.30), then we get following equation

$$\hat{C}_t = d\,b_1 + d\,b_2\,I_t + (1-d)\,C_{t-1} + U_t - (1-d)\,U_{t-1} \tag{6.31}$$

This is equivalent to

$$\hat{C}_t = d\,b_1 + d\,b_2\,I_t + (1-d)\,C_{t-1} + V_t \tag{6.32}$$

Where, $V_t = U_t - (1-d)\,U_{t-1}$

In equation (6.25), $b_2$ measures average response of $C_t$ to a unit change in $\overset{\bullet}{I_t}$. This is called long run elasticity. In equation (6.32), $b_2$ measures the average response of $C_t$ to a unit change in the actual or observed value of $I_t$.

If $d = 1$, then the short run elasticity and long run elasticity of $I_t$ are the same.

We estimate the equation (6.32) for the given data and estimate the coefficient of $C_{t-1}$, which is separated as 'X' and it is equal to (1-d).

$X = 1-d$
$-d = X - 1$
$d = -X + 1$
$d = 1 - X$

The coefficient of $I_t$ in equation (6.32) is equal to $d\,b_2$.

The coefficient of $I_t = \dfrac{db_2}{d} = \dfrac{(1-X)b_2}{(1-X)} = b_2$

## Numerical example

Let the estimated aggregate consumption function be given as

$$l_n\hat{C} = l_n\,1.263 + 0.3152\,l_n\,I_t + 0.6123\,l_n\,C_{t-1}$$

Here marginal propensity to consume (MPC) =0.3152 = 0.32. This means that one per cent increase in income in the current disposable income would increase consumption on the average by 0.32 percent, here other factors influencing the consumption are held constant. This is called short run elasticity of income for consumption.

$$\text{Long run elasticity} = \frac{0.3152}{1 - 0.6123} = 0.81$$

If the income is sustained in the long run, its long run elasticity would be 0.81%. This means if there is one percent change in permanent income, consumers would increase their consumption by 0.81 percent.

## Supply Response

In this method, we replace $A_{t-1}$ by $\hat{A}_{t-1}$ to remove the auto-correlation problem. Following are the steps involved.

1) First we regress $A_t$ on $P_{t-1}$ and obtain $\hat{A}_t$ by substituting observed value of $P_{t-1}$ in the estimated equation

$$A_t = a + b\, P_{t-1} + e_t \tag{6.33}$$

2) We lag $\hat{A}_t$ by one period to obtain $\hat{A}_{t-1}$ and this is included in the supply response model in place of $A_{t-1}$.

$$A_t = a + b\, P_{t-1} + C\, \hat{A}_{t-1} + e_t \tag{6.34}$$

Application of OLS to equation (6.34) yields biased estimates *i.e.*, estimates would have large variances especially for small samples. Hence it is desirable to use large sample size (N > 40) to get unbiased and consistent estimates from the equation (6.34). In addition to $P_{t-1}$ and $\hat{A}_{t-1}$ we can include all relevant variables which influence acreage under crop. These may be rainfall index in the current year and lagged output of the crop *i.e.*, $Y_{t-1}$ can be incorporated.

$$Y_{t-1} = \frac{\text{Lagged output of the crop}}{\text{Index of yields of alternative crops}} .$$

$P_{t-1}$ = Price of the crop in question deflated by index of post harvest prices of alternative crops.

$$A_t = a + b\, P_{t-1} + c\, R_t + d\, Y_{t-1} + c\, A_{t-1} + e_t \tag{6.35}$$

We should estimate equation (6.35) using logarithms to get elasticity coefficients directly.

Short run adjustment coefficient = $\delta = 1 - e$

We can estimate short run and long run elasticity coefficients for $P_{t-1}$, $Y_{t-1}$ and $R_t$ variable.

| Independent variable | Short run elasticity | Long run elasticity |
|---|---|---|
| $P_{t-1}$ | $b \times \delta$ | $B = b/\delta$ |
| $R_t$ | $c \times \delta$ | $C = c/\delta$ |
| $Y_{t-1}$ | $d \times \delta$ | $D = d/\delta$ |

For further details see Rajkrishna (1963)[*].

# ESTIMATION OF DEMAND AND SUPPLY THROUGH SIMULTANEOUS EQUATIONS MODEL

In a single equation model (Y = f(X), we predict the value of Y, given the fixed values of X variable. Here cause and effect relationship is unidirectional *i.e.*, here 'X' variable

---

[*] Rajkrishna "Farm supply response–A Punjab case study" Economic Journal, 1963.

determines the value of Y. This unidirectional cause and effect relationship is ruled out in most of the real world situations. For example, the price of a commodity at a point of time and place represents the intersection of aggregate demand and aggregate supply for that commodity. This means relationship between Y variable and X variables is not unidirectional but it is two-way directional or there is simultaneous relationship between Y and X variables (independent variables). In such a relationship Y is determined by some of X variables and some of X variables are in turn determined by Y. Under such situation it is better to lump together a set of variables, which are determined simultaneously by the remaining set of variables. This type of relationship is specified in the simultaneous equation model. We cannot estimate these simultaneous equation models by the method of ordinary least squares (OLS) technique. If we apply OLS method to estimate coefficients of each single equation model separately, such estimates are biased and inconsistent. In the demand and supply model, we assumed the linear demand and supply functions and they are given as:

$$Q_t^d = a_0 + a_1 P_t + U_{1t} \quad ..... \quad (a_1 < 0) \tag{6.36}$$

$$Q_t^s = b_0 + b_1 P_t + U_{2t} \quad ..... \quad (b_1 > 0) \tag{6.37}$$

$$Q_t^d = Q_t^s \tag{6.38}$$

Equation (6.36) is demand function and equation (6.37) is supply function and equation (6.38) denotes equilibrium condition of market.

$Q_t^d$ = Quantity of commodity demanded in the market in period 't'

$Q_t^s$ = Quantity of commodity supplied in the market in period 't'

$a_0$ and $a_1$ = Demand parameters
$\quad a_0$ = Constant term
$\quad a_1$ = Demand coefficient, the expected sign is negative because of law of demand.

$b_0$ and $b_1$ = Supply parameters i.e.,
$\quad b_0$ = Constant term
$\quad b_1$ = Supply coefficient, the expected sign is positive because of law of supply.

Here $a_0$, $a_1$, $b_0$ and $b_1$ are the structural coefficients of demand and supply equations. If we estimate the demand and supply equations separately we will not get the expected coefficients and even if we get expected signs, estimates are inconsistent[*] and biased[**].

To overcome such problems, an indirect least square (ILS) method is employed. In this method we will have to first assume, the exactly identified demand and supply functions. These are shown as follows.

## Demand Function

$$Q_t^d = a_0 + a_1 P_t + a_2 I_2 + U_{1t} (a_1 < 0, a_2 > 0) \tag{6.39}$$

---

[*] Inconsistent estimate means that it would not give a realistic picture about the true picture of parameter of population, even though sample size is increased.

[**] Baised estimate is one which is having large variance i.e., large standard error, finally becoming non-significant.

## Supply Function

$$Q_t^s = b_0 + b_1 P_t + b_2 P_{t-1} + U_{2t} \quad (b_1 > 0, b_2 > 0). \tag{6.40}$$

Demand equation is correctly identified with explanatory variables (*i.e.*, prices of the commodity in question ($P_t$) and income of the consumers ($I_t$). The supply function, tells us that quantity of commodity supplied depends on its current price ($P_t$) as well as its previous period's price ($P_{t-1}$).

Here the demand and supply functions satisfy the rules under identification conditions[***].

We have pre-determined additional variable *i.e.*, income variable in the demand function and lagged price variable in the supply function. Because of these two variables, we are now able to identify the demand and supply functions correctly. To estimate the parameters, we have to develop reduced form equations for these functions using the market clearing conditions. The equilibrium price equation is derived equating the parameters of demand and supply functions.

$$a_0 + a_1 P_t + a_2 I_t + U_{1t} = b_0 + b_1 P_t + b_2 P_{t-1} + U_2 t \tag{6.41}$$

$$a_1 P_t - b_1 P_t = b_0 - a_0 - a_2 I_t + b_2 P_{t-1} + U_{2t} - U_{1t} \tag{6.42}$$

$$P_t (a_1 - b_1) = b_0 - a_0 - a_2 I_t + b_2 P_{t-1} + U_{2t} - U_{1t} \tag{6.43}$$

$$P_t = \frac{b_0 - a_0}{a_1 - b_1} - \frac{a_2 I_t}{(a_1 - b_1)} + \frac{b_2 P_{t-1}}{(a_1 - b_1)} + \frac{U_{2t} - U_{1t}}{(a_1 - b_1)} \tag{6.44}$$

Now $P_t = C_0 + C_1 I_t + C_2 P_{t-1} + V_t$ \hfill (6.45)

$$\text{Where, } C_0 = \frac{b_0 - a_0}{a_1 - b_1}, \quad C_1 = \frac{-a_2}{(a_1 - b_1)}, \quad C_2 = \frac{b_2}{(a_1 - b_1)}, \quad V_t = \frac{U_{2t} - U_{1t}}{(a_1 - b_1)} \tag{6.46}$$

Equation (6.45) is reduced form equation.

Substituting equation (6.45) in equation (6.40), we get reduced form equation for equilibrium quantity.

$$Q_t = C_3 + C_4 I_t + C_5 P_{t-1} + W_t \tag{6.46}$$

---

[***] Identification: We have order and rank condition for determining the identification of an equation in the system of simultaneous equation. The order condition is specified as $K - k \geq m - 1$

If $K - k = m-1$, the equation is just identified, if $K - k > m-1$ the equation is said to be over identified. The notation 'm' refers to number of endogenous or dependent variables in the economic model. 'M' refers to the number of endogenous or dependent variables in the given equation, K refers to number of pre-determined or independent or exogenous variables in the model, k refers to number of predetermined or exogenous variables in the equation. In our market clearing model.

$M = 1$ similarly $k = 2$, $K = 2$

Hence $K-K = m-1$. The model is correctly identified.

This is only necessary condition.

The rank condition of identification is sufficient condition for identifying the 'M' equation with 'M' endogenous variables in the model. Here an equation in the model is identified if and only if at least one positive (non-zero) determinant of the order (M-1). (M-1) can be constructed from the coefficient of endogenous and exogenous variables excluded from that particular equation and included in the other equation of the model. For further details see Damodar N.R. Basic Econometrics, McGraw Hill Book Company, New York.

Where,

$$C_3 = \frac{a_1 b_0 - a_0 b_1}{a_1 - b_1}$$

$$C_4 = \frac{-a_2 b_1}{a_1 - b_1}$$

$$C_5 = \frac{a_1 b_2}{a_1 \cdot b_1}$$

$$W_t = \frac{a_1 U_{2t} - b_1 U_{1t}}{a_1 - b_1}$$

This is proved as follows:
Supply equation is

$$Q_t^s = b_0 + b_1 P_t + b_2 P_{t-1} + U_{2t} \tag{6.40}$$

Substitute value of $P_t$ in equation (6.45),

$$Q_t^s = b_0 + b_1 (C_0 + C_1 I_t + C_2 P_{t-1} + V_t) + b_2 P_{t-1} + U_{2t}$$

$$= b_0 + b_1 C_0 + b_1 C_1 I_t + b_1 C_2 P_{t-1} + b_1 V_t + b_2 P_{t-1} + U_{2t}$$

Now substitute the values $C_0$, $C_1$, $C_2$ and $V_t$ in the above equation

$$= b_0 + b_1 (\frac{b_0 - a_0}{a_1 - b_1}) + b_1 (\frac{-a_2}{a_1 - b_1}) I_t + b_1 (\frac{b_2}{a_1 - b_1}) P_{t-1} + b_1 (\frac{U_{2t} - U_{1t}}{a_1 - b_1}) + b_2 (P_{t-1}) + U_{2t}$$

$$= \frac{b_0 (a_1 - b_1) + (b_0 - a_0)}{(a_1 - b_1)} - \frac{b_1 \cdot a_2}{(a_1 - b_1)} I_t + \frac{b_1 \cdot b_2}{(a_1 - b_1)} P_{t-1} + b_2 (P_{t-1}) + \frac{b_1 (U_{2t} - U_{1t})}{(a_1 - b_1)} + U_{2t}$$

$$= \frac{a_1 b_0 - b_0 b_1 + b_0 b_1 - a_0 b_1}{(a_1 - b_1)} - \frac{b_1 \cdot a_2}{(a_1 - b_1)} + \frac{b_1 b_2 + b_2 a_1 - b_1 b_2}{(a_1 - b_1)} P_{t-1}$$

$$+ \frac{b_1 U_{2t} - b_1 U_{1t} + U_{2t} a_1 - U_{2t}}{a_1 - b_1}$$

$$= C_3 + C_4 I_t + C_5 P_{t-1} + W_t$$

Hence equation (6.46) is proved. Equation (6.45) and (6.46) are the reduced form equations and from these equations, we have to estimate structural coefficients of demand and supply equations.

$$C_5 = \frac{a_1 b_1}{a_1 - b_1} \tag{6.47}$$

$$C_2 = \frac{b_2}{a_1 - b_1} \tag{6.48}$$

Dividing equation (6.47) by (6.48), we get $a_1$

$$a_1 = \frac{C_5}{C_2} \qquad (6.49)$$

$$C_4 = \frac{-a_2 b_1}{a_1 - b_1} \qquad (6.50)$$

$$C_1 = \frac{-a_2}{a_1 - b_1} \qquad (6.51)$$

Dividing equation (6.50) by equation (6.51), we get

$$b_1 = \frac{C_4}{C_1} \qquad (6.52)$$

Substituting values of $a_1$, $b_1$ and $C_1$ in equation (6.51)

We get the values of $a_2 = -a_2 = C_1 a_1 - C_1 b_1$

$$a_2 = C_1 b_1 - C_1 a_1$$

$$C_2 = \frac{b_2}{a_1 - b_1} \qquad (6.48)$$

Substituting the values of $a_1$, $b_1$ and $C_2$ in equation (6.48), we get the value of $b_2$, therefore

$$b_2 = C_2 a_1 - C_2 b_1$$

$$C_0 = \frac{b_1 - a_0}{a_1 - b_1} \qquad (6.53)$$

$$C_3 = \frac{a_1 b_0 - a_0 b_1}{a_1 - b_1} \qquad (6.54)$$

Substitute values of $a_1$, $b_1$, $C_0$ and $C_3$ in equation (6.53) and (6.54) and solve the equation simultaneously to get the values of $a_0$ and $b_0$.

Now we have identified structural coefficients, $a_0$, $a_1$, $a_2$ for demand equation and $b_0$, $b_1$ and $b_2$ structural coefficients for supply equation. With these coefficients, we now designate demand and supply as

$$\hat{Q}_t^d = \hat{a}_0 + \hat{a}_1 P_t + \hat{a}_2 I_t + \hat{U}_{1t} \qquad (6.55)$$

$$\hat{Q}_t^s = \hat{b}_0 + \hat{b}_1 P_t + \hat{b}_2 P_{t-1} + \hat{U}_{2t} \qquad (6.56)$$

Using the market clearing conditions $i.e.,$

$Q_t^d = Q_t^s$, we get the equation for equilibrium price

$$\dot{P}_t = \hat{C}_0 + \hat{C}_1 I_t + \hat{C}_2 P_{t-1} + \hat{V}t \qquad (6.57)$$

Similarly substituting this equation (6.56) in equation (6.55) equation for equilibrium quantity is obtained.

$$\overset{*}{Q_t} = \hat{C}_3 + \hat{C}_4 I_t + \hat{C}_5 P_{t-1} + \hat{W}t \tag{6.58}$$

The parameters of equation (6.54) and (6.55) would satisfy the normal assumptions underlying the regression equation and would have the BLUE (Best–Linear Unbiased Estimate) properties. Direct estimation of parameters for equations (6.39) and (6.40) would not satisfy the normal assumption and the parameters do not have BLUE properties.

In estimating equilibrium price (equation 6.57) and equilibrium, quantity equation (6.58), we have to consider the price variables and crop output variable in terms of indices with references to particular base year. A normal year is selected as a base year and price and crop output variables are considered as 100. Indices for rest of the years are worked out based on the value of the variable in the base year. In the equation (6.57) price variable both as dependent variable and explanatory variable is considered. Direct estimation of this variable in the equation, through OLS would violate the normal assumptions underlying regressions resulting in biased and inconsistent estimates. To avoid this problem we have to choose the instrumental variable method. In this process first regress $P_t$ on $I_t$ (real per capita income or consumption expenditure real here means the nominal values of income variable are deflated by consumer price index) and then obtain $\hat{P}_t$.

Taking log of $P_t$ by one period.
We now estimate the equation (6.57) as

$$P_t = C_0 + C_1 I_t + C_2 \hat{P}_{t-1} + U_t \tag{6.59}$$

Equation (6.58) can be estimated taking the variables in indices with reference to a given base year, which is generally a normal year. Let us now assume the estimated equilibrium price equation and equilibrium quantity equation as:

$$\hat{P}_t = 381.92 + 0.1014\, I_t - 0.9638\, P_{t-1} \tag{6.60}$$

$$\hat{Q}_t = -10.382 + 0.0872\, I_t + 0.2447\, P_{t-1} \tag{6.61}$$

These equations (6.60) and (6.61) are reduced form equations now the reduced form coefficients are given as

$$
\begin{aligned}
C_0 &= \phantom{-}381.92 & C_3 &= -10.382 \\
C_1 &= \phantom{-}0.1014 & C_4 &= \phantom{-}0.872 \\
C_2 &= -0.9638 & C_5 &= \phantom{-}0.2447
\end{aligned}
$$

Now we can work out structural coefficients of demand and supply functions of a commodity from the estimated reduced form coefficients. The following procedure is adopted for this purpose.

$$a_1 = \frac{C_5}{C_2} = \frac{0.2447}{-0.9638} = -0.2539$$

Here $a_1$ = Price coefficient in demand equation

$$b_1 = \frac{C_4}{C_1} = \frac{0.872}{0.1014} = 0.860$$

$b_1$ = Price coefficient in supply equation

584

$-a_2 = C_1 (a_1-b_1)$

$\quad = 0.1014 (-0.2539 - 0.860)$

$-a_2 = -1.0125$

$\quad a_2 = 1.0125$

$\quad a_2$ = Coefficient of per capita income in the demand equation

$\quad b_2 = C_2 (a_1-b_1)$

$\quad = -0.9638 (-0.2539 - 0.860)$

$\quad = 1.0736$

$\quad b_2$ = Coefficient of lagged price in supply equation

Estimation of $a_0$ and $b_0$ (constant term)

$$b_0 - a_0 = C_0 a_1 - C_0 b_1 \tag{6.62}$$

$$a_1 b_0 - b_1 a_0 = C_3 a_1 - C_3 b_1 \tag{6.63}$$

$$b_0 - a_0 = 381.92 (-0.2539 - 0.860) \tag{6.64}$$

$$-0.2539 b_0 - 0.86 a_0 = -10.382 (-0.2539 - 0.860) \tag{6.65}$$

To cancel out $b_0$ multiply equation (6.64) by 0.2539 through out on both the sides.

$-0.2539 a_0 - 0.2539 a_0 = 0.2539 (-0.2539 - 0.860) (381.92)$

$-0.2539 b_0 - 0.86 a_0 = -10.382 (-0.2539 - 0.860)$

$\quad\quad\quad -1.1139 a_0 = -108.0143$

$$a_0 = \frac{-108.0143}{-1.1139}$$

$a_0$ = 96.97 (Intercept form for demand equation)

Substituting $a_0$ =96.97 in equation (6.64), we get $b_0$ value.

$b_0 - 96.97 = 381.92 (-0.2539 - -0.860)$

$\quad\quad b_0 = -0.4206 + 96.97$

$\quad\quad b_0 = -328.4502$ (Intercept term for supply equation)

Now the derived demand and supply functions are as follows:

$$Q_{Dt} = 96.97 - 0.2539 P_t + 1.0125 I_t + U_{t1} \tag{6.66}$$

$$Q_{st} = -328.4506 + 0.860 P_t + 1.0736 P_{t-1} + U_{t2} \tag{6.67}$$

Please note that equation (6.66) and (6.67) are correctly identified as demand and supply functions of the commodities respectively because for these equations we have expected signs and significant levels. In this process we have considered only two explanatory variables namely price variable and income variable. But in reality we have a host of variables that influence demand and supply of the commodity. The estimated equations would be correct if all the relevant variables with adequate number of observations are included in the model. Such demand and supply functions of the commodity can be used for prediction and planning purposes.

# CHAPTER 48

# Role of Government in Agricultural Marketing

## NATIONAL AGRICULTURAL CO-OPERATIVE MARKETING FEDERATION (NAFED)

It is the apex body constituted with co-operative principles existing at the national level. It deals with procurement, distribution, export and import of selected agricultural commodities. It promotes inter-state trade as well as foreign trade of farm commodities. The main export commodities are onions, chilies, ginger, garlic, cardamom, *etc.* Export of pulses, groundnut, onions and potatoes is mainly canalized through NAFED in the country. It further undertakes the movement of essential commodities from surplus areas to scarce areas. In the co-operative sector, NAFED is the central nodal agency for operating price support operations for pulses and oilseeds. It undertakes market intervention operations for horticultural products like onions, potatoes, grapes, black pepper, red chilies, *etc.* In 1995-96 its turnover was more than Rs. 870 crore. NAFED is exercising a healthy influence on market operations in stabilizing the market prices for favour of producers and consumers.

Other organizations working in co-operative sector at the national level are (1) National Co-operative Tobacco Growers Federation (NCTGF) (2) National Consumers Co-operative Federation (NCCF) (3) Tribal Co-operative Marketing Development Federation of India Ltd. (TRIFED).

## NATIONAL CO-OPERATIVE DEVELOPMENT CORPORATION (NCDC)

It was set up in 1973 with a prime aim of planning and promoting programmes for production, processing and marketing of agricultural produce through co-operative societies. Of course, these commodities should be notified by the corporation before they are being marketed. The corporation supports various marketing programmes for which financial assistance comes from the respective state Governments. Co-operative marketing and processing activities are being developed under the supervision of NCDC with a provision of requisite financial assistance and expertise and training. NCDC is now extending its activities to the various areas of co-operative dairies, fisheries, minor forest produce, *etc.*, which are basically aimed at economic development of the rural poor. Since its inception *i.e.*, from 1963 up to 1996, it has provided financial assistance to the tune of Rs. 2,800 crore besides striving hard for the development of co-operative marketing, storage, processing and distribution of consumer goods *viz.*, foodgrains, sugar, edible oils, kerosene, salt, soft drinks *etc.*, through service co-operatives.

# PROCUREMENT OF FOODGRAINS

The foodgrains are procured to make sure of their availability of foodgrains to the consumers at fair and reasonable prices. Central Government has been procuring substantial quantities of foodgrains and maintaining large buffer stock of foodgrains. The purchased foodgrains are wheat, rice, and coarse grains. Wheat and rice constitute 98 per cent of total foodgrains. Table 48.1 provides information on the procurement of foodgrains since 1965.

TABLE 48.1   Purchase of Foodgrains (in '000 tonnes).

| Year | Rice | Wheat | Coarse grains | Total quantity |
|------|------|-------|---------------|----------------|
| 1965 | 2,951 (73.21) | 375 (9.30) | 705 (17.49) | 4,031 (100) |
| 1975 | 5,042 (52.72) | 4,098 (42.85) | 423 (4.43) | 9,563 (100) |
| 1985 | 9,568 (47.59) | 10,355 (51.50) | 184 (0.91) | 20,107 (100) |
| 1995 | 9,997 (44.78) | 12,327 (55.22) | - | 22,324 (100) |

Figures in parentheses are percentages to total quantity.
Source: Government of India, Department of Food, New Delhi.

From the table it is clear that the Government of India through FCI started purchasing rice, wheat and coarse grains right from the year of its inception *i.e.*, 1965. The quantity of rice procured increased from 2,951 thousand tonnes to 9,997 thousand tonnes. This means that there is about three fold increase in the rice procurement. For wheat there is nearly four fold increase. Procurement of coarse grains constituted 17.49 per cent of the total procurement during 1965 while in 1995, there was no procurement of coarse grains. The procurement of wheat in 1995 increased by 33 fold from that of 1965.

## Buffer Stock Operations

It is the process of purchasing the requisite foodgrains at the announced procurement prices/levy prices in surplus areas and distribute the same in deficit regions through the public distribution system (PDS). This is meant to meet the emergency requirements and expected contingencies resulting from natural calamities like floods, wars, famines, cyclones, *etc*. The following are the advantages of maintaining buffer stocks in India.

1. To help stabilize prices of foodgrains.
2. To counteract the affects of marketing activities of speculators and hoarders in foodgrains.
3. To help maintain stability in the food economy of the country.
4. To protect the farmers against price hikes and price slumps.
5. To protect the consumers through fair price shops.

## Process of Procurement

Foodgrains are procured at the pre-announced procurement prices in the domestic market. Procurement prices are either higher or lower than the market prices. But whatever may the prevailing price, the Government makes a firm commitment to purchase the foodgrains at the announced prices. Announcement is generally done at the beginning of the season. The prices are announced by the Government on the recommendations of CACP. Up to 1990-91 procurement was confined to cereals by FCI only. Later other public agencies tried to procure necessary foodgrains at pre-determined prices, but the targets were not fulfilled by those public agencies. Hence inter-district movement restrictions on foodgrains were imposed with deliberate attempt to depress prices in surplus regions and procure foodgrains at announced prices. Other devises *viz.,* levy on producers, levy on traders and millers, pre-emptive purchases, open market purchases and monopoly procurement were also in vogue to achieve the targets in procurement. Let us know about these instruments of procurement.

### 1) Levy on Producers

Under this system the farmers are forced to sell part of their produce to the Government at the announced prices. Procurement is fixed at a flat rate and is being done directly by the Government or through the agents appointed by the Government. The appointed agencies *viz.,* co-operative marketing societies, Food Corporation of India and sometimes private traders are involved for this purpose. In this procurement process foodgrains are purchased from large number of farmers in the country.

### 2) Levy on Traders and Millers

Under this system, traders and millers are legally bound to deliver a fixed percentage of rice and wheat to the Government at the announced procurement prices. In Andhra Pradesh, millers are imposed a levy to an extent of 50 per cent on their stocks. Millers and traders are free to sell the remaining quantity of foodgrains in the open market at the prevailing market prices.

### 3) Pre-emptive Purchases

Here the Government purchases foodgrains from the open market at a price negotiated between the producers and traders.

### 4) Open Market Purchases

In this method, the Government or its agencies make their entry into the market as traders and purchase the produce like any trader in the open market. Here the purchase price is left to the supply and demand situations in the market.

### 5) Monopoly Procurement

It is a method, under which traders are not given permission to enter the market and the Government acquires monopoly rights in the market for the purchase of the entire lot of foodgrains. In the market, all the market infrastructure facilities are provided.

# PUBLIC DISTRIBUTION SYSTEM (PDS)

Public distribution system's (PDS) main aim is the distribution of foodgrains to the vulnerable sections of the society at fair prices during periods of scarcity. During periods of bumper production raising up the prices is the main aim through price policy instruments. Up to mid 1970s the public distribution system was fed to considerable extent by imported foodgrains. For example, the country imported ten million tonnes of foodgrains in 1966. After 1970s due to the increased production of foodgrains, procurement prices helped the nation to cut down imports of foodgrains. The quantity of foodgrains distributed is presented in the following Table 48.2.

TABLE 48.2 Distribution of Foodgrains in India (Million tones).

| Year | Rice | Wheat | Total foodgrains * |
|------|------|-------|--------------------|
| 1965 | 3,586(35.58) | 5,939(58.92) | 10,079(100) |
| 1975 | 3,211(19.65) | 7,545(67.05) | 11,253(100) |
| 1985 | 7,231(45.77) | 8,477(53.66) | 15,799 (100) |
| 1995 | 9,365(62.72) | 5,567(37.28) | 14,932 (100) |

*It includes other foodgrains *viz.*, jowar, maize, ragi, barly and small millets.

**Source:** Bulletin on food statistics, Directorate of Economics and Statistics, Ministry of Food and Civil Supplies, Govt. of India, New Delhi.

It is obvious from the table that, wheat distributed in 1965 was 5,939 million tonnes, which decreased to 5,567 million tonnes in 1995. We must note that wheat distribution increased to 7,545 million tonnes in 1975 and 8,477 million tonnes in 1985. On the contrary, rice distribution was continuously increasing from 3,586 million tonnes in 1965 to 9,365 million tonnes in 1995. Along with foodgrains, other items like edible oils, sugar, kerosene, cloth, *etc.*, are also provided.

# REVAMPED PUBLIC DISTRIBUTION SYSTEM (RPDS)

It was launched on January 1, 1992 but it came into functioning from June 1, 1992 in the selected blocks of drought prone, desert, tribal and hilly areas of the country. In addition to the essential commodities like rice and wheat included in PDS, some more commodities are added. These include pulses, tea, and iodized salt. Vigilance committees were constituted to keep an eye on the distribution of commodities at the village level. Items are delivered at the door of the fair price shop to minimize the leakage and pilferage. In the country, the RPDS was operating in 1775 blocks located in tribal, hilly and arid areas. For rice and wheat, there was a reduction in the issue price by Rs. 50 per quintal, under RPDS.

# THE TARGETED PUBLIC DISTRIBUTION SYSTEM

Government of India in 1997 took a decision to identify the vulnerable sections (below poverty line (BPL) families) on the basis of *Dr. Lakdowala Committee* and also to fix a uniform rate and scale for supply to these BPL families. Those who are not having access to the minimum per day energy requirements of 2100 calories per individual in rural areas and 2400 calories per individual in urban areas are treated as people living BPL. Accordingly, the Government of India identified 35.97 per cent of the population as BPL families and decided that they should be provided foodgrains at affordable prices in adequate quantities. Government of India had named this scheme as 'TARGETED PDS' and launched it with effect from 1-7-1997. According to this scheme all the families identified by the Government of India as BPL families are entitled to 10 kg of rice per family per month. State Governments also view that the TPDS as an important constituent of the strategy for poverty eradication to ensure food security to the population below poverty line to the extent possible. The Government is making its best efforts to implement the TPDS to the best advantage of the beneficiaries with utmost transparency and efficiency. Households belonging to the vulnerable sections of the society such as landless labourers, marginal farmers, artisans, craftsmen such as porters, tappers, weavers, blacksmiths, carpenters, etc., in rural areas and slum dwellers and persons earning their livelihood on daily basis in informal sectors viz., porters, coolies, rickshaw pullers, hand-cart pullers, fruit and vegetable sellers on pavement in urban areas, etc. are covered under BPL to get foodgrains at specially subsidized prices.

The vigilance cell wing of civil supplies department and revenue and civil supplies machinery is in charge of enforcing various control orders issued under the Essential Commodities Act and bringing to book the dealers, millers, etc., who violate the provisions of the order and indulge in malpractices like hoarding, black-marketing, profiteering, etc. The state commission and district consumer fora receive complaints against the unfair and restricted trade practices adopted by the trader and also the malpractices indulged in by the trade. The consumers can file complaints before these consumer fora for redressal of their grievances. A research and monitoring wing for essential commodities had been constituted in the civil supplies department to monitor the prices and the measures to be taken for market intervention to control the prices in the open market.

# DIRECTORATE OF MARKETING AND INSPECTION

Royal Commission on agriculture in 1928 identifying the need of efficient marketing system of agricultural commodities, recommended for the appointment of marketing experts in agricultural departments in all the major provinces, as no agency existed to look after this activity till then. Subsequently in 1931 accepting the realities of observation made by Royal Commission, the Central Banking Enquiry Committee made further recommendation of setting up a central agency, for coordinating the State activities in the field of agricultural marketing. Taking these recommendations into consideration, the Government of India had set up a central organization known as "the office of the Agricultural Marketing Adviser to the Government of India" in the year 1935. Later it came to be known as the Directorate of Marketing and Inspection (DMI). DMI is headed by the Agricultural Marketing Adviser to the Government of India. DMI maintains and controls 21 laboratories apart from Central AGMARK laboratory at Nagpur. It is also involved in activities like market research for development, training and extension.

### Functions of DMI

### Promotion of Grading and Quality Control

This was recognized as the principal function of DMI ever since it came into existence, because it helps the farmers, traders and consumers in their transaction. More specifically this function can be explained as follows.

## Framing of Grade Specifications

Grading of agricultural commodities is voluntary for internal trade under the Agricultural Produce (Grading and Marking) Act, 1937. This grading is endorsed by 'AGMARK' standards. This standard is given to the commodities at the request of the sellers.

Agricultural Produce (Grading and Marking) Act, 1937 gave powers to the central Government to prescribe standards for agricultural and livestock products in the form of quality specifications, packing, sealing, etc. (AGMARK standards). A standards division is established in the Directorate to look after revising the existing standards and formulating standards for new commodities to be covered.

Grading for internal trade is voluntary and it is done at two levels i.e., traders' level and at producers' level. Grading at traders' level, helps the consumers, to get quality commodities. At this level in internal trade, the commodities graded under 'AGMARK' are categorized into two groups called centralized commodities and decentralized commodities. In respect of centralized commodities, grading arrangements are made by the Directorate of Marketing and Inspection. DMI takes the function of quality testing in view of the requirement of elaborate testing arrangements. The commodities that are included are ghee, butter, vegetable oils, oilcakes, powdered spices, honey, etc. Regarding decentralized commodities, grading is done by the marketing departments of the respective State Governments, ofcourse, under the overall supervision of the DMI. Grading of decentralized commodities is relatively simple. Examples are rice, wheat, atta, pulses, etc. Producer level grading to ensure the farmers to receive fair returns, was introduced through State marketing departments in 1963-64. Under this process, the produce ready for sale is subjected to simple quality tests. This quality test is supposed to help the farmer to get the price in commensurate with the quality of the produce.

For the commodities, which are exported, compulsory quality control was introduced in 1942. The commodities are to be graded in accordance with the specifications prescribed under the Agricultural Produce (Grading and Marking) Act, 1937. The Export (Quality control and Inspection) Act was enacted in 1963 and the Export Inspection Council was set up to enforce compulsory quality control on commodities meant for export. The DMI is at present acting as an export inspection agency on behalf of Export Inspection Council.

## AGMARK Laboratories

The central AGMARK laboratory is an apex institution situated at Nagpur. This apex institution along with a network of AGMARK laboratories functioning at Mumbai, Kolkata, Chennai, Cochin, Guntur, Mangalore, Bhopal, Patna, Bangalore, Bhubaneswar, etc., provide technical guidance, necessary testing facilities, etc. The commodities that are graded under 'AGMARK' are ghee, vegetable oil, butter, spices, etc.

DMI also is enforcing the Meat Food Products Order, 1973 to exercise control over the production of meat products. This makes an obligation on the part of every businessman engaged in manufacturing or selling the packed meat products to obtain a license from the Agricultural Marketing Adviser to the Government of India. This order ensures hygienic standards prescribed under the order and at all stages of marketing through analysis of meat food by taking representative samples. Though the regulation of markets is a State subject, the DMI extends guidance and advice to all the State Governments in formulating market legislation and its enforcement.

The establishment of Market Planning and Design Centre (DMDC) in 1977 in the DMI with the assistance of the United Nations Development Programme (UNDP) and the Food and Agriculture Organization (FAO) was a major step to pay special care for marketing of fruits and vegetables which are highly perishable in nature. Though the assistance from UN ended, this activity is continuing under the activities of DMI.

DMI conducts market research and market surveys to study the problems of marketing of agricultural commodities so as to suggest suitable measures to solve them.

## BUREAU OF INDIAN STANDARDS (BIS)

Quality represents as a specification or set of specifications which are supposed to be found for the commodities within the given limits. This quality indicator needs to be specified in the form of national standards for the use by the industry and also consuming section. The Bureau of Indian Standards is the new name given to the Indian standards Institution (ISI) from April, 1 1987. The BIS operates a voluntary certification scheme under the BIS Act, 1986. The main objective of this scheme is to guarantee quality standards for the manufactured goods as per the standards set. To fulfill this objective, the certification scheme provides service to the manufacturers and consumers in the form of evaluation of products and production controls, the determination of the conformity with the established standards ensuring the same to the consumers and also providing an assurance that the products having BIS certification mark must be the recognized standards. Though BIS certification scheme is voluntary in use, but the Government made the scheme as compulsory for the number of items concerned with health and safety of the consumers, through Essential Commodities Act, *etc.*

The product certification scheme is basically voluntary in nature and aims at providing quality, safety and dependability to the ultimate customer. Presence of the certification mark ISI known as standard mark on a product is an assurance of conformity to the specification. Conformity is ensured by regular surveillance of the licensee's performance by surprise inspections and testing of samples, drawn both from the factory and the market. The Act also provides for:

1. Authorizing the foreign manufactures to use the standard mark.
2. Bringing articles under compulsory certification for reasons of safety, health, environmental protection, energy conservation; and
3. Penalty for spurious marketing and misuse of mark.

Considering the long experience of BIS in certification, Underwriters Laboratories (UL) of USA, South African Bureau of Standards (SABS) and Canadian Standards Association (CSA) have authorized BIS to operate their certification schemes in India.

### ECO Mark

The Government of India had instituted a scheme in February 1991 known as ECO Mark scheme for labelling environment friendly products. This scheme is administered by the Bureau of Indian Standards and provides for labeling of household and other consumer products which meet certain environmental criteria along with quality requirements prescribed in the relevant India standards.

The manufacturer can affix the BIS mark for his product only after he is granted a license. Before issuing BIS mark, not only that the final product is tested for quality but also strict enforcement is made regarding the quality of the inputs. Once the license is given, it is mandatory on the part of the manufacturers to observe strict implementation of quality standards laid down and maintain records concerned as prescribed by BIS. The license is initially given for a period of one year and renewed annually based on the performance of the business unit. To implement the standards, the Bureau is empowered to make surprise visits of the manufacturing units. During its visits it collects the samples of raw material as well as the final products for conducting the needed tests.

In agriculture, the items covered are foodgrain storage structures, fertilizers, pesticides, pumpsets, plastic pipes in irrigation, animal and tractor drawn implements and machinery, sprinkler and drip irrigation systems, *etc.*

The items that are related to food industry are cereal products, bakery and confectionery dairy products, processed fruits and vegetables, protein rich products, spices and condiments, tea, coffee, cocoa products, *etc.* Apart from the agricultural and food products, the items included are chemicals, electronics, textiles, products related to engineering field, *etc.*

In the formulation of standards, apart from BIS, a number of agencies, such as Central Committee for Food Standards, Department of Food, Directorate of Marketing and Inspection, Technical Standardization Committee of Defence, Directorate of Sugar, *etc.*, are currently involved, in maintaining standards for the same commodities. The standards prepared by these organizations reflect the different levels of quality. The bureau keeps a close liaison with all these organizations in its work.

## HACCP

Hazard Analysis and Critical Control Point (HACCP) is a process control system designed to identify and prevent microbial and other hazards in food production. It includes steps designed to prevent problems before they occur and to correct deviations as soon as they are detected. Such preventive control system with documentation and verification are widely recognized by scientific authorities and international organizations as the most effective approach available for producing safe food. As a member of WTO, India is signatory to the Sanitary and Phytosanitory (SPS) agreement and hence has to adopt for international trade, the standards, guidelines and recommendations issued by the food hygiene committee of the joint FAO/WHO Codex Alimentarius Commission which advocates the adoption of HACCP. Industries in the countries exporting to WTO member nations would now have to adopt HACCP, from dates specified by each importing country. For food industry in India, adoption of HACCP is becoming imperative to reach global standards, demonstrate compliance to regulations/customer requirements besides providing safer food to our millions.

## COMMODITY BOARDS

### Tobacco Board

India is the third largest producer of tobacco in the world. It is also one of the major exporting countries of tobacco along with USA, Brazil, Zimbabwe, Italy, Greece and Turkey. Of the various types of tobacco cultivated in the country, flue cured virginia (FCV) is the most important one in terms of exports and excise revenue earnings. Coming to the marketing aspects of tobacco, the Government of India had established the Indian Central Tobacco Committee to look after development, research and marketing of various types of tobacco cultivated in the country. This was later in the year 1965 was abolished by the Government. In the year 1956, the Government of India established the Tobacco Export Promotion Council under the Ministry of commerce to promote exports of Indian tobacco and tobacco products. The Directorate of Tobacco Development was established in 1966, under the Ministry of Agriculture to look after the development and marketing of tobacco. 'Subsequently experiencing the instances of erratic behaviour of market forces and the need to regulate production and promotion of overseas marketing, Tobacco Board was established in place of Tobacco Export Promotion Council under the Tobacco Board Act of 1975. The Board came into existence from 1976 with its headquarters at Guntur in Andhra Pradesh.

## Functions

It regulates the production and curing of FCV tobacco. To perform this function, the board weighs the factors like demand for and supply of tobacco (in estimation of total demand both domestic demand and demand overseas are taken into account). To measure the supply of tobacco, factors like availability of suitable land; and soil characteristics that influence both quality and quantity, *etc.*, are considered.

1. It regularly monitors the marketing of FCV tobacco in domestic market and international market and ensures that the farmers receive remunerative prices. It is always on the look out of new markets overseas, besides taking steps to sustain existing international markets.
2. It undertakes grading at the producers level.
3. It provides the needed information to growers, dealers, tobacco exporters, tobacco industry, *etc.*
4. It arranges auction platforms and acts as an auctioneer at auction platforms.
5. It recommends to the Government regarding fixation of minimum support prices for the exportable virginia tobacco.
6. It undertakes the purchase of virginia tobacco if the situation warrants so, to protect the interests of the growers.
7. It assists and coordinates in tobacco research for the development of tobacco industry, *etc.*

## Exports

The Board exports leaf tobacco and tobacco products (cigarettes, beedies, cut tobacco, chewing tobacco, snuffs, *etc.*). The total exports during 1995-96 were to the tune of Rs. 421.04 crore. Export earnings during the inception year (1974-75) were Rs. 82.22 croers. Tobacco and tobacco products are exported to countries like Sweden, UK., Germany, Denmark, Netherlands, Russia, Romania, Yugoslavia, Singapore, Hong Kong, Indonesia, Australia, New Zealand, *etc.*

## Spices Board

India holds 70 per cent of the share in the global spice market. International trade has grown tremendously since 1987. It has crossed Rs. 10,000 million during 1996-97 and touched an all time high of Rs. 13,521 million during 1997-98. The spices board (Ministry of Commerce, Government of India) is the apex organization established for the worldwide promotion of Indian spices. The list of spices numbering 52 including cardamom, pepper, chilly, ginger, turmeric, coriander, cumin, fennel, fenugreek, celery, aniseed, Bishop's weed, caraway, dill, cinnamon, cassia, garlic, curry leaf, kokum, mint mustard, parley, pomegranate, saffron, vanilla, teapot, pepper long, star anise, sweet flag, greater galanga, horse radish, caper, clove, asafoetida, combodge, hyssop, juniper berry, bay leaf, lovage, marjoram, nutmeg, mace, basil, poppy seed, all-spice, hose Mary, sage, savory, thyme, oregano, tarragon and tamarind fall under the purview of the spices board. It is the major link between the Indian exporters and the importers outside the country.

India holds 86 per cent of the share in the global spice market. During 2013-14 a total of 8,17,250 tonnes of spices and spice products valued at Rs.13,735.39 crore were exported from the country, against 7,26,613 tonnes valued at Rs.12,112.76 crore in 2012-13, registering an increase of 12 per cent in volume and 13 per cent in value terms.

## Major Functions of Spices Board

1. Formulation and implementation of programmes for better production.
2. Implementation of quality improvement methods through systematic research and development activities.
3. Imparting education and training to growers, processors, packers and exporters.
4. Providing registration and licensing.
5. Acting as a data bank and communication channel for importers and exporters, *etc.*

## Accreditation Systems for Quality Upgradation and Value Addition

Spices Board focusses its attention on quality upgradation and value addition through accreditation systems like issuing Indian spices logo, spice house certificate and brand name registration.

## Indian Spices Logo

To help the consumers to choose the best quality spices from among various brands, which often mislead and confuse international customers, spices board issues the Indian spices logo, which ensures quality of the product. The logo is displayed on all packs approved by the spices board. This logo is selectively given to exporters who follow certified processing and quality control measures. In essence the logo reflects the quality of Indian spices.

## Spices House Certificate

This is yet another measure which is very latest, aiming at the quality upgradation. Spices house certificate is given to the processors, who possess, their own processing facilities and infrastructure, with a competence and commitment to maintain quality. Quality should be maintained at all stages of processing right from the beginning to final shipping. Both Indian spices logo and Spice House Certificate have one objective in common *i.e.,* quality assurance but with a difference. Under logo programme, exporters of branded spices and spice products are covered, while Spice House certificate programme looks at exporters having processing facilities.

## Brand Name Registration

In the global market, wherein a large number of traders compete with each other to get a major market share, visual appeal is one, which, silently communicates the desired message to the consumers. Selection of package is therefore an impo.tant task. To assist the exporters in this regard, Spices Board has introduced a system to register their brand names with it and this registration is extended for those, which satisfy standards of packaging and labelling. The board with the help of Indian Institute of Packaging, Mumbai, specified norms for packaging.

## Various Schemes of Board

Following are the various schemes in operation by the Board for the benefit of registered exporters and registered manufacturer exporters.

### 1) Technology Transfer and Processing Upgradation

This scheme can be availed by all the registered exporters of spices and spices products to improve the processing technology. The assistance is limited to 33.33 per cent of the cost of developing/adopting technology, subject to a maximum Rs. 10 lakh.

### 2) Assistance for Research and Development Efforts

For those exporters who are interested to undertake product development research programmes, through private labs/research institute/agencies, approved by the department of science and technology, the financial assistance is made available by the Board. Under this programme assistance would be given to the extent of 50 per cent of cost of the research programme of Rs. 5 lakh each, whichever is less for the spices covered by the board.

### 3) Assistance for Stall Rent for Participation in International Food Fairs

Those exporters, who are registered with board are eligible to make use of the assistance under the scheme. The main purpose of introduction of this scheme is to facilitate the exporters to participate in international food fairs by setting up individual stalls. This facility is extended to the exporters to provide an opportunity to exhibit their products for promotion as well as obtaining orders for export. The financial assistance for individual under this scheme is fixed at 50 per cent of the stall rent or Rs. 1,00,000 which ever is less.

### 4) Setting up/Upgradation of Lab/Facilities for Monitoring Pesticide Residues

This scheme aims at assisting the exporters in quality upgradation processing undertaken on a continuous basis. The main idea is to enable the exporters to keep themselves upgraded in processing to the changing international quality standards.

### 5) Product Promotion Abroad

Exporters of branded consumer packs, whose packs are registered with board, are eligible for the financial assistance under the scheme. The assistance is fixed at 50 per cent of the cost of promotional activities undertaken abroad or Rs. 5,00,000 which ever is less for an individual.

### 6) Packaging Development

Exporters of branded consumer packed spices whose brands are registered with the board are eligible for assistance under the scheme. This assistance is meant for development of improved packaging. This assistance per individual exporter is limited to 50 per cent of the actual cost of packaging or Rs. 2,00,000 whichever is less.

## 7) Assistance for Sales Promotion Tours Abroad

Under this scheme, the exporters with logo/spice house certificate/registered brands are eligible to avail the assistance. This scheme encourages the exporters to undertake business promotion trips abroad and also for participation in international trade fairs. The assistance is limited to 50 per cent of the air fare for one trip in an year for one representative, subject to a maximum Rs. 60,000 for logo/spice house certificate holders and Rs. 40,000 for registered brand owners.

## 8) Reimbursement of Air Freight/Courier Charges for Sending Samples Abroad

Those exporters who are possessing logo or spice house certificate are entitled for assistance under the scheme. The amount of reimbursement is fixed at Rs. 50, 000 per exporter per year towards air freight/sending spices samples by courier to buyers abroad.

## 9) Brand Promotion Loan

Through this scheme, the board assists the logo exporters in undertaking long term promotional programmes to promote their brand in overseas markets. The exporters under this scheme get interest free loan to the extent of 50 per cent of the proposed expenditure on promotional programmes for a period of three years.

## 10) Modern Cleaning, Processing and Storage Facilities at Terminal Markets

This scheme is meant to assist association of exporters, farmers, agricultural produce market committees, agricultural marketing boards, *etc.*, in setting up infrastructure facilities for hygienic, cleaning, processing and storage for spices at assembling/ terminal markets. The board bears 50 per cent of cost of construction and the balance of 50 per cent is to be borne by the beneficiary. The fund will be released after the completion of structures only.

## Coffee Board

Coffee cultivation is mostly confined to the tropical and sub-tropical regions of Africa, Asia and Central/Latin America. In India its cultivation is mainly found in Karnataka, Kerala and Tamil Nadu, besides the North Eastern States, Andhra Pradesh and Orissa. Indian coffee has an international reputation for its blending quality. Indian coffee is exported to the USSR, UK, USA, Italy, Netherlands, France, Sweden, Australia, Japan, *etc*. India exports about 75 per cent of its production. The exports of coffee from India stood at 1.80 lakh tonnes in 1996-97 which increased to 2.98 lakh tonnes in 2013-14 and 3.01 lakh tonnes in 2014-15. During worldwide depression of the forties, the coffee industry made fervent appeals to the Government to interfere and look into its safety. The subsequent Coffee Control Conference in September 1940 recommended the prohibition of private marketing of coffee, which was incorporated in the Coffee Market Expansion Ordinance XIII of 1940. This Act provided the emergence of Indian Coffee Market Expansion Board. It was renamed as Indian Coffee Board in 1943, which was again changed as Coffee Board in 1955. Amendments were made in the Coffee Market Expansion (Amendment) Bill, 1945, incorporating which, a new Act came into force in 1955, which was called the Coffee Act.

The coffee board, a statutory organization under the Ministry of commerce, Government of India, looks after the development and growth of the Indian coffee industry. Increase in coffee production, internal marketing, export marketing, quality control, coffee research, extension, development, and promotion are all activities of the board. The following are the specific functions assigned to the board under the coffee act.

## Functions

1. Promotion of sale and consumption in India and elsewhere of the coffee produced in India.
2. Promotion of agricultural and technology research in the interest of the coffee industry.
3. Assistance to coffee estates for their development and
4. To provide better working conditions and improvement of amenities and incentives for workers.

## Marketing

One of the major functions of the coffee board is to undertake marketing of the entire produce of the coffee growers, through a common pool. The pool system of marketing is a unique feature of the coffee board. As per the provisions of the coffee Act, all the produce is pooled with coffee board. To perform the marketing function, it established unique dual auction system for domestic market and export market. The pooled coffee produce after curing and grading is kept open for auction for international market and domestic markets separately. Marketing expenses are deducted before the payments are made to the growers. Internal sales are operated through four channels of distribution, such as pool sales, dealers, local sales, permit holders, co-operative societies and board's promotional centres. Acting as a monopoly in the marketing of coffee, coffee board safeguards the interests of the farmers, curers, traders and exporters by providing them all the infrastructural support.

In export sales, exporters who have registered with board participate in the open auction and bid for the lots. Foreign firms are also allowed to participate in the export sales by registering with the coffee board, through their agents or branch offices in India. Besides sales through export auctions, coffee is also exported against direct sales made by the coffee board to buyers of bulk quantities. The share of direct sales in the total exports is around 50 per cent. The pricing of such direct sales is based on the prices prevalent in the export auctions. India is a member of the International Coffee Agreement, the main objective of which is to bring about a reasonable balance between supply and demand in the international market and ensure remunerative prices to producers and fair prices to consumers.

Among the other functions, coffee board is involved in conducting research and extension service to the growers. The research output of the board is conspicuous by its role in releasing high yielding and disease resistant varieties of coffee. Its role in this direction is also evident through the creation of a post-harvest technology division for quality upgradation and development of integrated pest and disease management measures.

## National Dairy Development Board (NDDB)

Dairy co-operatives have been established to strengthen the production and marketing of milk and milk products in the country. The network of dairy co-operatives consists of milk producers, co-operative societies at the village level, district milk co-operative unions at the district level, State co-operative dairy federations at the State level and the National Dairy Development Board (NDDB) at the national level. Objectives of NDDB are as follows:

1.  To promote, plan and organize programmes for the purpose of development of dairy and other agricultural based and allied industries.
2.  To finance co-operative federations and co-operative enterprises concerned to stimulate production, preservation and consumption of milk and milk products.
3.  To adopt and encourage co-operative ideologies for the development of dairy and other agricultural based and allied industries.
4.  To impart technical knowledge to such organizations engaged in co-operative or public sector, involved in production, procurement, marketing of milk and milk products, *etc.*
5.  To conduct research and promotional activities in the field of dairy, agriculture and horticulture.
6.  To involve in designing, planning, promoting, developing and setting up of dairy industries and undertaking any other related promotional activities.
7.  To provide consultancy and managerial services and provide storage, transportation, processing and distribution of milk and milk products and act as a lead institution regarding milk and milk products.
8.  To regulate dairy and allied industries and function as regulatory authority.
9.  To recommend to the Government to fix minimum and maximum prices for the purchase or sale of milk, as and when necessary and assisting the Government in enforcing it; and
10. To function as an agency in the marketing channel for the import and export of milk and milk products.

NDDB initiated the "restructuring edible oil and oil seeds production and marketing" project in 1979 to increase farmer investment in oilseeds productivity through farmer owned co-operatives. More than 9 lakh farmers have joined 5,331 Oilseeds Growers' Co-operative Societies affiliated, in turn, to 18 unions in Andhra Pradesh, Gujarat, Karnataka, Madhya Pradesh, Maharashtra, Orissa, Rajasthan and Tamilnadu.

## Other Activities

NDDB's pilot project on fruits and vegetables in Delhi provides a direct link between the fruit and vegetable growers in a number of States and consumers in Delhi by undertaking the entire range of operations from procurement to marketing.

Indian immunologicals, a unit of NDDB, is engaged in the production of veterinary vaccines, animal health products and vitamin premixes.

The National Tree Growers' Co-operative Federation (NTGCF) is implementing the Tree Growers' Co-operative Project (TGCP) at the instance of the National Wasteland Development Board (NWDB). The project is being implemented in the states of Andhra Pradesh, Gujarat, Karnataka, Orissa, Rajasthan and Uttar Pradesh. The major objective of the project is emphasizing on the coverage of wastelands with the plants that serve fuel wood and fodder needs of the villages.

# SECTION VII
## ECONOMIC PROBLEMS OF INDIAN AGRICULTURE

# CHAPTER 49

# Role of Agriculture in Indian Economy

From the evolution of human history, it is amply testified that economic development traces its origin to agriculture. Even the developed countries were basically agriculture in origin and through this sector, they could achieve industrial development. When a country progresses industrially, it is quite obvious that the share of agricultural sector in their economies declines. In the case of developing countries, the share of agricultural sector in the gross domestic product is high. However, notwithstanding the share of agricultural sector in the gross domestic product, this sector occupies a prominent place in all the economies on the globe. Now we shall examine the role of agriculture in the economic development of the country.

## Contribution of Agriculture to the National Income

Historically agriculture has contributed significantly to India's national income as indicated by the figures available right from the period 1950-51. The share of agriculture as a percentage to GDP was 55.40 in 1950-51, 30.90 in 1990-91, 26.40 per cent in 1997-1998 and 13.9 per cent in 2013-14. The share of agricultural sector in the national income has been declining over the years. This gives an indication that the share of industrial and service sectors is increasing, as natural outcome of a developing economy. The contribution of agriculture to GDP is 2 per cent in UK; 3 per cent in USA, 4 per cent in Canada and 5 per cent in Australia, the developed countries.

## Contribution of Agriculture to Employment

Agriculture is the main source of livelihood for more than 60 per cent of India's population. All those who live in rural areas directly or indirectly depend on agriculture for their livelihood. The owner farmers directly depend on agriculture, while those who are landless, work for the owner cultivators. They perform the operations right from land preparation to harvesting and threshing. In 1951 the working population dependent on agriculture was 69.90 per cent. In 1997, it has come down to 61 per cent. In 2012, working population dependent on agriculture was 51 per cent. Though in relative terms, the percentage of dependence seems to have declined, but in absolute terms, more number of people are found to engage in agriculture. The reason for this trend, is that other sectors *viz.*, secondary and tertiary sectors have not made much headway in providing employment to the rising population. When the comparative analysis of population dependent on agriculture between the developing and developed countries is made, it is found that 59 per cent in Bangladesh and 48 per cent in China as against 5 per cent in Japan, 4 per cent in France, 2 per cent in USA and UK were found to depend on agriculture.

## Agriculture as a Source of Food

Agriculture provides food and fibre. The role of agriculture with regard to food production has been remarkable, in that it has been able to provide sufficient quantities eventhough the population has increased tremendously. The foodgrain production has increased from 50.80 million tonnes to 203 million tonnes during the period from 1950-51 to 1998-1999. The per capita availability of cereals increased from 354.10 gms during 1950-51 and 1955-56 to 442 gms during 1996-97 and 1999-2000. However, in respect of pulses, the per capita availability declined from 64.70 gms per day during 1950-51 and 1955-56 to 35.90 gms during 1996-97 and 1999-2000. These figures reveal that India had succeeded in the production of cereals but in respect of pulses, similar strides could not be made keeping the requirements of the people in view by 2013, the per capita availability of cereals was 468.9 gms, while that of pulses stood at 41.9 gms.

## Agriculture as a Supplier of Raw Materials

There is a close interdependence between agriculture and industrial sectors. While agricultural sector supplies raw material and wage goods to the industrial sector, the latter provides materials for building up of social and economic overheads and also the basic consumption goods to the agricultural sector. Besides, agricultural sector provides basic items of consumption *viz.*, rice, wheat, pulses, oils, fruits, vegetables *etc.*, for the population. The growth and development of agro-based industries also rests on the development of agricultural sector. Agro-based industries are classified into four types *viz.*, agro-produce processing units, agro-produce manufacturing units, agro inputs manufacturing units and agro-service centre. Agro produce processing units deal with the processing of agricultural produce. The examples are rice mills, oil mills *etc.* Agro-produce manufacturing units help in bringing out altogether new products making use of the agricultural produce. The relevant examples to be cited are sugar factories, bakeries *etc.* Agro-input manufacturing units are involved in the production of farm supplies required in farming. The examples are fertilizers, pesticides, seed and feed industries, industries manufacturing machinery, equipment and implements *etc.* Agro-service centres are those service centres which provide service facilities to the machinery and implements used by the farmers. The agro-based industries meet domestic demand and at the same time contribute to the exports.

## Contribution of Agriculture to Capital Formation

Economic development is possible when the rate of capital formation increases to sufficiently high levels. Since agriculture is the largest industry in our country, this sector has a bigger role to play in capital formation. In agriculture, the productivity of the inputs directly depends on the capital resources *i.e.*, machinery, equipment irrigation and power projects. The main problem of the agricultural sector in the past was the inadequacy of capital resources. Mostly the agriculture practiced was labour intensive in nature. The methods of cultivation were traditional and hence did not help much in the capital formation. In the changed scenario, there should be greater public and private investment in agricultural sector to give a dimension, which will help to produce marketable surplus of the farm commodities. Along side, greater emphasis should be given for developing the marketing infrastructure without which any amount of efforts on production front would not yield the desired results.

## Agriculture and International Trade

Many farm commodities produced in the country are exported. India's agricultural exports are tea, coffee, tobacco, cashew kernels, spices (cloves, cardamom, chillies, pepper, *etc*), molasses, cotton, fruits, marine products, *etc*. The agricultural sector thus earns valuable foreign exchange. The percentage share of agricultural sector in total exports was 44.38 per cent in 1960-61, while the same during 1997-98, 1998-99, and 1999-2000 was 19.45, 18.50 and 14.43 per cent respectively in 2012-13 the same was 10.66 per cent. The share of agricultural exports has been declining as observed from this information. This is because of the faster increase in the export of manufactured products *viz.*, textile fabrics, cotton yarn fabric, readymade garments, leather and leather products and gems, and jewellery, chemicals and allied products, machinery *etc*. It should also be noticed that the agricultural sector has indirectly contributed to the export of some manufactured products by providing the basic raw materials. Infact, as the economy develops, the share of such indirect exports is expected to go up.

# CHAPTER 50

# Agriculture in Five-Year Plans

Agriculture has a long history of 10,000 years and it is the backbone of the Indian economy. At the time of independence, its contribution to the national income was 50 per cent, but in the year 2000 its contribution has come down to 25 per cent. Before independence there was stagnation in agriculture and it was crawling at an annual growth rate of 0.3 per cent, during the first half of the 20th century. Setting up of Royal Commission of Agriculture during 1928, under the chairmanship of Marquess of Linlithgow marked the first serious effort directed towards the improvement of agriculture. World economic depression of 1929-33 affected Indian market for raw and manufactured jute and demand for other Indian exports. There was a significant fall in the prices of agricultural commodities and this hit the rural people hard. Many farmers turned into agricultural labourers. So, the first task of Indian Government under the British regime was, therefore, to initiate growth in the process of agriculture.

Planning as an instrument of development was underlined by Sri. M. Visveswarayya in the year 1934. But the first attempt regarding planning was made by National Planning Committee under the chairmanship of Pandit Jawaharlal Nehru in 1938. However, the committee could bring out the reports on planning in 1948 only. Meanwhile, immediately after independence, the All India Congress Committee appointed the Economic Programme Committee with Pandit Jawaharlal Nehru as the chairman in November 1947. The committee submitted its reports recommending the institution of a permanent Planning Commission on January 25th, 1948. In March 1950, the Planning Commission was set up by the Government of India under the chairmanship of Pandit Jawaharlal Nehru. The first five-year plan draft outline was issued by the Planning Commission in 1951.

The first five-year plan in the country was started in the year 1951 and ended in 1956. It gave importance to agriculture and industry. Nearly Rs. 600 crore was allotted to agriculture, out of the actual public sector outlay of Rs. 1,960 crore. This constituted 30.61 per cent.

Second five year plan was implemented in 1956 and closed in 1961 with pubic outlay of Rs. 4,600 crore, out of which Rs. 950 crore was earmarked to agricultural sector. Though the actual outlay on agriculture increased to Rs. 950 crore in the second plan, but its percentage share in the public outlay decreased to 20.65 per cent.

Similarly, in the other subsequent plans though the outlay on agriculture was increasing but its percentage in total public outlay has remained more or less stable at 21 to 23 per cent. During sixth plan and ninth plan, the percentage share of

TABLE 50.1   Share of Agriculture in Five-Year Plans.

| Sl. No. | Plan | Actual public sector outlay (Rs. in crores) | Outlay on agriculture and irrigation (Rs. in crores) | Percentage share |
|---|---|---|---|---|
| 1 | First plan (1951-56) | 1,960 | 600 | 30.61 |
| 2 | Second plan (1956-61) | 4,600 | 950 | 20.65 |
| 3 | Third plan (1961-66) | 8,500 | 1,750 | 20.59 |
| 4 | Annual plans (1966-69) | 6,625 | 1,578 | 23.82 |
| 5 | Fourth plan (1969-74) | 15,900 | 3,810 | 23.96 |
| 6 | Fifth plan (1974-79) | 39,430 | 8,740 | 22.17 |
| 7 | Annual plans (1979-80) | — | — | — |
| 8 | Sixth plan (1980-85) | 1,09,290 | 26,130 | 23.91 |
| 9 | Seventh plan (1985-90) | 2,18,730 | 48,100 | 21.99 |
| 10 | Annual plans (1990-92) | — | — | — |
| 11 | Eighth plan (1992-97) | 4,95,670 | 1,02,730 | 20.73 |
| 12 | Ninth plan (1997-2002) | 7,26,000 | 1,66,583 | 22.95 |
| 13 | Tenth plan(2002 - 07) | 9,21,291 | 1,16,282.66 | 12.6 |
| 14 | Eleventh plan (2007 - 12) | 36,76,936 | 3,92,439 | 10.67 |
| 15 | Twelfth plan (2012 - 17) | 1,36,89,500 | 13,25,100 | 9.68 |

# FIRST FIVE-YEAR PLAN (1951-56)

At the time of independence the country was facing foodgrain crisis, hence the first five-year plan focussed on the increased production of foodgrains. Its target was fixed at 62.6 million tonnes. But due to popularization of Japanese method of culti-vation, the production of rice increased to 28.7 million tonnes. Another remarkable achievement made in the first plan was that all the actual production levels of rice, wheat, oilseeds, sugarcane, cotton *etc.*, were above their targeted levels. Community development programmes and extension services were given prime importance in this plan. The National Extension Service was initiated in 1953. In order to achieve targeted production level, emphasis was laid on improved tillage practices, use of improved seeds, fertilizers, water *etc.* Land reclamation and land development measures were also given importance during the first five-year plan.

# SECOND FIVE-YEAR PLAN (1956-61)

The target of foodgrain production was envisaged at 24 per cent higher in the second plan over that of the first plan. It wanted this increased production to be achieved through the adoption of improved technology and intensive cultivation. The second plan laid emphasis on diversified agriculture and development of livestock. The level of food production in the second plan was around 76 million tonnes. Similarly, the production of sugarcane also increased. The number of agricultural credit societies increased from 1,05,000 to 2,10,000. The total loan advanced by PACS was around Rs. 200 crore. The total net irrigated area increased from 51.5 million acres in the first five-year plan to 70 million acres in the second five-year plan.

# THIRD FIVE-YEAR PLAN (1961-66)

The third plan provided larger outlay for the development of large irrigation projects and small irrigation schemes, soil conservation programmes and development of cooperatives. The programmes aimed at the doubling the rate of agricultural production growth. The two specific priority goals envisaged were *viz.*, (1) to produce enough foodgrains and make the economy self-sufficient and (2) to produce commercial crops to meet the exports and industrial sector. With these goals in its framework, a target of increased agricultural production as a whole by 30 per cent was envisaged. Foodgrain production actually increased by 32 per cent over second plan. Rice, wheat, oilseeds, cotton, pulses *etc.*, had also increased by 30 to 50 per cent. Irrigation increased to 90 million acres, soil conservation measures were undertaken in an area of 11 million acres, dry farming technology in 22 million acres and land reclamation in 3.6 million acres. Extension and community development programmes were extended nearly to 360 million farmers. Co-operative farming was given a boost through pilot projects in 320 districts of the country. Intensive agricultural district programme (IADP) was started first in three districts during 1960-61 and later extended to 13 districts. In the remaining part of India, Intensive Agricultural Area Programme (IAAP) was started in 1964-65. Due to drought conditions, agricultural production had declined from 82.7 million tonnes in 1961-62 to 72.3 million tonnes in 1965-66. National Seeds Corporation was set up in 1963. Agro-Industries corporations were also started in different states in the country. National Co-operative Development Corporation (NCDC) and Agricultural Refinance Corporation (ARC) were also started in 1963. Co-operative marketing societies and co-operative processing societies were rigorous in their activities with a turnover of Rs. 583 crore. In sum, the third plan failed in agricultural front because only 10 per cent increase was obtained against the target of 30 per cent.

## ANNUAL PLANS (1966-69)

Due to the wars of 1962 and 1965, the fourth five year plan could not be immediately implemented in 1966. There was a steep fall in agricultural production during 1965 and 1966. Devaluation of rupee also occurred in 1966. These events necessitated readjustment in the planning process. As a result three annual plans were formulated for three consecutive years in the place of fourth plan. The total outlay for these three plans amounted to Rs. 6,625 crore out of which 23.82 per cent *i.e.*, Rs. 1,578 crore were allotted to agricultural sector. The High Yielding Varieties Programme (HYVP) was implemented during the year 1967. Food production rose to 95.1 million tonnes in 1967-68 due to HYV programme. Minor irrigation was given utmost importance in these plans. During the period 1966-69, 1,60,000 private tube wells and 7,00,000 pumpsets were added. Institutional source provided Rs. 200 crore for the development of Agro Industries Corporations, land development and irrigation development. Irrigated area increased by 18.6 million ha under major and medium irrigation projects and 8.1 million ha under minor irrigation schemes.

## FOURTH FIVE-YEAR PLAN (1969-74)

This plan had two major objectives for the development of agricultural sector. The first objective aimed at an annual growth of 5 per cent in agriculture and the second objective was aimed at facilitating economic development of large number of rural population including small cultivators, dry land farmers, landless agricultural labourers *etc.* Accordingly, priority programmes in agriculture looked into maximizing production from the rural areas and bridging the gaps between rural and urban areas eliminating imbalances in the region. The target of foodgrain production was kept at 125 million tonnes, oilseeds 10.7 million tonnes, cotton 8.5 million bales, jute 9 million bales *etc*. Special programmes in the selected area were implemented with improved package of practices of high yielding varieties of seeds. Emergency Food Production Programme was also started. Importance was given to the use of potential input supplies, agricultural research and extension programmes, incentives to farmers, efficient use of irrigation water, etc.

The target for rice production could not be realized and it was only 44.1 million tonnes. Oilseed production and pulse production were also affected very badly. Inspite of favourable price trends, centrally sponsored scheme of integrated dry land agricultural development was launched with 24 pilot projects in the selected areas for demonstrating dry land technology during 1970-71. All India coordinated research projects for dry land agriculture under the purview of Indian Council of Agricultural Research (ICAR) was also started. A central scheme on soil conservation was extended to eight more catchment areas to cover 80 million ha. under soil conservation programme. The total agricultural credit (ST and MT loans) disbursed to small and marginal farmers amounted to Rs. 757.9 crore and that of long-term credit was Rs. 628.3 crore in the year 1973-74. The community development programme covered nearly 5,123 blocks having 5,67,000 villages. On the whole, the plan achieved an annual growth of 2.8 per cent in agricultural production as against the targeted growth rate of 5 per cent.

## FIFTH FIVE-YEAR PLAN (1974-79)

At the time of implementing fifth plan the economy was facing severe inflation. The major objectives of this plan were (1) to achieve self-reliance and (2) to adopt suitable measures to raise standard of living of rural people below poverty line. The fifth plan strengthened the strategies of agricultural production of the previous plans. High yielding varieties of cereals, multiple cropping practices and water management methods were adopted mainly in the command areas of major and medium irrigation projects. Besides, seed technology, substantial development in the use of manures, and efficient distribution arrangement of chemical fertilizers were given prominence for overall agricultural development. The food grain production achieved during this plan was 125 million tonnes. Sugarcane, pulses and cotton production exceeded their targets, whereas, jute and mesta production was below their targets due to excess rains. The area under high yielding varieties was 35.2 million ha. Gross cropped area increased to 173.92 million ha. During 1977-78, Krishi Vignana Kendras and training centers were established for imparting advanced and potential agricultural technology to the farmers. 78,000 gobar gas plants were also installed during the period. National seeds projects were also launched in the states of Andhra Pradesh, Maharashtra, Bihar and Karnataka. The share of agriculture in the bank credit was 11.8 per cent in 1978. Short term loans advanced were Rs. 1,650 crore, medium term loans Rs. 235 crore, and long term loans Rs. 410 crore during the fifth plan period.

## SIXTH FIVE-YEAR PLAN (1980-85)

The sixth plan aimed at achieving an annual growth rate of 3.8 per cent in agriculture. To achieve this objective, it has devised strategies for crop production based on steady growth of foodgrain production, substantial increase in pulse production, self-sufficiency in oilseed production and increased production in tea, tobacco and spices.

Wheat and rice production was keeping pace with their targets during the plan, whereas rest of the crops did not achieve the target. The production of foodgrains at the end of the plan stood at 138.1 million tonnes. Rice and wheat recorded a production of 54.5 and 41.2 million tonnes respectively. The production of oilseeds was 10.4 million tonnes against its target of 12.7 million tonnes. A centrally sponsored scheme of community nurseries of rice and minikit distribution scheme of rice, wheat, bajra, maize and ragi were intensified in order to create a clear impact on production of cereal crops. NABARD was established in July 1982. The total agricultural credit advanced was Rs. 5,556 crore in 1984.

## SEVENTH FIVE-YEAR PLAN (1985-90)

The average annual production of foodgrains during seventh plan stood at 155 million tonnes. Production of cereals recorded 34.76 million tonnes in 1989. The recorded high level production of pulses and oilseeds was 14.06 million tonnes and 18.46 million tonnes respectively during 1990-91. Similarly, sugarcane production recorded a higher level of 240 million tonnes in 1990-91. Distribution of certified seeds was 57.04 lakh quintals during 1989-90. The area under high yielding varieties was 63.1 million ha. against the targeted area of 70 million ha. Agricultural credit sanctioned from institutional sources increased to Rs. 12,570 crore in 1989-90. The value of livestock sector rose to Rs. 27,700 crore in 1987-88. During this plan, dairy and fisheries development were given prominence, as a result, their production increased substantially. The plan aimed at direct attack on problems of rural poverty, rural unemployment and regional imbalances in the development. Rural employment programmes were launched during the seventh plan.

## EIGHTH FIVE-YEAR PLAN (1992-97)

The aim of the eighth plan was to generate surplus of food grains for export and attain self-sufficiency in pulses and oilseeds. The growth rate expected was 4 per cent in respect of gross value of output and 3 per cent with regard to the value added products. The targets set for 1996-97 were 88 million tonnes for rice, 66 million tonnes for wheat, and 17 million tonnes for pulses. Regarding oilseeds the targeted production was 23 million tonnes. The production of sugarcane was expected to be 275 million tonnes.

The strategy that was spelt out to achieve the targeted production levels were laying emphasis on dry land farming, spreading benefits of green revolution to the eastern region and improving efficiency of irrigation.

## NINTH FIVE-YEAR PLAN (1997-2002)

In order to achieve the objectives of removing the incidence of poverty and umemployment and ensuring food and nutritional security, ninth plan put a targeted annual growth rate of 4.7 per cent. The demand for foodgrain consumption would likely to increase from 197 million tonnes in 1996-97 to 298 million tonnes in the year 2011-2012. Hence, the production should increase from 199 million tonnes in 1996-97 to 337 million tonnes in 2011-2012. Similarly, demand for sugar and jaggery is expected to increase to 55 million tonnes in the year 2011 and that of milk and fish to 181 million tonnes and 11 million tonnes respectively. The supply of milk and fish production are put at 227 million tonnes and 15 million tonnes respectively in the year 2011-2012. The targeted growth rate of agricultural crops output is expected to increase at 3.82 per cent per annum and that of agriculture at 4.5 per cent. Annual targeted growth rate for livestock is 6.59 per cent, poultry 7.5 per cent, and fisheries 6.5 per cent.

## Strategies for Agricultural Development

We notice a significant shortfall in the investment on agriculture, from 24 to 21 per cent from sixth plan to eighth plan. This fall was not allowed in ninth plan and the public sector investment constituted nearly 23 per cent. Long gestation projects *viz.*, irrigation projects are given preference in the ninth Plan. Credit is considered as the most important factor of private investment and its availability to the people is ensured through interest rate reduction. Available minimum support policy is enunciated to provide adequate incentives to the farmers and at the same time keep the Indian agricultural commodities competitive in the markets. A clear cut input subsidy policy is framed to help the small farmers in the high potential areas. This plan envisages large expenditure on agricultural research, extension services, expansion of irrigation, and reducing the gap between irrigation potential and actual irrigation. Cropping intensity is expected to raise from 134.2 per cent to 140.4 per cent. Gross cropped area or gross sown area is expected to raise from 190.5 million ha. to 203 million ha. Fertilizer consumption is targeted to increase from 14.3 million tonnes to 20 million tonnes. To achieve high-targeted annual agricultural growth rate and exports, massive expansion and upgradation of agricultural marketing, storage and distribution infrastructure are given priority. Facilities for packing, grading, certification of agricultural commodities and development of future agricultural markets are given special attention with adequate funds.

## TENTH FIVE - YEAR PLAN (2002 - 07)

The targeted GDP growth rate was 8 per cent, while the actual growth achieved was 7.7 per cent. The foodgrain production target was 225mt to 243mt at the end of the tenth plan. Thrust areas for the tenth plan were: utilization of waste lands and un-utilised and under-utilised lands, reclamation/development of problem soils, rainwater harvesting and conservation for the development of rainfed areas, development of irrigation, especially minor irrigation, conservation and utilization of biological resources, diversification of high value crops, increasing cropping intensity, timely and adequate availability of inputs, strengthening of marketing, processing / value addition infrastructure, revamping and modernizing the extension systems and encouraging private sector to take up extension services, bridging the gap between research and farmer's yields, cost - effectiveness while increasing productivity, promotion of farming systems approach, promotion of organic farming, utilization of organic waste, etc. Continued emphasis will be placed on progressive institutionalisation for providing timely and adequate credit support to farmers with particular focus on small / marginal farmers and weaker sections of the society to enable them to adopt modern technology and improved practices for increasing agricultural production and productivity. An amount of Rs. 3,59,701 crore is estimated as production credit for distribution through institutional sources and Rs. 3,76,865 crore for investment credit, making a total of Rs. 7,36,566 crore for tenth plan.

## ELEVENTH FIVE - YEAR PLAN (2007 - 12)

The targeted GDP growth was 9 per cent, while the achievement was 7.9 per cent. The National Food Security Mission (NFSM) was launched starting with rabi 2007-08. The scheme has three components - NFSM - rice, NFSM - wheat and NFSM - pulses. Another objective was accelerating agri - growth through high - value segment (horticulture, livestock and fisheries). Incentivizing the states to ensure that APMCs are reformed and notified for direct buying from farmers and encourage clustering of farmers in groups through NGOs, be it in the form of cooperatives and farmer clubs, were the other areas highlighted. Planned to encourage NABARD refinance to SHGs at 7 per cent interest rate with the condition that they will not charge more than 11 per cent from farmers. National Land Records Modernization Programme (NLRMP) was launched in 2008. Massive programme of 'Jalayagnam' of creating 98.41 lakh ha of new irrigation potential was also planned.

## TWELFTH FIVE - YEAR PLAN (2012 - 17)

In the 12th five year plan 9 per cent GDP growth is expected. Agricultural and manufacturing sectors are expected to grow at 4 and 10 per cent respectively. Coming to agricultural research, the important issues are increasing funding to agricultural R & D, farming system based technology developments, national initiative on climate resilient agriculture, inter - departmental platforms for multi-disciplinary research in priority areas, national agricultural entrepreneurship project to nurture entrepreneurship through research and national agricultural education project to improve education quality in State Agricultural Universities. Rural economy growth has to be enhanced by sustained agricultural growth and development of rural areas by providing rural infrastructure and amenities. Forest economies and tribal societies need greater protection and promotion.

# CHAPTER 51

# Agricultural Technology

Technology is the knowledge applied by man to improve production or marketing process. It is seen in hybrid seeds, improved crop varieties, pesticides, machinery and fertilizers. The objective of technology is to provide more output from a given bundle of production inputs. It is vital to the economizing process. Technology is related to a specific activity. It connotes a way of completing a particular task. This is the reason why it is generally a suffix to the activity to which it is related. Hence, we have technologies like crop production technology, dairy technology *etc.*, with technology referring to a specific activity.

According to Moxcher, technology in farming is "simply the way things are done". A given technology implies a given set of inputs used in a production process.

Indian agriculture in the last three decades has witnessed growth and development. It started with the green revolution during mid sixties encompassing HYV seeds, fertilizer, irrigation and plant protection measures, coupled with substantial investment on research and development. The subsequent revolutions that followed in oilseeds (yellow revolution), milk (white revolution) and fish (blue revolution) contributed to the phenomenal increase in agriculture, dairy and fish production. India now stands as the largest producer of wheat, rice, fruits, vegetables, spices, milk, fish, eggs *etc.*

## Emerging Problems

Despite the significant achievements, there are problems causing concern as presented below:

1. The population of the country is growing at a higher rate of 1.8 per cent and in spite of the achievements on food front, it has over 200 million impoverished people.
2. It is estimated that the total food grain requirement would be about 220.5 million tonnes by the end of 2002. The Planning Commission has fixed a target of 4.5 per cent annual growth rate of agriculture during 1997-2002, so as to achieve a GDP growth rate of 6.5 to 7.5 per cent.
3. Providing food to every one is a problem confronting the country today. The role of sustainable agriculture is prominent here. We would have to produce more per unit of land with less of chemical inputs.
4. Declining per capita availability of arable land.
5. There is widespread mismatch between production and post-harvest technologies. Minimizing losses and making availability of required agricultural commodities is possible through improved post harvest technologies.

6. Inadequacy of infrastructural facilities in rural areas.
7. Emerging problem of declining capital investment in agriculture.
8. The existing institutions are in a state of reorientation, particularly following the WTO regime of liberalization and globalization. The role of state, co-operative sector, private sector, NGOs *etc.*, are to be redefined to face the new challenges relating to input use and input and output markets.

# TECHNOLOGICAL DEVELOPMENTS ACHIEVED

## Bio-technology

1. *Bio-diversity in Agriculture:* The gene map for the model experimental plant Arabidiopsis has been worked out. Scientists are of the opinion that mapping and sequencing should be given top priority, as they are the potential contributors of genetic improvement in crop plants and livestock. Perennial grasses along with others had been living over thousands of years without any cultural practices. However, the wild species are unsuitable for cultivation. Hence, an understanding of structure of the genes for these traits is essential for replacing them with genes of desirable qualities.

2. *Boll Guard Cotton:* Boll worm attack is the single major threat to the cultivation of the cotton. Efforts in the direction of controlling the boll worm challenge through spraying bacterial extract gave only temporary relief. Attempts were therefore made to transfer the genes responsible for the bacterial toxin to the cotton plant. The efforts were successful in 1981, but it took almost 15 years to complete field trials. In spite of this achievement, still cotton farmers are beset by the boll worm infestations. Therefore, it calls for studies on the suitability of boll guard cotton to our conditions and recommendations by the experts in the field.

3. *Genetic Enhancement of Plants:* The International Rice Research Institution (IRRI), Philippines, is in the forefront of the genetic engineering research in rice and had already transferred several genes targeted against stem borer insect, fungal sheath blight and bacterial blight. The new varieties have to be field tested before they are released for cultivation.

## Water Management

Since water being a scarce resource various techniques have been developed for increasing water use efficiency. Watershed management is the approach in the irrigation development particularly in rainfed agriculture. In scarcity regions the practice that is gaining momentum is protective/supplemental irrigation. As the watersheds are spatially laid from ridge to valley, they most efficiently conserve land and water resources and facilitate availability of water throughout the crop period. The objective of micro-irrigation technique like drip irrigation is basically to optimize water use, thereby covering larger area with the available water discharge for better plant growth, coupled with reduced operational costs and increased productivity.

## Agricultural Engineering

It deals with mechanization through efficient use of inputs to increase farm productivity, conserving natural resources, reduce crop losses, improve quality of agroproduce, *etc.* Mechanization is one of the measures of modernization of agriculture.

## Bio-intensive Integrated Pest Management

Considering the ill-effects of the indiscriminate use of chemical pesticides, bio-intensive integrated pest management (BIPM), a technically feasible and economically viable method of pest management has become popular, particularly among small and marginal farmers. BIPM relates to conservation and augmentation of natural enemies of crop pests and adoption of all compatible cultural, mechanical, physical, genetic, selective chemical pesticides, tolerant varieties and legal methods.

## Nutrient Management

The availability of nutrient responsive high yielding varieties of crops led to intensive nutrient application and improved farm management to derive full benefits from such varieties. The practices of integrated plant nutrient system and also integrated pest management (IPM), are an outcome of the apprehension that the application of large doses of fertilizers, pesticides, fungicides, weedicides *etc.*, would lead to deterioration of soil health and pollution hazards. The consumption of nutrients which is about 16.5 million tonnes realized from 40 million tonnes of nitrogenous, phosphatic and potassic fertilizers is also causing a serious strain on foreign exchange reserves of the country. Hence it calls for exploration of supplementary and indigenous renewable form of nutrient sources.

## Integrated Plant Nutrient System

According to FAO, integrated plant nutrient system is the "management and conservation of natural resources base and the orientation of the technological and institutional agencies, in such a manner as to ensure the attainment and continuing satisfaction of human needs for the present and future generations". The components of this system are:

(i) On-site resource generation (recycling of crop residues), (ii) mobilization of off-site nutrient resources (addition of chemical nutrients from outside), (iii) resource integration (integration of generation on-site resources with chemical nutrients) and (iv) resources management (optimal use of all the resources). On the whole, the integrated plant nutrient system reflects the management of the farming system as a whole.

## Post Harvest Technology

Substantial quantity of food crops is lost after harvest because of handling and storage problems. The loss is estimated at 9.33 per cent for foodgrains. Post harvest technology with emphasis on farm post-harvest handling and storage system for different commodities should receive suitable attention. Analogously post harvest technologies should be evolved for diary products, fish, eggs, meat *etc.*

## Remote Sensing in Agriculture

In view of the country's vast and various agro-climatic regions, remote sensing has special significance in agriculture. It has been emphasized by the planners about the need of accurate forecasting system on particulars of crop production. The estimates of crop production help the strategies for planning, distribution, procurement, price stability, storage of agricultural commodities, *etc.* To have the accurate forecasting of crop production estimates at the district, state and national level, we have forecasting agricultural output using space, agro-meteorology and land project (FASAL).

To sum up there is a need to expand and refine the successful models like Krishi Vigyan Kendras (KVKs) for linking research, extension and farmers. Losses at the farmers' level can be minimized through opening and strengthening of Agricultural Technology Information Centres (ATIC) in state agricultural universities. Further, the district level extension system needs to be strengthened for effective transfer of technology.

# CHAPTER 52

# Land Reforms

Agrarian reforms comprise reforms in respect of all such areas which have a bearing on agricultural production, marketing and income distribution in rural areas. Land reforms are a part of the agrarian reforms. According to the Food and Agriculture Organization (FAO) the term, land reforms means: *"More than redistribution of land either by breaking up large estates to improve consolidation of land holdings, it must include a number of measures to improve the relationship of the man who works the soil to the land he works, including opportunity for land ownership, improved conditions of tenancy, agricultural credit at reasonable rates of interest, reforms of exorbitant rents and taxes and facilities for obtaining agricultural supplies and marketing agricultural products with emphasis on cooperatives"*.

In the beginning, land pertained to those who cleared and cultivated it. Therefore, the land ownership was passed on to the next generation of sons in equal proportion. In the process of colonization, rights on land originated even by conquest. Different races, castes and communities participated in the process of colonization and naturally acquired rights in the ownership, cultivation and management of land[1]. By the end of 18th century and the beginning of nineteenth century, village communities constituted different castes, races and religious communities. If members of the castes to which the family of the original founders belonged were not sufficient for cultivating the land of a village, cooperation of persons of other castes was freely sought and they were allowed to occupy and cultivate the village land[2]. The most important characteristic of the village communities is their self-sufficiency. State enjoyed part of the agricultural produce, as revenue though there was no rigidity as far as the extent of share was concerned. With the growth of farming community and area under cultivation practical difficulties arose in the collection of revenue in the kind form. The first attempt of collecting revenue in money from the farmer was found during the period of Timur. The emergence of revenue farming in the history of land tenure is an important feature. The revenue farmers were vested with the power of collection of revenue from the other farmers and remit to the Government. The revenue farmers over the years developed their own strategies of cultivating large waste areas of the village in addition to their own land and became quite powerful by the time East India Company came to power.

---

[1] Maine, H.S. Village communities in the East and West, London, 1895 p. 176.
[2] Maine *Op Cit* p. 128.

# LAND TENURE SYSTEMS

Land tenure deals with the rights and pattern of control of the land resource. At the time of independence, there were three types of land tenure systems in India *viz.*, Zamindari system, Mahalwari system and Ryotwari system. We notice a basic difference in these systems with regard to mode of payment of land revenue. In Zamindari system land revenue was collected by the Zamindars from the farmers and given to the Government. In Mahalwari system, village head was responsible to collect the land revenue. Under ryotwari system, land revenue was paid by the farmers to the state Government. There were also three types of tenants via., occupancy tenants, tenants at will and sub-tenants. Occupancy tenants enjoy permanent and heritable rights on land. Tenants-at-will did not have any security of tenure on the land and could be evicted at any time whenever the landlord desired to do so. The position of sub-tenants was also similar. In addition to these classes of people, a big class of agricultural labourers existed with no rights on land but worked on land for wages. Let us now know more about these types of land tenures.

## Zamindari System

It is an intermediary agency system between the State and the actual cultivator, with ownership and control rights in land with the intermediary *i.e.*, the Zamindar entitling to collect rent or appropriate share of crop produce from the actual cultivator. The intermediaries were vernacularly known as *Zamindars, Jagirdars, Jahagirdars, Talukdars, Khot, Lambarder etc.*,

This system was created by the British Government in the year 1793. Lord Cornwallis made settlement with landlords (Zamindars) to collect the land revenue from the farmers. The landlords were declared as full proprietors of land and in return the task of collecting land revenue or land rent from the farmers was entrusted to them. They were acting as the intermediaries between the farmer and the government. This system was prevalent in India up to 1949, in the states of West Bengal, Orissa, Andhra Pradesh, Uttar Pradesh, and Madhya Pradesh.

## Defects of Zamindari System

Zamindari system exploited the farmers as the zamindars were conferred unlimited rights to extract as much rent as they wished. They also enjoyed 25 per cent of the collected rent. Rent constituted 25 per cent of the value of the produce. There was hindrance to the development of agriculture due to large share of rent. Cultivators were caught in poverty and misery. Zamindars exploited the farmers and forced them to offer gifts/nazrana.

## Mahalwari System

Mahal means a village. The mahalwari tenure refers to the whole village rather than the individuals. This system was introduced by William Bentinck in Agra and Oudh. Later it was extended to Madhya Pradesh and Punjab. Mahalwari was a joint village system. The ownership of property was joint or community based. The share of the state in rent varied from 40 to 70 per cent. Under this system the whole village was treated as an unit to collect the land revenue. The responsibility of collecting land revenue was given to village headman or a co-sharer appointed for the purpose.

Period of settlement on the land, fixation of land revenue *etc.*, were different in different Mahalwari systems.

The distinction between Zamindari and Mahalwari tenures lies in that, while the Zamindar as an individual enjoys rights of ownership on his estate and directly responsible to the States for land revenue, while in Mahalwari, the ownership rights are held by the villages. In ancient India, there was no individual ownership and the land was held by villagers in common and its use was determined by the village panchayat. Mahalwari system was based on that custom.

### Ryotwari System

A system wherein the cultivator had a direct relationship with the State and had the right to transfer the land by sale or through inheritance. This was first introduced in Tamil Nadu and later extended to Maharashtra, Assam, Punjab, Coorg *etc.* Under this system, the responsibility of paying the land revenue to the Government was given to the farmers. The ryot or the farmer had full rights on land regarding sale, transfer and lease as long as he pays land revenue to the Government. Due to borrowing of loans from money lenders, cultivators had lost their rights on the land. Later money lenders started giving land for lease and collected exorbitant rents from the cultivators. The situation was alarming in Punjab, where the tenants were evicted. In Gujarat at the time of independence, there were 60 per cent of tenants. The amount of rent extracted from them varied from one-half to one fourth of the value of the total produce per acre.

## OBJECTIVES OF LAND REFORMS

Planning Commission report stated the following objectives of land reforms.
1. To remove such impediments to increase agricultural production.
2. To eliminate all forms of exploitation and social injustice within the agrarian system.
3. To provide security to the tiller.
4. To assure equality of status and opportunity to all sections of rural people.

## MEASURES

The following measures were taken up to achieve the stated objectives of land reforms.

1. Abolition of intermediaries.
2. Tenancy reforms and
3. Reorganisation of agriculture.

Tenancy reforms include.

1. Regulation of rent.
2. Security of tenure.
3. Ownership rights.
4. Ceiling on land holdings.

Measures of reorganization of agriculture include.

1. Redistribution of land.
2. Consolidation of holdings and
3. Cooperative farming.

# ABOLITION OF INTERMEDIARIES

During the year 1949 land reforms committee under the chairmanship of **Kumarappa** recommended for abolition of intermediaries and the introduction of **land reforms**. By the end of first plan, Zamindari system was abolished and nearly **173 million acres** of land was acquired from intermediateries and 2 crore tenants were **brought into** direct relationship with the State. Tenants cultivating land for more than **6 years were** given right to purchase the land at fair prices fixed by land tribunal. They were protected from exorbitant rents and illegal extortions. Zamindars were given compensation for the land acquired from them based on the net income or net asset basis.

Though the official documents claim that Zamindari system was completely abolished, the fact is that it has changed only its garb. The previous Zamindars acquired large areas for their personal cultivation with the help of hired agricultural labourers. They are now called big land owners. In the States where land ceiling is in operation, the ceiling for them was kept very high. Flaws in the legislation enabled them to transfer the land in the name of their kith and kin to escape the land ceiling laws. Soon after independence, exploitation of land-lords and oppression of tenants and tillers of the soil had declined steeply and the rural feudal structure has crumbled down.

# TENANCY REFORMS

Due to the tenancy reforms, occupancy tenants enjoyed permanent rights like the owner of the land and did not face the fear of eviction. There were regulation of rent, security of tenure and conferment of ownership rights on tenants due to this tenancy reforms.

## Regulation of Rent

The first five year plan stated that rent should be fixed at one fouth or one fifth of the value of the total produce obtained from the land. Except the States of Andhra, Punjab, Haryana, Jammu and Kashmir and Madras, all the States accepted this limit. In Madras 40 per cent of the gross produce was fixed as a maximum rent. In Andhra 30 per cent of the gross produce of irrigated land and 25 per cent of gross produce of dry land was fixed as rent. In Bihar, this limit was raised to 50 per cent.

## Security of Tenure

Persons cultivating the land on payment of rent are treated as tenants. However, in some States like West Bengal and Uttar Pradesh, share croppers were not regarded as tenants. A limited right of resumption of land by land owners for personal cultivation was granted in all the States. Land owners were permitted to resume the land rights up to the ceiling limit. There were flaws in the definition of personal cultivation and this rendered all tenants insecure. Laws relating to the securities of tenure were not implemented fully due to the absence of land records. Many landlords compelled their tenants to give up their tenancies. Landlords applied various kinds of threats and pressures on tenants to surrender the land.

## Ownership Rights

It has been estimated that as a result of these laws nearly 40 lakh tenants were given ownership rights over 90 lakh acres. Some States *viz.,* Madras and Andhra had not adopted any legislation on conferment of ownership rights, while in other States, the laws fell short of expectation. Many tenants could not afford to pay the purchase price of land and in some States many tenants were unwilling to purchase the land. As a result many tenants had foregone the privilege of having ownership rights.

## Ceilings on Land Holdings

A measure of land reforms, aimed at restricting the maximum size of land area a family can own and redistributing the surplus land so generated among the landless or small and marginal farmers, came into force. The imposition of land ceilings has two prime aspects *viz.,* (1) ceiling on future acquisition of land and (2) ceiling on existing holdings. The first five year plan favoured only the first aspect *i.e.,* ceiling on future acquisition, whereas second five year plan categorically recommended ceiling on existing holdings. To implement the ceiling act we require, the definition of family holding. The family holding considers two aspects *viz.,* operational unit and certain area of land which yield certain average income to the family. The first five-year plan defined family holding as an *"area equivalent to work unit for a family of average size working with such assistance as is customary in agricultural operations"*. Second five-year plan allowed flexibility in the definition of family holding based on different types of areas in the State.

The following categories of farms were exempted during second five-year plan.

1. Tea, coffee, and rubber plantations.
2. Orchards.
3. Specialized farms *viz.,* cattle breeding farms, dairy farms, wool raising farms *etc.*
4. Sugarcane farms operated by sugar factories and
5. Efficiently managed farms.

National guidelines for fixing up ceilings were evolved by the Chief Ministers' Conference in 1972. The main features of these guidelines were as follows:

1. The unit of application of ceiling shall be a family of 5 members consisting of husband, wife and minor children.
2. The ceiling should range from 4.05 to 21.85 hectares. The limit for the best category of land with assured irrigation and capable of raising atleast two crops a year should be with in the range of 4.05 to 7.28 hectares. In the case of land having assured irrigation of only one crop in a year, ceiling should not exceed 10.93 hectares and maximum limit for any type of land should not exceed 21.85 hectares.
3. Compensation payable for surplus land should be well below the market value of land, preferably in multiples of land revenue payable for the land.
4. Landless agricultural labourers, particularly those belonging to scheduled castes and scheduled tribes should be given priority while distributing surplus land.
5. All the acts enacted and amended by the States should be placed beyond any challenge in the courts of law on the ground of alleged infringement of fundamental right.

The ceiling for country as a whole was fixed as 18 acres of wet land and 54 acres of dry land. This land ceiling is applicable to one family unit consisting of five members in the family. Benami transactions of land if detected would become null and void with retrospective date of application. However, in the 9<sup>th</sup> schedule of the constitution, a new policy was enunciated at the Chief Ministers' conference and in that most of the States were given permission to amend existing laws on ceilings. The levels of ceiling accepted by the states were varying depending upon the rental value of the land. For example Andhra Pradesh, Madhya Pradesh and Karnataka fixed the land ceiling ranging from 4.05 ha to 21.85 ha, in Tamil Nadu it ranged from 4.86 ha to 24.28 ha, in West Bengal it was from 5 ha to 7 ha, in Rajasthan from 7.25 ha to 21.85 ha *etc.*

## Social Rationality of Land Ceiling

It is argued that in a poor country like India where the land supply is limited and number of claimants is large and therefore it is socially unjust to allow small number of people to hold large part of land and subjugate the interests of millions of landless labourers. Such condition is against the justice, equality and prosperity of the majority of the people. Large number of landlords functioned as absentee landlords and the productivity of their lands continued to be very less and therefore agricultural production by far hampered. There were significant inequalities in the income levels of the people. Hence it is socially justifiable to impose ceilings on land and distribute it to the actual users of land, making the tenants as the owners. Therefore, it is better to distribute the land according to the ceiling laws in order to increase the agricultural production as a whole, thereby raising income and bringing prosperity to the toiling masses in the country.

## Economic Rationality

There are economists who support the view that the small farms are more efficient than the large farms. In this connection Prof. C.H. Hanumatha Rao, pleaded that small farms offer more opportunities for employment and they require less capital compared to the large farms. Further, the small farms would have more productivity, because of intensive use of inputs and prudent management. Therefore, the argument that large farms would have economies of scale, should not come in the way of implementing the land ceilings. He further added that small farms can be made into large farms through cooperative effort so as to have scale economies.

# MEASURES OF REORGANIZATION OF AGRICULTURE

1. *Redistribution of Land:* It has been decided that once surplus land is obtained after fixing ceiling limits by the State government, it can be distributed among the farmers and landless labourers who do not have land rights on their name. Further these landless labourers and petty peasants should be provided access to joint cultivation on one hand and enable them to realize the benefits of cooperative farming on the other hand.

2. *Consolidation of Holdings:* The average size of holdings in India in infact is very small. It was 2 ha. in 1976 and 1977 and at present it is 1.57 ha (1990-91). Nearly 72 per cent of holdings belonged to small and marginal farms *i.e.* less than 2 ha. during 1976-77. Large farms with more than 10 ha. constituted only 3 per cent. The size of the holdings is decreasing but number of holdings is increasing over time. This is due to the inheritance laws prevailing in the constitution. The laws of inheritance are such that all the children would have equal share in the property of their parents. The inevitable consequence of this inheritance law is that farms

are being subdivided and fragmented with every passing generation. The population growth in India is increasing at an annual rate of 2.1 per cent. But the land area under agriculture had increased marginally that too at the cost of other categories of land. Due to this fact large number of people have been forced back to choose agriculture as their source of living. Further there is a decline of joint family system which was prevailing in earlier period. Nucleus families or disjointed family system is now leading to sub-division and fragmentation of holdings. Sub-division and fragmentation result in the following disadvantages.

(i) Wastage of land
(ii) Difficulties in land management
(iii) Difficulty in modernization or adoption of new technology
(iv) Disputes over boundaries
(v) Disguised unemployment
(vi) Low productivity *etc.*

All the difficulties which are emanating from sub-division and fragmentation can be solved if appropriate measures for consolidation of holdings are taken up. Initially the programme of consolidation was taken up on voluntary basis in India. In some States later it was made compulsory. However, this programme had to face many problems due to the following factors.

Since the quality of the soil varies from land to land it became difficult to allot the same quality of land for consolidation purpose. It has posed lot of problems, hence many alternative methods of valuation came into vogue. These methods were based on the market value of land or rental value of land. Due to emotional and sentimental attachment, farmers in general are not wiling to part with their land, hence consolidation programme is not becoming successful. In many areas up-to-date records are not available and this leads to difficulties in fixing up the ownership of the land. In many States there is no adequate and trained staff to carry out the programme of consolidation successfully. All these factors are infact delaying the implementation of the programme. However, in West Bengal, Gujarat and Madhya Pradesh, voluntary programme of consolidation is in progress. Some States *viz.,* Kerala, Orissa, Tamil nadu and parts of Andhra Pradesh, took exemption from the programme. In rest of the States it is made compulsory. It was estimated that 45 million ha. (25%) of the cultivable land is consolidated. The work was completed in Punjab, Haryana and West Uttar Pradesh.

3. *Co-operative Farming:* It has been advocated to solve the problems of subdivision and fragmentation of holdings. In this reform, farmers pool their small holdings for the purpose of cultivation and reap benefits of large scale farming. Its rationale is that, it is an effective method of solving the problems of small and uneconomic holdings. Such problem holdings are pooled together through this method and joint cultivation is taken up by the members. Agricultural inputs like seeds, fertilizers, manures, labour and tractor services are purchased by the society in bulk to reduce the cost per unit of input. All the necessary inputs are made available to the members at a cheaper rate and in adequate quantities. The advantages of scientific farming, adopting the new potential technologies can be reaped. Adequate employment to the farmers, bullocks and other resources is often made possible through this programme. Marketable surplus of foodgrains and industrial raw material can also be obtained more easily from the cooperative farms and these can also be marketed on bulk basis to reduce the marketing

costs. Co-operative farming leads to lay the foundation of strong democracy, self-help and mutual help.

# PROGRESS OF LAND REFORMS

In sum, the progress of land reforms can be categorised into three different phases *viz.*, 1936-55, 1956-71 and 1972-95.

## First Phase (1936-55)

Top priority was given by the State Governments and Union territories during this phase in terms of bringing actual cultivators into direct contact with the State through the abolition of zamindars and other intermediaries. At the end of the second five-year plan, intermediaries covering about 40 per cent of the area were largely abolished and as many as 20 million tenants were brought into direct contact with the State. Nearly 57.7 lakh hectares were distributed to landless agricultural labourers under Zamindari/Jagirdari abolition acts in various States.

Regarding tenancy reforms which aimed at conferring the rights of ownership of tenants, the Government had taken measures *viz.*, declaring tenants as owners and making them to pay compensation to owners in suitable installments, acquisition of the right of ownership by the State for payment of compensation and transfer of ownership to tenants and State's acquisition of the landlord's rights bringing tenants into direct relationship with the State. About 30 million tenants had got rights on tenure-held land in the process, and most of them free of cost. A significant number of tenants in Assam, Kerala, Maharashtra, West Bengal, Gujarat and Karnataka had benefited from the tenancy reforms, while in other States it was marginal. In States in which ownership rights could not be conferred on tenants, they were protected through security of tenure by not allowing eviction of tenants and reducing the rent to 20 to 25 per cent of the yield. Out of an estimated 15 million protected tenants, West Bengal alone had given recorded rights of hereditary occupancy to 1.46 million tenants. Despite the progress made, tenancy reforms still suffer with shortcomings.

1. These reforms have excluded the share-croppers who form the bulk of the tenant cultivators.
2. Eviction of the tenants still takes place on several grounds.
3. The right of resumption given in the legislation has led to land grabbing by the unscrupulous.
4. Ownership rights could not be conferred on a large body of tenants because of non-recording of tenancy in land records.

## Second Phase (1956-71)

The salient feature of this phase regarding land reforms is the imposition of ceilings on agricultural land holdings. All States enacted ceiling laws by the end of 1961 but the States fixed up different ceiling limits. Up to December 1970, the ceilings could identify one million hectares of agricultural land, out of which only 50 per cent was distributed among the rural poor.

## Third Phase (1972-1995)

Comprehensive land reforms in the form of reduction in the existing land holding ceiling limit took place in the conference of chief ministers held in New Delhi on 23rd

July, 1972. As per the revised ceiling limits, the Government irrigated land should enjoy a lower ceiling limit ranging from 4 ha to 7 ha, privately irrigated land from 5 ha. to 7 ha. and for dry land and other kinds of agricultural land, the ceiling limit shall not exceed 22 ha. There was some progress in the implementation of ceiling laws during the sixth-five year plan period (1980-85) but during seventh and eighth plan periods, there was no progress and it had come to a standstill in a majority of the States. At the end of September 1996, the area declared surplus was 30.59 lakh hectares, of which 26.47 lakh hectares were taken possession of and 21.28 lakh hectares were distributed among 51.21 lakh beneficiaries in the country. About 8.78 lakh hectares of land was not available for distribution for reasons such as involvement of litigation in various courts, reserved or transferred surplus land for public purposes, unfit for cultivation *etc.* The areas distributed under ceiling laws accounted for only 1.5 per cent of the net cultivated area and beneficiaries covered about 3.50 per cent people below the poverty line in the country. Most of ceiling surplus lands were poor in quality and require heavy investment for bringing under the plough. In most cases, the beneficiaries received less than 0.5 ha. of land and hence such holding could not provide economic sustenance to the beneficiary farm households.

Consolidation of land holdings as a measure to save the labour costs and other inputs to increase the productivity initiated before independence, achieved some progress in States like U.P., Punjab, Maharashtra, Haryana, M.P. and Gujarat. At present the programme of consolidation of land holdings seems to have come to a halt.

# CHAPTER 53

# Agricultural Labour

According to second Agricultural Labour Enquiry Commission (1960), agricultural labour families constituted 25 per cent of the rural families and nearly 85 per cent of the agricultural labourers were casual labourers, and these emanated from socially backward sections of the rural area such as scheduled castes, scheduled tribes *etc.* Fifteen per cent of the labourers were attached to specific landlords as permanent labourers. More than 50 per cent of the workers did not possess any land and even those who possessed land did have very little land *i.e.,* less than 1 acre. The size of labour population is increasing over time. According to National Commission on Agriculture (1991) 33.30 million households were agricultural labourers among 108.3 million rural households. This amounts to 30.75 per cent. This implies that agricultural labourers are increasing in their population.

## WAGES AND EARNINGS OF AGRICULTURAL LABOURERS

It was pointed out by the various reports and Commissions that agricultural labourers in India did not receive notified minimum wages except in certain States of the country *viz.,* Kerala, Punjab, Haryana and Uttar Pradesh. Even in these States females did not get sufficient remuneration as notified in the minimum wages act. Due to the prevalence of labour unions and organizations in certain areas wages are closer to minimum wages. But in most of the areas agricultural labourers were disorganized and have weak bargaining power. Due to these factors there is a gap between actual wages and minimum wages fixed by the Government. Despite this the real wages[*] of agricultural labourers had increased in all the States during the period from 1971-72 to 1988-89. Two more trends are also apparent. First, there has been decline in the regional disparities in real wages. Secondly disparities between wages paid to male and female labourers were decreasing over time. National commission on Agricultural Rural Labour (NCRL) in 1981 explained that the decline in disparities in male-female wages were due to the factors like implementation of rural employment and afforestation programmes. In these programmes the minimum wages for rural poor women were stipulated.

In the periodical revision of minimum wages, provision was made to pay equal wages to all the labourers irrespective of the sex, caste and creed. The practice of serfdom was also abolished. The living conditions of agricultural labourers were continued to be pathetic and deplorable. The monthly per capita expenditure was less

---

[*] Real wage is the actual wage rate divided by general price index for that year.

than Rs. 100. This means that they were below the poverty line and caught in the vicious circle of utter poverty and misery.

### Causes for Poor Economic Status of Agricultural Labourers

1. The prime cause for poverty of agricultural labourers is their social and economic status. Most agricultural labourers belong to the suppressed and depressed sections of the society, who were neglected for ages together. These are the socially depressed and never had the courage to assert themselves.
2. Most agricultural labourers are illiterate and ignorant. They live in scattered villages and they do not have organizations and unions. However, in urban areas workers are having unions and organizations. Due to these factors, farm workers are not having enough bargaining power to plead with the farmers and land owners for securing remunerative wages.
3. Agricultural labourers don't have continuous work on farms. They are seasonally employed. In the large tracts of dry land areas, they work only for 1 to 2 months. Even in irrigated areas, they work only for 100 to 200 days per year. For rest of the year they are unemployed, idle and helpless. Apart from under-employment there is unemployment and disguised unemployment. In the disguised unemployment marginal value productivity (MVP) of agricultural labourers is nearer to zero. Consequently, agricultural labourers are in low economic position in India.
4. Paucity of non-agricultural occupations in Indian villages is another important cause of poverty of agricultural labourers. The population of landless labourers is increasing steadily. This is adding to the supply of labourers and hence the demand is less for agricultural labourers. Rural indebtedness is another cause for their poverty. The farm labourers are borrowing money from pawn brokers, money lenders, big farmers *etc.* They pledge their movable property and borrow the amount at exorbidant rates of interest. Sometimes in this process some of them are becoming bonded labourers accepting lower wages. Thus due to the factors beyond their control and partly due to their weak bargaining power, they are not getting adequate wages and hence forced to lead miserable life.

## PROGRAMMES FOR IMPROVING THE CONDITIONS OF AGRICULTURAL LABOURERS

### National Rural Employment Programme (NREP)

Central Government had launched food for work programme on 1-4-1977 to help the rural people particularly agricultural labourers by generating additional employment to them and create durable assets in the rural areas. Later the same programme was renamed as National Rural Employment Programme (NREP) in October 1980. The programme was part and parcel of VI plan and implemented on 50:50 share basis between the Central and State Governments. The allocation of resources to the States and union territories was made on the fixed formula based on 75 per cent weightage to the population of agricultural labourers and marginal farmers and 25 per cent weightage for incidence of poverty in the state. Payment of wages was 1kg of food grains per day per labourer and balance wage was paid in cash. Wages were paid according to the minimum wages act. 10 per cent of the resources was earmarked for the use on the programmes of direct benefit of SCs and STs. Similarly, 10 per cent was earmarked for social forestry and plantation programmes. Important objective of this programme was generation of 300 to 400 million mandays and creation of durable community assets.

## Training of Rural Youth for Self-employment (TRYSEM)

This national scheme was implemented in August 1979, with the prime aim of alleviating unemployment among the rural youth. The main thrust of this scheme is directed against imparting necessary skills and made the new technology adopted by the youth for enabling them to take up suitable vocations of self-employment. The target fixed for the scheme was 2 lakh rural youth every year under IRDP areas and non-IRDP areas. Separate funds were provided for non-IRDP areas but not for IRDP areas. This programme was expected to cover all the blocks of the country in a phased manner. In effect TRYSEM is a part and parcel of IRDP programme. Training is given to the rural youth between the age group of 18 to 35 years for self-employment avocations. Sixth plan had made a provision of Rs. 500 crore for strengthening TRYSEM.

## Small Farmers Development Agency (SFDA) and Marginal Farmers and Agricultural Labourers Development Agency (MFAL)

This project was started in the fourth-five year plan to improve the economic conditions of the rural poor and agricultural labourers. Initially 46 SFDAs and 41 MFALs were started to tackle the problems of small and marginal farmers and agricultural labourers and benefit about 50,000 small farmers, 15,000 marginal farmers and 25,000 agricultural labourers. The main thrust of this programme was laid on the crop husbandry which included several intensive multiple cropping practices, introduction of HYVs, water harvesting methods in rainfed areas such as land development, soil conservation, development of minor irrigation, development of horticulture and adoption of potential technology for dry farming. These projects implemented animal husbandry programmes *viz.*, supply of milch animals, poultry, piggery, sheep and goat rearing, fisheries *etc.* Small farmers were given 25 per cent subsidy under these projects. Agricultural labourers and marginal farmers were paid subsidy up to 33.33 per cent on their investment.

## Tribal Development Agencies (TDA)

About 6 tribal development projects were implemented in 1971 and 1972 with a view to bring about economic development and accelerate pace of agricultural development in tribal areas of the country. Later the number of projects was raised to 8 in the country in 1977. These agencies are operating on a pilot basis with the functions of strengthening co-operatives, providing relief indebtedness, and building up of rural industries based on agricultural and forest produce. A total area of 24,060 ha. was benefited under minor irrigation and 6,450 ha. of land was reclaimed in the country by the integrated tribal development authority (ITDA).

## Hill Area Development Agencies (HADA)

It was started in fifth-five year plan and is being continued in States of Assam (2 districts), Tamil Nadu (1 district), Uttar Pradesh (8 districts), West Bengal (3 districts), Western ghat region constituting the States of Maharashtra, Karnataka, Tamil Nadu, Keral and Goa (156 districts). The strategy of this programme is directed to develop economic status of agricultural labourers and rural poor by taking up the projects on watershed development, crop husbandry, development of horticulture, animal husbandry and infrastructure in rural areas.

## Drought Prone Area Programme (DPAP)

During the fourth plan 'rural works progamme' was initiated to provide assistance to drought prone areas through taking up of various developmental schemes in the drought prone areas and providing remunerative work to agricultural labourers in these areas. Schemes concerned with medium and minor irrigation, road laying, soil conservation, afforestation, *etc.*, were taken up on priority basis to provide employment in off-seasons. The working group of planting commission on integrated rural development observed that the programme was required to take up long term measure instead of short term measure. According to their recommendation, it was redesignated as Drought Prone Area Programme. The following are the main objectives of DPAP.

1.   Development of small and marginal farmers and agricultural labourers
2.   Development and management of irrigation sources
3.   Changes in the agronomic practices
4.   Restructuring the cropping pattern and pasture development
5.   Soil, water conservation and afforestation; and
6.   Livestock development

## 7. Desert Development Programme (DDP)

This was launched in the country in the year 1977-78, based on the recommendation of National Commission on Agriculture (NCA). About 132 blocks in the country are being benefited in the states of Rajasthan, Haryana, Gujarat, Jammu and Kashmir and Himachal Pradesh. The programme aims at the integrated development of desert areas and increasing the productivity, income levels and employment opportunities of the rural people. Its strategy is aimed at afforestation, ground water development, construction of water harvesting structures, rural electrification, development of agriculture, development of horticulture, development of sericulture, dairy, *etc.*

## 8. Minimum Wages Act for Agricultural Labourers

Minimum wages act was passed in 1948 in some States. Minimum wages in agriculture were not revised for a long period. There is a strong opinion which is contrary to the minimum wages act in rural areas. The basic argument is that at the peak agricultural seasons, labourers are demanding more wages than the wages stipulated in the minimum wages act. Therefore, the enforcement of minimum wages has no meaning, but on the other hand, during slack season due to excessive supply of labourers and paucity of work, labourers are being paid less wages than that indicated in the minimum wages act. The National Commission on Labour (NCL) gave due weightage to this view and strongly recommended for continuance of minimum wages act on the ground of rights given to labourer's quality and growth in the economy. As regards the enforcement of minimum wags act, the commission suggested that panchayats should be given adequate right to implement minimum wages act. We also notice considerable disparities in the payment of wages across regions, crops, States, sex, age *etc.* Labour Ministers' conference held in August 1981, decided that minimum wages fixed for labourers should not fall below the poverty line. Wages should be properly linked with consumer price index. In pursuance of their recommendations, general minimum wages were fixed, taking into necessities like food, fuel, and shelter in respect of workers.

# CHAPTER 54

# Farm Mechanization

Farm mechanization means the introduction and use of non-biological power in carrying out various operations. The mechanization in the farm sector has to take several factors into consideration as it facilitates the speedy completion of farm operations with much ease. Farm mechanization is a different area of high-tech agriculture, in which modern machines are being put to use for land preparation, land development, inter-cultural operations, sowing, transplanting, harvesting, threshing *etc.* Simply it means the use of machinery for farm operations in the place of human and cattle labour. Prior to 1960s, agricultural engineers focussed their attention on the improvement of manually operated farm tools, animal drawn equipment, chaff cutters, sugarcane crushers and irrigation pumpsets. Farm mechanization received a fillip in the wake of first green revolution, notably in the States of Punjab, Haryana, UP and AP manufacturing of tractors and tractor-drawn equipment was first taken on small scale after importing Ford tractors from USA. Power operated threshers for wheat, jowar, *etc.*, tractor drawn seed drills, disc harrows, cultivators, levelling blades, puddlers and a host of power operated sprayers are now being seen on Indian farms. Power operated centrifugal pumps and diesel engines are now available in large numbers. Electric motors and diesel engines are being used for a variety of farm operations *viz.*, drawing water from wells, pumping water from tanks, threshing, shelling, chaff cutting and also for operations of flour mills, rice mills, oil expellers, *etc.* Introduction of engines and electric motors paved the way for mechanization in farming. During 1970, India started producing tractors. Despite lack of concerted efforts on the part of planners, the use of tractors, power tillers, harvester combines and electric pumps has gone up significantly over the last 30 years, while in 1972, 1.48 lakh tractors were in use, the number had gone up to 24 lakhs by 1997. Power tillers had gone up from 17,000 to 1.4 lakhs. The number of electric pumps rose phenomenally from 16.18 lakh to 1.17 crores in the country during the period.

Different types of farm equipment and tools are now classified as high-tech agriculture. These are four-wheel tractors ranging from 20 to 100 hp. Power tillers ranging from 5 to 12 hp, self-propelled paddy transplanters, tractor operated potato planter, self-propelled combines, proclainers, power operated sprayers and dusters.

## CASE FOR MECHANIZATION

These high-tech agricultural tools raise the productivity of agriculture from 5 to 30 per cent. Seed-cum-fertilizer drill increased the productivity of the land by 10 to 15 per cent and plant protection equipment enhanced the production by 10 to 20 per cent. Similarly harvesting equipment and irrigation pumps raised the productivity of land by 5 to 10 per cent and 10 to 30 per cent respectively. Large scale farming is made possible and feasible by mechanization. Another obvious advantage of mechanization is, it brings down the labour costs per unit of operation and saves time. Mechanization, therefore, increases the productivity of labour. For example a pair of bullocks can plough the land in ten days but the tractor can plough the same piece of land efficiently within a day. Heavy jobs like digging, bunding, land development, land levelling *etc.*, are easily done with the help of machinery. Transportation of huge amount of produce is easily done with trucks. Large scale sowing, harvesting, threshing, raising of multiple crops *etc.*, are efficiently done with the appropriate machinery.

## CASE AGAINST MECHANIZATION

It has been argued that mechanization has no scope in India because of large number of small and fragmented holdings. Further, the small holdings are scattered hence machinery cannot be used efficiently. Infact, large farms should exist for viable mechanization. In USA, Canada, Australia, and USSR, where the average size of holding is very large, mechanization is having an edge in saving the labour and bringing down the costs. The second criticism is directed to the argument that mechanization in India displaces many agricultural workers out of jobs or works. Agriculture is the main source of livelihood for most of the people in India. Hence, mechanization is the most serious objection. Similarly, it would render the existing cattle population surplus and useless. Killing the cattle for beef purpose is against the ethics of Hinduism which is the major religion in India.

## SELECTIVE MECHANIZATION

The introduction of selective and need-based agricultural equipment suited to Indian soil and agro-climatic conditions was taking place and the new equipment that would find acceptance with the farmer are cotton pickers, sugarcane harvesters and combines, vegetable planters, orchard sprayers, forage harvesters, rotary and drum type movers, automatic planters and sunflower combines.

Hence, it is wrong to argue that there is no scope for mechanization in India under labour surplus situation. Machines suitable for small farms as well as large farms can also be introduced in India. Such machines would enhance the productivity of labour. Some machines are inevitable for certain operations *viz.*, reclamation of lands, bringing new soils under plough, conservation of soils, ploughing the barren lands, bunding, levelling *etc.* The policy of selective mechanization is applicable to large and specialized farms. India has registered spectacular progress in producing the tractors, pumpsets, power tillers, transplanters *etc.* Punjab and Haryana are using large number of machines for various farm operations.

# CHAPTER 55

# Capital Formation in Agriculture

Investments are being made in agriculture to improve the quality of rural assets and enhance their productivity. Such investment made on capital assets of agriculture by private and public enterprises is generally referred to as capital formation in agriculture. Public investments in agriculture include investments in irrigation, treatment and reclamation of land, watershed development, farm supplies, electricity, flood control, agricultural education and research *etc.* Investments by Government have been made on processing, creation of warehouses, afforestation, procurement and distribution agencies like Food Corporation of India and State Trading Corporation, agro-industries, markets, transportation, communication *etc.* Private investments in agriculture are being made on irrigation, farm mechanization, farm buildings, livestock, land development and land improvement, development of plantation and horticultural crops, processing, transport vehicles *etc.* Both public investment and private investment are essential for creating competition and quicken the process of capital formation in agriculture. Absence of any one of them leads to lopsided capital investment and development of the economy. Infact both investments are complimentary to each other due to the following facts.

For example, if capital investment is made on irrigation projects, a vast command area would facilitate raising of irrigated crops in the place of dry crops. This leads to diversification and use of technology. This shift further makes the farmers to make additional investment in agriculture in the form of seeds, fertilizers, land development activities *etc.* Similarly, dairy development investment by the Government would provide link with milk processing facilities and leads to the purchase of large number of milch animals by the farmers. Investments made by the Government on sugar factories provide incentives to the farmers to adopt new technology in raising the productivity of sugarcane farms. Thus we can see public investment is having complementary effect on private investment. Private investment in agriculture ultimately paves the way for higher income and growth of agricultural sector. Further, the confidence of the private entrepreneur in the local area enhances, leading to capital formation. Public investments not only create physical assets and infrastructure in agriculture, but also provide income and employment avenues to the people and improve their risk bearing and risk taking ability. Such aptitude and ability of the people is necessary for entrepreneurial development in agriculture. Once enterprises are developed, there would be increase in the income levels of the people and thus generate additional demand for a variety of products to be produced by the farmers, which in turn leads to further investment in agriculture through forward and backward linkages. In sum investments in agriculture generate demand and increased production. With the adequate market development for increased production, income levels of the producers, consumers, and middlemen increase. This inturn makes available the investable funds and resources in the hands of the people for the growth and development of the economy. All these occur in the cycle of prosperity. Table 55.1 provides information on gross capital formation in agriculture from the year 1960-61 to 1996-97.

TABLE 55.1   Gross Capital Formation* in Agriculture (Rs. in crores).

| Year | Total | Public | Private | Per cent share | |
| --- | --- | --- | --- | --- | --- |
| | | | | Public | Private |
| 1960-61 | 1,668 | 589 | 1,079 | 35.3 | 64.7 |
| 1970-71 | 2,758 | 789 | 1,969 | 28.6 | 71.4 |
| 1980-81 | 4,636 | 1,796 | 2,840 | 38.7 | 61.3 |
| 1990-91 | 4,594 | 1,154 | 3,440 | 25.1 | 74.9 |
| 1991-92 | 4,729 | 1,002 | 3,727 | 21.2 | 78.9 |
| 1992-93 | 5,372 | 1,061 | 4,311 | 19.7 | 80.3 |
| 1993-94 | 5,031 | 1,153 | 3,878 | 22.9 | 77.1 |
| 1994-95 | 6,256 | 1,316 | 4,940 | 21.0 | 79.0 |
| 1995-96 | 6,961 | 1,268 | 5,693 | 18.2 | 81.8 |
| 1996-97 | 6,999 | 1,132 | 5,867 | 16.2 | 83.8 |
| 1999-2000 | 50,151 | 8,670 | 41,481 | 17.29 | 82.71 |
| 2000-2001 | 45,480 | 8,084 | 37,396 | 17.77 | 82.23 |
| 2001-2002 | 56,977 | 9,711 | 47,266 | 17.04 | 82.96 |
| 2002-2003 | 55,668 | 8,734 | 46,934 | 15.69 | 84.31 |
| 2003-2004 | 53,541 | 10,804 | 42,737 | 20.18 | 79.82 |
| 2004-2005 | 76,096 | 16,187 | 59,909 | 21.27 | 78.73 |
| 2005-2006 | 86,604 | 19,940 | 66,664 | 23.02 | 76.98 |
| 2006-2007 | 92,057 | 22,987 | 69,070 | 24.97 | 75.03 |
| 2007-2008 | 1,05,741 | 23,257 | 82,484 | 21.99 | 78.01 |
| 2008-2009 | 1,27,127 | 20,572 | 1,06,555 | 16.18 | 83.82 |
| 2009-2010 | 1,33,162 | 22,693 | 1,10,469 | 17.04 | 82.96 |
| 2010-2011 | 1,32,734 | 19,854 | 1,12,880 | 14.96 | 85.04 |
| 2011-2012 | 1,57,172 | 21,184 | 1,35,988 | 13.48 | 86.52 |
| 2012-2013 | 1,62,083 | 23,886 | 1,38,197 | 14.74 | 85.26 |

Source: Economic survey, 1998-99

## Trends in Capital Formation

It is obvious from the table that the share of public investment in gross capital formation in agriculture is found to be declining from 1960-61 to 1996-97. On the contrary

---

* Gross capital formation: The process of adding to the net physical capital stock of an economy in an attempt to achieve greater total output is called capital formation. When the new physical capital stock of the given year is added to the existing physical stock, the value of the gross capital formation in the economy is arrived at. Net capital formation is a derivative of gross capital formation. It is known that the physical stock of buildings, machinery etc., is subject to depreciation. When this amount has been deducted from the gross capital formation, net capital formation is arrived at.

private investment is increasing. Another noteworthy point is that the share of private investment is also increasing over the given period. The Central Statistical Organization (CSO) did not include investments for agriculture in the gross capital formation. Infact this also should be included to arrive at the gross capital formation.

Investment in agriculture as percentage of GDP is presented in Table 55.2.

TABLE 55.2 Investment in Agriculture as Percentage of GDP.

| Year | Public | Private | Total |
|---|---|---|---|
| 1990-91 | 0.6 | 1.6 | 2.2 |
| 1996-97 | 0.4 | 1.1 | 1.5 |
| 1997-98 | 0.3 | 1.1 | 1.4 |
| 1998-99 | 0.3 | 1.1 | 1.4 |
| 1999-2000 | 0.4 | 1.1 | 1.5 |
| 2004-05 | 0.5 | 1.8 | 2.3 |
| 2005-06 | 0.6 | 1.9 | 2.5 |
| 2006-07 | 0.6 | 1.8 | 2.4 |
| 2007-08 | 0.5 | 1.9 | 2.4 |
| 2008-09 | 0.5 | 2.4 | 2.9 |
| 2009-10 | 0.5 | 2.3 | 2.8 |
| 2010-11 | 0.4 | 2.1 | 2.5 |
| 2011-12 | 0.4 | 2.2 | 2.6 |

*Source: Economic Survey, Ministry of Agriculture*

Public investment decreased from 0.6 per cent in 1990-91 to 0.4 per cent in 1999-2000. Barring 1990-91, in the remaining periods as indicated in the table, private investment stood constant at 1.1 per cent. Total investment decreased during the period from 1996-97 to 1999-2000. Private investments in agriculture depend on the following key factors.

(1) Returns to investment.
(2) Availability of investable resources such as institutional credit, and
(3) Favourable terms of trade to agriculture.

Infact agriculture suffers from implicit taxation but not explicit taxation on account of the protection given to the industries. There is a widespread feeling that the farmers are not really getting remunerative prices for their produce. Government is imposing movement controls on produce, resorting to centralized procurement, low support prices *etc.* Hence, there is low profitability to the resources used in agriculture. This makes the farmers reluctant to make further investments in agriculture at the requisite levels. The All India Debt and Investment Survey (1998) had brought out that the rural households used only 12 per cent of debt on capital expenditure in agriculture during the year 1991, whereas the same in 1971 was 31.2 per cent.

Another determinant of returns to capital investment in agriculture is the extent of risk in agriculture, which is the emanating form production and marketing. When the crops fail to give adequate returns due to the natural calamities and failure of technology, farmers do not get confidence to invest in agriculture. At times even after producing the crops to the expected level, the market prices would crash or slump down thereby bringing less returns to the farmers. Hence as long as such risks exist, it would be difficult to envisage private long-term investments.

Another important aspect is the availability of investable resources from the banking institutions. Though the financial institutions make available funds for investment in agriculture, the farmers are critical of quantum of loans and cost of credit thereof. The farmers are being affected by the bottlenecks in the flow of bank credit. Banks are reluctant to advance adequate loans for investments in agriculture on account of perceived high risks, long repayment period and low productivity per unit of capital. Non-availability of backward and forward linkages has diminished the enthusiasm of the banks to support well conceived agricultural projects. Therefore, private investment in agriculture is not progressing as per the requirements. It is here the government has to play a crucial role in providing capital investments for agriculture. In the years to come agricultural production should grow many fold in order to take advantage of increased access to world markets under WTO agreements. Doubling of present food production in the next ten years, calls for massive investments in agriculture. The Incremental Capital Output Ratio (ICOR) of 3.3 is required in the agricultural sector to achieve the target. Private entrepreneurs should be encouraged to invest in commercial infrastructure, especially in the areas of storage, processing, marketing, transport, irrigation, electricity, *etc.*

## FACTORS AFFECTING CAPITAL FORMATION

Factors having bearing on capital formation infact are innumerable. These are divided into two categories *viz.,* price factors and non-price factors. The dominant conventional view has accorded primary importance to the non-price factors than the price factors.

### Price Factors

Among the price factors, the important factors are prices of the farm products and farm inputs. For the purpose of aggregate investment analysis, all these price factors are assumed in a single factor called the 'terms of trade index'. Favourable terms of trade are infact the guiding forces to attract the farmers to invest more in agriculture.

### Non-price Factors

These include technology, institutions, Government investments made in infrastructure, Government investments made in agricultural sector, quality and quantity of Government services *etc.* All these factors have a bearing on capital formation in agriculture. Among the technologies, seed and fertilizer technologies are noteworthy. Bio-technology *i.e.,* bio-fertilizers, blue green algae, tissue culture *etc.* is holding immense promise in raising the productivity of farm resources. Technology improvements in India are coming from both private investment and public investment. Development of institutions *viz.,* banking institutions, marketing institutions, institutional reforms like land reforms, co-operative farming institutions, *etc.,* are playing key role in capital formation. Infrastructural institutions are providing roads and electricity. The Government in the development of watersheds, markets, production of seeds, processing industries *etc.,* is making lot of investment.

## Measures to Improve Capital Formation in Agriculture

The main factor influencing capital formation at the micro level is the motivation of a farmer. A farmer should be motivated to undertake capital investment with the basic idea of augmenting the productive capacity of scarce farm resources. Though motivation is a powerful factor to accelerate capital formation still suitable environment must be created by the Government so as to enable the farmer to undertake capital investment. These measures from the Government include agrarian reforms, public investment, farm credit facilities, *etc.*

# CHAPTER 56

# Irrigation

Water is one of the essential inputs to produce crops and livestock products. Proper application of water to the soil is very much essential for reaping the highest benefits from agriculture. It is also important to see the following additional functions of water in crop production.

1. Water supplies adequate moisture to the soil which is essential for germination of seeds, chemical and bacterial process during the plant growth. Water keeps the soil and its environs more favourable and conducive for plant growth.
2. It dilutes and drains out salts in the soil.
3. It softens the soil clods and helps in tillage operations.
4. It enhances the productivity of all farm resources, and
5. It reduces the hazards of frost action.

Water is also recognized as vital resource for life, human and societal development and environmental sustainability. In this context water should be treated as social and economic good. Its management throws very much concern to the planners, managers, users *etc.*

India is having the largest irrigated area. Its gross irrigated area had also increased from 22.6 million ha. in 1951 to 96.9 million ha in 1997. Irrigation water management is the most crucial and relevant policy objective in the 21st century. In order to feed the teeming population there is a need to increase the productivity of irrigated area by two to three times.

Water resources in the country were estimated in the past by various committees *viz.*, Irrigation Commission, National Commission on Agriculture *etc.* The total surface water flow including regenerated flow from ground water was estimated at about 187 M.HM (million ha meters), out of which only 70 M.HM could be used and rest flows into the sea. The total ground water recharge from all the sources was about 45 M.HM in 1997, out of which 36 M.HM is used for irrigation purpose.

## DEVELOPMENT OF IRRIGATION

Irrigation was given the highest priority in first, second and third five year plans. The share of irrigation in the total expenditure was more than 25 per cent during sixth, seventh and eighth five year plan periods. As a result there is an increase in the irrigated area by 2.4 million ha. per year, on an average, during these plan periods. But still we have lot of potential in irrigation development. The maximum potential is estimated around 140 million ha. which is the target for achievement by the year 2025.

Planning Commission broadly classified irrigation schemes into three types in the year 1978-79.

## Major Irrigation Projects

Projects with cultivable command area (CCA) of 10,000 ha and above.

## Medium Irrigation Projects

Projects with cultivable command area of 2000 ha to 10,000 ha.

## Minor Irrigation Projects

Projects with cultivable command area of less than 2,000 ha.

Based on the project cost, major irrigation projects require more than Rs. 5 crore for their completion. Projects costing Rs. 25 lakh to Rs. 5 crore come under medium irrigation projects. Projects costing less than Rs. 25 lakh fall under minor irrigation projects. This classification was made in 1951 and existed up to 1978.

Development of irrigation is to be viewed from the following four broad sub-divisions

1. Major and medium irrigation projects,
2. Minor irrigation projects,
3. Command area development and water management, and
4. Flood control.

## 1. Major and Medium Irrigation Projects

In the first two plan periods major irrigation projects *viz.*, Bhakranangal, Rajasthan canal, Gandhisagar dam, Nagarjuna sagar, Hirakud, Tungabhadra *etc.*, were taken up in the country. The performance of these projects was examined by the National Irrigation Commission in 1972 and National Commission on Agriculture in 1976 and several other committees thereafter. According to these committees, irrigation potential was more than 4 million ha. Large irrigated areas were deteriorated due to the excess use of irrigation water, resulting in water logging and salinity conditions.

The following causes were identified by the committees.

1. Lack of adequate drainage facilities in the irrigated areas
2. Absence of proper irrigation distribution system
3. Lack of non-introduction of rotational distribution system of water to the farmers
4. Lack of anticipatory research on the optimum irrigation water use
5. Lack of suitable infrastructure and extension services
6. Poor coordination among the concerned Government organizations in the command areas

The following are the measures required to be taken up for the development of major irrigation projects.

1. Expeditious completion of ongoing major projects, which are technically feasible and economically viable.
2. Quick implementation of the programmes by aiding, strengthening and monitoring the organizations of the projects at the State level.
3. Taking up of relevant work of modernization of irrigation systems in a phased and planned manner.
4. Proper advance planning for scarce materials of construction required for major irrigation projects

5. Optimization of benefits through better operation of the existing systems
6. Conjunctive use of surface and ground water
7. Introduction of *warabandi* on rotational distribution system on the existing and the new projects.
8. Efficient use of irrigation water through monitoring the irrigation programmes for different regions, and
9. Settling river water disputes.

## 2. Minor Irrigation Projects

In the case of minor irrigation projects the following measures are worthy for adoption.

1. Active involvement of the revenue staff and project staff in the collection of loan applications for minor irrigation development.
2. Giving maximum emphasis on organizing local campaigns (credit melas) for clearance of large number of applications for minor irrigation.
3. Simplification and streamlining the procedures underlying the loan application
4. Revitalizing the organization for collection of hydrological data and formulation of feasible schemes.
5. Adopting quick methods for ground water exploration through survey and investigation.

## 3. Command Area Development (CAD)

Various comprehensive programmes were systematically devised under the guidance and supervision of command area development authority (CADA). This department devises many systematic programmes of land consolidation, scientific land shaping, construction of water courses and field channels to carry surplus water away from the fields, and system of roads to enable the farmers to carry the produce from the farms to the market *etc*. Besides, these measures, adequate and timely supply of inputs is to be ensured and marketing and other infrastructural facilities created so that the farmers are able to derive maximum benefits from the land and water. Following major functions are entrusted to CADA.

1. Land shaping and levelling.
2. Modernization of irrigation system and the development of main drainage system.
3. Construction of field channels and field drains.
4. Lining of field channels.
5. Arranging adequate supply of inputs in right time.
6. Adoption and enforcement of suitable cropping pattern.
7. Enforcement of suitable rostering system of distribution among the farmers, and
8. Strengthening the existing extension system.

## 4. Flood Control

Government of India had set up Rashtriya Barh Ayog (RBA) in 1976 to develop long-term strategy of flood control and its management. During the sixth plan period the following measures were envisaged under the flood control programme. These include flood forecasting and warning system, new embankments, drainage improvement projects, soil conservation practices, watershed development *etc*.

## IRRIGATION POLICY AND STRATEGY

Considering the urgent need for evolving a National Water Policy, the Irrigation Commission has recommended for the creation of National water resource council. Prime Minister chairs this council and it is entrusted with the formulation of policies regarding irrigation. These pertain to water conservation, water use and inter basin transfer of water. Different strategies are infact needed to develop irrigation in different States based on water resources and irrigation problems. For example, the irrigation policy in Punjab and Haryana should aim at ameliorating the problems of drainage and reclamation of problem lands. Salinity and alkalinity problems are given top priority in the irrigation strategy. In Madhya Pradesh, the Government is introducing large-scale sprinkler irrigation system due to the paucity of irrigation water in command areas. In Maharashtra, farmers are using irrigation water under drip irrigation system.

Tube wells are prominent in North and Eastern parts of Uttar Pradesh, Bihar and Orissa. Of late the Government of Andhra Pradesh is giving top priority for water shed development concept, taking up renovating and desilting the tanks, wells, construction of percolation tanks *etc.*, for developing the underground water potential. Water management is the main concern of irrigation policy and strategy. Water delivery system should be strengthened according to the areas served and crops grown. In South India rice based cropping system is consuming lot of water leading to indiscriminate use of irrigation water. This results in salinity and water logging conditions. The *warabandi* system of distribution of water according to the time schedule is not fully successful in command areas due to lack of on farm development programmes. New technologies have to be developed to conserve, store, convey and apply irrigation water effectively, efficiently and sustainably. Various farm development progammes relevant here are providing separate irrigation and drainage channels, lining the irrigation channels, taking the water through pipes and measuring water through controlling and regulation devices.

## CONJUNCTIVE USE OF SURFACE AND GROUND WATER

When well water and canal water are used in accordance with need, such system is called conjunctive use of water. Under such system the drainage and the salinity problems are brought down. Large quantity of water is conserved in the reservoirs and productivity of water is also enhanced, but this system for its sustainability, requires water percolation measures.

### 1. Sprinkler Irrigation System

Under this system water is conveyed through sprinklers to the crop to prevent seepage losses and to have controlled irrigation. Many seed farms and nurseries are being successfully raised under this system. About 30 to 40 per cent of water is saved under this system. This system is gaining importance in dry areas of Madhya Pradesh, Andhra Pradesh and Karnataka.

### 2. Drip Irrigation System

In India, wells nearly contribute to 30 per cent of the total irrigated area in the country. In many States the ground water table is fast depleting. As a result many wells have become dry and abandoned. Drip irrigation system is becoming popular in Maharashtra and Madya Pradesh. Under this system banana, grapes, vegetables, sugarcane *etc.*, are being successfully grown. About 40 to 70 per cent of the water is saved under this system and yields of the crops can be enhanced by 10 to 100 per cent.

## 3. Tank Modernization

About 2 lakh tanks are existing in India, but majority of these tanks are associated with problems like encroachment, siltation, poor tank structures and lack of farmers' cooperation. Inspite of the problems, tank irrigation offers more scope for increasing irrigation potential in the country, compared to canals and wells. But what is required at this juncture is desilting of tanks and taking up of adequate percolation measures for increasing ground water. To reduce the conveyance losses, the irrigation strategy must include sluice modification, canal lining, provision of additional wells in ayacut (command area), catchments development and on farm development works. These are infact the important components of tank modernization.

## 4. Farmers' Participation

Many studies on irrigation development disclose that the farming community must be made active participants in all phases of planning, design, implementation, operation and maintenance activities, if good standards of water management are to be achieved. This can be achieved by creating water users' associations (Pani Panchayats) to solve the problems of water use. Farmers confront with various technical ills of irrigation management. They range from inadequate design and operation of the main irrigation system to those of poor irrigation layouts and practices at the farm level. These ills lead to considerable water losses and unequal distribution of irrigation water. Besides, there are adverse affects *viz.*, water logging and salinity, reduced benefits from agricultural inputs, low productivity of irrigation input and uneven farm incomes. All these problems should be solved effectively through Pani Panchayats.

## 5. Re-use of Waste Water

The demand for water is continuously increasing due to the need for increased agricultural production and its ever-increasing demand in industry and urban areas. Due to urbanization and industrialization quality of ground water as well as surface water is being degraded. However, it is also possible to reclaim the degraded water and reuse it for different purposes like cooling, land scape irrigation *etc.* Large quantities of industrial waste water and drainage water are not reclaimed at present. Therefore waste water reclamation and reusing the water is an urgent need especially in drought areas.

## 6. Inter-basin Transfers

The problem of water scarcity can be overcome by inter basin water transfer. This means all the rivers, which carry adequate water should be connected by constructing canals and reservoirs. According to the National Water Development Agency (NWDA) as conceived by Dr. K.L. Rao, the Mahanadhi, the Godavary, the Krishna, the Pennar, the Cauvery and the Vaigai are to be connected by a canal having a length of 3,716 km.

The agenda for research and development on irrigation should tackle the following thrust areas and problems in irrigation.

1. Develop appropriate irrigation strategies for ground water use in conjugation with surface water.
2. Develop improved water management practices for enhanced and sustainable agricultural production in the command areas.
3. Develop technologies for efficient drainage of agricultural land and use of drainage water for irrigation.
4. Appraisal of water pollution hazards and develop suitable methods for its control.
5. Develop suitable strategies for integrated watershed development including percolation measures for enhancing ground water potential.
6. Improving the suitable technologies for raising the productivity of water, and
7. Reclaiming and reusing the sewage water to minimize the water demand.

# CHAPTER 57

# Agricultural Price Policy

The basic aim of agricultural price policy is the intervention in the agricultural produce markets to influence the price levels and their fluctuations, particularly from farm gate to retail level. Price policy is directed to bring about growth and equity in the country, therefore it is occupying a prime place in economic and political debate. It involves conflicting objectives. The price policy and its instruments are constantly being reviewed because the majority of the population is affected by the price policy. At the time of independence the aim was to increase agricultural production and supply of foodgrains to consumers at reasonable prices. Up to mid sixties various forms of price controls mainly imports of foodgrains and their distribution at the prices below the market rate were followed. With the introduction of green revolution in 1960s, price policy was assigned a positive role to increase the domestic agricultural production. Agricultural Prices Commission (APC) was set up in January 1965 with a broad framework of price policies. It was renamed as Commission for Agricultural Costs and Prices (CACP) in the year 1985.

The need for agricultural price policy was clearly identified due to the following factors.

1. Agricultural prices fluctuate more violently than the prices of industrial products.
2. Price fluctuations bring disaster to producers as well as consumers. Middlemen take undue advantage and exploit the rest of the population through speculation.
3. Price fluctuations retard economic development of the nation.
4. Price fluctuations affect the welfare of the consumers.
5. During inflation the low-income group of consumers will be very much affected because most goods will not be with in the accessible reach of the consumers because of high prices.

Therefore major objectives of the price policy are to safeguard farmers' interest and giving necessary incentives to augment the agricultural production. Consumers' interests are safe guarded supplying the essential commodities through the Public Distribution System (PDS) at fair prices. Therefore, basic instruments of price policy should lay concentration on terms of trade favourable to agriculture. Such favourable terms of trade would influence the input-output price relationship on production side and consumer's real income on the consumption side. Relative prices of agricultural commodities will have bearing on the allocation of resources on the requisite inputs among the crops. Price policy is also directed towards the stabilization of prices. Following objectives of price policies are formulated by APC.

1. To achieve price stability without hampering the income of the farmers.
2. To provide price support to the farmers in the initial stages and remunerative prices at later stages.
3. To protect the interests of the consumers, supplying essential commodities at fair prices.
4. To maintain the reasonable relationship between the prices of agricultural commodities and manufactured products through appropriate parity prices, thereby terms of trade become favourable to the core sector.
5. To maintain appropriate relationship among the competing crops and
6. To provide favourable climate to the farmers to apply requisite new technology for augmenting agricultural production through remunerative prices for agricultural commodities.

The present price policy put an end to a series of Government measures that were initiated before independence. Procurement of foodgrains, inter regional movement restrictions on foodgrains, rationing and distribution of essential commodities through civil supplies corporation *etc.*, were in operation even before independence. These control measures were continued in varying degrees even after independence. After the establishment of APC, first price policy was the provision of price support given to the farmers as incentive for increasing agricultural production. Later the Government of India, related to agricultural price policy set up many committees.

## RECOMMENDATIONS ON AGRICULTURAL PRICE POLICY

Recommendations of the Expert committee headed by Dr. C.H. Hanumantha Rao on methodology of cost of production of crops.

### The Terms of Reference of this Committee are

1. To examine the design, content and methodology adopted with regard to generation of cost of production estimates under the comprehensive scheme for studying the cost of cultivation/production of various crops.
2. To review "the terms of trade" between agricultural and non-agricultural sectors and suggest methods to safeguard the interests of farmers.
3. To recommend any other measure to improve the remunerativeness of crop production.

The Committee examined all the issues and recommended the following:

1. The casual hired labour may continue to be evaluated on the basis of actual wages paid whether they are market wages or statutory minimum wages.
2. Family labour be valued on the basis of actual wage rates for casual labour.
3. Procurement/minimum support prices announced before the sowing season should always provide for the possible rise in the cost of production likely to occur during the cropping season. The Commission should also have a second look at the changes in input costs before the market arrival of the crops and adjust the procurement/minimum support prices in case, the observed rise in input costs turns out to be higher than the anticipated rise.
4. The commission should publish the methodology including weighting diagram and index numbers of input prices used by them in their reports.
5. In order to account for management input of the farmers, the paid out costs be raised by 10 per cent.

## Important Recommendations of High Power Committee on Agricultural Policies and Programme

This Committee headed by Sri Bhanu Pratap Singh recommended the following:

1. Farming should be given the formal status of industry if the bias against this sector could not be removed otherwise.
2. It has strongly recommended the adoption of parity principle to ensure equitable terms of trade between the farmers and the rest.
   Recommendations of special expert committee (Sen's committee):
   Following are the recommendations of the committee:
1. Introduction of a new cost classification including six cost concepts *viz.,* Cost A1, Cost A2, Cost B1 Cost B2, Cost C1 and Cost C2.
2. Use of index numbers of costs with a three year base instead of absolute base and a guide for price fixation.
3. Construction of index of prices received and prices paid (prices paid for production items only) by farmers to examine the relative movement of prices over time.
4. Costing of inputs on the basis of actual expenses and imputed values based on the principle of opportunity cost.

The following are the major functions of APC:

1. To advise of the Government on the price policy embodying fixation of minimum support prices for agricultural commodities *viz.,* paddy, wheat, jowar, bajra, maize, ragi, barley, gram, tur, mung, urad, groundnut, mustard, soyabean, sunflower seed, copra, cotton, jute and tobacco.
2. To recommend appropriate measures to make price policy effective.
3. To suggest suitable measures to reduce the cost of marketing and recommend fair price margins at various stages of agricultural marketing.
4. To advice Government of India on any problem relating to agricultural prices and agricultural production.
5. To fix various price policy instruments *viz.,* minimum support price (MSP) based on average cost of production, procurement price as a premium over and above the minimum support price, ceiling price, issue price, levy price *etc.*

# ADMINISTERED PRICES*

## Procurement Price

It is the price at which the Government procures commodities from farmers/ processors to feed public distribution system and maintain buffer stocks. However, the Government through the policy instrument of procurement prices is imposing an element of compulsion on the farmers to sell part of their produce to the Government at the announced procurement price.

---

*The administered price regime includes (i) minimum support prices (MSP) for 23 commodities (seven cereals, four pulses, eight oilseeds, copra, raw cotton, raw jute and VFC tobacco); (ii) statutory minimum price for sugarcane; (iii) levy prices for rice and sugar; (iv) central issue prices for rice, wheat and coarse cereals for sale under public distribution system. The public agencies like Cotton Corporation of India, National Agricultural Cooperative Markeging Federation (NAFED), Jute Corporation of India and state level federation undertake open market operations.

## Ceiling Price

Ceiling price is an upper price level of a commodity fixed by the Government to protect the consumers from unwarranted price rise. By fixing ceiling price Government checks the traders in charging a higher price than the maximum price. It does not apply to the farmers.

## Minimum Support Price (MSP)

It is fixed based on average cost of production. This is mainly aimed at protecting the farmers against the price fall during surplus production or market glut conditions. MSP is announced by the Government of India ahead of the agricultural seasons every year. The minimum support price is the price at which the Government of India makes a commitment to purchase all the quantities offered by the farmers. But in reality MSP is always lower than the market price for foodgrains, as such the farmers are not obliged to sell their products to the Government.

## Issue Price

It is the price at which the commodities are made available to the consumes at fair price shops. Issue price always remains higher than the procurement price.

## Levy Price

The levy price applies to both farmers and traders. It can be imposed on both the the farmers and traders depending on the market situation for the food grains. The levy system is a procurement system, which makes it obligatory for the farmers, and traders to sell a special quantity to the Government at the procurement price. They are permitted to sell the remaining part in the open market.

Due to the element of compulsion, the procurement price is becoming levy price and according to levy, the Government through Food Corporation of India is purchasing rice and wheat from the millers. The millers are forced to sell a part of their rice milled by them to the FCI at the levy price and the quantum of levy component is as high as 75 per cent in Punjab, Haryana and Uttar Pradesh and 50 per cent in Andhra Pradesh, West Bengal and Orissa. The system of levy price exists as long as the demand for rice remains higher than the supply and millers expect to make profit by selling the produce in the open market. The quantity of coarse cereals purchased at minimum support price continues to be considerably lower than that of rice and wheat.

# CHAPTER 58

# World Trade Organization (WTO)

International trade is not a new or recently started activity. Trade between countries has been an age-old practice. In reality, it was free and there were not many restrictions. Transportation and settlement systems were the problems of trade between countries. The mode of transportation was sea, encountering problems like loss of life and goods due to poor navigation and pirate menace. The settlement was through gold or barter system. As the communication and transportation facilities improved and new settlement systems evolved, it was natural to expect increased international trade. However, several restrictions on international trade started creeping in (1) to protect domestic industry (2) to protect the prices of the goods in local markets (3) due to balance of payment problems and (4) due to extraneous reasons like political, favouring one country while discriminating the other. The foreign trade was affected and led to scarcity of goods in some parts of the world and glut in other parts. Even before the second world war ended, the allied countries gave serious thought in developing a system that would end the chaotic conditions prevailing and pave the way for an orderly conduct of international trade and promote good monetary relations among the countries. They started working with the objective of finding a system which would:

1. Help to remove the restrictions on trade.
2. Ensure free convertibility of currencies (which was suspended during the inter war period due to exchange controls) and
3. Maintain stability in exchange rates among the currencies.

The representatives of 44 allied countries met at Brettonwoods, New Hampshire, USA in June 1944 to give a concrete shape to their ideas. The conference made recommendations to set up three international organizations *viz.*, (1) International Monetary Fund (IMF) to deal with the problems of balance of payments (2) International Bank for reconstruction and Development (IBRD) to deal with the problems of reconstruction and development of economies of countries shattered in the two world wars and (3) International trade organization (ITO) to deal with the problems of international trade. The first two were set up in 1945 but there were serious controversies about the third i.e. ITO and therefore could not be established.

An international conference on Trade and Employment was held in Havana in 1947-48. A charter was drawn and signed by 53 countries for setting up of an

International Trade Organization (ITO). ITO could not be established because the Havana charter was not ratified by the US Congress. However 23 countries had agreed to continue the exercise of negotiating for trade concessions, which were incorporated in the General Agreement on Tariff and Trade (GATT). This agreement was signed on 30 October 1947 and came into force with effect from 1 January 1948. India was the founder member of GATT. The main purpose of GATT was to ensure competition in commodity trade through the removal or reduction of trade barriers so as to bring about all round economic prosperity.

## OBJECTIVES OF GATT

The preamble of the GATT mentions the following as its important objectives.

1. Raising of standard of living of public world over.
2. Ensuring full employment and a large and steadily growing volume of real income and effective demand in all the countries.
3. Developing full use of resources of the world and
4. Expansion of production and international trade.

### Conventions and Principles of GATT

GATT embodies certain conventions and general principles governing international trade among the member countries that adhere to the agreement. The rules of conventions of GATT require that:

1. Any proposed change in the tariff or other type of commercial policy of a member country should not be undertaken without consultation of other parties to the agreement and
2. That the countries that adhere to GATT should work towards the reduction of tariffs and other barriers to international trade which should be negotiated within the frame work of GATT.

### GATT has adopted the following principles

1. *Non-discrimination:* The principle of non-discrimination requires that no member country shall discriminate between the members of GATT in the conduct of international trade. To ensure non-discrimination, the members of GATT agree to apply the principle of most favoured nation (MFN) to all member countries on import and export duties. As far as quantitative restrictions are permitted, they too are to be administered without favour to any country. Each nation shall be treated as well, as the most favoured nation.

2. *Prohibition of Quantitative Restrictions (QRs):* GATT rules seek to prohibit quantitative restrictions as far as possible and limit the restrictions on trade to the less rigid tariffs. However, certain exceptions to this prohibition are granted to countries confronted with balance of payments difficulties and to developing countries.

3. *Consultation:* By providing a forum for continuing consultations, GATT seeks to resolve disagreements through consultation.

On account of persistent follow up by member developing countries for removing or modifying the provisions which are perceived to be biased against them, eight rounds of trade negations were held during the period from 1947 to 1994. They were Geneva Round (1947), Anney Round (1949), Torquay Round (1950), Geneva Round

(1955-56), Dillon Round (1959-62), Kennedy Round (1963-67), Tokyo Round (1973-79) and Uruguay Round (1986-90).

## ORIGIN OF WTO

The eighth round of multilateral trade negotiations held under the auspices of the GATT is known as the Uruguay round because it was launched in punta del Esta in Uruguay in September 1986. Arthur Dunkel, the then Director General of GATT, presented a draft act embodying the negotiations of the Uruguay round. This was popularly known as Dunkel draft. This was replaced by an enlarged and modified final text which was approved by delegations from the member countries of the GATT on 15th December 1993. This final act was signed by Ministers of 125 Governments on 15th April 1994. The results of the Uruguay Round are to be implemented within ten years–different time periods are given for effecting the different agreements.

The most important decision at the Uruguay round was setting up of World Trade Organization (WTO) to replace GATT. The World Trade Organization located at Geneva (Switzerland) came into effect on 1.1.1995. WTO is the international organization dealing with the global rules of trade between nations. Its objective is to help trade flow smoothly, freely, fairly and predictably. It does this by:

1. Administering trade agreements.
2. Acting as a forum for trade negotiations.
3. Settling trade disputes.
4. Reviewing national trade policies.
5. Assisting developing countries in trade policy issues through technical assistance and training programmes, and
6. Co-operation with other international organizations.

### Objectives

The following are the objectives of WTO spelt out in the preamble to the WTO agreement.

1. In the field of trade and economic development, its relations will be guided by the following
   a. Raising the standards of living.
   b. Achieving full employment and increasing the volume of real income and effective demand and
   c. Expanding the production and trade in goods and services.
2. Towards the fulfillment of the overall objective of sustainable development to ensure the optimal use of world's resources by
   a. Promoting the protection and preservation of environment and
   b. Augmenting resources commensurate with the respective needs and concerns at different levels of economic development.
3. For ensuring that the developing countries, particularly the poorest countries, may get a share in the growth of international trade consistent with their needs of economic development.
4. For achievement of these objectives to enter into mutually beneficial arrangements aiming at reduction of barriers and elimination of discrimination in international trade.

5. Based on the past agreements and efforts and their results to develop a more viable and durable multilateral trading system.
6. For co-ordinating policies in the field of trade, environment and economic development, to take effective steps.

## Organizational Set-up

1. *Ministerial Conference:* It is the supreme authority and takes decisions on all matters falling with in the purview of an agreement. It consists of all the member countries. It meets alteast once in 2 years.

2. *General Council:* It comes next to the ministerial conference and consists of the representatives of all the member countries. It overseas the operations of the WTO agreement and ministerial decisions. Besides, it acts as a dispute settlement body (DSB) and a Trade Policy Review Body (TPRB). It meets once in a month on an average. The general council works through (a) council for trade in goods (b) council for trade in services and (c) council for trade related aspects of intellectual property rights (TRIPS).

There are three committees that perform the functions assigned to them under WTO agreement, the multilateral trade agreements and any other functions assigned by the general council. The three committees are
   1. Committee on trade and development.
   2. Committee on balance of payments restrictions.
   3. Committee on budget, finance and administration.

3. *Director General:* The head of the WTO secretarial is appointed by the general council. The term of the director general is four years.

## WTO AGREEMENTS

### Trade Related Investment Measures (TRIMS)

This agreement seeks to bring about multilateral disciplines on investment practices that distort trade flows. According to this agreement all measures obstructing foreign investment and distorting trade should be removed within two years by developed countries, five years by developing countries and seven years by the least developed countries. The important features of the TRIMS are as follows:

1. All restrictions on foreign capital/investors/companies should be scraped.
2. The foreign investor should be given the same rights in the matter of investment as a national investor.
3. Restrictions will not be placed on any area of investment.
4. There should not be any limitation on the extent of foreign investment.
5. There should not be restrictions on the imports of raw materials and components.
6. Foreign investor will not be obliged to use local products and materials.
7. It is not mandatory on the part of foreign investor to export a part of the output.
8. Elimination of restrictions on repatriation of dividend, interest and royalty.

### Trade Related Intellectual Property Rights (TRIPS)

The TRIPS agreement encompasses seven areas of intellectual property rights *viz.,* (i) copy right (ii) trade marks (iii) trade secrets (iv) industrial designs (v) geographical indications (vi) patents which also includes microorganisms and plant varieties (vii) layout designs of integrated circuits.

It allows a transition period of one year for developed countries, five years for developing countries and eleven years for the least developed counties to make their laws and practices to conform to the rules and conditions laid in the agreement.

## General Agreement on Trade in Services (GATS)

This agreement is based on the desire to establish a multilateral frame work of principles and rules for trade in services with a view to expanding such a trade under conditions of transparency and progressive liberalisation. The services refer to banking and insurance, transport, communication, mobility of labour *etc.*

Under GATS, Most Favoured Nation (MFN) principle will apply to all member countries *i.e.*, if a service of any kind is open to one member country, it should be extended to all member countries in equal measure. The second requirement is transparency *i.e.*, all agreements affecting trade in services shall be published. Further each country is required to make market access commitments on services.

## Agreement on Trade in Textiles and Clothing

The developed countries imposed most comprehensive quota restrictions under the Multi Fibre Agreement (MFA). This agreement on trade in textiles and clothing proposes to phase out quotas over ten year period beginning with 1 January 1995 and ending with 1 January 2005.

## Plurilateral Trade Agreement (PTA)

These agreements are: agreement on trade in civil aircraft, agreement on Government procurement, international dairy agreement and international bovine meat agreement.

## Dispute Settlement Body (DSB)

The dispute settlement process shall consist of consultations between the parties, failing which conciliation and mediation by the director general if agreed by the parties or a panel of three experts. It also provides for an Appellate review by a standing Appellate body, consisting of seven members set up by DSB to report to it within 90 days. The decision of the DSB will be binding on both the parties.

## Trade Policy Review Body (TPRB)

This has been set up to carryout trade policy reviews. Frequency of such reviews depends on the share of a member in world trade.

## Agreement on Agriculture (AOA)

The Agreement on Agriculture (AOA) forms a part of the final act of the Uruguay round of multilateral trade negotiations, which was signed by the member countries in April 1994 at Marrakesh, Morocco and came into force on 1st January 1995. The Uruguay round marked a significant turning point in world trade in agriculture, for the first time, agriculture featured in a major way in the GATT round of multilateral trade negotiations. The Uruguay round agreement sought to bring order and fair competition to this highly distorted sector of world trade by establishment of fair and market oriented agricultural trading sector.

The main cause of distortion of international trade in agriculture has been the massive domestic subsidies. (A subsidy is deemed to exist if there is a financial contribution by a Government on any public body and a benefit is conferred) given by the industrialized countries to their agricultural sector over many years. This in turn led to excessive production and it's dumping in international markets as well as import restrictions to keep out foreign farm products from their domestic market. Hence, the starting point for the establishment of a fair agricultural trade regime should be the reduction of domestic production subsidies given by the developed countries, reduction in the volume of subsidized exports and minimum market access opportunities for agricultural producers world wide.

The Agreement on Agriculture (AOA) has three main components; (i) Market access (ii) domestic subsidy or domestic support and (iii) export subsidies.

*i) Market Access:* On market access the agreement has two basic elements.

a. The market access commitment required all non-tariff barriers (NTBS) such as quantitative restrictions (QRs) and export, import licensing *etc.*, are to be replaced by tariffs to provide the same level of protection. Tariffs, resulting from tariffication process together with other tariffs on agricultural products are to be reduced by 36 per cent over six years in the case of developed countries and 24 per cent over 10 years in the case of developing countries.

b. The second element relates to setting up of a minimum level for imports of farm commodities by member countries as a share of domestic consumption. Countries are required to maintain current levels (1986-88) of access for each individual product. Where the current level of import is negligible, the minimum access should not be less than 3 per cent of the domestic consumption. This minimum level is to rise to 5 per cent by the year 2000 in the case of developed countries and by 2004 in the case of developing countries. However, special safeguard provisions allow for the application of additional duties when shipments are made at prices below certain reference levels or when there is a sudden import surge. The market access provision, however, does not apply when the commodity in question is a *'traditional staple'* of a developing country.

*ii) Domestic Support:* The provisions of the agreement regarding domestic support have two main objectives: (1) To identify the acceptable measure that support farmers and (2) to deny unacceptable trade distorting support to the farmers. These provisions are aimed largely at the developed countries where the levels of agricultural support have risen to extremely high levels in recent decades.

All domestic support is quantified through the mechanism of Aggregate Measure of Support (AMS). AMS is a means of quantifying the aggregate value of domestic support on subsidy given to each category of agricultural product. Each WTO member country has made calculations to determine its AMS wherever applicable. There was no requirement in the agreement for reduction commitment if the figure was below 5 per cent for developed countries and 10 per cent for developing countries. In other cases, member countries were required to reduce their total AMS by 20 per cent over 6 years (developed countries) or 13.30 per cent over 10 years (developing countries). The base period external reference price on which the reductions calculated was 1986-88.

AMS consists of two parts-product specific subsidies and non-product specific subsidies. Product specific subsidy refers to the total level of support provided for each individual agricultural commodity essentially signified by procurement price in India. Non-product specific subsidy, on the other, hand refers to the total level of support for the agricultural sector as a whole i.e. subsidies on inputs such as fertilizers, electricity, seeds, credit and irrigation, *etc.*

There are three categories of support measures that are not subject to reduction under the agreement and support within de-minimis (under the de-minimis provision of the agreement, there is no requirement to reduce trade distorting domestic support or subsidy where the aggregate value of support does not exceed a certain limit or ceiling. In case of developing countries the de-minimis ceiling is 10 per cent) level is allowed. The three categories of exempt support measures are:

a. Measure which have a minimum impact on trade and which meet the basic and policy specific criteria set out in the agreement (the so called Green Box measures in the terminology of WTO). These measures include Government assistance on general services like research, pest and disease control, training, extension and advisory services, public stock holding for food security purposes, domestic food aid and direct payments to producers like Governmental financial participation in income insurance, relief from natural disasters and payments under environmental assistance programmes.

b. Developing country measures other wise subject to reduction which meet the criteria set out in paragraph 2 of article 6 of the agreement (the so-called special and differential treatment or S & D Box). Examples of these are investment subsidies which are generally available to agriculture in developing countries and agricultural input services generally available to low income and resource poor producers in developing countries.

c. Direct payments under production limiting programmes (the so called Blue Box measures). These were relevant from the developed countries point of view only.

*iii) Export Subsidies:* Export subsidies also become a major factor in depressing or destabilizing world market prices for many farm products. The Uruguay round marked a radical departure from the earlier GATT disciplines in the areas of agricultural export subsidies. Members are required to reduce the value of direct export subsidies to a level of 36 per cent below the 1986-90 base period level over a six year implementation period. The quantity of subsidized export is to be reduced by 21 per cent over the same period. In the case of developing countries, the reductions are two thirds those of the developed countries over a ten year period and there are no reductions for least developed countries.

Export subsidies are defined as subsidies contingent on export performance and the list covers export subsidy practices such as direct export subsidies contingent. On export performance, sales of non-commercial stocks of agricultural products for export at prices lower than comparable prices for such goods in the domestic markets, producer financed subsidies such as government programmes which require a levy on production which is then used to subsidise the export of the product, cost reduction measures such as subsidies to reduce marketing costs for exports including handling costs and costs of international freight, internal transport subsidies applying only to exports *etc.* All such export subsidies are subject to reduction commitments in terms of both the volume of subsidized export and budgetary outlays for such subsidies. However, such measures are non-existent in India and hence the issue of reduction of export subsidies on agricultural products is not of particular relevance for India.

# References

---

## SECTION I
## Microeconomics

Ahuja, H.L., Advanced Economic Theory: Microeconomics Analysis, S. Chand & Company Ltd., New Delhi, 1979.

Barthwal, R.R., Microeconomic Analysis, Wiley Eastern Ltd. New Delhi, 1992.

Dewett, K.K. and Adarsh Chand., Modern Economic Theory, Shyam Lal Charitable Trust, New Delhi, 1997.

Dewett, K.K. and Varma, J.D., Elementary Economic Theory, S. Chand & Company Ltd., New Delhi, 1985.

Gail, L. Cramer and Clarence W. Jensen., Agricultural Economics and Agribusiness, John Wiley & Sons, Inc., New York, 1994.

Gupta, G.S., Managerial Economics, Tata McGraw-Hill Publishing Company Ltd, New Delhi, 1990.

Jhingan, M.L, Microeconomic Theory, Vikas Publishing House Pvt. Ltd., New Delhi, 1982.

Joshi, J.M. and Rajendra Joshi., Microeconomic Theory: An Analytical Approach, Wishwa Prakashan, New Delhi, 1994.

Koutsoyiannis, A., Modern Microeconomics, English Language Book Society/Macmillan, London, 1979.

Richard, W. Leftwich and Ross. D. Eckert., The Price System and Resource Allocation, Holt-Saunders, Japan, 1985.

Samuelson, Paul A. and William, D. Nordhaus., Economics, McGraw-Hill Book Company, New York, 1989.

Seth, M.L., Principles of Economics, Lakshminarayan Agarwal Educational Publishers, New Delhi, 1968.

Stonier, A.W. and Hague Douglas C., A Textbook of Economic Theory, The English Language Book Society and Longman's Green and Company Ltd., London, 1962.

Sundharam, K.P.M. and M.C. Vaish., Microeconomic Theory, S. Chand & Company Ltd., New Delhi, 1977.

Tewari, D.D. and Katar Singh., Principles of Microeconomics, New Age International (P) Ltd., Publishers, New Delhi, 1996.

# SECTION II
## Macroeconomics

Ackley Gardner., Macroeconomic Theory, the Macmillan Company, New York, 1961.

Dewett, K.K. and Adarsh Chand., Modern Economic Theory, Shyam Lal Charitable Trust, New Delhi, 1997.

Jhingan, M.L., Macroeconomic Theory, Vikas Publishing House Pvt. Ltd., New Delhi, 1983.

Seth, M.L., Macroeconomics, Lakshminarayan Agarwal Educational Publishers, Agra, 1981.

# SECTION III
## Agricultural Production Economics

Allan, N. Rae, Crop Management Economics, Granada Publishing Ltd., London, 1977.

Bishop, C.E. and W.D. Toussaint., Introduction to Agricultural Economic Analysis, John Wiley & Sons, Inc., New York, 1958.

Christopher Ritson, Agricultural Economics: Principles and Policy, Granada Publishing Ltd., London, 1977.

Heady, Earl O. and J.L. Dillon., Agricultural Production Functions, Kalyani Publishers, New Delhi, 1960.

Debertin David, L., Agricultural Production Economics, Macmillan Publishing Company, New York, 1986.

Handerson, James, M. and Richard E. Quandt., Microeconomic Theory: A Mathematical Approach, McGraw-Hill Book Company, New York, 1980.

Heady, Earl O., Economics of Agricultural Production and Resource Use, Prentice-Hall, Inc., Englewood Cliffs N.J., 1952.

Johnston, J., Econometric Methods, McGraw-Hill International Editions, 1984.

Julia Hebden., Applications of Econometrics, Heritage Publishers, New Delhi, 1988.

Klein, L.R., An Introduction to Econometrics, Prentice-Hall, Englewood Cliffs, N.J., 1962.

Maddala, G.S., Econometrics, McGraw-Hill Kogakusha Ltd., Tokyo, 1977.

Robertson, An Introduction to Agricultural Production Economics, Tata McGraw-Hill Publishing Company Ltd., New Delhi, 1971.

Shankayan, P.L., Introduction to the Economics of Agricultural Production, Prentice-Hall of Indian Pvt. Ltd., 1988.

# SECTION IV
## Farm Management

Adams, R.L., Farm Management, McGraw-Hill Book Company, Inc., New York, 1921.

Barnard, C.S. and J.S. Nix., Farm Planning and Control, Cambridge University Press, London, 1979.

Blagburn, C.H., Farm Planning and Management, Longman's Green & Co. Ltd., London, 1961.

Buckett, M., An Introduction to Farm Organization and Management, Pergamon Press New York, 1981.

Castle, E.N. and M.H. Becker., Farm Business Management: The Decision Making Process, The Macmillan Company, New York, 1965.

Dhondyal, S.P., Farm Management (an economic analysis), Friends Publications.

Efferson, J. Norman., Principles of Farm Management, McGraw-Hill Book Company, Inc., New York, 1953.

Forster, G.W., Farm Organization and Management, Prentice-Hall, Inc., New York, 1953.

Heady, Earl O. and Jensen Harald, R., Farm Management Economics, Prentice-Hall of India (Pvt. Ltd.), New Delhi, 1964.

Hopkins, John. A. and Murray William, G., Elements of Farm Management, Prentice-Hall Inc., Englewood cliffs, N.J., 1955.

Johl, S.S. and T.R. Kapur., Fundamentals of Farm Business Management, Kalyani Publishers, Ludhiana, 1987.

Kahlon, A.S. and Karam Singh., Economics of Farm Management in India: Theory and Practice, Allied Publishers Private Ltd., New Delhi, 1980.

Paul Loomba, N., Linear Programming: An Introductory Analysis, Tata McGraw-Hill Publishing Company Ltd., New Delhi, 1989.

Prem Kumar Gupta. and D.S. Hira., Operations Research, S. Chand & Company Ltd., New Delhi, 2000.

Robert O' Connor., Principles of Farm Business Analysis and Management, Irish University press, Shannon, Ireland, 1973.

Sankhayan, P.L., Introduction to Farm Management, Tata McGraw-Hill Publishing Company Ltd., New Delhi, 1983.

Sharma, A.N. and V.K. Sharma., Elements of Farm Management, Prentice-Hall of India Pvt. Ltd., New Delhi, 1981.

Singh, I.J., Elements of Farm Management Economics, Affiliated East-West Press Private Ltd., New Delhi, 1977.

Taha Hamdy, A., Operations Research: An Introduction, Macmillan Publishing Company, New York, 1989.

William J. Baumol., Economic Theory and Operations Analysis, Prentice-Hall of India Pvt. Ltd., New Delhi, 1978.

Yang, W.Y., Methods of Farm Management Investigations, Food and Agriculture Organisation of the United Nations, Rome, Italy, 1971.

# SECTION V
## Agricultural Finance

Ajit Singh., Rural Development and Banking in India: Theory and Practice, Deep and Deep Publications, New Delhi, 1989.

Choubey, B.N., Principles and Practice of Co-operative Banking in India, Asia Publishing House, Bombay, 1968.

Dwivedi, D.N., Managerial Economics, Vikas Publishing House, New Delhi, 1980.

Ghosal, S.N., Agricultural Financing in India, Asia Publishing House, Bombay, 1966.

Gittenger, J.C. Price., Economic Analysis of Agricultural Projects, A World Bank Publication, 1976.

Goel, B.B., Role of Co-operatives in Rural Development (with special reference to marketing co-operatives), Indian Co-operative Review, Volume XXXVIII (4), 2001.

Jain, M.K., Rural Banks and Rural Poor, Print Well Publishers, Jaipur, 1989.

James, C. Van Horn., Fundamentals of Financial Management, Prentice-Hall of India, New Delhi, 1964.

Johl, S.S. and C.V. Moore., Essentials of Farm Financial Management, Today and Tomorrow's Printers and Publishers, New Delhi, 1970.

John, J. Hamptron., Financial Decision Making: Concepts, Problems and Cases, Prentice-Hall of India, New Delhi, 1983.

Johnson, D.T., The Business of Farming: A Guide to Farm Business Management in the Tropic, Macmillan Publishers, London, 1982.

Kahlon, A.S. and Karam Singh., Managing Agricultural Finance. Theory and Practice, Allied Publishers Private Ltd., New Delhi 1984.

Kenneth, Duft D., Principles of Management in Agribusiness, Reston Publishing Company, Reston, 1979.

Mamoria, C.B. and R.D. Saksena., Co-operation in India, Kitab Mahal, Allahabad, 1973.

Mukhi, H.R., Co-operation in India and Abroad, New Heights Publishers, New Delhi, 1983.

Muniraj, R., Farm Finance for Development, Oxford & IBH Publishing Company Private Ltd., New Delhi, 1987.

Narendra Kumar, (Ed.)., Bank Nationalization in India, A Symposium, Lalvani Publishing House, Bombay, 1969.

Pandey, U.K., An Introduction to Agricultural Finance, Kalyani Publishers, New Delhi, 1990.

Puttaswamaiah, K., (Ed.)., Project Evaluation Criterion and Cost Benefit Analysis, Oxford & IBH Publishing Company, Private Ltd., New Delhi, 1989.

Report on Rural Debt and Investment Surveys, Bombay, 1962.

Sathya, M., Service Area Approach to Rural Lending, Kitab Mahal, New Delhi, 1993.

Subba Reddy, S. and P. Raghu Ram., Agricultural Finance and Management, Oxford & IBH Publishing Company Private Ltd., New Delhi, 1996.

Tokhi, M.R. and D.P. Sharma, (Ed.)., Rural Banking in India, Oxford & IBH Publishing Company Private Ltd., New Delhi, 1983.

Vasanth Desai., A Study of Rural Economics: A Systems Approach, Himalaya Publishing House, Bombay, 1983.

Vyas, M.R., Evolution and Management of Regional Rural Banks, Arihant Publishers, Jaipur, 1990.

William, G. Murray and Nelson Aaron, G., Agricultural Finance, The Iowa State University Press, Ames, Iowa, 1960.

# SECTION VI
## Agricultural Marketing

Acharya, S.S. and N.L. Agarwal., Agricultural Marketing in India, Oxford & IBH Publishing Company Pvt. Ltd., New Delhi, 1992.

Acharya, S.S. and N.L. Agarwal., Agricultural Prices-Analysis and Policy. Oxford & IBH Publishing Company Pvt. Ltd., New Delhi, 1994.

Edward W. Cundift, Richard R. Still and Norman A.P. Govoni., Fundamentals of Modern Marketing, Prentice-Hall of India Private Ltd., New Delhi, 1980.

Kahlon, A.S. and D.S. Tyagi., Agricultural Price Policy in India, Allied Publishers Pvt. Ltd., New Delhi, 1983.

Kulkarni, K.R., Agricultural Marketing in India, The Co-operators' Book Depot, Bombay, 1964.

Mamoria, C.B. and R.L. Joshi., Principles and Practice of Marketing in India, Kitab Mahal, Allahabad, 1995.

Ramaswamy, V.S. and S. Nama Kumari., Marketing Management: Planning, Implementation and Control, Macmillan India Ltd., New Delhi, 1995.

Thomsen, F.L. and Foote, R.J., Agricultural Prices, McGraw-Hill Book Company, Inc., New York, 1952.

# SECTION VII
## Economic Problems of Indian Agriculture

Agrawal, A.N., Indian Agriculture: Problems, Progress and Prospects, Vikas Publishing House Pvt. Ltd., New Delhi, 1981.

Bansil, P.C., Agricultural Problems of India, Vikas Publishing House Pvt. Ltd., New Delhi, 1977.

Dasgupta, Biplab., The New Agrarian Technology and India, The Macmillan Company of India Ltd., 1977.

Dhingra, I.C., Indian Economic Problems, Sultan Chand & Sons, New Delhi, 1995.

Misra, S.N., Capital Formation and Accumulation in Indian Agriculture Since Independence, Indian Journal of Agricultural Economics, Vol. I & II, 1996.

Patel, K.V. Shaah, A.C. and D'Mellow, L., Rural Economics, Himalaya Publishing House, New Delhi, 1984.

Raja Purohit., Land Reforms in India, Ashish Publishing House, New Delhi, 1984.

Raju, K.N., Land Reforms in India: Achievements and Failures, National Bank News Review, July-September, 1997.

Ruddar Datt, K.P.M. Sundaram., Indian Economy, S. Chand & Company Ltd., New Delhi, 2001.

Sadhu, A.N. and Mahajan, R.K., Technological Change and Agricultural Development in India, Himalaya Publishing House, New Delhi, 1985.

Satish Bhatia., Agricultural Price Policy and Production in India, Konark Publishers Pvt. Ltd., 1991.

The World Bank Annual Report 2000, The World Bank, Washington DC.

# Appendix A

| | | |
|---|---|---|
| 1. | ACD | Agricultural Credit Department |
| 2. | ADB | Agricultural Development Branch |
| 3. | AFC | Agricultural Finance Corporation |
| 4. | AFC | Average Fixed Cost |
| 5. | AGMARK | Agricultural Produce (Grading and Marking) Act |
| 6. | AIRCRC | All India Rural Credit Review Committee |
| 7. | AIRCSC | All India Rural Credit Survey Committee |
| 8. | AIRDISC | All India Rural Debt and Investment Survey Committee |
| 9. | AMS | Aggregate Measure of Support |
| 10. | AoA | Agreement on Agriculture |
| 11. | APC | Agricultural Prices Commission |
| 12. | APCCADB | Andhra Pradesh Central Co-operative Agricultural Development Bank |
| 13. | APCOBARD | Andhra Pradesh Co-operative Bank for Agriculture and Rural Development |
| 14. | APP | Average Physical Product |
| 15. | AR | Average Revenue |
| 16. | ARDC | Agricultural Refinance and Development Corporation |
| 17. | ATC | Agreement on Textiles and Clothing |
| 18. | ATC | Average Total Cost |
| 19. | AVC | Average Variable Cost |
| 20. | AVP | Average Value Product |
| 21. | BCR | Benefit-cost Ratio |
| 22. | BEP | Break-even Point |
| 23. | BIRD | Bankers Institute for Rural Development |
| 24. | BIS | Bureau of Indian Standards |
| 25. | BLBC | Block Level Bankers' Committee |
| 26. | BOP | Balance of Payments |
| 27. | BPL | Below Poverty Line |
| 28. | CACP | Commission on Agricultural Costs and Prices. |
| 29. | CADA | Command Area Development Authority |
| 30. | CALCOB | Committee on Agricultural Loans through Commercial Banks |
| 31. | CCA | Capital Consumption Allowance |
| 32. | CIRRS | Commercial Interest Reference Rates |
| 33. | CR | Current Ratio |

| 34. | CRAFICARD | Committee to Review Arrangements for Institutional Credit for Agriculture and Rural Development |
| 35. | CRR | Cash Reserve Ratio |
| 36. | CWC | Central Warehousing Corporation |
| 37. | CYI | Crop Yield Index |
| 38. | DBOD | Department of Banking Operations and Development |
| 39. | DCC | District Consultative Committee |
| 40. | DCCB | District Central Co-operative Bank |
| 41. | DCP | District Credit Plan |
| 42. | DI | Disposable Income |
| 43. | DICGC | Deposit Insurance and Credit Guarantee Corporation |
| 44. | DIRS | Differential Interest Rate Scheme |
| 45. | DISCOBARD | District Co-operative Bank for Agriculture and Rural Development |
| 46. | DMI | Directorate of Marketing and Inspection |
| 47. | DSB | Dispute Settlement Body |
| 48. | DTA | Domestic Tariff Area |
| 49. | DWCRA | Development of Women and Children in Rural Area |
| 50. | ECGC | Export Credit Guarantee Corporation |
| 51. | $E_p$ | Elasticity of Production |
| 52. | EPZ | Export Promotion Zones |
| 53. | EXIM Bank | Export Import Bank of India |
| 54. | FAO | Food and Agriculture Organization |
| 55. | FCI | Fertilizer Corporation of India |
| 56. | FCI | Food Corporation of India |
| 57. | FCR | Fixed Cost Ratio |
| 58. | FSS | Farmers Service Society |
| 59. | GATS | General Agreement on Trade in Services |
| 60. | GATT | General Agreement on Tariffs and Trade |
| 61. | GCR | Gross Cost Ratio |
| 62. | GDP | Gross Domestic Product |
| 63. | GI | Gross Income |
| 64. | GNP | Gross National Product |
| 65. | HADP | Hill Area Development Projects |
| 66. | HIPC | Heavily Indebted Poor Countries |
| 67. | IBRD | International Bank for Reconstruction and Development |
| 68. | IDA | International Development Association |
| 69. | IDADA | Integrated Dryland Agricultural Development Agency |
| 70. | IDBI | Industrial Development Bank of India |
| 71. | IFAD | International Fund for Agricultural Development |
| 72. | IFC | International Finance corporation |
| 73. | IMBP | Individual Maximum Borrowing Power |
| 74. | IMF | International Monetary Fund |
| 75. | IOU | I Owe You |
| 76. | IRDP | Integrated Rural Development Programme |
| 77. | IRR | Internal Rate of Return |
| 78. | ISI | Indian Standards Institute |
| 79. | IT | Income Tax |
| 80. | ITO | International Trade Organization |
| 81. | KVIB | Khadi and Village Industries Board |

| 82. | KVIC | Khadi and Village industries Commission |
| 83. | LDB | Land Development Bank |
| 84. | LP | Linear Programming |
| 85. | MC | Marginal Cost |
| 86. | MFALDA | Marginal Farmers and Agricultural Labourers Development Agency |
| 87. | MFC | Marginal Factor Cost |
| 88. | MFNS | Most Favoured Nations |
| 89. | MIGA | Multilateral Investment Guarantee Agency |
| 90. | MPCCS | Multipurpose Co-operative Credit Societies |
| 91. | MPP | Marginal Physical Product |
| 92. | MR | Marginal Revenue |
| 93. | MRS | Marginal Rate of Substitution |
| 94. | MRTS | Marginal Rate of Technical Substitution |
| 95. | MTNS | Multilateral Trade Negotiations |
| 96. | MVP | Marginal Value Product |
| 97. | NABARD | National Bank for Agriculture and Rural Development |
| 98. | NAFED | National Agricultural Co-operative Marketing Federation |
| 99. | NCA | National Commission on Agriculture |
| 100. | NCCF | National Consumers' Co-operative Federation |
| 101. | NCDC | National Co-operative Development Corporation |
| 102. | NCR | Net Capital Ratio |
| 103. | NDDB | National Dairy Development Board |
| 104. | NI | National Income |
| 105. | NIBM | National Institute of Bank Management |
| 106. | NNP | Net National Product |
| 107. | NPW | Net Present Worth |
| 108. | NREP | National Rural Employment Programme |
| 109. | NRY | Nehru Rojagar Yojana |
| 110. | NSS | National Sample Survey |
| 111. | NTB | Non-Tariff Barriers |
| 112. | NWDB | National Wasteland Development Board |
| 113. | OCR | Operational Cost Ratio |
| 114. | OECD | Organization for Economic Co-operation and Development |
| 115. | OGL | Open General License |
| 116. | PACS | Primary Agricultural Co-operative Credit Societies |
| 117. | PBR | Plant Breeders Rights |
| 118. | PBRS | Patents and Plant Breeders Rights |
| 119. | PDS | Public Distribution System |
| 120. | PI | Personal Income |
| 121. | PLDB | Primary Land Development Bank |
| 122. | PMWU | Productive Man Work Units |
| 123. | PR | Price Ratio |
| 124. | QRs | Quantitative Restrictions |
| 125. | RBI | Reserve Bank of India |
| 126. | REC | Rural Electrification Corporation |
| 127. | RLEGP | Rural Landless Employment Guarantee Programme |
| 128. | RPDS | Revamped Public Distribution System |
| 129. | SDR | Special Drawing Rights |
| 130. | SFDA | Small Farmers Development Agency |

| 131. | SHG | Self Help Group |
| 132. | SLBC | State Level Bankers Committee |
| 133. | SLR | Statutory Liquidity Ratio |
| 134. | SPS | Sanitary and Phytosanitary Measures |
| 135. | SWC | State Warehousing Corporation |
| 136. | TADP | Tribal Area Development Programme |
| 137. | TB | Tariff Barriers |
| 138. | TC | Total Cost |
| 139. | TFC | Total Fixed Cost |
| 140. | TPP | Total Physical Product |
| 141. | TPRM | Trade Policy Review Mechanism |
| 142. | TR | Total Revenue |
| 143. | TRIFED | Tribal Co-operative Marketing Development Federation |
| 144. | TRIMS | Trade Related Investment Measures |
| 145. | TRIPS | Trade Related Intellectual Property Rights |
| 146. | TRYSEM | Training of Rural Youth for Self Employment |
| 147. | TVC | Total Variable Cost |
| 148. | TVP | Total Value Product |
| 149. | UPOV | Union for the Protection of New Varieties |
| 150. | WR | Working Ratio |
| 151. | WTO | World Trade Organization |

# Appendix B

Committees/Study Groups Constituted with Regard to Rural Credit

| S. No. | Name of the Committee | Year | Chairman |
|---|---|---|---|
| 1. | All India Rural Credit Survey Committee (AIRCSC) | 1954 & 1969 | RBI |
| 2. | Study Group of National Credit Council for Organizational framework for the implementation of Social Objectives | 1969 | D.R. Gadgil |
| 3. | Committee of Bankers for the preparation of a coordinated programme for providing adequate banking facilities in the under banked district | 1969 | F.K.F. Nariman |
| 4. | Committee on Differential Rate of Interest | 1971-72 | Dr. R.K. Hazari |
| 5. | Committee to review the flow of the institutional credit especially to the weaker sections of the rural communities | 1975 | M. Narasimham |
| 6. | Study Group on the working of Lead Bank Schme in Gujarat and Maharashtra | 1975 | RBI |
| 7. | The Working Group on Rural Banks | 1975 | M. Narasimham |
| 8. | Working group on Multi-Agency Approach in Agricultural finance | 1976 | C.E. Kamath |
| 9. | Committee on Consumption credit | 1976 | B. Sivaraman |
| 10. | Working group on Block Level Panning | 1978 | M.L. Dantwala |
| 11. | Review Committee on RRB | 1977 | M.L. Dantwala |
| 12. | Expert Group on Agricultural Credit Schemes of Commercial Banks | 1978 | Gunvant Desal |
| 13. | Committee to study the Role of Banks in lending to priority sector advance and 20 point programme | 1980 | Dr. K.S. Krishnamurthy |
| 14. | Committee to Review Arrangements for Institutional Credit for Agricultural and Rural Development (CRAFICARD) | 1981 | B. Sivaraman |
| 15. | Working Group on Working of Lead Bank Scheme | 1982 | U.K. Sarma |

| S. No. | Name of the Committee | Year | Chairman |
|---|---|---|---|
| 16. | Committee to study the Role of Bank in the Implementation of New 20 point programme | 1982 | Amitaba Gosh |
| 17. | Committee to Review the Working of Monetary system | 1986 | S.M. Kelkar |
| 18. | The Working of Monetary System | 1986 | S.M. Kelkar |
| 19. | Agricultural Credit Review Committee (ACRC) | 1989 | A.M. Khusro |
| 20. | Committee to Examine certain Operational Aspects of Rural lending | 1989 | Dr. P.D. Ojha |
| 21. | Evaluation Study of SAA | 1991 | National Institute of Bank Management |
| 22. | Committee on Financial System | 1991 | M. Narasimham |
| 23. | Committee to Review IRDP | 1994 | D.R. Mehta |
| 24. | Committee on Restructuring of RRB | 1994-95 | Bhandari |
| 25. | Expert Group of Structuring RRB | June 1995 | Dr. N.K. Thingalaya |
| 26. | Committee for Revamping RRB (set up by NABARD) | Dec. 1995 | K. Basu |
| 27. | Working Group for studying the functioning of Self-Help Group (SHGs) and Non-Government Organisations (NGOs) | Apr. 1996 | S.K. Kalia |
| 28. | Working Group to evolve a detailed recommendation on the integration and rationalization of credit linked poverty alleviation programme and employment scheme | Sep. 1996 | RBI |
| 29. | Committee to examine the present structure of remuneration package of RRB | Nov. 1996 | S.C. Mahalik |

# Appendix C

## Rural Development Programmes

| S.No. | Programme | Year |
|---|---|---|
| 1. | Community Development Programme (CDP) | 1952 |
| 2. | National Extension Service | 1953 |
| 3. | Applied Nutrition Programme (ANP) | 1958 |
| 4. | Panchayti Raj | 1959 |
| 5. | Intensive Agriculture District Programme (IADP) | 1960 |
| 6. | Hill Area Development Programme (HADP) | 1960 |
| 7. | Tribal Area Development Programme (TADP) | 1962 |
| 8. | Intensive Agricultural Area programme (IAAP) | 1964 |
| 9. | High Yielding Variety Programme (HYVP) | 1965 |
| 10. | Intensive Area Development Scheme (IADS) | 1965 |
| 11. | Farmers' Training Education | 1966 |
| 12. | Wells Construction Programme | 1966 |
| 13. | Small Farmers Development Agency (SFDA) | 1969 |
| 14. | Marginal Farmers and Agricultural Labourers Development Agency (MFAL) | 1969 |
| 15. | Drought Prone Area Programme (DPAP) | 1970 |
| 16. | Rural Works Programme (RWP) | 1971 |
| 17. | Crash Scheme for Rural Employment (CSRE) | 1971 |
| 18. | Pilot Intensive Rural Employment Projects (PIREP) | 1972 |
| 19. | Pilot Project Tribal Development | 1972 |
| 20. | Employment Guarantee Scheme (EGS) | 1972 |
| 21. | Command Area Development Programme (CADP) | 1973 |
| 22. | Minimum Needs Programme (MNP) | 1974 |
| 23. | Food for Work Programme (FWP) | 1977 |
| 24. | Antyodaya | 1977 |
| 25. | Desert Development Programme | 1977 |
| 26. | Comprehensive Area Development Programme | 1978 |
| 27. | Integrated Rural Development Programme (IRDP) | 1978 |
| 28. | National Rural Employment Programme (NREP) | 1980 |
| 29. | Training of Rural Youth for Self-employment (TRYSEM) | 1980 |
| 30. | Whole Village Development Programme (WVDP) | 1980 |

| 31. | Bio-Gas Programme (BGP) | 1981 |
|-----|-------------------------|------|
| 32. | Integrated Rural Energy Programme (IREP) | 1981 |
| 33. | Self-Employment to Educated Unemployed Youth Programme (SEEUYP) | 1983 |
| 34 | Rural Landless Employment Guarantee Programme (RLEGP) | 1983 |
| 35 | Jawahar Rozgar Yojana (JRY) | 1984 |
| 36 | Special Foodgrains Production Programme (SFPP) | 1986 |
| 37 | Special Rice Production Programme (SRPP) | 1986 |
| 38 | Forest Farming for Rural Poor (FFRP) | 1986 |
| 39 | Indira Aawaas Yojana (JRY) | 1986 |
| 40 | Millions Well Scheme | 1989 |
| 41 | Rajiv Gandhi National Drinking Water Mission | 1991 |
| 42 | Employment Assurance Scheme | 1993 |
| 43 | National Social Assistance Programme | 1995 |
| 44 | Jawahar Gram Samridhi Yojana | 1999 |
| 45 | Swaran Jayanthi Gram Swarojgar Yojana | 1999 |
| 46 | Innovative Scheme for Rural Housing and Habitat Development Scheme | 1999 |

# Appendix D

## MICRO ECONOMICS AND MACRO ECONOMICS

### I. CHOOSE AND UNDERLINE THE CORRECT ANSWER

1. Public and Private sector co-exist under
   - (a) Capitalism
   - (b) Socialism
   - (c) Command economy
   - (d) Mixed economy

2. According Engels law of family expenditure, as income increases, the percentage expenditure on luxuries
   - (a) Decreases
   - (b) Remains the same
   - (c) Increases
   - (d) Increase or decrease

3. A consumer spends his income according to the
   - (a) Law of diminishing returns
   - b) Law of equimarginal utility
   - (c) Law of equimarginal returns
   - (d) Law of variable proportions

4. As we consume more of a commodity, its total utility
   - (a) Increases at increasing rate
   - (b) Increases at diminishing rate
   - (c) Increases at constant rate
   - (d) Decreases at increasing rate

5. A capitalist economy seeks to solve various problems with the help of
   - (a) Economic planning
   - (b) Price mechanism
   - (c) Efficient allocation of resources
   - (d) None of these

6. If cross elasticity coefficient is negative, the two goods are
   - (a) Substitutes
   - (b) Unrelated
   - (c) Compliments
   - (d) Perfect substitutes

7. One man, one vote is the principle of
   - (a) Partnership
   - (b) Joint stock company
   - (c) State enterprise
   - (d) Cooperatives

8. Number of firms in oligopoly is
   - (a) Many
   - (b) Few
   - (c) One
   - (d) Large

9. Demand curve slopes down from left to right because of the operation of
   - (a) Law of equimarginal utility
   - (b) Law of substitution
   - (c) Law of maximum satisfaction
   - (d) Law of diminishing marginal utility

10. The theory of rent is associated with the name of
    (a) J. S. Mill
    (b) J. M. Keynes
    (c) T. R. Malthus
    (d) David Ricardo

11. The concept of Quasi rent was introduced in economic theory by
    (a) David Ricardo
    (b) Alfred Marshall
    (c) J. S. Mill
    (d) Adam Smith

12. Those goods whose supply is scarce in relation to demand for them are called
    (a) Free goods
    (b) Nature's gifts
    (c) Economic goods
    (d) None of them

13. Human wants are
    (a) Unlimited
    (b) Complementary
    (c) Satiable
    (d) All of them

14. Superfluous consumption is defined as
    (a) Necessary
    (b) Luxury
    (c) Comfort
    (d) All of them

15. According to Engels law of consumption, as income increases, the percentage expenditure on necessaries of life
    (a) Increases
    (b) Remains the same
    (c) Decreases
    (d) May increase or decrease

16. Law of diminishing marginal utility fails to operate in the case of
    (a) Rare collections
    (b) Abnormal persons
    (c) Dissimilar units
    (d) All of them

17. A consumer stops consumption of a commodity at the point where
    (a) Initial utility is maximum
    (b) Marginal utility is maximum
    (c) Total utility is minimumd) Marginal utility is zero

18. The consumer is said to be in equilibrium when

    (a) $\dfrac{\text{MU of } x}{\text{Price of } x} = \dfrac{\text{MU of } y}{\text{Price of } y} = \dfrac{\text{MU of } z}{\text{Price of } z} = K$

    (b) $\dfrac{\text{MU of } x}{\text{Price of } x} = \dfrac{\text{MU of } y}{\text{Price of } y} = \dfrac{\text{Price of } z}{\text{MU of } z} = K$

    (c) MU of $y = Px$
    (d) None

19. The demand for a commodity that can be put to several uses is called
    (a) Joint demand
    (b) Derived demand
    (c) Composite demand
    (d) Direct demand

20. When the total amount of money spent increases with a fall in price or decreases with a rise in price, the elasticity of demand is said to be
    (a) Unity
    (b) Less than Unity
    (c) Greater than Unity
    (d) Zero

21. Consumer's surplus is large when the demand is
    (a) Perfectly elastic
    (b) Perfectly inelastic
    (c) Relatively inelastic
    (d) Relatively elastic

22. The demand for a commodity is said to be elastic, when it
    (a) Is a luxury
    (b) Has substitutes
    (c) Has several uses
    (d) All of them

23. Conversion of sugarcane into sugar is an example of
    (a) Possession utility
    (b) Form utility
    (c) Place utility
    (d) Time utility

24. The grain dealer stocks grain to supply it at the time of a shortage and creates
    (a) Place utility
    (b) Form utility
    (c) Time utility
    (d) Possession utility

25. Which of the following forms of business organization has limited liability
    (a) Individual proprietorship
    (b) Partnership
    (c) Joint stock company
    (d) State enterprise

26. Monopolist is a single seller of a commodity which has
    (a) Close substitutes
    (b) Close complements
    (c) No close substitutes
    (d) None of them

27. Under monopoly, price is
    (a) Higher than marginal revenue
    (b) Lower than marginal revenue
    (c) Equal to marginal revenue
    (d) None of them

28. The seller (firm) is price taker in the case of
    (a) Monopoly
    (b) Perfect competition
    (c) Monopolistic competition
    (d) Oligopoly

29. Dynamic theory of profits is associated with the name of
    (a) F. B. Hawley
    (b) J. B. Clark
    (c) J. B. Say
    (d) J. M. Keynes

30. Scarcity definition of economics is given by
    (a) Lionel Robbins
    (b) Adam Smith
    (c) Alfred Marshall
    (d) J. M. Keynes

31. Who introduced the terms micro and macro in economics
    (a) Alfred Marshall
    (b) Ragner Frisch
    (c) Adam Smith
    (d) J. M. Keynes

32. Supply of a commodity is influenced by
    (a) Number of firms
    (b) Price
    (c) Technology
    (d) All of them

33. Wage - price spiral leads to
    (a) Demand pull inflation
    (b) Stagflation
    (c) Cost push inflation
    (d) Walking inflation

34. There is no distinction between firm and industry in
    (a) Perfect competition
    (b) Monopsony
    (c) Monopolistic competition
    (d) Monopoly

35. Possession utility is created by
    (a) Storage
    (b) Buying and selling
    (c) Procuring
    (d) Transportation

36. Debts owed by individuals and nations are called
    (a) National wealth
    (b) Individual wealth
    (c) Public wealth
    (d) Negative wealth

37. The kinked demand curve is a feature of
    (a) Oligopoly
    (b) Monopoly
    (c) Duopoly
    (d) Monopsony

38. When the rate of inflation per annum ranges between 3 - 9 %, it is referred to as
    (a) Walking inflation
    (b) Running inflation
    (c) Creeping inflation
    (d) Hyper inflation

39. The distinction between stock and supply ceases to exist in the case of
    (a) Perishable goods
    (b) Semi durable goods
    (c) Durable goods
    (d) Intermediate goods

40. Personal income minus personal taxes is called as
    (a) Gross Domestic Product
    (b) Gross National Product
    (c) Disposable Income
    (d) National Income

41. Interest is the price paid for the use of
    (a) Land
    (b) Capital
    (c) Labour
    (d) Organisation

42. Modern theory of population was propounded by
    (a) Edwin Cannon
    (b) Alfred Marshal
    (c) T. R. Malthus
    (d) Adam Smith

43. 'One - Share, One - Vote' principle is followed in
    (a) State enterprise
    (b) Partnership
    (c) Cooperative organization
    (d) Joint Stock company

44. Consumer is said to be in equilibrium in the case of single commodity when
    (a) $MU_x < P_x$
    (b) $MU_x = P_x$
    (c) $MU_x > P_x$
    (d) $MVP = MFC$

45. Marshall defines economics as
    (a) The study of mankind in the ordinary business of life
    (b) The study of human behavior as a relationship between ends and scarce means
    (c) An enquiry into the nature and causes of the wealth of nations
    (d) A science of scarcity

46. The correct relationship between price, demand and supply of a commodity is that when
    (a) Price rises, supply rises and demand falls
    (b) Price rises, supply falls and demand rises
    (c) Price falls, supply falls and demand rises
    (d) Price rises, supply falls and demand falls

47. Net National Product at factor cost equals
    (a) NNP at market price - Indirect taxes + Subsidies
    (b) Disposable income - Personal tax
    (c) Gross National Product - Depreciation
    (d) All of these.

48. When several things are demanded for a joint purpose, it is a case of
    (a) Derived demand
    (b) Composite demand
    (c) Joint demand
    (d) Direct demand

49. Principle of progressive taxation is based on
    (a) Law of diminishing returns
    (b) Law of diminishing marginal utility
    (c) Law of equimarginal utility
    (d) Law of substitution

50. Who is considered to be the father of economics
    (a) Adam Smith
    (b) J. M. Keynes
    (c) Alfred Marshal
    (d) Lionel Robbins

## II. FILL IN THE BLANKS WITH APPROPRIATE WORDS

1. Human wants are _____ and the means to satisfy them are _____.

2. The reasoning proceeds from particulars to generals in _____ method of economic analysis.

3. The word economics is derived from the Greek word _____.

4. In deductive method of economic analysis, the reasoning proceeds from _____ to _____.

5. The _____ are the starting point of all economic activities whereas _____ is the end of all economic activities.

6. Anything that satisfies human want is called _____.

7. The power of commodity to command other things in exchange for itself is called _____.

8. Exchange of commodities for commodities is termed as _____ system.

9. Disguised unemployment is a sign of _____ population.

10. The price below which a seller refuses to sell a commodity is known as _____ price.

11. The price at which demand and supply are equal is called _____ price.

12. The compulsory payment made by the individuals and firms to the Govt. is called _____.

13. The difference between what we are prepared to pay and what we actually pay is termed as _____.

14. The theory of _____ is concerned with the evaluation of the services of the factors of production.

15. The method of measurement of national income from the output side is called _____ method.

16. The longer the period, the greater is the influence of _____ on the price of the commodity.

17. A business run by not less than two persons having unlimited liability is known as _____.

18. Any good or service which ais in production is called _____ of production.

19. According to Thomas Robert Malthus, population increases in _____ progression whereas food supply increases in _____ progression.

20. The size of the population at which per-capita income is the highest is _____ population.

21. If more quantity demanded at the same price or the same quantity at higher price, it is called _____ in demand.

22. A positive science is concerned with _____ whereas normative science tells _____.

23. Micro economics is also known as /_____.

24. Free goods have value in _____ but not value in _____.

25. The positive checks increase _____ rate whereas preventive checks decrease _____ rate.

26. The method which approaches national income from the distribution side is known as _____ method.

27. The demand is said to be _____ when small proportionate change in price is accompanied by big proportionate change in quantity demanded.

28. As the income of family increases, the proportion of income spent on food and other necessaries of life _____.

29. Economic laws are mere statements of _____.

30. According to the liquidity preference theory, the demand for money depends on _____, _____ and _____ motives.

31. The characteristic that any single want is satiable leads to the development of a law known as _____

32. Law of family expenditure was propounded by _____

33. If the rate of tax rises with rise in income, it is called a _____ tax.

34. The minimum price which must be paid to a factor of production to stay in the industry is called _____.

35. The demand for a factor of production is _____ demand.

36. The persistent and appreciable rise in general price level is known as _____.

37. Percentage change in quantity demanded of one commodity as a result of percentage change in the price of related commodity is called _____ elasticity of demand.

38. The market structure where there is only one buyer is termed as _____.

39. The total market value of all final goods and services produced in a year is called _____.

40. If the rate of tax falls with rise in income, it is called a _____ tax.

41. The supply of land is perfectly _____.

42. The extent of deviation of actual population of a country from the optimum population is termed as _____.

43. Increasing the stock of real capital in a country is called _____.

44. Law of equimarginal utility is also known as _____.

45. Modern economists divided subject of matter of economics into _____ and _____.

46. If the cross elasticity of demand is positive between the two goods, then they are said to be _____.

47. Expression of value in exchange of a good in terms of money is termed as _____.

48. Buying more quantity of a commodity at lesser price is called _____ of demand.

49. A tax which is really paid by the person on whom it is legally imposed is called _____.

50. The presence of one buyer and one seller in the market for a commodity is termed as _____.

## KEY

I. CHOOSE AND UNDERLINE THE CORRECT ANSWER

1. (d)  2. (c)  3. (b)  4. (b)  5. (b)  6. (c)  7. (d)
8. (b)  9. (d)  10. (d)  11. (b)  12. (c)  13. (d)  14. (b)
15. (c)  16. (d)  17. (c)  18. (a)  19. (c)  20. (c)  21. (c)
22. (d)  23. (b)  24. (c)  25. (c)  26. (c)  27. (a)  28. (b)
29. (b)  30. (a)  31. (b)  32. (d)  33. (c)  34. (d)  35. (b)
36. (d)  37. (a)  38. (a)  39. (a)  40. (c)  41. (b)  42. (a)
43. (d)  44. (b)  45. (a)  46. (a)  47. (a)  48. (c)  49. (b)
50. (a)

II. FILL IN THE BLANKS WITH APPROPRIATE WORDS

1. Unlimited, limited  2. Inductive  3. OIKONOMICAS
4. Generals, particulars  5. Wants, Consumption  6. Good or Service
7. Value  8. Barter  9. Over
10. Reserve  11. Equilibrium  12. Tax
13. Consumer surplus  14. Distribution  15. Product
16. Supply  17. Partnership  18. Factor
19. Geometric, Arithmetic  20. Optimum
21. Increase  22. What is, what ought to be
23. Price theory  24. Value in use, value in exchange
25. Death rate, Birth rate  26. Income  27. Elastic
28. Decrease  29. Tendencies

30. Transaction, Precautionary, Speculative
31. Law of diminishing marginal utility     32. Ernest Engel
33. Progressive       34. Transfer earnings     35. Derived
36. Inflation        37. Cross          38. Monopsony
39. Gross National Product                40. Regressive
41. Inelastic        42. Mal-adjustment     43. Capital formation
44. law of substitution or law of maximum satisfaction
45. Micro economics and Macro economics       46. Substitute
47. Price        48. Extension       49. Direct tax
50. Bilateral Monopoly

# AGRICULTURAL PRODUCTION ECONOMICS AND FARM MANAGEMENT

## I. CHOOSE AND UNDERLINE THE CORRECT ANSWER

1. Agricultural production economics is concerned with
   (a) Resource use efficiency
   (b) Resource substitution and combination
   (c) Resource allocation
   (d) All of them

2. The resources whose use varies with the level of production are called
   (a) Flow resources          (b) Fixed resources
   (c) Variable resources        (d) Poly period resources

3. All resources are variable in the
   (a) Very short period        (b) Short period
   (c) Long period          (d) Market period

4. Any good or service that helps in production is termed as
   (a) Input            (b) Resource
   (c) Factor of production       (d) All of them

5. The ratio of output to input is called
   (a) Average physical product     (b) Productivity
   (c) Technical efficiency       (d) All of them

6. An ideal condition in which costs are minimum or profits are maximum is known as
   (a) Sub optimality         (b) Optimality
   (c) Super optimality        (d) None of them

7. Single convenient unit of production for which technical coefficients are computed is
   (a) Plant            (b) Farm-firm
   (c) Technical unit         (d) Economic unit

8. Expenses incurred on inputs and input services in raising a crop on an unit area is termed as
   (a) Cost of production        (b) Average total cost
   (c) Replacement cost        (d) Cost of cultivation

9. The technical efficiency of variable input is indicated by
   (a) Average physical product     (b) Marginal physical product
   (c) Total physical product      (d) Marginal value product

10. The technical efficiency of fixed resource is indicated by
    (a) Average physical product
    (b) Marginal value product
    (c) Total physical product
    (d) All of them

11. Depreciation, interest, repairs and taxes are examples of
    (a) Variable costs
    (b) Explicit costs
    (c) Fixed costs
    (d) Fixed resources

12. Seeds, fertilizers, plant protection chemicals etc are examples of
    (a) Stock resources
    (b) Variable resources
    (c) Mono period resources
    (d) All of them

13. Labour and sunshine are examples of
    (a) Flow resources
    (b) Stock resources
    (c) Durable inputs
    (d) Semi durable inputs

14. Which of the following inputs is an example of fixed resources
    (a) Farm buildings
    (b) Land
    (c) Machinery and equipment
    (d) All of them

15. Marginal product is indicated by the slope of
    (a) Total product curve
    (b) Total cost curve
    (c) Iso-quant
    (d) Iso-cost line.

16. The inflection point is the point at which
    (a) APP is maximum
    (b) TPP is maximum
    (c) MPP is maximum
    (d) None of them

17. The ratio of MPP to APP is
    (a) Elasticity of substitution
    (b) Elasticity of production
    (c) Elasticity of demand
    (d) Elasticity of supply

18. Which of the following stages of production is irrational
    (a) I stage
    (b) II stage
    (c) III stage
    (d) I and III stages

19. Variable resource is in excess quantity in
    (a) I stage
    (b) III stage
    (c) II stage
    (d) Rational stage

20. Law of diminishing returns fails to operate when
    (a) New technology is adopted
    (b) Capital is scarce
    (c) New soil is brought under plough (d) All of them.

21. Cost minimization is the goal of
    (a) Factor - Factor relationship
    (b) Factor - Product relationship
    (c) Input - Output relationship
    (d) Product - Product relationship

22. The slope of iso-quant indicates
    (a) Marginal rate of product substitution
    (b) Marginal rate of technical substitution
    (c) Elasticity of substitution
    (d) None of them

23. Which of the following types of factor substitution is common in agriculture
    (a) Constant rate
    (b) Decreasing rate
    (c) Increasing rate
    (d) Fixed proportions

24. Within the ridge lines
    (a) MPP of both the inputs is zero
    (b) MPP of both the inputs increases
    (c) MPP of both the inputs decreases but remains positive
    (d) MPP of both the inputs is negative

25. Which of the following types of product substitution is common in agriculture
    (a) Increasing rate              (b) Constant rate
    (c) Decreasing rate              (d) Diminishing rate

26. Regional specialization in the production of commodities is explained by
    (a) Principle of comparative advantage (b) Law of equimarginal returns
    (c) Principle of factor substitution    (d) Principle of product substitution

27. Cost of production is indicated by
    (a) Total Cost (TC)              (b) Average Variable Cost (AVC)
    (c) Average Total Cost (ATC)     (d) Average Fixed Cost (AFC)

28. It is economical to substitute one input for another input as long as the
    (a) Cost of resource being added is less than the saving in the cost from the resource being replaced.
    (b) Cost of resource being added is more than the saving in the cost from the resource being replaced
    (c) Decrease in the returns from the product being replaced is less than the increase in the returns from the product being added.
    (d) Decrease in returns is less than increase in returns.

29. Which of the following economic principles guides the producer in deciding what to produce.
    (a) Principle of factor substitution   (b) Principle of product substitution
    (c) Law of diminishing returns         (d) Cost principle

30. If marginal rate of substitution is less than price ratio, profits can be increased by producing more of
    (a) Replaced product             (b) Added input
    (c) Added product                (d) Replaced product

31. The concept of opportunity cost is usually associated with
    (a) Law of diminishing returns         (b) Principle of factor substitution
    (c) Principle of product substitution   (d) Law of equimarginal return

32. Production of several products at the same time on the farm is known as
    (a) Specialised farming          (b) Diversified farming
    (c) Mixed farming                (d) Ranching

33. The practice of grazing of animals on public lands is termed as
    (a) Mixed farming                (b) Ranching
    (c) Specialised farming          (d) Dry farming

34. Which of the following types of farming is regarded as defence against risks
    (a) Dry farming                  (b) Ranching
    (c) Diversified farming          (d) Specialised farming

35. Which of the following systems of farming is common throughout the world
    (a) Peasant farming              (b) Collective farming
    (c) State farming                (d) Capitalistic farming

36. Which of the following systems of farming is owned by the Govt, and operated paid by management:
    (a) Capitalistic farming
    (b) Cooperative farming
    (c) Peasant farming
    (d) State farming

37. The type of ownership is individual and that of operatorship is collective in the case of cooperative
    (a) Tenant farming
    (b) Joint farming
    (c) Collective farming
    (d) Better farming

38. The type of ownership is collective and that of operatorship is individual in the case of cooperative
    (a) Tenant farming
    (b) Joint farming
    (c) Better farming
    (d) Collective farming

39. Partial budgeting is used to analyse
    (a) Enterprise substitution
    (b) Input substitution
    (c) Size or scale of operation
    (d) All of them

40. Linear programming was developed by
    (a) George B. Dantzig
    (b) T. R. Malthus
    (c) J. M. Keynes
    (d) Cobb-Douglass

41. Which of the following types of budgeting is used to analyse the expected change in profit for a proposed minor modification in the existing organization of farm business
    (a) Enterprise budget
    (b) Capital budget
    (c) Partial budget
    (d) Complete budget

42. Which of the following methods of reducing risk prevents large losses
    (a) Flexibility
    (b) Stable enterprises
    (c) Diversification
    (d) Hedging

43. Which of the following methods of reducing risk prevents the sacrifice of large gains
    (a) Flexibility
    (b) Diversification
    (c) Net-worth
    (d) Spreading sales

44. Which of the following cost concept reflects commercial cost of cultivation
    (a) Cost A1
    (b) Cost B
    (c) Cost C
    (d) Cost A2

45. Which of the following is not an assumption of L.P.
    (a) Non linearity
    (b) Divisibility
    (c) Additivity
    (d) Finiteness

46. Farm management is
    (a) Micro economic in its scope
    (b) Intra - farm study
    (c) Science of choice
    (d) All of them

47. Which of the following is a current liability
    (a) Crop loan
    (b) Accounts payable
    (c) Interest on all types of loans
    (d) All of them

48. Which of the following is the most commonly used method of computing depreciation
    (a) Annual revaluation method
    (b) Diminishing balance method
    (c) Sum of the years digits method
    (d) Straight line method

49. The difference between gross income and total variable costs is
    (a) Net income
    (b) Gross margin
    (c) Profit
    (d) Net working capital

50. Which of the following economic principles guides the producer in the allocation of resource when they are limited
    (a) Law of variable proportions
    (b) Principle of factor substitution
    (c) Principle product substitution
    (d) Law of equimarginal returns

## II. FILL IN THE BLANKS WITH APPROPRIATE WORDS

1. The process whereby some goods and services called inputs are transformed into other goods called products is known as _____.

2. The time required for a resource to be completely transformed into a product is called _____.

3. The relationship between different quantities of inputs and corresponding quantities of output is termed as _____.

4. An yardstick or criterion or index indicating which of the two or more alternatives will maximize a given end is referred to as _____.

5. The additional income received from producing an additional unit of output is _____.

6. Each successive unit of variable input when applied to fixed factors adds more and more to the total product in the case of _____ returns.

7. The planning period during which some inputs are fixed is called _____.

8. Factor - product relationship is explained by the law of _____ returns.

9. The profit maximizing or optimum input is where _____ equals with _____.

10. The additional cost incurred in using an additional unit of input is known as _____.

11. The marginal factor cost is always constant and is equal to the price per unit of _____.

12. The elasticity of production is the ratio of _____ to _____.

13. The profit maximizing or optimum output is where _____ equals with _____.

14. The marginal revenue is always constant and is equal to price per unit of _____.

15. Output is kept constant and inputs are varied in _____ relationship.

16. The absolute amount by which one input is reduced so as to gain another input by one unit is termed as _____.

17. A curve representing all possible combinations of two resources that produce same level of output is known as _____.

18. Inputs which are used together in fixed proportions is called _____.

19. A line or curve connecting the points of least cost combination for different levels of output is termed as _____.

20. A line that represents all possible combinations of two resources that can be purchased with given amount of funds is called _____.

21. The slope of iso cost line indicates the inverse price ratio of _____.

22. The least cost combination of resources is where _____ is tangent to _____.

23. In product - product relationship _____ are kept constant and _____ are varied.

24. A curve representing all possible combinations of two products that can be produced with given amount of resource is called _____.

25. The goal of product - product relationship is _____ maximization.

26. The optimum combination of enterprises or products is where _____ is tangent to _____.

27. A line representing all possible combinations of two products that yield same revenue is called _____.

28. The slope of _____ indicates inverse price ratio of products.

29. Two products are _____ if the increase or decrease in the output of one product is accompanied by increase or decrease in the output of another product.

30. Two enterprises are _____ if the increase in the output of one product necessitates sacrifice in the output of another product.

31. Two enterprises are _____ if the increase or decrease in the output of product does not influence the output of another product.

32. The method of finding present value of future sum is _____.

33. The method of finding future value of present sum is _____.

34. A producer has to stop the production temporarily if selling price fails to cover _____.

35. The total farm activity drawn up in advance by the farmer is called _____.

36. A method of estimating expected income, expenses and profits for the farm business is called _____.

37. List of all physical and financial assets owned by the farm business along with their values at a given point in time is called _____.

38. A systematic organization of assets and liabilities of a farm business at a given point in time is known as _____.

39. The owner's contribution into the business is termed as _____.

40. The difference between total assets and total liabilities is _____.

41. An obligation or debt owed to someone else is called _____.

42. Anything of value in the possession of the business or claim on something of value in the possession of others is defined as _____.

43. Decline in the value of an asset due to use or time obsolescence is referred to as _____.

44. A situation where all possible outcomes are known for a given management decisions and the probability associated with each possible outcome is also known, is termed as _____.

45. When a farm is organized to produce a single commodity which is the only source of income, then farm is said to be _____.

46. Prorating the original cost of an asset over its useful life is called _____.

47. Cost of self owned and self employed resources is known as _____.

48. Each unit increase in the output of one product requires larger and larger sacrifice in the output of another in the case of _____ rate of substitution.

49. The behavior of output when all the inputs are changed simultaneously in the same proportion is called _____.

50. In the short run, the goal of the producer is to recover _____ costs through generation of income.

## KEY

I. **CHOOSE AND UNDERLINE THE CORRECT ANSWER**

| | | | | | | |
|---|---|---|---|---|---|---|
| 1. (d) | 2. (c) | 3. (c) | 4. (d) | 5. (d) | 6. (b) | 7. (c) |
| 8. (d) | 9. (a) | 10. (c) | 11. (c) | 12. (d) | 13. (a) | 14. (d) |
| 15. (a) | 16. (c) | 17. (b) | 18. (d) | 19. (b) | 20. (d) | 21. (a) |
| 22. (b) | 23. (b) | 24. (c) | 25. (a) | 26. (a) | 27. (c) | 28. (a) |
| 29. (b) | 30. (c) | 31. (d) | 32. (b) | 33. (b) | 34. (c) | 35. (a) |
| 36. (d) | 37. (b) | 38. (a) | 39. (d) | 40. (a) | 41. (c) | 42. (c) |
| 43. (a) | 44. (c) | 45. (a) | 46. (d) | 47. (d) | 48. (d) | 49. (b) |
| 50. (d) | | | | | | |

II. **FILL IN THE BLANKS WITH APPROPRIATE WORDS**

1. Production
2. Transformation period or production period
3. Production function
4. Choice indicator
5. Marginal revenue
6. Increasing
7. Short run
8. Diminishing
9. Marginal value product, marginal input cost
10. Marginal factor cost or marginal input cost
11. Input
12. MPP, APP
13. Marginal Revenue, Marginal Cost
14. Output
15. Factor - Factor
16. Marginal rate of technical substitution.

17. Iso-quant18. Perfect complements      19. Expansion path
20. Iso-cost line          21. Inputs or factors or resources
22. Iso-quant, Iso-cost line
23. Inputs, outputs or products
24. Production possibility curve          25. Profit
26. Production possibility curve, Iso-revenue line      27. Iso-revenue line
28. Iso-revenue line      29. Complementary      30. Competitive
31. Supplementary      32. Discounting      33. Compounding
34. Average variable cost 35. Farm planning      36. Farm budgeting
37. Farm inventory      38. Balance sheet or net-worth statement
39. Net-worth or owner's equity
40. Net-worth or owner's equity          41. Liability
42. Asset          43. Depreciation      44. Risk
45. Specialised      46. Depreciation      47. Implicit costs
48. Increasing          49. Returns to scale      50. Total variable costs

# AGRICULTURAL FINANCE

## I. CHOOSE AND UNDERLINE THE CORRECT ANSWER.

1. According to the comprehensive crop insurance scheme indemnity is worked out through

   (a) $\dfrac{\text{Short fall in the yield of the crop}}{\text{Threshold yield of the crop}} \times \text{Sum insured}$

   (b) $\dfrac{\text{Threshold yield of the crop}}{\text{Short fall in the yield of the crop}} \times \text{Sum insured}$

   (c) $\dfrac{\text{Actual yield of the crop}}{\text{Threshold yield of the crop}} \times \text{Sum insured}$

   (d) $\dfrac{\text{Threshold yield of the crop}}{\text{Actual yield of the crop}} \times \text{Sum insured}$

2. One of the following is a discounted measure of project appraisal
   (a) Ranking by inspection
   (b) Proceeds per rupee of outlay
   (c) Average annual proceeds or rupee outlay
   (d) B-C Ratio

3. One of the following is an intangible benefit of a project
   (a) Increase in cropping intensity
   (b) Scale economies due to specialization
   (c) Income distribution
   (d) Improvement in cropping pattern

4. Pollution problems, health hazards etc., arising out of implementation of project indicate
   (a) Social costs          (b) Primary costs
   (c) Direct costs          (d) Real costs

5. BEP is estimated through

(a) $\dfrac{F}{P-V}$

(b) $\dfrac{VC}{P-V}$

(c) $\dfrac{F}{V-P}$

(d) $\dfrac{F}{P+V}$

6. Current ratio is worked out through

(a) $\dfrac{\text{Total current assets}}{\text{Total current liabilities}}$

(b) $\dfrac{\text{Total current liabilities}}{\text{Total current assets}}$

(c) $\dfrac{\text{Total assets}}{\text{Total liabilities}}$

(d) $\dfrac{\text{Total liabilities}}{\text{Total assets}}$

7. Even when each year of loan repayment period passes by, the principal component remains constant in the case of
   (a) Partial repayment plan
   (b) Balloon repayment plan
   (c) Amortized decreasing repayment plan
   (d) Amortized even repayment plan

8. Percentage of equity is worked out through

(a) $\dfrac{\text{Owned capitcal}}{\text{Borrowed capitcal}} \times 100$

(b) $\dfrac{\text{Owned capitcal}}{\text{Owned capitcal} + \text{Borrowed capitcal}} \times 100$

(c) $\dfrac{\text{Borrowed capitcal}}{\text{Owned capitcal} + \text{Borrowed capitcal}} \times 100$

(d) $\dfrac{\text{Borrowed capitcal}}{\text{Owned capitcal}} \times 100$

9. The measure that is adopted by institutional financial agencies like tie-up arrangement falls under one of the following Ps of credit
   (a) Principle of seasonality       (b) Principle of proper utilization
   (c) Principle of Protection        (d) Principle of payment

10. Debt - equity ratio is estimated using

(a) $\dfrac{\text{Owner's equity}}{\text{Total debts}}$

(b) $\dfrac{\text{Current liabilities}}{\text{Owner's equity}}$

(c) $\dfrac{\text{Total debts}}{\text{Owner's equity}}$

(d) $\dfrac{\text{Owner's equity}}{\text{Total liabilities}}$

11. NABARD came into existence in the year
    (a) 1982        (b) 1984
    (c) 1985        (d) 1990

12. RRBs were set up in the year
    (a) 1965        (b) 1975
    (c) 1978        (d) 1980

13. D. R. Gadgil and Sri. F.K.F. Nariman were connected with the introduction of
    (a) Branch expansion scheme
    (b) Lead bank scheme
    (c) Farm graduate scheme
    (d) Differential rate of interest scheme

14. Which one of the following commercial banks was not included in first spell of nationalization
    (a) Bank of India            (b) Andhra bank
    (c) Syndicate bank           (d) Indian bank

15. R. G. Saraiya was associated with
    (a) All India Rural Credit Survey Committee
    (b) All India Rural Credit Review Committee
    (c) Co-operative Planning committee
    (d) Committee on cooperative administration

16. FSSs were established on the recommendation of
    (a) NCA                      (b) Bawa Committee
    (c) Nariman Committee        (d) Venkatappaiah Committee

17. Crop loans are
    (a) Partially liquidating loans
    (b) Self - liquidating loans
    (c) Non - liquidating loans
    (d) Both partially and self - liquidating loans

18. Under IRDP, schedule tribes are eligible to receive a subsidy of
    (a) 25 %                     (b) 33.3 %
    (c) 50 %                     (d) 60 %

19. ACD, ARDC and RPCC of RBI were merged to establish
    (a) AFC                      (b) NABARD
    (c) IRDP                     (d) DRDA

20. The birth place of cooperative movement was
    (a) UK                       (b) Germany
    (c) France                   (d) USA

## II. FILL IN THE BLANKS WITH APPROPRIATE WORDS.

1. The loans advanced by the Government to the farmers for land improvement and distress relief are known as _____ loans.

2. The study pertaining to studying, examining and analysing the financial aspects pertaining to farm business is _____.

3. The financing agriculture at the aggregate level falls under the fold _____ finance, while the same for individual business units refers to _____ finance.

4. Term loans are _____ liquidating in nature.

5. Equitable mortgage is applicable in respect of _____ property

6. Financing fertilizer companies, construction of warehouses etc. fall under _____ loans.

7. Short term loans are called _____.

8. Short term loans are _____ liquidating in nature.

9. The rate of interest for DIR loans stands at _____ percent.

10. Three famous maxims of Sir Horace Plunkett with reference to cooperation are better farming, _____ and _____.

11. "One man and one vote" is the important principle of _____.

12. Frederick Nicholson's slogan on the formation of credit societies in India is popularly called _____.

13. The All Indian Rural Credit Survey Committee (AIRCSC) was appointed by RBI under the chairmanship of _____.

14. Sri. B. Venkatappaiah was the chairman of the _____.

15. District Central Co-operative Banks are the link between State - Cooperative Banks and _____.

16. Single window system in co-operative credit was introduced in 1987 in the state of _____ .

17. The first spell of nationalization of 14 commercial banks took place on _____ .

18. In second spell of nationalization, the number of banks that were nationalized stood at _____.

19. F.K.M. Nariman committee endorsed the views of the study groups on "area approach" and recommended the formulation of _____

20. The government of India recommended the establishment of Regional Rural Banks on the recommendation of working group headed by _____ .

21. The share capital of the Regional Rural Banks is shared by the _____, _____ and the _____ in the ratio of 50:15:35.

22. The DIR scheme was introduced on the recommendation of RBI committee under the chairmanship of _____.

23. The crop loan system in the country was introduced in _____.

24. IBRD, IDA, IFC and MIGA are collectively known as _____.

25. The Reserve Bank of India was established in _____ under the RBI act, 1934.

26. Sri. B. Sivaraman, Chairman of CRAFICARD was concerned with establishment of a national level institution called _____.

27. Returns, _____ and risk bearing ability are popularly known as three Rs of credit.

28. In estimation of risk bearing ability the gross returns are _____ .

29. Certificates such as 10 - 1 indicates _____ , while adangal shows _____.

30. In amortized even repayment plan, as the repayment period passes by, the principal portion _____ continuously.

31. In amortized decreasing repayment plan, the principal component _____ over the entire repayment period.

32. While presenting the balance sheet of a firm, cash is categorised under _____ asset.

33. The statement that presents a summary of receipts and gains minus expenses and losses during a specified period is called _____ or profit and loss statement.

34. Cash-flow statement is also known as _____ _____.

35. Cash-flow statement is prepared at the _____ of the agricultural year.

36. The point at which the total cost curve and total revenue curve intersect is called _____.

37. A firm is said to be at break - even point when contribution margin is exactly equal to _____.

38. The excess of production over the break-even point gives _____.

39. Project refers to a specific investment activity with specific _____ point and specific _____ point.

40. Project is treated as a cycle because each phase not only grows out of the _____ phase but also leads into _____ phase.

41. Costs with reference to a project at current market prices are _____ costs, while, the costs that are deflated by general price index are termed as _____ costs.

42. The technological externalities and technological spill over accrued to society due to the establishment of projects are _____ costs.

43. An increase in the productivity of crops due to the existence of projects refers to _____ benefits.

44. The timely collection and analysis of data on the progress of a project with the idea of identifying constraints is _____.

45. Investment analysis is otherwise called as _____ budgeting.

46. The present value of future money can be found out through _____ process.

47. Payback period is an _____ measure of project appraisal.

48. IRR is known as _____ efficiency of capital.

49. Warehouse receipt serves as _____ security in availing loans.

50. Based on the recommendations of the committee headed by _____, the cooperative credit societies act of 1904 was enacted.

51. The repayment of term loan in a series of installments is known as
_____.

## KEY

**I. CHOOSE AND UNDERLINE THE CORRECT WORD**

1. (a)  2. (d)  3. (c)  4. (a)  5. (a)  6. (a)  7. c
8. (b)  9. (c)  10. (c)  11. (a)  12. (b)  13. (b)  14. (b)
15. (c)  16. (a)  17. (b)  18. (c)  19. (b)  20. (b)

**II. FILL IN THE BLANKS WITH APPROPRIATE WORDS**

1. Taccavi
2. Agricultural finance
3. Macro, Micro
4. Partial
5. Self- acquired
6. Indirect
7. SAO loans or crop loans
8. Self
9. 4
10. Better business, better living
11. Cooperation
12. Find Raiffeessen
13. A. D. Gorwala
14. All India Rural Credit Review Committee
15. PACS
16. Andhra Pradesh
17. 19th July, 1969
18. 6
19. Lead bank scheme
20. M. Narasimham
21. Central Government, State Government, Sponsoring bank
22. B. K. Hazare
23. 1965
24. World bank group
25. 1935
26. NABARD
27. Repayment capacity
28. Deflated
29. Ownership of the land, cropping pattern
30. Increases
31. Remain constant
32. Current asset
33. Income statement
34. cash-flow summary, cash-flow budget or flow of funds statement
35. Beginning 36. Break-even point
37. Fixed costs
38. Margin of safety
39. Starting, end
40. Preceding, subsequent
41. Nominal, real
42. Social
43. Tangible
44. Monitoring
45. Capital
46. Discounting
47. Undiscounted
48. Marginal
49. Collateral
50. Sir Edward Law
51. Amortization

## AGRICULTURAL MARKETING

**I. CHOOSE AND UNDERLINE THE CORRECT ANSWER.**

1. Marketable surplus is influenced by one of the following factors
   (a) Level of production
   (b) Size of holding
   (c) Family consumption requirements
   (d) All the three factors

2. The most important method of buying and selling that takes place in regulated markets is
   (a) Morghum method
   (b) Sale under cover of a cloth
   (c) Sales through negotiations
   (d) Open auction sale

3. Wholesalers and retailers fall under the category of
   (a) Agent middlemen
   (b) Merchant middlemen
   (c) Speculative middlemen
   (d) Facilitative middlemen

4. In convergent form of Cob-Web cycle
   (a) Ed > Es
   (b) Ed < Es
   (c) Ed = Es
   (d) Es > Ed

5. Statutory minimum price is related to
   (a) Paddy
   (b) Groundnut
   (c) Sunflower
   (d) Sugarcane

6. The apex body in co-operative marketing is
   (a) MARKFED
   (b) NAFED
   (c) NCDC
   (d) APC

7. To protect the farmers against excess fall in price, the Government announces
   (a) Procurement price
   (b) Ceiling price
   (c) Minimen support Price
   (d) Issue price

8. AGMARK act was initiated in the year
   (a) 1930
   (b) 1935
   (c) 1937
   (d) 1947

9. Grading done at the farmers' level is
   (a) Kutcha
   (b) Pucca
   (c) Standard
   (d) Optional

10. Risk arising out of the changes in the policies of Government is
    (a) Physical risk
    (b) Technical risk
    (c) Institutional risk
    (d) Price risk

11. Vegetables markets fall under
    (a) General markets
    (b) Specialised markets
    (c) Terminal markets
    (d) None of the three

12. Open market is also called
    (a) Black market
    (b) Visible market
    (c) Invisible market
    (d) None of the three

13. Demand creation is the sub - function of
    (a) Transfer of ownership
    (b) Physical movement
    (c) Changing the form of the product
    (d) Facilitating

14. Agricultural produce (Development and Warehousing) Corporation act was passed in
    (a) 1956
    (b) 1966
    (c) 1970
    (d) 1976

15. Targeted PDS was launched in
    (a) 1997
    (b) 2000
    (c) 2001
    (d) 2002

16. Which of the following factors has negative relationship with marketable surplus.
    (a) Size of holding
    (b) Size of family
    (c) Price of commodity
    (d) Production

17. The act of holding and preserving farm commodities from the time of production till consumption is known as
    (a) Processing
    (b) Distribution
    (c) Storage
    (d) Assembling

18. The agent of the Government for purchase, sale, storage and distribution of farm products and farm inputs is
    (a) NAFED
    (b) FCI
    (c) SCB
    (d) PACS

19. The minimum support prices for farm commodities are recommended by
    (a) FCI
    (b) STC
    (c) CACP
    (d) NAFED

20. The market functionaries who take title of the products are
    (a) Merchant middlemen
    (b) Agent middlemen
    (c) Facilitative middlemen
    (d) Speculative middlemen

21. Ceiling price concept is to protect
    (a) Farmers
    (b) Consumers
    (c) Market intermediaries
    (d) Public agencies

22. Agricultural price commission was renamed as CACP is in the year
    (a) 1965
    (b) 1975
    (c) 1985
    (d) 1995

23. Market for coffee falls under the category of
    (a) Shandies
    (b) Hats
    (c) Primary markets
    (d) International markets

24. Warehousing corporation was established on the recommendation of
    (a) CRAFICARD
    (b) AIRCSC
    (c) AIRCRC
    (d) Gadgil committee

25. Adjustment of supply and demand on the basis of time, place and quantity is known as
    (a) Assembling
    (b) Concentration
    (c) Distribution
    (d) Equalisation

## II. FILL IN THE BLANKS WITH APPROPRIATE WORDS

1. Transportation service which is found in the marketing activity, creates _____ utility.

2. When large quantities of a commodity are bought and sold in the market, such a market is referred to as _____ market.

3. These markets are called _____ markets, in which future sales and purchases of commodities take place at current time.

4. In _____ markets, marketing practices are regulated and facilities created for smooth conduct.

5. The actual quantity of a commodity which a farmer sells in the market shows _____.

6. As the size of the holding increases _____ surplus also increases.

7. Invisible market can also be called as _____ market.

8. Assembling of the farm produce takes forms like _____ assembling and _____ assembling.

9. Time utility is added to the commodity by the _____ functioning of the marketing.

10. A single activity, which facilitates the movement of the product from the point of its production till it reaches the ultimate consumer, is _____.

11. Selling and _____ are the two important functions involved in the transfer of ownership.

12. Standardization precedes _____.

13. Distress sales can be averted if adequate marketing _____ is arranged for the farmers.

14. Marketing efficiency is the ratio of market output to market _____.

15. The agent middlemen that operate in the marketing of the agricultural commodities are specifically called _____ , _____.

16. By market _____, one means the description of number and nature of marketing participants in the marketing activities.

17. The execution of opposite sales or purchases in the future markets to effect the purchases or sales of physical products made in cash markets is _____.

18. The first legislation relating regulation of markets was enacted through _____ market act of 1897.

19. The idea of regulated market was imported into India from _____ in the pre-independence period.

20. The need for establishing licensed warehouse in India was early realized in the year _____ by Royal commission on agriculture.

21. The route through which the farm commodities pass through from producers to consumers is known as _____.

22. The remuneration that the intermediaries receive for their services in moving the commodity in the market channels is _____.

23. The extent of costs incurred in the movement of a commodity from the producer to consumer are _____.

24. Trend factors are those, which reflect movement in the economic variables over _____.

25. The components of time series data are trend, _____, seasonal and _____ factor.

26. The three forms of cob-web cycles in the prices of farm commodities are convergent , _____ and _____.

27. Nerlove model which is popularly known as _____ is used to estimate effect of prices on the acreage of crops.

28. The central AGMARK laboratory is an apex institution situated at _____.

29. The Bureau of Indian Standards is the name given to the _____, from April, 1987.

30. Secular markets are _____ in nature.

31. A market in which goods are exchanged for money immediately after the sale is called _____.

32. The loans that help the farmers to overcome distress sales are _____.

33. The reduction in price fluctuations within a certain range refers to _____.

34. The word marcatus means _____

35. Marketing adds _____ to the product but at the same time, it adds utility to the product.

36. The agency which decides parameters for the fixation of grade standards for farm commodities is _____.

37. FCI undertakes _____ operation in foodgrains to stabilize prices.

38. Price spread refers to price differential between consumer's price and _____ price

39. The markets with which consumers have direct access are called as _____ markets.

40. There is a _____ relationship between the size of the holding and the marketable surplus.

41. Packaging is a part of _____.

42. Greater the price spread _____ is the marketing efficiency.

43. Regulated markets narrow down the _____ spread between the producers and consumers

44. A market having a few buyers is known as _____ market.

# KEY

## I. CHOOSE AND UNDERLINE THE CORRECT WORD

| | | | | | | |
|---|---|---|---|---|---|---|
| 1. (d) | 2. (d) | 3. (b) | 4. (a) | 5. (d) | 6. (b) | 7. (c) |
| 8. (c) | 9. (a) | 10. (c) | 11. (b) | 12. (b) | 13. (a) | 14. (a) |
| 15. (a) | 16. (b) | 17. (c) | 18. (a) | 19. (c) | 20. (a) | 21. (b) |
| 22. (c) | 23. (d) | 24. (b) | 25. (d) | | | |

## II. FILL IN THE BLANKS WITH APPROPRIATE WORDS

1. Place
2. Wholesale
3. Forward
4. Regulated
5. Marketed surplus
6. Marketable
7. Black
8. Primary, Secondary
9. Storage
10. Marketing function
11. Buying
12. Grading
13. Finance
14. Input
15. Commission agents, brokers
16. Structure
17. Hedging
18. Berar cotton and grain
19. Great Britain
20. 1928
21. Marketing channel
22. Marketing margin
23. Market costs
24. time
25. Cyclical, Irregular
26. Divergent or Explosive, Regular or Uniform
27. Partial adjustment model
28. Nagpur
29. Indian Standards Institution
30. Permanent

31. Spot or cash market   32. Marketing loans   33. Price stabilization
34. Merchandise or trade              35. Cost
36. Directorate of Marketing and Inspection   37. Buffer-stock
38. Producer's          39. Retail              40. Positive
41. Packing            42. Lesser              43. Price
44. Oligopsony

# Index